W0111627

Platelets and Megakaryocytes

METHODS IN MOLECULAR BIOLOGY™

John M. Walker, SERIES EDITOR

METHODS IN MOLECULAR BIOLOGY™

Platelets and Megakaryocytes

Volume 2
Perspectives and Techniques

Edited by

Jonathan M. Gibbins

School of Animal and Microbial Sciences,
The University of Reading, Reading, UK

Martyn P. Mahaut-Smith

Department of Physiology,
University of Cambridge, Cambridge, UK

HUMANA PRESS ✳ TOTOWA, NEW JERSEY

© 2004 Humana Press Inc.
999 Riverview Drive, Suite 208
Totowa, New Jersey 07512

www.humanapress.com

All rights reserved. No part of this book may be reproduced, stored in a retrieval system, or transmitted in any form or by any means, electronic, mechanical, photocopying, microfilming, recording, or otherwise without written permission from the Publisher. Methods in Molecular Biology™ is a trademark of The Humana Press Inc.

All papers, comments, opinions, conclusions, or recommendations are those of the author(s), and do not necessarily reflect the views of the publisher.

This publication is printed on acid-free paper. ∞
ANSI Z39.48-1984 (American Standards Institute)
Permanence of Paper for Printed Library Materials.

Production Editor: Mark J. Breaugh.

Cover design by Patricia F. Cleary.

Cover illustration: Background images: Scanning electron microscopic images of human platelets at rest (upper panel) and after ADP-evoked aggregation (lower panel); from J-P Cazenave et al., chapter 2, volume 1. Foreground images: (left panel) Flow cytometric analysis of platelet integrin activation by thrombin; from B. Nieswandt, chapter 20, volume 1. (right panel) A megakaryocyte surrounded by smaller cells isolated from femoral marrow; M.Mahaut-Smith, chapter 17, volume 2.

For additional copies, pricing for bulk purchases, and/or information about other Humana titles, contact Humana at the above address or at any of the following numbers: Tel.: 973-256-1699; Fax: 973-256-8341; E-mail: humana@humanapr.com; or visit our Website: www.humanapress.com

Photocopy Authorization Policy:
Authorization to photocopy items for internal or personal use, or the internal or personal use of specific clients, is granted by Humana Press Inc., provided that the base fee of US $25.00 per copy is paid directly to the Copyright Clearance Center at 222 Rosewood Drive, Danvers, MA 01923. For those organizations that have been granted a photocopy license from the CCC, a separate system of payment has been arranged and is acceptable to Humana Press Inc. The fee code for users of the Transactional Reporting Service is: [1-58829-011-5/04 $25.00].

Printed in the United States of America. 10 9 8 7 6 5 4 3 2 1

E-ISBN: 1-59259-783-1

ISSN: 1064-3745

Library of Congress Cataloging in Publication Data

Platelets and megakaryocytes / edited by Jonathan M. Gibbins, Martyn P. Mahaut-Smith.
   p. ; cm. -- (Methods in molecular biology ; 272-273)
  Includes bibliographical references and index.
  ISBN 1-58829-101-4 (v. 1 : alk. paper) -- ISBN 1-58829-011-5 (v. 2 : alk. paper)
  1. Blood platelets--Laboratory manuals. 2. Megakaryocytes--Laboratory manuals.
  [DNLM: 1. Blood Platelets--physiology. 2. Histological Techniques. 3. Megakaryocytes--physiology.
WH 300 P71788 2004] I. Gibbins, Jonathan M. II. Mahaut-Smith, Martyn P. III. Methods in molecular biology (Clifton, N.J.) ; v. 272-273.
  QP97.P568 2004
  612.1'17--dc22
                             2004002112

Preface

The average human body has on the order of 10^{12} circulating platelets. They are crucial for hemostasis, and yet excessive platelet activation is a major cause of morbidity and mortality in Western societies. It is therefore not surprising that platelets have become one of the most extensively investigated biological cell types. We are, however, far from understanding precisely how platelets become activated under physiological and pathophysiological conditions. In addition, there are large gaps in our knowledge of platelet production from their giant precursor cell, the megakaryocyte. Understanding megakaryocyte biology will be crucial for the development of platelet gene targeting. The aim of *Platelets and Megakaryocytes* is therefore to bring together established and recently developed techniques to provide a comprehensive guide to the study of both the platelet and the megakaryocyte. It consists of five sections split between two volumes. The more functional assays appear in Volume 1, whereas Volume 2 includes signaling techniques, postgenomic methods, and a number of key perspectives chapters.

Part I of Volume 1, *Platelets and Megakaryocytes: Functional Assays*, describes many well-established approaches to the study of platelet function, including aggregometry, secretion, arachidonic acid metabolism, procoagulant responses, platelet adhesion under static or flow conditions, flow cytometry, and production of microparticles. Although one would ideally wish to perform experiments with human platelets, studies within the circulation using intravital microscopy require the use of animal models, which are described in Chapter 16, vol. 1. These approaches are becoming increasingly important in our understanding of how platelet responses contribute to the complex formation of thrombi within the circulation. Although naturally occurring genetic mutations can indicate the importance of specific proteins, these are limited in frequency and scope and thus many laboratories are using transgenic animals to delete or upregulate individual gene products (*see* Chapter 2, vol. 2). Consequently, the application of platelet techniques to murine models has become a focus of many labs in recent years (e.g., Chapters 2, 16, 20, vol. 1). In addition to basic and advanced approaches to study platelet function, several chapters in this section (particularly 1 and 2, vol. 1) focus on the long-standing issue of the effects of different anticoagulants and procedures to prepare platelets. The experimenter has a choice of studying platelets within the blood, in plasma, or in an artificial medium. In whole blood, potential interactions with other cell types and plasma proteins are included, which is in many ways the most physiological in vitro approach (*see* Chapter 6, vol. 1), however this is a complex situation and interpretation can be difficult. In studies within plasma, other cells are removed, but the clotting cascade is retained (*see* Chapter 5, vol. 1). Frequently, however, platelets are studied in isolation from other cells and plasma following their resuspension in an artificial medium. The preparation of platelets from human and other species is not a trivial matter and great care is required to ensure that the method of preparation does not adversely affect subsequent analysis (*see* Chapter 2, vol. 1).

Part II of Volume 1 focuses on approaches used to study megakaryocyte function, including the development of specialized structures for future production of platelets (e.g., the demarcation membrane system), the appearance of platelet-specific surface receptors, and the increase in ploidy. The source of megakaryocytes is often a complex issue facing many researchers owing to the extremely low density (<1%) of this cell type in its primary location, the marrow. Techniques to purify megakaryocytes from marrow based on their unique size and surface markers are described in Chapter 22, vol. 1, along with approaches to maintain these cells in culture and monitor formation of platelet-generating proplatelet structures. An alternative approach to generating megakaryocytes is to grow them in culture from precursor cells as detailed in Chapter 23, vol. 1. This requires the presence of thrombopoietin (Chapter 26, vol. 1) acting through its receptor, c-Mpl, and normally other cytokines. The availability of systems to generate megakaryocytes in vitro provides a promising avenue to generate genetically modified platelets. Although there is no doubt that continuous megakaryocyte cell lines are useful for some studies of signaling in these cells, they have their limitations and the pros and cons are discussed in Chapter 27, vol. 1.

Many basic and advanced techniques for the general study of cell signaling have been applied in studies to characterize the mechanisms of regulation of platelet function. These include ligand-binding assays, the study of protein and lipid kinases and phosphatases, the analysis of lipid rafts in the regulation of cell signaling, the measurement of intracellular calcium levels, electrophysiological techniques, nitric oxide signaling, the use of venom proteins, and the internalization of proteins into platelets through permeabilization. These techniques and more are presented in Part II of Volume 2. In many respects, the megakaryocyte is a giant platelet. Differences do occur in the arrangement of cellular organelles and cytoskeleton in the two cells, however, megakaryocytes respond to platelet agonists such as ADP with full downstream functional responses (discussed in Chapters 1 and 16, vol. 2). Therefore, despite differences in ultrastructure, the megakaryocyte has earned its place as a sufficient, if not comparable model of platelet signaling. Many of the signaling techniques are therefore beginning to be applied to the megakaryocyte, which, because of its size, is proving to be an extremely interesting model for platelet signaling, particularly using single cell approaches such as imaging and electrophysiology (*see* Chapters 16 and 17, vol. 2).

Part III of Volume 2 is dedicated to recent advances in molecular techniques and post-genomic techniques and how they may be applied to the study of platelets and megakaryocytes. This section includes descriptions of how retroviruses may be used to express genes in primary megakaryocytes, the use of GFP-fusion proteins to study signaling in live cells, two-dimensional electrophoresis for platelet proteomics, the production of platelet cDNA libraries and the use of gene array technology.

Although the main aim of the book is to include practical approaches to the study of platelets and megakaryocytes, a series of perspectives chapters are included (Part I, vol. 2). These chapters review the current understanding of platelet and megakaryocyte biology in addition to their discussions of important new developments and experimental strategies. Many of the methods chapters also include further discussion and background on specific techniques.

This book has only been made possible by the efforts of many international experts in the field. We are grateful to them for their willingness to contribute their knowledge, in particular their tricks of the trade, which have resulted from many years of dedicated hands-on work. We also wish to thank our colleagues within the Department of Physiology at Cambridge and the School of Animal and Microbial Sciences at Reading for helpful discussion during the course of the editing work, in particular Peter Wooding on electron microscopy and Gwen Tolhurst on molecular techniques. We are also grateful to Margaret Bardy and Karen Parr for considerable secretarial assistance. We are also grateful to the following companies for supporting the cost of color reproduction: Eli Lilly and Company, Cairn Research Ltd., Bio Rad Laboratories Ltd., and Sysmex UK Ltd.

Jonathan M. Gibbins
Martyn P. Mahaut-Smith

Contents of Volume 2

Perspectives and Techniques

CONTENTS OF THE COMPANION VOLUME

Volume 1: Functional Assays

xiii

Contributors

JAN WILLEM AKKERMAN • *Department of Haematology, University Medical Center Utrecht, The Netherlands*

ROBERT K. ANDREWS • *Department of Biochemistry and Molecular Biology, Monash University, Clayton, Victoria, Australia*

HAVA AVRAHAM • *Division of Experimental Medicine, Beth Israel Deaconess Medical Center, Boston, MA*

SHALOM AVRAHAM • *Division of Experimental Medicine, Beth Israel Deaconess Medical Center, Boston, MA*

MICHAEL C. BERNDT • *Department of Biochemistry and Molecular Biology, Monash University, Clayton, Victoria, Australia*

LAWRENCE F. BRASS • *Hematology-Oncology Division, University of Pennsylvania, Philadelphia, PA*

EMERY H. BRESNICK • *Molecular and Cellular Pharmacology, University of Wisconsin Medical School, Madison, WI*

K. RICHARD BRUCKDORFER • *Department of Biochemistry and Molecular Biology, University College London, London, UK*

JEAN-PIERRE CAZENAVE • *INSERM U.311, Etablissement Français du Sang-Alsace, Strasbourg, France*

DERMOT COX • *Department of Clinical Pharmacology, Royal College of Surgeons in Ireland, Dublin, Ireland*

GYLES E. COZIER • *Integrated Signalling Laboratories, Department of Biochemistry, University of Bristol, Bristol, UK*

PETER J. CULLEN • *Integrated Signalling Laboratories, Department of Biochemistry, University of Bristol, Bristol, UK*

NIKLA R. EMAMBOKUS • *Division of Hematology/Oncology, Children's Hospital, Boston, MA*

RICHARD W. FARNDALE • *Department of Biochemistry, University of Cambridge, Cambridge, UK*

ROBERT FLAUMENHAFT • *Center for Hemostasis and Thrombosis Research, Beth Israel Deaconess Medical Center and Harvard Medical School, Boston, MA*

RYAN FORTNA • *Hematology-Oncology Division, University of Pennsylvania, Philadelphia, PA*

JONATHAN FRAMPTON • *Department of Anatomy, Birmingham University Medical School, Birmingham, UK*

CHRISTIAN GACHET • *INSERM U.311, Etablissement Français du Sang-Alsace, Strasbourg, France*

ELIZABETH E. GARDINER • *Department of Biochemistry & Molecular Biology, Monash University, Clayton, Victoria, Australia*

MEENAKSHI GAUR • *Departments of Laboratory and Internal Medicine, University of California, San Francisco, CA*

JONATHAN M. GIBBINS • *School of Animal & Microbial Sciences, The University of Reading, Reading, Berkshire, UK*

GERTIE GORTER • *Department of Haematology, University Medical Center, Utrecht, The Netherlands*

KARINE GOUSSET • *Center for Biostabilization, University of California at Davis, Davis, CA*

STEVEN HEAD • *DNA Array Core Facility, Department of Research Resources, The Scripps Research Institute, La Jolla, CA*

BÉATRICE HECHLER • *INSERM U.311, Etablissement Français du Sang-Alsace, Strasbourg, France*

JEAN-MARC HERBERT • *Cardiovascular/Thrombosis Research Department, Sanofi-Synthelabo Recherche, Toulouse, France*

SUSANNA M. O. HOURANI • *School of Biomedical & Molecular Sciences, University of Surrey, Guildford, UK*

HONG JIANG • *Hematology-Oncology Division, University of Pennsylvania, Philadelphia, PA*

JIANGO JIN • *Department of Physiology, Temple University School of Medicine, Philadelphia, PA*

MATTHEW L. JONES • *Department of Pharmacology, School of Medical Sciences, University of Bristol, UK*

STEPHANIE M. JUNG • *Department of Protein Biochemistry, Institute of Life Science, Kurume University, Japan*

C. GRAHAM KNIGHT • *Department of Biochemistry, University of Cambridge, UK*

SATYA P. KUNAPULI • *Department of Physiology, and the Sol Sherry Thrombosis Research Center, Temple University School of Medicine, Philadelphia, PA*

THOMAS J. KUNICKI • *Department of Molecular and Experimental Medicine, The Scripps Research Institute, La Jolla, CA*

LLOYD T. LAM • *Center for Cancer Research, National Cancer Institute, Bethesda, MD*

ANDREW D. LEAVITT • *Departments of Laboratory and Internal Medicine, University of California, San Francisco, CA*

JOSÉ A. LÓPEZ • *Division of Thrombosis Research, Baylor College of Medicine and Houston VA Medical Center, Houston, TX*

SYLVIA Y. LOW • *Department of Biochemistry & Molecular Biology, University College London, UK*

MARÍA L. LOZANO • *Department of Haematology & Oncology, School of Medicine, University of Murcia and Centro Regional de Hemodonación, Spain*

MARTYN P. MAHAUT-SMITH • *Department of Physiology, University of Cambridge, Cambridge, UK*

KATRIN MARCUS • *Medical Proteome-Center, Ruhr-University Bochum, Germany*

MICHAEL J. MASON • *Department of Physiology, University of Cambridge, Cambridge, UK*

HELMUT E. MEYER • *Medical Proteome-Center, Ruhr-University Bochum, Germany*

MASAAKI MOROI • *Department of Protein Biochemistry, Institute of Life Science, Kurume University, Japan*

GEORGE J. MURPHY • *Department of Genetics, Children's Hospital, Harvard Medical School, Boston, MA*

KHALID M. NASEEM • *Department of Biomedical Sciences, University of Bradford, Bradford, UK*

PETER O'BRIEN • *Diagnostic and Experimental Medicine, Lilly Research Laboratories, Indianapolis, IN*

PHILIPPE OHLMANN • *INSERM U.311, Etablissement Français du Sang-Alsace, Strasbourg, France*

CATHERINE M. ONLEY • *Department of Biochemistry, University of Cambridge, Cambridge, UK*

BENJAMIN Z. S. PAUL • *Department of Pharmacology, Temple University School of Medicine, Philadelphia, PA*

BERNARD PAYRASTRE • *Inserm U563, Department of Oncogenesis and Signal Transduction in Hematopoietic Cells, Hopital Purpan, Toulouse, France*

ALASTAIR W. POOLE • *Department of Pharmacology, School of Medical Sciences, University of Bristol, Bristol, UK*

NICOLAS PREVOST • *Hematology-Oncology Division, University of Pennsylvania, Philadelphia, PA*

ROCIO RIBA • *Department of Biomedical Sciences, University of Bradford, Bradford, UK*

JOSÉ RIVERA • *Department of Haematology and Oncology, School of Medicine, University of Murcia and Centro Regional de Hemodonación, Spain*

DANIEL R. SALOMON • *Division of Experimental Hemostasis & Thrombosis, Department of Molecular & Experimental Medicine, The Scripps Research Institute, La Jolla, CA*

PIERRE SAVI • *Cardiovascular/Thrombosis Research Department, Sanofi-Synthélabo Recherche, Toulouse, France*

CORIE N. SHRIMPTON • *Division of Thrombosis Research, Baylor College of Medicine and Houston VA Medical Center, Houston, TX*

LOUIS F. STANCATO • *Eli Lilly & Company-Sphinx Laboratories, Research Triangle Park, NC*

FERN TABLIN • *Department of Anatomy, Physiology and Cell Biology, University of California, Davis, CA*

MASSIMILIANO TOGNOLINI • *Hematology-Oncology Division, University of Pennsylvania, Philadelphia, PA*

MAX TROXLER • *Vascular Surgical Unit, University of Leeds Medical School, Leeds, UK*

VICENTE VICENTE • *Department of Haematology and Oncology, School of Medicine, University of Murcia and Centro Regional de Hemodonación, Spain*

CHRIS J. VLAHOS • *Cardiovascular Research, Lilly Research Laboratories, Indianapolis, IN*

SIMON A. WALKER • *Inositide Group, Department of Biochemistry, School of Medical Sciences, University of Bristol, Bristol, UK*

DONNA WOULFE • *Hematology-Oncology Division, University of Pennsylvania, Philadelphia, PA*

JIE WU • *Hematology-Oncology Division, University of Pennsylvania, Philadelphia, PA*

JING YANG • *Centocor, Inc., Malvern, PA*

RADOSLAW ZAGOZDZON • *Division of Experimental Medicine, Beth Israel Deaconess Medical Center, Harvard Medical School, Boston, MA*

Color Plates

Color Plates 1–5 appear as an insert following p. 300.

PLATE 1   Schematic representation of basic retroviral structure and the process of infection. (See full caption on p. 37, Chapter 2.)

PLATE 2   Megakaryocyte-specific retroviral infection through the TVA receptor. (See full caption on p. 39, Chapter 2.)

PLATE 3   Ribbon diagram representation of the crystal structure of wild-type green fluorescent protein from *Aequoria victoria*. (See full caption on p. 408, Chapter 25.)

PLATE 4   Comparison of GFP images obtained with laser scanning and spinning disk confocal microscopes. (See full caption on p. 413, Chapter 25.)

PLATE 5   Summary of the microarray process. (See full caption on p. 481, Chapter 29.)

I

PERSPECTIVES

1

Signaling Receptors on Platelets and Megakaryocytes

Donna Woulfe, Jing Yang, Nicolas Prevost, Peter O'Brien, Ryan Fortna, Massimiliano Tognolini, Hong Jiang, Jie Wu, and Lawrence F. Brass

1. Introduction

Although the body of knowledge is far from complete, much has been learned about the receptors that enable circulating platelets to become activated at sites of vascular injury. A variety of approaches have been used to acquire this information, many of which are described elsewhere in this book. Far less is known about the receptors that are expressed on megakaryocytes, some of which play specific roles in megakaryocyte development and some of which mirror those that will subsequently be found on platelets. Our intent in this chapter is to provide a "receptor-centric" overview of platelet activation, followed by a very brief consideration of receptor function in megakaryocytes and megakaryoblastic cell lines. (Our apologies in advance to the authors of the many excellent studies that we have not cited in our effort to be brief.)

Formation of a platelet plug can be thought of as occurring in three phases: initiation, extension, and perpetuation, each of which involves a somewhat different group of receptors (**Fig. 1**). *Initiation* occurs when circulating platelets arrest on and are activated by exposed collagen and von Willebrand factor in the subendothelial matrix, allowing the accumulation of a platelet monolayer that will subsequently support thrombin generation and the formation of platelet aggregates. Key to this phase of platelet activation are the receptors that can bind to collagen (particularly the integrin $\alpha_2\beta_1$ and GPVI) and von Willebrand factor (the GPIb/IX/V complex and the integrin $\alpha_{IIb}\beta_3$) and initiate intracellular signaling. *Extension* occurs when additional platelets accumulate on the initial monolayer. Key to this phase is the presence on the platelet surface of receptors that can respond rapidly to soluble agonists such as thrombin, ADP, and thromboxane A_2 (TxA_2). Most of the receptors involved in these events are members of the superfamily of G protein-coupled receptors. *Perpetuation* refers to the late events of platelet plug formation, when direct interactions between platelets make contact-dependent signaling possible. These late events stabilize the platelet plug, prevent premature disaggregation, and regulate clot retraction. Perpetu-

From: *Methods in Molecular Biology, vol. 273:*
Platelets and Megakaryocytes, Vol. 2: Perspectives and Techniques
Edited by: J. M. Gibbins and M. P. Mahaut-Smith © Humana Press Inc., Totowa, NJ

1. Initiation

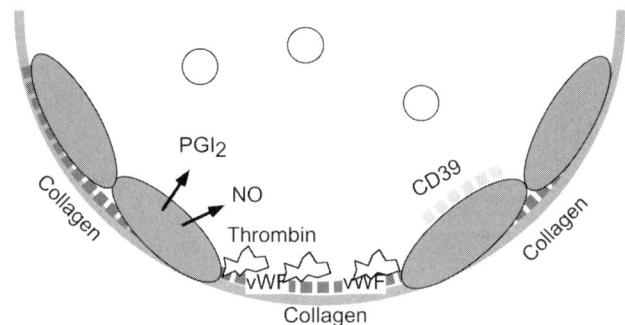

2. Extension

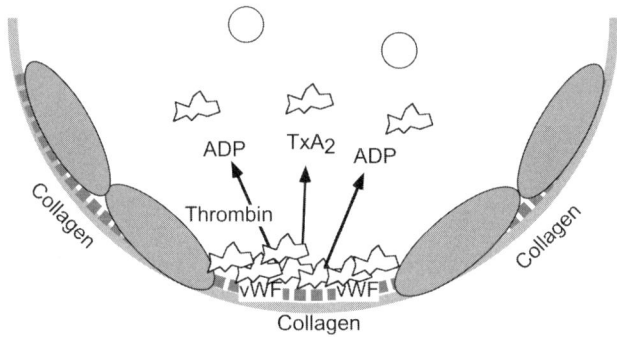

3. Perpetuation

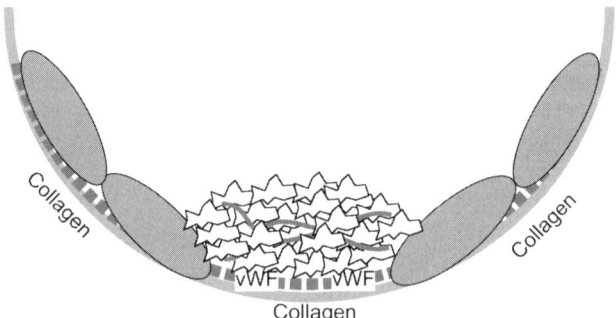

Fig. 1. Steps in platelet-plug formation. (1) *Initiation.* Prior to vascular injury, platelets are maintained in the resting state by a combination of inhibitory factors that generate a threshold to be crossed before platelets become activated. These factors include PGI_2 and NO released from endothelial cells, and CD39, an ADPase on the surface of endothelial cells that can hydrolyze small amounts of ADP to prevent

Table 1
G Protein-Coupled Receptors Expressed on Human Platelets

Agonist	Receptor	Gα protein families	Approximate no. of copies per platelet	Reference
Thrombin	PAR1	G_q, G_i, G_{12}	2000	*59,61,62,148,149*
	PAR4	G_q, G_{12}	?	*61,148*
ADP	$P2Y_1$	G_q, G_{12}	150	*33,34,151,152*
	$P2Y_{12}$	G_i (particularly G_{i2})	600	*38,39*
TxA_2	TPα and TPβ	G_q, G_{12}	1000	*84–86,153–155*
Epinephrine	α_{2A}-adrenergic	G_i (particularly G_z)	300	*75,76,156*
PGI_2	IP	G_s	?	*157*

This list is not intended to be exhaustive. Platelets also express G protein-coupled receptors for SDF-1 *(158,159)*, platelet-activating factor (PAF) *(160,161)*, and vasopressin (V1) *(162–165)*, among others.

ation is less well understood than initiation and extension, but recent studies point to an essential role for outside-in signaling through integrins and to the signals generated by receptor tyrosine kinases, including members of the Eph kinase family.

The surface of human platelets is crowded with receptors that support one or more of the phases of platelet plug formation. Those that are directly involved in binding to adhesive proteins such as collagen, von Willebrand factor, and fibrinogen are present in the greatest numbers. After recruitment of molecules that are initially hidden within the surface-connecting-membrane system, there are approx 80,000 copies of $\alpha_{IIb}\beta_3$ and 15,000 copies of GPIb on the surface of human platelets. In contrast, the number of receptors that respond to agonists such as thrombin, ADP, and epinephrine range from a few hundred to a few thousand (**Table 1**). Although these numbers are not large, when placed in the context of the relatively small size of human platelets the density is high.

2. Signaling During the Initiation Phase of Platelet Plug Formation

Under static conditions, collagen is able to activate platelets without the assistance of cofactors, but under the conditions of flow that exist in vivo, von Willebrand factor (vWF) plays an important role, particularly in arterial circulation. A number of integrins

---

Fig. 1. *(continued)* inappropriate platelet activation. The development of the platelet plug is initiated by the exposure of collagen in the subendothelial layers and the local generation of thrombin. This causes platelets to adhere and spread on the connective matrix, forming a monolayer. (2) *Extension.* The initial platelet monolayer is extended as additional platelets are activated via the release, secretion, or generation of TxA_2, ADP, thrombin, and other platelet agonists, most of which are ligands for G protein-coupled receptors on the platelet surface. (3) *Perpetuation.* Finally, close contacts between platelets in the growing hemostatic plug help to perpetuate and stabilize the platelet aggregate through mechanisms that include outside-in signaling and Eph/ephrin interactions.

have been shown to bind cells to collagen, but in platelets $\alpha_2\beta_1$ seems to be dominant. According to current models, the collagen:vWF complex binds to at least four molecules on the platelet surface: $\alpha_2\beta_1$ (collagen), $\alpha_{IIb}\beta_3$ (vWF), GPVI (collagen), and GPIb (vWF). These interactions are illustrated in **Fig. 2**. Of these, GPVI is arguably the most potent in terms of initiating signal generation, but the others are thought to contribute as well. GPVI was cloned in 1999 *(1)*. Its structure places it in the immunoglobulin domain superfamily and its ability to generate signals rests on its constitutive association with a second molecule, the Fc-receptor γ-chain. Platelets from mice that lack the γ-chain *(2)* have impaired responses to collagen. So do platelets from humans *(3,4)*, but possibly not mice *(5)* with reduced expression of $\alpha_2\beta_1$. The gene encoding GPVI has recently been knocked out in mice *(5a)*, and using blocking antibodies or a depletion strategy, Nieswandt et al. *(5,6)* concluded that GPVI is required for platelet responses to collagen. Given the requirement that the Fc-receptor γ-chain be present for GPVI to reach the platelet surface *(6)*, it is likely that the platelet phenotype of any future GPVI knockout will resemble the FcRγ knockout.

Studies using synthetic "collagen-related peptides" (CRP) allow stimulation of GPVI in isolation and suggest that activation of GPVI alone can be sufficient to activate platelets. Whether that is also true for collagen is still debated. According to current models, collagen causes clustering of GPVI and its associated γ-chain. This leads to the phosphorylation of the γ-chain by nonreceptor tyrosine kinases in the Src family, creating a tandem phosphotyrosine motif recognized by the SH2 domains of Syk. Association of Syk with GPVI/γ-chain activates Syk and leads to the phosphorylation and activation of the γ2 isoform of phospholipase C (PLCγ2) via the adaptor protein, SLP-76 *(7)*. Loss of Syk impairs collagen responses *(2)*. PLCγ, like other forms of phospholipase C in platelets, hydrolyzes $PI\text{-}4,5\text{-}P_2$ to form $1,4,5\text{-}IP_3$ and diacylglycerol (DG). IP_3 opens Ca^{2+} channels in the platelet dense tubular system, raising the cytosolic Ca^{2+} concentration and triggering Ca^{2+} influx across the platelet plasma membrane.

Although platelet activation by collagen involves some of the same signaling mechanisms used by soluble platelet agonists (described in **Subheading 3.1.**), it is slower in its rate of onset. Thus, it is common to observe a lag time between the addition of sol-

---

Fig. 2. *(see facing page)* Initiation of platelet-plug formation by collagen. Platelets appear to use several different molecular complexes to support platelet activation by collagen. These include (1) the vWF-mediated binding of collagen to the GPIb/IX/V complex and the $\alpha_{IIb}\beta_3$ integrin, (2) a direct interaction between collagen and the $\alpha_2\beta_1$ integrin, and (3) the GPVI/γ-chain complex. Clustering of GPVI results in the phosphorylation of tyrosine residues in the γ-chain by one or more members of the Src family of kinases (SFK), followed by the binding and activation of the tyrosine kinase, Syk. One consequence of Syk activation is the activation of phospholipase Cγ, leading to phosphoinositide hydrolysis, secretion of ADP and the production of TxA_2. ADP and TxA_2 bind to their own receptors in platelets, generating signals that support the more direct effects of collagen.

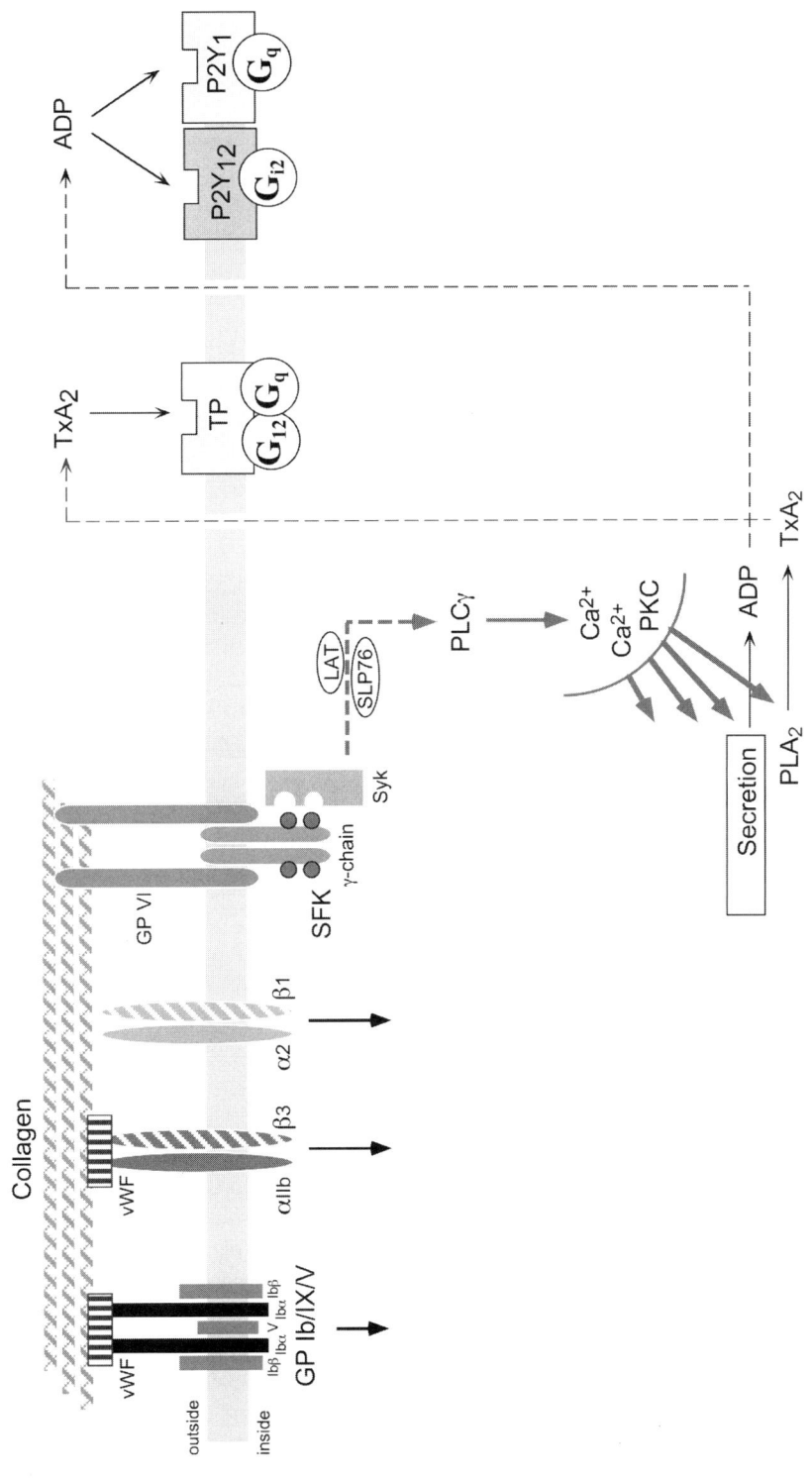

uble collagen to a platelet suspension and the time that aggregation is observed. This delay is attributable to the time required for soluble collagen to polymerize and for GPVI to form the clusters that are required for signal transduction. In vivo, platelets are activated by adhesion to polymerized collagen exposed beneath the broken cellular matrix. To allow extension of platelet plug formation, adhesion to the basement matrix must be maintained long enough to activate the release of soluble mediators described below. The coincident binding of vWF to $\alpha_{IIb}\beta_3$ and GPIbα strengthens these adhesive forces and makes it possible for this to occur.

3. Signaling During the Extension Phase of Platelet Plug Formation

The second step in the formation of the platelet plug is the recruitment of additional circulating platelets, which is made possible by the local accumulation of molecules such as thrombin, ADP, epinephrine, and TxA$_2$. Thrombin is generated locally from prothrombin once tissue factor has been exposed (a process facilitated by the negatively charged phospholipids on the surface of activated platelets). ADP is stored within platelet dense granules and secreted when platelets are activated. TxA$_2$ is synthesized within platelets once phospholipase A$_2$ has been activated and arachidonate released from membrane phospholipids. Epinephrine is available as a circulating catecholamine. Contacts between platelets are maintained by a variety of molecular interactions, of which the best-described is the binding of fibrinogen (and vWF) to activated $\alpha_{IIb}\beta_3$. Epinephrine, TxA$_2$, and serotonin cause vasoconstriction, but also potentiate the effects of thrombin and ADP on platelets. In contrast to collagen, platelet activation by most of the agonists involved in extension of the platelet plug is very rapid, with some responses occurring within a fraction of a second. Teleologically, this makes sense because circulating platelets would not be expected to linger at a wound site long enough for a slower process to occur.

Agonists that are involved in the extension phase of platelet plug formation typically bind to G protein-coupled receptors on the platelet surface. The design of these receptors makes them particularly well suited for this task. First, most of them bind their ligands with high affinity. Second, they consist of a single subunit that does not require oligomerization (although some G protein-coupled receptors may exist as homo- or heterodimers) *(8–10)*. Third, they are constitutively associated with G proteins, eliminating the time that would otherwise be required to recruit them into complexes *(8)*. Fourth, because they act as guanine nucleotide exchange factors (GEFs), each occupied receptor can activate multiple G proteins and, in some cases, more than one class of G proteins. This allows amplification of a signal that might begin with a relatively small number of receptors. It also potentially allows each receptor to signal via more than one effector pathway. Finally, because several generic mechanisms exist that can limit signaling by G protein-coupled receptors, platelet activation can be limited, a property that is essential when platelet activation is inappropriate and needs to be controlled *(11,12)*.

G protein-coupled receptors are comprised of a single polypeptide chain with an extracellular N-terminus and seven transmembrane domains. Binding sites for agonists can involve the N-terminus, the extracellular loops, or a pocket formed by the transmembrane domains *(8)*. The G proteins that act as mediators for these receptors are heterotrimers composed of a single α, β, and γ subunit. Within this complex the β subunit forms a propeller-like structure that is tightly, but not covalently, associated with the smaller γ subunit. The α subunit contains a guanine nucleotide-binding site that is normally occupied by GDP. Receptor activation causes the exchange of GTP for GDP, altering the conformation of the α subunit and exposing sites on both G_α and $G_{\beta\gamma}$ for interactions with downstream effectors *(13–15)*. Hydrolysis of the GTP by the intrinsic GTPase activity of the α subunit restores the resting conformation of the heterotrimer, preparing it to undergo another round of activation and signaling *(16)*. Regulators of G protein-signaling (RGS) proteins help to accelerate the hydrolysis of GTP by α subunits *(17)*. Prenylation of the γ chain and acylation (with palmitate or myristate) of the α subunit help to anchor the complete heterotrimer to the plasma membrane.

There are at least 22 mammalian genes encoding G_α, 5 encoding G_β, and 8 encoding G_γ. Human platelets express at least 10 of the known forms of G_α: one $G_{s\alpha}$ family member, four $G_{i\alpha}$ family members ($G_{i1\alpha}$, $G_{i2\alpha}$, $G_{i3\alpha}$, and $G_{z\alpha}$), three $G_{q\alpha}$ family members ($G_{q\alpha}$, $G_{11\alpha}$, and $G_{16\alpha}$) and two $G_{12\alpha}$ family members ($G_{12\alpha}$ and $G_{13\alpha}$). G_q family members are best known for their ability to activate phospholipase Cβ; G_s stimulates adenylyl cyclase; G_i family members inhibit adenylyl cyclase and activate PI 3-kinaseγ; G_{12} family members signal through Rho to regulate reorganization of the actin cytoskeleton. We have been particularly interested in the four G_i family members expressed in platelets. The α subunits of three of them ($G_{i1\alpha}$, $G_{i2\alpha}$, and $G_{i3\alpha}$) are approx 95% identical at the protein level, including a cysteine residue near the C-terminus that can be ADP-ribosylated by pertussis toxin. The fourth, $G_{z\alpha}$, is only about 60% identical to the others and is not a substrate for pertussis toxin. Recent studies on platelets from knockout mice have helped to define the receptors and effector pathways to which the G_i family members are linked. G_{i2} (the family member most highly expressed in platelets) couples P2Y$_{12}$ (ADP) receptors to adenylyl cyclase *(18,31)* (**Table 1**). G_z does the same for α_{2A}-adrenergic (epinephrine) receptors *(19)*. Both of these G proteins link their respective receptors to the activation of Rap1, in part through PI 3-kinase-γ *(20)*.

3.1. G Protein-Mediated Signaling in Platelets

Although the details vary, it is now generally accepted that agonists whose effects are mediated by G protein-coupled receptors must accomplish at least three tasks in order to activate platelets and recruit them into a growing platelet plug (**Fig. 3**). First, they must increase the cytosolic Ca^{2+} concentration and activate protein kinase C and PI 3-kinase isoforms. Typically this is accomplished by activating one or more isoforms of phospholipase Cβ, which in turn hydrolyzes PI-4,5-P$_2$ to 1,4,5-IP$_3$ and diacylglycerol. In resting platelets the cytosolic Ca^{2+} concentration is maintained at approximately 100 nM by limiting Ca^{2+} influx and by pumping Ca^{2+} out of the cytosol across the plasma membrane or into the dense tubular system. This creates a steep Ca^{2+} gradient across the plasma membrane. Once formed, 1,4,5-IP$_3$ releases Ca^{2+} from the dense tubular system, which in turn opens Ca^{2+} influx channels in the plasma membrane. Extracellular Ca^{2+} then pours in, following its concentration gradient. In activated platelets the cytosolic free concentration can exceed 1 μM with potent agonists like thrombin. Different phospholipase Cβ isoforms can be activated

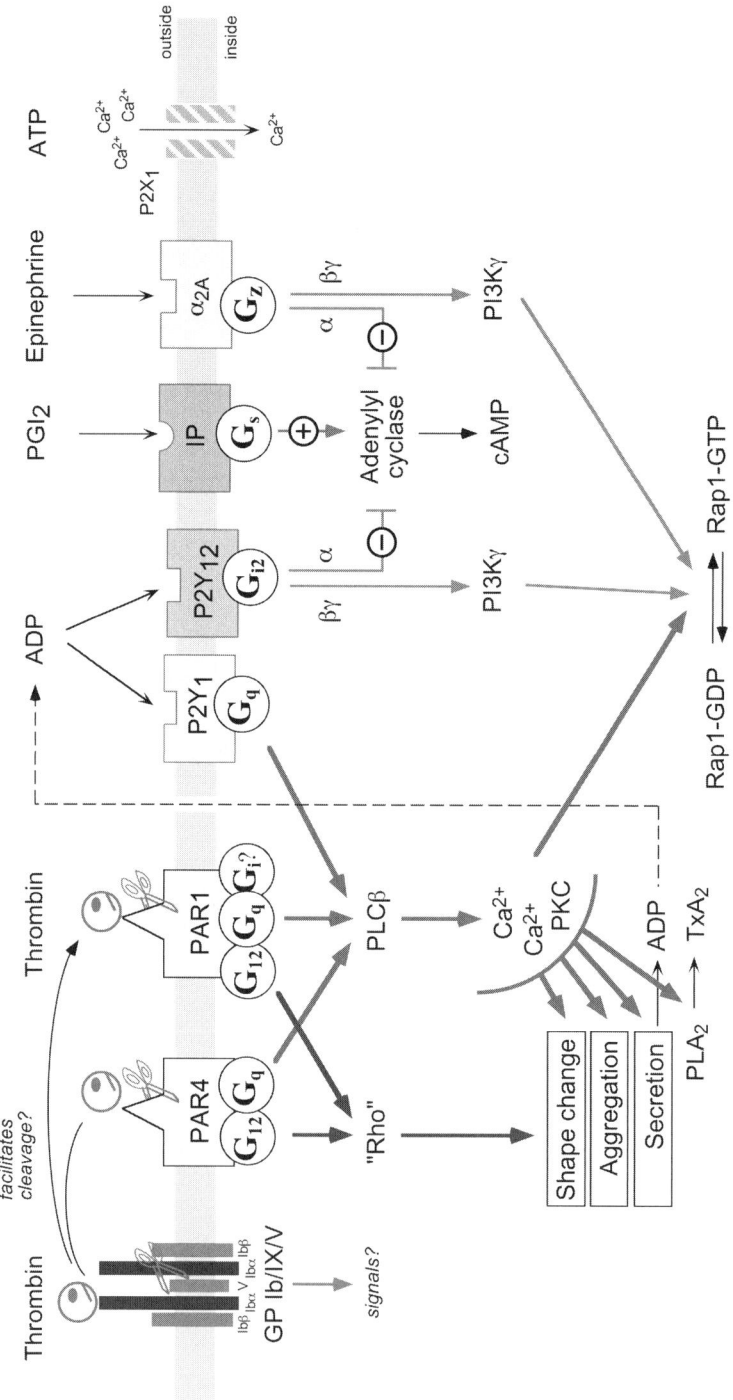

10

by either G_α or $G_{\beta\gamma}$ (or both). PLC-activating α subunits are typically members of the pertussis-toxin-resistant $G_{q\alpha}$ family. Phospholipase Cβ_2 can also be activated by $G_{\beta\gamma}$ derived from G_i-coupled receptors. However, in mouse platelets, loss of the gene encoding $G_{q\alpha}$ abolished the increase in Ca^{2+} otherwise caused by thrombin and ADP, despite continued signaling through G_i family members, leaving this an open issue—at least in mouse platelets *(21)*. Activation of phospholipase C can also release membrane-associated proteins that bind to PI-4,5-P$_2$ via their pleckstrin homology (PH) domains and lead to the PH-domain-dependent binding of other proteins as 3- and 5-phosphorylated polyphosphoinositides are rapidly synthesized by PI 3- and 5-kinases in response to platelet activation. These events bring molecules into new associations within signaling complexes, supporting the subsequent events of platelet activation.

The second task accomplished by platelet agonists via receptors coupled to G proteins is to trigger the reorganization of the actin cytoskeleton and microtubule network that not only underlies shape change, but may also be an important step in integrin activation. The details of this process are less well understood, but it appears that at least two potential effector pathways are involved: activation of myosin light-chain kinase downstream of G_q family members and activation of low-molecular-weight GTP-binding proteins in the Rho family, which occurs downstream of G_{12} family members *(22,23)*. Linking G_{12} family members to Rho family members may be one of several proteins that have both G_α-interacting domains and guanine nucleotide exchange factor (GEF) domains that bind to Rho and Rac. One example is p115RhoGEF, but others exist as well *(24)*. Their role in platelets has not been defined.

The third task accomplished by platelet agonists is to inhibit adenylyl cyclase if PGI$_2$ is present. Rising cAMP levels inhibit platelet responsiveness to agonists and an increase in cAMP synthesis is one of several mechanisms by which endothelial cells

---

Fig. 3. *(see opposite page)* Platelet activation via GPCRs and G proteins during recruitment of additional platelets into the platelet plug. A number of agonists that are capable of activating platelets do so via G protein-coupled receptors (GPCRs) on the platelet surface. Although the details vary, critical responses include G_q-mediated activation of phospholipase Cβ isoforms to allow an increase in cytosolic Ca^{2+}, activation of phospholipase A$_2$ and protein kinase C, and G_{12}-mediated guanine nucleotide exchange on Rho family members to support rearrangement of the platelet cytoskeleton (e.g., during shape change). Activated PGI$_2$ receptors (IP) stimulate adenylyl cyclase, raising platelet cAMP levels and causing a generalized inhibition of platelet responses to agonists. G_i family members are best known for their ability to support the suppression of adenylyl cyclase by platelet agonists, but also couple receptors to other effector pathways, including those that activate PI-3-kinaseγ and the Ras family member, Rap1B. Activation of human platelets by thrombin is shown to include the established role of two protease-activated receptor family members, PAR1 and PAR4, and the more speculative role of the GPIb-IX-V complex as a facilitator of PAR1 and PAR4 cleavage by thrombin or as a signaling complex in its own right. GPIb has been shown to have a high-affinity binding site for thrombin and GPV has been shown to be a substrate for thrombin.

prevent inappropriate platelet activation. PGI_2 released from endothelial cells causes a $G_{s\alpha}$-mediated increase in adenylyl cyclase activity. NO raises cGMP levels, inhibiting the hydrolysis of cAMP by phosphodiesterases *(25)*. When added to platelets in vitro, PGI_2 can cause as much as a 10-fold increase in the platelet cAMP concentration, but even small increases in basal cAMP levels can impair platelet activation by thrombin or ADP *(26–28)*. Therefore, it is perhaps not surprising that some platelet agonists inhibit PGI_2-stimulated cAMP synthesis. Chief among these are ADP and epinephrine, whose receptors are coupled to G_{i2} and G_z, respectively *(19,29)*. Loss of the genes encoding $G_{i2\alpha}$ or $G_{z\alpha}$ impairs responses to ADP and epinephrine. Conversely, loss of PGI_2 receptors (IP) predisposes mice to thrombosis in arterial injury models, ostensibly because of increased platelet responsiveness in vivo *(30)*. It is less clear that the ability to inhibit adenylyl cyclase is meaningful in the absence of PGI_2 *(31)*.

To summarize, platelet agonists working through G protein-coupled receptors activate phospholipase Cβ via $G_{q\alpha}$ and perhaps G_i-derived $G_{\beta\gamma}$; cause reorganization of the actin cytoskeleton via G_{12} family members and/or the products of G_q-mediated phosphoinositide hydrolysis; and suppress cAMP synthesis mediated by members of the G_i family **(Fig. 3)**. The full repertoire of events mediated by G_i family members is still being determined. It is clearly not limited to the inhibition of adenylyl cyclase: the impairment of platelet function that occurs in mice that are missing $G_{z\alpha}$ cannot be reversed by adding a membrane-permeable inhibitor of adenylyl cyclase *(31)*. Nor can such direct inhibitors restore ADP responses in the presence of $P2Y_{12}$ antagonists *(32)*. Therefore, it is likely that G_i family members regulate other effector pathways in platelets as well. Candidates for those effectors include the γ ($G_{\beta\gamma}$-activated) isoforms of PI 3-kinase, Src family members, and regulators of Rap1B **(Fig. 3)** *(20)*. As noted by others *(33,34)*, platelet responses to agonists such as thrombin, ADP, TxA_2, and epinephrine (including the three tasks of platelet activation noted above) can be mediated by a single class of receptors coupled to multiple different families of G proteins, by multiple classes of receptors each coupled to a single family of G proteins, or by the cumulative effect of two or more agonists, each of which evokes only a subset of the necessary G protein-mediated responses. The next section illustrates these principles and shows how gene deletions in mice have made it possible to dissect the relationships between receptors and G proteins in platelets.

3.2. ADP: Two G Protein-Coupled Receptors With Distinguishable Functions

ADP is stored in platelet dense (δ) granules and released upon platelet activation. It is also released from damaged red cells at sites of vascular injury. When added to platelets in vitro, ADP causes TxA_2 formation, protein phosphorylation, an increase in cytosolic Ca^{2+}, shape change, aggregation, and secretion. It also inhibits cAMP formation. However, ADP by itself is a weak activator of phospholipase C in human platelets *(35,36)*. According to current models, human platelets express two ADP-responsive purinergic G protein-coupled receptors, $P2Y_1$ and $P2Y_{12}$ (**Table 1, Fig. 3**) *(33,34,37–39)*. Optimal platelet activation by ADP requires activation of both. A third P2 purinergic receptor on platelets, $P2X_1$, is coupled directly to a nonselective cation channel (i.e., an

Table 2
Platelet Phenotypes of G Protein-Coupled Receptor Knockout Mice

Agonist	Receptor	Platelet phenotype	Reference
Thrombin	PAR1	None (not expressed on mouse platelets).	*166,167*
	PAR3	Higher concentrations of thrombin required to elicit responses, presumably because of the loss of the normal contribution of PAR3 to PAR4 activation.	*61*
	PAR4	Complete loss of thrombin responsiveness.	*66*
ADP	$P2Y_1$	Impaired response to ADP. Absent increase in cytosolic Ca^{2+} and shape change in response to ADP. Normal inhibition of cAMP formation by ADP.	*52,168*
	$P2Y_{12}$	Greatly diminished aggregation in response to ADP. Absent inhibition of adenylyl cyclase by ADP. Also impairs platelet responses to other agonists to the extent that they are dependent upon secreted ADP. Increased bleeding time.	*49*
TxA_2	$TP\alpha/\beta$	Prolonged bleeding time, absent aggregation in response to thromboxane A_2 agonists, and delayed aggregation with collagen.	*88*
PGI_2	IP	Increased carotid thrombosis following vascular injury with $FeCl_3$.	*30*

ionotropic rather than a G-protein-coupled receptor), and results in Ca^{2+} and Na^+ influx *(40–42)*. $P2X_1$ was initially identified as an ADP receptor, but is actually activated by ATP *(43)*. Its importance during platelet activation is unclear, but in theory could support platelet recruitment by responding to the ATP that is released from δ granules along with ADP *(44)*. $P2X_1$-deficient mice are viable, although difficult to breed due to reduced vas deferens contraction and thus low male fertility studies of $P2X_1$ transgenic mice suggest a role for this receptor in arterial thrombosis *(44a,b)*. Selective activation of $P2X_1$ receptors with exogenous α,β-meATP causes a transient increase in Ca^{2+} and shape change, but not detectable levels of aggregation *(45,46)*. $P2X_1$ receptors may also act by potentiating the signaling via other receptors, as proposed for $P2Y_1$ and collagen receptors *(47,170)*.

The genes encoding $P2Y_1$, $P2Y_{12}$, and $P2X_1$ have been deleted in mice (**Table 2**). In the absence of $P2Y_1$, ADP is still able to inhibit cAMP formation, but its ability to cause an increase in cytosolic Ca^{2+}, shape change, and aggregation is greatly impaired or eliminated—not unlike that which is observed in platelets from mice that lack G_q *(21)*. $P2Y_1(-/-)$ mice have a minimal increase in bleeding times and show some resistance to thromboembolic mortality following injection of ADP, but have no predisposition to spontaneous hemorrhage. Primary responses to platelet agonists other than ADP are unaffected and when combined with serotonin, which is a weak stimulus for phospholipase C in platelets, ADP causes the $P2Y_1(-/-)$ platelets to aggregate. Thus the results with the $P2Y_1$ knockout agree with earlier studies using antagonists *(33,48)* and suggest that platelet $P2Y_1$ receptors are coupled to G_q to activate phospholipase C.

Whether they also can couple to G_{12} family members in platelets is an open question because of the absence of ADP-induced shape change in platelets from $G_{q\alpha}(-/-)$ mice. The second platelet ADP receptor, $P2Y_{12}$, was independently cloned by two groups *(38,39)* and then knocked out in mice *(49)*. As predicted by inhibitor studies and the phenotype of a patient lacking functional $P2Y_{12}$ *(50)*, platelets from $P2Y_{12}(-/-)$ mice do not aggregate normally in response to ADP *(49)*. They retain $P2Y_1$-associated responses, but lack the ability to inhibit cAMP formation. The G_i family member associated with $P2Y_{12}$ responses appears to be primarily G_{i2}, since $G_{i2\alpha}(-/-)$ platelets have an impaired ability to aggregate and to inhibit PGI_2-stimulated cAMP formation *(29,31)*, while those lacking G_z or G_{i3} respond normally *(19,31)*.

The identification of the receptors that mediate platelet responses to ADP, the development of antagonists that target each of the known receptors, and the successful knockouts of the genes encoding $P2Y_1$, $P2Y_{12}$, $G_{q\alpha}$, and $G_{i2\alpha}$ have brought an increased appreciation of the contribution of ADP to platelet plug formation in vivo. Absence of $P2Y_{12}$ produces a hemorrhagic phenotype in humans, albeit a relatively mild one *(38,50,51)* (**Table 2**) and drugs that block $P2Y_{12}$ have proved to be useful as antiplatelet agents. Deletion of either $P2Y_1$ or $P2Y_{12}$ in mice prolongs the bleeding time and impairs platelet responses not only to ADP, but also to thrombin and TxA_2 *(49,52,53)*. Since platelet TxA_2 receptors do not couple directly to G_i family members, platelet aggregation induced by TxA_2 requires the secretion of ADP to inhibit adenylyl cyclase *(54)*. Lack of $P2Y_1$ improves survival in mouse models of disseminated thrombosis following the injection of platelet agonists *(52)* and tissue factor *(55)*.

3.3. Thrombin: (At Least) Two Receptors with Overlapping Functions

Thrombin is able to activate platelets at concentrations as low as 0.1 nM (approx 0.01 U/mL). Within seconds of the addition of thrombin, the cytosolic Ca^{2+} concentration increases 10-fold to approx 1 μM, triggering downstream Ca^{2+}-dependent events, including the activation of phospholipase A_2. Of these responses, aggregation, phosphoinositide hydrolysis, and the increase in cytosolic Ca^{2+}, but not shape change, have been shown to be abolished in platelets from mice lacking $G_{q\alpha}$ *(21)*. Thrombin also activates Rac and Rho in platelets, leading to rearrangement of the actin cytoskeleton and shape change—presumably mediated by G_{12} family members. Finally, thrombin is able to inhibit adenylyl cyclase activity in human platelets, either directly (via a G_i family member coupled to a thrombin receptor) *(56)* or indirectly (via released ADP) *(57)*.

Platelet responses to thrombin are largely mediated by protease-activated receptor (PAR) family members, although other proteins on the platelet surface may play a role as well (reviewed in **ref. 58**). Three of the four known PAR family members can be activated by thrombin. Of these, PAR1 and PAR4 are expressed on human platelets, while mouse platelets express PAR3 and PAR4 (**Table 1**). PAR1 and PAR4 activation occurs when thrombin binds to the extended N-terminus of the receptor, cleaving it and exposing a new N-terminus that serves as a tethered ligand *(59)*. Synthetic peptides based on the sequence of the tethered ligand domain of PAR1 and PAR4 are able to activate the receptors, mimicking many of the cellular actions of thrombin. When heterologously overexpressed, human PAR3 can respond to thrombin. However, on mouse platelets, PAR3 appears to serve primarily to promote cleavage of PAR4, as deletion of the gene encoding PAR3 causes a rightward shift in the thrombin dose/response curve *(60)*. PAR4 is expressed on human and mouse platelets and accounts for the continued ability of platelets from PAR3 knockout mice to respond to thrombin *(61,62)*. Activation of PAR4 requires 10- to 100-fold higher concentrations of thrombin than PAR1, in part because it lacks the hirudin-like sequences that can interact with thrombin's anion-binding exosite and facilitate receptor cleavage *(60–63)*. Kinetic studies suggest that on human platelets thrombin signals first through PAR1 and subsequently through PAR4, prolonging the duration of the thrombin response *(64)*.

Given what is known about them, are the PAR family members sufficient to account for platelet responses to thrombin? Clearly they are necessary: simultaneous blockade of PAR1 and PAR4 abolishes the response of human platelets to thrombin *(65)*, as does deletion of PAR4 in mice *(65)*. Both receptors appear to be coupled to G_{12} to cause shape change, while G_q and possibly G_i-derived $G_{\beta\gamma}$ activate phospholipase Cβ. So is there a reason to make room for another thrombin receptor on platelets? One recurring candidate is the GP Ib/IX/V complex. GP Ib is a heterodimer comprised of an α and a β subunit that are disulfide-linked. It forms a complex with GPIX and GPV in a $2:2:1$ (Ib:IX:V) ratio that serves as a binding site for vWF and as an anchor for the platelet cytoskeleton (reviewed in **ref. *67***). There is a high-affinity binding site for thrombin located at approx residues 268–287 on GPIbα, which is thought to interact with domains other than the active site *(68)*. Deletion of the sialic-acid-rich "glycocalicin" domain of GPIbα removes the thrombin-binding site and decreases platelet responses to thrombin, as does blockade of the binding site with antibodies and lectins *(69,72)*. Dörmann et al. *(73)* have suggested that the binding of thrombin to GPIb is needed for platelets to fully express procoagulant activity. The binding of thrombin to GPIbα could facilitate the cleavage of a PAR family member on human platelets, much as the binding of thrombin to PAR3 is thought to facilitate cleavage of PAR4 on mouse platelets. Consistent with that hypothesis, De Candia et al. *(72)* have shown that blockade of the interaction between thrombin and GPIb impairs the cleavage of PAR1 on human platelets. Thus, it remains possible that interactions with one or more members of the GPIb/IX/V complex may facilitate cleavage/activation of PAR1 or otherwise regulate platelet activation by thrombin.

3.4. Epinephrine: G_i-Mediated Responses That Potentiate Effects of Other Agonists

In contrast to ADP and thrombin, epinephrine is a weak activator of human platelets; when added to mouse platelets, it fails to cause aggregation at all. Platelet responses to epinephrine are mediated by α_{2A}-adrenergic receptors (**Table 1**) *(74–76)*. In both mice and humans, epinephrine is able to potentiate the effects of other agonists so that the combination of epinephrine with a suboptimal concentration of ADP or thrombin is a stronger stimulus for platelet aggregation than ADP or thrombin alone. Potentiation is usually attributed to the ability of epinephrine to inhibit cAMP formation, but this may be irrelevant unless PGI_2 is present *(31)*. Epinephrine has no detectable direct effect on phospholipase C in platelets and does not cause shape change, although it can trigger phosphoinositide hydrolysis indirectly by stimulating TxA_2 formation *(77)*. Taken together, these results suggest that platelet α_{2A}-adrenergic receptors are coupled to G_i family members, but not to G_q or G_{12} family members (**Fig. 3**).

Since platelets express four different members of the $G_{i\alpha}$ family, we have recently asked whether α_{2A}-adrenergic receptors are able to couple to all of them or only to a subset. The four $G_{i\alpha}$ family members in platelets are not expressed to the same extent. Based on immunoblotting, the relative levels of expression are $G_{i2\alpha} > G_{i3\alpha} >> G_{i1\alpha}$, with the fourth family member, $G_{z\alpha}$, expressed at levels intermediate between $G_{i2\alpha}$ and $G_{i3\alpha}$ *(78)*. $G_{z\alpha}$ is only about 60% identical to the others; it diverges particularly at the C-terminus, which is a domain critical for receptor interactions. $G_{z\alpha}$ is also notable for having a limited tissue distribution and a relatively slow rate of intrinsic GTP hydrolysis *(79)*, and for being a substrate for protein kinase C *(80)* and p21-activated kinase (PAK) *(81)*. When overexpressed in cells other than platelets, $G_{z\alpha}$ interacts with a variety of receptors, including α_{2A}-adrenergic receptors, serotonin receptors, and PAR1 *(56,82)*. To see whether these results predict what happens in platelets, we used homologous recombination to replace $G_{z\alpha}$ with green fluorescent protein *(19)*. Studies on these mice show normal responses to ADP and a PAR4-agonist peptide, but the ability of epinephrine to potentiate responses to other agonists is absent at concentrations likely to be encountered in vivo. The decreased aggregation is accompanied by a loss of epinephrine's normal ability to inhibit PGI_2-stimulated cAMP formation and by resistance to fatal thromboembolism following tail-vein injection of epinephrine plus collagen, but not ADP plus collagen *(19)*. In more recent studies, we have examined the effects of epinephrine on platelets from mice lacking $G_{i2\alpha}$ or $G_{i3\alpha}$ and found no decrease in the ability of epinephrine to potentiate aggregation or inhibit cAMP formation *(31)*. Therefore, it appears that in mouse platelets α_{2A}-adrenergic receptors are coupled to G_z, but not to G_{i2} or G_{i3} (**Fig. 3**).

3.5. Thromboxane Receptor(s)

TxA_2 is produced from arachidonate in platelets by the aspirin-sensitive cyclooxygenase pathway. When added to platelets in vitro, the TxA_2 analog U46619 causes shape change, aggregation, secretion, phosphoinositide hydrolysis, protein phosphorylation, and an increase in cytosolic Ca^{2+}. Once formed, TxA_2 can diffuse across the plasma membrane and activate other platelets (**Figs. 1–3**) *(83)*. This process is effective locally, but is limited by the brief (~30 s) half-life of TxA_2 in solution, helping to confine the spread of platelet activation to the original area of injury. Only one gene encoding a TxA_2 receptor (TP) has been identified, but there are two splice variants (TPα and TPβ) that differ in their cytoplasmic tails (**Table 1**) *(84)*. Human platelets express both *(84)*. Biochemical studies show that platelet TxA_2 receptors associate with $G_{q\alpha}$ *(85)* and $G_{13\alpha}$ *(86)*, and are able to activate G_{12} family members *(87)*. Loss of $G_{q\alpha}$ abolishes IP_3 formation, but does not prevent platelet shape change or guanine nucleotide exchange on $G_{12\alpha}/G_{13\alpha}$ *(21,22)*. The inhibitory effects of U46619 on PGI_2-stimulated cAMP formation appear to be mediated by secreted ADP. This suggests that platelet TP receptors are coupled to G_q and G_{12} family members, but not to G_i family members (**Fig. 3**). Platelets from mice lacking the gene encoding TP have a prolonged bleeding time and are unable to aggregate in response to thromboxane A_2 agonists (**Table 2**). They also show delayed aggregation with collagen, presumably reflecting the role of TxA_2 in platelet responses to collagen *(88)*.

4. Perpetuation of Platelet Signaling Within the Platelet Plug

Intracellular signaling does not cease when the hemostatic plug forms. Instead, the close proximity of one platelet to another makes possible a third wave of signaling that is dependent on stable contacts between platelets. Increasing evidence suggests that this helps to stabilize the platelet plug and perpetuate its growth. The last section of this chapter will briefly consider two examples of contact-dependent signaling: outside-in signaling through integrins and the signaling that arises through the interactions of Eph kinases and ephrins. Both of these involve tyrosine kinases.

4.1. Outside-In Signaling

In the context of platelets, outside-in signaling refers to the intracellular signaling events that occur downstream of activated $\alpha_{IIb}\beta_3$ once aggregation has begun. There is now ample evidence that outside-in signaling supports irreversible platelet aggregation and clot retraction. Since this topic has been reviewed in depth recently by others *(89–91)*, only a few points will be made here. Platelet aggregation results in the formation of large signaling and structural complexes that include $\alpha_{IIb}\beta_3$. One of the proteins in these complexes is the kinase FAK, which, when phosphorylated, can provide a binding site for (among other proteins) the p85 form of PI 3-kinase *(92)*. A number of other proteins have been identified that are capable of direct binding to the cytoplasmic domains of $\alpha_{IIb}\beta_3$, including β_3-endonexin *(93)*, CIB *(94,95)*, talin *(96)*, Syk *(97,98)*, myosin *(99)*, and Shc *(100)*. Shc binding requires phosphorylation of Y759 in the β_3 cytoplasmic domain *(100)*. Myosin binding requires phosphorylation of both Y747 and Y759 *(99)*. The binding of talin, β_3-endonexin, and CIB is thought to occur independent of β_3 tyrosine phosphorylation. Some of the protein:protein interactions that involve the cytoplasmic domains of $\alpha_{IIb}\beta_3$ help to regulate integrin activation; others are thought to participate in outside-in signaling and clot retraction. Tyrosine phosphorylation of β_3 is thought to be integral to the latter two events.

Tyrosines 747 and 759 of β_3 become phosphorylated following platelet aggregation *(99)*. Phosphorylation requires both activation of the integrin and its engagement with an adhesive protein. Inhibition of aggregation with an RGD-containing peptide or a peptidomimetic inhibits phosphorylation of β_3 *(99)*. According to the model proposed by Phillips and coworkers *(90,91)*, platelet activation and aggregation leads to the activation of the Src family member, Fyn, and the subsequent phosphorylation of β_3 by Fyn. Consistent with this, phosphorylation is diminished in Fyn(–/–) mouse platelets *(91)*. Mutation of both Y747 and Y759 to phenylalanine produces mice whose platelets tend to disaggregate and that show reduced clot retraction and a tendency to rebleed from tail bleeding time sites *(101)*. Clot retraction is dependent on platelets in general and a direct or indirect interaction of $\alpha_{IIb}\beta_3$ with myosin in particular. Therefore, the diminished clot retraction in the Y747F/Y759F mice is consistent with a reduction in phosphorylation-dependent binding of myosin to β_3. Failure to phosphorylate β_3 in the Y747F/Y759F mice would also prevent the binding of the adaptor protein Shc *(100)*, abolishing any signaling pathways that lie downstream of Shc. Although all the events and participants in outside-in signaling have not been identified, enough is known to justify its inclusion as an important participant in the late events of platelet-plug formation.

4.2. Eph Kinases and Ephrins

Eph kinases are a large family of cell-surface-receptor tyrosine kinases whose ligands (known as ephrins) are themselves held to the cell surface by either a glycophosphatidylinositol (GPI) anchor (the ephrin A family) or a transmembrane domain (the ephrin B family). The kinases are also divided into two groups and, with certain exceptions, the "A" ligands bind promiscuously to the "A" kinases and the "B" ligands bind to the "B" kinases *(102,103)*. Initially identified as orphan receptors, Eph kinases have been shown to play a critical role in neuronal patterning and axonal guidance during development *(104)* and to modulate adhesive interactions *(105–107)*. In addition, the expression of a subset of Eph kinases and ephrins in the developing vasculature has been shown to be an early marker that differentiates arteries from veins *(108–110)*. The recently reported crystal structure of ephrin B2 bound to EphB2 *(111)* confirms what was already suspected: The binding of an ephrin on one cell to an Eph kinase on another cell favors formation of receptor/ligand dimers (and higher-order structures) *(112)*. Clustering triggers signaling and responses in both the receptor-expressing cells and the ligand-expressing cells *(113–119)*. These events can be dependent on the phosphorylation of the Eph and ephrin (at least for the B ephrins), but phosphorylation-independent interactions have also been described, including those mediated by the PDZ target domains at the C-terminus of the Eph kinases and the B ephrins *(114–117,120–122)*. Phosphorylation of ephrin B family members occurs on conserved tyrosine residues within the cytoplasmic domain and can be mediated by members of the Src family *(123,124)* or by growth factor receptors *(125,126)*. Eph kinases in both the A and B families become phosphorylated on tyrosine residues when clustering occurs. Although this provides a clear role for the kinase activity of the Eph, examples have now been reported in which kinase gene knockout phenotypes can be partially reversed with kinase-dead receptor variants *(120,127,128)*. Eph kinases and the ephrin B family members have SAM domains that support lateral clustering *(129–132)*. Taken together, phosphorylated tyrosine residues, PDZ domains, and SAM domains make possible the regulated formation of large signaling complexes around clustered Eph kinases and ephrins.

Several years ago we proposed that, if expressed on platelets in suitable combinations, Eph kinases and ephrins could interact *in trans* (the term "*in trans*" being used to describe an interaction in which a ligand on one cell binds to a receptor on another cell, rather than an interaction "*in cis*," which would imply that a surface-bound ligand could somehow bind to and activate a receptor on the surface of the same cell) once platelet aggregation had occurred and provide a novel mechanism for contact-dependent signaling that might contribute to the stability of the platelet plug. Our observations to date suggest that this is likely the case. Human platelets express two Eph kinases (A4 and B1) and at least one ephrin (B1) that can serve as a ligand for both *(133)*. Forced clustering of either EphA4 or ephrin B1 causes platelets to adhere to immobilized fibrinogen and secrete the contents of their α- and δ-granules, although ephrin or Eph kinase activation by itself does not cause an increase in cytosolic Ca^{2+} and does not put $α_{IIb}β_3$ into the high-affinity state needed to support platelet aggregation. Activation of platelets by ADP results in the formation of signaling complexes with EphA4 that include two tyrosine kinases, Fyn and Lyn, and the cell adhesion molecule, L1/Ng-CAM *(133)*. Since L1 can in turn bind *in trans* to activated $α_{IIb}β_3$, this provides an additional mechanism to reinforce platelet:platelet interactions *(134)*. Finally, interruption of Eph/ephrin interactions causes platelets activated by ADP to disaggregate *(133)* and inhibits clot retraction (N. Prevost et al., unpublished observation). Taken together, these results suggest that once sustained platelet:platelet contacts have occurred, Eph/ephrin interactions, along with outside-in signaling, help to perpetuate platelet aggregation and stabilize the hemostatic plug **(Fig. 4)**.

5. Signaling in Megakaryocytes

Receptors on the surface of megakaryocytes can potentially perform two roles. Some, such as thrombopoietin receptors, play a clear role in megakaryocyte growth and development *(135)*. Others, including thrombin receptors, do not have an obvious role in megakaryocytes, but may appear on the megakaryocyte surface as a prelude to incorporation into what will become the platelet plasma membrane *(136,137)*. Since relatively little protein synthesis occurs in platelets, most if not all of the receptors present on the surface of platelets were synthesized by megakaryocytes. Integrins, particularly $α_{IIb}β_3$, appear on the surface of megakaryocytes relatively early and are sometimes used as a marker of megakaryocyte development. Based on limited information, it appears that receptor function is acquired as a late event *(138)*. Human and murine megakaryocytes will respond to thrombin, which can cause activation of $α_{IIb}β_3$ *(139,140)*. Shattil and co-workers *(140)* have shown that signaling through Rap plays a role in this response. Studies on a number of megakaryoblastic cell lines, such as HEL, CHRF-288, and Meg-01, show that these cells also express $α_{IIb}β_3$ as well as at least some of the agonist receptors present on platelets (e.g., **refs.** *141–144*). Thrombin, for example, can activate PAR1 on some of the cell lines and cause phosphoinositide hydrolysis and an increase in cytosolic Ca^{2+} *(141,142)*. In general, the megakaryoblastic cell lines that are most widely available can bind to immobilized fibrinogen, but lack the ability to activate $α_{IIb}β_3$ and bind soluble fibrinogen *(143)*. This may mean that key intermediates required for inside-out signaling are either not expressed or not fully functional. One advantage that the megakaryoblastic cell lines have over platelets is that they can be transfected, but although studies with them can be useful, caution needs to be applied in extrapolating the results to platelets. For this reason, many laboratories have put considerable effort into the development of in vitro cell cultures of megakaryocyte progenitor cells, which exhibit more characteristics of primary megakaryocytes and in a few instances have been shown to manufacture functional platelets *(145)*. These culture systems provide enormous scope for in vitro genetic modification of proteins to study megakaryocyte and platelet signaling pathways *(146,147)*.

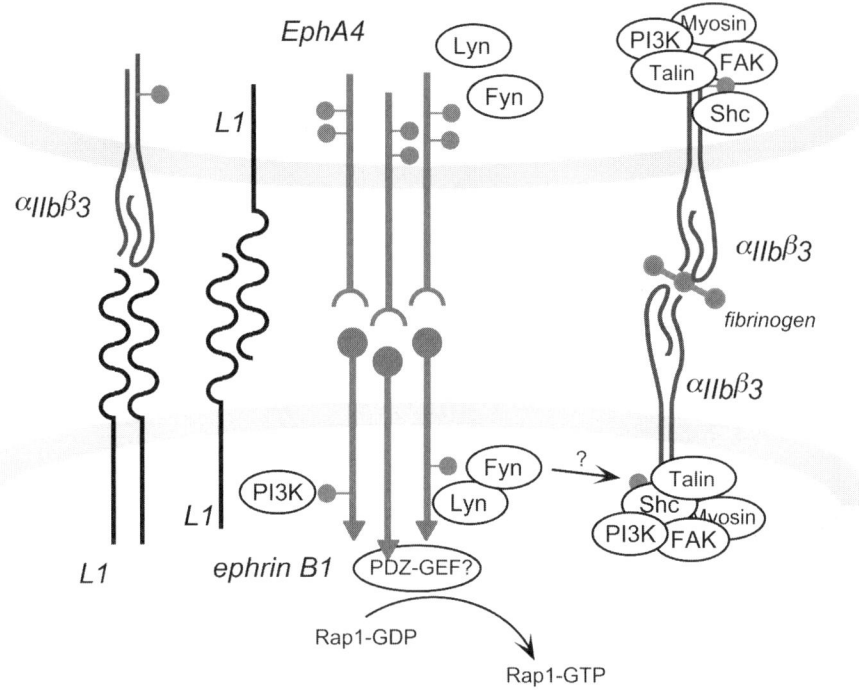

Fig. 4. Inside-out and Eph/ephrin signaling help to perpetuate platelet plug formation. Once platelets are activated and incorporated into a growing platelet plug, close
cell-to-cell contact appears to make possible additional types of signaling mechanisms.
One of these involves outside-in signaling via the $\alpha_{IIb}\beta_3$ integrin (illustrated on the
right side of the figure). Another involves signaling through the family of Eph kinases
and ephrins, both of which are represented on the surface of platelets. In each case, outside-in signaling and Eph/ephrin interactions result in the formation of signaling complexes and the recruitment of other membrane molecules. Outside-in signaling
complexes include Shc, talin, myosin, and FAK, some of which are regulated by the
phosphorylation of tyrosine residues within the β3 cytoplasmic domain. Molecules that
become associated with EphA4 during platelet aggregation include nonreceptor tyrosine
kinases, Fyn and Lyn, and the cell adhesion molecule, L1. PI 3-kinase and PDZ-GEF
(an exchange factor for ras family members including Rap1) have been shown to associate with ephrin B family members in cells other than platelets, but not yet shown to
do so in platelets. L1 has binding sites for heterotypic interactions with $\alpha_{IIb}\beta_3$ or $\alpha_v\beta_3$,
as well as for homotypic interactions *in trans* with L1 on adjacent platelets. The impact
of these latter interactions for the perpetuation of platelet aggregation remain to be
demonstrated.

6. Conclusion

In summary, the intracellular signaling that underlies platelet activation is a dynamic process with different receptor and effector pathways dominant at different phases in the initiation, extension, and perpetuation of platelet plugs. The essential problem throughout is to activate $\alpha_{IIb}\beta_3$ so that it can bind adhesive proteins, and then maintain it in an active (bound) state so that platelet plugs will remain stable long enough for wound healing to occur. In general, this requires the activation of phospholipase C and PI 3-kinase-dependent pathways and, perhaps, the rearrangement of the actin cytoskeleton. It also involves the suppression of inhibitory mechanisms normally designed to prevent platelet activation, including the formation of cAMP by adenylyl cyclase. If the initial stimulus for platelet activation is the exposure of collagen and von Willebrand factor, then $\alpha_{IIb}\beta_3$ activation is accomplished by a process that involves activation of PLCγ. If the initial stimulus is the generation of thrombin, as can be the case in pathological states, then a G protein-dependent mechanism results in a more rapid and robust activation of PLCβ—the same signaling mechanism that supports the recruitment of circulating platelets by secreted ADP and released TxA$_2$. Once $\alpha_{IIb}\beta_3$ has been activated and platelet aggregation has occurred, then close cell-to-cell contacts between platelets make possible a third wave of signaling, which includes outside-in signaling through integrins and the binding of ephrins to Eph kinases. All these events require proteins on the platelet surface that can serve as receptors for agonists or adhesive ligands.

References

1. Clemetson, J. M., Polgar, J., Magnenat, E., Wells, T. N. C., and Clemetson, K. J. (1999) The platelet collagen receptor glycoprotein VI is a member of the immunoglobulin superfamily closely related to FcαR and the natural killer receptors. *J. Biol. Chem.* **274,** 29,019–29,024.
2. Poole, A., Gibbins, J. M., Turner, M., Van Vugt, M. J., Van de Winkel, J. G. J., Saito, T., et al. (1997) The Fc receptor gamma-chain and the tyrosine kinase Syk are essential for activation of mouse platelets by collagen. *EMBO J.* **16,** 2333–2341.
3. Nieuwenhuis, H. K., Akkerman, J. W. N., Houdijk, W. P. M., and Sixma, J. J. (1985) Human blood platelets showing no response to collagen fail to express glycoprotein Ia. *Nature* **318,** 470–472.
4. Sixma, J. J., Van Zanten, G. H., Huizinga E. G., Van der Plas, R. M., Verkley, M., Wu, Y. P., et al. (1997) Platelet adhesion to collagen: An update. *Thromb. Haemost.* **78,** 434–438.
5. Nieswandt, B., Brakebusch, C., Bergmeier, W., Schulte, V., Bouvard, D., Mokhtari-Nejad, R., et al. (2001) Glycoprotein VI but not α2β1 integrin is essential for platelet interaction with collagen. *EMBO J.* **20,** 2120–2130.
5a. Kato, K., Kanaji, T., Russell, S., Kunicki, T. J., Furihata, K., Kanaji, S., et al. (2003) The contribution of glycoprotein VI to stable platelet adhesion and thrombus formation illustrated by targeted gene deletion. *Blood* **102,** 1701–1707.
6. Nieswandt, B., Bergmeier, W., Schulte, V., Rackebrandt, K., Gessner, J. E., and Zirngibl, H. (2000) Expression and function of the mouse collagen receptor glycoprotein VI is strictly dependent on its association with the FcRγ chain. *J. Biol. Chem.* **275,** 23,998–24,002.
7. Gross, B. S., Lee, J. R., Clements, J. L., Turner, M., Tybulewicz, V. L. J., Findell, P. R., et al. (1999) Tyrosine phosphorylation of SLP-76 is downstream of Syk following stimulation of the collagen receptor in platelets. *J. Biol. Chem.* **274,** 5963–5971.

8. Hamm, H. E. (2001) How activated receptors couple to G proteins. *Proc. Natl. Acad. Sci. USA* **98,** 4819–4821.

9. Gomes, I., Jordan, B. A., Gupta, A., Rios, C., Trapaidze, N., and Devi, L. A. (2001) G protein coupled receptor dimerization: implications in modulating receptor function. *J. Mol. Med.* **79,** 226–242.

10. Jordan, B. A. and Devi, L. A. (1999) G-protein-coupled receptor heterodimerization modulates receptor function. *Nature* **399,** 697–700.

11. Bünemann, M., Lee, K. B., Pals-Rylaarsdam, R., Roseberry, A. G., and Hosey, M. M. (1999) Desensitization of G-protein-coupled receptors in the cardiovascular system. *Annu. Rev. Physiol.* **61,** 169–192.

12. Penn, R. B., Pronin, A. N., and Benovic, J. L. (2000) Regulation of G protein-coupled receptor kinases. *Trends Cardiovasc. Med.* **10,** 81–89.

13. Lambright, D. G., Sondek, J., Bohm, A., Skiba, N. P., Hamm, H. E., and Sigler, P. B. (1996) The 2.0 Å crystal structure of a heterotrimeric G protein. *Nature* **379,** 311–319.

14. Ford, C. E., Skiba, N. P., Bae, H. S., Daaka, Y. H., Reuveny, E., Shekter, L. R., et al. (1998) Molecular basis for interactions of G protein βγ subunits with effectors. *Science* **280,** 1271–1274.

15. Hamm, H. E. (1998) The many faces of G protein signaling. *J. Biol. Chem.* **273,** 669–672.

16. Gilman, A. G. (1987) G proteins: transducers of receptor-generated signals. *Ann. Rev. Biochem.* **56,** 615–649.

17. Ross, E. M. and Wilkie, T. M. (2000) GTPase-activating proteins for heterotrimeric G proteins: Regulators of G protein signaling (RGS) and RGS-like proteins. *Annu. Rev. Biochem.* **69,** 795–827.

18. Jantzen, H. M., Milstone, D. S., Gousset, L., Conley, P. B., and Mortensen, R. M. (2001) Impaired activation of murine platelets lacking $G\alpha_{i2}$. *J. Clin. Invest.* **108,** 477–483.

19. Yang, J., Wu, J., Kowalska, M. A., Dalvi, A., Prevost, N., O'Brien, P. J., et al. (2000) Loss of signaling through the G protein, G_z, results in abnormal platelet activation and altered responses to psychoactive drugs. *Proc. Natl. Acad. Sci. USA* **97,** 9984–9989.

20. Woulfe, D., Jiang, H., Mortensen, R., Yang, J., and Brass, L. F. (2002) Activation of Rap1B by Gi family members in platelets. *J. Biol. Chem.* **277,** 23,382–23,390.

21. Offermanns, S., Toombs, C. F., Hu, Y. H., and Simon, M. I. (1997) Defective platelet activation in $G\alpha_q$-deficient mice. *Nature* **389,** 183–186.

22. Klages, B., Brandt, U., Simon, M. I., Schultz, G., and Offermanns, S. (1999) Activation of G_{12}/G_{13} results in shape change and Rho/Rho-kinase-mediated myosin light chain phosphorylation in mouse platelets. *J. Cell Biol.* **144,** 745–754.

23. Offermanns, S. (2001) In vivo functions of heterotrimeric G-proteins: studies in Gα-deficient mice. *Oncogene* **20,** 1635–1642.

24. Fukuhara, S., Chikumi, H., and Gutkind, J. S. (2001) RGS-containing RhoGEFs: the missing link between transforming G proteins and Rho? *Oncogene* **20,** 1661–1668.

25. Haslam, R. J., Dickinson, N. T., and Jang, E. K. (1999) Cyclic nucleotides and phosphodiesterases in platelets. *Thromb. Haemost.* **82,** 412–423.

26. Mills, D. C. B. and Smith, J. B. (1971) The influence on platelet aggregation of drugs that affect the accumulation of adenosine 3′:5′ cyclic monophosphate in platelets. *Biochem. J.* **121,** 185.

27. Eigenthaler, M., Nolte, C., Halbrugge, M., and Walter, U. (1992) Concentration and regulation of cyclic nucleotides, cyclic-nucleotide-dependent protein kinases and one of their

major substrates in human platelets. Estimating the rate of cAMP-regulated and cGMP-regulated protein phosphorylation in intact cells. *Eur. J. Biochem.* **205,** 471–481.

28. Keularts, I. M. L. W., Van Gorp, R. M. A., Feijge, M. A. H., Vuist, W. M. J., and Heemskerk, J. W. M. (2000) α_{2A}-adrenergic receptor stimulation potentiates calcium release in platelets by modulating cAMP levels. *J. Biol. Chem.* **275,** 1763–1772.

29. Jantzen, H.-M., Milstone, D. S., Gousset, L., Conley, P. B., and Mortensen, R. M. (2001) Impaired activation of murine platelets lacking $G_{\alpha i2}$. *J. Clin. Invest.* **108,** 477–483.

30. Murata, T., Ushikubi, F., Matsuoka, T., Hirata, M., Yamasaki, A., Sugimoto, Y., et al. (1997) Altered pain perception and inflammatory response in mice lacking prostacyclin receptor. *Nature* **388,** 678–682.

31. Yang, J., Wu, J., Mortensen, R., Austin, S., Manning, D. R., Woulfe, D., et al. (2002) Signaling through G_i family members in platelets: redundancy and specificity in the regulation of adenylyl cyclase and other effectors. *J. Biol. Chem.* **277,** 46,035–46,042.

32. Daniel, J. L., Dangelmaier, C., Jin, J. G., Kim, Y. B., and Kunapuli, S. P. (1999) Role of intracellular signaling events in ADP-induced platelet aggregation. *Thromb. Haemost.* **82,** 1322–1326.

33. Daniel, J. L., Dangelmaier, C., Jin, J. G., Ashby, B., Smith, J. B., and Kunapuli, S. P. (1998) Molecular basis for ADP-induced platelet activation I. Evidence for three distinct ADP receptors on human platelets. *J. Biol. Chem.* **273,** 2024–2029.

34. Jin, J. G., Daniel, J. L., and Kunapuli, S. P. (1998) Molecular basis for ADP-induced platelet activation II. The P2Y1 receptor mediates ADP-induced intracellular calcium mobilization and shape change in platelets. *J. Biol. Chem.* **273,** 2030–2034.

35. Fisher, G. J., Bakshian, S., and Baldassare, J. J. (1985) Activation of human platelets by ADP causes a rapid rise in cytosolic free calcium without hydrolysis of phosphatidylinositol-4,5-bisphosphate. *Biochem. Biophys. Res. Commun.* **129,** 958–964.

36. Daniel, J. L., Dangelmaier, C. A., Selak, M., and Smith, J. B. (1986) ADP stimulates IP_3 formation in human platelets. *FEBS Lett.* **206,** 299–303.

37. Léon, C., Hechler, B., Vial, C., Leray, C., Cazenave, J. P., and Gachet, C. (1997) The $P2Y_1$ receptor is an ADP receptor antagonized by ATP and expressed in platelets and megakaryoblastic cells. *FEBS Lett.* **403,** 26–30.

38. Hollopeter, G., Jantzen, H. M., Vincent, D., Li, G., England, L., Ramakrishnan, V., et al. (2001) Identification of the platelet ADP receptor targeted by antithrombotic drugs. *Nature* **409,** 202–207.

39. Zhang, F. L., Luo, L., Gustafson, E., Lachowicz, J., Smith, M., Qiao, X. D., et al. (2001) ADP is the cognate ligand for the orphan G protein-coupled receptor SP1999. *J. Biol. Chem.* **276,** 8608–8615.

40. MacKenzie, A. B., Mahaut-Smith, M. P., and Sage, S. O. (1996) Activation of receptor-operated channels via P_{2X1} not P_{2T} purinoreceptors in human platelets. *J. Biol. Chem.* **271,** 2879–2881.

41. Vial, C., Hechler, B., Léon, C., Cazenave, J. P., and Gachet, C. (1997) Presence of $P2X_1$ purinoceptors in human platelets and megakaryoblastic cell lines. *Thromb. Haemost.* **78,** 1500–1504.

42. Sun, B., Li, J., Okahara, K., and Kambayashi, J. (1998) P2X1 purinoceptor in human platelets—Molecular cloning and functional characterization after heterologous expression. *J. Biol. Chem.* **273,** 11,544–11,547.

43. Mahaut-Smith, M. P., Ennion, S. J., Rolf, M. G., and Evans, R. J. (2000) ADP is not an agonist at $P2X_1$ receptors: evidence for separate receptors stimulated by ATP and ADP on human platelets. *Br. J. Pharmacol.* **131,** 108–114.

44. Oury, C., Toth-Zsamboki, E., Thys, C., Tytgat, J., Vermylen, J., and Hoylaerts, M. F. (2001) The ATP-gated $P2X_1$ ion channel acts as a positive regulator of platelet responses to collagen. *Thromb. Haemost.* **86,** 1264–1271.

44a. Hechler, B., Lenain, N., Marchese, P., Vial, C., Heim, V., Freund, M., et al. (2003). A role of the fast ATP-gated $P2X_1$ cation channel in thrombosis of small arteries in vivo. *J. Exp. Med.* **198,** 661–667.

44b. Oury, C., Kuijpers, M. J., Toth-Zsamboki, E., Bonnefoy, A., Danloy, S., Vreys, I., et al. (2003). Overexpression of the platelet $P2X_1$ ion channel in transgenic mice generates a novel prothrombotic phenotype. *Blood.* **101,** 3969–3976.

45. Rolf, M. G., Brearley, C. A., and Mahaut-Smith, M. P. (2001) Platelet shape change evoked by selective activation of $P2X_1$ purinoceptors with α,β-methylene ATP. *Thromb. Haemost.* **85,** 303–308.

46. Rolf, M. G. and Mahaut-Smith, M. P. (2002) Effects of enhanced $P2X_1$ receptor Ca^{2+} influx on functional responses in human platelets. *Thromb. Haemost.* **88,** 495–503.

47. Oury, C., Toth-Zsamboki, E., Thys, C., Tytgat, J., Vermylen, J., and Hoylaerts, M. F. (2001) The ATP-gated $P2X_1$ ion channel acts as a positive regulator of platelet responses to collagen. *Thromb. Haemost.* **86,** 1264–1271.

48. Jin, J. G. and Kunapuli, S. P. (1998) Coactivation of two different G protein-coupled receptors is essential for ADP-induced platelet aggregation. *Proc. Natl. Acad. Sci. USA* **95,** 8070–8074.

49. Foster, C. J. (2001) Molecular identification and characterization of the platelet ADP receptor targeted by thienopyridine drugs using P2Yac-null mice. *J. Clin. Invest.* **107,** 1591–1598.

50. Nurden, P., Savi, P., Heilmann, E., Bihour, C., Herbert, J-M., Maffrand, J.-P., et al. (1995) An inherited bleeding disorder linked to a defective interaction between ADP and its receptor on platelets. Its influence on glycoprotein IIb-IIIa complex function. *J. Clin. Invest.* **95,** 1612–1622.

51. Cattaneo, M. and Gachet, C. (1999) ADP receptors and clinical bleeding disorders. *Arterioscler. Thromb. Vasc. Biol.* **19,** 2281–2285.

52. Léon, C., Hechler, B., Freund, M., Eckly, A., Vial, C., Ohlmann, P., et al. (1999) Defective platelet aggregation and increased resistance to thrombosis in purinergic $P2Y_1$ receptor-null mice. *J. Clin. Invest.* **104,** 1731–1737.

53. Fabre, J. E., Nguyen, M. T., Latour, A., Keifer, J. A., Audoly, L. P., Coffman, T. M., et al. (1999) Decreased platelet aggregation, increased bleeding time and resistance to thrombo-embolism in $P2Y_1$-deficient mice. *Nature Med.* **5,** 1199–1202.

54. Paul, B. Z. S., Jin, J. G., and Kunapuli, S. P. (1999) Molecular mechanism of thromboxane A_2-induced platelet aggregation—Essential role for $P2T_{AC}$ and $α_{2A}$ receptors. *J. Biol. Chem.* **274,** 29,108–29,114.

55. Léon, C., Freund, M., Ravanat, C., Baurand, A., Cazenave, J. P., and Gachet, C. (2001) Key role of the $P2Y_1$ receptor in tissue factor-induced thrombin-dependent acute thrombo-embolism: Studies in P2Y1-knockout mice and mice treated with a $P2Y_1$ antagonist. *Circulation* **103,** 718–723.

56. Barr, A. J., Brass, L. F., and Manning, D. R. (1997) Reconstitution of receptors and GTP-binding regulatory proteins (G proteins) in Sf9 cells—A direct evaluation of selectivity in receptor. G protein coupling. *J. Biol. Chem.* **272,** 2223–2229.

57. Kim, S., Quinton, T. M., Cattaneo, M., and Kunapuli, S. P. (2000) Evidence for diverse signal transduction pathways in thrombin receptor activating peptide (SFLLRN) and other agonist-induced fibrinogen receptor activation in human platelets. *Blood* **96,** 242a(Abstract).

58. O'Brien, P. J., Molino, M., Kahn, M., and Brass, L. F. (2001) Protease activated receptors: theme and variations. *Oncogene* **20,** 1570–1581.
59. Vu, T.-K. H., Hung, D. T., Wheaton, V. I., and Coughlin, S. R. (1991) Molecular cloning of a functional thrombin receptor reveals a novel proteolytic mechanism of receptor activation. *Cell* **64,** 1057–1068.
60. Nakanishi-Matsui, M., Zheng, Y. W., Sulciner, D. J., Weiss, E. J., Ludeman, M. J., and Coughlin, S. R. (2000) PAR3 is a cofactor for PAR4 activation by thrombin. *Nature* **404,** 609–610.
61. Kahn, M. L., Zheng, Y. W., Huang, W., Bigornia, V., Zeng, D. W., Moff, S., et al. (1998) A dual thrombin receptor system for platelet activation. *Nature* **394,** 690–694.
62. Xu, W.-F., Andersen, H., Whitmore, T. E., Presnell, S. R., Yee, D. P., Ching, A. C., et al. (1998) Cloning and characterization of human protease-activated receptor 4. *Proc. Natl. Acad. Sci. USA* **95,** 6642–6646.
63. Ishii, K., Gerszten, R., Zheng, Y. W., Welsh, J. B., Turck, C. W., and Coughlin, S. R. (1995) Determinants of thrombin receptor cleavage. Receptor domains involved, specificity, and role of the P3 aspartate. *J. Biol. Chem.* **270,** 16,435–16,440.
64. Covic, L., Gresser, A. L., and Kuliopulos, A. (2000) Biphasic kinetics of activation and signaling for PAR1 and PAR4 thrombin receptors in platelets. *Biochemistry* **39,** 5458–5467.
65. Kahn, M. L., Nakanishi-Matsui, M., Shapiro, M. J., Ishihara, H., and Coughlin, S. R. (1999) Protease-activated receptors 1 and 4 mediate activation of human platelets by thrombin. *J. Clin. Invest.* **103,** 879–887.
66. Sambrano, G. R., Weiss, E. J., Zheng, Y.-W., Huang, W., and Coughlin, S. R. (2001) Role of thrombin signaling in platelets in hemostasis and thrombosis. *Nature* **413,** 74–78.
67. Lopez, J. A., Andrews, R. K., Afshar-Khargan, V., and Berndt, M. C. (1998) Bernard-Soulier syndrome. *Blood* **91,** 4397–4418.
68. De Cristofaro, R., De Candia, E., Rutella, S., and Weitz, J. I. (2000) The Asp^{272}-Glu^{282} region of platelet glycoprotein Ibα interacts with the heparin-binding site of α-thrombin and protects the enzyme from the heparin-catalyzed inhibition by antithrombin III. *J. Biol. Chem.* **275,** 3887–3895.
69. De Marco, L., Mazzucato, M., Masotti, A., Fenton, J. W. II, and Ruggeri, Z. M. (1991) Function of glycoprotein Ibα in platelet activation induced by α-thrombin. *J. Biol. Chem.* **266,** 23,776–23,783.
70. Harmon, J. T. and Jamieson, G. A. (1988) Platelet activation by thrombin in the absence of the high affinity thrombin receptor. *Biochemistry* **27,** 2151–2157.
71. Mazzucato, M., De Marco, L., Masotti, A., Pradella, P., Bahou, W. F., and Ruggeri, Z. M. (1998) Characterization of the initial α-thrombin interaction with glycoprotein Ibα in relation to platelet activation. *J. Biol. Chem.* **273,** 1880–1887.
72. De Candia, E., Hall, S. W., Rutella, S., Landolfi, R., Andrews, R. K., and De Cristofaro, R. (2001) Binding of thrombin to glycoprotein Ib accelerates hydrolysis of PAR1 on intact platelets. *J. Biol. Chem.* **276,** 4692–4698.
73. Dörmann, D., Clemetson, K. J., and Kehrel, B. E. (2000) The GPIb thrombin-binding site is essential for thrombin-induced platelet procoagulant activity. *Blood* **96,** 2469–2478.
74. Newman, K. D., Williams, L. T., Bishopric, N. H., and Lefkowitz, R. J. (1978) Identification of α-adrenergic receptors in human platelets by ^{3}H-dihydroergocryptine binding. *J. Clin. Invest.* **61,** 395–402.
75. Kaywin, P., McDonough, M., Insel, P. A., and Shattil, S. J. (1978) Platelet function in essential thrombocythemia: decreased epinephrine responsivenesss associated with a deficiency of platelet alpha-adrenergic receptors. *N. Engl. J. Med.* **299,** 505–509.

76. Motulsky, H. J. and Insel, P. A. (1982) [³H]Dihydroergocryptine binding to alpha-adrenergic receptors of human platelets. A reassessment using the selective radioligands [³H]prazosin, [³H]yohimbine, and [³H]rauwolscine. *Biochem. Pharmacol.* **31,** 2591–2597.

77. Siess, W., Weber, P. C., and Lapetina, E. G. (1984) Activation of phospholipase C is dissociated from arachidonate metabolism during platelet shape change induced by thrombin or platelet-activating factor. Epinephrine does not induce phospholipase C activation or platelet shape change. *J. Biol. Chem.* **259,** 8286–8292.

78. Williams, A., Woolkalis, M. J., Poncz, M., Manning, D. R., Gewirtz, A., and Brass, L. F. (1990) Identification of the pertussis toxin-sensitive G proteins in platelets, megakaryocytes and HEL cells. *Blood* **76,** 721–730.

79. Casey, P. J., Fong, H. K. W., Simon, M. I., and Gilman, A. G. (1990) G_z, a guanine nucleotide-binding protein with unique biochemical properties. *J. Biol. Chem.* **265,** 2383–2390.

80. Lounsbury, K. M., Casey, P. J., Brass, L. F., and Manning, D. R. (1991) Phosphorylation of G_z in human platelets: selectivity and site of modification. *J. Biol. Chem.* **266,** 22,051–22,056.

81. Wang, J., Frost, J. A., and Ross, E. M. (1999) Reciprocal signaling between heterotrimeric G proteins and the p21-stimulated protein kinase. *J. Biol. Chem.* **274,** 31,641–31,647.

82. Ho, M. K. C. and Wong, Y. H. (2001) G_z signaling: emerging divergence from G_i signaling. *Oncogene* **20,** 1615–1625.

83. FitzGerald, G. A. (1991) Mechanisms of platelet activation: Thromboxane A_2 as an amplifying signal for other agonists. *Am. J. Cardiol.* **68,** 11B–15B.

84. Hirata, T., Ushikubi, F., Kakizuka, A., Okuma, M., and Narumiya, S. (1996) Two thromboxane A_2 receptor isoforms in human platelets—Opposite coupling to adenylyl cyclase with different sensitivity to Arg[60] to Leu mutation. *J. Clin. Invest.* **97,** 949–956.

85. Knezevic, I., Borg, C., and Le Breton, G. C. (1993) Identification of G_q as one of the G-proteins which copurify with human platelet thromboxane A_2/prostaglandin H_2 receptors. *J. Biol. Chem.* **268,** 26,011–26,017.

86. Djellas, Y., Manganello, J. M., Antonakis, K., and Le Breton, G. C. (1999) Identification of $G\alpha_{13}$ as one of the G-proteins that couple to human platelet thromboxane A_2 receptors. *J. Biol. Chem.* **274,** 14,325–14,330.

87. Offermanns, S., Laugwitz, K.-L., Spicher, K., and Schultz, G. (1994) G proteins of the G_{12} family are activated via thromboxane A_2 and thrombin receptors in human platelets. *Proc. Natl. Acad. Sci. USA* **91,** 504–508.

88. Thomas, D. W., Mannon, R. B., Mannon, P. J., Latour, A., Oliver, J. A., Hoffman, M., et al. (1998) Coagulation defects and altered hemodynamic responses in mice lacking receptors for thromboxane A_2. *J. Clin. Invest.* **102,** 1994–2001.

89. Payrastre, B., Missy, K., Trumel, C., Bodin, S., Plantavid, M., and Chap, H. (2000) The integrin α_{IIb}/α_3 in human platelet signal transduction. *Biochem. Pharmacol.* **60,** 1069–1074.

90. Philips, D. R., Prasad, K. S. S., Manganello, J., Bao, M., and Nannizzi-Alaimo, L. (2001) Integrin tyrosine phosphorylation in platelet signaling. *Curr. Opin. Cell Biol.* **13,** 546–554.

91. Phillips, D. R., Nannizzi-Alamio, L., and Prasad, K. S. S. (2001) β3 tyrosine phosphorylation in $\alpha_{IIb}\beta_3$ (platelet membrane GP IIb-IIIa) outside-in integrin signaling. *Thromb. Haemost.* **86,** 246–258.

92. Guinebault, C., Payrastre, B., Racaud-Sultan, C., Mazarguil, H., Breton, M., Mauco, G., et al. (1995) Integrin-dependent translocation of phosphoinositide 3-kinase to the cytoskeleton of thrombin-activated platelets involves specific interactions of p85α with actin filaments and focal adhesion kinase. *J. Cell Biol.* **129,** 831–842.

93. Shattil, S. J., O'Toole, T., Eigenthaler, M., Thon, V., Williams, M., Babior, B. M., et al. (1995) β_3-Endonexin, a novel polypeptide that interacts specifically with the cytoplasmic tail of the integrin β_3 subunit. *J. Cell Biol.* **131,** 807–816.

94. Naik, U. P., Patel, P. M., and Parise, L. V. (1997) Identification of a novel calcium-binding protein that interacts with the integrin α_{IIb} cytoplasmic domain. *J. Biol. Chem.* **272,** 4651–4654.

95. Shock, D. D., Naik, U. P., Brittain, J. E., Alahari, S. K., Sondek, J., and Parise, L. V. (1999) Calcium-dependent properties of CIB binding to the integrin αIIb cytoplasmic domain and translocation to the platelet cytoskeleton. *Biochem. J.* **342,** 729–735.

96. Calderwood, D. A., Zent, R., Grant, R., Rees, D. J., Hynes, R. O., and Ginsberg, M. H. (1999) The talin head domain binds to integrin beta subunit cytoplasmic tails and regulates integrin activation. *J. Biol. Chem.* **274,** 28,071–28,074.

97. Gao, J., Zoller, K. E., Ginsberg, M. H., Brugge, J. S., and Shattil, S. J. (1997) Regulation of the pp72^{syk} protein tyrosine kinase by platelet integrin $\alpha_{IIb}\beta_3$. *EMBO J.* **16,** 6414–6425.

98. Woodside, D. G., Obergfell, A., Leng, L., Wilsbacher, J. L., Miranti, C. K., Brugge, J. S., et al. (2001) Activation of Syk protein tyrosine kinase through interaction with integrin β cytoplasmic domains. *Curr. Biol.* **11,** 1799–1804.

99. Jenkins, A. L., Nannizzi-Alaimo, L., Silver, D., Sellers, J. R., Ginsberg, M. H., Law, D. A., et al. (1998) Tyrosine phosphorylation of the β_3 cytoplasmic domain mediates integrin-cytoskeletal interactions. *J. Biol. Chem.* **273,** 13,878–13,885.

100. Cowan, K. J., Law, D. A., and Phillips, D. R. (2000) Identification of Shc as the primary protein binding to the tyrosine-phosphorylated β_3 subunit of $\alpha_{IIb}\beta_3$ during outside-in integrin platelet signaling. *J. Biol. Chem.* **275,** 29,113–29,107.

101. Law, D. A., DeGuzman, F. R., Heiser, P., Ministri-Madrid, K., Killeen, N., and Phillips, D. R. (1999) Integrin cytoplasmic tyrosine motif is required for outside-in $\alpha_{IIb}\beta_3$ signalling and platelet function. *Nature* **401,** 808–811.

102. Gale, N. W., Holland, S. J., Valenzuela, D. M., Flenniken, A., Pan, L., Ryan, T. E., et al. (1996) Eph receptors and ligands comprise two major specificty subclasses and are reciprocally compartmentalized during embryogenesis. *Neuron* **17,** 9–19.

103. Klein, R. (2001) Excitatory Eph receptors and adhesive ephrin ligands. *Curr. Opin. Cell Biol.* **13,** 196–203.

104. Dodelet, V. C. and Pasquale, E. B. (2000) Eph receptors and ephrin ligands: embryogenesis to tumorigenesis. *Oncogene* **19,** 5614–5619.

105. Huai, J. and Drescher, U. (2001) An ephrin-A-dependent signaling pathway controls integrin function and is linked to the tyrosine phosphorylation of a 120 kDa protein. *J. Biol. Chem.* **276,** 6689–6694.

106. Zou, J. X., Wang, B., Kalo, M. S., Zisch, A. H., Pasquale, E. B., and Ruoslahti, E. (1999) An Eph receptor regulates integrin activity through R-Ras. *Proc. Natl. Acad. Sci. USA* **96,** 13,813–13,818.

107. Davy, A. and Robbins, S. M. (2000) Ephrin-A5 modulates cell adhesion and morphology in an integrin-dependent manner. *EMBO J.* **19,** 5396–5405.

108. Gerety, S. S., Wang, H. U., Chen, Z.-F., and Anderson, D. J. (1999) Symmetrical mutant phenotypes of the receptor EphB4 and its specific transmembrane ligand ephrin-B2 in cardiovascular development. *Molec. Cell* **4,** 403–414.

109. Adams, R. H., Wilkinson, G. A., Weiss, C., Diella, F., Gale, N. W., Deutsch, U., et al. (1999) Roles of ephrinB ligands and EphB receptors in cardiovascular development: demarcation of arterial/venous domains, vascular morphogenesis and sprouting angiogenesis. *Genes Dev.* **13,** 295–306.

110. Adams, R. H. and Klein, R. (2000) Eph receptors and ephrin ligands: Essential mediators of vascular development. *Trends Cardiovasc. Med.* **10,** 183–188.

111. Himanen, J. P., Rajashankar, K. R., Lackmann, M., Cowan, C. A., Henkemeyer, M., and Nikolov, D. B. (2001) Crystal structure of an Eph receptor-ephrin complex. *Nature* **414,** 933–938.

112. Huynh-Do, U., Stein, E., Lane, A. A., Liu, H., Cerretti, D. P., and Daniel, T. O. (1999) Surface densities of ephrin-B1 determine EphB1-coupled activation if cell attachment through $\alpha_v\beta_3$ and $\alpha_5\beta_1$ integrins. *EMBO J.* **18,** 2165–2173.

113. Holland, S. J., Gale, N. W., Gish, G. D., Roth, R. A., Zhou, S. Y., Cantley, L. C., et al. (1997) Juxtamembrane tyrosine residues couple the Eph family receptor EphB2/Nuk to specific SH2 domain proteins in neuronal cells. *EMBO J.* **16,** 3877–3888.

114. Hock, B., Bohme, B., Karn, T., Yamamoto, T., Kaibuchi, K., Holtrich, U., et al. (1998) PDZ-domain-mediated interaction of the Eph-related receptor tyrosine kinase EphB3 and the ras-binding protein AF6 depends on the kinase activity of the receptor. *Proc. Natl. Acad. Sci. USA* **95,** 9779–9784.

115. Torres, R., Firestein, B. L., Dong, H. L., Staudinger, J., Olson, E. N., Huganir, R. L., et al. (1998) PDZ proteins bind, cluster, and synaptically colocalize with Eph receptors and their ephrin ligands. *Neuron* **21,** 1453–1463.

116. Lin, D., Gish, G. D., Songyang, Z., and Pawson, T. (1999) The carboxyl terminus of B class ephrins constitutes a PDZ binding motif. *J. Biol. Chem.* **274,** 3726–3733.

117. Dodelet, V. C., Pazzagli, C., Zisch, A. H., Hauser, C. A., and Pasquale, E. B. (1999) A novel signaling intermediate, SHEP1, directly couples Eph receptors to R-Ras and Rap1A. *J. Biol. Chem.* **274,** 31,941–31,946.

118. Pandey, A., Duan, H., and Dixit, V. M. (1995) Characterization of a novel src-like adapter protein that associates with the Eck receptor tyrosine kinase. *J. Biol. Chem.* **270,** 19,201–19,204.

119. Cowan, C. A. and Henkemeyer, M. (2001) The SH2/SH3 adaptor Grb4 transduces B-ephrin reverse signals. *Nature* **413,** 174–179.

120. Birgbauer, E., Cowan, C. A., Sretavan, D. W., and Henkemeyer, M. (2000) Kinase independent function of EphB receptors in retinal axon pathfinding to the optic disc from dorsal but not ventral retina. *Development* **127,** 1231–1241.

121. Buchert, M., Schneider, S., Meskenaite, V., Adams, M. T., Canaani, E., Baechi, T., et al. (1999) The junction-associated protein AF-6 interacts and clusters with specific Eph receptor tyrosine kinases at specialized sites of cell-cell contact in the brain. *J. Cell Biol.* **144,** 361–371.

122. Bruckner, K., Pablo Labrador, J., Scheiffele, P., Herb, A., Seeburg, P. H., and Klein, R. (1999) EphrinB ligands recruit GRIP family PDZ adaptor proteins into raft membrane microdomains. *Neuron* **22,** 511–524.

123. Holland, S. J., Gale, N. W., Mbamalu, G., Vancopoulos, G. D., Henkemeyer, M., and Pawson, T. (1996) Bidirectional signaling through the the EPH-family receptor Nuk and its transmembrane ligands. *Nature* **383,** 722–725.

124. Palmer, A., Zimer, M., Erdmann, K. S., Eulenberg, V., Porthin, A., Heumann, R., et al. (2002) Ephrin B phosphorylation and reverse signaling: regulation by Src kinases and PTP-BL phosphatase. *Mol. Cell* **9,** 725–737.

125. Bruckner, K., Pasquale, E. B., and Klein, R. (1997) Tyrosine phosphorylation of transmembrane ligands for Eph receptors. *Science* **275,** 1640–1643.

126. Chong, L. D., Park, E. K., Latimer, E., Friesel, R., and Daar, I. O. (2000) Fibroblast growth factor receptor-mediated rescue of x-ephrin B1-induced cell dissociation in Xenopus embryos. *Mol. Cell Biol.* **20,** 724–734.

127. Kullander, K., Mather, N. K., Diella, F., Dottori, M., Boyd, A. W., and Klein, R. (2001) Kinase-dependent and kinase-independent functions of EphA4 receptors in major tract formation in vivo. *Neuron* **29,** 73–84.

128. Grunwald, I. C., Korte, M., Wolfer, D., Wilkinson, G. A., Unsicker, K., Lipp, H.-P., et al. (2001) Kinase-independent requirement of EphB2 receptors in hippocampal synaptic plasticity. *Neuron* **32,** 1027–1040.

129. Schultz, J., Ponting, C. P., Hofmann, K., and Bork, P. (1997) SAM as a protein interaction domain involved in developmental regulation. *Protein Sci.* **6,** 249–253.

130. Thanos, C. D., Goodwill, K. E., and Bowie, J. U. (1999) Oligomeric structure of the human Ephb2 receptor SAM domain. *Science* **283,** 833–836.

131. Stapleton, D., Balan, L., Pawson, T., and Sicheri, F. (1999) The crystal structure of an Eph receptor SAM domain reveals a mechanism for modular dimerization. *Nature Struct. Biol.* **6,** 44–49.

132. Smalla, M., Schmieder, P., Kelly, M., Ter Laak, A., Krause, G., Ball, L., et al. (1999) Solution structure of the receptor tyrosine kinase EphB2 SAM domain and identification of two distinct homotypic interaction sites. *Protein Sci.* **8,** 1954–1961.

133. Prevost, N., Woulfe, D., Tanaka, T., and Brass, L. F. (2002) Interactions between Eph kinases and ephrins provide a novel mechanism to support platelet aggregation once cell-to-cell contact has occured. *Proc. Natl. Acad. Sci. USA* **99,** 9219–9224.

134. Felding-Habermann, B., Silletti, S., Mei, F., Siu, C. H., Yip, P. M., Brooks, P. C., et al. (1997) A single immunoglobulin-like domain of the human neural cell adhesion molecule L1 supports adhesion by multiple vascular and platelet integrins. *J. Cell Biol.* **139,** 1567–1581.

135. Geddis, A. E., Linden, H. M., and Kaushansky, K. (2002) Thrombopoietin: a pan-hematopoietic cytokine. *Cytokine & Growth Factor Rev.* **13,** 61–73.

136. Shivdasani, R. A. (2001) Molecular and transcriptional regulation of megakaryocyte differentiation. *Stem Cell* **19,** 397–407.

137. Shattil, S. J. and Leavitt, A. D. (2001) All in the family: Primary megakaryocytes for studies of platelet $\alpha_{IIb}\beta_3$ signaling. *Thromb. Haemost.* **86,** 259–265.

138. Shiraga, M., Ritchie, A., Aidoudi, S., Baron, V., Wilcox, D., White, G., et al. (1999) Primary megakaryocytes reveal a role for transcription factor NF-E2 in integrin $\alpha_{IIb}\beta_3$ signaling. *J. Cell Biol.* **147,** 1419–1429.

139. Faraday, N., Rade, J. J., Johns, D. C., Khetawat, G., Noga, S. J., DiPersio, J. F., et al. (1999) Ex vivo cultured megakaryocytes express functional glycoprotein IIb-IIIa receptors and are capable of adenovirus-mediated transgene expression. *Blood* **94,** 4084–4092.

140. Bertoni, A., Tadokoro, S., Eto, K., Pampori, N., Parisi, L. V., White, G. C., et al. (2002) Relationships between Rap1b, affinity modulation of integrin $\alpha_{IIb}\beta_3$ and the actin cytoskeleton. *J. Biol. Chem.* **277,** 25,715–25,721.

141. Brass, L. F., Manning, D. R., Williams, A., Woolkalis, M. J., and Poncz, M. (1991) Receptor and G protein-mediated responses to thrombin in HEL cells. *J. Biol. Chem.* **266,** 958–965.

142. Van der Vuurst, H., Van Willigen, G., Van Spronsen, A., Hendriks, M., Donath, J., and Akkerman, J. W. N. (1997) Signal transduction through trimeric G proteins in megakaryoblastic cell lines. *Arterioscler. Thromb. Vasc. Biol.* **17,** 1830–1836.

143. Cichowski, K., Orsini, M. J., and Brass, L. F. (1999) PAR1 activation initiates integrin engagement and outside-in signaling in megakaryoblastic CHRF-288 cells. *Biochim. Biophys. Acta* **145,** 265–276.

144. Brass, L. F., Pizarro, S., Ahuja, M., Belmonte, E., Blanchard, N., Stadel, J. M., et al. (1994) Changes in the structure and function of the human thrombin receptor during receptor activation, internalization and recycling. *J. Biol. Chem.* **269**, 2943–2952.

145. Choi, E. S., Nichol, J. L., Hokom, M. M., Hornkohl, A. C., and Hunt, P. (1995) Platelets generated in vitro from proplatelet-displaying human megakaryocytes are functional. *Blood* **85**, 402–413

146. Faraday, N., Rade, J. J., Johns, D. C., Khetawat, G., Noga, S. J., DiPersio, J. F., et al. (1999) Ex vivo cultured megakaryocytes express functional glycoprotein IIb-IIIa receptors and are capable of adenovirus-mediated transgene expression. *Blood* **94**, 4084–4092.

147. Eto, K., Murphy, R., Kerrigan, S. W., Bertoni, A., Stuhlmann, H., Nakano, T., et al. (2002) Megakaryocytes derived from embryonic stem cells implicate CalDAG-GEFI in integrin signaling. *Proc. Natl. Acad. Sci. USA* **99**, 12,819–12,824.

148. Brass, L. F., Vassallo, R. R. Jr., Belmonte, E., Ahuja, M., Cichowski, K., and Hoxie, J. A. (1992) Structure and function of the human platelet thrombin receptor: studies using monoclonal antibodies against a defined epitope within the receptor N-terminus. *J. Biol. Chem.* **267**, 13,795–13,798.

149. Ishihara, H., Connolly, A. J., Zeng, D., Kahn, M. L., Zheng, Y. W., Timmons, C., et al. (1997) Protease-activated receptor 3 is a second thrombin receptor in humans. *Nature* **386**, 502–508.

150. Faruqi, T. R., Weiss, E. J., Shapiro, M. J., Huang, W., and Coughlin, S. R. (2000) Structure-function analysis of protease-activated receptor 4 tethered ligand peptides—Determinants of specificity and utility in assays of receptor function. *J. Biol. Chem.* **275**, 19,728–19,734.

151. Mills, D. C. B. (1996) ADP receptors on platelets. *Thromb. Haemost.* **76**, 835–856.

152. Gachet, C., Hechler, B., Léon, C., Vial, C., Leray, C., Ohlmann, P., et al. (1997) Activation of ADP receptors and platelet function. *Thromb. Haemost.* **78**, 271–275.

153. Hirata, M., Hayashi, Y., Ushikubi, F., Nakanishi, S., and Narumiya, S. (1991) Cloning and expression of cDNA for a human thromboxane A_2 receptor. *Nature* **349**, 617–620.

154. Hanasaki, K. and Arita, H. (1988) Characterization of thromboxane A_2/prostaglandin H_2 (TXA_2/PGH_2) receptors of rat platelets and their interaction with TXA_2/PGH_2 receptor antagonists. *Biochem. Pharmacol.* **37**, 3923–3929.

155. Furci, L., Fitzgerald, D. J., and FitzGerald, G. A. (1991) Heterogeneity of prostaglandin H_2/thromboxane A_2 receptors: Distinct subtypes mediate vascular smooth muscle contraction and platelet aggregation. *J. Pharmacol. Exp. Ther.* **258**, 74–81.

156. Kobilka, B. K., Matsui, H., Kobilka, T. S., Yang Feng, T. L., Francke, U., Caron, M. G., et al. (1987) Cloning, sequencing, and expression of the gene coding for the human platelet α_2-adrenergic receptor. *Science* **238**, 650–656.

157. Vane, J. R. and Botting, R. M. (1995) Pharmacodynamic profile of prostacyclin. *Am. J. Cardiol.* **75**, 3A–10A.

158. Kowalska, M. A., Ratajczak, J., Hoxie, J., Brass, L. F., Gewirtz, A., Poncz, M., et al. (1999) Megakaryocyte precursors, megakaryocytes and platelets express the HIV co-receptor CXCR4 on their surface: determination of response to stromal-derived factor-1 by megakaryocytes and platelets. *Br. J. Haematol.* **104**, 220–229.

159. Kowalska, M. A., Ratajczak, M. Z., Majka, M., Jin, J. G., Kunapuli, S., Brass, L., et al. (2000) Stromal cell-derived factor-1 and macrophage-derived chemokine: 2 chemokines that activate platelets. *Blood* **96**, 50–57.

160. Chao, W. and Olson, M. S. (1993) Platelet-activating factor: Receptors and signal transduction. *Biochem. J.* **292**, 617–629.

161. Honda, Z., Nakamura, M., Miki, I., Minami, M., Watanabe, T., Seyama, Y., et al. (1991) Cloning by functional expression of platelet-activating factor receptor from guinea-pig lung. *Nature* **349,** 342–346.

162. Bichet, D. G., Arthus, M.-F., Barjon, J. N., Lonergan, M., and Kortas, C. (1987) Human platelet fraction arginine-vasopressin: potential physiological role. *J. Clin. Invest.* **79,** 881–887.

163. Inaba, K., Umeda, Y., Yamane, Y., Urakami, M., and Inada, M. (1988) Characterization of human platelet vasopressin receptor and the relation between vasopressin-induced platelet aggregation and vasopressin binding to platelets. *Clin. Endocrinol. (Oxf.)* **29,** 377–386.

164. Siess, W., Stifel, M., Binder, H., and Weber, P. (1986) Activation of V1-receptors by vasopressin stimulates inositol phospholipid hydrolysis and arachidonate metabolism in human platelets. *Biochem. J.* **233,** 83–91.

165. Vittet, D., Cantau, B., Mathieu, M.-N., and Chevillard, C. (1988) Properties of vasopressin-activated human platelet high affinity GTPase. *Biochem. Biophys. Res. Commun.* **154,** 213–218.

166. Connolly, A. J., Ishihara, H., Kahn, M. L., Farese, R. V. Jr., and Coughlin, S. R. (1996) Role of the thrombin receptor in development and evidence for a second receptor. *Nature* **381,** 516–519.

167. Darrow, A. L., Fung-Leung, W. P., Ye, R. D., Santulli, R. J., Cheung, W. M., Derian, C. K., et al. (1996) Biological consequences of thrombin receptor deficiency in mice. *Thromb. Haemost.* **76,** 860–866.

168. Fabre, J.-E., Nguyen, M., Latour, A., Kiefer, J. A., Audoly, A. P., Coffman, T. M., et al. (1999) Decreased platelet aggregation, increased bleeding time and resistance to thromboembolism in P2Y$_1$-deficient mice. *Nature Med.* **5,** 1199–1202.

169. Mulryan, K., Gitterman, D. P., Lewis, C. J., Vial, C., Leckie, B. J., Cobb, A. L., et al. (2000) Reduced vas deferens contraction and male infertility in mice lacking P2X$_1$ receptors. *Nature* **403,** 86–89.

170. Vial, C., Rolf, M. G., Mahaut-Smith, M. P., and Evans, R. J. (2002) A study of P2X$_1$ receptor function in murine megakaryocytes and human platelets reveals synergy with P2Y receptors. *Br. J. Pharmacol.* **135,** 363–372.

2

Manipulation of Gene Expression in Megakaryocytes

Nikla R. Emambokus, George J. Murphy, and Jonathan Frampton

1. Introduction

The process of megakaryocyte generation in the bone marrow and subsequent differentiation leading to platelet production is little understood. Known as *megakaryopoiesis*, it involves a number of unique biological features, including an increase in the nuclear DNA content (endoreplication) and partitioning of cytoplasm and membranes into platelets. Abnormalities of thrombosis and hemostasis can occur as a result of alterations in the number of platelets generated and maintained in the circulation or through aberrations in the functional behavior of the platelet itself. Although some of these abnormalities are most likely indirect in their etiology (for example, immune thrombocytopenias), many are directly linked to an inherited or acquired genetic effect operating at some point during megakaryopoiesis. Inherited mutations affecting megakaryocytes and platelets are relatively rare, although there is increasing interest in the association between genetic polymorphisms in platelet proteins that have no profound phenotype but may be linked to an increased thrombotic risk. Specific genetic changes leading to deregulation of proliferation and/or differentiation at some point during megakaryopoiesis are also implicated in several acquired conditions, including leukemias, preleukemic states, and dysplasias.

Experimentation directed at understanding normal or disease-related biological processes has been restricted in different ways for megakaryocytes and platelets. Classically, studies of megakaryocyte function have involved work with cell lines that have features in common with megakaryoblasts and can often be induced to undergo some aspects of terminal differentiation. Although such cells lend themselves to manipulation, they are immortalized, often deriving from leukemic patients, and therefore have aberrant proliferation and differentiation characteristics. Although it is preferable to use primary cells, the scarcity of megakaryocytic cells in the bone marrow has limited their usage. In recent years, improvements in cell purification and culture conditions, especially through use of the specific growth factor thrombopoietin, has increased the range of research that can be conducted ex vivo. The study of abnormal human

From: *Methods in Molecular Biology, vol. 273:*
Platelets and Megakaryocytes, Vol. 2: Perspectives and Techniques
Edited by: J. M. Gibbins and M. P. Mahaut-Smith © Humana Press Inc., Totowa, NJ

megakaryopoiesis is further restricted either because of the small number of affected individuals or because it is impractical or clinically undesirable to collect bone marrow. Although such work has been, and will continue to be, of great value, it would also be advantageous to be able to study normal or mutated gene products in an in vivo context. Studies on platelets have been less affected by restraints on material availability, except in relation to that deriving from patients with rare specific genetic defects. Biochemical approaches aimed at examining affector and effector responses and the signaling pathways linking the two are well worked out. The major limitation, however, is with respect to the manipulation of gene products within the platelet. This is most obviously a consequence of the absence of nuclear DNA, and the difficulties associated with expression of exogenous proteins.

The clearest solution to many of these limitations is to work with cells derived from individuals or animals carrying specific genetic changes in the proteins of interest. Some, but not all, genetic diseases affecting human megakaryopoiesis or platelet function have been linked to mutation in a specific gene. In a few cases a corresponding spontaneous mutant mouse strain has been identified. However, most genes of interest to those studying megakaryopoiesis and platelet function do not have a naturally occurring mutant form in either humans or mice.

In this perspective we will highlight the advantages of using the mouse for the investigation of megakaryocyte and platelet biology through modification or ablation of a gene product of interest although some of the methods, especially those used on cells ex vivo, are equally suited to work with human cells. Broadly speaking the discussion will be divided into two areas: (1) introduction of exogenous genetic material, and (2) ablation or modification of endogenous genes. Within each of these areas we will consider techniques which are suited to either ex vivo or in vivo studies and we will attempt to speculate on future directions.

2. Introduction of Exogenous Material

Several strategies are available for the expression of exogenous proteins in cells of the megakaryocyte lineage, although the possibilities for applying such methods directly to platelets are more limited. These methods can be categorized into those suitable for ex vivo use and those that enable exogenous gene expression in vivo. Methods utilized ex vivo are largely dependent on purification and expansion of megakaryocytic cells, usually bone marrow- or umbilical cord-derived, or cells that can become committed to megakaryopoiesis under appropriate culture conditions. Such cell enrichment protocols are outside the scope of this chapter and the reader is instead referred to recent reviews on the topic *(1)*. Of the three principal routes for introducing exogenous material, the most commonly used is infection with viral vectors. Transfection of expression vectors, although strictly speaking a possible means of introducing genes of interest, is not generally considered for work with primary megakaryocytes and will not be discussed here. Recently, though, there has been considerable interest in the prospect of direct introduction of proteins as fusions to small basic peptide sequences, and this method will be considered in particular because of its potential for use with platelets as well as megakaryocytes. Expression of exogenous sequences in vivo can also be achieved through viral infection; however, there are considerable difficulties with respect to the efficiency and specificity of this route (*see* **Subheading 2.1.3.** for a possible way around some of these limitations). The generic method of choice for in vivo expression of protein-coding sequences in mice is transgenesis, and a number of permutations will be discussed, especially in regard to how expression of the exogenous gene product can be limited to megakaryocytes and platelets.

2.1. Infection

Vector systems derived from retroviruses, lentiviruses, adenoviruses, and herpes-related viruses are available for use. None of these viruses has an inherent capacity for restricted expression in megakaryocytic cells; however, as suggested above, "selectivity" can be achieved by expansion of the desired cells in culture prior to infection. A few (as yet little-used) alternative means will be discussed that can be used to obtain megakaryocyte-specific infection or expression from a viral vector.

2.1.1. Retroviral Infection

Delivery of exogenous sequences as retroviruses has the advantage that not only can a large proportion of cells be infected, but also the conversion of the retroviral RNA genome into an integrated DNA intermediate (the provirus) results in permanent genetic modification (**Fig. 1A**). Classical retroviral vectors usually yield proviruses that constitutively express the exogenous gene from the viral LTR promoter/enhancer sequences (**Fig. 1B**). This relatively simple approach has been used effectively to infect CD34+ progenitors derived from human peripheral blood or bone marrow. For example, Burstein et al. *(2)* were able to culture infected progenitors under conditions favoring megakaryocyte development and demonstrate that 60% of the platelets contained the exogenous gene product. The murine stem cell virus (MSCV, *[3]*) is ideal for infection of mouse hemopoietic cells and is frequently used in a form that co-expresses the exogenous gene of interest together with GFP to enable tracking of those cells that have been infected. Such a viral vector (MIGR, **Fig. 1C**) was used, for example, by Baccini et al. *(4)* to introduce the cell-cycle inhibitor p21 into ex vivo cultured megakaryocytes.

2.1.2. Regulation of Retroviral Gene Expression

If it is desirable to infect megakaryocytic cells selectively in a mixed population (perhaps to avoid preselection and culturing) or to restrict expression to a defined stage of differentiation, then one possible solution is to use retroviral vectors that contain an internal promoter (**Fig. 1D**). Full effectiveness of the specificity of the internal promoter is ensured by an additional modification to the vector in the form of a deletion of the enhancer sequences from the 3′LTR. The mechanism of proviral integration of such so-called self-inactivating (SIN) retroviruses results in no LTR-driven transcription. That this approach can be utilized to restrict expression to megakaryocytes was shown by Wilcox et al. *(5)*, who infected human CD34+ cells with a retrovirus in which the protein of interest was driven from approx 900 bp of the promoter of the *gpIIb* (α_{IIb} integrin) gene. Increasing understanding of megakaryocyte-specific

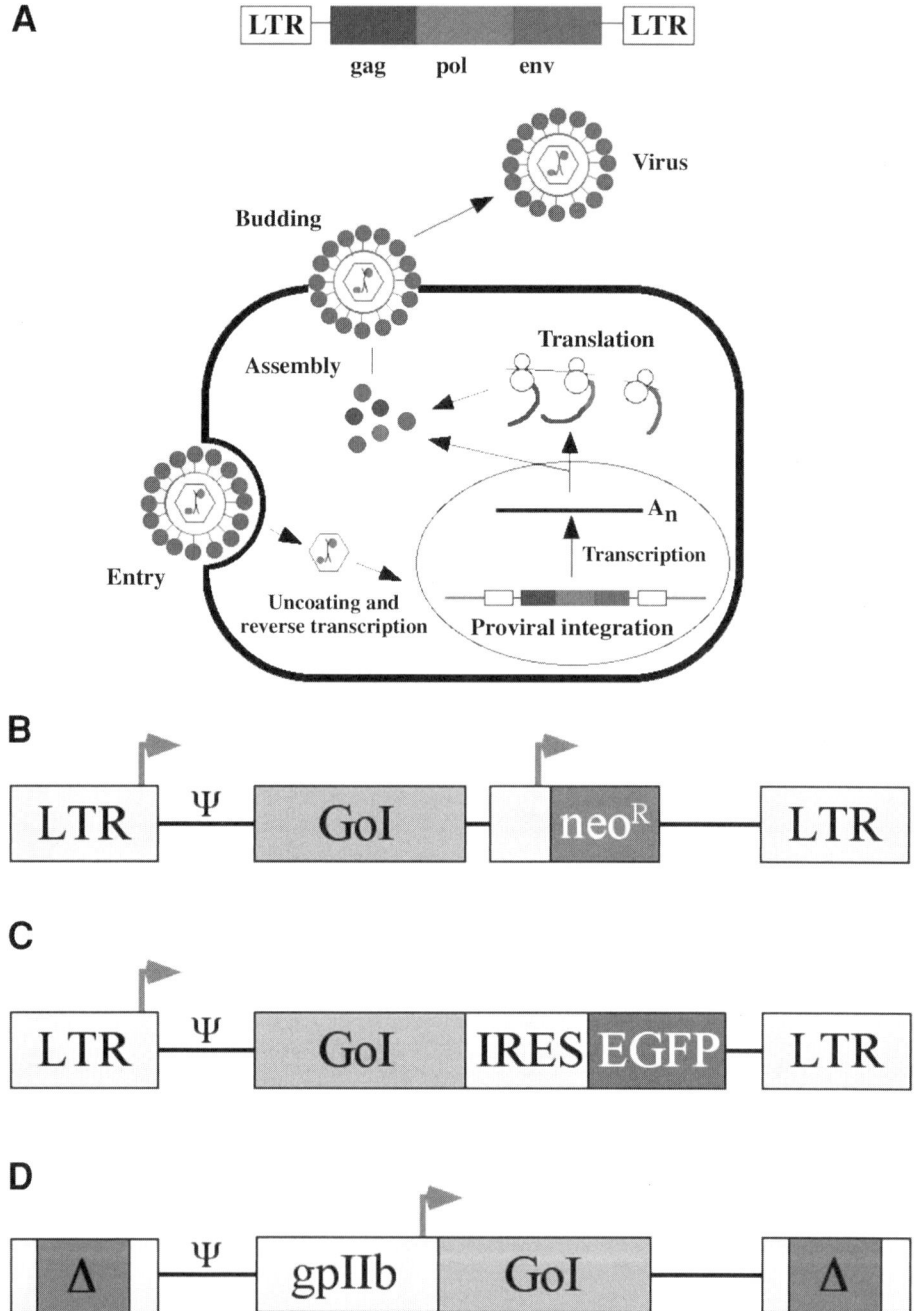

gene regulation may enable this approach to be extended if it is possible to identify promoter sequences that act during particular stages of megakaryocyte differentiation (**Table 1**).

2.1.3. Targeting of Retroviral Infection

Unlike their mammalian counterparts, avian retroviruses cannot infect and replicate in murine cells due to the lack of a cell surface receptor and incomplete intracellular virus assembly *(6)*. However, engineered expression of the subgroup A avian leukosis virus (ALV-A) receptor, TVA, on the surface of murine cells confers susceptibility to infection by ALV-A viruses and by murine retroviruses packaged with the ALV-A envelope protein (EnvA). By expressing TVA from a transgene (*see* **Subheading 2.3.**) driven by megakaryocyte-specific promoters, it is possible to restrict retroviral infection (**Fig. 2** and *[7]*). Both avian (e.g., RCAS, *[6]*) and murine (e.g., MuLV- or MSCV-based) vectors can be employed, although the latter have to be packaged into virions containing the EnvA protein ("pseudotyping") and cannot generally be produced in such high titers as the avian viruses.

In addition to the specificity of infection, this system has two other major advantages. First, multiple sequential infections can be performed on the target cells using

---

Fig. 1. *(see facing page)* (**A**) Schematic representation of basic retroviral structure and the process of infection. A retroviral genome consists of a single-stranded RNA molecule encoding the viral core proteins (gag), reverse transcriptase (pol) and the envelope glycoprotein (env). These genes are flanked by sequences (LTRs—long terminal repeats) involved in the conversion of the RNA genome into a double-stranded DNA form allowing integration into the host genome and subsequent transcription from the resultant so-called provirus. Retroviral vectors are modified versions of this basic structure, in which coding sequences for the gene of interest (GoI) are inserted between the LTRs, allowing their stable integration into the genome. Such vectors are generally unable to provide all the components for retroviral infection because of removal of all or part of the gag, pol, and env genes (replication-incompetent). In this case, packaging of the vector genome into viral particles is achieved by supplying these gene products on a "helper" retrovirus or through use of a "packaging" cell line. Many variations can be made to the basic retroviral structure to generate a suitable vector. The simplest type (**B**) contains the gene of interest which is transcribed from the LTR promoter elements and a selection cassette (e.g., neoR) driven by a second internal promoter (usually a strong constitutive promoter such as that from the PGK gene or the SV40 virus). It is often desirable to track infected cells; this can be achieved by co-expression of a fluorescent protein. In the case illustrated (**C**, the so-called MIGR vector), the fluorescent protein (EGFP) is produced from a bicistronic RNA through use of an internal ribosome entry site (IRES). If it is desirable to restrict expression of the gene of interest, one possible solution is to use a lineage-specific internal promoter, such as that from the *gpIIb* gene (**D**). In this case, a deletion is included in the LTR sequences to prevent these from driving strong constitutive expression. (*See* color insert following p. 300.)

Table 1
Examples of Genes Expressed Predominantly or Exclusively in the MK Lineage

Gene	Other tissues where expression has been detected	Used to drive transgene expression
gpIbα	Endothelium	+
gpIbβ	Endothelium	
gpIIb	Hemopoietic progenitors, mast cells	+
gpV	Endothelium	
gpVI		
gpIX	Endothelium	+
von Willebrand factor	Endothelium	
P-selectin		
PCLP1 (thrombomucin)	Endothelium, kidney podocytes	
AA4.1	Endothelium, lung epithelium	
c-Mpl	Hemopoietic progenitors	+
PF4		+
PBP		
β1 tubulin		

viruses encoding different genes. This is possible because, unlike infections with mammalian viruses, mouse cells infected with EnvA-packaged retroviruses do not express sufficient EnvA to elicit the phenomenon of resistance to superinfection (*6*). Second, a single transgenic mouse strain can be used as a means to introduce many different genes of interest, thereby avoiding the need to generate transgenic lines for each gene product or variant to be analysed. TVA expression driven by *gpIbα* or *gpIIb* gene promoter elements has been used very effectively to reconstitute protein expression in megakaryocytes derived from both the NF-E2 and c-*mpl* knockout mice (*8* and *9*, respectively). Future extensions of this methodology will doubtless include the generation of TVA-expressing strains in which the transgene is driven by alternative megakaryocyte-specific promoters. Indeed, additional strains expressing TVA in megakaryocytes have been created using promoters from the *PF4, gpIX,* and c-*mpl* genes (G. J. M. and Andrew Leavitt, unpublished). The possibility of infection and exogenous gene expression in progenitors prior to their commitment to megakaryopoiesis has also been made possible with a mouse strain in which TVA is expressed from the SCL gene 3′ enhancer (*9a*).

2.1.4. Lentiviruses as a Means of Infecting Nonreplicating Cells

Retroviruses fail to integrate in nonreplicating target cells due to a block that occurs before entry into the nucleus of the infected cell. This problem may be relevant to the targeting of megakaryocytic cells once they have commenced endoreplication. Hence, it is known that simple retroviruses are not effective as a means of transducing mature

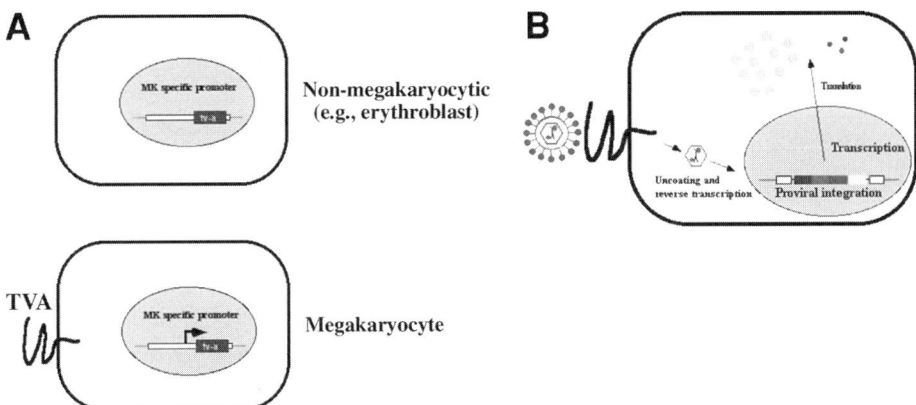

Fig. 2. Megakaryocyte-specific retroviral infection through the TVA receptor. (**A**) The *tv-a* transgene is transcribed from a specific promoter so that the TVA receptor is only expressed on the surface of megakaryocytic cells. (**B**) Retroviral particles coated with the EnvA glycoprotein bind to TVA, allowing viral entry and proviral integration. Following LTR-driven transcription, the protein encoded by the gene of interest (yellow) is expressed efficiently compared to the avian retroviral proteins *gag* (blue), *pol* (red), and *env* (green). (*See* color insert following p. 300.)

megakaryocytes, but it is unclear whether this is due to an absence of nuclear membrane breakdown that may be associated with endoreplication. However, as for other nonreplicating cells, this limitation can be overcome through use of lentiviral vectors *(10)*. The relatedness of lentiviruses to retroviruses allows for the possibility of using them in combination with the TVA system to facilitate entry into megakaryocytic cells. Pseudotyping of a modified human immunodeficiency virus-based lentiviral vector with EnvA has been shown to be an effective way of using lentiviral vectors in conjunction with TVA systems *(11)*. Indeed, recently we have been able to utilize a feline immunodeficiency virus-based lentiviral vector in a similar way to efficiently infect fully mature polyploid megakaryocytes (G. J. M., J. F., and Andrew Leavitt, unpublished observations).

2.1.5. Other Viral Delivery Systems

Sindbis viral vectors have been utilized once, to our knowledge, to provide transient gene expression in terminally differentiated megakaryocytes *(8)*. A much-investigated and exploited vector for transfer of exogenous sequences to mammalian cells, particularly for human gene therapy, is based on adenovirus. The feasibility of using recombinant adenoviruses to infect megakaryocytes has been shown, for example, by Faraday et al. *(12)*, although there seems to be no obvious advantage to match the efficiency, specificity, and heritable integration that can be achieved with retroviral and lentiviral vectors. Finally, if it is necessary to introduce very large segments of

DNA (20–100 kbp), then herpes simplex virus type I (HSV-1) could be a possible solution (*see*, for example, **ref. *13***), although we know of no examples yet of application to studies on megakaryocytes.

2.2. Direct Introduction of Protein

Recently, protein transduction has been shown to be a highly efficient method to introduce proteins into mammalian cells *(14)*. By fusing a basic 11-amino-acid peptide derived from HIV-TAT to a protein of interest, it is possible to render it into a cell-permeable form. A wide variety of cell types are able to incorporate the fusion proteins, although there is no specific description yet of the efficiency of entry into megakaryocytic cells. However, this method clearly should have useful applications in the study of megakaryocytes and it will be interesting to determine whether platelets are also amenable to protein uptake.

2.3. Transgenesis

Transgenesis involves the integration of copies of DNA sequences encoding the gene of interest (the "transgene") randomly into the genome. These genetic modifications are inherited and can be bred into appropriate genetic backgrounds. The number of copies may vary widely and the site of integration can determine the specificity and level of transgene expression. The basic method of transgenesis involves injection of the transgene DNA into 2-d-old embryos that are then transplanted back into a pseudopregnant female. Live pups (potential founders) can then be screened by Southern blotting or PCR of a small sample from the tip of the tail to determine the presence and copy number of integrated DNA *(15,16)*. The essential features of the transgene are a promoter, an intron element to mimic normal gene structure, the coding sequences of the molecule under investigation, and a polyadenylation signal **(Fig. 3A)**.

2.3.1. Conventional Transgenes

The choice of promoter used to drive expression of the sequences encoded in the transgene is of crucial importance. Promoters that give rise to widespread expression of the exogenous gene are of limited use because of the likelihood of problems associated with inappropriate expression. Usually, the transgene is designed so that it should be expressed in specific cell types. Limiting expression of transgenes to the megakaryocyte lineage can be achieved by harnessing the control elements of genes that are exclusively, or at least predominantly, expressed in megakaryocytes. Examples of gene promoters that have been, or have the potential to be, used in this way are listed in **Table 1**.

2.3.2. Regulated Expression from a Transgene

An alternative way of limiting transgene expression is to make use of a promoter that is able to be expressed in all cell types but is under the control of a regulator. The activity of the regulator is controlled by the in vivo administration of a small molecule that can be taken up by all cells. There are a number of systems of this type (for review, *see* **ref. *17***), but by far the most widely exploited one makes use of the tetracycline-dependent

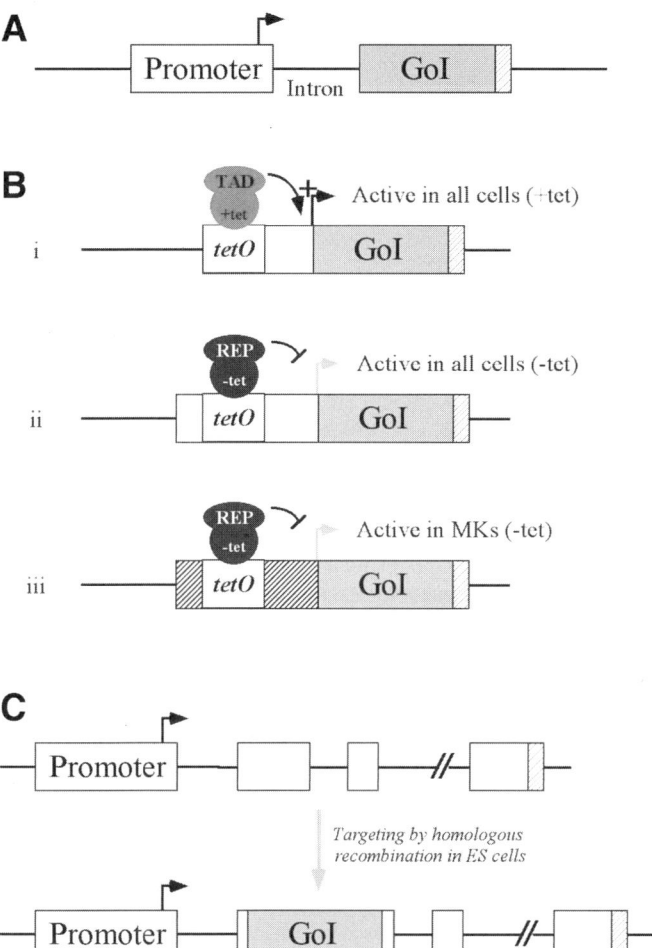

Fig. 3. Some strategies for transgenic expression of the gene of interest. (**A**) A conventional transgene consists of a promoter and the coding sequences (GoI) separated by a small intron. A polyadenylation signal sequence (hatched box) ensures correct transcriptional termination. (**B**) Tetracyline (tet)-regulated systems for the control of transgene expression. In all examples shown, the tetracycline-dependent regulator binds to its recognition motif *(tetO)* either in the presence *(i)* or absence *(ii and iii)* of tetracycline. In *(i)* the promoter element is essentially inactive until tet binding to the regulator brings the transcription activation domain (TAD) into proximity of the initiation site (arrow). In contrast, in *(ii)* and *(iii)* the promoter driving the transgene is dominantly repressed by a repressor domain (REP) linked to the regulator. When tetracycline is added, the regulator no longer binds to *tetO* and the gene of interest can be transcribed from either a constitutive *(ii)* or a megakaryocyte-specific promoter *(iii)*. (**C**) Precise lineage and stage-specific expression can often be achieved by introduction of the gene of interest into the genomic locus of a gene which has the desired expression pattern. In the example shown, targeted insertion is into a coding exon.

reversal of the binding of the bacterial tet repressor (TetR) to its DNA-recognition motif, the tet operator (*tetO*, *[18]*). Although there are several possible permutations of this system using different engineered variants of TetR *(19)*, we will consider three basic strategies that could be employed in megakaryocytes (**Fig. 3B**). Fusion of TetR to the transactivation domain of HSV-1 VP16 creates a transactivator (tTA) that binds *tetO* in the absence of tetracycline *(18)*. Wider applicability in vivo has been achieved by mutation of tTA so that it binds to *tetO* only in the presence of tetracycline. Therefore, by linking the gene of interest to an essentially inactive promoter element containing multimers of *tetO*, a transgene is created that should be expressed only when cells encounter tetracycline in vivo or ex vivo (**Fig. 3Bi**). However, there are two drawbacks to this approach. First, induced expression will be in all cell types, perhaps leading to some of the same problems that might be encountered with a constitutive transgene. Second, it is difficult to limit the basal expression of the transgene in the absence of tetracycline. An alternative, but as yet little-utilized, strategy involves fusion of TetR to a dominant suppressor domain (tTS, *[20]*). tTS will dominantly repress expression from a promoter containing *tetO* unless tetracycline is added to bring about its removal from DNA (**Fig. 3Bii**). In the context of restricting transgene expression to megakaryocytes, it is likely that *tetO* could be incorporated into a transgene driven by a megakaryocyte-lineage-specific promoter. If this transgenic line is bred together with a strain expressing tTS from a constitutive promoter, then megakaryocytic cells in the resultant offspring will express the gene of interest only when the animals are fed tetracycline (**Fig. 3Biii**).

2.3.3. Knock-In

The random integration of multiple copies of transgenic DNA into the genome often leads to expression that is not ideal, either in terms of cell type restriction or the level of transcription. A possible solution is to insert the transgenic cassette directly into a gene locus that is expressed specifically or predominantly in megakaryocytes (**Fig. 3C**). This so-called knock-in approach involves gene targeting in mouse embryonic stem (ES) cells. Practically, this is achieved by generating a targeting vector that consists of DNA isogenic with the ES cells to be modified and designed to carry the required modifications as well as flanking arms of unmodified DNA that are necessary for homologous recombination *(21)*. The vector bears a constitutively active antibiotic resistance gene (usually *neo^R* conferring resistance to G418) for selection of targeted cells and a negative selection cassette (usually the HSV-1 *tk* gene conferring sensitivity to gangcyclovir) for elimination of those integration events that have occurred through nonhomologous recombination (**Fig. 3C**). The targeting vector is introduced into ES cells by transfection and positive and negative selections are applied. Potential clones are expanded and analyzed by Southern blotting and PCR to determine whether homologous recombination has occurred. A correctly targeted ES clone is then injected into blastocysts, which, after reimplantation, develop to produce a chimeric founder. If the ES clone has retained totipotent potential and contributes significantly to the chimera, then the modified gene should be transferable to the next generation and a line will have been created (*see* **Fig. 6**). This is a labor-intensive strategy and should perhaps be undertaken only if transgenesis using a specific promoter fails. Another possible disadvantage is that the sequences being "knocked in" can sufficiently disrupt the target gene so that a null allele is generated. In this case modified animals could only be used as heterozygotes. An example of such a situation is described by Tronick-LeRoux et al. *(22)* who inserted a thymidine kinase transgene into the *gpIIb* gene.

3. Gene Ablation or Modification
3.1. mRNA Ablation

A number of methods exist for reducing the level of mRNAs that could be applied to expanded and purified megakaryocytes ex vivo, although there are few examples in the literature of their use on such cells. The common theme to these techniques is that they rely on RNA or DNA sequence complementarity to the target mRNA as a means to interrupt its translation or to initiate its destruction.

3.1.1. Antisense

The simplest approach of this type involves the use of a short single-strand oligo-nucleotide that, upon formation of a hybrid RNA:DNA or RNA:RNA duplex, initiates destruction of the mRNA by recruitment of RNAseH. Delivery need not be directly in the form of an oligonucleotide but can be as RNA transcribed from an introduced expression construct or virus (*see* **Subheading 2.1.**) or from a transgene (*see* **Subheading 2.3.**). One adaptation to the use of oligonucleotides, which is not amenable to introduction on a vector but which has become popular recently, uses synthetic DNA analogs called morpholino phosphorodiamidates. Morpholinos, as they are usually known, have highly favorable properties and are being widely used for functional genomic applications (see *[23]* for review). Usually morpholinos are employed to elicit translational inhibition, although through interruption of splicing events (**Fig. 4**) they also have the capacity to modify splicing reactions, a property that could be useful in defining the role of alternatively spliced mRNAs in megakaryocytes. A drawback is the means of delivery since the preferred option is microinjection; this is obviously not ideal for megakaryocytes, although electroporation might be feasible.

3.1.2. RNA Interference

A method that is stimulating tremendous interest at the moment is that of so-called RNA interference (RNAi). This differs from antisense approaches in that although it involves complementarity between a small RNA molecule and the target mRNA, the active molecule is double-stranded RNA of a defined length (21–22 bp). Specific recog-nition of the target mRNA elicits its cleavage through an evolutionarily conserved RNase III ("Dicer," *[24]*). The most effective RNAi molecules are those in which the two strands are linked by a short loop sequence that mimics the structure of an unprocessed intron. Until recently, use of this technique involved injection or uptake of synthetic RNAi molecules into cells. However, the range of possibilities has been increased since it has been shown by several groups (*see*, for example, **refs. *25,26***) that stable expression of an RNAi molecule can be achieved from a plasmid vector using the RNA pol III-dependent promoter from the U6 RNA gene (**Fig. 5**). This raises

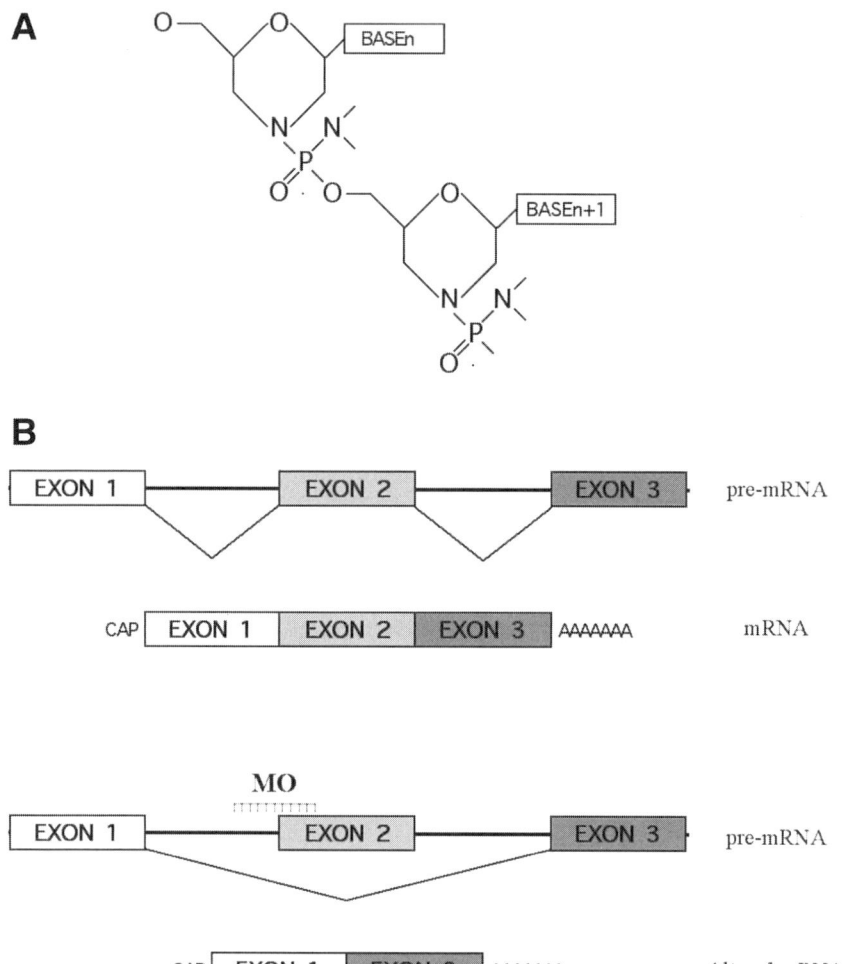

Fig. 4. (**A**) Diagrammatic representation of the basic structural unit of a morpholino oligonucleotide. (**B**) Possible application of morpholino oligonucleotides to investigate alternative spliced mRNAs. The morpholino (MO) is shown annealing specifically at the junction between intron 1 and exon 2 in the pre-mRNA thereby preventing the preferred splicing and leading to a mature mRNA containing only exons 1 and 3.

the possibility that this strategy could be employed in a transgenic context, providing a convenient means to make "knockouts" of genes of interest. Unless a mechanism can be designed to allow imposition of control on such a transgene (perhaps along the lines suggested in **Subheading 2.3.2.**), this approach would have to be limited to genes whose loss of function is not lethal.

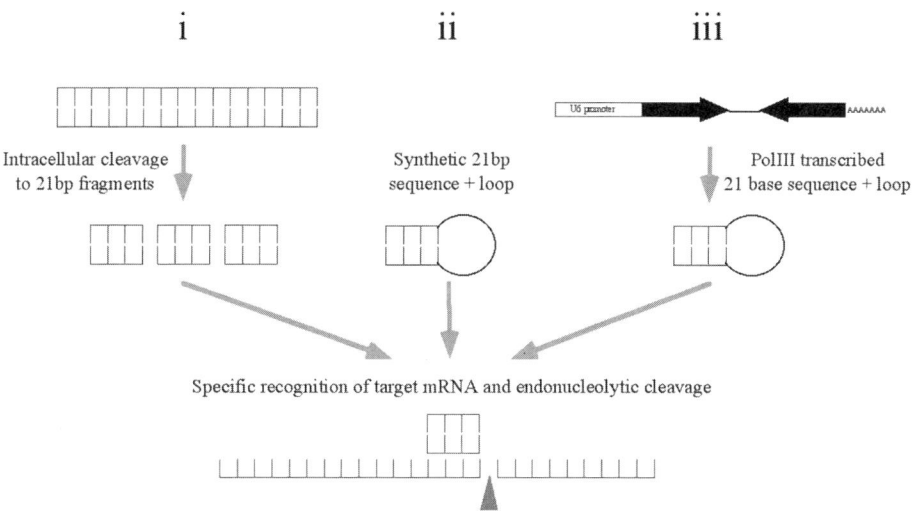

Fig. 5. RNA interference (RNAi). Three routes for the generation of 21-bp double stranded RNAi molecules are illustrated. *(i)* Large double-stranded RNAs are introduced by injection or transfection and are cleaved into fragments by a nonspecific endonuclease. *(ii)* Synthetic oligonucleotides can be introduced by injection or transfection. These are most efficient if the two strands are connected by a small loop. *(iii)* The 21-base sequence and its complement separated by a small "intron" are transcribed from plasmid DNA utilizing a polIII promoter (e.g., from the U6 RNA gene). A short run of Ts at the end of the construct ensures correct termination.

3.2. Removal or Mutation of a Gene

3.2.1. Constitutive Knockout

The most widely adopted approach to removing function of a specific gene involves the irreversible modification, or "knockout," of the chromosomal locus *(27)*. Ways to achieve a knockout are through the removal of key exons or promoter elements or by insertion of a block that prevents correct transcription, splicing, or translation of the RNA. Careful design of the targeting construct is very important to ensure that gene ablation is complete and that alternative transcription or splicing does not lead to an aberrant product from the modified locus. As described above (**Subheading 2.3.3.**), targeting by homologous recombination involves insertion of vector sequences into ES cells followed by selection for clones carrying the knockout allele (**Fig. 6**).

Quite a large number of knockouts have been described that have a megakaryocyte/ platelet phenotype (**Table 2**). These knockouts fall into roughly two categories: (1) those that confirmed previously inferred roles for the gene in megakaryocytes and platelets, and (2) those that revealed an unexpected role for the ablated gene product in the megakaryocytic lineage. Examples in the first group are classic megakaryocyte and platelet molecules such as the *c-mpl* gene encoding the receptor for TPO *(28)* and the *gpIIb* gene *(22)*, while

Emambokus et al.

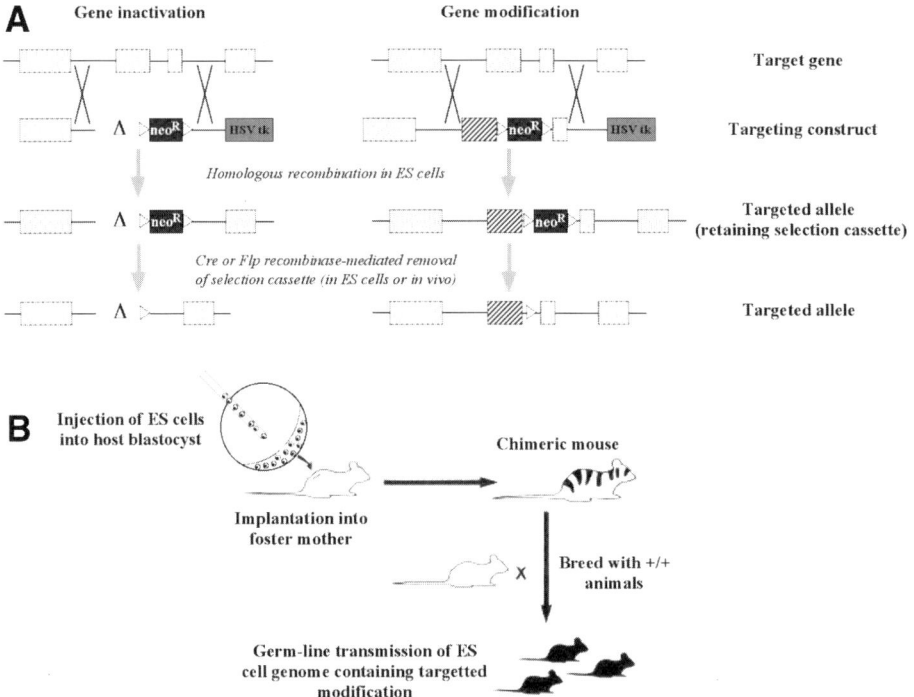

Fig. 6. Gene targeting by homologous recombination in ES cells. (**A**) Two possible strategies are illustrated for modification of a target gene of interest. Coding exons are depicted as light shaded boxes. Homologous recombination between the target gene and the targeting construct is indicated by crosses. On the left, gene inactivation is achieved by removal of sequences including a crucial exon, while on the right, no sequences are deleted but the coding sequence is altered in a particular exon (cross-hatched). Positive and negative selection cassettes (neo^R and HSV-tk) are shown as darker shaded boxes. The open arrows represent recognition sites for either Cre or Flp recombinase. (**B**) Schematic representation of the production of a stable mouse line from a genetically modified ES cell clone. Successful incorporation of the modified ES cells (donor) into the host blastocysts is determined by appearance of the donor coat color (black in this illustration) in the initial chimaeras and following demonstration of germ line transfer.

examples of a gene whose importance in megakaryocyte development was serendipitously revealed is that coding for the NF-E2 transcription factor *(29)*.

The number of genes successfully knocked out is growing rapidly. The best way to keep up with these is through one of the Web sites that are regularly updated (for example, http://tbase.jax.org or http://research.bmn.com/mkmd).

Table 2
Gene Knockouts With a Megakaryocyte/Platelet Phenotype

Gene	Nature of knockout	MK/platelet phenotype	Reference
FOG-1	Constitutive	Failure of commitment to megakaryopoiesis	*44,46*
Gfi-1b	Constitutive	Failure of megakaryocyte development	*46*
ICAM	Constitutive	Defective megakaryocyte progenitor	*47*
GATA-1	Lineage-specific (enhancer deletion)	Decreased platelets, increased immature megakaryocytes	*40*
Fli-1	Constitutive	Decreased platelets, dysmegakaryopoiesis	*48*
MafG	Constitutive	Decreased platelets, increased megakaryocytes	*49,50*
NF-E2	Constitutive	Decreased platelets	*40*
c-Mpl	Constitutive	Decreased megakaryocytes and platelets	*28,51*
TPO	Constitutive	Decreased platelets	*52*
Bcl-x	Lineage-specific (E/MK-specific Cre-loxP)	Decreased platelets, increased immature megakaryocytes	*37*
VASP	Constitutive	Megakaryocyte hyperplasia and platelet dysfunction	*53*
c-cbl	Constitutive	Megakaryocyte hyperplasia and thrombocytosis	*54*
WASP	Constitutive	Decreased and dysfunctional platelets	*55*
gpIbα	Constitutive	Dysfunctional platelets	*56*
α_{IIb} integrin	Constitutive	Dysfunctional platelets	*22*
β_3 integrin	Constitutive	Dysfunctional platelets	*57*
GPV	Constitutive	Dysfunctional platelets	*58*
PAR 3	Constitutive	Dysfunctional platelets	*59*
PAR 4	Constitutive	Dysfunctional platelets	*60*
PECAM	Constitutive	Dysfunctional platelets	*61*
CD39	Constitutive	Dysfunctional platelets	*62*
LAT	Constitutive	Dysfunctional platelets	*63*
FcγRII	Constitutive	Dysfunctional platelets	*64*
Syk	Constitutive	Dysfunctional platelets	*30,64*
SLP-76	Constitutive	Dysfunctional platelets	*65*
TGF-β1	Constitutive	Dysfunctional platelets	*66*
CGMP kinase I	Constitutive	Dysfunctional platelets	*67*

3.2.2. Conditional Knockout

One of the major drawbacks of conventional knockouts is that if the gene product is essential in many tissues in addition to the cell type of interest, then it is quite likely that the consequence of homozygosity for the mutated allele will be lethality, often at some point during development. Several experimental strategies are employed or have the potential to overcome the problem of lethality.

A possible, although little-explored, solution is to render the mutant allele homozygous in a genetic background in which the gene product is expressed from a transgene only in those cells that are thought to be responsible for the lethal phenotype. As an example, Syk kinase is an important component of signaling in megakaryocytes/platelets; however, the knockout is a perinatal lethal *(30)* possibly because defects in vessel endothelial cells lead to serious bleeding problems. If the endothelial defect could be selectively rescued using a *syk* transgene driven by an endothelial-specific promoter (for example, from the *flk-1* gene) then mice might survive adulthood and the influence of the absence of Syk in platelets could be examined.

Another solution, which has been applied in a few cases, is to use hemopoietic cells from the fetal liver of knockout embryos to reconstitute the hemopoietic system of lethally irradiated wild-type animals (so-called radiation chimeras). It is important to ensure that the genetic background of the donor and host are closely matched, if necessary by crossing the knockout strain to the host wild-type strain for at least five generations. Application to the *syk* knockout again provides a good example of the use of this technique *(31)*.

The solution to the problem of knockout lethality that is rapidly becoming almost standard is to restrict the cell type in which deletion occurs using the Cre-LoxP technology based on bacteriophage recombination systems *(32)*. Cre recombinase specifically recombines DNA sequences flanked by LoxP sites ("floxed") as illustrated in **Fig. 7A** *(33)*. The gene of interest is modified in ES cells by homologous recombination as described above. A large number of genes have been modified in this way and many of these will probably be of interest to those studying various aspects of megakaryocyte/platelet biology (*see* **ref. 34** for review of modified genes). Lineage- or stage-specific deletion is then achieved by crossing the floxed gene to a transgenic or knock-in strain that expresses Cre recombinase in a specific cell type **(Fig. 7B)**. Strains expressing a CreER fusion protein can be used for direct activation of Cre activity after feeding animals with estrogen *(35)*, while a Cre transgene driven from the interferon-responsive Mx promoter can be used to control expression at the transcriptional level by injecting animals with interferon or double-stranded RNA, which induces an interferon response *(36)*. As more strains become available,

---

Fig. 7. *(see opposite page)* Conditional gene ablation using Cre-loxP technology. **(A)** The gene of interest or a critical domain within it is flanked by Cre recombinase recognition sites (loxP), which are indicated by the arrows. When bound by Cre, two sites in the same orientation are recombined resulting in the deletion of the intervening sequences. **(B)** Strategy for the generation of a conditional allele. This is very similar to the scheme

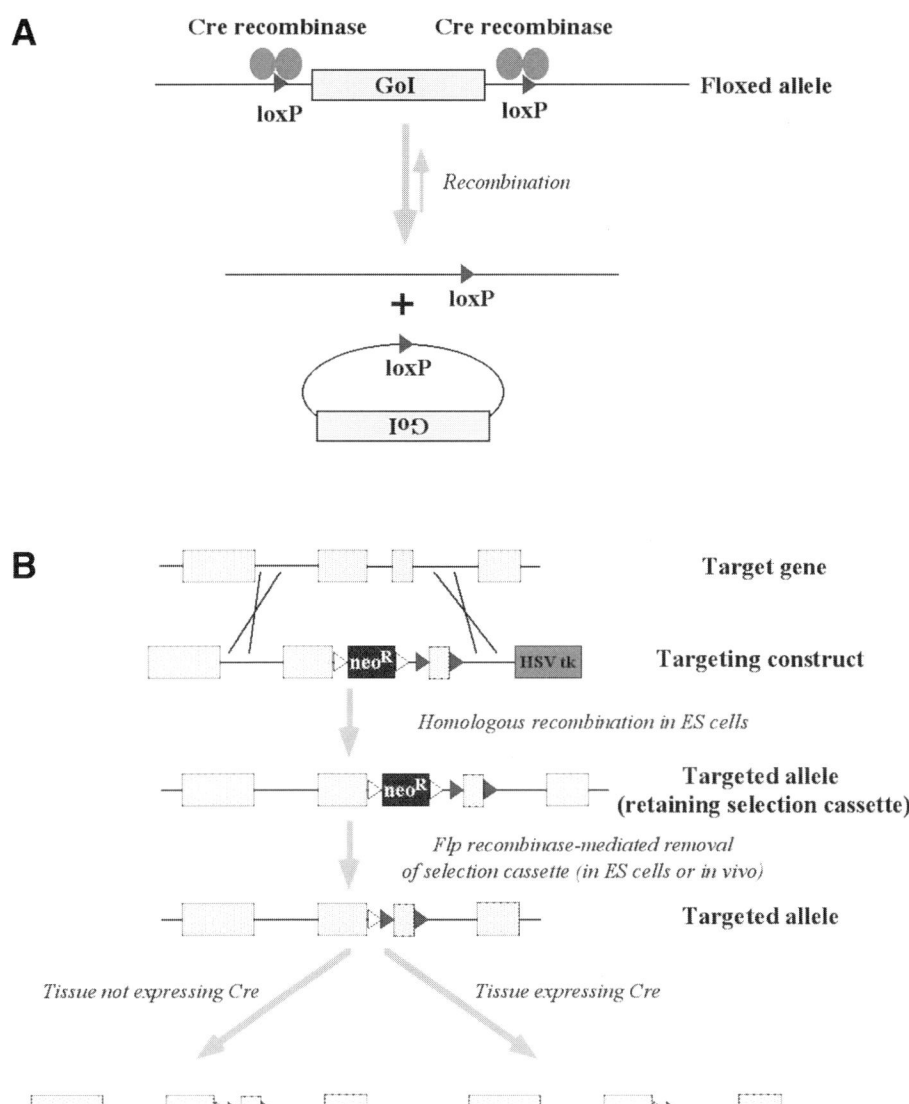

Fig. 7. *(continued)* described in **Fig. 6A**, except that two sets of recombinase recognition sites are introduced. In this example, Flp recognition sites (open arrows) are placed around the neo[R] selection cassette to enable its subsequent removal by Flp either in the ES cells or once transmitted to the mouse germ line. Cre recognition sites (shaded arrows) surround the sequences to be conditionally deleted.

lineage-specific expression of Cre recombinase from a transgene or knock-in gene is growing in popularity as a means to control gene deletion. Available strains expressing Cre recombinase can be found in a database compiled by Andreas Nagy (http://www.mshri.on.ca/nagy/Cre.htm).

Cre-LoxP technology shows great promise; nevertheless, there are inherent problems associated with it, the most challenging one remaining the specificity of Cre-recombinase expression. Although knock-ins are conceptually more likely to give the desired pattern of expression, they may be limited in that expression from a single copy might not always generate sufficient Cre protein to elicit efficient recombination. To date, there is no published megakaryocyte-specific Cre-recombinase strain, although a number of laboratories are actively trying to create one. To our knowledge attempts are being made using the *PF4* gene promoter, the *gpVI* gene as a knock-in and, in our own laboratory, the gpIbα gene promoter (J. F., unpublished). In the absence of a perfectly lineage-specific Cre strain, some success has been obtained in deletion of genes in megakaryocytes using mice that express Cre in a limited number of cell types. Hence, floxed alleles of the *bcl-x* and *piga* genes have been specifically recombined in megakaryocytes by making use of the MMTV-Cre *(37)* and GATA-Cre *(38)* strains, respectively.

As we learn more about the regulatory elements that control lineage-specific expression, it may become possible to achieve lineage-specific gene knockouts by promoter-element deletion rather than deletion of sequences encoding functional domains. An example of such an approach has been described relating to gene ablation in the megakaryocytic lineage. The GATA-1 transcription factor, when deleted constitutively, results in embryonic lethality from anemia due to blocked erythropoiesis *(39)*. Serendipitously, Shivdasani et al. *(40)* found that deletion of a particular upstream regulatory element of the GATA-1 gene caused specific loss of expression in the megakaryocytic lineage and a consequent block to differentiation.

3.2.3. Knockdown

Serendipity has often also led to the identification of gene modifications that produce a reduction in gene expression rather than a complete ablation. In some cases this may have experimental value in determining the function of a gene product. It is difficult at the moment to predict exactly which genetic modifications will produce such a "knockdown" phenotype, although the insertion of the neo[R] selection cassette within an intron can often have such an effect. Alternatively, modification of certain promoter elements might produce a reduction in expression levels, as has been seen for the GATA-1 gene knockdown mutation in a promoter proximal regulatory element *(41)*. Again, prediction of which strategies are likely to be effective in megakaryocytes will become easier as we learn more from this technology and discover more about regulatory mechanisms that control megakaryocyte-specific gene expression.

3.2.4. Knock-In Mutations

Ablation or reduction of gene expression is not the only way in which modification of a gene can be used to explore the function of the encoded product. Homologous recombination in ES cells can also be used to modify specific residues in the coding sequence so that alternative amino acids are introduced into the protein (**Fig. 6A**). Production of mutated proteins in this way will be particularly useful as a way of producing animal models for polymorphic variants of proteins thought to be linked to disease susceptibility. More subtle mutations of this sort also provide a potential means for unraveling the function of individual protein domains. For instance, a mouse model was generated to test the role of the cytoplasmic tyrosine motifs in the β_3 chain of the $\alpha_{IIb}\beta_3$ integrin receptor in mediating outside-in signaling by replacing them with phenylalanines *(31)*.

4. Future Advances

We have tried to summarize many of the most up-to-date advances in technologies that enable manipulation of gene expression in mammalian cells, and which might be particularly useful in relation to work on megakaryocytes and platelets. Most of these techniques are undergoing rapid improvements and we can only wonder at the sophistication that lies around the corner. Certainly, many of the methodologies will become more streamlined and more rapid. Dissemination of the wealth of methods and reagents—for example, available transgenic or floxed gene strains and mice expressing Cre recombinase in specific cells—is bound to facilitate advances. Combination of manipulated gene expression in megakaryocytes with analysis of the transcriptosome or proteome is also likely to become common practice. Our discussion of methodologies has centered on application to already partially characterized genes, but of course there are bound to be many more genes relevant to megakaryocyte/platelet biology that are yet to be identified. Candidates will emerge from microarray screening of expressed sequences and proteomic analyses, but there must also be a huge potential for screening for mouse mutations, probably recessive, that influence various aspects of megakaryopoiesis and platelet function. Many such large-scale projects are either planned or under way *(42,43)* and it is hoped that the phenotypic screening strategies that are employed will also include examination for effects at the level of platelet numbers and function.

Acknowledgments

N. R. E. and J. F. are supported by the Wellcome Trust as part of the Senior Fellowship award to J. F. G. J. M. was supported by the British Heart Foundation.

References

1. Maurer, A. M., Liu, Y., Caen, J. P., and Han, Z. C. (2000) Ex vivo expansion of megakaryocytic cells. *Int. J. Hematol.* **71**, 203–210.
2. Burstein, S. A., Dubart, A., Norol, F., Debili, N., Friese, P., Downs, T., et al. (1999) Expression of a foreign protein in human megakaryocytes and platelets by retrovirally medicated gene transfer. *Exp. Hematol.* **27**, 110–116.
3. Hawley, R. G., Lieu, F. H., Fong, A. Z., and Hawley, T. S. (1994) Versatile retroviral vectors for potential use in gene therapy. *Gene Ther.* **1**, 136–138.
4. Baccini, V., Roy, L., Vitrat, N., Chagraoui, H., Sabri, S., Le Couedic, J.-P., et al. (2001) Role of p21 Cip1/Waf1 in cell-cycle exit of endomitotic megakaryocytes. *Blood* **98**, 3274–3282.

5. Wilcox, D. A., Olsen, J. C., Ishizawa, L., Griffith, M., and White, G. C. (1999) Integrin αIIb promoter-targeted expression of gene products in megakaryocytes derived from retrovirus-transduced human hematopoietic cells. *Proc. Natl. Acad. Sci. USA* **96**, 9654–9659.

6. Federspiel, M. J. and Hughes, S. H. (1997) Retroviral gene delivery. *Methods Cell Biol.* **52**, 179–214.

7. Murphy, G. J. and Leavitt, A. D. (1999) A model for studying megakaryocyte development and biology. *Proc. Natl. Acad. Sci. USA* **96**, 3065–3070.

8. Shiraga, M., Ritchie, A., Aidoudi, S., Baron, V., Wilcox, D., White, G., et al. (1999) Primary megakaryocytes reveal a role for transcription factor NF-E2 in integrin alpha IIb beta 3 signaling. *J. Cell Biol.* **147**, 1419–1430.

9. Gaur, M., Murphy, G. J., deSauvage, F. J., and Leavitt, A. D. (2001) Characterization of MPL mutants using primary megakaryocyte-lineage cells from mpl$^{-/-}$ mice: a new system for Mpl structure-function studies. *Blood* **97**, 1653–1661.

9a. Murphy G. J., Göttgens, B., Vegiopoulos, A., Sanchez, M-J., Leavitt A., Watson, S. P., et al. Manipulation of mouse hematopoietic progenitors by specific retroviral infection. *J. Biol. Chem.* **278**, 43,556–43,563.

10. Curran, M. A., Kaiser, S. M., Achacoso, P. L., and Nolan, G. P. (2000) Efficient transduction of nondividing cells by optimized feline immunodeficiency virus vectors. *Mol. Ther.* **1**, 31–38.

11. Lewis, B. C., Chinnasamy, N., Morgan, R. A., and Varmus, H. E. (2001) Development of an avian leukosis-sarcoma virus subgroup A pseudotyped lentiviral vector. *J. Virol.* **75**, 9339–9344.

12. Faraday, N., Rade, J. J., Johns, D. C., Khetawat, G., Noga, S. J., DiPersio, J. F., et al. (1999) Ex vivo cultured megakaryocytes express functional glycoprotein IIb-IIIa receptors and are capable of adenovirus-mediated transgene expression. *Blood* **94**, 4084–4092.

13. Wade-Martins, R., Smith, E. R., Tyminski, E., Chiocca, E. A., and Saeki, Y. (2001) An infectious transfer and expression system for genomic DNA loci in human and mouse cells. *Nat. Biotech.* **19**, 1067–1070.

14. Schwarze, S. R., Ho, A., Hruska, K. A., and Dowdy, S. F. (2000) Protein transduction: unrestricted delivery into all cells? *Trends Cell Biol.* **10**, 290–295.

15. Gordon, J. W. and Ruddle, F. H. (1981) Integration and stable germ line transmission of genes injected into mouse pronuclei. *Science* **214**, 1244–1246.

16. Gordon, J. W., Scangos, G. A., Plotkin, D. J., Barbosa, J. A., and Ruddle, F. H. (1980) Genetic transformation of mouse embryos by microinjection of purified DNA. *Proc. Natl. Acad. Sci. USA* **77**, 7380–7384.

17. Mills, A. A. (2001) Changing colors in mice: an inducible system that delivers. *Genes & Dev.* **15**, 1461–1467.

18. Gossen, M. and Bujard, H. (1992) Tight control of gene expression in mammalian cells by tetracycline-responsive promoters. *Proc. Natl. Acad. Sci. USA* **89**, 5547–5551.

19. Bujard, H. (1999) Controlling genes with tetracyclines. *J. Gene Med.* **1**, 372–374.

20. Deuschle, U., Meyer, W. K.-H., and Thiesen, H.-J. (1995) Tetracycline-reversible silencing of eukaryotic promoters. *Mol. Cell. Biol.* **15**, 1907–1914.

21. Capecchi, M. R. (1989) Altering the genome by homologous recombination. *Science* **244**, 1288–1292.

22. Tronick-Le Roux, D., Roullot, V., Poujol, C., Kortulewski, T., Nurden, P., and Marguerie, G. (2000) Thrombasthenic mice generated by replacement of the integrin α$_{IIb}$ gene: demonstration that transcriptional activation of this megakaryocytic locus precedes lineage commitment. *Blood* **96**, 1399–1408.

23. Ekker, S. C. (2000) Morphants: a new systematic vertebrate functional genomics approach. *Yeast* **17,** 302–306.
24. Fire, A., Xu, S., Montgomery, M. K., Kostas, S. A., Driver, S. E., and Mello, C. C. (1998) Potent and specific genetic interference by double-stranded RNA in *Caenorhabditis elegans. Nature* **391,** 806–811.
25. Paul, C. P., Good, P. D., Winer, I., and Engelke, D. R. (2002) Effective expression of small interfering RNA in human cells. *Nat. Biotech.* **20,** 505–508.
26. Sui, G., Soohoo, C., Affar, E. B., Gay, F., Shi, Y., Forrester, W. C., et al. (2002) A DNA vector-based RNAi technology to suppress gene expression in mammalian cells. *Proc. Natl. Acad. Sci. USA* **99,** 5515–5520.
27. Thomas, K. R. and Capecchi, M. R. (1987) Site-directed mutagenesis by gene targeting in mouse embryo-derived stem cells. *Cell* **51,** 503–512.
28. Gurney, A. L., Carver-Moore, K., de Sauvage, F. J., and Moore, M. W. (1994) Thrombo-cytopenia in c-mpl-deficient mice. *Science* **265,** 1445–1447.
29. Shivdasani, R. A., Rosenblatt, M. F., Zucker-Franklin, D., Jackson, C. W., Hunt, P., Saris, C. J., et al. (1995) Transcription factor NF-E2 is required for platelet formation independent of the actions of thrombopoietin/MGDF in megakaryocyte development. *Cell* **81,** 695–704.
30. Turner, M., Mee, P. J., Costello, P. S., Williams, O., Price, A. A., Duddy, L. P., et al. (1995) Perinatal lethality and blocked B-cell development in mice lacking the tyrosine kinase Syk. *Nature* **378,** 298–302.
31. Law, D. A., DeGuzman, F. R., Heiser, P., Ministri-Madrid, K., Killeen, N., and Phillips, D. R. (1999) Integrin cytoplasmic tyrosine motif is required for outside-in $\alpha_{IIb}\beta_3$ signalling and platelet function. *Nature* **401,** 808–811.
32. Gu, H., Marth, J. D., Orban, P. C., Mossmann, H., and Rajewsky, K. (1994) Deletion of a DNA polymerase beta gene segment in T cells using cell type-specific gene targeting. *Science* **265,** 103–106.
33. Nagy, A. (2000) Cre recombinase: the universal reagent for genome tailoring. *Genesis* **26,** 99–109.
34. Kwan, K.-M. (2002) Conditional alleles in mice: practical consideratioins for tissue-specific knockouts. *Genesis* **32,** 49–62.
35. Schwenk, F., Kuhn, R., Angrand, P. O., Rajewsky, K., and Stewart, A. F. (1998) Temporally and spatially regulated somatic mutagenesis in mice. *Nucl. Acids Res.* **26,** 1427–1432.
36. Kuhn, R., Schwenk, F., Aguet, M., and Rajewsky, K. (1995) Inducible gene targeting in mice. *Science* **269,** 1427–1429.
37. Wagner, K.-U., Claudio, E., Rucker, E. B., Riedlinger, G., Broussard, C., Schwartzberg, P. L., et al. (2000) Conditional deletion of the Bcl-x gene from erythroid cells results in hemolytic anemia and profound splenomegaly. *Development* **127,** 4949–4958.
38. Jasinski, M., Keller, P., Fujiwara, Y., Orkin, S. H., and Bessler, M. (2001) GATA1-Cre mediates Piga gene inactivation in the erythroid/megakaryocytic lineage and leads to circulating red cells with a partial deficiency in glycosyl phosphatidylinositol-linked proteins (paroxysmal nocturnal hemoglobinuria type II cells). *Blood* **98,** 2248–2255.
39. Pevny, L., Simon, M. C., Robertson, E., Klein, W. H., Tsai, S. F., D'Agati, V., et al. (1991) Erythroid differentiation in chimaeric mice blocked by a targeted mutation in the gene for transcription factor GATA-1. *Nature* **349,** 257–260.
40. Shivdasani, R. A., Fujiwara, Y., McDevitt, M. A., and Orkin, S. H. (1997) A lineage-selective knockout establishes the critical role of transcription factor GATA-1 in megakaryocyte growth and platelet development. *EMBO J.* **16,** 3965–3973.

41. Harigae, H., Takahashi, S., Suwabe, N., Ohtsu, H., Gu, L., Yang, Z., et al. (1998) Differential roles of GATA-1 and GATA-2 in growth and differentiation of mast cells. *Genes Cells* **3,** 39–50.
42. Nolan, P. M., Peters, J., Strivens, M., Rogers, D., Hagan, J., Spurr, N., et al. (2000) A systematic, genome-wide, phenotype-driven mutagenesis programme for gene function studies in the mouse. *Nat. Genet.* **25,** 440–443.
43. de Angelis, M. H., Flaswinkel, H., Fuchs, H., Rathkolb, B., Soewarto, D., Marschall, S., et al. (2000) Genome-wide, large-scale production of mutant mice by ENU mutagenesis. *Nat. Genet.* **25,** 444–447.
44. Tsang, A. P., Visvader, J. E., Turner, C. A., Fujiwara, Y., Yu, C., Weiss, M. J., et al. (1997) FOG, a multiple zinc finger protein, acts as a cofactor for transcription factor GATA-1 in erythroid and megakaryocytic differentiation. *Cell* **90,** 109–119.
45. Tsang, A. P., Fujiwara, Y., Hom, D. B., and Orkin, S. H. (1998) Failure of megakaryopoiesis and arrested erythropoiesis in mice lacking the GATA-1 transcriptional cofactor FOG. *Genes & Dev.* **12,** 1176–1188.
46. Saleque, S., Cameron, S., and Orkin, S. H. (2002) The zinc-finger proto-oncogene Gfi-1b is essential for development of the erythroid and megakaryocytic lineages. *Genes & Dev.* **16,** 301–306.
47. Gerwin, N., Gonzalo, J. A., Lloyd, C., Coyle, A. J., Reiss, Y., Banu, N., et al. (1999) Prolonged accumulation in allergic lung interstitium of ICAM-2 deficient mice results in extended hyperresponsiveness. *Immunity* **10,** 9–19.
48. Hart, A., Melet, F., Grossfeld, P., Chien, K., Jones, C., Tunnacliffe, A., et al. (2000) Fli-1 is required for murine vascular and megakaryocytic development and is hemizygously deleted in patients with thrombocytopenia. *Immunity* **13,** 167–177.
49. Shavit, J. A., Motohashi, H., Onodera, K., Akasaka, J., Yamamoto, M., and Engel, J. D. (1998) Impaired megakaryopoiesis and behavioral defects in *mafG*-null mutant mice. *Genes & Dev.* **12,** 2164–2174.
50. Onodera, K., Shavit, J. A., Motohashi, H., Yamamoto, M., and Engel, J. D. (2000) Perinatal synthetic lethality and hematopoietic defects in compount *mafG* : :*mafK* mutant mice. *EMBO J.* **19,** 1335–1345.
51. Alexander, W. S., Roberts, A. W., Nicola, N. A., Li, R., and Metcalf, D. (1996) Deficiencies in progenitor cells of multiple hematopoietic lineages and defective megakaryocytopoiesis in mice lacking the thrombopoietin receptor c-mpl. *Blood* **87,** 2162–2170.
52. Bunting, S., Widmer, R., Lipari, T., Rangell, L., Steinmetz, H., Carver-Moore, K., et al. (1997) Normal platelets and megakaryocytes are produced in vivo in the absence of thrombopoietin. *Blood* **90,** 3423–3429.
53. Hauser, W., Knobeloch, K.-P., Eigenthaler, M., Gambaryan, S., Krenn, V., Geiger, J., et al. (1999) Megakaryocyte hyperplasia and enhanced agonist-induced platelet activation in vasodilator-stimulated phosphoprotein knockout mice. *Proc. Natl. Acad. Sci. USA* **96,** 8120–8125.
54. Murphy, M. A., Schnall, R. G., Venter, D. J., Barnett, L., Bertoncello, I., Thien, C. B. F., et al. (1998) Tissue hyperplasia and enhanced T-cell signalling via ZAP-70 in c-Cbl-deficient mice. *Mol. Cell. Biol.* **18,** 4872–4882.
55. Snapper, S. B., Rosen, F. S., Mizoguchi, E., Cohen, P., Khan, W., Liu, C. H., et al. (1998) Wiskott-Aldrich syndrome protein-deficient mice reveal a role for WASP in T but not B cell activation. *Immunity* **9,** 81–91.
56. Ware, J., Russell, S., and Ruggeri, Z. M. (1999) Generation and rescue of a murine model of platelet dysfunction: the Bernard-Soulier syndrome. *Proc. Natl. Acad. Sci. USA* **97,** 2803–2808.

57. Hodivala-Dilke, K. M., McHugh, K. P., Tsakiris, D. A., Rayburn, H., Crowley, D., Ullman-Culleré, M., et al. (1999) β3-integrin-deficient mice are a model for Glanzmann thrombasthenia showing placental defects and reduced survival. *J. Clin. Invest.* **103**, 229–238.
58. Ramakrishnan, V., Reeves, P. S., DeGuzman, F., Deshpande, U., Ministri-Madrid, K., DuBridge, R. B., et al. (1999) Increased thrombin responsiveness in platelets from mice lacking glycoprotein V. *Proc. Natl. Acad. Sci. USA* **96**, 13,336–13,341.
59. Kahn, M. L., Zheng, Y. W., Huang, W., Bigornia, V., Zeng, D., Moff, S., et al. (1998) A dual thrombin receptor system for platelet activation. *Nature* **394**, 690–694.
60. Sambrano, G. R., Weiss, E. J., Zheng, Y. W., Huang, W., and Coughlin, S. R. (2001) Role of thrombin signalling in platelets in hemostasis and thrombosis. *Nature* **413**, 26–27.
61. Duncan, G. S., Andrew, D. P., Takimoto, H., Kaufman, S. A., Yoshida, H., Spellberg, J., et al. (1999) Genetic evidence for functional redundancy of platelet/endothelial cell adhesion molecule-1 (PECAM-1): CD31-deficient mice reveal PECAM-1-dependent and PECAM-1-independent functions. *J. Immunol.* **162**, 3022–3030.
62. Enjyoji, K., Sevigny, J., Lin, Y., Frenette, P. S., Christie, P. D., Esch, J. S. 2nd, et al. (1999) Targeted disruption of CD39/ATP diphosphohydrolase results in disordered hemostasis and thromboregulation. *Nat. Med.* **5**, 1010–1017.
63. Zhang, W., Sommers, C. L., Burshtyn, D. N., Stebbins, C. C., DeJarnette, J. B., Trible, R. P., et al. (1999) Essential role of LAT in T cell development. *Immunity* **10**, 323–332.
64. Poole, A., Gibbins, J. M., Turner, M., van Vugt, M. J., van de Winkel, J. G. J., Saito, T., et al. (1997) The Fc receptor g-chain and the tyrosine kinase Syk are essential for activation of mouse platelets by collagen. *EMBO J.* **16**, 2333–2341.
65. Clements, J. L., Lee, J. R., Gross, B., Yang, B., Olson, J. D., Sandra, A., et al. (1999) Fetal hemorrhage and platelet dysfunction in SLP-76-deficient mice. *J. Clin. Invest.* **103**, 19–25.
66. Hoying, J. B., Yin, M., Diebold, R., Ormsby, I., Becker, A., and Doetschman, T. (1999) Transforming growth factor beta 1 enhancers platelet aggregation through a non-transcriptional effect on the fibrogen receptor. *J. Biol. Chem.* **274**, 31,008–31,013.
67. Massberg, S., Sausbier, M., Klatt, P., Bauer, M., Pfeifer, A., Siess, W., et al. (1999) Increased adhesion and aggregation of platelets lacking cyclic guanosine 3′,5′-monophosphate kinase I. *J. Exp. Med.* **189**, 1255–1263.

3

In Vitro Changes of Platelet Parameters

Lessons From Blood Banking

José Rivera, María L. Lozano, and Vicente Vicente

1. Introduction

Platelets, the smallest cellular components of blood, are critically involved at each step of the hemostatic response, from the initial sealing of damaged endothelium to supporting coagulation reactions, and finally, in the retraction of the fibrin clot that enhances fibrinolysis and wound healing. Consequently, if platelet concentration is decreased and/or platelet function is abnormal, the risk of hemorrhage is increased. Ever since Duke's 1910 report of the first use of platelet-containing fresh whole blood to treat three bleeding thrombocytopenic patients, platelet transfusions have gained value in medicine. Nowadays they are essential for the management of patients with primary thrombocytopenia, or for support of those treated with intensive chemo/radiotherapeutic regimens associated with prolonged periods of bone marrow aplasia *(1,2)*.

The collection, processing, and storage of platelets for clinical use have undergone significant changes over the last few decades, and novel approaches are being investigated to develop improved platelet products. These include research on new anticoagulants and additive solutions, modified storage containers and collection sets, procedures for virus and/or bacterial inactivation, chemical and/or physical methods to extend the shelf-life, and development of artificial or pseudoartificial platelet products *(3,4)*. Platelet concentrates (PCs) are routinely prepared during blood banking from donated whole blood collected into multiple plastic bags. The blood is collected in a primary bag that most frequently contains citrate-phosphate-dextrose (CPD) or CPD-adenine (CPD-A) as the anticoagulant (not normally acid-citrate-dextrose, as maintenance of a pH above 6.2 is a widely accepted requirement for PCs) and then separated into its components using either the platelet-rich plasma (PRP) or the buffy-coat (BC) procedure. These both use a two-stage centrifugation procedure, but differ in the speed and duration of each step (summarized in **Fig. 1**). In addition, PCs can be obtained directly from donors by apheresis *(5)*. The main feature that differentiates the use of apheresis from whole blood

From: *Methods in Molecular Biology, vol. 273:*
Platelets and Megakaryocytes, Vol. 2: Perspectives and Techniques
Edited by: J. M. Gibbins and M. P. Mahaut-Smith © Humana Press Inc., Totowa, NJ

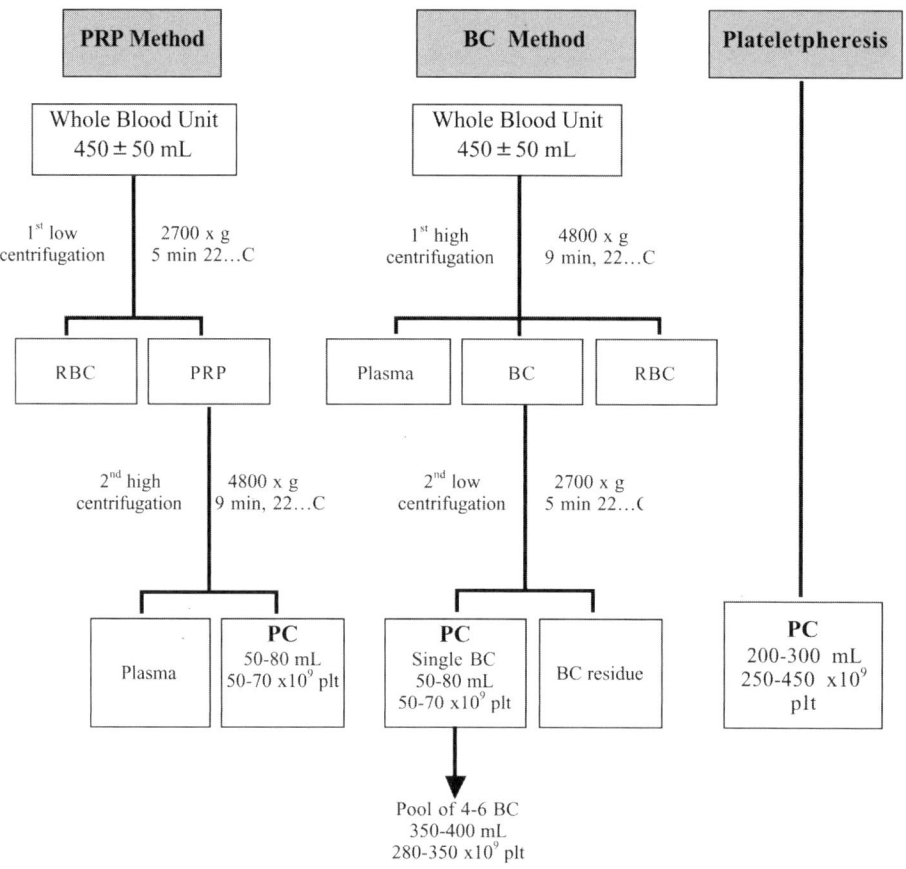

Fig. 1. Current methods for preparation of platelet concentrates. BC: buffy coat; PRP, platelet-rich plasma; RBC, red blood cells; PC, platelet concentrate; plt, platelets.

units to prepare PCs is that apheresis reduces the number of donors to whom it is necessary for the recipient to be exposed, although the theoretical advantages related to this have not been conclusively documented.

Despite the basic differences in the three preparative approaches, PCs prepared by all methods share major biochemical features, and potential differences in the activation state of platelets depend more on the production performance, storage modalities, and PC age, rather than in the preparative approach (PRP, BC, or apheresis) *(5–7)*. Currently, all conventional liquid PCs are stored in highly permeable plastic bags at a controlled temperature of 22 ± 2°C, and with constant mild agitation to facilitate gas exchange through the plastic container. Under these conditions, PCs for clinical use can be stored for only 5 d, a limitation that results in the frequent outdating of units and complicates the management of platelet inventories in blood banking. The primary reason for the 5-d shelf life of current PCs is concern about unacceptable increase in the risk of bacterial growth in PCs associated with extended storage at 22°C that may trigger septic reactions in recipients. Because of this, there is now considerable interest in developing methods to achieve bacterial inactivation of PCs as a strategy to produce safer PCs and to lengthen their shelf life *(8,9)*.

It is well known that when platelets are removed from circulation and stored under blood-bank conditions, exposure to a variety of mechanical and chemical influences results in a series of changes, collectively referred to as the platelet storage lesion (PSL) *(10–12)*. In addition, during storage of PCs, a small but growing number of platelets reach the end of their normal life span, and programmed cell death (apoptosis) begins. Recently it has been shown that platelets, despite being enucleated cells, have retained the memory of the parental megakaryocytes for programmed cell death, and that platelet apoptosis and PSL may be related to each other *(13,14)*. In this chapter we highlight some of the key features of the PSL, placing emphasis on several available tests for its monitoring.

2. Platelet Storage Lesion

The storage lesion developing in PCs under standard blood-bank conditions can be defined as a complex process encompassing all the deleterious changes in platelet morphology, structure, and function occurring from the act of venipuncture to the moment of administration of PCs to patients. We know that the rate of appearance of PSL in PCs can be influenced to a variable degree by several physical, chemical, and metabolic factors related to platelet withdrawal, preparation, and storage. These factors include blood flow rate during donation, type and ratio of anticoagulant, the characteristic of the plastic storage bags, the preparation procedure (PRP, BC, or apheresis), the overall storage conditions, and any kind of supplementary treatments of PCs **(Table 1)**. The questions that remain unresolved are what type and extent of storage-promoted changes may be tolerable without unacceptable losses of platelet viability and hemostatic efficacy upon transfusion.

The current gold standard of clinical platelet efficacy evaluation is in vivo survival of transfused radiolabeled platelets, based on the assumption that only viable circulating platelets can participate in the responses. In this sense, several studies have shown that increasing platelet storage time results in decreased post-transfusion platelet count increments *(15–18)*, this being more obvious if patients have an adverse clinical condition or are given specific drugs at the time of transfusion *(19)*. Unfortunately, when the quality of PCs is evaluated on a routine basis, the use of in vivo assays is clearly not practical, and in vitro tests are required *(20)*. Over the years no single in vitro test has stood out as a direct surrogate for platelet efficacy; however, it has now been recognized that examination of a number of different aspects of platelet physiology, metabolism, and morphology can produce a reasonable estimate of in vivo platelet efficacy.

3. Platelet Morphology Changes During Blood Banking

Marked changes in platelet morphology are associated with the development of PSL. Human platelets collected in citrate solution initially retain the basic resting discoid form, although they may display a few tendrils. However, within a few days of storage

Table 1
Factors Influencing Development of PSL

Collection and preparation features
  Blood-flow rate
  Type and ratio of anticoagulant
  Temperature and length of storage of blood before and during PC processing
  Whole-blood fractionation scheme (PRP or BC method)
  Centrifugation parameters (temperature, speed, acceleration/deceleration times)
  Resting period before resuspension
Characteristics of storage containers
  Composition of plastic bags
  Bag size and thickness of container wall
  Gas transfer capability of plastic bag
Conditions of the blood-banking storage
  Volume and cell content (platelets, WBC, RBC) of PCs
  Storage in plasma, supplemented plasma, or artificial media
  Type and rate of agitation of PCs
  Temperature and length of storage
Supplementary treatments
  Leukodepletion by filtration before or after PC obtention
  UV-B or gamma irradiation
  Pathogen inactivation
  Pooling, cryopreservation, lyophilization

under blood-banking conditions, most platelets have lost the resting discoid shape, becoming spherical and exhibiting protrusions and long pseudopods. At the same time, separation of tendrils and fragmentation of the cellular membrane lead to shedding of an increasing number of microvesicles. The platelet shape transformation, which is also an early event in the activation of platelets by agonists during the in vivo clotting process, is mainly due to cytoskeleton rearrangements within the platelet. The platelet cytoskeleton is an intricately woven network arranged in three major structures: a cytoplasmic actin network, a rim of membrane-associated cytoskeleton, and a marginal band consisting of a microtubule coil. Together, these support the platelet plasma membrane and give shape to both resting and activated platelets. During platelet activation, an orchestrated signaling cascade occurs, regulated in part by changes in phosphoinositides, that has a profound impact on actin assembly and cytoskeletal reorganization. Similarly, during preparation and storage of PCs, platelets are exposed to a variety of mechanical and chemical influences that may promote changes in membrane lipids, phosphoinositides, and second messenger levels that lead to actin assembly and hence to shape change.

Evaluation of the disk-to-sphere transformation status can be measured using a variety of techniques, including observation of the swirling phenomenon, assessment of platelet size distribution pattern in terms of volume, evaluation of the changes in platelet volume in response to either agonist stimulation (extent of shape change, ESC) or osmotic stress (hypotonic shock response, HSR), and direct assessment of platelet morphology by phase microscopy. Several correlation studies between in vivo and in vitro variables support the observation that assessment of platelet discoid shape status, by any of the available tests, is a useful predictor of in vivo platelet viability *(12)*.

3.1. Visual Inspection and the Determination of Swirling Phenomenon

The so-called "swirling" or "shimmering" phenomenon can be defined as the property of discoid platelets to reflect light when exposed to a light source. In contrast, nondiscoid platelets lack the property of swirling *(21,22)*. Although subjective, evaluation of swirling in PCs by gently rotating the container in front of a light source is a simple, noninvasive procedure for routine quality control of PCs. The swirling characteristics are described as follows:

- *Positive:* This degree of swirling is found in 83% of regular PCs stored for 1 d, and is still present in 65% of PCs after 5 d of storage at 22°C.
- *Intermediate:* A noticeable reduction in swirling is observed, as occurs when PCs are subjected to a decline in pH to the range of 6.4 to 6.7, or a rise above 7.5 during storage. It is also typical of PCs stored for 7 to 10 d at 22°C.
- *Negative:* Swirling is absent or only slightly detectable. This is typical of PCs subjected to a decline in pH below 6.3 during storage, PCs stored at 4°C for >24 h, and PCs stored for 10 to 14 d at 22°C.

3.2. Assessment of Platelet Size Distribution Pattern and/or Mean Platelet Volume Using Automatic Cell Counters

Platelet shape change results in about a 25% apparent increase in mean platelet volume (MPV) *(23)*. Throughout storage of PCs this transformation can be easily detected by assessing changes in the pattern of platelet-size distribution provided by particle counters based on impedance. This test is simple and rapid; changes in this variable during storage have been related to metabolic alterations in PCs, i.e., decreased pH, bicarbonate, and oxygen consumption *(24)*. However, it may provide limited information unless measurement is carried out at least twice during storage, i.e., immediately after preparation and sometime thereafter. Alternatively, the particle counter can be used to evaluate platelet discoid-shape status in PCs at a single point during storage, by inducing shape change in a PC sample with cold or prolonged exposure to EDTA, conditions known to cause this transformation *(12)*. Thus, if most platelets have already undergone a disc-to-sphere transformation, no further changes in size distribution would be observed. To perform the assay, PC samples are diluted $1:3$ with saline (3×10^5 platelets/μL) and MPV is measured before and 2.5 h after the addition of the diluted samples to K_2EDTA (4.7 mM final concentration) in a commercially available vacuum tube (Terumo Europe NV, Leuven, Belgium). The extent of the difference in MPV (dMPV) is calculated as (EDTA MPV-CPD-A MPV/CPD-A MPV) × 100. Throughout storage, a significant decrease in dMPV is observed, from 15.3% on day 1, to 7.6% on day 3, and 0.7% on day 7 *(25)*. These changes correlate well with those observed for pH (*see* **Subheading 6.**) and swirling (*see* **Subheading 3.1.**) *(12)*. In addition, an evaluation of platelet counts in PC samples with and without EDTA could also be informative of the aggregation state of platelets (*see* **Subheading 3.3.**).

3.3. Extent of Shape Change Induced by Agonist Stimulation

Another useful in vitro parameter for evaluating loss/maintenance of the platelet discoid shape in PCs during storage is the extent of shape change (ESC) caused by exposure to platelet agonists, often ADP. The ESC is defined as the ADP-induced decrease in the light transmittance (increase in optical density, OD) of the platelet suspension measured by means of an aggregometer or a photometer with stirring capability (*see* Chapter 6, vol. 1 for further discussion of light transmittance and OD measurements).

One method to carry out this assay is to test the ESC response of platelets in PRP to 20 μM ADP under conditions where aggregation is blocked—for example, using external EDTA. Briefly, platelets are adjusted to a density of 3×10^8 cells/mL (using platelet-poor plasma [PPP]) and 20 µL of 0.1 M EDTA is added to a 0.5-mL aliquot of a platelet suspension in a standard aggregometer using PPP as a reference, followed immediately by 10 µL of 1 mM ADP. The deflection from the baseline is measured at its peak and the ESC is determined by the percentage increase in optical density *(26)*. For fresh discoid platelets, the increase in ESC has been found to be at 20 to 30%, while PCs stored for 5 d have an increase in ESC of 15 to 25% *(27)*.

3.4. Hypotonic Shock Response (HSR)

In response to a hypotonic stress, human platelets initially take up water and swell *(28)*. As platelets swell, their refractive index decreases, resulting in increased light transmittance (decreased OD), which can be measured photometrically. In platelets with preserved functionality, regulatory volume responses then lead to extrusion of salts, and thus also water, resulting in a reduction of volume and a measurable decrease in transmitted light (increase in OD).

To perform the assay *(26)*, adjust PCs in PRP to a concentration of 3×10^8 cells with PPP, stir gently in a spectrophotometer cuvet and add water (1.5:1 volume ratio of PRP:water). The optical density at 610 nm is recorded immediately after water addition (highest transmittance or lowest optical density) and following incubation of cells for 15 min. An identical optical density recording is made in parallel in a platelet suspension mixed with isotonic PBS instead of water. The percentage HSR is determined by applying the formula:

[(OD post 15 min in water – lowest OD)/(OD in PBS – lowest OD)] × 100.

The HSR has long been used for evaluation of the quality of platelet products. In a multilaboratory evaluation of in vitro platelet assays, HSR was found to perform satisfactorily in terms of accuracy and precision, and to correlate reasonably well with in vivo platelet viability. Commonly, fresh units of platelets display an HSR ranging from 50 to 90%, while a unit stored for 5 d under current optimal conditions will have an HSR level ranging from 40 to 80% *(27)*.

4. Changes Reflecting Platelet Secretion and Activation During Storage

In addition to shape transformation, storage-promoted activation triggers a release reaction in platelets measurable by determining the concentration of granule contents in the supernatant of stored PCs. Platelets have three major organelles whose content is released upon activation: α-granules, dense granules (also called dense bodies), and lysosomes. The α-granules are the largest and most abundant, numbering about 50–80 per platelet. At rest, α-granules contain a variety of different proteins, some of which are platelet-specific, such as platelet factor 4 (PF4), platelet-derived growth factor (PDGF), and beta thromboglobulin (β-TG). α-Granules also contain small amounts of other plasma proteins (vWF, fibrinogen, FV, albumin, immunoglobulins, etc.). Dense granules are the second-largest storage compartment within the platelet, and the first to fuse with channel of the open canalicular system (OCS) during storage-promoted or physiological platelet activation. Dense granules contain calcium, serotonin, ADP, and ATP, as well as pyrophosphate, epinephrine, dopamine, and histamine. Platelets also have lysosomal granules that contain acid hydrolases such as β-glucoronidase, cathepsins, elastase, and collagenase. When platelets undergo secretion, lysosomal contents are more slowly and incompletely released than are the contents of α-granules and dense granules.

An increase in the concentration of granule-released compounds in PCs may be easily measured by means of commercially available kits, and used as an index of storage-promoted platelet activation. For instance, the level PDGF and β-TG within α-granules progressively diminishes during storage (17% and 19% reduction on day 1, and 31% and 35% reduction on day 5, respectively) *(29)*. PDGF is a potent mitogen and acts as a chemotactic factor for fibroblasts, thus promoting wound healing, although it may also have a negative feedback effect on the activation of platelets. Overall, it is reasonable to assume that platelets having lost a significant proportion of PDGF are less efficient at stimulating tissue restoration, by virtue of a reduction in growth-promoting activity.

In addition to the secretion of granule proteins, the release reaction during storage of PCs also increases expression on the platelet surface of glycoproteins from the granule membranes. Thus, indices widely used for the evaluation of quality of PCs include the enhancement of platelet surface expression of P-selectin (CD62P or GMP 140, from α-granules and dense granules), granulophysin (CD63, from dense granules and lysosomes), and LAMP 1 (CD107a) and LAMP-2 (CD107b) (lysosome-associated membrane proteins). Particularly, the surface expression of P-selectin has been postulated to associate with increased platelet clearance by the reticuloendothelial system (and thus with shortening of platelet survival within circulation) and with an increased formation of platelet-leukocyte aggregates through the PSGL-1 receptor *(30)*. The exact role of P-selectin in platelet function and platelet survival remains unknown, although it has been shown in vivo that the surface-expressed P-selectin is rapidly cleaved from the platelet membrane, releasing a 100-kDa P-selectin fragment to the plasma while platelets remains in circulation *(31)*. These data suggest that while the mechanism(s) involved in the recognition and clearance of senescent or damaged platelets is unknown, CD62P expression does not play a role or uniformly predict platelet viability.

Flow-cytometric analysis of platelets *(32,33)* with fluorochrome-labeled specific antibodies is currently the standard technique to evaluate surface expression of CD62P and CD63 in granule release assessment of PCs. The platelet surface expression of CD62P and CD63 increase during storage, from 19% and 15% positive cells on day 1, to 31% and 14% on day 5, and 67% and 30% on day 10, respectively *(34)*. In addition, levels of P-selectin in plasma can be quantified by ELISA *(35)*. The amount of soluble

P-selectin seems to be a more sensitive marker for platelet activation than the percentage of P-selectin positive cells, especially during the first days of storage. Values range from around 300 ng/mL on day 1, to 400 ng/mL on day 3, and 500 ng/mL on day 6.

The fusing of granules with the OCS as a result of storage-promoted activation also favors the recycling of major adhesive receptors such as the GPIb and GPIIb/IIIa between the external plasma membrane and granular membranes. The level of the immunoreactive platelet complexes can be investigated by flow cytometry using monoclonal antibodies labeled with a fluorophore, and measurements of changes in the mean fluorescence intensity (MFI). In our experience, 5-d blood banking of PCs has no apparent effect on the surface expression of GPIb/IX and GPIIb/IIIa. However, after prolonged storage for up to 9 d there is a 50% decrease in the levels of the surface GPIb/IX expression, while those of GPIIb/IIIa remain remarkably stable *(34,36)*. The reduction in the GPIb/IX immunoreactive levels upon storage is consistent with a proteolytic cleavage of the complex, generating glycocalicin, the soluble extracellular domain of GPIbα containing the vWF- and thrombin-binding sites. By means of a competitive assay and specific monoclonal antibodies, we have shown that glycocalicin levels in the plasma of stored PCs are within normal values (<5 µg/mL) in units stored for up to 5 d, whereas a significant increase can be observed upon prolonged storage (day 9, 35 µg/mL) *(37)*.

Since the hemostatic effectiveness of platelets is related to the ability of the membrane glycoprotein Ib/IX and IIb/IIIa complexes to interact with their respective physiological ligands (mainly vWF and thrombin for GPIb/IX and fibrinogen for GPIIb/IIIa), an evaluation of the functional integrity of these receptors in in vitro stored platelets is of great significance. This can be performed by assessing the ability of platelets to bind radio- or fluorochrome-labeled specific ligands in response to appropriate stimuli. Binding assays are performed by incubating washed platelets with increasing concentrations of ^{125}I-radiolabeled vWF or fibrinogen in the presence of ristocetin, and α-thrombin, respectively. Platelet-bound and free ligands are separated by centrifugation and counted, and the parameters of binding are analyzed *(37)* (*see* also Chapters 6, 7, 8, vol. 2). Prolonged storage of PCs, for up to 9 d, results in a 50% decrease both in the functional GPIb/IX content of platelets (associated with halved GPIb/IX immunoreactive levels) and in the level of ligand binding of GPIIb/IIIa (without quantitative changes of the complex) *(37)*. A similar binding assay approach demonstrates a storage-promoted loss of high-affinity thrombin receptors (GPIb/IX), which leads to impairment of the reactivity of stored platelets to thrombin *(36)*.

In addition to binding assays, the interaction of ligands with these major receptors can also be quantified by flow cytometry using fluorochrome-labeled fibrinogen or vWF. Cytometric analysis of FITC calibration beads allows the generation of standard curves relating bead-fluorescence intensity to number of fluorescent equivalents per bead (*see* Chapter 8, vol. 2; 19 in vol. 1). With this information, the binding parameters (dissociation constant and molecules per platelet) can be calculated *(38)*. Bound ligands (vWF, fibrinogen, or fibronectin), which are present at low levels on unstimulated platelets but appear at significant levels on stored, activated platelets, can be recognized by incubating washed samples with polyclonal FITC-labeled anti-ligand antibodies, followed by flow-cytometric analysis to obtain the percentage of fluorescence-positive platelets *(39)*.

Flow cytometry also allows estimation of storage-promoted conformational changes in GPIIb/IIIa-generating neoepitopes detectable by using monoclonal antibodies such as PAC-1 or ligand-induced binding sites *(32)*.

5. Changes During PC Storage Reflecting Platelet Membrane Damage and Cell Death

Platelet storage lesion is also associated with an alteration of lipids in the plasma membrane. These changes include loss of membrane-lipid asymmetry, resembling that occurring following agonist-induced activation, and loss of a significant amount of lipids, which most likely results from microvesiculation. The former results in surface exposure of negatively charged phospholipids, mainly phosphatidyl serine (PS), which are required for the assembly of Ca^{2+}-dependent procoagulant complexes: tenase (FIXa and FVIIIa) and prothrombinase (FXa and FVa). Enzymatic assays can be used to directly evaluate platelet procoagulant activity, classically referred to as PF3 activity. However, measurement of fluorochrome-labeled annexin V binding to platelets by flow cytometry is an alternative, simple method *(40,41)* to evaluate platelet PS exposure during blood banking storage. (For further discussion of measuring procoagulant activity, *see* Chapter 12, in vol. 1) Annexin V binding to stored platelets is at low levels on day 1 of storage (5–10%), but rises steadily throughout storage (15–20% on day 3, and 20–40% on day 5) *(42,43)*.

Together with the increase in platelet-associated procoagulant activity, several studies have demonstrated that procoagulant microvesicles of different sizes appear in the supernatant of standard PC during storage. Few methods are available for the quantitation and characterization of these platelet microparticles, but in practice, flow cytometry is the current method of choice (*see* also Chapter 21 in vol. 1). Briefly, PC samples are labeled with a fluorochrome-conjugated anti-platelet glycoprotein antigen (GPA) (e.g., anti-CD61 or anti-CD41), and only GPA-positive events are analyzed. The GPA particle populations are divided into subpopulations according to their forward light scatter properties, and thus their size. The smallest events (lowest forward-scatter region) are defined as microparticles, with a diameter of approx 0.5 μm, the largest events correspond to platelet aggregates (>5 μm), and the remaining majority of particles are free intact platelets (0.5–5 μm). During storage, the proportion of platelet-derived microparticles increases from about 6% at day 1 to 20% at the end of their shelf life *(44)*. As these microparticles are found to be procoagulant and to bind to and activate neutrophils in vivo, their levels in PCs may be relevant upon transfusion *(45)*.

In parallel to the expanding activation causing lipid rearrangement and microvesiculation, the inevitable aging and the progressive metabolic depletion of PCs during storage impairs the barrier function of the plasma membrane, leading to swelling of the platelets and eventually to lysis. Assessing cell death is not straightforward in platelets because their small size precludes the use of vital dyes such as trypan blue. Similarly, the absence of a nucleus obviates the use of DNA-staining vital dyes such as Hoechst and propidium iodide. A flow-cytometric analysis for platelet membrane integrity has been developed using actin-binding phalloidin-FITC as a vital dye *(46)*. In addition, assays with the methyltiazoletetrazolium dyes MTT and MTS have been modified to measure platelet viability *(13)*. However, most studies use the supernatant level of lactate dehydrogenase (LDH) activity as marker of cell lysis in PCs. The levels of LDH released into the medium can be measured directly on biochemical multianalyzers, or spectrophotometrically. In the latter, the supernatants from PCs are incubated with NADH and pyruvate, and LDH activity is measured as the time-dependent and pyruvate-dependent decrease in the absorbance of NADH at 339 nm *(46)*. We have found that during standard storage of PCs, supernatant LDH levels rise from 246 U/L on day 1, to 280 U/L on day 5, and 335 U/L on day 9, suggestive of cell lysis *(34)*. As yet, no clear correlation has been found between LDH levels in PCs and in vivo platelet viability.

6. Changes in the Metabolic Activity of PCs During Blood-Banking Storage

As stated above, the complex set of platelet changes associated with PSL is similar to that occurring following activation of platelets with physiological agonists. However, a significant difference exists between these two processes, as physiological activation changes occur in seconds to minutes, while the PSL develops over several days, during which metabolic exhaustion may happen. Indeed, assessment of metabolic status is an important aspect in quality assurance of PCs. Platelets, as other cells, utilize energy for a variety of biochemical events associated with activation, and their principal source of energy is the hydrolysis of ATP. In platelets, adenine nucleotides are partitioned into a storage and a metabolic pool. The former contains equal amounts of ADP and ATP and is released upon stimulation with agonists. The second contains predominantly ATP and provides energy for both platelet activation changes and maintenance of cell integrity in the circulation and, most likely, during in vitro storage. Thus, reduction of ATP concentration or change in ATP/ADP ratio is an index of PSL, which parallels the decrease in supernatant glucose and pH *(48,49)*. ATP levels can be determined by the luciferin-luciferase assay using a luminometer, and ADP is measured by the same assay after enzymatic conversion to ATP with phosphoenol pyruvate and pyruvate kinase *(49)* (*see* also Chapters 7 and 9 in vol. 1).

During storage of PCs, both glycolysis and oxidative phosphorylation fuel regeneration of ATP to avoid energy depletion. Platelets have a prominent glycolytic rate, significantly exceeding that of erythrocytes and skeletal muscle, resulting in substantial lactate production. Since hydrogen cation production accompanies lactate generation, a fall in pH would occur unless the storage medium of the PCs provides a mechanism to buffer the hydrogen ions. In practice, the presence of plasma (at least 0.65 mL/10^9 platelets) containing 20 mM bicarbonate provides sufficient buffering capacity for standard PCs stored for 1 wk. The oxidative pathway, carried out by few small-size mitochondria, is much more efficient than glycolysis, regenerating six ATP molecules for each consumed oxygen without production of hydrogen ions. Surprisingly, the main fuel for this oxidative phosphorylation in platelets is not pyruvic acid, readily available from glycolysis, but rather free fatty acids, amino acids, and/or organic anions such as acetate. The greater efficiency of ATP generation of the oxidative pathway, and its lack of association with a pH decrease, justify the need to store platelets in containers that allow adequate oxygen exchange to meet the platelets' needs. Without sufficient oxygen, the glycolytic rate would increase as a compensatory energy mechanism, eventually leading to glucose exhaustion, accumulation of lactate and acidosis, effects often referred to as the Pasteur effect. A fall of pH below 6.0 has been associated with the release of components from both alpha and dense granules, irreversible changes in platelets, and loss of viability. The pO_2 and the pCO_2 levels in PC during storage are currently used as markers of the respiratory activity of stored platelets. In addition, determination of pH, glucose, and lactate levels are informative of the metabolic status and the extent of PSL. The pH, pO_2, pCO_2, and plasma bicarbonate levels are measured under anaerobic conditions immediately after plasma is drawn from PCs in a blood-gas analyzer. Glucose and lactate can be measured enzymatically on a spectrophotometer in the supernatant of the PCs *(50)*. In our experience, blood-banking storage of standard PCs promotes a time-dependent significant change in all these metabolic parameters *(32)*. The ongoing metabolic activity during storage may also alter the platelet balance in second messengers, the natural regulators of platelet activation. Cyclic nucleotides, mainly cAMP and cGMP, are inhibitors of this process, whereas messengers such as calcium, diacylglycerol, inositol 1,4,5-trisphosphate, or thromboxane A_2 are activators. Work by our group and others has evaluated the potential of cAMP to protect against the activation of platelets during storage of PCs *(51,52)*. The cAMP levels in PCs can be determined using a commercial enzyme immunoassay (cAMP EIA system RPN 225, Amersham BioSciences, Buckinghamshire, UK). Thromboxane B_2 (stable thromboxane A_2 metabolite) levels can be measured in PPP samples using a dextran-coated charcoal radioimmunoassay with ^3H-TxB2. Storage promotes a decrease in the cAMP content of PCs while increasing the levels of the activating metabolite TxA_2. Thus the TxB2:cAMP ratio rises 1.5- and 3-fold at days 5 and 9, respectively, as compared to day 1 *(34)*.

7. Measurements Reflecting Potential Changes in Overall Hemostatic Capacity of Platelets During Blood-Banking Storage

Quality assessment of PCs often attempts to reflect the overall platelet status, applying the same criteria that are used in the evaluation of platelet function in patient studies.

A simple classical test available for this purpose is the measurement of platelet aggregation following stimulation with different concentrations and combinations of physiological agonists (e.g., ADP, collagen, epinephrine, thrombin). In functional platelets, a sufficient stimulus results in exposure of fibrinogen binding sites on GPIIb/IIIa receptors and, in the presence of external fibrinogen, an aggregation process which can be followed optically as an increase in light transmission (*see* Chapter 6, vol. 1 for further discussion of aggregation measurements). Platelets stored as concentrates respond poorly to the action of weak agonists such as epinephrine, ADP, and collagen *(53)*. Thrombin-induced aggregation of stored platelets compared to fresh units is also impaired, albeit to a lesser extent *(34)*. Nevertheless, significant aggregation to agonists has been demonstrated when autologous plasma in unresponsive stored PCs is replaced by fresh allogeneic plasma just before testing *(54)*. The improved functional response of stored PCs elicited by the addition of plasma, a situation mimicking the infusion of the product into the bloodstream, suggests a role for some unknown plasma factors in potentiating the weak agonist-mediated activation.

Measurement of platelet interaction with thrombogenic substrata using perfusion devices that simulate conditions in the bloodstream has been applied to gain more objective information on the quality and function of stored platelets *(55)*. In these assays, samples drawn from PCs at selected times during storage are added to in vitro platelet-depleted blood. The blood containing stored platelets is recirculated through annular chambers mounted with everted aorta segment or an alternative thrombogenic substance. At the end of the perfusion, the segments are morphometrically evaluated for the percentage of surface covered by attached but not spread platelets (adhesion), groups of platelets that form small (<5 μm) aggregates (aggregation), and thrombus platelets that form large aggregates (>5 μm). The general impression from these studies is that platelets undergo a slight, yet progressive, impairment in their adhesive and aggregating function throughout storage *(39,55)*. Parameters influencing the reduction in platelet adhesion (10–50%) in these assays include the extent of PC storage (1–8 d), the type of thrombogenic matrix, the shear rate, or the anticoagulant used *(39,55–57)*.

Despite their great potential for research, widespread use of perfusion procedures in quality assessment of PCs is limited because processing of vascular (thrombogenic) segments and morphometric evaluations are rather time-consuming and require dedicated personnel and equipment. In recent years, a number of reliable platelet-function analyzers have been developed that can provide useful information on global or particular aspects of platelet function *(58)*. Among these new hemostatic tests are the platelet-function analyzer (PFA-100, a modified version of the former Thrombostat 4000), the laser platelet aggregometer (PA-200), the ultegra(RPFA)-rapid platelet function analyzer, the thrombotic-status analyzer (TSA), the clot-signature analyzer, the cone and plate analyzer, the hemodyne, and the hemostasis device. A detailed explanation of these new tests is beyond the scope of this chapter, but *see* Chapter 18 in vol. 1 for further details on the PFA-100). In general, these new devices offer reliability, simplicity, and the use of small amounts of samples. However, they remain expensive and rather inflexible tests, and there is little widespread experience in their use in routine assessment of the quality of stored platelets. The most used system in studies of stored platelets is probably the PFA-100 *(59,60)*. This consists of a microprocessor-controlled instrument and disposable test cartridge containing a membrane that has been coated with collagen and either epinephrine or ADP as agonists. Samples of citrated blood are perfused (shear rate 5000–6000/s) through a 150-μm aperture cut into the membrane. For use in functional studies of PCs, thrombocytopenic blood is reconstituted with stored platelet samples. The hemostatic capacity of platelets is indicated by the time required for the platelet plug to occlude the aperture (closure time, CT, or in vitro bleeding time, IVBT), expressed in seconds up to a maximum of 300, the flow rate (μL/min), and the total flow volume (μL). Using these data a PFA-predictive index (PFA-PI) can be obtained *(60)*. According to these assays, closure times tend to increase after the third day of storage and may exceed 300 s (the limit of the device) when PC units have been stored for more than 6 d. Of the two available agonist cartridges, epinephrine appears to be particularly sensitive for the detection of initial platelet hyporeactivity (PFA-PI on fresh platelets and 3-, 5-, or 7-d-old platelets of 3.97, 6.74, 11.28, and 13.00, respectively). ADP, on the other hand, is more useful for measuring the residual platelet reactivity (PFA-PI on fresh platelets and 3-, 5-, or 7-d-old platelets of 2.43, 2.36, 5.03, and 13.74, respectively) *(60)*. Although one is restricted to rigidly designed cartridges, the low coefficient of variation with repeated analysis of platelet function and the apparent close correlation of the IVBT with in vivo platelet function suggest that the PFA-100 may be useful as a simple and standardized protocol for quality control of PCs.

8. Conclusions

Platelet transfusion is still a major clinical tool in the modern management of severe thrombocytopenic patients. Our increasing knowledge of morphological, biochemical, and functional defects involving the development of PSL may help to improve the conditions for the preparation and storage of PCs in blood banking, and aid design and evaluation of novel platelet products. A wide number of in vitro tests are available for routine monitoring of the extent of PSL, although a universally accepted and standardized protocol is still lacking. Thus, the selection of test(s) for PSL often relies on the experience and technical abilities of individual laboratories. Furthermore, there are still doubts that available tests can predict the survival and performance of stored platelets upon transfusion; thus, further studies of cross-evaluating in vitro parameters of platelet quality with objective clinical parameters are required.

References

1. Rebulla, P. (2000) Trigger for platelet transfusion. *Vox. Sang.* **78,** 179–182.
2. Clark, P. and Mintz, P. D. (2001) Transfusion triggers for blood components. *Curr. Opin. Hematol.* **8,** 387–391.
3. Rivera, J. and Vicente, V. (1999) Potential and current development of platelet products. *Haematologica* **84 (EHA-4 educational book),** 115–119.
4. Lee, D. H. and Blajchman, M. A. (2001) Novel treatment modalities: new platelet preparations and substitutes. *Br. J. Haematol.* **114,** 496–505.
5. Rebulla, P. (1998) In vitro and in vivo properties of various types of platelets. *Vox. Sang.* **74(Suppl. 2),** 217–222.
6. Heaton, W. A., Rebulla, P., Pappalettera, M., and Dzik, W. H. (1997) A comparative analysis of different methods for routine blood component preparation. *Transfus. Med. Rev.* **11,** 116–129.
7. Ness, P. M. and Campbell-Lee, S. A. (2001) Single donor versus pooled random donor platelet concentrates. *Curr. Opin. Hematol.* **8,** 392–396.
8. Corash, L. (2000) Inactivation of viruses, bacteria, protozoa and leukocytes in platelet and red cell concentrates. *Vox. Sang.* **78(Suppl. 2),** 205–210.
9. Blajchman, M. A. and Goldman, M. (2001) Bacterial contamination of platelet concentrates: incidence, significance, and prevention. *Semin. Hematol.* **38(Suppl. 11),** 20–26.
10. Chernoff, A. and Snyder, E. L. (1992) The cellular and molecular basis of the platelet storage lesion: a symposium summary. *Transfusion* **32,** 386–390.
11. Klinger, M. H. F. (1996) The storage lesion of platelets: ultrastructural and functional aspects. *Ann. Hematol.* **73,** 103–112.
12. Seghatchian, J. and Krailadsiri, P. (1997) The platelet storage lesion. *Transfus. Med. Rev.* **11,** 130–144.
13. Li, J., Xia, Y., Berino, A. M., Coburn, J. P., and Kuter, D. (2000) The mechanism of apoptosis in human platelets during storage. *Transfusion* **40,** 1320–1329.

14. Seghatchian, J. and Krailadsiri, P. (2001) Platelet storage lesion and apoptosis: are they related? *Transfus. Aphe. Sci.* **24,** 103–105.
15. Schiffer, C. A., Lee, E. J., Ness, P. M., and Reilly, J. (1986) Clinical evaluation of platelet concentrates stored for one to five days. *Blood* **67,** 1591–1594.
16. Peter-Salonen, K., Bucher, U., and Nydegger, U. E. (1987) Comparison of postransfusion recoveries achieved with either fresh or stored platelet concentrates. *Blut* **54,** 207–212.
17. Duguid, J. K. M., Carr, R., Jenkins, J. A., Hutton, J. L., Lucas, G. F., and Davies, J. M. (1991) Clinical evaluation of the effects of storage time and irradiation on transfused platelets. *Vox. Sang.* **60,** 151–154.
18. Owens, M., Holme, S., Heaton, A., Sawyer, S., and Cardinali, S. (1992) Post-transfusion recovery of function of 5-day stored platelet concentrates. *Br. J. Haematol.* **80,** 539–544.
19. Norol, F., Kuentz, M., Cordonnier, C., Beaujean, F., Haioun, C., Vernant, J. P., et al. (1994) Influence of clinical status on the efficacy of stored platelet transfusion. *Br. J. Haematol.* **86,** 125–129.
20. Murphy, S., Rebulla, P., Bertolini, F., Holme, S., Moroff, G., Snyder, E., et al. (1994) In vitro assessment of the quality of stored platelet concentrates. *Transfus. Med. Rev.* **8,** 29–36.
21. Bertolini, F. and Murphy, S. (1994) A multicenter evaluation of reproducibility of swirling in platelet concentrates. *Transfusion* **34,** 796–801.
22. Holme, S., Sawyer, S., Heaton, A., and Sweeney, J. D. (1997) Studies on platelets exposed to or stored at temperatures below 20°C or above 24°C. *Transfusion* **37,** 5–11.
23. Gear, A. R. L. and Polanowska-Grabowska. The platelet shape change, in *Platelets* (Gresele, P., Page, C. P., Fuster, V., and Vermylen, J., eds.), Cambridge University Press, pp. 319–337.
24. Järemo, P. (1997) Some correlations between light transmission changes and some commonly used in vitro assays for the assessment of platelet concentrates. *Eur. J. Haematol.* **58,** 181–185.
25. Vagace, J. M., Alonso, N., Groiss, J., Cabanillas, Y., Casado, M. S., Rincón, R., et al. (2001) Nuevo método para evaluar la calidad de los concentrados plaquetarios (New method to evaluate the quality of platelet concentrates). *Haematologica* **86(2),** a-C-080.
26. Connor, J., Currie, L. M., Allan, H., and Livesey, S. A. (1996) Recovery of in vitro functional activity of platelet concentrates stored at 4°C and treated with second-messenger effectors. *Transfusion* **36,** 691–698.
27. Holme, S., Moroff, G., and Murphy, S. (1998) A multi-laboratory evaluation of in vitro platelet assays: the tests for extent of shape change and response to hypotonic shock. *Transfusion* **38,** 31–40.
28. Fantl, P. (1968) Osmotic stability of blood platelets. *J. Phsysiol.* **198,** 1–16.
29. Ledent, E., Wasteson, A., and Berlin, G. (1995) Growth factor release during preparation and storage of platelet concentrates. *Vox. Sang.* **68,** 205–209.
30. Mehta, P., Patel, K. D, Laue, T. M., Erikson, H. P., and McEver, R. P. (1997) Soluble monomeric P-selectin containing only the lectin and epidermal growth factor domains bind to P-selectin glycoprotein ligand-1 on leukocytes. *Blood* **90,** 2381–2389.
31. Berger, G., Hartwell, D., and Wagner, D. D. (1998) P-selectin and platelet clearance. *Blood* **92,** 4446–4452.
32. Schimitz, G., Rothe, G., Ruf, A., Barlage, S., Tschöpe, D., Clemetson, K. J., et al. (1998) European working group on clinical cell analysis: Consensus protocol for the flow cytometric characterization of platelet function. *Thromb. Haemost.* **79,** 885–896.
33. Michelson, A. D. (2002) Flow cytometric analysis of platelets. *Vox. Sang.* **78(Suppl. 2),** 137–142.

34. Rivera, J., Lozano, M. L., Corral, J., Connor, J., González-Conejero, R., Ferrer, F., et al. (1999) Quality assessment of platelet concentrates supplemented with second-messenger effectors. *Transfusion* **39,** 135–143.
35. Kostelijk, E. H., Fijnheer, R., Nieuwenhuis, H. K., Gouwerok, C. W. N., and de Korte, D. (1996) Soluble P-selectin as parameter for platelet activation during storage. *Thromb. Haemost.* **76,** 1086–1089.
36. Lozano, M. L., Rivera, J., González-Conejero, R., Moraleda, J. M., and Vicente, V. (1997) Loss of high-affinity thrombin receptors during platelet concentrate storage impairs the reactivity of platelets to thrombin. *Transfusion* **37,** 368–375.
37. Rivera, J., Sánchez-Roig, M. J., Rosillo, M. C., Moraleda, J. M., and Vicente, V. (1994) Stability of glycoproteins Ib/IX and IIb/IIIa during preparation and storage of platelet concentrates: detection by binding assays with epitope-defined monoclonal antibodies and physiological ligands. *Vox. Sang.* **67,** 166–171.
38. Faraday, N., Goldsmidt-Clermont, P., Dise, K., and Bray, P. (1994) Quantitation of soluble fibrinogen binding to platelets by fluorescence-activated flow cytometry. *J. Lab. Clin. Med.* **123,** 728–740.
39. Lozano, M., Estebanell, E., Cid, J., Diaz-Ricart, M., Mazzara, R., Ordinas, A., et al. (1999) Platelet concentrates prepared and stored under currently optimal conditions: minor impact on platelet adhesive and cohesive functions after storage. *Transfusion* **39,** 951–959.
40. Corral, J., Gonzalez-Conejero, R., Martinez, C., Rivera, J., Lozano, M. L., and Vicente, V. (2002) Platelet aggregation through prothombinase activation induced by non-aggregant doses of platelet agonist. *Blood. Coag. Fibrinolysis* **13,** 95–103.
41. Xiao, H. Y., Matsubayashi, H., Bonderman, D. P., Borderman, P. W., Reid, T., Miraglia, C. C., et al. (2002) Generation of annexin V-positive platelets and shedding of microparticles with stimulus-dependent procoagulant activity during storage of platelets at 4°C. *Transfusion* **40,** 420–427.
42. Metcalfe, P., Williamson, L. M., Reutelingsperger, C. P. M., Swann, I., Ouwehand, W. H., and Goodall, A. H. (1997) Activation during preparation of therapeutic platelets affects deterioration during storage: a comparative flow cytometric study of different production methods. *Br. J. Haematol.* **98,** 86–95.
43. Shapira, S., Friedman, Z., Shapiro, H., Pressezen, K., Radnay, J., and Ellis, M. H. (2000) The effect of storage on the expression of platelet membrane phosphatidylserine and the subsequent impact on the coagulant function of stored platelets. *Transfusion* **40,** 1257–1263.
44. Scharf, R. E. and Hanfland, P. (1993) Platelet storage lesions: analysis of platelet membrane glycoproteins and platelet-derived microparticles by fluorescence-activated flow cytometry. *Transfus. Sci.* **14,** 189–94.
45. Owens, M. R. (1994) The role of platelet microparticles in hemostasis. *Transfus. Med. Rev.* **8,** 37–44.
46. Brown, S. B., Clarke, M. C. H., Magowan, L., Sanderson, H., and Savill, J. (2000) Constitutive death of platelets leading to scavenger receptor-mediated phagocytosis. *J. Biol. Chem.* **275,** 5987–5996.
47. Savage, B. (1982) Platelet adenine nucleotide levels during room temperature storage of platelet concentrates. *Transfusion* **22,** 288–291.
48. Murphy, S. (1994) Metabolic patterns of platelets—impact on storage for transfusion. *Vox. Sang.* **67(S3),** 271–273.
49. Summerfield, G. P., Keenan, J. P., Brodie, N. J., and Bellingham, A. J. (1981) Bioluminescent assay of adenine nucleotides: rapid analysis of ATP and ADP in red cells and platelets using the LKB luminometer. *Clin. Lab. Haematol.* **3,** 257–271.

50. Dohlhofer, R. (1976) Experiences with the glucose-dehydrogenase UV method for determination of blood glucose. *J. Clin. Chem. Clin. Biochem.* **4**, 415–417.
51. Bode, A. P. and Norris, H. T. (1994) Sustained elevation of intracellular cyclic 3′-5′ adenosine monophosphate is necessary for preservation of platelet integrity during long-term storage at 22°C. *Blood* **83**, 1235–1243.
52. Lozano, M. L., Rivera, J., Bermejo, E., Corral, J., Pérez, E., and Vicente, V. (2000) In vitro analysis of platelet concentrates stored in the presence of modulators of 3′,5′ adenosine monophosphate, and organic anions. *Transfus. Sci.* **22**, 3–11.
53. Rao, G. H. R., Escolar, G., and White, J. G. (1993) Biochemistry, physiology, and function of platelets stored as concentrates. *Transfusion* **33**, 766–778.
54. McShine, R. L., Weggemans, M., Das, P. C., Smit Sibinga, C. Th., and Brozovic, B. (1993) The use of fresh plasma and a plasma-free medium to enhance the aggregation response of stored platelets. *Platelets* **4**, 338–340.
55. Escolar, G., Galán, A. N., Mazzara, R., Castillo, R., and Ordinas, A. (2001) Measurement of platelet interaction with subendothelial substrata: Relevance to transfusion medicine. *Transfus. Med. Rev.* **15**, 144–156.
56. Boomgaard, M. N., Gouwerok, C. W. N., Homburg, C. H. E., de Groot, Ph. G., Ijsseldijk, M. J. W., and de Korte, D. (1994) The platelet adhesion capacity to subendothelial matrix and collagen in a flow model during storage of platelet concentrates for 7 days. *Thromb. Haemost.* **72**, 611–616.
57. Boomgaard, M. N., Gouwerok, C. W. N., Palfenier, C. H., Pankalla-Blandeau, I. E., Veldman, H. A., de Korte, D., et al. (1995) Pooled platelet concentrates prepared by the platelet-rich plasma method and filtered with three different filters and stored for 8 days. *Vox. Sang.* **68**, 82–89.
58. Harrison, P. (2000) Progress in the assessment of platelet function. *Br. J. Haematol.* **111**, 733–744.
59. Böck, M., Groh, J., Glaser, A., Storck, K., Kratzer, M. A. A., and Heim, M. U. (1995) Quality control of platelet concentrates by the Thrombostat 4000. *Semin. Thromb. Haemost.* **21(Suppl. 2)**, 91–95.
60. Borzini, P., Lazzaro, A., and Mazzuco, L. (1999) Evaluation of the hemostatic function of stored platelet concentrates using the platelet function analyzer (PFA-100™). *Haematologica* **84**, 1104–1109.

4

Pharmacological Approaches to Studying Platelet Function

An Overview

Susanna M. O. Hourani

1. Theoretical Considerations

Although pharmacology may be strictly viewed as the study of drugs (see, for example, the British Pharmacological Society Web site at http://www.bps.ac.uk/BPS.html), for many of us it also involves the use of drugs to study receptors. Indeed, the fundamental principle behind a pharmacological approach to the study of any biological system is that quantitative analysis of the effects of drugs on the responses of cells, tissues, or whole animals can provide useful information on the receptors involved. The approach is well-suited to situations where it is relatively easy to measure a response that is closely related to receptor activation, and although originally developed using isolated smooth-muscle preparations, platelets are also a very suitable system. Indeed, platelets have some advantages over smooth-muscle preparations in this respect, as the problem of diffusion of drugs through solid tissues is largely avoided in a suspension of cells. In addition, because in general each measurement is made on a separate sample of cells, there is no risk of desensitization due to repetitive addition of drugs to a single tissue, although obviously there may be time-dependent changes in the responses, which have to be taken into account. An important consideration is which response to measure: for platelets, one has the choice of overall functional responses such as shape change, aggregation, and the release of granule contents, which are more complex but may be of pathophysiological relevance, or of simpler biochemical responses such as changes in the levels of intracellular cyclic AMP or calcium. In theory any quantifiable response can be analyzed, although in some cases the results may differ depending on the response measured. A pharmacological approach can also be applied to cells in culture containing the receptors of interest, and in this context the importance of receptor density and the efficacy of receptor coupling (discussed in **Subheading 1.1.**) must be borne in mind.

The predominant theory of drug action, occupation theory, assumes that the actions of drugs are due to their ability to bind with high affinity and high specificity to some

From: *Methods in Molecular Biology, vol. 273:*
Platelets and Megakaryocytes, Vol. 2: Perspectives and Techniques
Edited by: J. M. Gibbins and M. P. Mahaut-Smith © Humana Press Inc., Totowa, NJ

component of the cell that we call the receptor. Although we now know a great deal about the structure and function of many receptors, these were a concept rather than a physical reality at the time when models describing the effects of drugs were developed. The term "receptor" was originally applied to any drug target, be it a channel, an enzyme, an uptake system, or a true receptor, whereas the word is now normally restricted to those proteins, usually within the plasma membrane, whose function it is to recognize chemical messengers and transmit the information to the cell. A drug that mimics the endogenous chemical messenger binds to the receptor and causes some change in the receptor, which in turn triggers a response in the cell: this drug is termed an *agonist*. A drug that simply binds to the receptor and does not trigger a response competes for binding with agonists and is termed an *antagonist*. From these simple ideas comes a very important concept, that drugs have two properties: *affinity*, or the ability to bind; and *efficacy*, the ability to activate the receptor. An agonist has both affinity and efficacy, whereas an antagonist has affinity but no efficacy.

1.1. Agonist Concentration-Response Curves

When the concentration of agonist used is plotted against the response measured, the graph usually has a hyperbolic shape (*see* **Fig. 1A**). To allow the study of a wider concentration range and to facilitate analysis, such data are usually plotted in a semilogarithmic format, as the response against the $\log_{10}$ of the agonist concentration. When plotted in this way the graph usually has a sigmoid or S-shaped appearance, which plateaus to give a clear maximal response (*see* **Fig. 1B**). From such a graph it is possible to calculate the potency of an agonist, a rather flexible concept that reflects how much of the drug is required to cause a response. It is best given in molar units as the EC_{50} (also called $[A]_{50}$, with [A] standing for the concentration of agonist), the concentration of the drug required to cause 50% of the maximal response. However, other measures, such as the concentration required to produce some defined response, may also be used. The EC_{50} is a useful measure particularly when comparing two agonists, but what does it really mean? To further discuss this issue, it is important to understand the basic theory of drug-receptor interactions.

The binding of any drug to a receptor can be described by the Langmuir equation:

$$[A] + [R] \underset{k_{-1}}{\overset{k_{+1}}{\rightleftharpoons}} [AR] \tag{1}$$

where [A] is the free drug concentration, [R] the concentration of unoccupied receptors, [AR] the concentration of occupied receptors, and k_{+1} and k_{-1} the association and dissociation rate constants, respectively. Making the assumption that binding to the receptors does not significantly affect [A], at equilibrium this can be reduced to:

$$\frac{[AR]}{[R]_T} = \frac{[A]}{[A] + K_A} \tag{2}$$

where $[R]_T$ is the total concentration of receptors and K_A is k_{-1}/k_{+1}, or the dissociation constant. If the fractional receptor occupancy $[AR]/[R]_T$ is plotted against [A], the graph is hyperbolic, or if plotted against $\log_{10}[A]$ it is sigmoidal. From such a graph

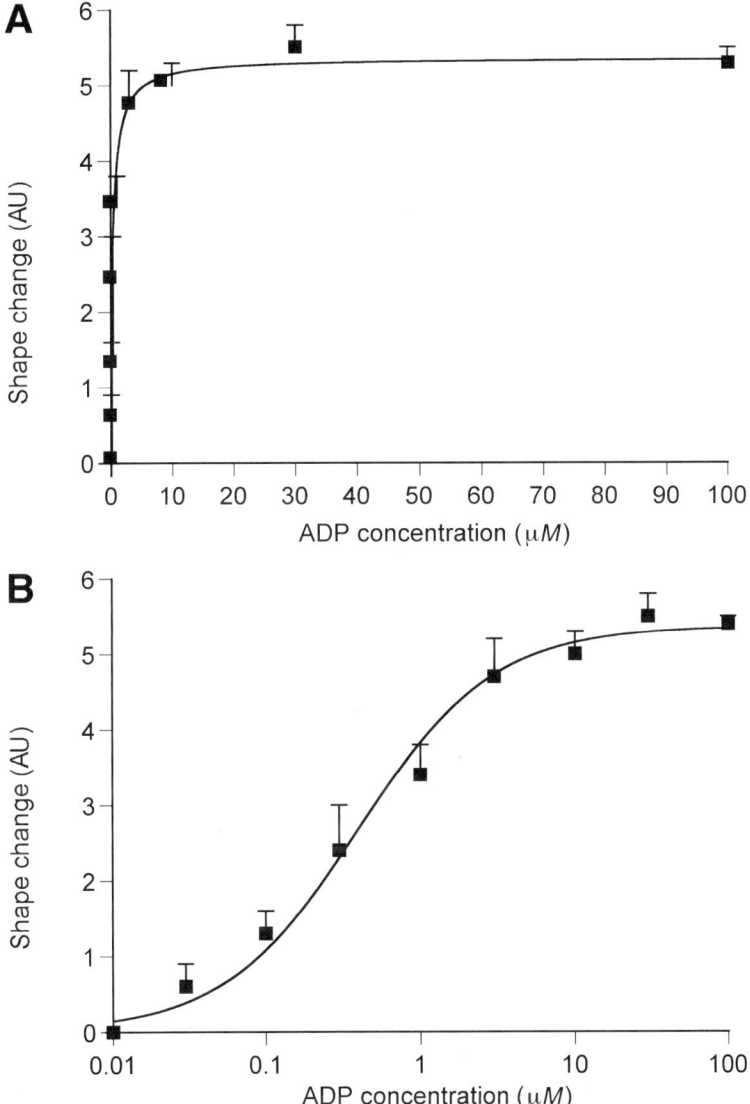

Fig. 1. Experimental concentration-response curves. A typical experimental concentration-response curve plotted using (A) a linear agonist concentration scale and (B) a logarithmic agonist concentration scale. Results are for the induction of shape change of human washed platelets by ADP (mean ± s.e. mean, $n = 7$–8), and the lines are fitted to the following equation (equivalent to Eq. 9; see text for further details):

$$E = \frac{E_{max} \cdot [ADP]}{[ADP] + EC_{50}}$$

Shape change was measured in arbitrary units (AU) as the maximum decrease in light transmission through a stirred sample of washed platelets at 37°C in HEPES buffer (see ref. 7 for details).

$[R]_T$ can be calculated and so can the K_A, which is mathematically equal to the concentration of drug required to occupy half the total number of receptors (*see* **Fig. 2**). There is an obvious similarity between this binding curve and a normal experimental concentration-response curve (e.g., as shown in **Fig. 1**), which also has a hyperbolic shape. It is very tempting, therefore, to make the assumption that the measured EC_{50} is in some simple way related to the K_A—in other words, that the potency of an agonist is simply related to its affinity. Indeed, as originally proposed *(1)*, occupation theory assumed that receptor occupancy for an agonist is linearly related to response, so that a maximal response would be observed at 100% receptor occupancy, and 50% of the maximal response would be observed when 50% of the receptors are occupied, so that EC_{50} would be equal to K_A. However, there is now abundant evidence from many systems that this is too simple, and indeed if one considers the many steps that are likely to occur between binding of an agonist and even the simplest response, with amplification and feedback built into the system, it becomes very unlikely that such a simple relationship could ever hold true. The major complication, however, is the concept of efficacy—poorly understood but clearly crucial. Efficacy is not an all-or-none property of drugs, but instead covers a complete range, from powerful agonist to true antagonist (and even beyond, to "inverse agonists" in some situations—*see* **refs. 2,3**). In between come drugs known as *partial agonists*, which produce a response but do not achieve the same maximal response as a full agonist.

A simple way of thinking about efficacy is that proposed by Stephenson *(4)*, who introduced the idea that a drug, by binding to a receptor, delivered a stimulus to the tissue. The stimulus was defined as:

$$S = \frac{e \cdot [A]}{[A] + K_A} \tag{3}$$

where e reflects the ability of the drug to induce a response (i.e., its efficacy), and the response is some nonlinear function of the stimulus. The importance of e in determining the observed potency of a drug is demonstrated in **Fig. 3**, which shows the predicted concentration-response curves for a series of drugs with the same affinity (K_A 30 nM, as in **Fig. 2**) but varying efficacies, assuming a hyperbolic relationship between stimulus S and the response (*see* legend to **Fig. 3**). This nonlinear, saturable function allows the response to plateau as e increases even though the stimulus is still increasing, to take account of the fact that there is always a limit to the response of any cell or tissue. Rather than the maximal response increasing as e increases, the curve shifts to the left and the potency increases. Clearly for the drugs in **Fig. 3** the measured EC_{50} is not equal to the K_A. For a drug with a high efficacy (e.g., Drug A in **Fig. 3**), only a very small percentage of the receptors need to be occupied to produce a maximal response, giving rise to the concept of "spare receptors." For these drugs, inactivation of a large percentage of receptors will not depress the maximal response observed, but will instead only reduce the observed potency. For the partial agonists Drugs D and E, with a low efficacy, even at 100% receptor occupancy only a submaximal response can be achieved, and reduction of the receptor number will reduce the maximum response.

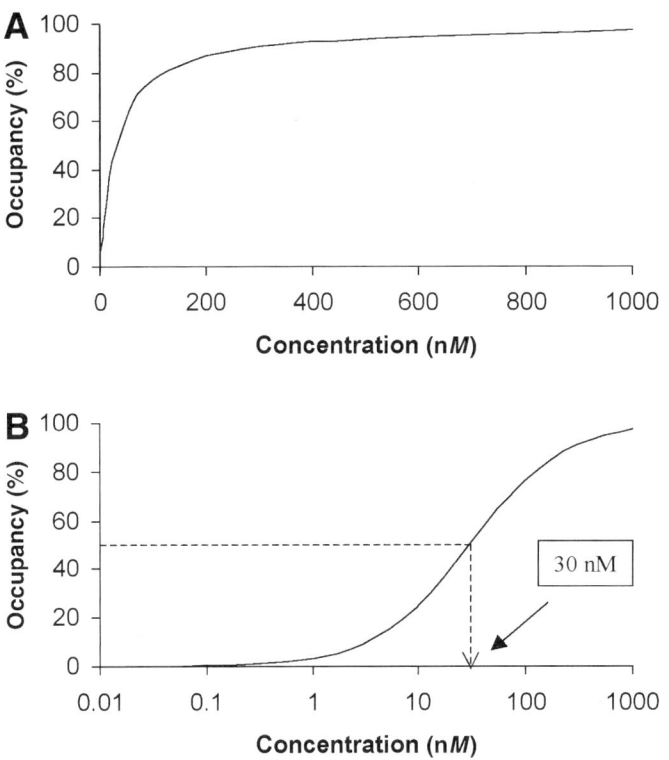

Fig. 2. Theoretical binding curves. Binding curves for a model drug with a K_A of 30 n*M*, plotted from **Eq. 2** using **(A)** a linear drug concentration scale or **(B)** a logarithmic drug concentration scale. The K_A is shown as the concentration that occupies 50% of the total receptors (R_T).

These drugs may also behave as antagonists since they bind the receptors but do not activate them fully, but by binding may compete with the action of a full agonist **(Fig. 4)**.

The value for *e* as used above includes both drug and tissue components, so a drug could have different *e* values in different tissues. A refinement of this analysis was introduced by Furchgott *(5)*, who proposed that *e* was the product of the total receptor concentration ($[R]_T$) and a drug-specific parameter ε, the intrinsic efficacy, such that

$$e = \varepsilon \cdot [R]_T \qquad (4)$$

Thus the family of curves in **Fig. 3**, rather than representing the effects of a number of different drugs in the same tissue, each with different values for ε, could instead represent the effects of a single drug in a number of tissues each with different receptor numbers. Black and Leff *(6)* defined this more explicitly in their operational model of receptor activation, where they introduced the term τ to represent efficacy as $[R]_T/K_E$,

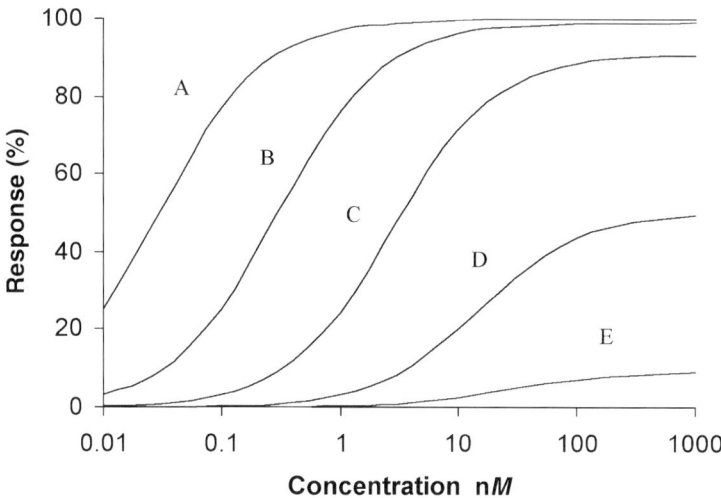

Fig. 3. Effect of varying agonist efficacy on concentration-response curves. Predicted concentration-response curves for a series of drugs with the same affinity ($K_A = 30$ nM) but varying efficacies (A; $e = 1000$; B, $e = 100$; C, $e = 10$; D, $e = 1$; E, $e = 0.1$). The curves were modeled using Stephenson's equation to derive the stimulus S (*see* text, **Eq. 3**), and assuming a hyperbolic relationship between S and response such that:

$$\text{Response} = \frac{100 \cdot S}{S + 1}$$

where K_E is a parameter (equivalent to K_A) describing the hyperbolic relationship between receptor occupancy and response. This parameter is both agonist- and tissue-dependent.

The influence of R_T or receptor density is an important consideration when attempting to compare the potencies of a drug in native tissues or cells with its potency at a cloned receptor expressed (often at very high levels) in a cell, where the receptor densities may be very different. Indeed, a drug could appear to be a full agonist like Drug A in a tissue with a large number of receptors, but a partial agonist like Drug E in one with a small number. In the context of platelets, this could easily explain the apparent paradox that nucleotide triphosphates, such as ATP and its analogs, act as competitive antagonists at $P2Y_1$ and $P2Y_{12}$ receptors in platelets (where the receptor number is low) as indicated by their ability to antagonize the effects of ADP on shape change and calcium mobilization ($P2Y_1$) and on inhibition of adenylate cyclase ($P2Y_{12}$) (as well as on aggregation itself, a composite response) *(7–9)*, whereas they can act as agonists at the cloned $P2Y_1$ receptor *(10)* and the cloned $P2Y_{12}$ receptor *(11)* expressed at a higher density. It is currently unclear why ATP has been reported to be an agonist at the P2Y receptor evoking Ca^{2+} mobilization in primary megakaryocytes *(12–14)*; however, this may also be explained by a higher density of $P2Y_1$ receptors in the precursor cell.

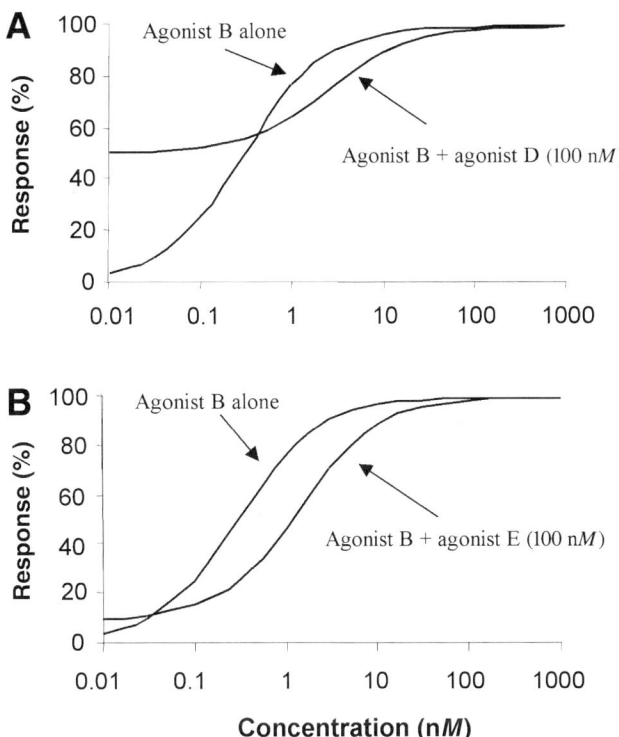

Fig. 4. Effect of combining drugs of different efficacies on concentration-response curves. Predicted concentration-response curves for Drug B ($e = 100$) in the presence of 100 n*M* of either (**A**) Drug D ($e = 1$) or (**B**) Drug E ($e = 0.1$). The curves were modeled assuming competition between B and D or E, adding together the stimulus *S* produced by the two drugs, and assuming a hyperbolic relationship between *S* and response as in **Fig. 3**.

1.2. Antagonist Affinity

As a consequence of the above theoretical considerations concerning the use of agonists, antagonists are generally regarded as more reliable pharmacological tools. The equation *(15)* describing the interaction of an agonist with the receptors in the presence of a fixed concentration [B] of an antagonist with dissociation constant K_B is as follows:

$$\frac{[AR]}{[R]_T} = \frac{[A]}{[A] + K_A \cdot (1 + [B]/K_B)} \tag{5}$$

Although the relationship between receptor occupancy by an agonist and the observed response is complex and usually unknown (see discussion in **Subheading 1.1.**), the assumption can be made that if the same number of receptors are occupied by an agonist in the presence and absence of an antagonist this will give rise to an equal response. If in the absence of an antagonist the concentration of agonist required to give any particular response (e.g., 50%) response is [A] and in the presence of antagonist it is [A′], the ratio between these (e.g., for the 50% response $[A]'_{50}/[A]_{50}$) is known as the dose ratio (dr). Assuming equal receptor occupancy, it can be assumed that:

$$\frac{[A]}{[A] + K_A} = \frac{[A']}{[A'] + KA \cdot (1 + [B]/K_B)} \tag{6}$$

This equation can be transformed into:

$$K_B = \frac{[B]}{(dr - 1)} \tag{7}$$

This is an extremely useful equation, as by constructing concentration-response curves to an agonist in the presence and absence of a fixed concentration of antagonist and calculating the dose-ratio, the dissociation constant of the antagonist can be measured directly. If a series of different concentrations of antagonist are used, then this analysis can be taken further and a Schild plot can be constructed (as shown in **Fig. 5**) that gives a very accurate measure of antagonist affinity as well as providing a check that the antagonism is truly competitive *(16)*. The Schild plot uses a logarithmic manipulation of this equation to transform it into a linear relationship:

$$\log(dr - 1) = \log([B]) - \log(K_B) \tag{8}$$

The slope of a plot of $\log(dr - 1)$ against $\log[B]$ should be unity; if this is the case, the intercept gives a measure of $-\log(K_B)$, called here the pA_2 (the negative logarithm of the concentration of antagonist required to produce a dose-ratio of 2). The K_B measured from this functional study is independent of the agonist used and should match the K_D of the antagonist as measured in a binding study, so this approach can be used to define the receptor subtype involved in any particular response. If the slope of the line is not unity, then an explanation must be sought: either the antagonism is not purely competitive, or there are other confounding factors *(see* **ref. 17**). When considering the ADP receptor(s) on platelets, an example of the former is suramin, which presents with a Schild plot slope of around 2 for both aggregation and an increase in intracellular calcium level ($P2Y_1$) and clearly has a nonspecific component to its action. However, as an inhibitor of the effect of ADP on adenylate cyclase ($P2Y_{12}$), suramin has a Schild plot slope of unity and appears to be more specific *(18,19)*. An example of antagonism with other confounding factors is ATP. When purified it has a Schild slope of 1 for aggregation, increasing calcium and inhibiting adenylate cyclase *(8,20)*, but if used unpurified the Schild slope is significantly lower (S. Hourani, unpublished observation), presumably due to the presence of ADP as an impurity.

2. Practical Considerations

2.1. Agonist Concentration-Response Curves

To construct a full concentration-response curve for an agonist, a concentration range spanning at least three (preferably four) log cycles is usually required, with the

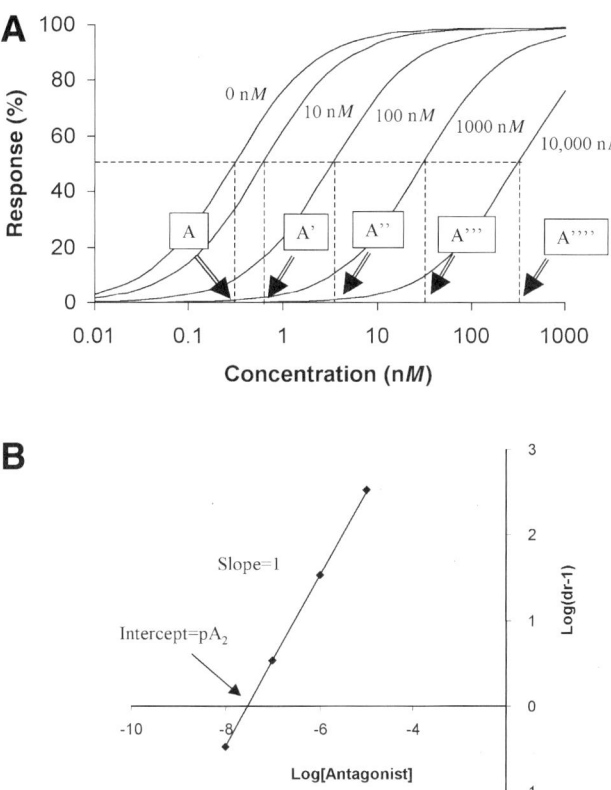

Fig. 5. Pharmacological analysis of antagonism. (**A**) Predicted concentration-response curves for Drug B ($e = 100$) in the presence of increasing concentrations of a competitive antagonist with a K_D of 30 nM. The curves were modeled using **Eq. 5** to calculate the receptor occupancy by the agonist, then using Stephenson's equation (**Eq. 3**) to derive the stimulus S, and assuming a hyperbolic relationship between S and response as for **Fig. 3**. The dose-ratio (dr) for 10 nM antagonist is given by A$'$/A, for 100 nM antagonist is A$''$/A, for 1000 nM antagonist is A$'''$/A and for 10,000 nM antagonist is A$''''$/A. (**B**) Schild plot of data in (**A**) (log(dr -1) against the log of the concentration of antagonist (M)). This has a slope of 1 and an intercept with the x axis at pA$_2$ [$= -\log(K_B)$]; *see* **Eq. 8**.

concentrations chosen so that they are evenly spaced when plotted on a logarithmic scale. For example, if the expected EC$_{50}$ is around 10 nM, a suitable set of concentrations would be 0.1, 0.3, 1, 3, 10, 30, 100, 300, and 1000 nM. If more points on the curve are required, they should be spaced at intervals of 1, 2, 5, 10 nM, etc. In platelet studies each concentration is usually tested on a different aliquot of the platelet suspension which avoids the problem of receptor desensitization, but there may well be time-dependent changes in responsiveness that need to be considered. For example, the response to a "standard" (usually maximal) concentration could be measured at regular intervals throughout the experiment and used to assess response decay versus time and generate an algorithm for normalization of all data. Ideally the concentrations should be tested in random order, but in practice it is often more convenient (and less likely to result in errors) to test them in increasing or decreasing order, which may theoretically introduce a systematic error into the results. The data from each individual experiment (agonist alone with and without antagonist) should be analyzed separately and the results of these analyses of several experiments (using, for example, blood from different donors) should be combined for statistical evaluation. In practice, when using ADP as agonist we have found that experiments carried out in duplicate or triplicate on the blood of three donors are usually sufficient to generate acceptable results. For comparison of EC_{50} or K_B values, remember that it is only the logarithms of these values that are normally distributed and therefore can be compared using the usual statistical tests such as t-tests or analysis of variance *(21)*. The data from a concentration-response curve can be fitted using the general hyperbolic ("logistic") function to relate the response E to the agonist concentration [A] *(22)*:

$$E = \frac{E_{max} \cdot [A]^n}{[A]^n + [A]_{50}{}^n} \tag{9}$$

where $[A]_{50}$ is the EC_{50}, E_{max} is the maximal response to the agonist, and n is a slope constant that allows for the different gradients that experimental concentration-response curves can display. This slope parameter, although chiefly a fitting parameter, can give useful information; for example, it has been used to indicate an interaction of an agonist with more than one receptor *(23)*. For an agonist, as discussed above, it is important to avoid the assumption that the $[A]_{50}$ is related to the K_D in any simple way, but these values can of course be used to calculate dose-ratios for use in the estimation of antagonist K_B.

The curve-fitting can be achieved using computer programs such as Prism (Graph-Pad Software Inc.) or Deltagraph (SPSS Inc.), which allow the correct equation to be defined. It should be noted however that some packages include both E_{max} and E_{min} as parameters, but of course by definition in the absence of agonist there is no response so E_{min} must be set at zero. If the curves do not plateau, an accurate estimate of the fitting parameters including $[A]_{50}$ may not be possible. However, in practice, because of the shape of a concentration-response curve, it is usually possible to fit a straight line to the central portion of the curve and calculate an $[A]_{50}$ (or the concentration that gives any other fixed level of response) from this. Values obtained in either way can then be used to compare agonist potency or to calculate antagonist affinity, as described below.

2.2. Antagonist Affinity

To calculate antagonist affinity using the Schild equation, concentration-response curves to an agonist (as above) are constructed in the presence and absence of one or more concentrations of antagonist. Again, care should be taken to try to reduce errors introduced by changes in the responsiveness of platelets with time or to implement a normalization procedure. If a Schild plot is being constructed, ideally the antagonist concentrations, like the agonist concentrations, should be chosen so that they are evenly spaced when plotted on a logarithmic scale. The $[A]_{50}$ of each agonist concentration-response curve should be calculated; if this is not possible, then the concentration that causes some fixed response should be determined. If only one concentration of antagonist is used, a K_B can be estimated using **Eq. 7** above, or if several concentrations have been used a Schild plot can be constructed and a more accurate measure of the K_B derived from the pA_2 value using **Eq. 8**. This method also provides a check for the competitivity of the antagonist and the validity of the approach in the form of the Schild plot slope, which should not be significantly different from unity. The points should then be fitted to the Schild equation using linear regression with the slope constrained to unity, and the value of K_B obtained will then accurately reflect the affinity of the antagonist for the receptor in question. It should be noted that this is independent of the agonist used, so if different K_B estimates are obtained using different agonists this is strong evidence that more than one receptor is involved.

Ideally in a single experiment a full set of concentration-response curves in the absence and presence of antagonist should be generated and a pA_2 or K_B value obtained. This can then be repeated several times to give a mean and standard error for statistical analysis if required, remembering that as with agonist potency it is only the log of the affinity (i.e., the pA_2 but not the K_B) that is normally distributed.

It is quite common to see results reported in which a single concentration of agonist is used and the inhibition of this response in response to various antagonist concentration is measured—a kind of antagonist concentration-response curve. While this may give useful information, the IC_{50} value for the antagonist (the concentration required to give 50% inhibition of the response to the agonist) is not the same as the K_B, and is dependent on the concentration of agonist and its potency. It is tempting but incorrect to apply a modified form of the Cheng-Prusoff equation *(24)*:

$$K_B = \frac{IC_{50}}{[A]/EC_{50} + 1} \tag{10}$$

This gives an estimate of the K_B from the IC_{50} value in the same way as is commonly done for binding studies, but it is theoretically invalid for functional studies even though it may in some cases give the correct result *(25)*.

3. Discussion

The approaches described above for the estimation of agonist potency and antagonist affinity are very straightforward and robust, and can be used to analyze data obtained from any kind of experiment, from measurement of signaling pathways through platelet aggregation or release to models of thrombosis. They are equally applicable to cells, tissues, or whole animals, and the only requirement is that quantitative data must be obtained relating the concentration of a drug to some measurable response. Problems usually arise only if the data are not reliable or if attempts are made to apply the equations to situations that do not conform to the requirements of the simple model. For example, attempts to carry out Schild analysis on antagonists that are not simply competitive will yield meaningless or misleading results, as will attempts to relate in a simple way the potency of an agonist to its affinity. Other confounding factors arise where the drugs used have more than one effect and do not act only on the receptor concerned.

In this short chapter it is obviously impossible to do more than cover the basic principles behind quantitative pharmacology, but for further information the most comprehensive source is the book by Terry Kenakin *(26)* that covers not only the theories and equations behind the various models of receptor mechanisms, but also gives many examples of how these can be applied to real data. The multiauthor volume edited by Kenakin and Angus *(27)* has more detailed chapters on various aspects of receptor pharmacology; another useful source of reference is the IUPHAR recommendations on terms and symbols in quantitative pharmacology *(28)*. In addition, there are frequent articles in *Trends in Pharmacological Sciences* devoted to specific aspects of receptor theory.

References

1. Clark, A. J. (1937) General pharmacology, in *Heffter's Handbuch der Experimentallen Pharmakologie* (Clark, A. J., ed.), Vol. 4, Springer-Verlag, Berlin, pp. 1–223.
2. Leff, P. (1995) The 2-state model of receptor activation. *Trends Pharmacol. Sci.* **16**, 89–97.
3. Bond, R. A., Leff, P., Johnson, T. D., Milano, C. A., Rockman, H. A., McMinn, T. R., et al. (1995) Physiological effects of inverse agonists in transgenic mice with myocardial overexpression of the beta(2)-adrenoceptor. *Nature* **374**, 272–276.
4. Stephenson, R. P. (1956) A modification of receptor theory. *Br. J. Pharmacol.* **11**, 379–393.
5. Furchgott, R. F. (1966) The use of β-haloalkylamines in the differentiation of receptors and in the determination of dissociation constants of receptor-agonist complexes, in *Advances in Drug Research* (Harper, N. J. and Simmonds, A. B., eds.), Vol. 3, Academic Press, New York, pp. 21–55.
6. Black, J. W. and Leff, P. (1983) Operational models of pharmacological agonism. *Proc. R. Soc. Lond.* **220**, 141–162.
7. Park, H. S. and Hourani, S. M. O. (1999) Differential effects of adenine nucleotide analogues on shape change and aggregation induced by adenosine 5′-diphospbate (ADP) in human platelets. *Br. J. Pharmacol.* **127**, 1359–1366.
8. Hall, D. A. and Hourani, S. M. O. (1993) Effects of analogs of adenine nucleotides on increases in intracellular calcium mediated by P2T-purinoceptors on human blood platelets. *Br. J. Pharmacol.* **108**, 728–733.
9. Hourani, S. M. O. (2001) Discovery and recognition of purine receptor subtypes on platelets. *Drug Devel. Res.* **52**, 140–149.
10. Palmer, R. K., Boyer, J. L., Schachter, J. B., Nicholas, R. A. and Harden, T. K. (1998) Agonist action of adenosine triphosphates at the human P2Y(1) receptor. *Mol. Pharmacol.* **54**, 1118–1123.
11. Zhang, F. L., Luo, L., Gustafson, E., Lachowicz, J., Smith, M., Qiao, X. D., et al. (2001) ADP is the cognate ligand for the orphan G protein-coupled receptor SP1999. *J. Biol. Chem.* **276**, 8608–8615.
12. Uneyama, C., Uneyama, H., and Akaike, N. (1993) Cytoplasmic Ca^{2+} oscillation in rat megakaryocytes evoked by a novel type of purinoceptor. *J. Physiol.* **470**, 731–749.
13. Somasundaram, B. and Mahaut-Smith, M. P. (1994) Three cation influx currents activated by purinergic receptor stimulation in rat megakaryocytes. *J. Physiol.* **480**, 225–231.

14. Mahaut-Smith, M. P., Ennion, S. J., Rolf, M. G., and Evans, R. J. (2000) ADP is not an agonist at $P2X_1$ receptors: evidence for separate receptors stimulated by ATP and ADP on human platelets. *Br. J. Pharmacol.* **131,** 108–114.
15. Gaddum, J. H. (1937) The quantitative effects of antagonistic drugs. *J. Physiol.* **89,** 7P–9P.
16. Arunlakshana, O. and Schild, H. O. (1959) Some quantitative uses of drug antagonists. *Br. J. Pharmacol.* **14,** 48–58.
17. Kenakin, T. P. (1982) The Schild regression in the process of receptor classification. *Can. J. Physiol. Pharmacol.* **60,** 249–265.
18. Hall, D. A. and Hourani, S. M. O. (1994) Effects of suramin on increases in cytosolic calcium and on inhibition of adenylate cyclase induced by adenosine 5′-diphosphate in human platelets. *Biochem. Pharmacol.* **47,** 1013–1018.
19. Hourani, S. M. O., Hall, D. A., and Nieman, C. J. (1992) Effects of the P2-purinoceptor antagonist, suramin, on human platelet-aggregation induced by adenosine 5′-diphosphate. *Br. J. Pharmacol.* **105,** 453–457.
20. Cusack, N. J. and Hourani, S. M. O. (1982) Adenosine 5′-diphosphate antagonists and human platelets - no evidence that aggregation and inhibition of stimulated adenylate cyclase are mediated by different receptors. *Br. J. Pharmacol.* **76,** 221–227.
21. Gaddum, J. H. (1945) Lognormal distributions. *Nature* **156,** 463–466.
22. Leff, P., Prentice, D. J., Giles, H., Martin, G. R., and Wood, J. (1990) Estimation of agonist affinity and efficacy by direct, operational model-fitting. *J. Pharm. Methods* **23,** 225–237.
23. Prentice, D. J. and Hourani, S. M. O. (1997) Information in agonist curve shape for receptor classification, in *Ann. N.Y. Acad. Sci.*, Vol. 812, pp. 234–235.
24. Cheng, Y. C. and Prusoff, W. H. (1973) Relationship between the inhibition constant (K_I) and the concentration of inhibitor which causes 50 per cent inhibition (I_{50}) of an enzymatic reaction. *Biochem. Pharmacol.* **22,** 3099–3108.
25. Lazareno, S. and Birdsall, N. J. M. (1993) Estimation of competitive antagonist affinity from functional inhibition curves using the Gaddum, Schild and Cheng-Prusoff equations. *Br. J. Pharmacol.* **109,** 1110–1119.
26. Kenakin, T. P. (1997) *Pharmacologic Analysis of Drug-Receptor Interaction*, 3rd ed., Raven Press, New York.
27. Kenakin, T. P. and Angus, J. A. (2000) *The Pharmacology of Functional, Biochemical, and Recombinant Receptor Systems.* Handbook of Experimental Pharmacology, 148, Springer-Verlag, Berlin Heidelberg.
28. Neubio, R.R., Spedding, M., Kenakin, T., and Christopoulos, A. (2003) International Union of Pharmacology Committee on Receptor Nomenclature and Drug Classification. XXXVIII Update on Terms and Symbols in Quantitative Pharmacology. *Pharmacol. Rev.* **55,** 597–606.

5

Inhibitors of Cellular Signaling Targets
Designs and Limitations

Chris J. Vlahos and Louis F. Stancato

1. Introduction

Kinases carry out the reversible phosphorylation of proteins and lipids, and are responsible for direct or indirect control of almost every signaling pathway in cells, leading to responses such as proliferation, differentiation, metabolism, transport, and gene expression. It is estimated that the human genome may contain about 1000 kinases *(1)*. To date, approx 500 protein kinases have been identified, of which about 400 are serine/threonine kinases and approx 100 are tyrosine kinases. Because of their central role in cellular signaling, as well as their primary role in disease progression, protein kinases are attractive therapeutic targets *(2,3)*. Seven molecules targeting protein kinases, imatinib mesylate (Gleevec) and trastuzumab (Herceptin), an inhibitor of BCR-abl and an antibody against ERB-2, respectively, are currently marketed as anti-oncolytics *(4–7)*. In addition, six molecules that block kinase activity are currently being investigated in Phase III clinical trials, whereas as many as 30 candidates targeting kinases are in early clinical development (Phase I or Phase II) *(8,9)*.

Many kinases are activated by more than one stimulus and many proteins are phosphorylated by multiple kinases; therefore the complexity and redundancy of these reactions imposes major limitations on the development of "selective inhibitors" of kinases. The objective of this chapter is to provide an understanding of the process to evaluate potentially useful therapeutic compounds directed toward kinases. Areas that will be highlighted include target validation (demonstration that the target kinase is involved in disease), target enablement (the design of a systematic screening approach to evaluate kinase inhibitors through a series of cell-free and cell-based assays), and hit-to-lead determination (the iterative process of compound evaluation culminating in the selection of a clinical candidate). In addition, limitations are discussed in order to give the investigator guidance in developing, selecting and/or using pharmacological kinase inhibitors.

From: *Methods in Molecular Biology, vol. 273:*
Platelets and Megakaryocytes, Vol. 2: Perspectives and Techniques
Edited by: J. M. Gibbins and M. P. Mahaut-Smith © Humana Press Inc., Totowa, NJ

2. Determination and Use of Kinase Inhibitors

Hanks and Hunter *(10)* showed that eukaryotic protein kinases form a large superfamily of homologous proteins that share common features in their catalytic domains. There are two main subdivisions within the superfamily: the protein-serine/threonine kinases and the protein-tyrosine kinases; these kinases can be further classified based on related substrate specificity and mode of regulation. In general, the catalytic region of protein kinases consists of an ATP-binding site and a substrate-binding site. Although both sites are potentially amenable to the binding of small-molecule inhibitors, the nucleotide-binding site is most often targeted for drug development because ATP binds in a discrete pocket. The three-dimensional crystal structure has been solved for a number of protein kinases, revealing the shape and topography of the ATP binding cleft, and how that cleft is affected by kinase activation. Due to the unique structural differences in ATP binding domains among the various kinase superfamily members, it is possible to develop clinically useful therapeutics with a high degree of specificity for the target of interest via competition for ATP binding *(11–15)*. Another potentially useful site for small-molecule intervention is the pocket involved in the transfer of phosphate to the kinase substrate. While most of the compounds currently in clinical evaluation were developed against the ATP binding site (including the BCR-abl inhibitor Gleevec *[16]*), there is a report of a Raf-1 inhibitor currently in Phase I/II clinical trials (BAY 43-9006) that is capable of interacting with both the ATP site as well as the phospho-transfer site *(17–19)*. For the remainder of this chapter we will focus on small molecule inhibition at the ATP site.

Many inhibitors of selected protein and lipid kinases were originally developed during drug discovery efforts, and regardless of whether or not they were successful in the clinic, they have proven to be valuable pharmacological tools in elucidating cellular signaling pathways. However, determination of cellular mechanisms has been complicated by the fact that many of the more commonly used kinase inhibitors show limited selectivity, inhibiting many isoforms of a given kinase family, as well as inhibiting multiple unrelated protein kinases. For example, staurosporine, an indocarbazole, is a very potent inhibitor of several PKC isoforms, but it also inhibits many other ser/thr and tyrosine kinases with remarkable potency. Unfortunately, there is a great deal of literature in which nonselective kinase inhibitors such as staurosporine have been reported to specifically inhibit certain enzymes, thus improperly defining a role for the kinase in question in specific cell signaling pathways. This leads to further use of these compounds, as well as "supported" claims in catalogs touting the "selectivity" of these compounds. It is important that the investigator take into account several considerations when designing an experiment using pharmacological kinase inhibitors (**Table 1**). A recent paper by Davies et al. investigated the specificity and mechanism of action of several commonly used protein kinase inhibitors and found that in most cases, these compounds inhibited multiple protein kinases, sometimes with a higher level of potency than against their presumed targets *(20)*. Thus, the development of selective kinase inhibitors must not only be directed against the kinase of interest, but should also include testing against related and unrelated kinases to determine selectivity.

Table 1
Considerations for Using Protein Kinase Inhibitors in Cell-Based Assays

1. Since most kinase inhibitors are competitive inhibitors against the ATP-binding site, higher concentrations are often needed in cell-based assays versus the effective concentration in a cell-free system. A good starting concentration is 10X higher than the IC_{50} from the cell-free assay.

2. Direct measurement of the product of the kinase phosphorylation reaction is preferred over measurement of a biological process (such as protein synthesis, cell proliferation, etc.). Since competitive inhibitors can be washed out, restoration of kinase activity and biological function is a useful way to establish that a kinase inhibitor is not toxic to the cell.

3. It is best that selectivity of the inhibitor be established against as many kinases as possible. Unfortunately, many commercially available inhibitors are marketed as "specific" or "selective." If possible, they should be tested against several related and unrelated kinases in cell-free assays prior to use in cell-based assays.

4. It is important that results obtained with a given inhibitor be confirmed by at least one other method. At a minimum, a second, structurally unrelated inhibitor against the same kinase target should be used. If available, a structurally related but inactive compound against the kinase (identified in a cell-free assay) can also be used to demonstrate target validity.

5. Antisense oligonucleotides or RNAi can be useful, nonpharmacological tools to confirm target-mediated effects in a signaling pathway. These are especially useful in eliminating a specific isoform of a kinase without affecting other members of the signaling family. However, one limitation to consider when interpreting the data is that if a kinase is part of a multiprotein signaling complex, complete removal of the kinase can affect the signaling pathway by disrupting the complex independent of effects on enzyme activity.

6. Cell systems expressing catalytically inactive and/or drug-resistant mutants of the target kinase are useful tools to examine nonspecific effects of inhibitors.

3. Target Validation

The cellular significance of many kinases in platelet function and adhesion is well-established *(21–25)*. From a drug discovery perspective, however, it is important to establish a link between a given kinase and the disease of interest. This is especially true when dealing with novel kinases identified through genomics-based efforts such as microarray analysis comparing mRNA level from diseased versus nondiseased tissue. An important limitation to consider is that, in certain diseases, the mRNA or protein level of the kinase may not be different from the nondiseased state, but rather the activation state of the kinase may be significantly higher. Analysis of mRNA should therefore be coupled with a proteomic approach that can identify post-translational modifications such as phosphorylation of either the kinase of interest or its substrate(s). Often, these analyses yield significantly more data, and therefore potential targets, than can be adequately resourced. Therefore, some guidelines should be established that could help prioritize the targets (kinases as well as other proteins):

1. *Is the target amenable to small-molecule inhibition?* In most cases, kinases are good targets for small-molecule inhibitors based on their discrete ATP- and substrate-binding pockets.

2. *Is the target novel?* This can be determined by comparing the sequence of the target to other kinase sequences in databases. In many cases, homology searches can help identify related kinases that, in turn, can yield clues in terms of function, downstream signaling targets, or modulators.

3. *Is the target predominantly found in the tissue of interest?* Again, this is important in developing therapeutic agents against novel kinases or selected isoforms for a particular disease. For example, if one is attempting to develop an inhibitor of a novel kinase for use in cardiac disease, then it is important that the target be expressed in the heart, or at least restricted to muscle lineages.

4. *Is there an association between the target and disease?* Sometimes, an increase in the expression of a given protein is observed when comparing tissue homogenates from diseased tissue versus nondiseased tissue *(26)*. Similarly, one can also show increases in protein or message in a primary tumor cell culture versus a nontransformed cell *(27)*. It should be recognized that overexpression of the target is not the only means leading to its increased activation; loss of upstream regulators such as phosphatases can lead to aberrant signaling and subsequent manifestation of disease (e.g., PTEN) *(28)*. In some cases, it is possible to create transgenic mice overexpressing the protein, and the use of organ- or cell-specific promotors (such as α-MHC for cardiac expression *[29]* or CD-2 promotor for T-cell expression *[30]*) can show that overexpression leads to expression of the disease phenotype. Further support of the role of the kinase in disease can be confirmed by lack of disease phenotype in a transgenic mouse expressing the catalytically inactive mutant kinase.

4. Target Enablement

Once a kinase target has been clearly defined, it is important to develop an efficient, sequential methodology that can rapidly screen several thousand compounds in order to identify potent and selective inhibitors. These assays must not only be amenable to developing a structure-activity relationship, but must relate directly to biological function. An ideal screening paradigm, illustrated in **Fig. 1**, consists of a primary screen, which is usually an isolated enzyme assay under cell-free conditions; a cell-based assay in which compounds are tested for the ability to block kinase activity in a cell; a functional assay in cells, in which inhibition of the kinase is shown to be directly responsible for blocking a cellular function; and, finally, in vivo testing in an animal model of disease.

High-throughput screening involves the testing of several thousand compounds against the target kinase using a rapid means of testing and analysis. Compounds often tested include traditional organic chemicals, natural product extracts from plants, microorganisms, or marine species, combinatorial chemistry mixtures, and peptides and peptide mimetics. There are a number of commercially available "biased" chemical libraries for a number of classes of targets, including kinases. These libraries are relatively small collections of compounds that have a relatively high success rate at inhibiting kinases, such as bisindolylmaleimides, flavonoids, and indoles. Initially, compounds are tested at a single concentration, with active compounds defined as those inhibiting >50% of the enzyme activity. Actives are then retested using multipoint concentration ranges to determine IC_{50} values (the concentration in which 50% of the enzyme activity is inhibited). A hit is defined as an active molecule that is confirmed by an IC_{50} measurement in the primary assay, in which the chemical structure, purity, and stability are confirmed.

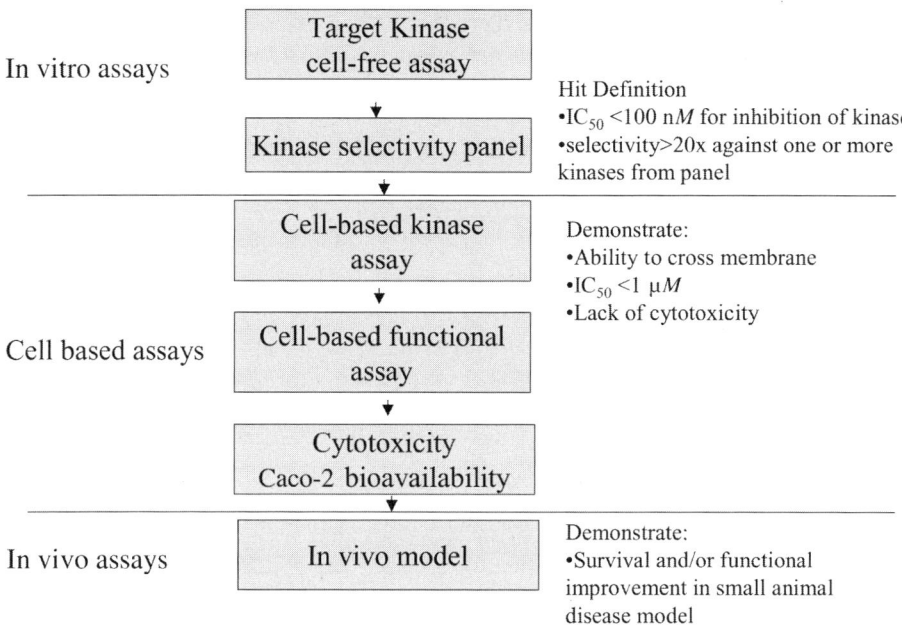

Fig. 1. Representative testing scheme for kinase inhibitors. An effective screen consists of a primary enzyme assay against the kinase of interest in a high-throughput format, a selectivity panel consisting of several representative kinases from various subfamilies, a secondary, cell-based assay measuring kinase activity and/or cellular function directly mediated by the kinase, assays to assess cytotoxicity and compound bioavailability, and an in vivo disease model to determine compound efficacy. Critical success factors describing minimal potency, selectivity, and efficacy are defined based on benchmark compounds and the output from the initial screen.

There are several important considerations to take into account when designing a kinase screen. The quantity and quality of enzyme are critical to the success of the assay, and enough enzyme must be on hand to allow for the testing of potentially tens of thousands of compounds (assuming that a biochemical assay, and not a cell-based approach, is the preferred method of choice for lead generation). Often, commercial protein vendors represent a useful source of protein suitable for kinase lead generation, often providing well-characterized enzymes from major signaling pathways. In some cases the enzymes are packaged as components of kinase assay kits designed to give the researcher an out-of-the-box solution for kinase drug discovery. Our experience dictates, however, that commercially available protein kinase preparations, while often as effective as advertised, can sometimes fail to meet established criteria for the level of purity and validation necessary for initiating drug-discovery projects. It is important, therefore, to verify the sequence and purity of the procured kinase prior to initiating the drug screen.

For many of these reasons, researchers may opt to express and isolate kinases in their laboratories. The complexity of the target, vis-à-vis mechanisms of activation, cofactors, and potential protein-protein interactions, are also critical factors in the decision to express and purify a given kinase. Insect cell lines such as *Sf9* and *Sf21* infected with baculovirus often represent the expression method of choice due to their ability to perform post-translational modifications and their ease of scale-up. However, because these cells contain endogenous kinases and phosphatases, care must be taken to ensure purity. *E. coli* represents an alternative source for protein expression due to its ability for high expression, large-scale production, and relatively minimal lack of contaminating kinase and phosphatase activities. In most cases, kinases are expressed as fusion proteins to GST or with affinity tags such as $(His)_6$, myc, or flg, which are added to aid in the purification and in most instances do not affect enzymatic activity.

There are some situations whereby expression of the target kinase may not be sufficient to generate an active form of the enzyme. For those targets known to play a role in discrete signaling pathways, *a priori* knowledge of kinase activation mechanisms may be important in determining not only the cDNA library from which an enzyme is cloned, but also the appropriate state for expression. As discussed above, the kinase domain most commonly targeted for drug development is the ATP-binding cleft. There are also several cellular and biochemical issues surrounding the activation of a particular kinase, leading to a series of questions important for determining how a kinase should be expressed and purified:

1. *Should the active or inactive form of an enzyme be screened?* For example, the catalytic domain of p38/SAPK family members is known to undergo a profound conformational change upon phosphorylation *(31,32)*. Literature reports cite this difference between the active and inactive forms of the enzyme as the basis for the discovery of relatively specific p38 and SAPK inhibitors that target the inactive form of these kinases *(33)*. Likewise, Gleevec, the BCR-Abl inhibitor useful in the treatment of chronic myeloid leukemia, was discovered through a drug screen using an inactive form of the BCR-Abl kinase fusion protein *(16)*.

2. *Is full-length enzyme required, or will expression of the kinase catalytic domain be sufficient?* In the case of large molecular weight and receptor protein kinases, expression of the catalytic domain provides a way to generate sufficient enzyme for the high-throughput screen *(34)*.

3. *Should various regulatory sites, or even whole regulatory domains, be modified in such a way as to artificially increase activity?* Many of the so-called AGC subfamily of kinases *(10)* such as AKT/PKB, SGK, and PKC can be expressed in a constitutively active state by expressing an src myristoylation signal *(35)* or by mutating a consensus PDK1 phosphorylation site within the activation loop of the catalytic domain to an aspartic acid *(36)*. Of course, mutation of residues near those that bind ATP may artificially alter the pocket's topography and thus adversely affect small-molecule binding *(37)*. In the case of Src family members, deletion of the regulatory tyrosine at position 527, significantly away from the ATP binding site, can dramatically increase kinase activity and generate an enzyme suitable for small-molecule targeting *(38)*.

Once enzyme is obtained it must be demonstrated that the kinase preparation is enzymatically pure, i.e., that any associated kinase activity is directly attributed to the target of interest, as active kinase contaminants may slip through purification procedures undetected. If there are contaminating kinases in the reaction mixture, they may not be inhibited by test compounds and can therefore phosphorylate the substrate (especially if a general kinase substrate such as myelin basic protein or histone is used), resulting in false negatives. One way to test this is to determine autokinase activity present in an enzyme reaction mixture by incubating the kinase plus radioactive ATP in an in vitro kinase reaction. Following electrophoresis and autoradiography, one phosphoenzyme band corresponding to the molecular weight of the kinase would be observed in the case of pure material.

Finally, after all questions regarding enzyme expression, purity and activity are answered, it is necessary to establish a systematic screening format in order to identify inhibitors of the kinase. In virtually any drug discovery effort, the underlying goal is to identify active molecules (ultimately specific for the target of interest) from tens of thousands of potentially unrelated chemical compounds. To increase the likelihood that only high-quality leads are identified, each active compound must overcome a series of assays designed to test for efficacy and potency against the target in question (**Fig. 1**). These assays must be amenable to developing a structure-activity relationship (SAR), which is an iterative process of chemical synthesis and biological testing in which information is derived relating molecular structure to changes in experimental test results. In addition, the assays must also relate directly to biological function. In nearly all cases a combination of cell-free and cell-based assays are used to generate leads.

4.1. Primary Assay

Generally, the simplest methods for rapidly screening kinase inhibitors are any of a number of biochemical assays that detect changes in the phosphorylation state of peptide or protein substrates. All assay formats share one common and essential characteristic: each is capable of functioning in microtiter plates, the density of which ranges from 96 wells/plate to 1536 wells/plate. The biochemical assay formats are broadly split into two categories, depending on the method of detection: (1) radioactive, which includes filter-binding and scintillation-proximity assays (SPAs), and (2) nonradioactive, which includes fluorescence polarization (FP), homogenous time-resolved fluorescence (HTRF) and mobility shift/microfluidics. The most common assay format to measure kinase activity is filter binding, in which $^{33}P\gamma$-ATP is incubated with the isolated enzyme and a substrate protein or peptide. The radioactive product is trapped onto a membrane and directly quantitated by use of a scintillation counter or Top-Count. An alternative radioactive format is SPA, in which phosphorylation of a substrate attached to a bead or plate embedded with a scintillant generates a quantifiable signal. Each assay format has its own strengths and limitations (**Table 2**). While a detailed examination of each format is beyond the scope of this chapter, it is important to keep in mind that they all share one very important characteristic: each is useful in screening for kinase inhibitors from large chemical libraries. While radioactive assays do pose potential safety and waste concerns, the filter-binding format remains the most widely used format in many lead generation projects. Some investigators have used in-gel kinase assays or SDS-PAGE of an in vitro kinase reaction

Table 2
Comparison of Radioactive and Nonradioactive Kinase Assay Formats

---

Radioactive Formats

Filter Binding

Phosphate detection: ^{33}P-labeled substrate bound to filter

Format: nonhomogeneous

Strengths: direct measurement of phosphate incorporation into substrate; easiest of all formats to design; industry "gold standard"

Limitations: requires use of radioisotope; may require large amounts of enzyme and substrate; 96-well only; labor-intensive

Scintillation Proximity Assay

Phosphate detection: immobilized ^{33}P-labeled substrate in proximity to scintillant-impregnated bead

Format: homogeneous

Strengths: ease of operation; miniaturizable beyond 96-well

Limitations: requires use of radioisotope; potential for large number of false positives; may require use of custom reagents that are not readily produced

Nonradioactive Formats

Fluorescence Polarization, HTRF, and Chemiluminescence

Measurement: antibody-based detection of phosphate incorporation into peptide substrate

Format: homogeneous

Strengths: nonradioactive; miniaturizable beyond 96-well; ease of operation; very low protein and substrate requirement

Limitations: requires phosphospecific antibodies; can only use peptide substrates; potential compound interference with optical output

Mobility Shift/Microfluidic Chip

Measurement: separation of fluorescently tagged product from substrate via change in charge-to-mass ratio

Format: homogeneous

Strengths: direct measurement of product formation; highly sensitive; low protein and substrate requirement; high degree of data reproducibility

Limitations: requires fluorescently tagged peptides with a near-neutral charge; high instrument expense; instrument reliability

---

mixture to assess kinase activity, but these techniques are not suitable for high-throughput screening due to labor intensivity, reduced sensitivity, difficulty in quantitation, and length of time to perform the analysis.

More recently, nonradioactive technologies are being employed to measure kinase activity. One such format is fluorescence polarization, which uses a substrate peptide conjugated to a fluorescent reporter group. Upon phosphorylation, the fluorescent product is recognized by a phosphospecific antibody, which causes it to rotate more slowly in solution. The fluorophore is excited in both parallel and perpendicular planes, and the slower rotation of the conjugated pair enables it to emit light in a given plane rather than randomly scattered. Fluoresence polarization is extremely sensitive and amenable to 384-well plate densities (and beyond), although it is possible that some test compounds could interfere with the assay by acting as quenchers or due to their own inherent fluorescence. The use of this format, the technology for which has existed for decades, has increased dramatically due to the availability of phosphospecific antibodies as well as novel phosphate-derivatizing detection agents *(39)*. Regardless of the assay format, it is often most efficient to initially assay test compounds at a single concentration and to then confirm the actives by determining IC_{50}s for any compounds that show >50% inhibition at the fixed concentration due to the large number of compounds screened (*see* **Subheading 5.**). This will also help to eliminate false positives that seem to be inherent in assays regardless of the screening format.

The outcome of a kinase screen can be greatly influenced by the choice of substrate used in the enzymatic assay. Ideally, the substrate should be the relevant cellular substrate for the kinase. In many cases, however, the physiologic substrate is not known, so it is possible to develop enzyme assays using general kinase substrates such as histone H4 or myelin basic protein. Since some kinases will autophosphorylate, it is also possible to develop an assay either directly measuring this response or using a substrate peptide containing the autophosphorylation site (if known). Oriented peptide libraries have been used to identify the substrates for several kinases, including Akt, PKA, and ZAP-70 *(40–42)*. This approach can utilize combinatorial peptide libraries or expression of single peptides containing known kinase phosphorylation sites. In either case, kinase activity can be detected by measuring incorporation of radiolabeled phosphate or by using fluorescence-labeled phosphospecific antibodies on the peptide microarrays. This promising technology can be especially valuable in identifying substrate sequences for novel kinases, with minimal reagent requirement and rapid analysis *(43,44)*.

Finally, kinetics of the assay are critical for identifying inhibitors. It is important that the concentration for compounds to be tested in the assay should be at or near the K_m for ATP when screening for competitive inhibitors of ATP binding. In addition, the reaction should be run at initial rate conditions to minimize any product inhibition or potential loss of assay sensitivity. While it is important to assess effects of metal ions, detergent, pH, ionic strength, and other assay conditions, it is also crucial to assess the sensitivity of the enzyme to DMSO, since most compound libraries are stocked as DMSO solutions.

4.2. Secondary Cell-Based Assays

Hits from the primary enzyme screen are next investigated for their ability to inhibit the kinase in a cell-based assay where the readout is directly associated with kinase activity. Novel tools such as antisense oligonucleotides or the expression of an inactive mutant of the kinase can be used to validate the target in intact cells. Ideally, the most relevant screen is to monitor changes in substrate phosphorylation as a direct result of the target kinase. For example, phosphorylation of MAPKAP-K2 or MNK-1, substrates for activated p38, can be detected in anisomycin-treated cells *(45,46)*. Therefore, it is possible to evaluate inhibitors of p38 such as SB203580 in a cell-based assay measuring phosphorylation of these enzymes. If substrates for the target kinase are not known, or if appropriate tools to measure kinase activity have not been developed, it may be necessary to develop a functional cell-based assay. Examples include calcium mobilization, glucose uptake (PI3K), glycogen synthesis (GSK3β), reporter assay (Jnk/cJun), cell proliferation, protein synthesis, contraction, actin reorganization, motility, and nuclear or membrane translocation. There are several complicating factors, however, in dealing with functional assays. For example, if an endpoint is dependent on the action of multiple kinases (e.g., protein synthesis or cell proliferation), these assays may not distinguish between selected kinases. Another potential problem results when the measured response is several kinase steps away from the kinase of interest (e.g., using a MEKKK inhibitor to monitor c-jun phosphorylation). Care must be taken to ensure that the candidate inhibitor is acting on the desired enzyme rather than another kinase downstream in the pathway. Cell-based assays are also useful in that they can eliminate compounds from consideration that cannot cross a cell membrane. In many cases, cell-based assays can be performed in 96-well formats. These assays are of particular use in driving a structure-activity relationship, as they will provide a means for simultaneously evaluating several compounds with appropriate concentration curves in relatively short times. Finally, cell-based assays can also provide an initial assessment of cytotoxicity, although this may dependend on the cell type.

4.3. In Vivo Analysis of Kinase Inhibitors

Compounds that have been successfully identified in the primary screens and in the cell-based assays are then tested in animal models of disease. Due to the specialized nature of in vivo models, it is beyond the scope of this chapter to comprehensively review the animal models currently available. In some cases, however, transgenic models overexpressing the kinase of interest (as well as parallel models using the kinase inactive mutants) not only have been useful in target validation, but have also been employed in demonstrating the utility of a kinase inhibitor *(47–50)*. Systemic knockout models can also assist in target validation, as well as give an indication of the effects of blocking kinase activity *(51–53)*. While transgenic models have utility, the ideal animal model directly shows that a kinase inhibitor blocks (or restores) physiological function in an animal model of disease.

5. Hit and Lead Determination

A typical screen can yield hundreds to thousands of actives and several hundred hits, so it is important to design a testing scheme that will enable the declaration of a lead compound. A lead compound is defined as a hit that has been validated by appropriate primary and secondary assays. The testing scheme leading to lead identification depicted in **Fig. 1** includes selectivity (testing against other kinases) as well as secondary cell-based assays (described in **Subheading 4.2.**) that are used to demonstrate that the activity measured relates to the target of interest. These assays therefore enable the medicinal chemist to construct a structure-activity relationship around the lead compound. It is important that these secondary assays be reproducible and robust enough to rapidly evaluate all screen hits. Other considerations in lead declaration include the potency of the compound (affinity and/or functional activity), the emergence of a proprietary intellectual property strategy, and some early indication of a structure-activity relationship. At the point of lead declaration, in vivo activity is not necessarily required, although some limited activity in animal models can certainly help in validating the target. In addition, cytotoxicity (XTT, MTT) *(54)* or surrogate bioavailability assays (CaC02, HERG) *(55,56)* can also provide useful information in prioritizing candidate molecules to proceed into in vivo testing.

As compounds are evaluated through the hit-to-lead and lead-to-clinical candidate stages, selectivity is of consideration in order to maximize efficacy and minimize toxicity. Early in the discovery process, an indication of kinase selectivity is important in the identification and elimination of broad-acting kinase inhibitors. Selectivity assays can also enable selection of certain chemical scaffolds on which to focus drug development. Testing confirmed screen hits against a panel of other kinases, consisting of both closely related kinases as well as unrelated kinases, can accomplish this. A comparison of the IC_{50} values for the target kinase versus these other kinases will give an indication of the relative potency of compounds. However, due to the relative similarities in ATP binding sites among kinase subfamilies, it may be very difficult to achieve a large degree of separation of a target kinase over its family members. However, when comparing kinases across other subfamilies with diverse substrate binding sites, a wider range of compound potency will be observed. It is naïve to envision a molecule that will selectively inhibit only one kinase. For example, even the marketed antioncolytic kinase inhibitor Gleevec and the antiangiogenesis clinical candidate SU6668 have been demonstrated to inhibit multiple kinases (BCR-abl, c-kit, PDGF-R, VEGF-R) *(57,58)*. However, it is possible to select compounds for further development in which a large selectivity profile is observed among groups of unrelated kinases. In addition, panel kinases should also include one or more kinases in which inhibition would be undesirable. For example, PKA and CamKII are both key enzymes involved in regulating cardiac contraction and relaxation, so it would likely be important to ensure that inhibitors of the target kinase do not significantly affect these enzymes.

In concert with the classical screening approach, there is also an increased focus on designing kinase inhibitors based on enzyme mechanism and structure. Most protein kinases form a dissociative transition state between the gamma phosphate of ATP and the substrate protein. In many cases, X-ray crystal structure of the kinase complexed to an inhibitor has yielded useful information about mechanism. For example, co-crystallization of PI3Kγ with ATP or various inhibitors (wortmannin, LY294002, quercetin) reveals conformational changes in the protein, as well as different binding interactions that may be useful in the design of isoform-selective inhibitors *(59)*. Although not a rapid or efficient method for supporting inhibitor structure-activity relationship, structure determination can be a useful tool in refining the design of an inhibitor or for understanding mechanism of action.

Another useful tool to refine a structure-activity relationship is virtual screening, which is the process of estimating the biological activity of compounds selected from a database (i.e., a chemical library) with unknown activity against a given target

(60–63). Virtual screening results in a prioritized list of compounds, from which these agents can be added to a screening list in order to potentially increase the number of actives. Not only is virtual screening useful to support an existing screen, but it can also be used as an alternative to screening. Using the structure of the enzyme as well as the structure of known chemical entities, it is possible to generate a mathematical model of the ligand-receptor complex. These techniques can support inhibitor discovery efforts ranging from prediction of binding, lead optimization, and evaluation of biased libraries. X-ray structures from various public databases are readily available and can also be used in target evaluation.

6. Conclusion

Protein kinases are therapeutic targets based on their essential role as regulators of cellular signaling pathways. Selective, cell-permeable inhibitors of protein kinases have been useful in the elucidation of signal transduction pathways and have been extremely valuable in defining the function of specific kinases. More recently, and certainly more significantly, protein kinase inhibitors are beginning to make their way into the clinic and into the marketplace as effective therapeutic agents. High-throughput screens have been used to develop selective inhibitors against target kinases, although the format of these assays is not usually consistent with the subcellular milieu in which these kinases normally operate. However, proper design of a screening paradigm to evaluate a target kinase, which includes both cell-free and cell-based assays to evaluate enzyme activity and cellular function, and the inclusion of selectivity assays, can minimize limitations associated with primary assay design and lead to the development of novel pharmacologic agents.

Acknowledgments

We wish to thank Dr. Susan McDowell for her critical reading of the manuscript and helpful suggestions.

References

1. Venter, J. C., et al. (2001) The sequence of the human genome. *Science* **291,** 1304–1351.
2. Cohen, P. (2002) Protein kinases—the major drug targets of the twenty-first century? *Nat. Rev. Drug Discov.* **1,** 309–315.
3. Vlahos, C. J., McDowell, S. A., and Clerk, A. (2003) Kinases as therapeutic targets for heart failure. *Nat. Rev. Drug Discov.* **2,** 99–113.
4. Thiesing, J. T., Ohno-Jones, S., Kolibaba, K. S., and Druker, B. J. (2000) Efficacy of STI571, an abl tyrosine kinase inhibitor, in conjunction with other antileukemic agents against bcr-abl-positive cells. *Blood* **96,** 3195–3199.
5. Druker, B. J., Talpaz, M., Resta, D. J., Peng, B., Buchdunger, E., Ford, J. M., et al. (2001) Efficacy and safety of a specific inhibitor of the BCR-ABL tyrosine kinase in chronic myeloid leukemia. *N. Engl. J. Med.* **344,** 1031–1037.
6. Vogel, C., Cobleigh, M. A., Tripathy, D., Gutheil, J. C., Harris, L. N., Fehrenbacher, L., et al. (2001) First-line, single-agent Herceptin (trastuzumab) in metastatic breast cancer: a preliminary report. *Eur. J. Cancer* **37(Suppl. 1),** S25–S29.
7. Stebbing, J., Copson, E., and O'Reilly, S. (2000) Herceptin (trastuzamab) in advanced breast cancer. *Cancer Treat. Rev.* **26,** 287–290.

8. Traxler, P., Bold, G., Buchdunger, E., Caravatti, G., Furet, P., Manley, P., et al. (2001) Tyrosine kinase inhibitors: from rational design to clinical trials. *Med. Res. Rev.* **21**, 499–512.

9. Zwick, E., Bange, J., and Ullrich, A. (2002) Receptor tyrosine kinases as targets for anticancer drugs. *Trends Mol. Med.* **8**, 17–23.

10. Hanks, S. K. and Hunter, T. (1995) Protein kinases 6. The eukaryotic protein kinase superfamily: kinase (catalytic) domain structure and classification. *Faseb J.* **9**, 576–596.

11. Vlahos, C. J., Matter, W. F., Hui, K. Y., and Brown, R. F. (1994) A specific inhibitor of phosphatidylinositol 3-kinase, 2-(4- morpholinyl)-8-phenyl-4H-1-benzopyran-4-one (LY294002). *J. Biol. Chem.* **269**, 5241–5248.

12. Jirousek, M. R., Gillig, J. R., Gonzalez, C. M., Heath, W. F., McDonald, J. H. 3rd, Neel, D. A., et al. (1996) (S)-13-[(dimethylamino)methyl]-10,11,14,15-tetrahydro-4,9:16, 21-dimetheno-1H, 13H-dibenzo[e,k]pyrrolo[3,4-h][1,4,13]oxadiazacyclohexadecene-1,3(2H)-dione (LY333531) and related analogues: isozyme selective inhibitors of protein kinase C beta. *J. Med. Chem.* **39**, 2664–2671.

13. Buchdunger, E., Zimmermann, J., Mett, H., Meyer, T., Muller, M., Druker, B. J., et al. (1996) Inhibition of the Abl protein-tyrosine kinase in vitro and in vivo by a 2-phenylaminopyrimidine derivative. *Cancer Res.* **56**, 100–104.

14. Pollack, V. A., Savage, D. M., Baker, D. A., Tsaparikos, K. E., Sloan, D. E., Moyer, J. D., et al. (1999) Inhibition of epidermal growth factor receptor-associated tyrosine phosphorylation in human carcinomas with CP-358,774: dynamics of receptor inhibition in situ and antitumor effects in athymic mice. *J. Pharmacol. Exp. Ther.* **291**, 739–748.

15. Mendel, D. B., Laird, A. D., Smolich, B. D., Blake, R. A., Liang, C., Hannah, A. L., et al. (2000) Development of SU5416, a selective small molecule inhibitor of VEGF receptor tyrosine kinase activity, as an anti-angiogenesis agent. *Anticancer Drug Des.* **15**, 29–41.

16. Schindler, T., Bornmann, W., Pellicena, P., Miller, W. T., Clarkson, B., and Kuriyan, J. (2000) Structural mechanism for STI-571 inhibition of abelson tyrosine kinase. *Science* **289**, 1938–1942.

17. Lyons, J. F., Wilhelm, S., Hibner, B., and Bollag, G. (2001) Discovery of a novel Raf kinase inhibitor. *Endocr. Relat. Cancer* **8**, 219–225.

18. Lowinger, T. B., Riedl, B., Dumus, J., and Smith, R. A. (2002) Design and discovery of small molecules targeting Raf-1 kinase. *Curr. Pharm. Des.* **8**, 2269–2278.

19. Wilhelm, S. M., Housley, T., Kennure, N., Rong, H., Carlson, R., Hibner, B., et al. (2001) A novel diphenylurea Raf-1 kinase inhibitor (RKI) blocks the Raf/MEK/ERK pathway in tumor cells. *Proceedings of the AACR* **42**, Abstract 4956.

20. Davies, S. P., Reddy, H., Caivano, M., and Cohen, P. (2000) Specificity and mechanism of action of some commonly used protein kinase inhibitors. *Biochem. J.* **351**, 95–105.

21. Kralisz, U. and Cierniewski, C. S. (1997) Differential effects of the tyrosine kinase inhibitors on collagen type 1-induced platelet aggregation and adhesion to this protein. *Thromb. Res.* **86**, 287–299.

22. Law, D. A., Nannizzi-Alaimo, L., Ministri, K., Hughes, P. E., Forsyth, J., Turner, M., et al. (1999) Genetic and pharmacological analyses of Syk function in $\alpha_{IIb}\beta_3$ signaling in platelets. *Blood* **93**, 2645–2652.

23. Koziak, K., Kaczmarek, E., Park, S. Y., Fu, Y., Avraham, S., and Avraham, H. (2001) RAFTK/Pyk2 involvement in platelet activation is mediated by phosphoinositide 3-kinase. *Br. J. Haematol.* **114**, 134–140.

24. Crosby, D. and Poole, A. W. (2002) Interaction of Bruton's tyrosine kinase and protein kinase Ctheta in platelets. Cross-talk between tyrosine and serine/threonine kinases. *J. Biol. Chem.* **277**, 9958–9965.

25. Yap, C. L., Anderson, K. E., Hughan, S. C., Dopheide, S. M., Salem, H. H., and Jackson, S. P. (2002) Essential role for phosphoinositide 3-kinase in shear-dependent signaling between platelet glycoprotein Ib/V/IX and integrin alpha(IIb)beta(3). *Blood* **99,** 151–158.

26. Bowling, N., Walsh, R. A., Song, G., Estridge, T., Sandusky, G. E., Fouts, R. L., et al. (1999) Increased protein kinase C activity and expression of Ca2+-sensitive isoforms in the failing human heart. *Circulation* **99,** 384–391.

27. Shayesteh, L., Lu, Y., Kuo, W. L., Baldocchi, R., Godfrey, T., Collins, C., et al. (1999) PIK3CA is implicated as an oncogene in ovarian cancer. *Nat. Genet.* **21,** 99–102.

28. Lu, Y., Lin, Y. Z., LaPushin, R., Cuevas, B., Fang, X., Yu, S. X., et al. (1999) The PTEN/MMAC1/TEP tumor suppressor gene decreases cell growth and induces apoptosis and anoikis in breast cancer cells. *Oncogene* **18,** 7034–7045.

29. Gulick, J., Subramaniam, A., Neumann, J., and Robbins, J. (1991) Isolation and characterization of the mouse cardiac myosin heavy chain genes. *J. Biol. Chem.* **266,** 9180–9185.

30. Parsons, M. J., Jones, R. G., Tsao, M. S., Odermatt, B., Ohashi, P. S., and Woodgett, J. R. (2001) Expression of active protein kinase B in T cells perturbs both T and B cell homeostasis and promotes inflammation. *J. Immunol.* **167,** 42–48.

31. Zhang, J., Zhang, F., Ebert, D., Cobb, M. H., and Goldsmith, E. J. (1995) Activity of the MAP kinase ERK2 is controlled by a flexible surface loop. *Structure* **3,** 299–307.

32. Wang, Z., Harkins, P. C., Ulevitch, R. J., Han, J., Cobb, M. H., and Goldsmith, E. J. (1997) The structure of mitogen-activated protein kinase p38 at 2.1-A resolution. *Proc. Natl. Acad. Sci. USA* **94,** 2327–2332.

33. Wang, Z., Canagarajah, B. J., Boehm, J. C., Kassisa, S., Cobb, M. H., Young, P. R., et al. (1998) Structural basis of inhibitor selectivity in MAP kinases. *Structure* **6,** 1117–1128.

34. Parast, C. V., Mroczkowski, B., Pinko, C., Misialek, S., Khambatta, G., and Appelt, K. (1998) Characterization and kinetic mechanism of catalytic domain of human vascular endothelial growth factor receptor-2 tyrosine kinase (VEGFR2 TK), a key enzyme in angiogenesis. *Biochemistry* **37,** 16,788–16,801.

35. Kohn, A. D., Summers, S. A., Birnbaum, M. J., and Roth, R. A. (1996) Expression of a constitutively active Akt Ser/Thr kinase in 3T3-L1 adipocytes stimulates glucose uptake and glucose transporter 4 translocation. *J. Biol. Chem.* **271,** 31,372–31,378.

36. Sable, C. L., Filippa, N., Hemmings, B., and Van Obberghen, E. (1997) cAMP stimulates protein kinase B in a Wortmannin-insensitive manner. *FEBS Lett.* **409,** 253–257.

37. Bishop, A. C., Buzko, O., and Shokat, K. M. (2001) Magic bullets for protein kinases. *Trends Cell Biol.* **11,** 167–172.

38. Liu, X. and Pawson, T. (1994) Biochemistry of the Src protein-tyrosine kinase: regulation by SH2 and SH3 domains. *Recent Prog. Horm. Res.* **49,** 149–160.

39. Parker, G. J., Law, T. L., Lenoch, F. J., and Bolger, R. E. (2000) Development of high throughput screening assays using fluorescence polarization: nuclear receptor-ligand-binding and kinase/phosphatase assays. *J. Biomol. Screen.* **5,** 77–88.

40. Obata, T., Yaffe, M. B., Leparc, G. G., Piro, E. T., Maegawa, H., Kashiwagi, A., et al. (2000) Peptide and protein library screening defines optimal substrate motifs for AKT/PKB. *J. Biol. Chem.* **275,** 36,108–36,115.

41. Dostmann, W. R., Nickl, C., Thiel, S., Tsigelny, I., Frank, R., and Tegge, W. J. (1999) Delineation of selective cyclic GMP-dependent protein kinase Ialpha substrate and inhibitor peptides based on combinatorial peptide libraries on paper. *Pharmacol. Ther.* **82,** 373–387.

42. Nishikawa, K., Sawasdikosol, S., Fruman, D. A., Lai, J., Songyang, Z., Burakoff, S. J., et al. (2000) A peptide library approach identifies a specific inhibitor for the ZAP-70 protein tyrosine kinase. *Mol. Cell.* **6,** 969–974.

43. Reineke, U., Volkmer-Engert, R., and Schneider-Mergener, J. (2001) Applications of peptide arrays prepared by the SPOT-technology. *Curr. Opin. Biotechnol.* **12,** 59–64.

44. Gast, R., Glokler, J., Hoxter, M., Kiess, M., Frank, R., and Tegge, W. (1999) Method for determining protein kinase substrate specificities by the phosphorylation of peptide libraries on beads, phosphate-specific staining, automated sorting, and sequencing. *Anal. Biochem.* **276,** 227–241.

45. Fukunaga, R. and Hunter, T. (1997) MNK1, a new MAP kinase-activated protein kinase, isolated by a novel expression screening method for identifying protein kinase substrates. *EMBO J.* **16,** 1921–1933.

46. Neininger, A., Thielemann, H., and Gaestel, M. (2001) FRET-based detection of different conformations of MK2. *EMBO Rep.* **2,** 703–708.

47. Wakasaki, H., Koya, D., Schoen, F. J., Jirousek, M. R., Ways, D. K., Hoit, B. D., et al. (1997) Targeted overexpression of protein kinase C beta2 isoform in myocardium causes cardiomyopathy. *Proc. Natl. Acad. Sci. USA* **94,** 9320–9325.

48. Esposito, G., Prasad, S. V., Rapacciuolo, A., Mao, L., Koch, W. J., and Rockman, H. A. (2001) Cardiac overexpression of a G(q) inhibitor blocks induction of extracellular signal-regulated kinase and c-Jun NH(2)-terminal kinase activity in in vivo pressure overload. *Circulation* **103,** 1453–1458.

49. Liao, P., Georgakopoulos, D., Kovacs, A., Zheng, M., Lerner, D., Pu, H., et al. (2001) The in vivo role of p38 MAP kinases in cardiac remodeling and restrictive cardiomyopathy. *Proc. Natl. Acad. Sci. USA* **98,** 12,283–12,288.

50. Liao, P., Wang, S. Q., Wang, S., Zheng, M., Zhang, S. J., Cheng, H., et al. (2002) p38 Mito-gen-activated protein kinase mediates a negative inotropic effect in cardiac myocytes. *Circ. Res.* **90,** 190–196.

51. Standaert, M. L., Bandyopadhyay, G., Galloway, L., Soto, J., Ono, Y., Kikkawa, U., et al. (1999) Effects of knockout of the protein kinase C beta gene on glucose transport and glucose homeostasis. *Endocrinology* **140,** 4470–4477.

52. Fruman, D. A., Snapper, S. B., Yballe, C. M., Alt, F. W., and Cantley, L. C. (1999) Phospho-inositide 3-kinase knockout mice: role of p85alpha in B cell development and proliferation. *Biochem. Soc. Trans.* **27,** 624–629.

53. Hirsch, E., Katanaev, V. L., Garlanda, C., Azzolino, O., Pirola, L., Silengo, L., et al. (2000) Central role for G protein-coupled phosphoinositide 3-kinase γ in inflammation. *Science* **287,** 1049–1053.

54. Sieuwerts, A. M., Klijn, J. G., Peters, H. A., and Foekens, J. A. (1995) The MTT tetra-zolium salt assay scrutinized: how to use this assay reliably to measure metabolic activity of cell cultures in vitro for the assessment of growth characteristics, IC50-values and cell survival. *Eur. J. Clin. Chem. Clin. Biochem.* **33,** 813–823.

55. Tannergren, C., Langguth, P., and Hoffmann, K. J. (2001) Compound mixtures in Caco2 cell permeability screens as a means to increase screening capacity. *Pharmazie* **56,** 337–342.

56. Finlayson, K., Turnbull, L., January, C. T., Sharkey, J., and Kelly, J. S. (2001) [^{3}H]dofetilide binding to HERG transfected membranes: a potential high throughput preclinical screen. *Eur. J. Pharmacol.* **430,** 147–148.

57. Mauro, M. J. and Druker, B. J. (2001) STI571: targeting BCR-ABL as therapy for CML. *Oncologist* **6,** 233–238.

58. Hoekman, K. (2001) SU6668, a multitargeted angiogenesis inhibitor. *Cancer J.* **7(Suppl. 3),** S134–S138.

59. Walker, E. H., Pacold, M. E., Perisic, O., Stephens, L., Hawkins, P. T., Wymann, M. P., et al. (2000) Structural determinants of phosphoinositide 3-kinase inhibition by wortmannin, LY294002, quercetin, myricetin, and staurosporine. *Mol. Cell.* **6,** 909–919.

60. Godden, J. W., Stahura, F., and Bajorath, J. (1998) Evaluation of docking strategies for virtual screening of compound databases: cAMP-dependent serine/threonine kinase as an example. *J. Mol. Graph. Model.* **16,** 139–143, 165.
61. Bissantz, C., Folkers, G., and Rognan, D. (2000) Protein-based virtual screening of chemical databases. 1. Evaluation of different docking/scoring combinations. *J. Med. Chem.* **43,** 4759–4767.
62. Langer, T. and Hoffmann, R. D. (2001) Virtual screening: an effective tool for lead structure discovery? *Curr. Pharm. Des.* **7,** 509–527.
63. Good, A. C., Krystek, S. R., and Mason, J. S. (2000) High-throughput and virtual screening: core lead discovery technologies move towards integration. *Drug Discov. Today* **5,** 61–69.

II

SIGNALING TECHNIQUES

6

Ligand-Binding Assays for Collagen

Stephanie M. Jung and Masaaki Moroi

1. Introduction

Collagen is one of the most physiologically important agonists for platelet function. Two collagen-specific receptors, integrin $\alpha_2\beta_1$ and glycoprotein (GP) VI, have been identified on the platelet surface from studying patient's platelets deficient in one of these proteins (*1,2*). Integrin $\alpha_2\beta_1$ was found to be the major receptor responsible for platelet adhesion to collagen (*3,4*). The other collagen receptor, GPVI, was indicated to contribute to platelet activation and aggregation by its involvement in collagen-induced signaling pathways (*5,6*).

In the past, analyzing the interaction between collagen and platelets has been hampered because platelets react only with fibrous collagen (*7*), an insoluble, and therefore nonideal, ligand for binding studies. Consequently, collagen–platelet interactions have been measured mostly by platelet adhesion to immobilized collagen (*4,8*). However, platelet adhesion is different from the binding reaction itself; adhesion takes a long time and platelets become activated while being in contact with the collagen surface. A major disadvantage of this type of analysis is that kinetic parameters cannot to be calculated from adhesion data. Although typical binding measurements have been made through soluble-ligand binding to cells, soluble collagen was indicated not to bind to platelets (*7*), making it difficult to apply the conventional binding assay to the platelet–collagen interaction. Although soluble-collagen binding to platelets in the presence of Arg has been reported (*9*), this method is not reproducible. The measurement of fibrous-collagen binding to platelets was also attempted (*10,11*), but it was not clearly indicated that the observed binding was specific.

However, several years ago, we found that although soluble collagen does not bind to resting platelets, it does bind specifically to activated platelets (*12*). Utilizing this property, we have developed a ligand-binding assay that can be used to study the collagen-platelet interaction and the mechanism of integrin activation in platelets. Soluble collagen was shown to bind to activated integrin $\alpha_2\beta_1$ and fibrous collagen was indicated to bind to GPVI and activate platelets. Although the binding technique is a classical

From: *Methods in Molecular Biology, vol. 273:*
Platelets and Megakaryocytes, Vol. 2: Perspectives and Techniques
Edited by: J. M. Gibbins and M. P. Mahaut-Smith © Humana Press Inc., Totowa, NJ

procedure, this is a newly found phenomenon with many potential applications remaining to be explored. Notably, we found that the activation mechanism of integrin $\alpha_2\beta_1$ was quite similar to that of integrin $\alpha_{IIb}\beta_3$ *(13)*. Because integrin $\alpha\beta_1$ was indicated to show no outside-in signaling, it would be a good system for analyzing the integrin-activation mechanism. An assay for ligand binding to integrin $\alpha_2\beta_1$ is the subject of this chapter.

In this section, two basic procedures will be described: (1) preparing ^{125}I-labeled soluble collagen and (2) performing the binding assay. The binding assay consists of the following steps: pre-incubation of the washed platelets with any reagents to be tested; incubation of the platelets with an activator and ^{125}I-labeled soluble collagen; separation of the platelet-bound and unbound ^{125}I-labeled soluble collagen by centrifugation; and measuring the platelet-bound soluble collagen by γ-counting. In our binding system, nonspecific binding can be determined in the presence of 5 mM ethylene diamine tetraacetic acid (EDTA), instead of using excess ligand, because integrin-dependent binding is cation (e.g., Mg^{2+})-dependent; this enables us to obtain binding curves with an adequate range of concentrations to accurately determine binding parameters, since at saturating concentrations, it would be impossible to obtain the excess of cold ligand necessary due to the concentration limits of soluble collagen and the physical characteristics of soluble collagen itself at high concentrations.

2. Materials

2.1. Equipment for Binding Assays

1. Adjustable-volume repetitive micropipet; e.g., Eppendorf Multipipette 4780 with 0.5-, 1.0-, and 2.5-mL Combitips-plus (Eppendorf, Hamburg, Germany).
2. 0.5-mL plastic microcentrifuge tubes.
3. 0.3-mL narrow-tipped, polypropylene microcentrifuge tubes (cat. no. 72.702; Sarstedt, Germany).
4. High-speed micro-refrigerated centrifuge with drum-type rotor; e.g., Tomy MRX-150 with TMH-2 rotor (maximum speed of 12,000 rpm) that holds 4 sample blocks with the capacity of 20 tubes/block (Tomy, Japan).
5. Polystyrene counting tubes.
6. Racks for 0.5-mL centrifuge tubes (10×10 samples capacity is convenient).
7. Rack for the narrow microcentrifuge tubes (commercial or homemade from a rectangular piece of polystyrene, on which you can punch holes at regular intervals to fit the tubes).
8. Deep freezer (-80 to $-40°C$) or dry ice (practical only for a small number of samples) for freezing the gradients after the centrifugation separation of the binding mixtures.
9. Multisample γ counter (e.g., Aloka Auto Well Gamma System ARC-1000M; Aloka, Japan).

2.2. Reagents[1]

1. Bovine type III collagen (1 mg/mL solution of monomeric collagen; Koken, Tokyo, Japan).
2. NaI[125].
3. Iodobeads (Pierce, Rockford, IL).
4. Prostaglandin I$_2$ (PGI$_2$) dissolved in ethanol (0.1 mg/mL) and stored under nitrogen gas.

---

[1]Where no specific source is indicated, reagents can be obtained from any manufactuer.

2.3. Buffers and Solutions

2.3.1. Radiolabeling Collagen and Platelet Preparation

1. HEPES/Tyrode's solution: 136 mM NaCl, 2.7 mM KCl, 0.42 mM NaH$_2$PO$_4$, 12 mM NaHCO$_3$, 5.5 mM glucose, 5 mM HEPES, pH 7.4.
2. Platelet washing buffer 1: 6.85 mM citrate, 130 mM NaCl, 4 mM KCl, 5.5 mM glucose, pH 6.5.

2.3.2. Buffers and Solutions for the Binding Assay[2]

1. 20% sucrose solution: 20% (w/v) sucrose, 0.2% (w/v) BSA in HEPES/Tyrode's solution. After preparation, the solution should be filtered under vacuum.
2. Total binding buffer: 14 mM MgCl$_2$, 7% (w/v) BSA in HEPES/Tyrode's solution, pH 7.4.[3]
3. Nonspecific binding buffer: 35 mM EDTA and 7% (w/v) BSA in HEPES/Tyrode's solution, pH 7.4.[3]

3. Methods

3.1. Preparation of ^{125}I-Labeled Collagen (12) (see Note 1)

3.1.1. Preparation of Soluble Collagen

1. Dialyze 20 mL of bovine type III collagen (1 mg/mL) in standard cellulose dialysis tubing (14,000 MW cutoff) against two one-liter changes of HEPES-Tyrode's buffer at 4°C overnight.
2. After dialysis, incubate the collagen at 37°C for 1 h to allow fibrous collagen formation.
3. Transfer the resultant collagen gel to a centrifuge tube and centrifuge it at 8000g at 25°C for 20 min; separate the supernatant and pellet.
4. Centrifuge the supernatant at 100,000g for 1 h at 25°C. The supernatant obtained is the soluble collagen preparation (concentration of about 0.1 mg/mL) for the iodination procedure (*see* **Subheading 3.1.2.**).

3.1.2. ^{125}I-Labeling of Soluble Collagen

1. 20 mL soluble collagen is radiolabeled with 74 MBq of NaI[125] by using eight pieces of Iodobeads at room temperatures for 2–4 h according to the manufacturer's instructions. NaI[125] should be used in accordance with local guidelines for the safe usage and disposal of radioactivity. NaI[125] should be used in a suitable fume hood.
2. The obtained sample is dialyzed against HEPES-Tyrode's solution at 4°C with several changes of the buffer solution to remove free [125]I. The dialysis solutions will contain the bulk of the radioactivity; the minimum amount of buffer necessary for thorough dialysis should be used and carefully disposed according to local rules for the safe use of radio-

---

[2]All these buffers should be divided into convenient aliquots, and stored at –20°C. They can be kept indefinitely in the deep freezer and then thawed out as needed.

[3]These buffers contain MgCl$_2$ and BSA or EDTA and BSA at 7 times the final concentrations in the binding mixtures. In the binding assay mixture (70 μL total volume), 10 μL of either binding buffer will be used, so the final concentrations of MgCl$_2$ will be 2 mM and that of EDTA will be 5 mM, with 1% (w/v) BSA.

isotopes. The specific radioactivity of this obtained collagen is about 3000–9000 cpm/μL with 4–8% yield of radioactivity. The radiolabeled collagen can be stored at 4°C (do not freeze) for up to 2 mo.

3.1.3. Determining Concentration of the ^{125}I-Labeled Soluble Collagen

The concentration of soluble collagen is determined by assaying the protein by the bicinchoninic protein assay *(14)*, with collagen (bovine collagen type III, as supplied by the manufacturer) as the standard protein. The molar concentration of soluble collagen is calculated with 3×10^5 as the molecular weight of collagen.

3.2. Selecting the Assay Conditions (see Notes 2–4)

1. *Reaction volume:* Volumes of 50–80 μL can be used and are adequately separated by the 250-μL sucrose gradient. The step-by-step procedure described in **Subheading 3.6.** is based on a binding assay volume of 70 μL.
2. *Platelet concentration:* Under the assay conditions, an adequate number of counts without the problems of platelet aggregation and ligand depletion from too many receptors can be obtained by using 4–8×10^8 cells/mL, in assay volumes of 50–80 μL.
3. *Concentration of ^{125}I-labeled collagen:* The actual concentration of the labeled collagen will differ each time the preparation is incubated and precentrifuged prior to each binding experiment. It will not be practical to determine the concentration at the time of the experiment; you can estimate the amount of collagen to use by counting an aliquot and determining how much is needed to give adequate cpm per binding reaction for accurate data. The number of bound counts in the total binding reactions is about 0.5–6.0% of the added counts. Nonspecific binding is about 5–8% of the total bound counts, depending on the platelet preparation and whether there is any activation during the washing procedures. Practically speaking, the labeled soluble collagen is usually diluted one- to twofold and 20 μL is used per 70-μL assay. As indicated in the labeling procedure of soluble collagen, the ^{125}I-labeled soluble collagen usually has an activity of about 3000–9000 cpm/μL, with a protein concentration of 0.1 mg/mL, giving a specific activity of 3–9×10^4 cpm/μg.
4. *Concentration of agonists:* The platelets must be activated before they can bind soluble collagen, since only the activated form of integrin $\alpha_2\beta_1$ can bind this ligand. Some examples of agonists and the concentrations employed are given below. However, note that platelets from different individuals show different sensitivities to agonists. Thus, experiments comparing the effect of reagents on activation induced by different agonists should be performed on the same day with the same platelet preparation to minimize differences in platelet reactivity from different individuals and platelet preparations.
   a. **ADP:** This is a weak agonist that can be used between 2 μ*M* (low agonist concentration) and 20 μ*M* (high agonist concentration, for full activation by this agonist).
   b. **Thrombin:** The working range of concentrations is 0.05–0.2 NIH U/mL; the binding is approximately twice that induced by ADP; thrombin solutions should be made up in 0.2–0.5% BSA-containing HEPES/Tyrode's to avoid absorption of the thrombin on the walls of the container. Maximum activation can be achieved at 0.1 NIH U/mL.
   c. **Collagen-related peptide (CRP):** CRP (cross-linked preparation) *(15)* is used to study activation induced by the collagen receptor glycoprotein VI, as actual fibrous collagen would compete with soluble collagen to activated integrin $\alpha_2\beta_1$. The working concentration range of this strong agonist is 40–400 ng/mL. Full activation by this agonist

can be achieved with 300–400 ng/mL; avoid using CRP at higher concentrations, where it is capable of displacing soluble collagen binding from activated integrin $\alpha_2\beta_1$ *(12)*.

5. *Platelet-sedimenting conditions:* The time required to sediment the platelets will depend on the centrifuge you are using; you may have to determine this empirically. Five minutes at 12,000 rpm is sufficient for complete platelet sedimentation with the Tomy MRX-150 with TMH-2 rotor.

6. *Stability of the activating response:* ADP-induced activation is the least stable, so these assays should be performed first if possible. However, there should be no problems if all the experiments are completed within a 5-h period, from the completion of making washed platelets.

3.3. Preparing Sucrose "Gradients" for Sample Separations

1. For each assay, you will need one sucrose gradient; label the narrow-tipped centrifuge tubes and set them in the polystyrene tube rack.

2. Thaw the 20% sucrose/0.2% BSA/HEPES-Tyrode's and warm to room temperature in a water bath, to avoid bubble formation in the gradients.

3. With the repetitive micropipet, pipet 250 or 300 μL of the sucrose solution into each centrifuge tube. It is easier to make the gradients if you attach an additional pipet tip to the tip of the Multitip-plus, long and narrow enough to reach the bottom of the centrifuge tube. Insert the pipet tip to the bottom of the centrifuge tube and slowly draw it up as you are dispensing the aliquot of 20% sucrose. If any bubbles form on the bottom, tap the tube tip to remove them. Leave the gradients at room temperature until you use them in the binding assay.

3.4. Preparing the ^{125}I-Labeled Soluble Collagen for the Binding Assay

This procedure *must* be performed prior to using the labeled collagen each time the binding experiments are performed.

1. Incubate the labeled soluble collagen solution (amount sufficient for the binding assays + 0.5 mL extra) at 37°C for 1 h.

2. Centrifuge the incubated sample at 34,000*g* for 1 h at 25°C using an ultracentrifuge (e.g., TL-100, Beckman Instruments, Palo Alto, CA) to remove any fine particles of insoluble collagen.

3. Being careful not to disturb any pelleted material, remove the supernatant, the soluble collagen. Usually, there is no precipitate in this step. If there is a large precipitate, this may indicate that the collagen preparation has denatured or has formed insoluble particles, and this cannot be used for the binding study.

4. Take an aliquot of this collagen (about 50 μL) for protein determination (as described in **Subheading 3.1.3.**) at a later time when the radioactivity has sufficiently decayed (several weeks to a month) for safer handling of the sample.

3.5. Preparing Washed Platelets for the Binding Assay

3.5.1. Isolation and Washing of Platelets

1. Whole blood is obtained from the cubital veins of healthy, drug-free volunteers with sodium citrate as the anticoagulant. For 400 binding assays, 150 mL of whole blood is usually sufficient.

2. Platelets are then sedimented from PRP (platelet-rich plasma) by centrifugation at 900g for 12 min after the addition of sodium prostaglandin I_2 (PGI$_2$) to a final concentration of 0.1 μg/mL.

3. The centrifuged platelets are washed once with platelet washing solution 1, and the washed platelets are finally suspended with HEPES-Tyrode's buffer containing 2% (w/v) bovine serum albumin at the appropriate concentration; this is usually 3.5-fold the desired final concentration of platelets in the binding assay (usually 4×10^8 cells/mL). For example, a typical binding assay uses 20 μL of platelet suspension (1.4×10^9 cells/mL) per 70-μL assay.

3.5.2. Pretreatment of Platelets With Inhibitors and Drugs

Any drugs or inhibitors to be tested should be preincubated with the washed platelet suspension just before adding the platelets to the assay mixture described in the next section. The concentration of drug or inhibitor is calculated on the basis of the volume of the washed platelet suspension. For control samples, platelets are incubated with HEPES/Tyrode's buffer instead of drugs or inhibitors. The preincubation time will depend on which reagent is to be used, but 5 min is usually sufficient. This is preferably done just prior to initiating the binding reaction with the platelets to avoid prolonged contact of the platelets with the drugs or inhibitors, which would increase the probability of nonspecific effects.

3.6. Step-by-Step Description of the Binding Procedure

1. For each assay point, five replicates each are run for total binding and nonspecific binding. After labeling of tubes and placing them in order in the racks, perform the following procedures for each tube.

2. The repetitive micropipet, set to an appropriate volume, is used for all the following reagent additions and the addition of platelets to initiate the binding reaction. The total volume of each binding mixture will be 70 μL, so use agonist solutions of the appropriate concentration to yield the desired final concentration (e.g., for a 70-μL final volume, use agonist at seven times the final concentration). Adding a regular narrow-tipped pipet to the end of the combitip-plus of the repetitive pipet will facilitate adding the sample into the sample tubes and avoid "splashing" due to the force of expelling the solution.

3. Pipet 10 μL of total binding buffer or nonspecific binding buffer into the sample tube.

4. Add 10 μL of the solution of stock activating agonist (e.g., ADP, thrombin, CRP), which should be seven times the final desired concentration in the binding reaction.

5. Add 30 μL of the appropriately diluted ^{125}I-labeled soluble collagen (precentrifuged as indicated in **Subheading 3.4.**).

6. Dispense five 30-μL aliquots of the ^{125}I-labeled soluble collagen into separate counting tubes (to be counted later to determine the total number of added counts).

7. Vortex each binding mixture for 1 s on the high setting to mix the reagents.

8. Inititate binding by adding 20 μL of the washed platelet suspension. Touch the tube to the vortex mixer (at high setting) to rapidly mix. It is particularly important to avoid over-mixing, especially when using thrombin or CRP as the activating agonist.

9. Let the assay mixtures stand at room temperature for 80 min, with no mixing.

10. For each assay, layer the entire 70-μL mixture onto a sucrose gradient by using a micropipet without disturbing the gradient. This can be done easily if the sample is allowed to slide down the side of the gradient tube onto the sucrose layer in one smooth, continuous motion. To economize, you can use the same pipet tip for the total binding or nonspecific binding replicates of each point.

11. Centrifuge the gradients at 12,000 rpm for 5 min in the refrigerated microcentrifuge set at a temperature of about 10°C.
12. Remove the tubes from the centrifuge and place them in the deep freezer until you are ready to count the samples. You should let the samples freeze for at least 30 min.

Counting the samples: Using a pair of scissors, cut off the tips of each tube, placing them in individual counting tubes. This can be done easily if you hold the gradient tube directly above the counting tube while you are cutting the tip off. Count these samples along with the 30-µL aliquots of the ^{125}I-labeled soluble collagen (*see* **step 6**) in the gamma counter. Because the platelets form a tight pellet on the bottom of the narrow-tipped centrifuge tube, the tips can be cut off at about 2 mm from the bottom; it is not important how much you cut off of the tip, as long it it contains the pellet; the unbound labeled soluble collagen is clearly separated from the pellet, being mostly on the top portion of the sucrose pad. If you need to count the supernatants, put the rest of the centrifuge tube, which contains the bulk of the sucrose and unbound labeled collagen, into a separate counting tube.

3.7. Data Analyses (see Note 5)

1. Specific binding is determined by subtracting the mean of the nonspecific binding (determined in the presence of 5 m*M* EDTA) from the mean of the total binding data, five replicates for each data point.
2. We analyze the data using the computer software PRISM (Version 3; GraphPad Software, La Jolla, CA); nonlinear regression analysis is used to find the best fit curves for the kinetics data.
3. For the binding kinetics of soluble binding to ADP- and thrombin-induced platelets, we compare the data fit to both the "one-site" and "two-site" binding models by using PRISM, from which we concluded that the soluble collagen binding to ADP- and thrombin-induced platelets follows a one-site binding model with the thrombin-induced platelets showing an activated integrin $\alpha_2\beta_1$ with 3.7–12.7-fold higher than the form in ADP-induced platelets. A complete description of the software and analysis method can be found in the manual for PRISM supplied by GraphPad Software. For a further description of binding methods in general (theoretical and practical), refer to the book by E. C. Hulme (ed.), *Receptor-Ligand Interactions: A Practical Approach (16)*, particularly chapters 4 and 5. For typical binding curves and binding parameters (the dissociation constant K_d and the number of binding sites B_{max}) and the accuracy of this binding method, refer to our previously reported data *(13)*.

3.8. Application of Basic Binding Technique to Fibrinogen Binding

With a few modifications, this binding technique is directly applicable to the binding of fibrinogen to integrin $\alpha_{IIb}\beta_3$ (glycoprotein IIb/IIIa), which is a measure of the activation of this integrin.

1. Fibrinogen can be ^{125}I-labeled by the technique used for soluble collagen directly. Use fibrinogen in the labeling reaction after dialysis against HEPES/Tyrode's. A substantially higher specific activity can be obtained for fibrinogen since it contains many more Tyr residues than collagen.
2. Fibrinogen, unlike soluble collagen, can be divided into aliquots and frozen if storage for a longer time is desired; however, avoid repeated thawing and freezing of the preparation.

3. After thawing, the labeled fibrinogen should be warmed at 37°C for about 30 min and then precentrifuged as described for collagen to remove any precipitate is present.
4. $CaCl_2$ should be substituted for the Mg^{2+} in the total binding buffer.
5. For fibrinogen binding, because of the much higher number of binding sites and higher specific activity of the ^{125}I-labeled fibrinogen, the binding assays can be performed with one quarter to one half of the cell concentration used for the soluble collagen binding assay.
6. Fibrous collagen can be used as an agonist instead of CRP to study the mechanism of collagen-induced activation. This is a major advantage because fibrous collagen is the physiological agonist of platelets and there is evidence that CRP does not have the same properties as collagen as an agonist even though they both act through the collagen receptor GPVI.
7. If thrombin is used as an agonist, it is necessary to activate the platelets with thrombin first, followed by adding an inhibitor such as PPACK (D-Phe-Pro-Arg-chloromethylketone) (use at 50–100 μM final concentration) or hiruidin to inhibit the thrombin activity, to avoid cleavage of the fibrinogen to fibrin and resultant gel formation.

4. Notes

1. Because the content of tyrosine residues in collagens of all types is very low, the efficiency for radiolabeling collagen by the Iodobeads method is also very low. If you need to obtain higher specific radioactivity, another method, such as the Bolton-Hunter reagent, must be used. The Bolton-Hunter reagent incorporates the ^{125}I into the amino residues of collagen, which are more abundant than the tyrosine residues.
2. Is it possible to use type I collagen? This type of collagen has even lower Tyr content than Type III collagen, so we were unable to obtain a similar specific radioactivity for type I. For this reason, and because of problems with solubility and monomer yield, we do not recommend the use of type I.
3. What is the maximum concentration of soluble collagen obtainable? Because the conversion from soluble collagen to insoluble collagen is an equilibrium, you cannot obtain a high concentration of soluble collagen even if you use a high concentration of the original monomeric collagen. To avoid unwanted precipitation (polymerization) of collagen, it is important to treat the collagen properly. The most important factor is the temperature: keep the collagen on ice so that the collagen will remain monomeric (soluble form) until the actual step where you want it to polymerize.
4. Can collagen from different commercial sources be used? From our experience, we have found that the collagen from various commercial sources can markedly differ in purity, solubility properties, and stated protein concentration, even when it is suppose to be the same type of collagen. We have had the most success with the collagen provided by the manufacturer listed above.
5. *Erratic data:* Sometimes this happens if the collagen preparation is "too old," stored improperly, or has not by been incubated at 37°C and precentrifuged as indicated in the methods. We must emphasize that the preparation used in these binding studies is soluble collagen, probably consisting of mainly monomeric collagen but also containing some dimers and the like. If the preparation is stored for a long period of time or becomes denatured, it is possible that larger multimers may form; however, the size of these multimers may still be too small to be precipitated by ultracentrifugation. This would result in some samples having a much larger number of counts, while most of the samples will show a typical amount of binding for the amount of soluble collagen added. The only solution is to make up a new batch of collagen.

6. No binding or very low binding can be due to some of the following problems:

   a. *Denatured soluble collagen preparation:* Storage for about 2 mo is about the limit for the soluble collagen preparation to still be "active," aside from the obvious loss of specific activity due to decay of the isotope.

   b. *Problem with the agonist:*

      i. ADP: It is best to store the ADP as a stock solution of lower concentration, e.g., 1 mM. We found that ADP loses its activity much more rapidly (due to hydrolysis) when it is stored as a concentrated solution like 10 mM.

      ii. Thrombin: When using low thrombin concentrations, it is necessary to make up low thrombin stocks in the presence of BSA (0.2–0.5%), since absorption of thrombin onto the walls of the container become significant at low concentrations, resulting lower than expected concentrations.

      iii. CRP: As stated above, avoid using excessively high concentrations of this agonist (greater than 400 ng/mL) because at such concentrations, CRP can compete with soluble collagen binding to activated integrin $\alpha_2\beta_1$.

   c. *Platelet preparation is "too old":* Use the washed platelets within about 5 h after you prepare them. The response to ADP and other weak agonists should be determined first, as these responses are the most labile.

   d. *Ambient temperature is too low:* The binding should be done at a room temperature no lower than about 20°C because we found that at temperatures lower than this, the time it takes for the binding to come to equilibrium is higher and/or the amount of binding is lower.

7. *Aggregation of the platelets:* This would normally not occur under the conditions of the binding assay. This will not occur when using the weak agonist ADP. However, this might occur if the sample is mixed too long after addition of the platelets to initiate the binding reaction or if you stir or unnecessarily agitate the binding mixtures during the incubation period required for binding to reach equilibrium; this is particularly true when higher concentrations of thrombin or CRP are used to activate the platelets. When you are using high agonist concentrations, sometimes you may see some microaggregates, but this should not affect the results if they are small enough to not "trap" unbound labeled collagen. With high thrombin, you may also see a small amount of a clear, gel-like precipitate, which also should not affect the results.

References

1. Nieuwenhuis, H. K., Akkerman, J. W. N., Houdijk, W. P. M., and Sixma, J. J. (1985) Human blood platelets showing no response to collagen fail to express surface glycoprotein Ia. *Nature* **318,** 470–472.

2. Moroi, M., Jung, S. M., Okuma, M., and Shinmyozu, K. (1989) A patient with platelets deficient in glycoprotein VI that lack both collagen-induced aggregation and adhesion. *J. Clin. Invest.* **84,** 1440–1445.

3. Saelman, E. U. M., Nieuwenhuis, H. K., Hese, K. M., De Groot, P. G., Heijnen, H. F. G., Sage, E. H., et al. (1994) Platelet adhesion to collagen types I through VIII under conditions of stasis and flow is mediated by GP Ia/IIa ($\alpha_2\beta_1$-integrin). *Blood* **83,** 1244–1250.

4. Santoro, S. A. (1986) Identification of a 160,000 dalton platelet membrane protein that mediates the initial divalent cation-dependent adhesion of platelets to collagen. *Cell* **46,** 913–920.

5. Ichinohe, T., Takayama, H., Ezumi, Y., Yanagi, S., Yamamura, H., and Okuma, M. (1995) Cyclic AMP-insensitive activation of c-Src and Syk protein-tyrosine kinases through platelet membrane glycoprotein VI. *J. Biol. Chem.* **270,** 28,029–28,036.

6. Watson, S. P., Berlanga, O., Best, D., and Frampton, J. (2000) Update on collagen receptor interactions in platelets: is the two-state model still valid? *Platelets* **11**, 252–258.

7. Gordon, J. L. and Dingle, J. T. (1974) Binding of radiolabelled collagen to blood platelet. *J. Cell Sci.* **16**, 157–166.

8. Moroi, M., Okuma, M., and Jung, S. M. (1992) Platelet adhesion to collagen-coated wells: Analysis of this complex process and a comparison with the adhesion to matrigel-coated wells. *Biochim. Biophys. Acta* **1137**, 1–9.

9. Misselwitz, F., Domogatsky, S. P., Leytin, V. L., and Repin, V. S. (1987) Binding of human monomeric type I collagen to platelets. *Biochim. Biophys. Acta* **923**, 436–442.

10. Mazurov, A. V., Idel'son, G. L., Khachikyan, M. V., and Domogatskii, S. P. (1988) Interaction of platelets with ^{125}I-labeled type III collagen. Necessity of formation of fibrillar structures. *Biokhimiya* **54**, 1280–1289.

11. Brass, L. F., Faile, D., and Bensusan, H. B. (1976) Direct measurement of the platelet: collagen interaction by affinity chromatography on collagen/Sepharose. *J. Lab. Clin. Med.* **87**, 525–534.

12. Jung, S. M. and Moroi, M. (1998) Platelets interact with soluble and insoluble collagens through characteristically different reactions. *J. Biol. Chem.* **273**, 14,827–14,837.

13. Jung, S. M. and Moroi, M. (2000) Signal-transducing mechanisms involved in activation of the platelet collagen receptor integrin $\alpha_2\beta_1$. *J. Biol. Chem.* **275**, 8016–8026.

14. Smith, P. K., Krohn, R. I., Hermanson, G. T., Mallia, A. K., Gartner, F. H., Provenzano, M. D., et al. (1985) Measurement of protein using bicinchoninic acid. *Anal. Biochem.* **150**, 76–85.

15. Morton, L. F., Hargreaves, P. G., Farndale, R. W., Young, R. D., and Barnes, M. J. (1995) Integrin $\alpha_2\beta_1$-independent activation of platelets by simple collagen-like peptides: collagen tertiary (triple-helical) and quaternary (polymeric) structures are sufficient alone for $\alpha_2\beta_1$-independent platelet reactivity. *Biochem. J.* **306**, 337–344.

16. Hume, E. C., ed. (1992) *Receptor-Ligand Interactions: A Practical Approach.* Oxford University Press, Oxford.

7

Use of Radiolabeled 2-Methylthio-ADP to Study P2Y Receptors on Platelets and Cell Lines

Pierre Savi and Jean-Marc Herbert

1. Introduction

The use of radioactive 2-methylthio-ADP (2MeS-ADP) to study ADP receptors on platelets was first developed by MacFarlane et al. *(1)* in 1983. However, this technique may not have initially been pursued by many groups because of the relative paucity of studies into platelet ADP receptors and the absence of a commercially available radiolabeled ligand. From 1992 to 1994, a few papers described the use of 2MeS-ADP to demonstrate the antagonistic activity of clopidogrel on the platelet ADP receptor *(2–4)* and the presence of ADP receptors on megakaryocytoblastic cells *(5)*. Since these initial reports, several groups have published studies describing the binding characteristics of 2MeS-ADP on platelets *(6,7)*. Two different teams provided evidence for a defect of ADP receptor numbers in platelets from patients with a low sensitivity to ADP *(8,9)*. Consequently, a variety of antiplatelet compounds have been evaluated as antagonists on platelet ADP receptors *(10–13)*. A recent study has demonstrated the ability of $P2Y_{12}$ to bind 2MeS-ADP, supporting its identity as the previously named $P2T_{AC}$ or low-affinity platelet ADP receptor *(14)*.

2MeS-ADP was initially labeled with ^{32}P in the beta position, by Mc Farlane et al. *(1)*. The compound has also been tritiated on the C8 position of the adenine ring by Moravek (Brea, CA) although its synthesis has been discontinued. This ligand had a low specific activity (6 Ci/mmol), but could be used for several years, due to its stability and the relatively long half-life of tritium. ^{33}P-labeled 2MeS-ADP is now commercially available (NEN PerkinElmer, Boston, MA) and relatively easy to use. It is water-soluble and exhibits a low level of binding to tubes or filters, allowing a low background to be achieved. Due to its strong specific activity (>600 Ci/mmol), ^{33}P-labeled 2MeS-ADP binding does not require a high quantity of platelets or cells, even if the number of binding sites is rather low. Moreover, 2MeS-ADP is relatively

From: *Methods in Molecular Biology, vol. 273:*
Platelets and Megakaryocytes, Vol. 2: Perspectives and Techniques
Edited by: J. M. Gibbins and M. P. Mahaut-Smith © Humana Press Inc., Totowa, NJ

stable in biological media, which is a great advantage when one wants to work with platelets in plasma. The method we have developed uses 96-well microplates for incubation and glass-fiber filters for filtration.

2. Materials

1. Anticoagulant: 3.8% sodium citrate solution.
2. Citric acid and dextrose solution: 22 g trisodium citrate dihydrate, 8 g citric acid monohydrate, 25 g dextrose in 1 L sterile water.
3. PGI_2 (from Sigma, St. Louis, MO): A stock solution is made at 1 mg/mL in methanol and stored at –30°C.
4. Resuspension buffer: 134 mM NaCl, 12 mM NaHCO$_3$, 0.34 mM Na$_2$HPO$_4$, 2.9 mM KCl, 1 mM MgCl$_2$, 5 mM glucose, 5 mM HEPES, 0.35% bovine serum albumin (BSA), titrated to pH 7.4 with HCl.
5. 96-well microplates: Flat-bottom 300-µL wells, made of clear polystyrene (Nunc, BD Falcon).
6. Incubation buffer: For platelets, binding is conducted in 145 mM NaCl, 5 mM KCl, 0.1 mM MgCl$_2$, 5.5 mM glucose, 15 mM HEPES, titrated to pH 7.4 with NaOH. This has yielded satisfactory results on platelets from different sources (humans, rabbits, or rats). Other buffers, such as those routinely used for platelet washing, can also be employed for binding provided they do not contain any ADP-consuming reagent such as apyrase or creatine phosphate/creatine phosphokinase. In order to avoid any aggregation of platelets by reagents or during incubation, 5 mM Na$_2$EDTA can be added, which blocks aggregation but not the binding of 2MeS-ADP.
7. Washing buffer: Washing of filters can be performed with incubation buffer or, more simply, with either phosphate-buffered saline (PBS) or 0.9% NaCl. The buffers are stored at 4°C. Incubation buffer is warmed by aliquot; the pH is verified and, if necessary, corrected immediately before use. Washing buffer is used at 4°C. All the buffers are renewed each week. (For cultured cells, the medium used for culture, without serum, can be used for washing and resuspension.)
8. Radiolabeled ligand: We currently use [33]P-2MeS-ADP with a specific activity of 1500–2000 Ci/mmol. The concentration commercially available is between 250 and 500 nM and is stored at –20°C. For binding studies, it is diluted immediately before use in incubation buffer or cell buffer at concentrations ranging from 0.1 to 10 nM, depending on the study. The dilutions are not stored.
9. Cold ligand: e.g., 2-MeS ADP or ADP (*see* **Note 1**).
10. Instrumentation: Apart from the usual laboratory instruments such as centrifuges and cell counters, binding experiments require filtration devices (Skatron, Tomtec, Brandel). We use the automated Skatron filtration device (Molecular Devices, St. Gregoire, France), which harvests cells suspended in 96-well plates and presents a relatively large filtration area. This is an advantage if the binding is low and a cell-rich sample is required—for example, when binding is measured on platelets from clopidogrel-treated patients or subjects presenting a defect in the number of $P2Y_{12}$ receptors. In some cases, we have also used a dot blotter, a filtration box usually devoted to titration of antibodies (Minifold, Schleicher & Schüll, Dassel, Germany), connected to a water pump, with all the filling and washing steps being performed manually. This very simple device has been of considerable help when a portable system was required for experiments outside our laboratory.
11. Filters: Filters need to exhibit a high capacity for cell trapping and a low nonspecific binding. Good cell trapping can be achieved by choosing a thick filter. Nonspecific binding

can be achieved with glass-fiber-based filters, as 2MeS-ADP is highly water-soluble and a short washing with an aqueous buffer (saline or PBS) is enough to remove the background labeling. We routinely use 1-μm retention glass-fiber filters (Filtermats 11734, Molecular Devices) on the Skatron filtration device.

3. Methods

The following protocol is used to measure the binding of ^{33}P-2MeS-ADP to human or animal platelets. A complete binding experiment involves five different steps: sample preparation, incubation with the ligand, separation of bound ligand from free, counting, and calculation. These are each described in **Subheadings 3.1.** to **3.5. Subheading 3.6.** outlines an experiment designed to study the concentration-inhibition curve for a test compound on 2MeS-ADP binding. **Subheading 3.7.** provides information for adaptation of the method to platelets in plasma and other cells. Authorization for use of radioactive compounds and human material must be obtained and local standard operating procedures followed (*see* **Note 2**).

3.1. Preparation of Platelets

The method used to prepare platelets is particularly important, as some types of ADP receptors are easily desensitized. Several authors have described suitable platelet-preparation protocols; an existing approach within in your laboratory can be attempted if it produces platelets that are sensitive to ADP. The method we currently use is presented here.

1. After taking blood by a firm and rapid puncture, mix gently with a 3.8% sodium citrate solution (9:1 v/v, blood:citrate) into a plastic tube. Blood is kept at room temperature and used as soon as possible, ideally within 20 to 30 min after withdrawal.
2. Obtain platelet-rich plasma (PRP) by centrifuging whole blood for 5 min at 100g at room temperature. Remove the PRP (upper layer), taking care to avoid the red cells, which contain a significant level of ADP, and acidify with a solution of citric acid and dextrose (120 μL for 880 μL of PRP).
3. Add PGI$_2$ (12 ng/mL final), centrifuge at 500g for 10 min and resuspend the platelet pellet in the same volume of resuspension buffer.
4. After 15 min incubation at room temperature, centrifuge again (500g, 10 min) and resuspend platelets in the incubation buffer at between 0.3 and 0.7×10^6 platelets/μL. The suspension must have a shimmering aspect when the tube is gently shaken (*see* **Note 3**).

3.2. Incubation With Ligand

The binding is measured in a total volume of 200 μL, and all the reagents are warmed to room temperature immediately before use.

1. Fill the wells of a microplate with 50 μL of incubation buffer containing either (a) no added drug, for total binding, (b) cold 2MeS-ADP (40 μ*M*) (or ADP, 4 m*M*) for determination of the nonspecific binding, or (c) the drug to be tested, at a concentration fourfold higher than the required final concentration.
2. Add 50 μL of radioactive ^{33}P-2MeS-ADP solution to each well. The concentration of the ligand depends on the type of study you want to perform. Saturation experiments

need various concentrations, from 0.05 to 10 n*M*, but for kinetics and competition, it is recommended to choose a low concentration, below the K_D value. For example, for competition binding studies on P2Y$_{12}$ in our laboratory, concentrations range between 0.2 and 0.6 n*M*.

3. Finally, add 100 μL of the platelet suspension into each well. Shaking is not needed and incubation can be done at room temperature. For platelets, the kinetics of binding are rapid and a maximum is reached at 5 min. This binding is stable for at least 2 h, but, since the stability of tested compounds may vary and because platelet integrity can be modified, we prefer the filtration to occur between 5 and 20 min of incubation. If one wants to measure the kinetics of association, various incubation times ranging from 1 min to several hours will be required.

3.3. Separation of Bound From Free Ligand

1. Pre-wet filters with the washing buffer.
2. Filter platelet suspensions under vacuum and quickly rinse three times with 500 μL of ice-cold washing buffer. This operation should not take more than 6 s. Filters are then air-dried. This can be done with an automatic cell harvester, by programming a 10-s sequence of aspiration without refilling the microplate.

3.4. Counting

1. Cut the filter into individual disks (as defined by the filtration device) and place each disk in a scintillation vial with 5 mL of a water-compatible liquid scintillation cocktail.
2. After a brief vortexing, samples are kept in the dark overnight then counted in a beta counter, in the same channel as used for ^{14}C. A quenching curve, made from a standard solution of ^{33}P radionucleotide (it could be ^{33}P-2MeS-ADP, but it is not mandatory), will allow disintegrations per minute (DPM) to be obtained, as DPM is used for saturation experiments. This consists of measuring the number of CPM in the window corresponding to ^{33}P and the emission spectrum of a known DPM quantity of radioactivity (obtained from a standard solution) in solution with scintillation liquid and increasing volumes of a quenching reagent (for example, carbon tetrachloride [CCl_4]). The counting efficacy, calculated as CPM/DPM, is plotted as a function of the quenching, expressed as the shift of emission spectrum. Each sample is counted and the counting efficacy is deduced from the emission spectrum. The count in DPM is then calculated, correcting the CPM count by the counting efficacy. Today, all these calculations are integrated in the counter, but you need to make standard samples to establish a quenching curve for each isotope and each type of sample (scintillation liquid, filter, counting vial, etc.). If you wish to determine the effect of a drug against the binding, counts per minute (CPM) are sufficient.

3.5. Calculations

The exploitation of results from ligand-binding studies follows classical methods of calculation whose details have been described elsewhere (*16–18*). Software packages (such as Kell, from Biosoft, Cambridge, UK) contain several tools devoted to the exploitation of results from kinetic, saturation, and competition experiments. They allow the estimation of the affinity and of the number of binding sites for the ligand, and the affinity of competitors. When different classes of binding sites are present, these can also be detected if their affinities are significantly different. Another way to provide evidence for different binding sites is to use specific ligands for these sites as competitors. For instance, $P2Y_1$ and $P2Y_{12}$, which are both receptors for 2MeS-ADP, have been differentiated in rabbit platelets, by measuring the effect of A3P5PS (an antagonist of $P2Y_1$) or the influence of a treatment with clopidogrel (whose in vivo metabolite acts as an antagonist of $P2Y_{12}$) *(19)*.

3.6. Sample Standard Experiment: Effect of Test Compound on Binding of 2MeS-ADP to Platelets

The following protocol is aimed at establishing the concentration:inhibition curve for a test compound on platelet P2Y receptors. **Steps 1–8** are illustrated schematically in **Fig. 1**. Each concentration is tested in duplicate for each run. Usually, a concentration: inhibition curve is an average of three independent experiments.

1. Prepare 200 µL of the test compound in incubation buffer at a concentration fourfold higher than the highest test concentration (stock solution).
2. In incubation buffer, prepare five tubes of serial dilutions from the initial stock solution (usually, the range of dilutions is 0.3, 0.1, 0.03, 0.01, 0.003), one tube with buffer alone for control, and one tube containing a solution of 4 m*M* ADP for nonspecific binding (200 µL of each).
3. Prepare 1 mL of ^{33}P-2MeS-ADP at the concentration of 0.6 n*M* in incubation buffer.
4. Prepare 2 mL of a suspension of washed platelets (0.5×10^6/µL) in incubation buffer containing 5 m*M* Na$_2$EDTA, as described in **Subheading 3.1., steps 1–4**.
5. To individual wells of a 96-well microplate, add (a) 50 µL of a specific concentration of the test compound (stock solution or one of each serial dilution), or buffer alone, or ADP, (b) 50 µL of radiolabeled ligand, and (c) 100 µL of platelets.
6. Incubate for 15 min at room temperature.
7. Aspirate each well of the microplate with the filtration device and rinse three times with 500 µL of ice-cold washing buffer.
8. Dry the filters and count them by scintillation to yield counts per minute (CPM) for each well.
9. Calculate the mean of radioactivity for each duplicate.
10. Calculate the percentage inhibition using the following formula:

$$\% \text{ inhibition} = 100 \times [1 - (S - NS) / (C - NS)]$$

where S is the CPM count in the sample, NS is the CPM count for ADP (nonspecific binding), and C is the CPM count in the control (total binding). **Figure 2** shows the competition curves for 2 MeS-ATP (an antagonist of platelet P2Y receptors) and GTP (a triphosphate nucleotide, not recognized by platelet P2Y receptors), obtained on human platelets, according to this protocol (*see* **Note 4**).

3.7. Variations and Alternatives

The method described above can be easily modified and adapted for different conditions.

3.7.1. Platelet-Rich Plasma

Platelets can be used in platelet-rich plasma. Under these conditions, the addition of 5 m*M* Na$_2$EDTA to the PRP is recommended. Alternatively, heparin (0.01 U/mL) or hirudin can also be used to prevent coagulation.

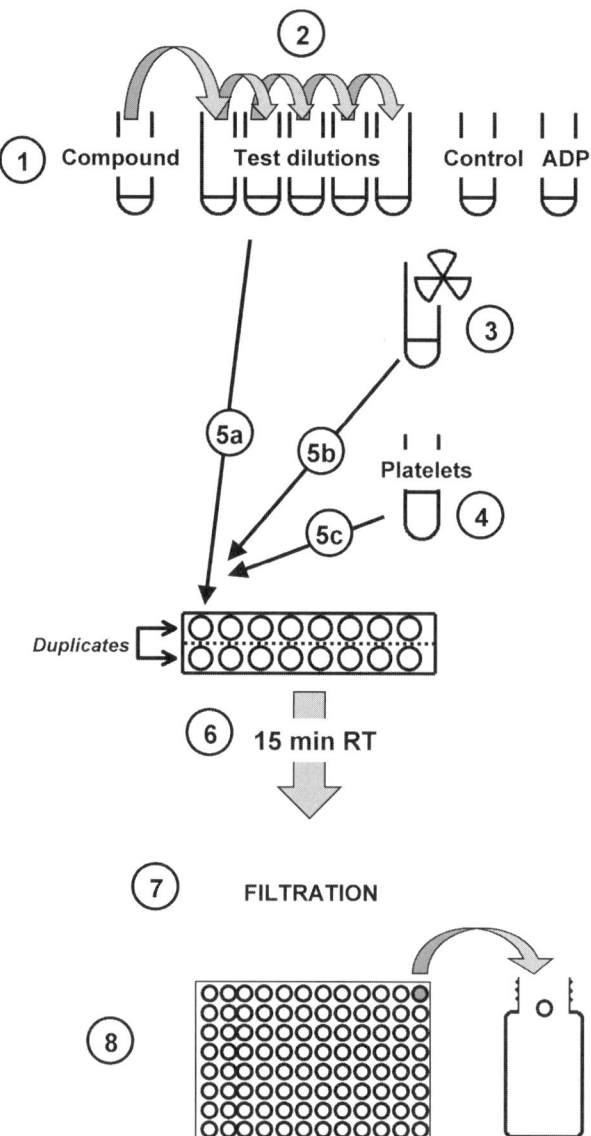

Fig. 1. Schematic representation of the steps involved in an experiment to assess the effect of a test compound on the binding of 2MeS-ADP to platelets. The numbers refer to **steps 1–8** of **Subheading 3.6.** Values from scintillation counting in **step 8** are used to generate (*see* **steps 9,10, Subheading 3.6.**) a concentration-inhibition relationship as shown in **Fig. 2**.

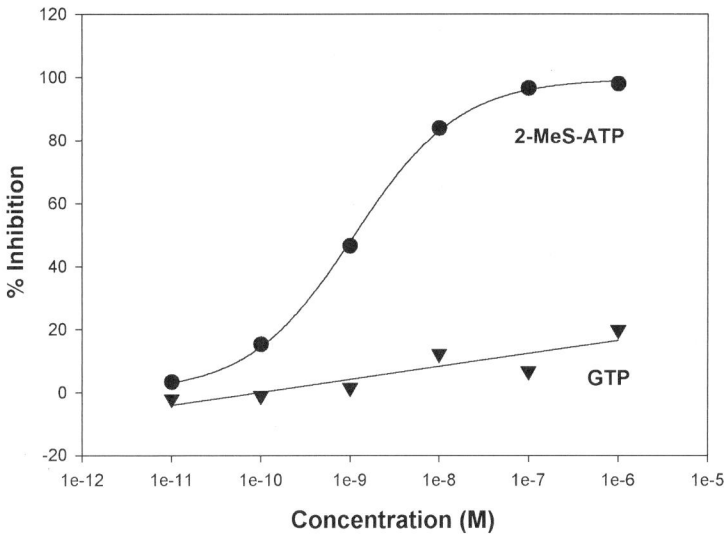

Fig. 2. Concentration-inhibition relationship of 2MeS-ATP on 2MeS-ADP binding to platelets. Data from a typical experiment (**Subheading 3.6.**) designed to assess the effects of 2MeS-ATP (an antagonist of platelet P2Y receptors) on 2MeS-ADP binding to platelets. GTP, a triphosphate nucleotide not recognized by platelet P2Y receptors, was included as a control.

3.7.2. Cultured Cells

Immediately prior to use, wash cells in serum-free medium, in order to remove secreted nucleotides. When cells are adherent, they are harvested after a short incubation time in EDTA (5 mM) containing PBS. If you are uncertain whether the ADP receptor being studied is trypsin-sensitive, trypsin should be avoided. If you suspect nucleotide release, apyrase (0.1 U/mL) can be added to the washing medium in order to destroy all the released ATP and ADP and protect P2Y against desensitization. Cells are finally suspended into the washing medium without apyrase. We have successfully used this method to measure the binding on CHO cells stably transfected with P2Y$_{12}$ (*14*). A preliminary experiment was performed to give an indication of the kinetics of association (5–120 min), using 15,000 cells per well and a low ligand concentration (0.1–0.3 nM ^{33}P-2MeS-ADP). From this first experiment, a period of incubation was selected that gave maximum and stable binding. We have also observed that, with adhesive cells such as CHO, it is better to use a short period of incubation. The binding was measured with increasing cell densities in order to determine a density giving a signal at least 100-fold higher than the background and less than 10% of the total radioactivity, and to verify that we were in the linear part of the curve. At a fixed incubation time and cell density, we performed saturation experiments by incubating increasing concentrations of ligand (0.001 to 3 nM). The K$_D$ was calculated and a concentration corresponding to 50% of the K$_D$ was chosen for competition

Table 1
Troubleshooting Chart

Problem	Cause	Solution
Suspension difficult to aspirate from the microwell	Coagulation or formation of platelet aggregates	Add EDTA (5 mM final)
Suspension difficult to filtrate, plugging	Inadequate type of filter	Check if your filter is really made with glass fiber
	Platelet suspension too concentrated	Dilute the platelet suspension
	Vacuum source too weak	Increase the aspiration pressure
Damaged filters	Vacuum source too strong	Reduce the aspiration pressure
Nonspecific counts elevated (>2.5% of the total ligand)	Concentration of ADP too low	Check the final concentration of ADP (1 mM)
	Inadequate filter	Check if your filter is really made with glass fiber
	Washing volume too low	Increase the washing volume to 3 mL
Total binding in controls too high (>10% of the total ligand)	Too many binding sites in your sample	Reduce the number of platelets
Total binding in controls too low (<3X nonspecific binding)	Concentration of platelets too low	Increase the number of platelets
	Problems in the preparation of the platelet sample	Check the ADP sensitivity by aggregometry (in the absence of EDTA)
	Washing time too long	Reduce the washing time under 6 s
	Vacuum source too strong	Reduce the aspiration pressure

experiments. The kinetic experiments were repeated in order to obtain more precise time constants at this selected concentration.

4. Notes

1. Specific reagents for studying P2 receptors are mainly nucleotides. Their purity must be as high as possible, and storage instructions have to be strictly respected. The majority of solutions are made immediately prior to use in water and storage is avoided. 2MeS-ADP can be stored as an aqueous solution at 0.01 M, in aliquots at –30°C. Under these conditions, it has been found to be stable for up to 4 mo. It should be recognized that a number of nucleotides are known to contain contaminants. Commercial 2MeS-ATP, for instance, has been found to contain 2MeS-ADP; an enzymatic method has been described by Hechler et al. *(15)* to limit the effects of this contaminant.

2. Laboratory safety and security rules must be observed, including appropriate authorization for use of radioactive compounds, and established procedures for biohazard and radioprotection. In our institute, ^{33}P is not manipulated behind a screen, its radiation being in the same energetic category as ^{14}C; however, local regulations should always be followed. Latex gloves are worn throughout the procedure.

3. It is also advisable to check the integrity of the ADP-dependent platelet response in a standard aggregometer. Add 2 mM CaCl$_2$ and 1 mg/mL fibrinogen to a sample of the platelet suspension, place an aliquot into the aggregometer, and stimulate with 5 µM ADP (*see* Chapter 5, vol. 1).

4. The recent discovery of P2Y$_{12}$ has highlighted the importance of the platelet ADP receptors in hemostasis and thrombosis. Binding studies performed with 2MeS-ADP as a ligand have demonstrated the effect of several compounds developed as antithrombotic drugs. However, caution must be taken when interpreting binding studies using 2MeS-ADP since this agonist is known to bind to other P2Y receptors, including, for example, P2Y$_1$ on platelets.

5. *Troubleshooting:* If you are not satisfied with your results, **Table 1** may be useful in identifying the problem.

References:

1. MacFarlane, D. E., Srivastava, P. C., and Mills, D. C. B. (1983) 2-Methylthioadenosine [γ-^{32}P] diphosphate. An agonist and radioligand for the receptor that inhibits the accumulation of cyclic AMP in intact blood platelets. *J. Clin. Invest.* **71,** 420–428.

2. Mills, D. C. B., Puri, R., Hu, C-J., Minniti, C., Grana, G., Freedman, M. D., Colman, R. F., and Colman, R. W. (1992) Clopidogrel inhibits the binding of ADP analogues to the receptor mediating inhibition of platelet adenylate cyclase. *Arteriosclerosis Thrombosis* **12,** 430–436.

3. Savi, P., Laplace, M. C., Maffrand, J. P., and Herbert, J. M. (1994) Binding of [^{3}H]-2-methylthio ADP to rat platelets: effect of clopidogrel and ticlopidine. *J. Pharm. Exp. Ther.* **269,** 772–777.

4. Savi, P., Laplace, M. C., and Herbert, J. M. (1994) Evidence for the existence of two different ADP-binding sites on rat platelets. *Thromb. Res.* **76,** 157–169.

5. Savi, P., Troussard, A. and Herbert, J. M. (1994) Characterization of specific binding sites for [^{3}H]-2-MeS-ADP on megakaryocytoblastic cell lines in culture. *Biochem. Pharmacol.* **48,** 83–86.

6. Gachet, C., Cattaneo, M., Ohlmann, P., Hechler, B., Lecchi, A., Chevalier, J., et al. (1995) Purinoceptors on blood platelets: further pharmacological and clinical evidence to suggest the presence of two ADP receptors. *Brit. J. Haematol.* **91,** 434–444.

7. Jantzen, H. M., Gousset, L., Bhaskar, V., Vincent, D., Tai, A., Reynolds, E. E., et al. (1999) Evidence for two distinct G-protein-coupled ADP receptors mediating platelet activation. *Thromb. Haemost.* **81,** 111–117.

8. Nurden, P., Savi, P., Heilmann, E., Bihour, C., Herbert, J. M., Maffrand, J. P., et al. (1995) An inherited bleeding disorder linked to a defective interaction between ADP and its receptor on platelets. Its influence on glycoprotein IIb-IIIa complex function. *J. Clin. Invest.* **95,** 1612–1622.

9. Cattaneo, M., Lombardi, R., Zighetti, M. L., Gachet, C., Ohlmann, P., Cazenave, J. P., et al. (1997) Deficiency of ^{33}P-2MeS-ADP binding sites on platelets with secretion defect, normal granule stores and normal thromboxane A2 production. Evidence that ADP potentiates platelet secretion independently of the formation of large platelet aggregates and thromboxane A$_2$ production. *Thromb. Haemost.* **77,** 986–990.

10. Sugidachi, A., Asai, F., Yoneda, K., Iwamura, R., Ogawa, T., Otsuguro, K., et al. (2001) Antiplatelet action of R-99224, an active metabolite of a novel thienopyridine-type G_i-linked P2T antagonist, CS-747. *Brit. J. Pharmacol.* **132**, 47–54.

11. Baurand, A., Raboisson, P., Freund, M., Leon, C., Cazenave, J. P., Bourguignon, J. J., et al. (2001) Inhibition of platelet function by administration of MRS2179, a P2Y1 receptor antagonist. *Eur. J. Pharmacol.* **412**, 213–221.

12. Scarborough, R. M., Laibelman, A. M., Clizbe, L. A., Fretto, L. J., Conley, P. B., Reynolds, E. E., et al. (2001) Novel tricyclic benzothiazolo[2,3-c]thiadiazine antagonists of the platelet ADP receptor (P2Y$_{12}$). *Bioorg. Med. Chem. Lett.* **11**, 1805–1808.

13. Savi, P., Pereillo, J. M., Uzabiaga, M. F., Combalbert, J., Picard, C., Maffrand, J. P., et al. (2000) Identification and biological activity of the active metabolite of clopidogrel. *Thromb. Haemost.* **84**, 891–896.

14. Savi, P., Labouret, C., Delesque, N., Guette, F., Lupker, J., and Herbert, J. M. (2001) P2Y$_{12}$, a new platelet ADP receptor, target of clopidogrel. *Biochem. Biophys. Res. Commun.* **283**, 379–383.

15. Hechler, B., Vigne, P., Léon, C., Breittmayer, J. P., Gachet, C., and Frelin, C. (1998) ATP derivatives are antagonists of the P2Y$_1$ receptor: Similarities to the platelet ADP receptor. *Mol. Pharmacol.* **59**, 727–733.

16. Munson, P. J. and Rodbard, D. (1980) Ligand: a versatile computerized approach for the characterization of ligand binding systems. *Anal. Biochem.* **107**, 220–239.

17. Scatchard, G. (1949) The attractions of proteins for small molecules and ions. *Ann. NY Acad. Sci.* **51**, 660–672.

18. Cheng, Y. C. and Prusoff, W. H. (1973) Relationship between the inhibition constant (Ki) and the concentration of inhibitors which causes 50% inhibition (IC50) of an enzymatic reaction. *Biochem. Pharmacol.* **22**, 3099–3108.

19. Savi, P., Beauverger, P., Labouret, C., Delfaud, M., Salel, V., Kaghad, M., et al. (1998) Role of P2Y$_1$ purinoceptor in ADP-induced platelet activation. *FEBS Lett.* **422**, 291–295.

8

Ligand-Binding Assays

Fibrinogen

Dermot Cox

1. Introduction

The fibrinogen receptor on platelets is glycoprotein (GP) IIb/IIIa, also known as $\alpha_{IIb}\beta_3$. This receptor can bind to immobilized fibrinogen in its resting state. Binding to soluble fibrinogen, however, requires activation of the receptor. This can occur due to inside-out signaling in response to a platelet agonist such as ADP, or directly by agents such as manganese. The interaction between GPIIb/IIIa and fibrinogen is essential in the formation of platelet aggregates *(1,2)*.

[125]I-fibrinogen binding to platelets has been widely used to study the activity of GPIIb/IIIa *(3–6)*. While ^{125}I-fibrinogen binding is a straightforward assay, there are difficulties with interpretation and the use of radiolabels. One issue involved is the low affinity of fibrinogen for GPIIb/IIIa, such that high levels of fibrinogen are required to saturate binding sites at the receptor. The original study, which identified a fibrinogen receptor on platelets, found the Scatchard plot to be linear *(4)* while subsequent studies using higher concentrations of fibrinogen found this relationship to be curvilinear *(3,7,8)*. The curvilinear nature of the Scatchard plot was thought to suggest the existence of multiple binding sites *(7,8)* but analysis of the binding data using a nonlinear regression model showed the existence of only one binding site *(3)*. These studies illustrate the problems with using the Scatchard plot to interpret binding data.

Peerschke et al. *(7)* estimated the number of binding sites for fibrinogen to be $12,896 \pm 2456$, while Cox and Seki *(3)*, using higher concentrations of fibrinogen, estimated the number of receptors to be $39,000 \pm 11,000$. The latter figure is more in line with studies using antibodies that identified $54,162 \pm 2,455$ GPIIb/IIIa receptors per platelet *(9)*. This difference may be due to the failure to activate all receptors in the fibrinogen-binding studies. Another problem is that of variation in the number of receptors on platelets from different donors *(3)*. In our hands we can often find normal

From: *Methods in Molecular Biology, vol. 273:*
Platelets and Megakaryocytes, Vol. 2: Perspectives and Techniques
Edited by: J. M. Gibbins and M. P. Mahaut-Smith © Humana Press Inc., Totowa, NJ

platelets with between 30,000 and 60,000 receptors per platelet using an antibody-based method. In ligand-binding studies this problem of interdonor variability was overcome by using a nonlinear model with a correction factor for each experiment, which greatly increased the fit of the model *(3)*. The association constant (K) for ^{125}I-fibrinogen binding was estimated as $4.19 \pm 1.3 \times 10^6/M$ *(3)*. A major source of error in binding studies is depletion of the radiolabel, which leads to free concentrations much lower than expected. Using the ^{125}I-fibrinogen-binding method described below, the depletion due to binding to the platelets was estimated as no more than 1% and even when depletion due to binding to tubes and the like was considered, it was never more than 10%.

As an alternative to radioligand binding, fluorescently labeled fibrinogen can be used in a flow-cytometric assay. Either FITC-fibrinogen can be prepared *(10)* or a commercial Oregon Green fibrinogen purchased from Molecular Probes (Leiden, The Netherlands). This assay is easy to perform and useful for detecting inhibition of fibrinogen binding. However, it is not quantitative and cannot be used for determining the receptor number.

Depending on the data required, there are alternative assays of GPIIb/IIIa receptors. If a diagnosis of Glanzmann's thrombasthenia *(11)* is required, the best method is the use of antibodies in flow cytometry. In fact, the platelet glycoprotein screen from BioCytex (Marseille, France) is ideally suited to this. It contains antibodies to GPIIIa, GPIb, and GPIa. This has the advantage of allowing a diagnosis of Bernard-Soulier disease (absence of GPIb) and deficiencies in the collagen receptor as well. The kit uses quantitative immunofluorescence to allow accurate determination of the number of receptors per platelet. The use of multiple markers also controls for disorders of platelet size and the assay requires only a few microliters of blood. The main drawback of the assay is that it cannot discriminate between functional and nonfunctional GPIIb/IIIa receptors. Thus, patients with normal levels of GPIIb/IIIa but reduced capacity to bind fibrinogen will not be detected.

The recent use of GPIIb/IIIa antagonists *(12)* in the treatment of cardiovascular disease *(13)* has led to an interest in measuring the percent occupancy of GPIIb/IIIa by inhibitors *(14)*. This is of particular interest, as this class of drugs has a steep dose-response curve; high levels of drug lead to bleeding complications and low levels lead to an increased risk of myocardial infarction. A number of assays have been developed to utilize the functional response of platelets to infer this information. The GPIIb/IIIa receptor occupancy assay (BioCytex) is particularly useful in that it allows determination of the total number of receptors and the number of occupied receptors. This assay uses two monoclonal antibodies to GPIIIa. The binding of one is inhibited by abciximab and the other is inhibited by eptifibatide and tirofiban. Thus, the combination of these two antibodies allows the determination of free receptors and total receptors *(9)*. The kit is based on quantitative flow cytometry.

Another useful method to measure inhibition of GPIIb/III binding to fibrinogen is an adhesion assay. While the IC$_{50}$ values are higher than in other assays, it is specific for GPIIb/IIIa, uses a 96-well plate format that is good for high-throughput assays, and uses an inexpensive assay of an intracellular enzyme (acid phosphatase) to detect platelet adhesion *(15)*.

One of the problems with many fibrinogen-binding assays is the need to remove the unlabeled fibrinogen from the system before adding the labeled fibrinogen, which makes it impossible to study binding in patient samples ex vivo. However, when fibrinogen binds to GPIIb/IIIa it undergoes a conformational change and exposes neo-epitopes known as receptor-induced binding sites (RIBS). This makes it possible to monitor fibrinogen binding in the presence of plasma and to measure fibrinogen binding in blood samples ex vivo using anti-RIBS antibodies. One such antibody is 9F9 *(16)*. The platelet fibrinogen kit from BioCytex uses an anti-RIBS antibody to measure fibrinogen bound to platelets. This kit allows direct quantification of bound fibrinogen in whole blood by quantitative immunofluorescent flow cytometry.

2. Materials

1. HEPES-Tyrode's buffer: 137 mM NaCl, 2.7 mM KCl, 0.5 mM MgCl$_2$, 3 mM NaH$_2$PO$_4$, 3.5 mM HEPES, 5.5 mM glucose, and 0.35% BSA, adjusted to pH 7.5 with NaOH/HCl (*see* **Note 1**).
2. Human fibrinogen (Calbiochem-Novabiochem, San Diego, CA).
3. Bovine serum albumin (Sigma, Poole, UK).
4. Sepharose Cl2B (Sigma) column equilibrated in HEPES-Tyrode's buffer.
5. Na^{125}I (Amersham Biosciences, Buckinghamshire, UK; PerkinElmer Life Sciences, Boston, MA).
6. Acid-citrate dextrose: 38 mM citric acid, 75 mM sodium citrate, 124 mM dextrose.
7. PGE$_1$ (Sigma).
8. Apyrase Grade VII (Sigma).
9. 0.1 M sodium phosphate buffer: mix together 0.1 M NaH$_2$PO$_4$, 0.1 M Na$_2$HPO$_4$ (pH 7.2, adjustment not required).
10. Paraformaldehyde fixative: 4% w/v paraformaldehyde in 0.1 M sodium phosphate buffer.
11. Neutralizing buffer: 20 mM NH$_4$Cl, 150 mM NaCl, and 300 mM Tris-HCl, titrated to pH 7.2 with NaOH/HCl.
12. Iodobeads/Iodo-Gen (Pierce, Rockford, IL).
13. Oregon Green fibrinogen (Molecular Probes).
14. Thrombin-receptor-activating peptide (TRAP 14; Sigma, Cat. no. S 7152) (*see* **Note 2**).
15. ADP (Sigma).
16. GPIIb/IIIa antagonist: abciximab (ReoPro; Lilly, Indianapolis, IN), tirofiban (Aggrastat; Merck), or eptifibatide (Integrelin; Millennium, San Francisco, CA).
17. Platelet GP Screen (Biocytex, Marseille, France).
18. Platelet fibrinogen kit (Biocytex).
19. GPIIb/IIIa receptor occupancy kit (Biocytex).
20. 0.129 M trisodium citrate.
21. 96-well plates.
22. Phosphate-buffered saline (PBS): 10 mM Na$_2$HPO$_4$ and 10 mM NaH$_2$PO$_4$, mixed to give pH 7.4, and 120 mM NaCl.
23. Substrate buffer: 0.1 M sodium acetate, pH 5.5, containing 0.1% Triton X-100 and 10 mM *p*-nitrophenyl phosphate (Sigma).

3. Methods

3.1. Preparation of Plasma-Free Platelets

To perform labeled-fibrinogen binding studies, it is necessary to remove all the free fibrinogen. The three most frequently used methods are described in **Subheadings 3.1.3.–3.1.5.**

3.1.1. Collection of Blood

Donors should be free of aspirin for two weeks. Fill syringes with 1/10 volume of acid citrate dextrose (ACD) prior to collection (*see* **Note 3**).

3.1.2. Preparation of Platelet-Rich Plasma

1. Adjust blood to pH 6.5 by further addition of ACD.
2. Centrifuge at $850g$ for 3 min or $150g$ for 10 min.
3. Remove the platelet-rich plasma (PRP) layer using a plastic Pasteur pipet without disturbing the red blood cells.

3.1.3. Albumin Gradient Centrifugation

The removal of all the plasma proteins associated with platelets is difficult. The best method is probably to centrifuge the platelets through an albumin gradient. This also has the benefit of not activating the platelets *(17)*.

1. Dissolve BSA in water at a concentration of 50% w/v (*see* **Note 4**).
2. Use this stock solution to make up solutions of 25%, 17%, 12%, and 10% w/v.
3. Add 1.5 mL of the 50% BSA to a 15-mL centrifuge tube and layer 1.5 mL of 25% BSA on top. Repeat this procedure for other concentrations of BSA.
4. After the gradients are prepared, layer 1.5 mL of PRP on top of each tube and centrifuge at $1200g$ for 15 min.
5. The platelets form a visible band on top of the 50% BSA layer. Remove these with a plastic Pasteur pipet.

3.1.4. Gel Filtration

1. Add PGE_1 (1 µg/mL) and apyrase (1 U/mL) to the PRP (*see* **Note 5**).
2. Centrifuge PRP at $600g$ for 20 min to form a platelet pellet.
3. Remove the plasma and resuspend the pellet in a small volume (~0.5 mL) of HEPES-Tyrode's buffer.
4. Layer the platelets onto a Sepharose Cl2B column equilibrated in HEPES-Tyrode's buffer and collect fractions.
5. Pool fractions with a visible platelet component.

3.1.5. Doubly Washed Platelets

1. Add PGE_1 (1 µg/mL) and apyrase (1 U/mL) to PRP (*see* **Note 5**).
2. Centrifuge the PRP at $600g$ for 20 min to form a platelet pellet.
3. Remove the supernatant and resuspend the pellet in HEPES-Tyrode's buffer containing PGE_1 and apyrase.
4. Repeat **steps 2** and **3**.

After obtaining a platelet preparation free of plasma using one of the three afore-mentioned methods, the platelet count should be determined by hemocytometer or particle counter (*see* **Note 6**) (*see* Chapter 3, vol. 1 for details of platelet counting). It is also useful to add a small amount of apyrase (<0.1 U/mL) to preserve the sensitivity to ADP.

3.2. ^{125}I-Fibrinogen-Binding Studies

Activated GPIIb/IIIa returns to the resting state if it does not bind fibrinogen; therefore, it is necessary to fix the platelets after stimulation, prior to use in the binding assay. Free fibrinogen is separated from bound fibrinogen by centrifugation through sucrose.

3.2.1. Platelet Activation and Fixation

1. Add agonist to the platelets for 5 min at room temperature (*see* **Note 7**).
2. Add fixative solution so the final concentration of paraformaldehyde is 0.8%, and leave for 30 min at room temperature (*see* **Note 8**).
3. Add an equal volume of neutralizing buffer and centrifuge the platelets at 600g for 20 min.
4. Resuspend the pellet in HEPES-Tyrode's buffer and wash three times.
5. Finally, determine the platelet concentration as before.

3.2.2. Fibrinogen Labeling

1. Add Na^{125}I solution to Iodobeads in a tube. The number of beads used is dependent on the amount of fibrinogen used and is determined according to the manufacturer's instructions (*see* **Note 9**).
2. After 5 min add fibrinogen (*see* **Note 10**).
3. After incubation for 2–15 min, remove the beads with forceps or a Pasteur pipet.

3.2.3. Binding Studies

1. Incubate washed, fixed platelets (2 × 10^8 platelets/mL) in HEPES-Tyrode's buffer containing 2 mM CaCl$_2$, ^{125}I-fibrinogen, and inhibitor at room temperature for 30 min (*see* **Note 11**).
2. Layer an aliquot of the reaction mixture onto 200 μL of HEPES-Tyrode's buffer containing 20% w/v sucrose in a 0.4-mL microcentrifuge tube in triplicate for each tube (*see* **Note 12**).
3. Centrifuge the tubes at high speed (≈10,000 rpm) in a microcentrifuge for 5 min.
4. Using a sharp razor blade, cut the tip off the tube and count in a γ-counter (*see* **Note 13**).

3.2.4. Data Analysis

For saturation-binding studies it is important to perform the appropriate analysis. As the Scatchard plot is curvilinear with fibrinogen binding, it is necessary to use a nonlinear regression analysis program such as Prism (www.GraphPad.com) (*see* **Note 14**). It is important to have a suitable estimate of the background. This can be obtained using a 50-fold excess of unlabeled fibrinogen. However, with higher levels of fibrinogen this is unsuitable and the best method is to use an excess of a GPIIb/IIIa antagonist: abciximab (10 μg/mL), eptifibatide (10 μM), or tirofiban (10 μM). This should be

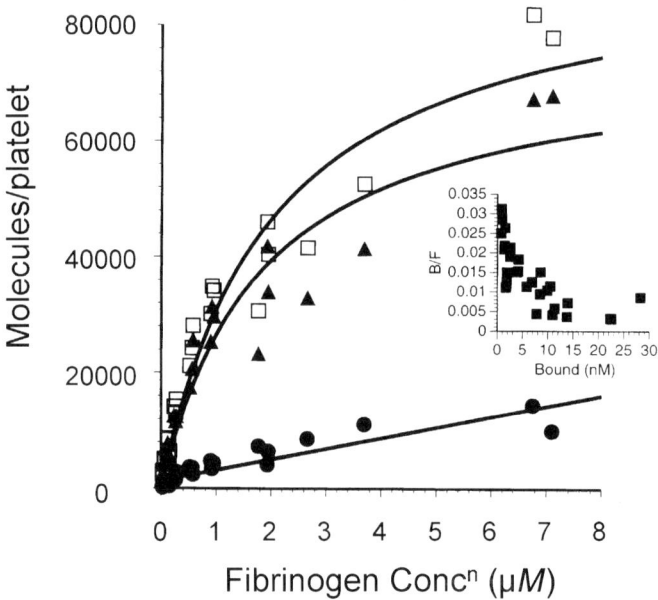

Fig. 1. Analysis of radiolabeled fibrinogen binding. Binding isotherm for ^{125}I-fibrinogen binding to platelets with the Scatchard plot showing bound/free against bound inset. Total binding (□), specific binding (▲), and nonspecific binding (●) are shown. Binding is determined by measuring dpm from ^{125}I-fibrinogen and converting it into molecules per platelet using the specific activity, fibrinogen concentration, and platelet concentration. Bound fibrinogen was separated from free fibrinogen by centrifugation through 20% sucrose. From **ref. 3** with permission.

performed for every concentration of fibrinogen used. Specific binding is determined by subtracting the binding in the presence of excess antagonist from the binding in the absence of antagonist (*see* **Fig. 1**).

3.3. Oregon Green Fibrinogen-Binding Studies

An alternative to ^{125}I-fibrinogen is fluorescently labeled fibrinogen. Binding is detected by flow cytometry. This has the advantage that there is no need for radiolabels and it is a quick assay to perform. The major disadvantage is that it is not possible to quantify the binding, only to obtain data on relative binding. There are two possibilities for fluorescent labels. One is to directly label fibrinogen with FITC *(10)* and the other is to use the commercially available Oregon Green fibrinogen *(19)*.

3.3.1. Oregon Green Fibrinogen Binding

1. Collect blood and prepare a fibrinogen-free platelet preparation as described in **Subheading 3.1.**

2. Incubate platelets with Oregon Green fibrinogen (up to 1 mg/mL) for 30 min at room temperature in the presence of 2 mM CaCl$_2$ and agonist (ADP [20 μM] or TRAP [5 μM]) (*see* **Note 15**). Dilute the samples with 1 mL of buffer prior to analysis.
3. Run the negative and positive controls through the cytometer and adjust the cytometer parameters (*see* **Note 16**). Read the samples using these parameters.

3.4. GPIIb/IIIa Receptor Occupancy Studies

This assay is a monoclonal antibody-based flow cytometric assay from BioCytex. mAb 1 is a monoclonal antibody that directly binds to GPIIIa and competes with abciximab for its site. mAb 2 is a monoclonal antibody that binds to a site on GPIIIa that is distinct from the abciximab site. However, both tirofiban and eptifibatide cause conformational changes in GPIIIa that prevent mAb 2 from binding. This binding site is known as a ligand-attenuated binding site (LABS). This assay also uses calibration beads that contain known numbers of molecules of antibody, which makes it possible to determine the number of molecules of antibody bound. By using both antibodies it is possible to determine the total number of receptors and the number of receptors occupied by the antagonists *(9)*.

3.4.1. GPIIb/IIIa Receptor Occupancy Assay

1. Collect a blood sample in citrate (9:1 v/v blood:0.129 M trisodium citrate) (heparin, 20 U/mL, can also be used as an anticoagulant) (*see* **Note 17**).
2. Place 100 μL of blood in a tube with 300 μL platelet-poor plasma (PPP) (*see* **Note 18**).
3. Add 20 μL of the PPP:blood mixture to three flow-cytometry tubes along with 20 μL of either isotype control antibody, mAb 1 or mAb 2. Then briefly vortex the samples and incubate at room temperature for 20 min. If in vitro studies of GPIIb/IIIa antagonists are being performed, then they are included at this stage.
4. Add 20 μL of labeled antibody to each tube and also to a fourth tube. Add 40 μL of calibration beads to the fourth tube. Briefly vortex all tubes and allow to incubate at room temperature for 20 min.
5. Add 1 mL of buffer (supplied with the kit) to each tube and then analyze by flow cytometry (*see* **Note 19** and **Fig. 2**).
6. Using the mean fluorescence for each of the four peaks from flow cytometry (*see* **Note 20**) and the molecules of antibody per bead from the kit, draw a standard curve (log-log) of fluorescence vs number of molecules. Use this to calculate the number of antibodies bound in the sample (*see* **Note 21**).

3.5. Direct Measurement of Bound Fibrinogen

One of the biggest difficulties in measuring binding of fibrinogen to platelets is the need to remove all the free plasma fibrinogen. However, just as ligands induce conformational changes in GPIIb/IIIa when they bind (LIBS and LABS), GPIIb/IIIa induces a conformational change in fibrinogen upon binding. This conformational change is known as a receptor-induced binding site (RIBS) and can be recognized by the antibody 9F9 *(16)*. Thus, this antibody can distinguish between contaminating unbound fibrinogen and bound fibrinogen. This is the basis for the Platelet Fibrinogen flow cytometry kit from BioCytex.

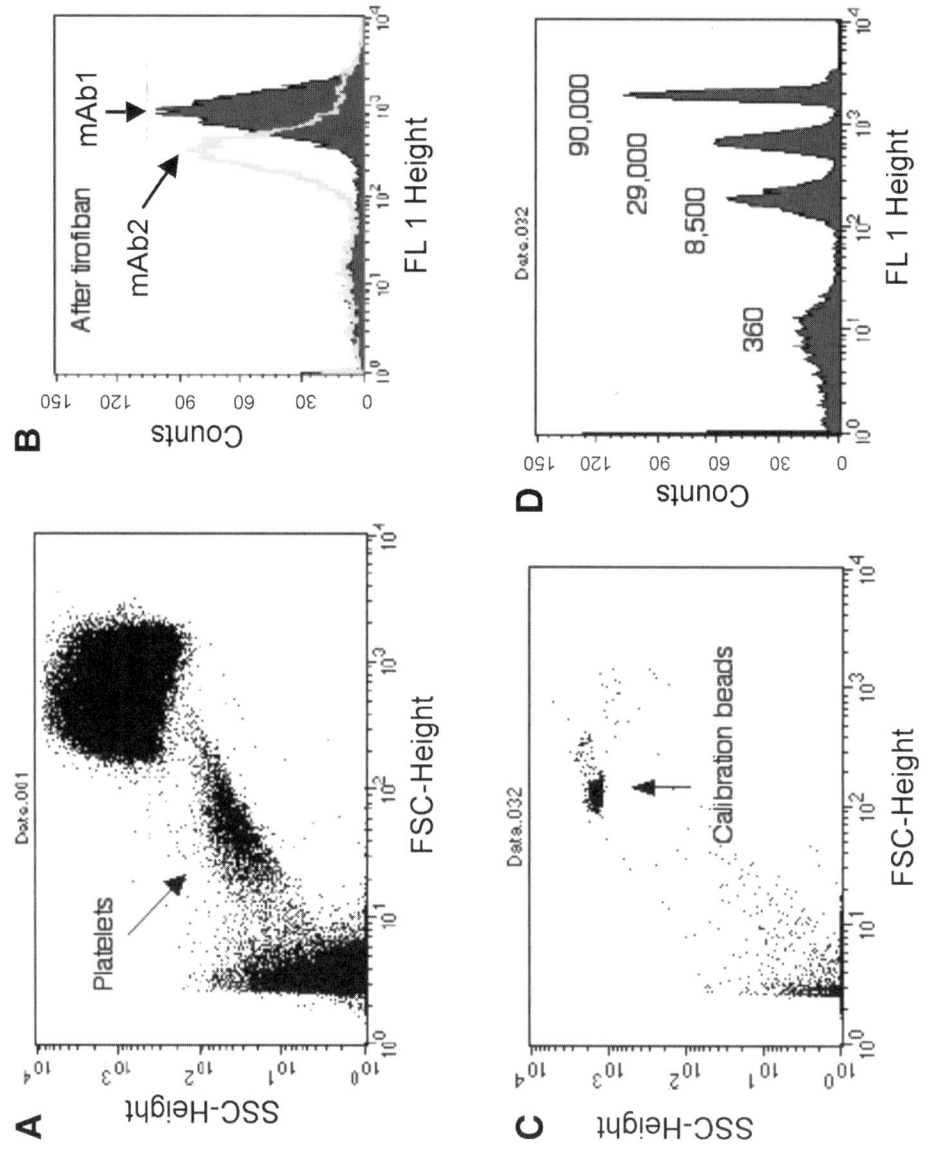

132

3.5.1. RIBS Assay

1. Collect blood in citrate or heparin.
2. Set up tubes containing 20 μL of whole blood and 40 μL of buffer (supplied), and 20 μL of buffer or inhibitory reagent (supplied) or activator (TRAP or ADP, not supplied).
3. After 2 min at room temperature, add 20 μL of each of these samples to 20 μL of antifibrinogen antibody.
4. After 10 min at room temperature, add 2 mL of buffer (supplied) to each tube; after 30 min at 4°C, analyze the samples by flow cytometry (*see* **Note 22**).

3.6. Platelet-Adhesion Assays

While platelet activation is required for soluble-fibrinogen binding to platelets, resting platelets can bind to fibrinogen if it is immobilized on a surface. There is evidence that different epitopes on fibrinogen are used for adhesion and aggregation *(20)*. As the following adhesion assay can be performed in a 96-well plate, it is particularly useful for high-throughput screening *(15)*. However, it is important to note that the IC_{50} values for GPIIb/IIIa antagonists are much higher in this assay compared with those obtained in ^{125}I-fibrinogen binding *(21)*. An agonist is not necessary in this assay, although platelet adhesion is increased if an agonist is used.

3.6.1. Platelet Adhesion

1. Collect blood in ACD and prepare plasma-free platelets as described in **Subheading 3.1.**
2. Coat 96-well plates with fibrinogen (20 μg/mL in PBS for 1 h at 37°C or overnight at 4°C).
3. Block the plates with 1% BSA for 1 h at 37°C.
4. Wash the plates three times in PBS before use.
5. Add 50 μL of platelets (8×10^7 platelets/mL) and 50 μL of antagonist (or buffer, as control) and incubate for 30 min at 37°C.
6. Empty the plate and wash three times in PBS.
7. Add 100 μL of substrate buffer are then added and incubate the plates for 2 h at 37°C (*see* **Note 23**).
8. Stop the reaction by addition of 20 μL of 1 *M* NaOH and read the plates at 410 nm.

4. Notes

1. The following buffer (JNL buffer) can also be used: 130 m*M* NaCl, 9 m*M* NaHCO$_3$, 10 m*M* sodium citrate, 10 m*M* Tris, 3 m*M* KCl, 0.8 m*M* KH$_2$PO$_4$, 6 m*M* dextrose, 0.9 m*M* MgCl$_2$, pH 7.35 with NaOH/HCl.

Fig. 2. *(see opposite page)* Flow cytometric analysis of platelet samples using GPIIb/IIIa receptor occupancy kit. **(A)** Side scatter (SSC) against forward scatter (FSC) dot plot of whole blood showing the platelet population. **(B)** Histogram of gated populations showing a sample from a patient treated with the GPIIb/IIIa antagonist tirofiban and stained with mAb1 and mAb 2, which are two antibodies that recognize GPIIIa. The presence of tirofiban inhibits mAb2 binding but not mAb1 binding. **(C)** FSC vs SSC dot plot showing the calibration beads. **(D)** Histogram of the mean fluorescence of the calibration beads showing the four populations of beads and the number of antibody molecules bound to each bead. This information can be used to produce a standard curve of molecules bond against mean fluorescence.

2. TRAP 6 (SFLLRN) is equally effective and more convenient if it is being custom-synthesized.

3. It is important that blood be collected in a manner that does not lead to platelet activation. This is best achieved using a wide-bore needle (19G) and to release the tourniquet prior to collecting the blood. The use of a syringe and butterfly cannula is preferable to that of a Vacutainer®.

4. It is important to use high-grade BSA that yields a pH near 7.0 when dissolved in water. To dissolve the BSA it should be sprinkled on water in a wide beaker and left at 4°C overnight without stirring.

5. PGE_1 is essential to prevent activation of the platelets and thus also aggregation. Apyrase is used to break down ADP that can be released during the centrifugation process. Released ADP is responsible for the desensitization of platelet ADP receptors and this is the main reason for the loss of sensitivity to ADP in washed platelets. Apyrase has both ATPase and ADPase activity and a preparation with high ADPase activity should be used at a concentration of 1 U/mL ADPase activity.

6. The platelet count should be as high as possible with platelets from a 60-mL blood donation being resuspended in 2–3 mL of HEPES-Tyrode's buffer.

7. Since resting platelets do not bind soluble fibrinogen, it is necessary to activate them prior to the binding study. GPIIb/IIIa is found on the platelet surface as well as in internal granules. Strong platelet agonists will activate GPIIb/IIIa as well as exposing the internal pool of receptors, while weak agonists activate only the surface receptors. If it is important to quantify surface receptors only, a weak agonist such as ADP (5 µM) should be used. Epinephrine often does not work with washed platelets. To cause maximum release, the best agonist is thrombin-receptor-activating peptide (TRAP) at 5–20 µM (either TRAP 14 or TRAP 6 [SFLLRN]), which activates the protease-activated receptor 1 (PAR1) on platelets. This is a very strong agonist and can be used on platelet preparations that fail to respond to other agonists. Thrombin (0.2 U/mL) can also be used but it must be neutralized (using PPACK) before addition of fibrinogen; otherwise, clot formation occurs.

8. Paraformaldehyde will require heating to 60°C to get it to dissolve, which should be done in a fume hood.

9. Fibrinogen can be labeled using the chloramine-T method *(18)*. However, Iodobeads is a more convenient method; alternatively, Iodo-Gen reagent can also be used. ^{125}I-fibrinogen of specific activity 0.3–1 mCi/mg can easily be obtained.

10. It is important to use fibrinogen of high purity and high solubility. Concentrations of 10–20 mg/mL fibrinogen are required.

11. The final concentration of fibrinogen should range from 10 µg/mL to 2 mg/mL. For the higher concentrations of fibrinogen the ^{125}I-fibrinogen can be diluted with unlabeled fibrinogen. An aliquot (5 µL) of the ^{125}I-fibrinogen stock solution should be counted. Since the volume and concentration are known, the number of molecules can be calculated. This then allows the relationship between DPM and molecules of fibrinogen to be determined. If mixtures of cold and hot fibrinogen are used, aliquots of these should be counted as well to determine the relationship between dpm and fibrinogen molecules in those samples. To determine nonspecific binding an excess (50–100-fold) of unlabeled fibrinogen (5 mg/mL) is used. Alternatively a GPIIb/IIIa antagonist such as tirofiban can be used, especially with high concentrations of ^{125}I-fibrinogen. If nonfixed platelets are used, an agonist (e.g., 20 µM ADP) should be included at this stage. If ADP is used it is important that the dilution factor of the platelets is such that the final apyrase levels are less than 0.05 U/mL. If effects of inhibitors of GPIIb/IIIa are to be investigated, these inhibitors should be added at this stage.

12. To ensure sufficient counts, 100 µL of reaction mixture is used for ^{125}I-fibrinogen concentrations of 10–100 µg/mL, 50 µL with 200–300 µg/mL and 25 µL with 0.6 mg/mL to 2.5 mg/mL.

13. The pellet should be visible in the tube. The tube should be kept tightly sealed while cutting to prevent loss of liquid from the tube. Cutting should be done behind a lead screen, as the tip can fly off when cut.

14. It is best not to pool data, as there are significant differences in maximum binding between individuals.

15. The platelet concentration is not critical for flow cytometry. Because 10,000 platelets are usually collected, 1×10^5 platelets per tube is usually sufficient.

16. The negative control uses an excess of unlabeled fibrinogen or a GPIIb/IIIa antagonist. The positive control uses TRAP-stimulated platelets. The forward scatter (FSC), side scatter (SSC), and channel 1 mean fluorescence (FL1) parameters are set to a log scale. The FSC and SSC PMT voltages are set so that the platelets appear in the center of the window. The FL1 voltage is set so that the negative events are centered on the first decade of the log scale for the FL1 histogram.

17. As little as 1 mL is all that is required for this assay.

18. The kit recommends using the supplied buffer. However, as this results in a 1-in-4 dilution of the drug concentration, I recommend using PPP from the sample. This can be achieved by spinning the remaining blood in an Eppendorf centrifuge at top speed for 10 min.

19. It is important that all tubes be set up at the same time and the same instrument settings used for each sample. The forward scatter (FSC), side scatter (SSC), and FL1 parameters must be set to a log scale. A sample of unstained, anticoagulated blood can be used to set the parameters. It should be possible to see three populations of cells **(Fig. 2A)**. On the extreme right of the FSC will be a circular population of all the red and white blood cells. In the middle, dominating the dot plot will be a long oval population of platelets. At the left of the FSC scale is a population of microparticles. A region should be drawn around the platelet population. The FL1 histogram will be set up using this region as a gate and 10,000 events from this region collected. The FL1 voltage is adjusted so that the isotype control is in the first decade of the log FL1 histogram. When samples are run there should be a tight marker placed over the positive histogram and the mean (or geometric mean) fluorescence of this peak is noted (*see* **Fig. 2B**).

20. When the beads are being analyzed it is important to identify the singlet population in the FSC-against-SSC dot plot **(Fig. 2C)**. A tight region around this should be drawn and used as a gate for collecting the sample. Four peaks should appear in the FL1 histogram; *see* **Fig. 2D** (if there are more it is probably due to the region including some double beads). A tight marker is drawn over each histogram and the mean fluorescence for each peak noted.

21. The mAb 1 value tends to be a little higher than the mAb 2 value. Also, while abciximab will completely inhibit mAb 1, there appears to be a residual mAb 2 binding that cannot be displaced by tirofiban or eptifibatide. The effect of a drug can be quantified by determining the ratio of mAb 2 to mAb 1 or, in the case of abciximab, the ratio of mAb 1 to mAb 2. Alternatively the binding levels from a pre-treatment and post-treatment sample can be compared.

22. A log FSC against log SSC dot plot is set up and the platelet population identified (a long population of events between the red and white blood cells on the right and microparticles on the left). A region is drawn around this population and this is used as a gate. There are two possible methods for analyzing the fibrinogen bound to platelets—mean fluorescence of the platelets or the percentage of platelets that have fibrinogen bound. Using the sample

with the supplied inhibitory reagent to set the background, the mean fluorescence of each sample is then determined. Alternatively, using the sample containing the supplied inhibitor, a histogram marker is set up from the extreme right of the FL1 range to the edge of the histogram so that only 1% of cells fall into this marker. The samples are then analysed using this marker. Any platelet that falls within the region defined by this marker is considered to be positive for fibrinogen. The percentage of platelets in the sample that fall into this region is then determined from the histogram statistics. These are the positive platelets since platelets in the general circulation should not have any bound fibrinogen.

23. To prepare a standard curve, known numbers of cells can be added to blank wells containing substrate buffer at this stage. The stock platelet suspension can be used. The volume of platelets added to the wells should be no more than 10–15 µL. This allows absorbance values to be converted to number of platelets adhered.

References

1. Parise, L. (1999) Integrin $\alpha_{IIb}\beta_3$ signaling in platelet adhesion and aggregation. *Curr. Opin. Cell. Biol.* **11,** 597–601.
2. Calvete, J. (1999) Platelet integrin GPIIb/IIIa: structure-function correlations. An update and lessons from other integrins. *Proc. Soc. Exp. Biol. Med.* **222,** 29–38.
3. Cox, D. and Seki, J. (1998) Characterization of the binding of FK633 to the platelet fibrinogen receptor. *Thromb. Res.* **91,** 129–136.
4. Marguerie, G., Plow, E., and Edington, T. (1979) Human platelets possess an inducible and saturable receptor specific for fibrinogen. *J. Biol. Chem.* **254,** 5357–5363.
5. Nicholson, N. S., Panzer-Knodle, S. G., Salyers, A. K., Taite, B. B., Szalony, J. A., Haas, N. F., et al. (1995) SC-54684A: an orally active inhibitor of platelet aggregation. *Circulation* **91,** 403–410.
6. Mousa, S., Bozarth, J., Naik, U., and Slee, A. (2001) Platelet GPIIb/IIIa binding characteristics of small molecule RGD mimetic: distinct binding profile for Roxifiban. *Br. J. Pharmacol.* **133,** 331–336.
7. Peerschke, E. I., Zucker, M. B., Grant, R. A., Egan, J. J., and Johnson, M. M. (1980) Correlation between fibrinogen binding to human platelets and platelet aggregability. *Blood* **55,** 841–847.
8. Peerschke, E. I. (1982) Evidence for interaction between platelet fibrinogen receptors. *Blood* **60,** 973–978.
9. Quinn, M., Deering, A., Stewart, M., Cox, D., Foley, B., and Fitzgerald, D. (1999) Quantifying GPIIb/IIIa receptor binding using 2 monoclonal antibodies: discriminating abciximab and small molecular weight antagonists. *Circulation* **99,** 2231–2238.
10. Xia, Z., Wong, T., Liu, Q., Kasirer-Friede, A., Brown, E., and Frojmovic, M. (1996) Optimally functional fluorescein isothiocyanate-labelled fibrinogen for quantitative studies of binding to activated platelets and platelet aggregation. *Br. J. Haematol.* **93,** 204–214.
11. French, D. and Seligsohn, U. (2000) Platelet glycoprotein IIb/IIIa receptors and Glanzmann's thrombasthenia. *Arterioscler. Thromb. Vasc. Biol.* **20,** 607–610.
12. Coller, B. S. (2001) Anti-GPIIb/IIIa drugs: current strategies and future directions. *Thromb. Haemost.* **86,** 427–443.
13. Bosch, X. and Marrugat, J. (2001) Platelet glycoprotein IIb/IIIa blockers for percutaneous coronary revascularization, and unstable angina and non-ST-segment elevation myocardial infarction. *Cochrane Database Syst. Rev.* CD002130.
14. Cox, D. and Fitzgerald, D. (2002) Monitoring antiplatelet therapy, in *Platelets 2000* (Greesele, P., Page, C., Fuster, V., and Vermylen, J., eds.), Cambridge University Press, pp. 471–484.

15. Cox, D., Aoki, T., Seki, J., Motoyama, Y., and Yoshida, K. (1996) Pentamidine is a specific, non-peptide, GPIIb/IIIa antagonist. *Thromb. Haemost.* **75,** 503–509.

16. Ugarova, T. P., Budzynski, A. Z., Shattil, S. J., Ruggeri, Z. M., Ginsberg, M. H., and Plow, E. F. (1993) Conformational changes in fibrinogen elicited by its interaction with platelet membrane glycoprotein GPIIb-IIIa. *J. Biol. Chem.* **268,** 21,080–21,087.

17. Timmons, S. and Hawiger, J. (1989) Isolation of human platelets by albumin gradient and gel filtration. *Methods Enzymol.* **169,** 11–21.

18. Greenwood, F., Hunter, W., and Glover, J. (1963) The preparation of ^{131}I-labelled human growth hormone of high specific radioactivity. *Biochem. J.* **89,** 114–123.

19. Fullard, J., Murphy, R., O'Neill, S., Moran, N., Ottridge, B., and Fitzgerald, D. (2001) A Val193Met mutation in GPIIIa results in a GPIIb/IIIa receptor with a constitutively high affinity for a small ligand. *Br. J. Haematol.* **115,** 131–139.

20. Rooney, M., Farrell, D., van Hemel, B., de Groot, P., and Lord, S. (1998) The contribution of the three hypothesized integrin-binding sites in fibrinogen to platelet-mediated clot retraction. *Blood* **92,** 2374–2381.

21. Cox, D., Aoki, T., Seki, J., Motoyama, Y., and Yoshida, K. (1994) The pharmacology of the integrins. *Med. Res. Rev.* **14,** 195–228.

9

Techniques for Analysis of Proteins by SDS-Polyacrylamide Gel Electrophoresis and Western Blotting

Jonathan M. Gibbins

1. Introduction

The separation of mixtures of proteins by SDS-polyacrylamide gel electrophoresis (SDS-PAGE) is a technique that is widely used—and, indeed, this technique underlies many of the assays and analyses that are described in this book. While SDS-PAGE is routine in many labs, a number of issues require consideration before embarking on it for the first time. We felt, therefore, that in the interest of completeness of this volume, a brief chapter describing the basics of SDS-PAGE would be helpful. Also included in this chapter are protocols for the staining of SDS-PAGE gels to visualize separated proteins, and for the electrotransfer of proteins to a membrane support (Western blotting) to enable immunoblotting, for example. This chapter is intended to complement the chapters in this book that require these techniques to be performed. Therefore, detailed examples of why and when these techniques could be used will not be discussed here.

1.1. Principles of SDS-PAGE

The basic principle of SDS-PAGE is the separation of proteins in accordance with their molecular masses by electrophoretic migration through a polyacrylamide gel. Polyacrylamide gels are formed through the polymerization of acrylamide monomers into long polymers that are cross-linked together by bisacrylamide. This gives rise to a gel matrix, the porosity of which is directly related to the acrylamide concentration. Therefore, as the acrylamide concentration increases the pore size decreases. Poly-acrylamide gels are used as a medium through which proteins are allowed to migrate according to charge. For SDS-PAGE, however, the intrinsic charge of a protein is of little consequence, as proteins gain a net negative charge due to interaction with SDS. Proteins are denatured and dissociated by boiling in SDS, and optionally a reducing agent, and then applied to the sample wells of a gel. Treated proteins are loaded into wells in the top of the gel. When a current is supplied through the gel, the proteins

From: *Methods in Molecular Biology, vol. 273:*
Platelets and Megakaryocytes, Vol. 2: Perspectives and Techniques
Edited by: J. M. Gibbins and M. P. Mahaut-Smith © Humana Press Inc., Totowa, NJ

migrate toward the anode. The relative mobility of each protein is determined by the porosity of the gel that acts as a molecular sieve, and therefore low-molecular-mass proteins migrate more quickly than high-molecular-mass species. The inclusion of molecular mass standards allows the apparent molecular-mass of a protein separated by SDS-PAGE to be determined. As discussed later in this chapter, the selection of an appropriate acrylamide concentration is therefore essential for optimal separation of proteins of interest. This brief description of SDS-PAGE does not do justice to the complexities and variations that are possible. For an excellent description of the principles of gel electrophoresis of proteins, and the many variations that are possible to suite all manner of experimental requirements, readers are directed to **ref. *1***.

1.2. Gel Staining

It is frequently necessary to stain proteins within a polyacrylamide gel so that they may be visualized. There is an enormous number of types of stain available, with the sensitivity of commercial stains continually increasing. With their application in proteomics techniques, numerous highly sensitive fluorescent stains have been developed. The visualization of these, however, requires specialized equipment. In this chapter three basic stains are described that require no specialist equipment, and the chapter also includes a reliable and highly sensitive method for the staining of proteins with silver.

1.3. Western Blotting

A number of protocols described in this book require the transfer of proteins separated by SDS-PAGE onto a membrane support. These include immunoblotting and the development of in vitro kinase assays. Two main methods are usually employed to achieve this: semi-dry and wet Western blotting. This chapter includes detailed protocols for each of these approaches, which we have utilized in the study of platelets and megakaryocytes for a number of years. As with most techniques of this type, a number of variations and refinements are possible to suite specific experiments. For further details on different protocols, and particularly alternative buffer compositions, readers are directed to **ref. *2***.

2. Materials

2.1. Preparation of Gels

1. SDS-PAGE system and electrophoresis power supply. There are systems available from a number of suppliers, including Bio-Rad (Hercules, CA) and Amersham Biosciences (Buckinghamshire, UK).
2. 70% (v/v) ethanol.
3. 30% (w/v) acrylamide: 0.8% (w/v) bisacrylamide solution. Due to the hazards associated with handling acrylamide it is recommended to use a prepared solution (e.g., Protogel, supplied by National Diagnostics, Hull, UK).
4. Resolving gel buffer: 3.0 *M* Tris-HCl, pH 8.8.
5. Stacking gel buffer: 0.5 *M* Tris-HCl, pH 6.8.
6. 10% (w/v) sodium dodecyl sulfate (SDS, Sigma, St. Louis, MO).
7. 1.5% (w/v) ammonium persulfate (Sigma) solution, prepared immediately before use.
8. *N,N,N',N'*-tetramethylethylenediamine (TEMED) (Sigma).

9. Water-saturated butan-1-ol. Prepared by mixing together equal proportions of water and butan-1-ol. Allow for phases to separate and use the upper phase.
10. Running buffer stock solution (10X concentrate): 250 mM Tris-HCl, 1.92 M glycine, 1% (w/v) SDS, pH 8.3. Dilute 10-fold before use. We find that it is cheaper to buy ready-made 10X concentrate (e.g., from National Diagnostics).

2.1.1. Gradient Gels

1. Linear gradient former (available from Bio-Rad and Amersham Biosciences).
2. Peristaltic pump and tubing.
3. Glycerol.

2.2. Sample Preparation, and Loading and Running Gels

1. Laemmli sample treatment buffer (STB, 2X concentrate): 50 mM Tris-HCl, 4% (w/v) SDS, 10% (v/v) 2-mercaptoethanol, 20% (v/v) glycerol, a trace of bromophenol blue (or Coomasie brilliant blue R), pH 6.8.
2. Heating block to heat samples to 100°C, or a boiling water bath with floats to hold microfuge tubes.
3. Prestained molecular mass markers.
4. Gel-loading tips.

2.3. Coomassie Staining

1. Coomassie brilliant blue R stain solution: 50% (v/v) methanol, 10% (v/v) glacial acetic acid, 0.1% (w/v) Coomassie brilliant blue R (Sigma), prepared in deionized water.
2. Destaining solution: 50% (v/v) methanol, 10% (v/v) glacial acetic acid, prepared in deionized water.
3. Staining trays.
4. Coomassie brilliant blue G colloidal stain concentrate: 2% (v/v) phosphoric acid, 15% (w/v) ammonium sulfate, 0.1% (w/v) Coomassie brilliant blue G (Sigma), prepared in deionized water.
5. Methanol.

2.4. Silver Staining

1. Acid alcohol fixing solution: 50% (v/v) ethanol, 10% (v/v) glacial acetic acid, prepared in deionized water.
2. 10% (v/v) glutaraldehyde solution (cross-linking solution).
3. 90 mM NaOH, freshly filtered.
4. 14.8 M ammonia solution.
5. 1.14 M AgNO$_3$, filtered and stored in the dark.
6. 47.6 mM citric acid.
7. 37% formaldehyde solution.
8. Magnetic stirrer.
9. Staining trays.

2.5. Semi-Dry Western Blotting

1. Western blotting electrotransfer cell (e.g., Trans Blot SD, Bio-Rad) and power supply.
2. Whatman 3MM paper (Maidstone, UK) cut to the dimensions of the gel to be blotted.
3. Polyvinylidene difluoride (PVDF) (Bio-Rad, Amersham Biosciences, PALL, VWR international, Poole, UK) or nitrocellulose (Bio-Rad, Amersham Biosciences) blotting membranes.

4. Anode buffer 1: 0.3 *M* Tris-HCl, 20% (v/v) methanol, pH 10.4.
5. Anode buffer 2: 25 m*M* Tris-HCl, 20% (v/v) methanol, pH 10.4.
6. Cathode buffer: 25 m*M* Tris-HCl, 40 m*M* 6-amino-*n*-hexanoic acid, 20% (v/v) methanol, pH 9.4.
7. Large test tube.
8. Methanol.

2.6. Wet Western Blotting

1. Wet Western blotting system (e.g., Mini Trans-Blot Cell, Bio-Rad) and power supply.
2. Large tray for blotter cell assembly submerged in buffer.
3. CAPS transfer buffer: 10 m*M* 3-[cyclohexyamino]-1-propanesulphonic acid (CAPS), 10% (v/v) methanol, pH 11.
4. Polyvinylidene difluoride (PVDF) (Bio-Rad, Amersham Biosciences, Pall) or nitrocellulose (Bio-Rad, Amersham Biosciences) blotting membranes.
5. Methanol.

3. Methods

3.1. Choice and Preparation of Gels

This discontinuous system is suitable for most protein separations. Before embarking on casting a gel, it is important to consider the percentage acrylamide gel to use to provide optimal separation of proteins of interest. For separation of a wide range of proteins, for example a whole cell lysate, a 10% resolving gel will usually be suitable. However, if your focus is on proteins of low molecular mass (e.g., 10–30 kDa), then higher-percentage acrylamide gels will be necessary, such as 12, 15, or 18%. Conversely, the separation of high-molecular-mass proteins will require lower-percentage acrylamide gels, such as 8%. Gels of below 8% acrylamide are very difficult to handle without damage. There are no hard and fast rules concerning the choice of acrylamide concentration, and in many cases this comes down to personal choice. This is, however, an element that may require optimization for some experiments. If the following method does not provide a high enough level of resolution, particularly between lower-molecular-mass proteins, readers are recommended to use the Tris-Tricine SDS-PAGE method as described by Schagger and von Jagow *(3)*.

Acrylamide is a carcinogen and a neurotoxin, and therefore suitable care should be taken in its use. Latex gloves should be worn throughout this process to prevent skin contact with acrylamide (and to prevent contamination of gels and blots with proteins present on the skin), and care must be taken to prevent the formation of aerosols.

1. Ensure that the glass plates, spacers and combs are scrupulously clean. Wash in warm water with detergent and rinse with distilled water and then 70% (v/v) ethanol. Dry thoroughly and assemble following manufacturer's instructions (*see* **Note 1**).
2. In clean beakers, prepare the resolving and stacking gel mixtures according to the proportions in **Table 1**, but do not add the ammonium persulfate or TEMED at this stage. The volumes given in **Table 1** may require adjustment to account for the size format and number of gels that are to be prepared. Care should be taken in the disposal or cleaning of apparatus that has been in contact with nonpolymerized acrylamide.

Table 1
Gel Preparation for Various Percentage Acrylamide Gels

	4% (Stacking gel)		8%		10%		12%		15%		18%	
	Mini	Maxi	Mini	Maxi	Mini	Maxi	Mini	Maxi	Mini	Maxi	Mini	Maxi
Acrylamide:bisacrylamide (30:0.8% [w/v]) (mL)	0.666	3.333	2.667	18.667	3.333	23.333	4.000	28.000	5.000	35.000	6.000	42.000
Resolving gel buffer (mL)	—	—	1.250	8.750	1.250	8.750	1.250	8.750	1.250	8.750	1.250	8.750
Stacking gel buffer (mL)	1.250	6.250	—	—	—	—	—	—	—	—	—	—
10% (w/v) SDS (mL)	0.050	0.250	0.100	0.700	0.100	0.700	0.100	0.700	0.100	0.700	0.100	0.700
Water (mL)	2.780	13.897	5.478	38.348	4.812	33.682	4.145	29.015	3.145	22.015	2.145	15.015
1.5% (w/v) ammonium persulfate (mL)	0.250	1.250	0.500	3.500	0.500	3.500	0.500	3.500	0.500	3.500	0.500	3.500
TEMED (mL)	0.004	0.020	0.005	0.035	0.005	0.035	0.005	0.035	0.005	0.035	0.005	0.035

Quantities are given that are sufficient to prepare two mini or two maxi gels. Mini gel dimensions approx $80 \times 60 \times 0.75$ mm. Maxi gel dimensions approx $160 \times 120 \times 1.5$ mm. Resolving gel buffer: 3.0 M Tris-HCl (pH 8.8). Stacking gel buffer: 0.5 M Tris-HCl (pH 6.8).

3. To determine the depth of stacking and resolving gels to be cast, ensure that the stacking gel is at least twice as deep as the wells themselves. As a guide, place the sample well comb in the top of assembled plates, and mark one of the glass plates with marker pen in the appropriate position. This is the level to fill to, when pouring the resolving gel. The casting stand should be checked using a spirit level to ensure that it is exactly level and adjusted if necessary.

4. Add the ammonium persulfate and TEMED to the resolving gel mixture and swirl gently to mix. This ensures even polymerization, but be careful not to introduce air bubbles. Transfer the gel mixture to the prepared glass plates on the casting stand, using a Pasteur pipet. Fill to just above the level marked on the glass (to allow for slight shrinkage of the gel on polymerization) (*see* **Note 2**).

5. Gently layer over the acrylamide some water-saturated butan-1-ol (about 2 to 3 mm depth). This prevents the gel from drying out after polymerization and also ensures that the gel is exactly flat at the edges. Allow to polymerize at room temperature for about 30 min (*see* **Note 3**).

6. Decant the butan-1-ol and wash the inside of the plates and the top of the gel with a large volume of deionized water (a wash bottle may make this easier), until the smell of butan-1-ol can no longer be detected. Dry the insides of the plates with filter paper down to the level of the gel, taking care not to touch the gel.

7. Add the appropriate volumes of ammonium persulfate and TEMED to the stacking gel mixture, swirl to mix thoroughly, and fill the gel assembly to the top. Using a Pasteur pipet, remove any air bubbles.

8. Lower the sample well comb into the stacking gel, slanting at an angle to avoid the trapping of any air. As the comb is lowered, the gel will begin to overflow. This may be collected and used to top up the gel when the comb is fully in place.

9. When the stacking gel has polymerized, gently remove the comb. Rinse the top of the gel with deionized water and shake the gel to remove this prior to assembly into the running tank. If well dividers are not straight, they may be repositioned using a gel-loading tip (*see* **Note 4**).

10. Add the running buffer to the upper and lower reservoirs, ensuring that any bubbles that become lodged under the gel are displaced. Follow the apparatus manufacturer's instructions for the appropriate volumes of buffer to be used. If you are running large-format gels, the tank will probably incorporate a water cooling system. Ensure that this is filled and remains turned on.

3.1.1. Gradient Gels

On occasions where a sample contains proteins of wide-ranging molecular mass, yet a high degree of resolution is required throughout the gel, the use of a gradient gel should be considered. With such gels, the percentage acrylamide increases as proteins migrate from the top of the gel to the bottom. Low-percentage acrylamide gel at the top enables resolution of high-molecular-mass proteins, and higher-percentage gel at the bottom ensures resolution of low-molecular-mass species. The gradient is formed from two acrylamide solutions as the gel is poured using a linear gradient former and a peristaltic pump. The choice of gradient depends on the range of sizes of proteins to be separated. We routinely use 10 to 18% gradients to separate proteins in whole-platelet lysates.

Table 2
Gel Preparation for 10 to 18% Gradient Gel

	4% (Stacking gel)	10%	18%
Acrylamide : bisacrylamide (30 : 0.8% [w/v]) (mL)	1.667	5.833	10.500
Resolving gel buffer (mL)	—	2.190	2.190
Stacking gel buffer (mL)	3.125	—	—
10% (w/v) SDS (mL)	0.125	0.175	0.175
Water (mL)	6.948	8.200	2.753
Glycerol (mL)	—	0.200	1.000
1.5% (w/v) ammonium persulfate (mL)	0.625	0.875	0.875
TEMED (mL)	0.010	0.007	0.007

Quantities are given that are sufficient to prepare maxi gel ($160 \times 120 \times 1.5$ mm). Resolving gel buffer: 3.0 M Tris-HCl (pH 8.8). Stacking gel buffer: 0.5 M Tris-HCl (pH 6.8).

1. Prepare the gel mixtures as presented in **Table 2**, but do not add the ammonium persulfate or TEMED. This table indicates quantities required for one Maxi gel using the Bio-Rad Protean II xi system, with gel dimensions of approx $160 \times 120 \times 1.5$ mm. If using different dimension gels, calculate the required quantity for the resolving gel and divide by 2 to give the volume of each of the two gels to be prepared. Note that the gels contain different quantities of glycerol. If different percentage gels are required use the same proportion of glycerol in the high- and low-percentage acrylamide mixes and adjust the water volume accordingly.
2. Prepare glass plates and spacers as described in **Subheading 3.1.** and place the whole apparatus, including the prepared acrylamide solutions and the gradient former, into a cold room at 4°C. This slows down polymerization and thereby prevents premature polymerization before the pouring is complete.
3. Using peristaltic pump tubing (ID of around 1 mm), connect the pump to the gradient former. To the other end of the tubing attach a plastic pipet tip and prise between the glass plates of the gel apparatus in a central position. The pump should be adjusted prior to use with the chosen tubing to allow a flow rate of about 5 mL/min.
4. Add the ammonium persulfate solution and TEMED to the two acrylamide solutions and mix. Place the higher-percentage gel in the chamber of the gradient former from which the pump draws, and the lower-percentage mixture in the other chamber. If the gradient former has a tap between the two chambers, open this to allow flow between the chambers and turn on the pump.
5. When all of the acrylamide solution has been poured, layer water-saturated butan-1-ol over the gel and allow to polymerize. Note that this will take considerably longer than at room temperature. Cast the stacking gel and run as described below.

3.2. Sample Preparation, and Loading and Running Gels

1. Samples should be treated with an equal volume of Laemmli 2X sample treatment buffer (STB). If adding to a solid sample (e.g., following immunoprecipitation, or a precipitated protein), the buffer should be diluted twofold with water. If added to megakaryocytes or

cell lines directly, it will be necessary to shear the DNA by repeated drawing in and out of a 1-mL syringe using a narrow-gauge hypodermic needle. If the sample turns yellow, it is too acidic. This may be adjusted by the addition of a very small volume of 1 *M* NaOH until a blue color is restored. The volume to add will depend on the volume that can be added to each well of the gel system that is utilized (*see* **Note 5**).

2. Heat the samples to 100°C for 5 min, using either a heating block or a boiling water bath. This ensures that proteins are effectively coated in SDS and that even relatively insoluble proteins are dissolved (*see* **Note 6**). Stained or unstained molecular-mass markers should be treated in an identical manner to the samples.
3. Centrifuge tubes to ensure that the entire sample may be collected using a pipet fitted with a gel-loading tip.
4. Pipet the samples into the bottom of the wells of the stacking gel (the sample will fall to the bottom of the well due to the inclusion of glycerol in the STB). For even running, try to keep the volume in each well the same, and fill unused lanes with an equivalent volume of 1X STB. Avoid spillover into the next lane or excessive mixing of the sample with the tank buffer (*see* **Note 7**).
5. Attach the tank to a power supply. Select the appropriate current and voltage and turn on. As an SDS-PAGE gel runs the electrical resistance will increase, and therefore the current will change relative to voltage. Therefore for even separation of proteins without the gel slowing down as time proceeds, it is recommended to use a power supply designed for gel electrophoresis where the current or voltage may be fixed. The current or voltage selected will depend on the dimensions of the gel and whether the two gels in one tank are in series or parallel. It is recommended to refer to the manufacturer's instructions for guidance on this (*see* **Notes 8** and **9**).
6. When the blue dye in the STB has reached the bottom of the gel, or has just eluted, turn off the power supply. It is important to proceed immediately to the next part of the experiment (e.g., staining or Western blotting) to avoid the diffusion of proteins. This would be more noticeable for low-molecular-mass proteins where lots of resolution may be noticed. If you are unready to continue, maintain very low power to the gel until ready (*see* **Note 10**).

3.3. Coomassie Staining

3.3.1. Coomassie Brilliant Blue R Stain

This stain is suitable for most purposes. For observation of very low quantities of protein, Coomassie brilliant blue G colloidal stain or silver stain is recommended (*see* below).

1. Pry apart the glass plates using one of the spacers and cut away the stacking gel from the resolving gel using the edge of one of the glass plates.
2. Handling only at the edges, lift the gel from the glass plate and submerge in the stain solution. Incubate with gentle rocking at room temperature for around 2 h, or preferably overnight.
3. Transfer to destain solution and mix. Replace the destain solution with fresh reagent when it becomes difficult to see through, until the background is as clear as possible and protein bands are visible.

3.3.2. Coomassie Brilliant Blue G Colloidal Stain

This stain is reported to be of similar sensitivity to silver stain. In our experience it does not quite live up to this expectation; however, it is very simple and does not require destaining.

1. Prepare stain prior to use by mixing four parts Coomassie brilliant blue G colloidal concentrate with one part methanol.
2. Submerge the gel in the stain and gently mix at room temperature for several hours (at least overnight). With this stain, greater sensitivity is achieved by incubation for longer periods (up to days if necessary). For extended incubation, wrap the tray in aluminium foil to minimize the evaporation of the methanol.
3. Destaining is not necessary; however, background coloration can be removed by extended rinsing in deionized water.

3.4. Silver Staining

A number of kits are available for silver staining of protein gels. Our experience with these is not particularly good. While probably more time-consuming, the following technique that was originally described by Giulan et al. *(4)* produces reliable results if the method is followed closely and fresh reagents are prepared. This method is described for the staining of one minigel and should be scaled up as appropriate for more or larger gels.

1. Transfer SDS-PAGE gel to acid-alcohol fixing solution and agitate gently for 30 min.
2. Transfer the gel to 10% (v/v) glutaraldehyde cross-linking solution and incubate at room temperature for 30 min. If you need to subsequently extract the protein from the gel for any reason, this step must be omitted. Decant the glutaraldehyde and wash in many changes of deionized water until the smell of glutaraldehyde can no longer be detected.
3. Preparation of the stain (this should be performed in a scrupulously clean 50-mL beaker, stirred continuously using a magnetic stirrer): Add 420 µL 14.8 M ammonia solution. Add 1.2 mL 1.14 M AgNO$_3$ dropwise, ensuring that the precipitate that forms on addition returns into solution before addition of the next drop (if the precipitate does not disappear, add more ammonia solution dropwise until the solution is clear). Add 22 mL deionized water. The stain is now ready for incubation with the gel.
4. Preparation of developer (100 mL—sufficient for the development of one minigel): To 500 µL 47.6 mM citric acid solution add 50 µL 37% (v/v) formaldehyde solution and make up to 100 mL with deionized water.
5. Transfer the gel to the ammonical silver stain solution and agitate vigorously (200 s for 0.75-mm-thick gel and 300 s for a 1-mm-thick gel).
6. Wash gels twice for 1 min in a large volume of deionized water and transfer to about 30 mL developer in a fresh container. Agitate vigorously. As soon as bands begin to appear, discard the stain and replace with the remainder of the fresh stain.
7. When sufficient development of bands has appeared, terminate the development by equilibration of gel in deionized water (*see* **Note 11**).

3.5. Semi-Dry Western Blotting

A sample setup for a semi-dry Western blotting system is shown in **Fig. 1A**. The following transfer system involves the use of filter papers soaked in three different buffer solutions. We find this very effective for the Western blotting of most proteins, and it works well for samples that contain proteins of a wide range of molecular masses. If a good level of transfer of a specific protein is not achieved, then alternative buffer systems maybe tried (*see* **ref. 2** for excellent suggestions for this).

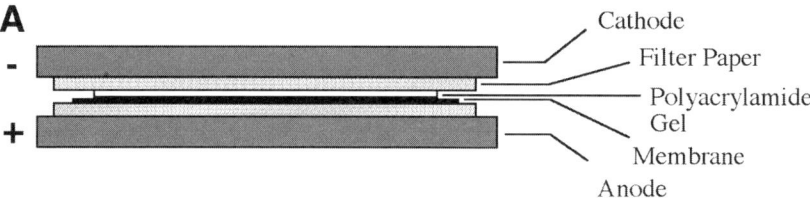

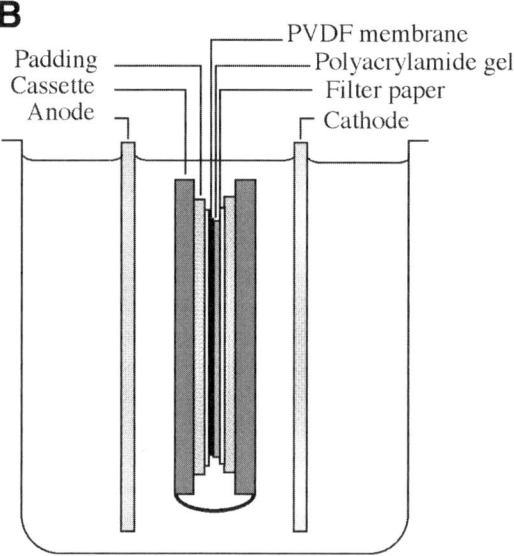

Fig. 1. Schematic representation of typical apparatus for Western blotting. (**A**) Semi-dry blotting is performed in a transfer cell comprising two large plate electrodes. Buffer-soaked filter papers are placed on the anode, followed by a PVDF or nitrocellulose membrane, the SDS-PAGE gel, and more pieces of soaked filter paper. Following the placement of the cathode, electrophoretic transfer is performed by supplying power to the unit. Proteins, which are coated in negatively charged SDS, migrate toward the anode and are captured on the membrane. (**B**) The same principle applies to wet Western blotting, except that the sandwich of filter papers, gel, and membrane are submerged in a large tank of buffer between the anode and the cathode, and power supplied.

Note that latex gloves should be worn throughout this procedure, even while cutting filter papers, to prevent contamination of blots with proteins from the skin. In addition, all trays and implements must be scrupulously clean.

1. Soak four pieces of Whatman 3MM paper in anode buffer 1, two pieces in anode buffer 2, and six pieces in cathode buffer. Soak PVDF membrane in analytical-grade methanol, or alternatively, if transferring proteins to a nitrocellulose membrane, soak this in anode buffer 2 (*see* **Notes 12** and **13**).
2. Clean the electroblotter electrode surfaces with deionized water.

3. Place the anode buffer 1-soaked paper on the anode of the blotter (on most systems this is the lower plate [e.g., the Bio-Rad Trans-Blot SD], but this must be checked) and roll out any trapped air bubbles using a large test tube. Some buffer will be released from the filter papers, but this does not matter.

4. Place the two sheets of anode buffer 2-soaked filter paper on top of the other sheets, and again roll flat.

5. Handling around the edges only, place the PVDF membrane on top of the filter papers and expel any air that is trapped beneath it using the test tube.

6. Pry apart the glass plates to release the polyacrylamide gel. Handling only around the edges (preferably at the end where the molecular mass markers are), pick up the gel and lower onto the PVDF membrane. Do this slowly and avoid lifting the gel on and off the membrane until the correct position is found. If required, the surface of the membrane may be flooded with cathode buffer, which will allow the gel to be gently slid into position. Remove any trapped air bubbles below the gel by rolling with the test tube, taking great care not to put too much pressure on the gel and cause stretching or tearing (*see* **Note 14**).

7. Place the 6 pieces of paper soaked in cathode buffer over the gel, aligned exactly over the rest of the layers, and roll out any air bubbles. Do not use too much pressure or the gel will become damaged or slide around. Mop up any excess buffer from around the gel sandwich.

8. Assemble the upper electrode (cathode) and plug into power supply. Apply 15 V constant voltage for 2 h (*see* **Notes 15** and **16**).

9. Disassemble the apparatus. It is recommended to Coomassie stain the gel following transfer (*see* **Subheading 3.3.1.**) to ensure that the transfer was successful and even (if using prestained marker proteins, these should be completely transferred to the membrane). In addition, it is useful to cut a corner from the blot so that its orientation can be determined in the future. Remember that the proteins isolated on a PVDF membrane will be on the surface that was in contact with the gel.

10. Proceed as required for the specific experiment described elsewhere—for example, blocking to enable subsequent immunoblotting or radioactivity detection.

3.6. Wet Western Blotting

As the name suggests, this system utilizes considerably more buffer, although transfer may be achieved more quickly than with semi-dry blotting. The following protocol is described as used with the Bio-Rad Mini Trans-Blot Cell (*see* **Fig. 1B**). Subtle differences are to be found with other systems, but the principles are identical.

1. Remove the SDS-PAGE gel from its glass plates and place in CAPS transfer buffer for 20 min to equilibrate.

2. In a tray, open the transfer cassette and place a piece of Whatman number 1 paper on each piece of padding (cut to the same size as the padding). Fill the tray with CAPS buffer to the level of the padding.

3. Cut some PVDF membrane the size of the gel to be transferred and soak in methanol before placing in CAPS buffer for 5 min to equilibrate.

4. Place the gel on the filter paper on the cathode side of the cassette (gray on this system), ensuring that no air bubbles are trapped beneath it. The gel should be covered in CAPS buffer to prevent drying out.

5. Position the PVDF membrane over the gel, ensuring that no air bubbles become trapped, followed by the remaining piece of filter paper. Close the cassette, taking care that the cassette is completely soaked in buffer, and transfer to the running tank.

6. Apply 200 mA of constant current for 20 to 30 min (*see* **Note 17**).
7. Disassemble the apparatus. It is recommended to Coomassie stain the gel following transfer (*see* **Subheading 3.3.1.**) to ensure that the transfer was successful and even (if using prestained marker proteins, these should be completely transferred to the membrane). In addition, it is useful to cut a corner from the blot so that its orientation can be determined in the future. Remember that the proteins isolated on a PVDF membrane will be on the surface that was in contact with the gel.

4. Notes

1. Check that the bottom edges of the glass plates are not cracked or chipped, and ensure that the bottom edge of the spacers are exactly flush with the glass plates. This will prevent leakage of the acrylamide solution during gel casting, since a perfect seal will be created along the bottom of the plates in the casting apparatus.
2. Buffer may be degassed in a vacuum to minimize the occurrence of bubbles in the solutions.
3. Avoid moving or tipping the apparatus to check for polymerization. This should be apparent from checking any remainder acrylamide solution in the beaker and also by observing the top of the gel. An interface will appear where a small amount of buffer is excluded from the top of the polymerized gel. At this stage allow another 5 to 10 min to ensure that polymerization is complete.
4. Cast gels may be stored overnight at 4°C wrapped in cling film with a layer of water or gel buffer placed in the wells. Longer storage is not recommended.
5. The sample treatment buffer described includes a reducing agent (mercaptoethanol, although this maybe replaced with dithiothreitol) to ensure complete protein denaturaton through disruption of disulfide bonds. In some experiments it may be preferable to run a gel in the absence of a reducing agent (for example, immunoblotting of some proteins will work only on nonreduced samples), where the reducing agent should simply be omitted from the buffer and replaced with water.
6. Samples are ideally heated in microfuge tubes. Ensure that a cap-lock is used or the lid is pierced to prevent the lids from popping off and the sample being lost.
7. While loading sample, keep the tip of the pipette below the level of the sample at the bottom of the well. This reduces swirling and mixing of the sample into the tank buffer.
8. It is advisable to run the first part of the gel more slowly at a lower current and voltage until all of the samples are focused in the stacking gel.
9. Checking the progress of the gel is achieved through looking at the progress of stained molecular mass markers and the blue solvent front. If the solvent front is not straight, the gel is too hot and the voltage and current should be reduced. Uneven running in small hotspots may be caused by a very high concentration of a single protein. Again, results may be improved by reducing the power. It is possible to run quite severely overloaded gels, but this should be done very slowly.
10. If using large-format gels, large volumes of tank buffer may have been used. It is possible to reuse tank buffer once or twice.
11. Because termination of staining is by equilibration in water, this does not occur immediately. The addition of water shortly before optimal development has occurred will reduce the risk of overstained gels with high levels of background staining. This will probably require practice.
12. Following the wetting of PVDF membrane in methanol, the membrane may also be pre-equilibrated in anode buffer 2. We have found, however, that this step is not necessary.

13. We routinely use PVDF membrane for Western blotting since it is easier to handle and has a greater protein binding capacity. For most applications for Western-blotted proteins, nitrocellulose may also be used. PVDF is a hydrophobic substance that traps proteins on its surface due to hydrophobic interactions. Proteins may pass through nitrocellulose during blotting, and therefore far more care is required to ensure that electrotransfer is not performed for too long. This may become an important issue if one protein in a mixture migrates from the gel very slowly. Care in the handling of nitrocellulose is essential, because it is explosive. If proteins are to be visualized using fluorescent stains, PVDF must be used, since nitrocellulose must not be placed near a UV light source.

14. If a gel does tear during assembly of the blotting procedure, then it may be pieced back together providing it will lie exactly flat. Transfer should not be too adversely affected.

15. Some researchers recommend calculating the voltage to be applied based on the surface area of the gel or gels that are transferred. We have found that this is not necessary when using PDVF and 15 V for 2 h is suitable for all our uses. Note that even if several gels are to be transferred together in one blotter, 15 V is still fine.

16. If the power supply has a timer facility, blotting may be performed at the end of the day for 2 h and be left until the following morning before disassembly.

17. For more lengthy transfer, cooling of the tank may be necessary.

References

1. Hames, B. D. (ed.) (1998) *Gel electrophoresis of proteins—A practical approach (3rd Ed.).* Oxford University Press, Oxford, UK.
2. Dunbar, B. S. (ed.) (1994) *Protein Blotting—A practical approach.* Oxford University Press, Oxford, UK.
3. Schagger, H. and von Jagow, G. (1987) Tricine-sodium dodecyl sulphate-polyacrylamide gel electrophoresis for the separation of proteins in the range from 1 to 100 kDa. *Anal. Biochem.* **166,** 368–379.
4. Guilan, G. G., Moss, R. L., and Greaser, M. (1983) Improved methodology for analysis and quantitation of proteins on one-dimensional silver-stained slab gels. *Anal. Biochem.* **129,** 277–287.

10

Study of Tyrosine Kinases and Protein Tyrosine Phosphorylation

Jonathan M. Gibbins

1. Introduction

In recent years, our increased understanding of the complex signal transduction mechanisms that regulate cellular function has fueled huge advances in all aspects of biomedical science and cell biology. Platelet and megakaryocyte function is no exception to this. In the last 10 yr our understanding of the receptor biochemistry and the systems that they control has been pivotal in the development of new strategies to inhibit platelet function and thereby prevent thrombosis. Experimental techniques have become more and more elegant, however; the basic toolbox that a researcher requires to study signaling in platelets and megakaryoctes is described in this and several subsequent chapters.

What do we mean by signal transduction, or cell signaling? I think of it as a jigsaw puzzle on which is pictured a detailed diagram of the engine that drives all aspects of cellular function. The catch is that the pieces are so small that you cannot see the picture or the even shape of the pieces. The challenge is determine how all the pieces fit and function together to orchestrate an appropriate cellular response to a given stimulus.

Cell signaling is not platelet- or megakaryocyte-specific, and many of the techniques that are described in this and the following chapters may be applied to any cell type or tissue. I have, however, endeavored to include a few tips and tricks that we have picked up along the way, working with these cells.

A central feature of many aspects of signaling is protein tyrosine phosphorylation. The phosphorylation of specific tyrosine residues in an array of signaling proteins acts as a molecular switch, regulating critical protein–protein interactions and enzymatic activity. Indeed, tyrosine phosphorylation is often involved in the regulation of tyrosine kinases themselves, the enzymes that are responsible for protein tyrosine phosphorylation. In recent years the work of many researchers has established that tyrosine phosphorylation is an essential aspect of many platelet and megakaryocyte

From: *Methods in Molecular Biology, vol. 273:*
Platelets and Megakaryocytes, Vol. 2: Perspectives and Techniques
Edited by: J. M. Gibbins and M. P. Mahaut-Smith © Humana Press Inc., Totowa, NJ

signaling pathways. A range of both activatory and inhibitory platelet receptors have been shown to become tyrosine phosphorylated on ligand binding or cell activation, such as the GPVI-Fc receptor γ-chain complex collagen receptor *(1–4)*, FcγRIIA *(5,6)*, integrin $\alpha_{IIb}\beta_3$ *(7)*, and platelet endothelial cell adhesion molecule-1 *(8–10)*. The tyrosine kinases responsible for phosphorylation of some of these receptors has been identified as members of the Src-family of kinases, such as Fyn, Lyn, Yes, Src, and Lck *(9,11,12)*. A number of tyrosine kinases have been implicated as acting further downstream in the stimulation of platelet aggregation, including Bruton's tyrosine kinase (BTK) *(13)* and focal adhesion kinase (FAK) *(14,15)*. Phosphorylation of receptors such as the GPVI-Fc receptor γ-chain complex results in the recruitment and/or phosphorylation of a range of molecules such as adaptor proteins (e.g., linker for activation of T-cells [LAT] *[16]*, Vav *[17]*, Grb2 *[6]*, SLP-76 *[18]*) and other enzymes such as phospholipase Cγ2 *(18,19)* and phosphoinositide 3-kinase *(16,20)*, leading to the activation of multiple signaling cascades and ultimately platelet activation. Tyrosine kinases have also been shown to be involved in the response of platelets to G protein-coupled receptor agonists such as thrombin *(21)*. Many of the receptor proximal tyrosine-kinase-dependent signaling events in these cells are becoming apparent, but the role of tyrosine phosphorylation further downstream—for example, in the regulation of integrins and secretion—are far less clear.

This chapter is divided into several sections, which describe different techniques that are routinely used to characterize tyrosine-kinase-dependent signaling in platelets, megakaryocytes, and other cell types. Methods are described to measure the level of protein tyrosine phosphorylation of a cell extract or specific protein, the measurement of tyrosine kinase activity in vitro, the analysis of tyrosine-phosphorylation-dependent protein–protein interaction, and considerations for the experimental use of inhibitors of specific tyrosine kinases.

By the very nature of this area of research, no description of this type could be exhaustive. It is hoped, however, that this will provide the newcomer to cell signaling in platelets and megakaryocytes with some of the essential basics.

2. Materials

2.1. Preparation of Platelets, Megakaryocytes, and Cell Lines

1. 4% (w/v) sodium citrate solution.
2. Acid-citrate-dextrose (ACD): 85 mM sodium citrate, 70 mM citric acid, and 110 mM glucose.
3. Modified Tyrode's-HEPES buffer: 134 mM NaCl, 0.34 mM Na$_2$HPO$_4$, 2.9 mM KCl, 12 mM NaHCO$_3$, 20 mM HEPES, 5 mM glucose, and 1 mM MgCl$_2$, pH 7.3.
4. Prostacyclin (PGI$_2$) Prostaglandin E$_1$ (PGE$_1$). Stock solution of 125 µg/mL dissolved in ethanol.
5. Ethylene glycol-bis(β-aminoethyl ethylether)-*N,N,N′,N′*-tetraacetic acid (EGTA). Prepare 100 mM stock in modified Tyrode's-HEPES buffer and dilute 100-fold into platelet suspension *(optional)*.
6. Indomethacin: prepare 10 mM stock dissolved in dimethyl sulfoxide and dilute 1000-fold into platelet suspension *(optional)*.
7. Apyrase: prepare 2000 U/mL stock solution prepared in modified Tyrode's-HEPES buffer. Dilute 1000-fold into platelet suspension *(optional)*.

2.2. Measurement of Protein Tyrosine Phosphorylation

1. Platelet aggregometer.
2. Reagents and equipment to cast and run SDS-PAGE gels (refer to Chapter 9, vol. 2).
3. Reagents and equipment to electrotransfer proteins from an SDS-PAGE to a membrane support (Western blotting; refer to Chapter 9).
4. Polyvinylidene difluoride blotting membrane or nitrocellulose.
5. Bovine serum albumin (BSA).
6. Tris-buffered saline containing 0.1% (v/v) Tween-20 (TBST): 20 mM Tris-HCl, 137 mM NaCl, 0.1% (v/v) Tween-20, pH 7.6.
7. Sodium azide.
8. Anti-phosphotyrosine antibody; e.g., mAb4G10 (Upstate Biotechnology, Milton Keynes, UK) or PY20 (several suppliers).
9. Thermal food-bag sealer and polyethylene sheets.
10. Horseradish peroxidase-conjugated anti-mouse IgG secondary antibody (e.g., Cat. no. NA931, Amersham Biosciences, Buckinghamshire, UK).
11. Enhanced chemiluminescence (ECL) detection reagents (several are on the market, such as ECL—Amersham Biosciences).
12. X-ray film or film specifically designed for ECL detection, such as Hyperfilm ECL (Amersham Biosciences).
13. X-ray film developer (this is preferable to developing by hand, due to the fixed time for development).
14. X-ray film scanning system and analysis software.
15. Camera-based system for the detection of ECL signals and analysis software *(optional)*.

2.3. Measurement of Kinase Activity In Vitro

1. Modified Tyrode's-HEPES buffer: 134 mM NaCl, 0.34 mM Na$_2$HPO$_4$, 2.9 mM KCl, 12 mM NaHCO$_3$, 20 mM HEPES, 5 mM glucose, and 1 mM MgCl$_2$, pH 7.3.
2. Ethylene glycol-bis(β-aminoethyl ethylether)-N,N,N',N'-tetraacetic acid (EGTA). Prepare 100 mM stock in modified Tyrode's-HEPES buffer and dilute 100-fold into platelet suspension.
3. Indomethacin (Sigma): prepare 10 mM stock dissolved in dimethyl sulfoxide and dilute 1000-fold into platelet suspension *(optional)*.
4. Apyrase (Sigma): prepare 2000 U/mL stock solution prepared in modified Tyrode's-HEPES buffer. Dilute 1000-fold into platelet suspension *(optional)*.
5. 2X NP40 lysis buffer with protease and phosphatase inhibitors: 2% (v/v) NP40 (IGEPAL CA-630, Sigma), 300 mM NaCl, 10 mM EDTA, 20 mM Tris-HCl, pH 7.3, 1 mM PMSF, 4 mM NaF, 2 mM Na$_3$VO$_4$, 10 µg/mL aprotinin, 10 µg/mL leupeptin, and 1 µg/mL pepstatin A.
6. Protein A- or protein G-sepharose (Sigma, St. Louis, MO).
7. Antibody to kinase of interest (preferably monoclonal).
8. Tris-buffered saline containing 0.1% (v/v/) Tween-20 (TBST): 20 mM Tris-HCl, 137 mM NaCl, 0.1% (v/v) Tween-20, pH 7.6.
9. Kinase-assay buffer: 105 mM NaCl, 20 mM HEPES, pH 7.4, 5 mM MnCl$_2$, 5 mM MgCl$_2$, 2 mM NaF, 1 mM Na$_3$VO$_4$, 10 mM ATP, and 200 µCi/mL of [γ-^{32}P] ATP.
10. Kinase substrate, if required.
11. 50 mM ethylenediaminetetraacetic acid (EDTA).
12. Laemmli SDS-PAGE sample treatment buffer (refer to Chapter 9, vol. 2).

13. Reagents and equipment to cast and run SDS-PAGE gels (refer to Chapter 9, vol. 2).
14. Reagents and equipment to electrotransfer proteins from an SDS-PAGE to a membrane support (Western blotting: refer to Chapter 9, vol. 2).
15. Polyvinylidene difluoride blotting membrane or nitrocellulose.
16. Bovine serum albumin (BSA).
17. Perspex screen suitable for use with ^{32}P.
18. Autoradiography film cassette, X-ray film, and film development system or storage phosphor screen and phosphorimager.

2.4. Study of Tyrosine Phosphorylation-Induced Protein–Protein Interactions

The following reagents will be required, in addition to those itemized above, for in vitro kinase assays:

1. Antibodies specific for the proteins of interest.
2. Immunoblot stripping buffer: TBST containing 2% (w/v) SDS and 5% (v/v) 2-mercaptoethanol.

2.5. Use of Tyrosine Kinase Inhibitors

Inhibitors of a wide range of signaling enzymes are available from a range of suppliers such as Calbiochem (Nottingham, UK), Sigma (Poole, UK), Biomol (Plymouth Meeting, PA), Tocris (Avonmouth, UK), and Alexis (Nottingham, UK). Pharmaceutical companies are often able to supply specific inhibitors for research purposes.

3. Methods

3.1. Preparation of Platelets, Megakaryocytes, and Cell Lines

Megakaryocytes and cell lines may be suspended in a range of physiological buffers for experimentation. For ease of comparison with platelet experiments, one recommendation would be to use modified Tyrode's-HEPES buffer. Every platelet lab has its favorite method for the preparation of platelets, with each method having advantages and disadvantages. The critical factor when examining signaling in these cells is to ensure that they are maintained in their resting state throughout preparation, since platelets are exquisitely sensitive to the stimulation of tyrosine phosphorylation, and our techniques for detecting tyrosine phosphorylation are also very sensitive. I will briefly describe the system that we have successfully used for a number of years for this type of analysis using human platelets (a scaled-down version of this protocol has also been used with mouse platelets). Detailed methods for the preparation of platelets are, however, described in Chapters 1 and 2, vol. 1, and readers are encouraged to refer to that chapter for further discussion on this matter.

1. Draw blood into a 50-mL syringe containing 2 mL sodium citrate (4% [w/v]) using a 21 g butterfly needle.
2. To 50 mL blood add 7.5 mL ACD and centrifuge at 200g for 20 min at room temperature.
3. Collect platelet-rich plasma into a 50-mL tube and add 50 ng/mL PGI$_2$ or PGE$_1$ immediately before centrifugation at 1000g for 10 min at room temperature.

4. Remove plasma with a pipet and resuspend platelets in 1 mL modified Tyrode's-HEPES buffer. Dilute further to 25 mL with modified Tyrode's-HEPES buffer; add 3 mL ACD and 50 ng/mL PGI$_2$ or PGE$_1$. A sample may be taken here to estimate platelet count (refer to Chapter 3, vol. 1). Centrifuge at 1000g for 10 min at room temperature and decant buffer.

5. Immediately resuspend the platelets in 1 mL modified Tyrode's-HEPES buffer and dilute to the desired cell density in the same buffer. Rest platelets for at least 30 min at 30°C prior to use.

6. For several of the experiments below it may be desirable to prevent aggregation—for example, to ensure equivalent levels of protein immunoprecipitated from resting and activated platelets, or to prevent secondary signaling stimulated by aggregation. In these cases, 1 mM EGTA may be added to the suspension for this purpose. Other secondary signaling effects may be inhibited by the addition of inhibitors to the platelet suspension such as 10 µM indomethacin (cyclooxygenase inhibitor—to prevent thromboxane A$_2$ production) and 2 U/mL apyrase (rapidly metabolizes secreted ADP).

3.2. Measurement of Protein Tyrosine Phosphorylation

This technique involves the immunodetection of tyrosine phosphorylated proteins that have been separated by SDS-polyacrylamide gel electrophoresis (SDS-PAGE) and transferred onto a membrane support by Western blotting. The technique described utilizes an enhanced chemiluminescence (ECL) detection system that is recorded on blue-sensitive X-ray film. The methods may be easily adapted to allow fluorescence detection techniques (which allow more accurate quantitation) or colorimetric detection. This approach may be applied to whole cell extracts or specific isolated proteins.

1. Prepare cells as described in **Subheading 3.1.** If required, stimulate with agonist, preferably while stirred in a platelet aggregometer. Terminate stimulation with an equal volume of 2X Laemmli SDS-PAGE sample treatment buffer and heat to 100°C for 5 min. For isolated proteins, add sample treatment buffer directly and heat as above. Separate proteins on SDS-PAGE gel of choice (for further details of SDS-PAGE techniques please refer to Chapter 9, vol. 2). The amount of protein to add to each lane of the gel may require optimization, and in some cases it is advisable to assay the protein content of samples prior to loading a gel to ensure that equal amounts of protein are added to each lane. On minigels, proteins from 8×10^6 platelets or 8×10^3 megakaryocytes is a recommended starting point for anti-phosphotyrosine immunoblots, although considerably greater levels of protein may be loaded, particularly if running large-format gels.

2. Electrotransfer (Western blot) the entire gel onto a membrane support. We recommend polyvinylidene difluoride (PVDF) for this purpose. (For further details of Western blotting techniques, please refer to Chapter 9, vol. 2.)

3. On completion of the transfer, mark the positions of stained molecular mass markers with a permanent marker pen, and place the membrane into a 10% (w/v) solution of bovine serum albumin (BSA) in Tris-buffer saline solution with Tween-20 (TBST) to block remaining protein binding surface on the membrane. Incubate with gentle rocking for 30–60 min at room temperature (*see* **Note 1**).

4. Transfer the blot to a tray containing fresh TBST and rinse with agitation for 5 min.

5. Prepare a 1 µg/mL solution of anti-phosphotyrosine antibody (e.g., mAb4G10) in TBST containing 2% (w/v) BSA. Small blots (e.g., 6 × 8 cm) require approx 2 mL, where blots from larger-format gels will require between 5 and 10 mL (*see* **Notes 2** and **3**).

6. Use a food-bag sealer to seal the blot between two pieces of polyethylene, sealing three sides before the addition of the mAb4G10 solution on the sample-protein-bound side of the blot.

Remove as many air bubbles as possible before sealing the last side. Incubate for 1 h at room temperature, either with gentle mixing (e.g., by rotation) or flat on the bench (*see* **Note 4**).

7. Remove the blot from the bag and wash in a large volume (at least 200 mL) of TBST for 2 h with mixing, and change the buffer every 30 min (*see* **Note 5**).

8. As described in **step 6** incubate the blot with anti-mouse IgG horseradish peroxidase (HRP)-conjugated antibody. We use Cat. no. NA931 (Amersham Biosciences) for this, at a dilution of 1 in 10,000 prepared in TBST containing 2% (w/v) BSA and *no* sodium azide. Other similar secondary antibodies should be optimised for the appropriate concentration to be used in this assay. The presence of sodium azide either in the secondary antibody preparation or carried over due to insufficient washing will cause the poisoning of the HRP and therefore no signal.

9. Discard the secondary antibody and wash the blot as described in **step 7** (*see* **Note 5**).

10. Incubate the blot with ECL reagent following the manufacturer's instructions. Shake off excess reagent and sandwich between Saran™ wrap.

11. Expose X-ray film to the blot and develop using standard autoradiography techniques to detect tyrosine phosphorylated proteins. Adjust the exposure time for subsequent exposures in order to obtain maximum sensitivity without overexposure and saturation of the film (*see* **Note 6**). Transfer the positions of the molecular mass markers onto the blot using a permanent marker pen.

12. Film may be used for relative quantitation of tyrosine phosphorylation using one of a number of film-scanning systems and software (*see* **Note 7**).

3.3. Measurement of Kinase Activity In Vitro

A useful approach that has been applied to the study of the regulation of tyrosine kinases is the in vitro measurement of the activity of a given enzyme (either in isolation or in a cell extract). In order for an assay to be designed, it is important that a substrate is known for the enzyme system under scrutiny. This need not necessarily be an endogenous substrate, although clearly this is ideal. It is not uncommon for tyrosine kinases to be substrates for their own enzymic activity (for example, Syk *[5]*), and in these circumstances it is not necessary to add an exogenous substrate. In many cases the addition is required of a substrate such as whole protein, relevant peptide derived from a known substrate, or an "artificial" substrate known to be phosphorylated by the enzyme (for example, enolase phosphorylation by Src-family kinases *[11]*). The following method details how in vitro kinase assays may be used on a specific tyrosine kinase that is isolated from a cell extract by immunoprecipitation. In this system the incorporation of radioactive phosphate into the substrate is measured. This technique is not restricted to tyrosine kinases and may also be applied to the study of serine and threonine kinases (*see* Chapter 12, vol. 2). It should be noted that a number of commercially available nonradioactive alternatives to this type of assay are becoming available, most utilizing phospho-substrate-specific antibodies and detection by immunoblotting.

The possibilities and permutations for this type of assay are enormous, so the following method is provided as a guideline and should be adapted by the reader as is appropriate. This section describes the isolation of a protein from a cell lysate, and the in vitro kinase assay.

1. Prepare platelets, megakaryocytes or cell lines for example as described in Chapters 2, 22, 23, and 27, vol. 1 and if required, stimulate the cells with agonist, preferably in a platelet aggregometer with stirring. To ensure equal recovery of kinase from activated and nonactivated platelets, it is advisable to prevent platelet aggregation by the inclusion of 1 mM EGTA in the platelet suspension. The number of cells required depends on the expression levels of the kinase of interest and the affinity and quantity of the kinase-specific antibody that is available. This may therefore require optimization (*see* **Note 8**).

2. Terminate stimulation by the addition of an equal volume of ice-cold 2X NP40 lysis buffer containing protease and phosphatase inhibitors. Transfer lysate to microfuge tubes, vortex, and incubate on ice for at least 15 min.

3. Centrifuge samples at 14,000g at 4°C for 10 min to pellet insoluble debris and transfer as much of the supernatant to fresh microfuge tubes as is possible, but ensure an equal volume is collected from each sample.

4. Prior to immunoprecipitation, pre-clear the sample of proteins that may bind to protein A/protein G-Sepharose (PAS/PGS). Prepare a 50% (v/v) slurry of PAS or PGS in TBST and add 20 µL to each sample. Rotate at 4°C for 30 to 60 min (*see* **Note 9**).

5. Pellet the Sepharose by brief centrifugation and carefully remove the supernatant to fresh microfuge tubes (*see* **Note 9**).

6. To each tube add an antibody that specifically binds to the protein of interest. We would recommend the use of monoclonal rather than polyclonal antibodies for this, and would suggest using approx 1 µg/sample. Depending on the affinity of the antibody and the amount of the respective antigen in the sample, this amount of antibody may be reduced (*see* **Note 10**). Mix by rotation for 1 h at 4°C.

7. Add 20 µL PAS/PGS and mix by rotation for 1 h at 4°C (*see* **Note 11**).

8. Pellet the Sepharose by centrifugation for a few seconds at 14,000g and remove supernatant (*see* **Note 12**). Wash each sample pellet twice by resuspension with 1 mL ice-cold 1X NP40 lysis buffer (prepared by 1:1 dilution of 2X buffer with water) followed by centrifugation as described above.

9. Perform one final wash in TBST (no inhibitors present), remove as much buffer as possible, and replace with 25 µL kinase assay buffer (this contains 5 µCi of [γ-^{32}P]ATP). If an exogenous substrate is to be used, then this should also be added to the kinase buffer mix. The assay time begins for each sample upon the addition of buffer. Mix and incubate in a Perspex-shielded water bath at 30°C for 20 min (*see* **Notes 13** and **14**).

10. Termination of the assay: If the assay if designed to measure autophosphorylation, then unbound [γ-^{32}P]ATP may be removed from the sample. For this, terminate the assay by the addition of 500 µL 50 mM EDTA at 4°C. Pellet the Sepharose by brief centrifugation and remove the supernatant, which should be disposed of in an appropriate manner. Add Laemmli SDS-PAGE sample buffer to the pellet, mix, and heat to 100°C for 5 min. For assays that include an added subtrate, the unreacted radioactivity cannot be removed. In these cases add 25 µL Laemmli sample treatment buffer, mix, and heat to 100°C for 5 min.

11. Separate proteins by SDS-PAGE. Extreme care should be taken because samples are radioactive. All buffer, gel, glass plates, and sample tubes must be treated in an appropriate manner and in accordance with local rules. If necessary, the gel electrophoresis tank should be placed behind a suitable Perspex shield. Note that for assays that utilize an added substrate, unreacted [γ-^{32}P]ATP will elute into the tank buffer.

12. Results may be obtained directly from a polyacrylamide gel either dried or hydrated; however in our experience better results are obtained if proteins are first transferred to a PVDF membrane. This may be achieved as described in Chapter 9, vol. 2 (*see* **Note 15**).

We also routinely block such membranes with BSA as described in **Subheading 3.2., step 3**, to allow the identification of proteins present in the blot, and the confirmation of successful immuniprecipition and equal levels of kinases in each sample, by subsequent immunoblotting. For immunoblotting techniques refer to **Subheading 3.2.**

13. Expose X-ray film to the blot and develop by conventional autoradiography techniques. Alternatively, expose a storage phosphor screen to the blot and develop using a phosphor-imager. The latter method is preferable because it enables more reliable quantitation of ^{32}P incorporation/kinase activity.

3.4. Study of Tyrosine Phosphorylation-Induced Protein–Protein Interactions

Many of the regulatory roles of tyrosine phosphorylation may be attributed to the ability of specific protein tyrosine residues to support protein–protein interactions when phosphorylated. This is enabled through structurally conserved domains known as Src homology 2 (SH2) domains or phosphotyrosine binding (PTB) domains. Such interactions are responsible for the recruitment of signaling proteins to an activated receptor (e.g., the binding of Syk to the collagen receptor GPVI-FcR γ-chain complex when activated), the translocation of proteins to different cellular compartments (for example, to the location of a substrate), and the direct regulation of enzymic activity. In this way, tyrosine phosphorylation affects all aspects of cell signaling. The ability to assess how protein tyrosine phosphorylation affects protein-protein interactions has proven vital in the successful characterization of a number of platelet and megakaryocyte signaling pathways *(1,16,18,19)*.

There are several ways in which this may be studied. These include two examples outlined below, namely co-immunoprecipitation techniques and the precipitation of proteins with recombinant proteins. There are many examples in the literature where this approach has been successfully applied to the study of signaling pathways of platelets and megakaryocytes. While these are the workhorse techniques employed by many researchers, a number of new technologies, such as fluorescence energy tranfer (FRET), surface plasmon resonance (e.g., Biacore AB, Stevenage, UK), and protein arrays are also being used more frequently to approach these questions.

3.4.1. Coimmunoprecipitation

The principle behind this technique is that specific proteins are isolated from cell lysates produced using buffers containing mild detergents (*see* **Note 16**), and the proteins are separated by SDS-PAGE and immunoblotted or stained to detect proteins that may co-isolate with the protein of interest. Analysis of co-association under different states of cellular activity, and comparison with the level of tyrosine phosphorylation of the proteins involved, can provide important clues as to the mechanism of protein interaction, how it is regulated, and what effect this has on a given signaling pathway or cellular function.

1. The immunoprecipitation of proteins from platelets, megaokaryocytes or cell lines should be performed as described in **Subheading 3.3., steps 1–8**. For in vitro kinase assays it is strongly recommended to use, if possible, monoclonal antibodies. Co-immuoprecipitation relies on the maintenance of endogenous protein–protein interactions in detergent lysates

with antibodies bound to the protein of interest. Some antibodies may displace such an interaction, and this is more likely with polyclonal preparations that contain a mixure of antibodies that potentially bind multiple sites on the target molecule. If the experiment has involved the use of antibodies, for example to stimulate or block specific receptors, it is advised to omit the preclearing step during immunoprecipitation since this may result in removal of proteins of interest from the sample before immunoprecipitation.

2. Perform a final wash step in TBST. Centrifuge briefly and remove as much buffer as possible. Add Laemmli sample treatment buffer (the volume added should be chosen in accordance with the volume that is possible to run on a single lane in SDS-PAGE system of choice) and heat to 100°C for 5 min. Following centrifugation remove all the solution until the Sepharose pellet is dry, and load an SDS-PAGE gel. It does not matter if some beads are transferred to the wells of the gel.

3. Separate proteins by SDS-PAGE and electrotransfer to a PVDF membrane as described in Chapter 9, vol. 2.

4. Block membranes with BSA as described in **Subheading 3.2., step 3** (*see* **Note 17**).

5. Following the protocol described in **Subheading 3.2., steps 4–8** to immunoblot to detect co-immunprecipated proteins. Clearly, in order for this to work, it is important to have some idea of what may be present. If potential interacting proteins are not known (clues are often to be found from the literature on different cell types), a useful approach is to first probe the blot for phosphotyrosine residues or stain an SDS-PAGE gel with silver (*see* Chapter 9, vol. 2). If proteins that are tyrosine-phosphorylated are present, the apparent molecular mass determined from the blot will be useful in beginning the search for its identity. The level of tyrosine phosphorylation of an interacting protein or the protein immunoprecipitated, or the levels of co-immunoprecipitated proteins, may be analyzed to determine whether the interaction is due to regulated protein tyrosine phosphorylation.

6. It will be frequently necessary to reprobe immunoblots to determine the presence of other proteins, check levels of tyrosine phosphorylation, and particularly to confirm that, for each sample, equivalent levels of protein were immunoprecipitated. Blots should be stripped of bound antibodies before reprobing, unless the new signal is expected in a different region of the blot from the previous results. Reagents specifically designed for this are widely available. Alternatively, incubate a blot with immunoblot stripping buffer (*see* **Subheading 2.4.**) at 80°C for 20–40 min with agitation, and then wash the blot in multiple changes of TBST until the smell of mercaptoethanol is no longer detectable. Whichever method is used, blots should be reblocked in BSA before reprobing (*see* **Note 18**).

3.4.2. Precipitation With Recombinant Fusion Proteins

The principle behind this technique is to use recombinant proteins containing defined regions of a protein of interest (or the entire protein) to identify proteins from cell extracts with which they can interact. Fusion proteins (e.g., glutathione-*S*-transferase [GST] proteins as described below) are frequently used to provide an effective tag to enable the isolation of the protein during the assay. The production of GST fusion constructs is simplified through the use of commercially available vector systems such as the pGex range from Amersham Biosciences. These systems usually incorporate a protease-cleavage site should the removal of the fusion partner be necessary. A description of the methods required to construct a fusion protein expression vector and the production and purification of such proteins is not possible here. Detailed information is, however, available from http://www.amershambiosciences.com.

The following technique has been successfully used by this and other labs for the use of GST fusion proteins to investigate platelet tyrosine-kinase-dependent signal transduction. Many examples are available in the literature where this approach has been successfully used to investigate the binding properties of SH2 domains from specific proteins in different cellular contexts. This protocol assumes that a suitable purified GST fusion protein is available for use.

1. Platelets, megakaryocytes, or cells lines should be prepared, stimulated, and lysed using NP40 lysis buffer containing protease inhibitors and phosphatase inhibitors as described in **Subheading 3.3., steps 1–3**.
2. Rehydrate and wash approx 80 mg of glutathione agarose in 50 mL deionized water. Centrifuge the suspension to pellet the agarose, remove the supernatant, and resuspend in 1 mL PBS containing 0.1% (v/v) Triton X-100 (PBST).
3. Ideally the GST-fusion protein should also be diluted in PBST. If necessary, exchange protein into the buffer by dialysis or a desalting column. Add fusion protein to the glutathione-agarose bead suspension in proportions of approx 200 µg to 1 mL of suspension.
4. Mix by rotation at 4°C for 2 h. Centrifuge for a few seconds at 14,000*g* to pellet the agarose and resuspend in 1 mL PBST. Centrifuge again, remove the supernatant, and repeat this washing step twice more, once using PBST and then once in 50 m*M* Tris-HCl, pH 7.3.
5. Resuspend in 50 m*M* Tris-HCl, pH 7.3. The preparation can usually be stored at 4°C for a number of days with the addition of sodium azide (0.05% [w/v]). Successful immobilization may be checked by running a sample of the fusion protein solution before and after incubation with glutathione agarose together with a sample of the immobilized protein (boiled in Laemmli sample treatment buffer) on SDS-PAGE and staining of the gel (*see* Chapter 9, vol. 2).
6. Add agarose-immobilized protein to the lysate samples (corresponding to around 1–5 µg protein) and mix by rotation for 1–2 h at 4°C. To control for nonspecific binding to the agarose or GST component of the fusion protein, replicate samples using glutathione-agarose-immobilized GST alone should be performed. GST-agarose should be used at a molar equivalent to the fusion protein used, and prepared as described above for GST fusion proteins (*see* **Note 20**).
7. Pellet agarose by brief centrifugation, and wash the pellet in 2X 1 mL NP40 lysis buffer with protease and phosphatase inhibitors (1X strength) followed by a final wash step using 1 mL TBST. Remove as much buffer as possible, add Laemmlli SDS-PAGE sample treatment buffer, mix, and heat to 100°C for 5 min.
8. Pellet the beads by centrifugation and remove all the sample buffer, until the pellet is dry. Load onto an SDS-PAGE gel for separation of proteins. It does not matter if some beads are transferred to the wells of the gel. (For further details on SDS-PAGE refer to Chapter 9, vol. 2.)
9. Transfer the proteins to a membrane such as PVDF (for further details on Western blotting refer to Chapter 9, vol. 2) block the membrane by incubation for 30–60 min in TBST containing 10% (w/v) BSA, and process for the identification of fusion-protein-binding proteins that are isolated from the cell extract by immunoblotting as described for immunoprecipitation experiments in **Subheading 3.4.1., steps 4** and **5** (*see* **Notes 20** and **21**).

3.5. Use of Specific Tyrosine Kinase Inhibitors

The are a number of ways in which the function of a specific kinase may be determined in a given cell type. Many of these approaches involve some form of genetic modification of cells in vitro. While this type of approach is possible with megakaryocytic cell lines and is becoming possible with primary megakaryocytes maintained in culture, this approach is not currently feasible with platelets. The use of transgenic animals and kinase inhibitors has therefore become a popular approach to address the function of kinases in platelets. Strategies for the use of transgenic animals are described in Chapters 2 and 23, vol. 2. In addition, Chapter 5, vol. 2 describes the designs and limitations of inhibitors of cell signalling.

Although kinase inhibitors have been widely and successfully used in the study of signal transduction, a degree of care is required in their use and the interpretation of results. The main issue is possibly the specificity of an inhibitor. Obtaining this type of information is largely dependent on how extensively an inhibitor has been studied, and whether all its potential targets are known. It is becoming increasingly apparent that some inhibitors are far less selective than was once thought, casting doubt on the interpretation of some previous results *(22)*. The use of inhibitors is valid and they are valuable tools, although they should be considered as complementary approaches to other techniques that are available.

In most cases, the use of a kinase inhibitor would involve its incubation with cells, with or without other forms of stimulation, and the assay of a given cellular function or detailed scrutiny of a specific signaling pathway. Great care is required in the choice of inhibitor and the experimental design for its use. Rather than providing a list of detailed methodology, the following are points that should be considered in the choice and use of a kinase inhibitor.

1. According to the literature, to what degree is the inhibitor selective toward your protein of interest? This will depend on the concentration at which it is used.
2. What is the IC_{50} concentration? This will differ between use on whole cells and isolated proteins, where substantially higher concentrations are usually required for whole cells. It is important to relate the concentration planned for use to the available data on specificity.
3. How soluble is the inhibitor, and what solvent is appropriate? Inhibitors that gain access to the cytosol are frequently required to be dissolved in solvents such as dimethyl sulfoxide or ethanol. If these solvents are to be used, they should not be added to platelets or megakaryocytes at final concentrations greater then 0.5% (v/v). Above this concentration, these solvents will begin to affect cell signaling and function. Experiments using such solvents must contain a solvent-only control.
4. Is the inhibitor membrane-permeable (i.e., is it suitable for use on whole cells)? How quickly does it gain access to the cytosol? This will have implications on incubation times required.
5. Is the compound toxic? This will be important if an inhibitor is to be used on cells that are to be maintained in culture following treatment.
6. How may data obtained using an inhibitor be confirmed? For example, can the activity of the inhibitor target be measured? Are alternative and structurally unrelated inhibitors available for the molecule? Can similar confirmatory experiments be performed in cells where molecular modification is possible?

4. Notes

1. Blots may be left in blocking solution at 4°C overnight. For this reason, it is essential that the solution contain a bacteriostatic agent such as 0.05% (w/v) sodium azide. The use of high-quality BSA is essential to minimize the presence of phosphatases. Alternative blocking

solutions may be used, but for antiphosphotyrosine blots it is recommended to avoid the use of skimmed milk powder due to the presence of phosphatases, unless heat-inactivated and used in the presence of phosphatase inhibitors.

2. A number of other antiphosphotyrosine antibodies are commercially available, such as mAb PY20. While several antibodies are suitable for these types of studies, the experimenter should be aware that these reagents do not produce identical results. It is therefore unwise to switch between different antibody types.

3. If 0.05% (w/v) sodium azide is included in the mAb4G10 solution it may be reused numerous times over a period of months. For this, the solution must be stored at 4°C, and centrifuged prior to each use to remove debris.

4. Small-format blots may be incubated with antibodies in a 50-mL polypropylene tube, with the blot placed around the inner surface of the tube, and mixed on a tube roller.

5. Blot washing times may be reduced if necessary. In our experience, however, extensive washing when using this antibody produces results with extremely low background signal levels.

6. For most purposes normal autoradiography film is perfectly acceptable. However, several manufacturers supply film designed for ECL. This has higher sensitivity for the wavelength of light produced, and develops clear, thereby enhancing densitometry analysis.

7. We successfully utilize a CCD camera-based system to record and quantify tyrosine phosphorylation. This avoids the use of X-ray film.

8. A suggested starting point for cell number would be to stimulate 500 μL platelets suspended at a density of 8×10^8 cells/mL; due to the substantially greater size of megakaryocytes, try 1000–2000-fold lower number of cells (i.e., 1 to 2×10^5).

9. The pellet of PAS/PGS is not easy to see, yet this reagent is expensive; therefore it is not a good idea to increase dramatically the volume of slurry added to samples, although this should be increased if the pellet is not visible. We have found that the use of colored microfuge tubes can make the pellet easier to see.

10. The kinase assay is performed while the kinase is coupled to antibody. There is no guarantee that antibody binding will not affect the activity of the kinase of interest. The use of polyclonal antibodies will substantially increase the likelihood of such interference, and therefore monoclonal antibodies are recommended if possible. It may be necessary to try more that one antibody clone in order to successfully assay kinase activity.

11. Some isotypes of IgG (e.g., mouse IgG_1, rat IgG_1, IgG_{2a}, IgG_{2b}, and all goat and sheep IgGs) bind poorly to protein A. Where such antibodies are to be used it is recommended to use PGS (which binds most IgGs) in place of PAS, or include a secondary antibody that is able to bind PAS, such as rabbit anti-mouse IgG, which may be added along with the Sepharose.

12. It may be useful to save a small volume of cell extract from before and after immunoprecipitation. These may be used for immunoblot analysis to determine the level of depletion of the respective kinase during the procedure.

13. The amount of substrate to be added will depend on its ability to become phosphorylated and detected in the assay. This will therefore require optimization for each assay system. A suggested starting concentration for a peptide substrate would be 1–5 μg per assay, although larger proteins may require more.

14. During optimization of a new assay, the length of the assay and the temperature at which it is performed may be adjusted as required to maximize sensitivity and dynamic range within a time period that is practicable.

15. Electrotransfer of proteins using a semi-dry transfer cell will result in radioactive contamination of the anode.

16. The cell lysis buffer used in this protocol is NP40, a relatively mild detergent in which many protein-protein interactions will be maintained (e.g., mediated by SH2 domains). The maintenance of interactions upon cell lysis may be dependent on the choice and concentration of the detergent used. It may be necessary, therefore, to experiment with the use of different detergents. Examples of some detergents that may be considered include Triton X-100, digitonin, CHAPS, and Octyl β-D glucoside.

17. For immunoblotting, the protein used to block a membrane is critical. The combination of some antibodies and blocking solutions sometimes produces high levels of nonspecific binding or a patchy background signal. If this occurs it is recommended to try an alternative blocking reagent such as 5% (w/v) skimmed milk powder in TBST or a proprietary reagent such as Blotto (Santa Cruz Biotechnology, Santa Cruz, CA). If blotting for phosphotyrosine residues and using milk protein to block the membrane, ensure that the blocking solution is heated to inactivate phosphatases, and preferably include a phosphatase inhibitor.

18. Effective stripping may be confirmed by probing the blot with secondary antibody alone, where no signal should be detected.

19. If nonspecific binding to agarose is a problem, samples may be precleared using glutathione agarose exactly as described for pre-clearing in immunoprecipitation procedures using PAS or PGS (*see* **Subheading 3.3., steps 4** and **5**).

20. This type of analysis does not provide information on whether the interaction between proteins is direct, or whether other unidentified molecules are involved in holding the complex together. This type of question may be addressed by extending studies using additional techniques that are beyond the scope of this chapter, such as Far Western blotting (probing a blot with the one of the interacting proteins in place of an antibody; GST fusion proteins are often used for this), surface plasmon resonance combined with mass-spectrometric identification of associated proteins, and yeast/mammalian two-hybrid analysis.

21. Interaction of proteins with recombinant proteins does not confirm that such interactions actually occur with native protein within the cell, expressed at normal levels. Interactions discovered in this way will require confirmation using an alternative approach such as coimmunoprecipitation.

References

1. Gibbins, J., Asselin, J., Farndale, R., Barnes, M., Law, C. L., and Watson, S. P. (1996) Tyrosine phosphorylation of the Fc receptor gamma-chain in collagen-stimulated platelets. *J. Biol. Chem.* **271,** 18,095–18,099.

2. Gibbins, J. M., Okuma, M., Farndale, R., Barnes, M., and Watson, S. P. (1997) Glycoprotein VI is the collagen receptor in platelets which underlies tyrosine phosphorylation of the Fc receptor γ-chain. *FEBS Lett.* **413,** 255–259.

3. Poole, A., Gibbins, J. M., Turner, M., van Vugt, M. J., van de Winkel, J. G. J., Saito, T., et al. (1997) The Fc receptor γ-chain and the tyrosine kinase Syk are essential for activation of mouse platelets by collagen. *EMBO J.* **16,** 2333–2341.

4. Tsuji, M., Ezumi, Y., Arai, M., and Takayama, H. (1997) A novel association of Fc receptor γ-chain with glycoprotein VI and their co-expression as a collagen receptor in human platelets. *J. Biol. Chem.* **272,** 23,528–23,531.

5. Yanaga, F., Poole, A., Asselin, J., Blake, R., Schieven, G., Clark, E. A., et al. (1995) Syk interacts with tyrosine-phosphorylated proteins in human platelets activated by collagen and cross-linking of the Fcg-IIA receptor. *Biochem. J.* **311,** 471–478.

6. Robinson, A., Gibbins, J., Rodriguez-Linares, B., Finan, P. M., Wilson, L., Kellie, S., et al. (1996) Characterization of Grb2-binding proteins in human platelets activated by Fc gamma RIIA cross-linking. *Blood* **88,** 522–530.

7. Law, D. A., DeGuzman, F. R., Heiser, P., Ministri-Madrid, K., Killeen, N., and Phillips, D. R. (1999) Integrin cytoplasmic tyrosine motif is required for outside-in alpha IIb beta 3 signalling and platelet function. *Nature* **401,** 808–811.

8. Cicmil, M., Thomas, J. M., Leduc, M., Bon, C., and Gibbins, J. M. (2002) PECAM-1 signalling inhibits the activation of human platelets. *Blood* **99,** 137–144.

9. Cicmil, M., Thomas, J. M., Sage, T., Barry, F. A., Leduc, M., Bon, C., et al. (2000) Collagen, convulxin, and thrombin stimulate aggregation-independent tyrosine phosphorylation of CD31 in platelets. Evidence for the involvement of Src family kinases. *J. Biol. Chem.* **275,** 27,339–27,347.

10. Jones, K. L., Hughan, S. C., Dopheide, S. M., Farndale, R. W., Jackson, S. P., and Jackson, D. E. (2001) Platelet endothelial cell adhesion molecule-1 is a negative regulator of platelet-collagen interactions. *Blood* **98,** 1456–1463.

11. Briddon, S. J. and Watson, S. P. (1999) Evidence for the involvement of p59(fyn) and p53/56(lyn) in collagen receptor signalling in human platelets. *Biochem. J.* **338,** 203–209.

12. Quek, L. S., Pasquet, J. M., Hers, I., Cornall, R., Knight, G., Barnes, M., et al. (2000) Fyn and Lyn phosphorylate the Fc receptor gamma chain downstream of glycoprotein VI in murine platelets, and Lyn regulates a novel feedback pathway. *Blood* **96,** 4246–4253.

13. Quek, L. S., Bolen, J., and Watson, S. P. (1998) A role for Bruton's tyrosine kinase (Btk) in platlet activation by collagen. *Curr. Biol.* **8,** 1137–1140.

14. Achison, M., Knight, G. G., Barnes, M. J., and Farndale, R. W. (1997) Activation by collagen and a collagen-related peptide of the intracellular tyrosine kinase, P125(fak), in human platelets which is independent of the integrin $\alpha_2\beta_1$. *Thromb. Haemost.* PD637–PD637.

15. Achison, M., Elton, C. M., Hargreaves, P. G., Knight, C. G., Barnes, M. J., and Farndale, R. W. (2001) Integrin-independent tyrosine phosphorylation of p125(fak) in human platelets stimulated by collagen. *J. Biol. Chem.* **276,** 3167–3174.

16. Gibbins, J. M., Briddon, S., Shutes, A., van Vugt, M. J., van de Winkel, J. G., Saito, T., et al. (1998) The p85 subunit of phosphatidylinositol 3-kinase associates with the Fc receptor gamma-chain and linker for activitor of T cells (LAT) in platelets stimulated by collagen and convulxin. *J. Biol. Chem.* **273,** 34,437–34,443.

17. Cichowski, K., Brugge, J. S., and Brass, L. F. (1996) Thrombin receptor activation and integrin engagement stimulate tyrosine phosphorylation of the proto-oncogene product, p95(vav), in platelets. *J. Biol. Chem.* **271,** 7544–7550.

18. Gross, B. S., Melford, S. K., and Watson, S. P. (1999) Evidence that phospholipase C-γ2 interacts with SLP-76, syk, lyn, LAT and the Fc receptor γ-chain after stimulation of the collagen receptor glycoprotein VI in human platelets. *Eur. J. Biochem.* **263,** 612–623.

19. Pasquet, J. M., Gross, B., Quek, L., Asazuma, N., Zhang, W. G., Sommers, C. L., et al. (1999) LAT is required for tyrosine phosphorylation of phospholipase C gamma 2 and platelet activation by the collagen receptor GPVI. *Mol. Cell. Biol.* **19,** 8326–8334.

20. Zhang, J., Zhang, J., Shattil, S. J., Cunningham, M. C., and Rittenhouse, S. E. (1996) Phosphoinositide 3-kinase gamma and p85/phosphoinositide 3-kinase in platelets. Relative activation by thrombin receptor or beta-phorbol myristate acetate and roles in promoting the ligand-binding function of alphaIIbbeta3 integrin. *J. Biol. Chem.* **271,** 6265–6572.

21. Cho, M. J., Pestina, T. I., Steward, S. A., Lowell, C. A., Jackson, C. W., and Gartner, T. K. (2002) Role of the Src family kinase Lyn in TxA2 production, adenosine diphosphate secretion, Akt phosphorylation, and irreversible aggregation in platelets stimulated with γ-thrombin. *Blood* **99,** 2442–2447.
22. Davies, S. P., Reddy, H., Caivano, M., and Cohen, P. (2000) Specificity and mechanism of action of some commonly used protein kinase inhibitors. *Biochem. J.* **351,** 95–105.

11

Protein Tyrosine Phosphatases

Matthew L. Jones and Alastair W. Poole

1. Introduction

The reversible covalent modification of proteins by the addition and removal of a phosphate group is an important theme in signal transduction in mammalian cells. There is a large superfamily of protein kinases that catalyse the addition of a phosphate group to hydroxyl residues on the side chains of the amino acids serine, threonine, or tyrosine. Similarly there is a large family of protein phosphatases that catalyze the removal of phosphate groups from phosphoserine, phosphothreonine, and phosphotyrosine residues. A dynamic interplay between protein kinases and protein phosphatases exists in vivo whereby the phosphorylation status of the target protein is determined by the opposing actions of these kinases and phosphatases.

Protein tyrosine phosphatases comprise a large superfamily of related enzymes, as shown in **Fig. 1**. Included in this superfamily are "classical" PTPs, which dephosphory-late phosphotyrosine alone; dual specificity phosphatases, which also dephosphorylate phosphoserine/phosphothreonine; and PTEN, a phosphoinositide phosphatase. The classical PTPs are divided into two categories, receptor and nonreceptor; to date only three nonreceptor PTPs have been shown to be expressed in human platelets: PTP1B, SHP-1, and SHP-2 *(1,2)*. These three phosphatases all contain a single catalytic subunit, but have different regulatory domains. PTP1B is constitutively localized to the endoplasmic reticulum, and is released from its anchoring site by proteolytic cleavage by calpains *(3)*. SHP-1 and SHP-2, in contrast, are regulated and localized by two N-terminal SH2 domains that bind phosphotyrosine-containing sequences within associated regulatory proteins *(4)*.

This chapter will first lead the reader through the steps necessary to measure the activity of individual phosphatases present in platelets. Identification of phosphatase substrates in platelets is also a key question, and the details of an approach using substrate-trapping mutant phosphatases will then be described. This is a popular approach for identification of tyrosine phosphatase substrates and exploits the two-

From: *Methods in Molecular Biology, vol. 273:*
Platelets and Megakaryocytes, Vol. 2: Perspectives and Techniques
Edited by: J. M. Gibbins and M. P. Mahaut-Smith © Humana Press Inc., Totowa, NJ

Non-receptor PTPs   ## Receptor PTPs

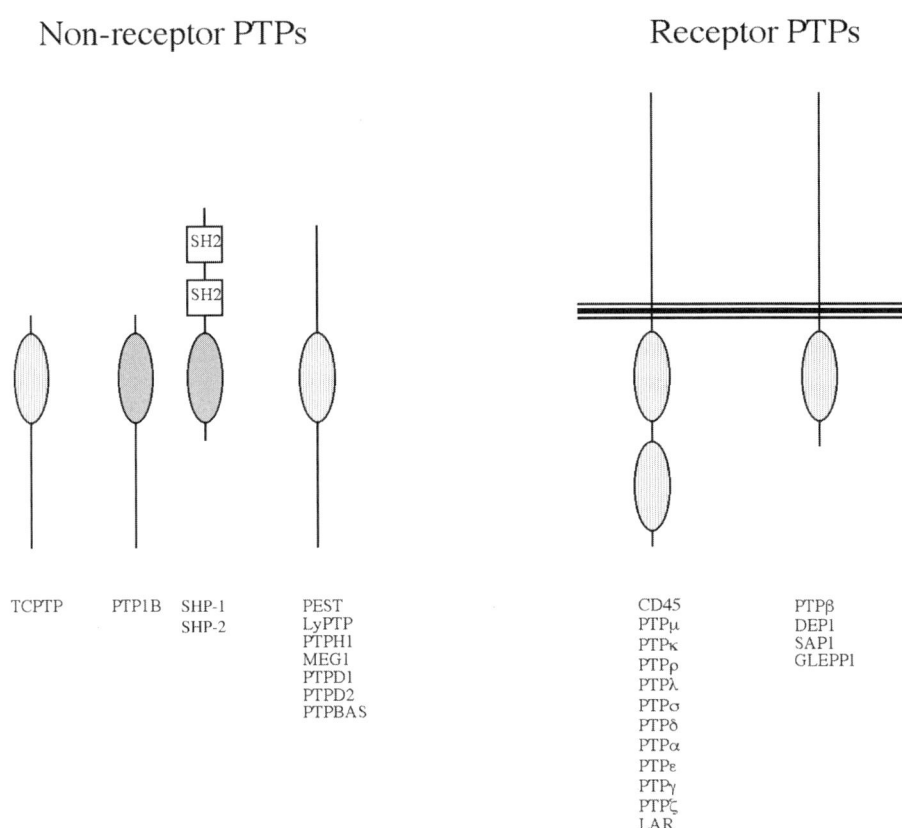

		SH2 SH2			CD45	PTPβ

TCPTP   PTP1B   SHP-1
                      SHP-2

PEST
LyPTP
PTPH1
MEG1
PTPD1
PTPD2
PTPBAS

CD45
PTPμ
PTPκ
PTPρ
PTPλ
PTPσ
PTPδ
PTPα
PTPε
PTPγ
PTPζ
LAR

PTPβ
DEP1
SAP1
GLEPP1

Fig. 1. The protein tyrosine phosphatase family. Shown are the "classical" (phosphotyrosine-specific) PTPs, which are divided into nonreceptor and receptor PTPs. All members of this class of PTPs contain the PTP recognition sequence (HCSAGxGRxG) within their catalytic domain. Not included in this diagram are the dual-specificity phosphatases, which are capable of dephosphorylating phospho-tyrosines and phosphothreonines/phosphoserines. Only PTP1B, SHP-1, and SHP-2 have so far been shown to be expressed in human platelets, and are therefore shown with solid shading.

step mechanism used by these enzymes to catalyze dephosphorylation. Inactivation of the phosphatase domain by point mutagenesis leaves the enzyme able to bind with high affinity to the substrate, thus trapping it and enabling the identification of the trapped substrate protein. Substrates may then be trapped using the mutant phosphatase domain as a GST fusion protein using an immunoprecipitation approach, or more definitively using a Far-Western blotting approach, as described.

2. Materials

2.1. Radioactive Phosphopeptide Assay for Protein Tyrosine Phosphatase Activity

1. Raytide peptide (Calbiochem, Nottingham, UK).
2. A construct encoding the catalytic domain of p60[c-src] fused to the C-terminus of GST; it is possible to use other tyrosine kinase catalytic domains, e.g., Abl tyrosine kinase.
3. [γ^{32}P]-ATP (*see* **Note 1**).
4. Whatman P81 chromatography paper.
5. Phosphorylation buffer: 50 mM HEPES, pH 7.5, 100 μM EDTA, 0.2% (v/v) 2-mercaptoethanol, and 10 mM MgCl$_2$.
6. 0.5% (v/v) phosphoric acid.
7. 1 M ammonium bicarbonate.
8. Assay buffer: 50 mM HEPES, pH 7.5, and 0.2% (v/v) 2-mercaptoethanol.
9. Termination solution: 900 mM HCl, 90 mM sodium pyrophosphate, 2 mM Na$_2$H$_2$PO$_4$, and 4% (v/v) activated charcoal Norit A (Sigma, Dorset, UK).

2.2. Substrate-Trapping Mutants as Tools for Identifying Substrates of Protein Tyrosine Phosphatases

1. Expression construct: catalytic domain in pGEX vector (Amersham Biosciences, Chalfont, UK) fused to the C-terminus of GST.
2. Transformation-competent *Escherichia coli* host strain for protein expression, e.g., BL21 strain.
3. LB-agar plates (1% [w/v] bactotryptone, 0.5% [w/v] sodium chloride, 0.5% [w/v] yeast extract, and 1.5% [w/v] bactoagar).
4. LB broth (1% [w/v] bactotryptone, 0.5% [w/v] sodium chloride, and 0.5% [w/v] yeast extract).
5. 1 M stock of filter-sterilized (0.2-μm filter) glucose in deionized water; store at 4°C.
6. 100 mg/mL stock solution of ampicillin in deionized water; store in aliquots at –20C.
7. 17.5 mg/mL stock solution of chloramphenicol in absolute ethanol (for BL21 Rosetta strain only); store at –20°C.
8. 1 M stock solution of IPTG (isopropyl-β-D-thiogalactopyranoside) in water, filter-sterilized (0.2-μm filter) and stored in 1-mL aliquots at –20°C.
9. Lysis buffer: 50 mM sodium phosphate, 100 mM sodium chloride, 1 mM EDTA, 0.1% (v/v) NP-40, 1 mM DTT, 0.1 mM PMSF, and 0.1 μM leupeptin, pH 7.0.
10. 10% (v/v) solution of rehydrated GSH-Sepharose/agarose beads (Sigma) in lysis buffer. Large volumes can be made and stored at 4°C.
11. PBST: 50 mM sodium phosphate, 100 mM sodium chloride and 0.05% (v/v) Tween-20. Store at 4°C.
12. 2X SDS-PAGE sample buffer: 24 mM Tris-HCl, pH 6.8, 10% (v/v) glycerol, 0.8% (w/v) SDS, and 6 mM DTT. Store in small aliquots at –20°C.
13. 2X NP40 lysis buffer for lysis of platelets: 2% (v/v) NP40, 300 mM NaCl, 20 mM Tris-HCl, 1 mM PMSF, 10 mM EDTA, 2 mM Na$_3$VO$_4$, 10 μg/mL leupeptin, 10 μg/mL aprotinin, 1 μg/mL pepstatin A, pH 7.3.

2.3. Far-Western Detection of PTPase Substrates

For GST fusion protein expression, as above for substrate-trapping mutant, plus 20 mM reduced glutathione solution in 20 mM Tris-HCl buffer, pH 8.0.

3. Methods

3.1. Radioactive Phosphopeptide Assay for Protein Tyrosine Phosphatase Activity

[^{32}P]-Phosphopeptides are commonly used as substrates for phosphatases. They are often prepared by incubating an appropriate peptide with [γ-^{32}P]ATP and a suitable kinase followed by separation of the [^{32}P]phosphopeptides from [γ-^{32}P]ATP. It is likely that the [^{32}P]phosphopeptides will be a mixture of phosphorylated and unphosphorylated peptides, but in subsequent analysis it is assumed that the phosphopeptide concentration corresponds to that of the [^{32}P]phosphate incorporated into the peptide.

The method detailed here describes the phosphorylation of a commercial peptide (Raytide) by a recombinant GST-fusion protein of the catalytic domain of the tyrosine kinase p60$^{c\text{-}src}$.

3.1.1. Preparation of [^{32}P]Phosphopeptides

1. Incubate 40 µg of the Raytide peptide with GST-catalytic domain of p60$^{c\text{-}src}$, 30 µM ATP, and 40 µCi of [γ^{32}-P]-ATP in a total volume of 120 µL of phosphorylation buffer; incubate at 30°C for 30 min (*see* **Note 2**).
2. Separate the [^{32}P]phosphopeptides from free [γ^{32}P]-ATP by isolation on Whatman P81 paper. Allow the phosphorylation reaction mixture to absorb to the paper and then wash the paper with 200 mL of 0.5% phosphoric acid. Repeat the wash four times. Elute the [^{32}P]phosphopeptides from the paper by addition of 1 mL of 1 M ammonium bicarbonate.
3. Lyophilize the peptides and resuspend in water. Take a small aliquot of the [^{32}P]phosphopeptides and determine the concentration of the peptide by counting in a scintillation counter.

3.1.2. Phosphatase Assay

1. In a total volume of 50 µL of assay buffer, mix 6 µM of [^{32}P]Raytide and enzyme and incubate for 10 min at 30°C (*see* **Notes 3–5**).
2. Terminate the reaction by adding 750 µL of termination solution, vortex, and centrifuge at top speed in a microcentrifuge for 10 min. Count the radioactivity in the supernatant.

3.2. Substrate-Trapping Mutants as Tools for Identifying Substrates of Protein Tyrosine Phosphatases

One of the aims of studying protein tyrosine phosphatases is to identify physiological substrates; this can be achieved by using so-called substrate-trapping mutants. The 200–300 residue catalytic domain of protein tyrosine phosphatases is characterized by the signature motif HCxxGxxRS, located at the base of the active site. The catalytic mechanism of PTPs have been well studied and can be summarized as (1) nucleophilic attack on the phosphorus atom by the thiolate group of the active site Cys to form a phospho-enzyme intermediate, (2) release of the dephosphorylated product and (3) hydrolysis of the thiol-phosphate intermediate and release of phosphate *(5)*.

The starting point for substrate-trapping experiments involves generation of appropriate mutations in the cDNA encoding the phosphatase, heterologous expression in an appropriate host, and subsequent purification of the expressed protein *(6)*.

One method for preparing a substrate-trapping mutant involves subcloning the cDNA encoding the mutant into a pGEX bacterial expression vector (APS Biotec), which results in the production of recombinant protein tagged at the N-terminus with GST, thus enabling a relatively simple one-step purification of protein from bacterial lysate by affinity chromatography using GSH-agarose/sepharose.

Described below is a starting protocol for the expression of GST-fusion proteins in *Escherichia coli*, and subsequent use of a GST-substrate trapping mutant for the purification of trapped substrates of tyrosine phosphatases.

1. Transform an appropriate *E. coli* host strain with the DNA construct and spread an aliquot of the transformation mixture on a LB-agar plate containing the appropriate antibiotic (*see* **Notes 6–8**).
2. Select a colony and inoculate 5 mL of LB culture supplemented with 10 mM glucose, containing the appropriate antibiotic for the vector and also where necessary, for the host strain; grow overnight at 37°C with shaking (200 rpm).
3. Dilute the overnight culture 1:100 into 100 mL of LB in a 500-mL flask, prewarmed to 37°C, containing the appropriate antibiotic(s), and incubate at 37°C until the density of the culture reaches mid-log phase ($A_{600nm} = 0.4$ to 0.6).
4. Lower the temperature of incubation to 30°C and add IPTG to a final concentration of 0.1 mM. Incubate at 30°C for 3 h. After 3 h recover the cells by centrifugation; the pelleted cells can be stored at –20°C until required or used immediately (*see* **Note 9**).
5. Resuspend the bacterial cell pellet in 10 volumes of lysis buffer and sonicate at medium power in 30-s bursts. Centrifuge the lysed bacteria at 10,000g for 10 min; the supernatant is the clarified cell lysate.
6. Immobilize the GST-fusion proteins on GSH-Sepharose (100 µL of 10% (v/v) GSH-beads per 1.5 mL of clarified bacterial lysate) by incubating at 4°C with constant rotation for 1 h (*see* **Notes 10** and **11**).
7. After GST-fusion proteins have been immobilized, pellet the beads by centrifugation at 1000g for 3 min and remove the supernatant. Resuspend the beads in 1 mL of lysis buffer, briefly vortex, and centrifuge again at 1000g for 3 min. Repeat this wash three times (*see* **Note 12**).
8. Having generated a purified GST-fusion of point mutant phosphatase domains with a substrate-trapping capability, it will then be important to use these proteins to fish for substrates from lysates of platelets. Since the substrate-trapping mutant relies on the substrate being phosphorylated on tyrosine in order to trap it, platelet lysates may need to be prepared from activated platelets. Lysates may be made using a variety of detergents, but usually addition of an equal volume of 2X (2%) NP40 buffer is the buffer of choice. After preclearing of the cell lysate by incubation of the lysate with glutathione-agarose beads for 30 min at 4°C, pellet the beads by centrifugation at 1000g for 3 min. The supernatant is then added to the washed glutathione agarose-beads containing the immobilized GST-fusion protein and incubated at 4°C with constant rotation for 1 h (*see* **Note 13**).
9. After 1 h pellet the beads by centrifugation at 1000g for 3 min and the remove the supernatant. Resuspend the beads in 200 µL of PBST, vortex and pellet by centrifugation at 1000g for 3 min. Repeat this wash 3 times.
10. After the final wash, resuspend the beads in 50 µL of 2X SDS-PAGE sample buffer and heat at 90°C for 5 min. The samples are now ready for SDS-PAGE.

It is advisable to subject the samples to electrophoresis on a gradient, e.g., 4–20%, polyacrylamide gel in the first instance. This is useful for identifying novel substrates of unknown molecular weight since a gradient gel will enable separation of proteins over a broad range of molecular weights on the same gel.

Proteins can be visualized by staining the polyacrylamide gel after electrophoresis with Coomassie blue stain or silver stain. A variety of gel staining reagents and kits based around Coomassie blue stain and silver stain are available from a number of suppliers with varying degrees of detection sensitivity. We would recommend using a staining reagent based on a Coomassie blue stain, e.g., the EZBlue™ gel staining reagent (Sigma) or Bio-Safe™ Coomassie stain (Bio-Rad, Hercules, CA), which are capable of detecting a few nanograms of protein. Any novel proteins found to interact with the substrate-trapping mutant can be identified by amino acid sequencing using standard methodologies (*see* **Note 14**).

3.3. Far-Western Detection of PTPase Substrates

GST fusions of mutant phosphatase domains can be used to co-precipitate trapped substrates from lysates of basal or stimulated platelets in standard 1% NP40 or equivalent detergent conditions. In order to more definitively demonstrate a direct interaction between the substrate-trapping mutant phosphatase domain and a tyrosine-phosphorylated substrate protein, however, it may be necessary to use a Far-Western blotting approach. This will help to rule out phosphoproteins that indirectly bind to the substrate trap through a multimolecular protein–protein complex. This approach has been used successfully to isolate SHP-1 substrates from T lymphocytes *(7)*. The principle behind Far-Western blotting is that membranes are probed directly with a solution of the recombinant substrate-trapping mutant, so as to detect a direct interaction with a putative substrate. Any bound recombinant protein may be detected by standard blotting using an antibody directed against the recombinant protein tag—in this case, GST—followed by an appropriate secondary antibody. The process is therefore very similar to standard Western blotting but includes an additional blotting step. Far-Western detection may be performed on whole cell lysates or immunoprecipitates blotted onto standard PVDF membranes.

1. By SDS-PAGE, run the whole cell lysate or immunoprecipitate of the putative substrate from samples prepared from basal and stimulated cells. Electrotransfer the proteins to a PVDF blotting membrane. Block the membrane in the standard way using 10% (w/v) BSA in TBS-T (for further details on blotting refer to Chapter 9, vol. 2).

2. Prepare GST fusion protein of substrate-trapping mutant phosphatase domain, as described above. Elute the fusion protein from 175 μL glutathione-agarose beads by incubation with 100 μL of 20 m*M* glutathione. Incubate in rotating tubes for 1 h at 4°C. Spin the beads and take off the supernatant. Check the protein concentration, and adjust to 0.5 mg/mL. Take 80 μL of this protein solution and add to 2 mL block solution (10% (w/v) BSA in TBS-T), making a 20 μg/mL solution (*see* **Notes 15** and **16**).

3. Incubate the PVDF membrane with the 2 mL of GST-fusion protein in block solution for 1 h at room temperature. Because of the small volume of solution used, it will probably be necessary to use a sealed-bag method for blotting.

4. Wash the membrane in TBS-T and probe for bound GST-fusion protein using an anti-GST antibody, following standard immunoblotting protocols (for further details refer to Chapter 10, vol. 2) (*see* **Note 17**).

4. Notes

1. The specific activity of the $[\gamma^{32}P]ATP$ has to be in the range of 2000–5000 cpm/pmol and *must* be precisely known because the concentration of the phosphopeptide is assumed to be equal to the amount of $[^{32}P]$phosphate incorporated into it.

2. The parameters used to phosphorylate the peptide given here should be considered as a starting point. In some instances, it may be necessary to increase the time period of phosphorylation of the peptide to several hours to ensure a more extensive phosphorylation of the peptide.

3. Careful consideration of how you are going to prepare the phosphatase is essential; make sure phosphatase inhibitors are absent from the preparation.

4. The parameters of the dephosphorylation reaction should be optimized to ensure that the reaction rate is linear. Perform a dilution of the enzyme preparation to ensure that the rate of dephosphorylation is proportional to the concentration of enzyme.

5. It may be necessary to calculate kinetic constants using this assay. In brief, for the calculation of kinetic constants obtain an estimate of the K_m by varying the peptide concentration and plotting velocity against peptide concentration. To determine K_m (peptide) and V_{max}, vary the peptide concentration in the following multiples of the estimated K_m: 0.125 K_m, 0.25 K_m, 0.5 K_m, 1.0 K_m, 2.0 K_m, 4.0 K_m, and 8.0 K_m. Obtain the kinetic values V_{max} and K_m by plotting a Lineweaver-Burk plot. More detailed descriptions of how to obtain kinetic constants can be found in standard enzyme kinetics texts.

6. To avoid retransforming *E. coli* every time one wants to express protein, a glycerol stock can be made from the starter culture. Take a volume of culture and add an equal volume of filter-sterilized 50% v)/v) glycerol; glycerol stocks can be stored at –70°C. It is not recommended to inoculate a starter culture directly from a glycerol stock, since loss of protein expression can occur overtime. To avoid this, inoculate a 1.0-mL LB culture, containing the appropriate selective antibiotic(s), with approx 2 µL of glycerol stock and incubate for 6 h at 37°C; spread 100 µL of this miniculture on a selective LB-agar plate, invert, and incubate overnight at 37°C. Pick a single colony from this plate as described in **Subheading 3.2., step 2.**

7. In our laboratory we have started to use the BL21 Rosetta strain (Novagen), a host strain for expression that contains a plasmid encoding genes for tRNAs that are rare in *E. coli* and thus overcoming "hungry-codon" syndrome. However, it may not be necessary to use this particular BL21 strain in all cases (see company catalogs for advice on choosing an appropriate host strain). Transformation-competent BL21 cells, and glycerol stocks of noncompetent BL21 cells, are commercially available (Novagen). Transformation-competent BL21 cells can be made using the calcium-chloride method *(8)*.

8. The ampicillin resistance gene encodes β-lactamase, a secreted protein, which degrades ampicillin in the culture medium. In saturated cultures the amount of soluble β-lactamase can be quite high and thus it is important to dilute the starter culture appropriately; otherwise a reduced level of protein expression may occur. We use 100 µg/mL ampicillin in all cultures, and it is possible to increase the concentration to 200 µg/mL. Alternatively, you can substitute carbenicillin for ampicillin; carbenicillin is more stable than ampicillin.

9. In some cases proteins expressed in *E. coli* are insoluble. Lowering the incubation temperature after induction with IPTG can improve solubility. In the protocol described here we lower the temperature to 30°C and incubate for 3 h but one may lower the incubation temperature further to 18°C, for example, and incubate for a longer period of time. However, if this does not work then a change of expression system might be considered.

10. In addition to expression as a GST-fusion in *E. coli*, one could express the protein with a variety of different tags to enable purification. The pET system (Novagen) has an extensive array of different vectors encoding the protein with myriad different tags or combinations of tags. In these cases the protocol described above could easily be adapted to accommodate these vectors. However, the pET system is based on a T7-driven promoter, and thus for expression of protein constructs, must be transformed into a BL21 host bearing the T7 RNA polymerase gene, i.e., a λDE3 lysogen. In addition to expressing proteins in *E. coli*, other hosts, e.g., mammalian cells, yeast, and insect cells, may be used to express GST-fusion proteins if expression in *E. coli* is of concern.

11. Although this protocol describes the purification of GST-fusion proteins from bacterial cell lysates, it can be adapted for the purification of GST-fusion proteins from other host cells. For example, if purifying GST-fusion proteins from insect cells or mammalian cells, effective lysis could be achieved by raising the concentration of NP40 to 1% (v/v). When using sonication to lyse cells it is important to keep the lysate on ice at all times and sonicate in short bursts. This should help to prevent overheating of the lysate and concomitant degradation of your GST-fusion protein.

12. After immobilizing your fusion protein on GSH-beads it is important to wash the beads thoroughly with lysis buffer to remove any contaminating proteins before incubating with the lysate.

13. When designing the experiment, include a GST-only negative control to ascertain if any observed interaction actually occurs between your immobilized protein and protein in the platelet lysate or if the observed interaction occurs with the GST moiety. The recommended preclearing of the platelet lysate with GSH-beads should help to remove most of the proteins that would interact with the GSH-beads, thus leaving only a potential interaction with the GST moiety as a spurious interaction.

14. Any observed interaction between the immobilized fusion protein and a protein present in the platelet lysate may be indirect, i.e., the interaction occurs through a multimeric complex. It is possible to validate whether this is a direct interaction by obtaining pure protein and using this in a GST-pulldown assay to see if the pure proteins are able to interact.

15. It is important to dissolve the glutathione solution to a pH of 8.0. If the solution is neutral then the elution efficiency for many proteins is significantly reduced.

16. It is usually sufficient to use 1 mL of GST-fusion-expressing bacterial lysate to generate enough protein for 2 mL of purified blotting protein at 20 μg/mL. The exact amount to use, however, will depend on the expression levels achieved.

17. Proteins run on SDS-PAGE gels in this manner will not appear in a structurally natured state. It is possible, therefore, that the phosphatase domain will not recognize the substrate in this form on the membrane. There are protocols used for renaturing the substrate on the membrane prior to blotting, but in the authors' hands these are generally not required.

References

1. Jackson, J. P., Schoenwaelder, S. M., Yuan, Y., Salem, H. H., and Cooray, P. (1996) Non-receptor protein tyrosine kinases and phosphatases in human platelets. *Thromb. Haemost.* **76**, 640–650.

2. Edmead, C. E., Crosby, D. A., Southcott, M., and Poole, A. W. (1999) Thrombin-induced association of SHP-2 with multiple tyrosine-phosphorylated proteins in human platelets. *FEBS Lett.* **459(1)**, 27–32.

3. Pasquet, J. M., Dachary-Prigent, J., and Nurden, A. T. (1998) Microvesicle release is associated with extensive tyrosine dephosphorylation in platelets stimulated by A23187 or a mixture of thrombin and collagen. *Biochem. J.* **333,** 591–599.

4. Barford, D., Das, A. K., and Egloff, M.-P. (1998) The structure and mechanism of protein phosphatases: insights into catalysis and regulation. *Ann. Rev. Biophys.* **27,** 133–164.

5. Pannifer, A. D. B., Flint, A. J., Tonks, N. K., and Barford, D. (1998) Visualization of the cysteinyl-phosphate intermediate of a protein-tyrosine phosphate by X-ray crystallography. *J. Biol. Chem.* **273(17),** 10,454–10,462.

6. Flint, A. J., Tiganis, T., Barford, D., and Tonks, N. J. (1997) Development of "substrate-trapping" mutants to identify physiological substrates of protein tyrosine phosphatases. *Proc. Natl. Acad. Sci. USA* **94,** 1680–1685.

7. Sathish, J. G., Johnson, K. G., Fuller, K. J., LeRoy, F. G., Meyaard, L., Sims, M. J., and Matthews, R. J. (2001) Constitutive association of SHP-1 with leukocyte-associated Ig-like receptor-1 in human T cells. *J. Immunol.* **166,** 1763–1770.

8. Sambrook, J. A. R., D. W. (2001) *Molecular cloning. A laboratory manual.* Third ed. Vol. 1. Cold Spring Harbor Laboratory Press, New York, pp. 1.116–1.118.

12

The Study of Serine-Threonine Kinases

Gertie Gorter and Jan Willem Akkerman

1. Introduction

Serine-threonine kinases play crucial roles in activating and inhibitory signaling pathways that control platelet activation. Diversity in isozyme composition, catalytic and regulatory subunits, and subcellular localization provide a multitude of signaling properties through which the platelet initiates its functional responses or prevents them. This chapter describes a number of techniques and assays that may be used to study various specific serine-threonine kinases in platelets and their regulation during the platelet-activation process.

2. Materials

Where concentrations of antibodies for immunoblotting are indicated in the methods section, these are for the specific reagents recommended. Other reagents may be equally suitable, but optimization of concentrations to be used may be required before use.

2.1. Isolation of Platelets

1. Acid citrate dextrose (ACD): 85.0 mM trisodium citrate, 71.4 mM citric acid, 111 mM glucose in distilled water.
2. HEPES-Tyrode's buffer: 145 mM NaCl, 5 mM KCl, 0.5 mM Na$_2$HPO$_4$, 1 mM MgSO$_4$, 10 mM HEPES, 5.5 mM glucose per liter distilled water; adjust to pH 6.5 or 7.25, as defined below.
3. Prostacyclin or PGI$_2$ (Cayman Chemical, Ann Arbor, MI).

2.2. Protein Kinase A

1. For the detection of protein kinase A catalytic subunits: Primary antibody, rabbit polyclonal antiprotein kinase A-C IgG, Cat. no. SC-903 (Santa Cruz Biotechnology, Santa Cruz, CA) in combination with a secondary antibody horseradish peroxidase (HRP)-labeled anti-rabbit IgG, Cat. no. 7071-1 (Cell Signaling, Beverly, MA).
2. For the detection of protein kinase A regulatory subunits R-I, R-IIα and R-IIβ, the following primary antibodies may be used: mouse anti-R-I IgG2b, Cat. no. P19920, mouse anti-R-IIα

From: *Methods in Molecular Biology, vol. 273:*
Platelets and Megakaryocytes, Vol. 2: Perspectives and Techniques
Edited by: J. M. Gibbins and M. P. Mahaut-Smith © Humana Press Inc., Totowa, NJ

IgG1, Cat. no. P55120 and mouse anti-R-IIβ IgG1, Cat. no. P54720 respectively (all from Transduction Laboratories, Lexington, KY) in combination with a secondary antibody HRP-conjugated goat anti-mouse IgG (e.g., GAMPO; DAKO A/S, Glostrup, Denmark).

3. VASP is detected using the primary antibody: goat anti-VASP Cat. no. SC-1950 (Santa Cruz Biotechnology) in combination with a secondary antibody HRP-conjugated rabbit anti-goat IgG (e.g., RAGPO, Cat. no. P0449, DAKO A/S).

4. Laemmli buffer 3X: 0.003% (w/v) bromophenol blue, 15% (v/v) β-mercaptoethanol, 30% (v/v) glycerol, 6% (w/v) SDS in 0.18 M Tris-HCl, pH 6.8.

5. Transfer buffer: 25 mM Tris-HCl, 192 mM glycine, 20% (v/v) methanol (pH of approx 9.0, although this does not require to be adjusted).

6. Blocking buffer: Tris-buffered saline (10 mM Tris-HCl, 150 mM NaCl, pH 7.4) containing 2% (w/v) fat-free dried milk , 0.5% (w/v) acid-free bovine serum albumin, and 0.1% (v/v) Tween-20.

7. Wash buffers: Tris-buffered saline containing 0.1% (v/v) Tween-20 and Tris-buffered saline without detergent.

8. Prestained protein molecular mass standards—broad range (e.g., from Bio-Rad Laboratories, Hercules, CA).

9. Nitrocellulose Western blot membrane (e.g., Optitran BA-S85, reinforced NL, Schleicher & Schuell, Dassel, Germany).

2.3. Protein·Kinase B

1. For the detection of protein kinase Bα: primary antibody, goat polyclonal anti-Akt1 IgG Cat. no. SC-7126 (Santa Cruz Biotechnology) in combination with the secondary antibody HRP-conjugated rabbit anti-goat IgG RAGPO (e.g., Cat. no. P0449, DAKO A/S).

2. For the detection of protein kinase Bβ: primary antibody, sheep polyclonal anti-Akt2 IgG 06-606 (Upstate) in combination with a secondary antibody HRP-labeled rabbit anti-sheep IgG (e.g., Cat. no. SC-1618, Santa Cruz Biotechnology).

3. For the detection of phosphorylated Thr[308] in protein kinase Bα: primary antibody, rabbit polyclonal ant-phosphoAkt Thr308 (Cat. no. 9275, Cell Signaling Technology) in combination with a secondary antibody HRP-labeled goat anti-rabbit IgG Cat. no. 7071-1 (Cell Signaling Technology).

4. For the detection of phosphorylated Ser[473] in protein kinase Bα: primary antibody, phospho-Akt Ser[473] (Cat. no. 9271, Cell Signaling Technology) in combination with a secondary antibody HRP-labeled anti-rabbit IgG (e.g., Cat. no. 7071-1, Cell Signaling Technology).

5. For the detection of total protein kinase B: primary antibody, goat polyclonal anti-Akt IgG (Cat. no. SC-1618, Santa Cruz Biotechnology) in combination with a secondary antibody rabbit anti-goat HRP conjugated RAGPO (e.g., Cat. no. P0449, DAKO A/S).

6. Laemmli buffer 3X: 0.003% (w/v) bromophenol blue, 15% (v/v) β-mercaptoethanol, 30% (v/v) glycerol, 6% (w/v) SDS in 0.18 M Tris-HCl, pH 6.8.

7. Transfer buffer: 25 mM Tris-HCl, 192 mM glycine, 20% (v/v) methanol (pH of approx 9.0, although this does not require to be adjusted).

8. Blocking buffer: Tris-buffered saline containing 5% fat-free dried milk, 0.1% (v/v) Tween-20.

9. Tris-buffered saline containing 1% (w/v) fat-free dried milk and 0.1% (v/v) Tween-20 and Tris-buffered saline with no detergent.

10. Nitrocellulose blot membrane (e.g., Optitran BA-S85, reinforced NL, Schleicher & Schuell).

11. For the detection of protein kinase Bα catalytic activity: 2X concentrated lysis buffer: 100 mM Tris-HCl, pH 7.5, 2 mM EDTA, 2 mM EGTA, 1 mM Na$_3$VO$_4$, 100 mM sodium fluoride, 10 mM sodium pyrophosphate, 20 mM sodium β-glycerophosphate, 0.1 mM

phenylmethylsulfonyl fluoride, 1 µg/mL aprotinin, 1 µg/mL pepstatin, 1 µg/mL leupeptin, 1 µM microcystin.

12. Wash buffer: 50 mM Tris-HCl, pH 7.5, 0.03% (w/v) Brij35, 0.1 mM EGTA, 0.1% (v/v) 2-mercaptoethanol.

13. Assay buffer: 10 mM MOPS, pH 7.2, 25 mM β-glycerolphosphate, 5 mM sodium orthovanadate, 1 mM dithiothreitol.

14. cAMP-dependent protein kinase A inhibitor peptide (TYADFIASGRTGRRNAI-NH₂) (custom-synthesized).

15. Protein kinase B substrate peptide (RPRAATF) (custom-synthesized).

2.4. Protein Kinase C

1. For the detection of protein kinase C-α: primary antibody, rabbit polyclonal IgG, SC-208 (Santa Cruz Biotechnology) in combination with a secondary antibody HRP-conjugated swine anti-rabbit IgG (e.g., SWARPO, DAKO A/S).

2. For the detection of protein kinase C-βII: primary antibody, rabbit polyclonal IgG SC-210 (Santa Cruz Biotechnology) in combination with a secondary antibody HRP-conjugated swine anti-rabbit IgG (e.g., DAKO A/S).

3. For the detection of phosphorylated protein kinase Cα/β: primary antibody, polyclonal rabbit anti-phospho-protein kinase Cα/β IgG (Thr^{638} on protein kinase Cα/Thr^{641} on protein kinase CβII) Cat. no. 9375 (Cell Signaling Technology) in combination with a secondary HRP-labeled goat anti-rabbit IgG (e.g., Cat. no. 7071-1, Cell Signaling Technology).

4. Laemmli buffer 3X: 0.003% (w/v) bromophenol blue, 15% (v/v) β-mercaptoethanol, 30% (v/v) glycerol, 6% (w/v) SDS in 0.18 M Tris-HCl, pH 6.8.

5. Transfer buffer: 25 mM Tris-HCl, 192 mM glycine, 20% (v/v) methanol.

6. Blocking buffer: Tris-buffered saline containing 5% (w/v) fat-free dried milk, 0.1% (v/v) Tween-20.Wash buffers: Tris-buffered saline containing 1% (w/v) fat-free dried milk and 0.1% (v/v) Tween-20 and Tris-buffered saline without added detergent.

7. For the detection of the phosphorylation of pleckstrin: [^{32}P]orthophosphoric acid Cat. no. NEX 053 (NEN Life Science Products, Boston, MA); polyacrylamide (Bio-Rad, Hercules, CA).

8. Staining solution: 0.1% (w/v) Coomassie brilliant blue R250, 45% (v/v) methanol, 10% (v/v) acetic acid, prepared in deionized water.

9. Destaining solution-1: 30% (v/v) acetic acid, 20% (v/v) methanol.

10. Destaining solution-2: 15% (v/v) acetic acid, 10% (v/v) methanol.

11. Destaining solution-3: 7.5% (v/v) acetic acid, 5% (v/v) methanol.

2.5. Protein Kinase G

See **Subheading 2.2.** for reagents to measure VASP phosphorylation.

2.6. p38-MAPKinase

1. For analysis of total p38^{MAPK}: primary antibody, polyclonal rabbit anti-p38^{MAPK} (Cat. no. 9212, Cell Signaling Technology) together with a secondary HRP-labeled goat anti-rabbit IgG (e.g., Cat. no. 7071-1, Cell Signaling Technology).

2. For analysis of phosphorylated p38^{MAPK}: primary antibody polyclonal rabbit anti-phospho-p38^{MAPK} (Thr^{180}/Tyr^{182}) (Cat. no. 9211, Cell Signaling Technology) together with a secondaryHRP-labeled goat anti-rabbit IgG (Cat. no. 7071-1, Cell Signaling Technology).

3. Laemmli buffer 3X: 0.003% (w/v) bromophenol blue, 15% (v/v) β-mercaptoethanol, 30% (v/v) glycerol, 6% (w/v) SDS in 0.18 M Tris-HCl, pH 6.8.

4. Transfer buffer: 25 mM Tris-HCl, 192 mM glycine, 20% (v/v) methanol.

5. Blocking buffer: 5% (w/v) fat-free dried milk and 0.1% (v/v) Tween-20 in phosphate buffered saline (140.4 mM NaCl, 2.1 mM NaH$_2$PO$_4$, 8.4 mM Na$_2$HPO$_4$).
6. Wash buffer: phosphate-buffered saline containing 1% (w/v) fat-free dried milk and 0.1% (v/v) Tween-20.

2.7. p42-MAPKinase

1. Rabbit polyclonal anti-p42^{MAPK} antibody Cat. no. SC-154 (Santa Cruz Biotechnology).
2. Protein A, HRP-linked (NA 9120, Amersham Biosciences, Lille Chalfont, Buckinghamshire, UK).
3. Laemmli buffer 3X: 0.003% (w/v) bromophenol blue, 15% (v/v) β-mercaptoethanol, 30% (v/v) glycerol, 6% (w/v) SDS in 0.18 M Tris-HCl, pH 6.8.
4. Transfer buffer: 25 mM Tris-HCl, 192 mM glycine, 20% (v/v) methanol (pH of approx 9.0, although this does not require to be adjusted).
5. Blocking buffer: 2% (w/v) fat-free dried milk, 0.5% (w/v) bovine serum albumin, 0.1% (v/v) Tween-20 in phosphate-buffered saline (140.4 mM NaCl, 2.1 mM NaH$_2$PO$_4$, 8.4 mM Na$_2$HPO$_4$).
6. Wash buffer: Tris-buffered saline containing 0.1% (v/v) Tween-20.
7. Prestained molecular mass protein standards (e.g., Kaleidoscope, Bio-Rad).

3. Methods

3.1. Isolation of Platelets

A major challenge for studies on serine-threonine kinases in platelets is to start with platelets that have preserved a resting, dormant state during blood collection and further handling, while at the same time being maximally responsive to platelet-activating agents. This is not an easy task. It demands special precautions with respect to donor selection, blood collection, and platelet-isolation procedures. Even when these steps have been performed with the greatest care, the researcher will occasionally find the enzymes in a preactivated state before platelets have made contact with an agonist. Further changes during platelet activation are then minor or virtually absent, and such experiments should be abandoned. For a more detailed discussion of the issues regarding the collection of blood and preparation of platelets, please refer to Chapters 1 and 2, vol. 1.

Blood should be obtained (with informed consent) from healthy, medication-free donors. Donors who take acetylsalicylic acid should be avoided, since the formation of thromboxane A$_2$ via an aspirin-sensitive step (via the activity of cyclo-oxygenase, COX1) serves as an important positive feedback loop in platelet activation. As discussed below, this loop may be either left intact when the feedback of this pathway on the regulation of serine-threonine kinases is a point of interest, or completely blocked through the addition of aspirin or indomethacin, for example, when the coupling between agonist-receptors and kinases is studied without interference by other pathways. It is recommended that donors avoid lipid-rich meals prior to blood collection to avoid interference by chylomicrons. This is especially important when lipophilic agonists—e.g., platelet-activating factor—are subject of study. In practice, this is often a problem.

Blood should be collected by means of a free-flow system, and the use of vacuum-driven systems for blood collection should be avoided, as these are certain means to activate intracellular kinases. Collect blood into trisodiumcitrate (13 mM, final concen-

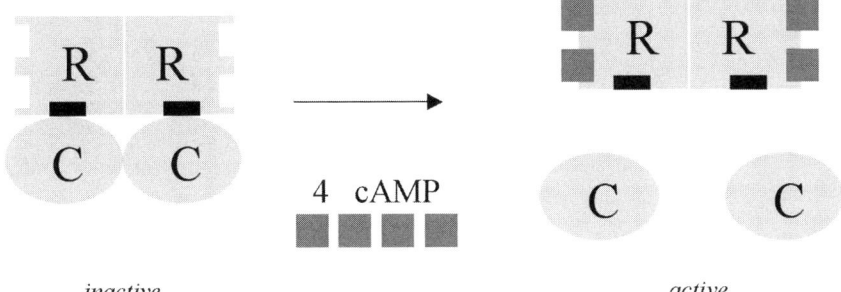

Fig. 1. Activation of protein kinase A. The protein kinase A complex consists of two regulatory subunits (R) and two catalytic subunits (C). In this configuration the complex is inactive. Binding of four molecules of cAMP to the regulatory domains releases and activates the catalytic subunits.

tration) as anticoagulant; avoid the use of heparin for this purpose, since this often leads to platelet activation. Do not use EDTA, as this will deplete intracellular Ca^{2+} stores and interfere with ligand binding to integrin $\alpha_{IIb}\beta_3$ (glycoprotein IIb-IIIa), which is a bivalent cation-dependent process.

1. Donors should be healthy and free from factors that might pose a risk to the investigators; donors should be medication-free.
2. Collect venous blood by free flow directly into 0.1 volume trisodiumcitrate (130 m*M*) in plastic tubes; mix gently.
3. Prepare platelet-rich plasma (PRP) by centrifugation of the blood for 15 min at 150*g* at 20°C.
4. Carefully collect the supernatant (PRP) and take a sample for platelet count; use plastic pipets.
5. Add 0.1 volume ACD to the PRP and mix gently.
6. Centrifuge the PRP at 330*g* for 15 min at 20°C.
7. Remove the complete supernatant.
8. Carefully resuspend the pellet in HEPES-Tyrode's solution, pH 7.25, to a density of 2×10^{11} cells/L.
9. Let the platelets recover for 30 min at 20°C to ensure a resting state before stimulation with agonists, or other treatment.
10. If a washing step is required, carefully resuspend the platelet pellet obtained after centrifugation of PRP in 10 mL HEPES-Tyrode's buffer, pH 6.5. Add prostacyclin (PGI$_2$, not the stable analog Iloprost) to a final concentration of 10 ng/mL.
11. Centrifuge at 330*g* for 15 min at 20°C.
12. Remove supernatant and carefully resuspend the platelets in HEPES-Tyrode's buffer, pH 7.25, to a density of of 2×10^{11}/L (*see* **Notes 1–5**).

3.2. Protein Kinase A

The protein kinase A complex consists of two regulatory subunits (R) and two catalytic subunits (C). In this configuration the complex is inactive. Binding of four molecules of cAMP to the regulatory domains releases and activates the catalytic subunits (**Fig. 1**). Megakaryocytic cell lines MEG-01, DAMI, and CHRF-288-11,

time (min)   0   5   10   15

Sp-5,6-DCl-cBIMPS 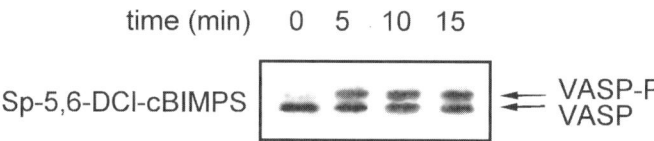 ← VASP-P
                                       ← VASP

Fig. 2. Phosphorylation of VASP by protein kinase A. Phosphorylation of VASP by Sp-5,6-DCl-cBIMPS, an activator of protein kinase A (from **ref. 6**).

and also human hematopoietic stem cells, megakaryocytes, and platelets, contain the different isotypes of the catalytic subunit and the regulatory subunits -Iα/β, -IIα, and -IIβ. Expression depends on the maturation stage of the cells *(1)*.

The activity of protein kinase A or cAMP-dependent kinase is deduced from the phosphorylation of one of its major substrates, the vasodilator-stimulated phospho-protein or VASP (49 kDa). This protein contains three phosphorylation sites (Ser[157], Ser[239], and Thr[278]). These sites can also be phosphorylated by protein kinase G. Thus, phosphorylation of VASP is not specific for protein kinase A activation and should be accompanied by incubation with a protein kinase A inhibitor such as H89, to discriminate between protein kinase A- and protein kinase G-mediated phosphorylations. Analysis of VASP phosphorylation is based on detection of the mobility shift on a polyacrylamide gel, which accompanies the phosphorylation of VASP **(Fig. 2)**.

3.2.1. Measurement of Protein Kinase A Subunit Expression

These methods are based on detection of the catalytic and regulatory subunits by Western blotting.

1. Collect a 100-μL sample from the platelet suspension and mix with 50 μL 3X concentrated Laemmli buffer. The use of screwcap or safe-lock microtubes is recommended.
2. Heat the samples for 5 min at 100°C, cool to 20°C, and centrifuge at 10,000g for 1 min at 20°C in a microcentrifuge.
3. Collect the supernatant for analysis by polyacrylamide gel electrophoresis (PAGE), for example using the Bio-Rad Miniprotean II system (Bio-Rad). (For further details regarding SDS-PAGE, refer to Chapter 9, vol. 2.)
4. Prepare a 10% SDS polyacrylamide gel with a thickness of 1.5 mm; acrylamide/bisacrylamide 37.5/1 (w/w).
5. Apply samples of 30 or 40 μL and apply prestained molecular mass protein standards (broad range). Separate the proteins by electrophoresis (30 mA/gel, maximal voltage; 20°C using the stated system).
6. Collect the gels and transfer the proteins electrophoretically to a nitrocellulose blot membrane for example using a Transblot system (Bio-Rad) for 1 h at 125 V. The transfer procedure is carried out at 4°C and the system is placed on a magnetic stirrer to prevent local heating of the transfer buffer. (For further details regarding Western blotting, please refer to Chapter 9, vol. 2.)
7. Collect the membrane and place it against the inner surface of a 50-mL tube with the proteins exposed.

8. Block the proteins by adding 10 mL blocking buffer to diminish nonspecific binding of the antibody used later for detection of protein kinase A subunits; incubate 1–2 h at 20°C with constant agitation.

3.2.1.1. ANALYSIS OF PROTEIN KINASE A CATALYTIC SUBUNITS

1. Incubate the membrane with a 2000-fold dilution of the anti-protein kinase A catalytic subunits IgG in Tris-buffered saline containing 0.1% (v/v) Tween-20 for 4 h or, if more convenient, overnight with constant agitation at 4°C.
2. Continue with 6 washing steps, each consisting of a 10-min incubation with 10 mL Tris-buffered saline containing 0.1% (v/v) Tween-20 with constant agitation at 4°C.
3. Incubate the membrane with a 3000-fold dilution of horseradish peroxidase (HRP)-labeled anti-rabbit IgG in Tris-buffered saline containing 0.1% (v/v) Tween-20 for 1–2 h with constant agitation at 4°C.
4. Continue with five washing steps, each consisting of a 10-min incubation with 10 mL Tris-buffered saline containing 0.1% (v/v) Tween-20 with constant agitation at 4°C.
5. Carry out a final washing step with 10 mL Tris-buffered saline without Tween-20 for 10 min at 4°C.
6. Visualize the protein band by chemiluminescence using the Western blot chemiluminescence reagent (e.g., from PerkinElmer Life Sciences, Boston, MA) and suitable X-ray film (e.g., Kodak XOmat blue XB-1 film). Record more than one exposure time.

3.2.1.2. ANALYSIS OF PROTEIN KINASE A REGULATORY SUBUNITS

1. Incubate the membrane with a 250-fold dilution of the anti-protein kinase A R-I IgG, a 500-fold dilution of the anti-protein kinase A R-IIα IgG, or a 2000-fold dilution of the anti-protein kinase A R-IIβ IgG (depending on the subunit under investigation) in Tris-buffered saline containing 0.1% Tween-20 for 4 h or—if more convenient—overnight with constant agitation at 4°C.
2. Continue with six washing steps, each consisting of a 10-min incubation with 10 mL Tris-buffered saline containing 0.1% (v/v) Tween-20 with constant agitation at 4°C.
3. Incubate the membrane with a 5000-fold dilution of HRP-conjugated goat anti-mouse IgG (GAMPO) in Tris-buffered saline containing 0.1% (v/v) Tween-20 for 1–2 h with constant agitation at 4°C.
4. Wash membranes and visualize and record protein bands using a chemiluminescent reagent as decribed in **Subheading 3.2.1.1., steps 4–6**.

3.2.2. Measurement of VASP Phosphorylation

The measurement is based on the mobility shift on SDS-PAGE that occurs when VASP is phosphorylated, shifting the apparent molecular weight from 48 to 50 kDa.

1. Prepare and separate platelet proteins by SDS-PAGE, and transfer onto nitrocellulose membranes in preparation for immunoblotting as described in **Subheading 3.2.1., steps 1–8**.
2. Incubate the membrane with a 250-fold dilution of the anti-VASP IgG in Tris-buffered saline containing 4% (w/v) fat-free dried milk and 0.1% (v/v) Tween-20 for 4 h or, if more convenient, overnight with constant agitation at 4°C.
3. Continue with six washing steps, each consisting of a 10-min incubation with 10 mL Tris-buffered saline containing 0.1% (v/v) Tween-20 with constant agitation at 4°C.

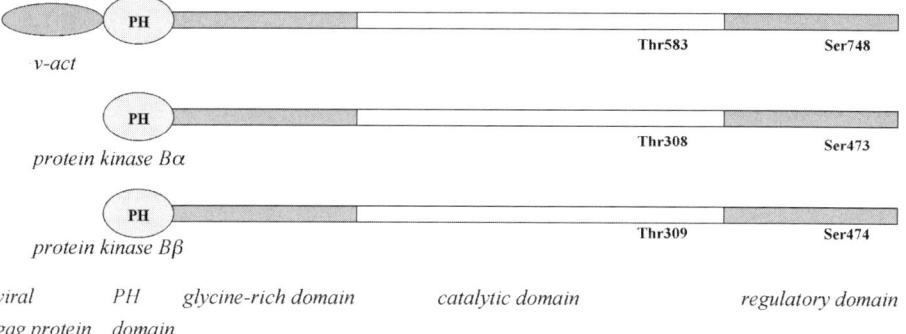

Fig. 3. Structure of protein kinase B. Platelets contain the protein kinase Bα and -β isoforms differing slightly in the position of the phosphorylation sites. Both contain a pleckstrin homology (PH) domain for binding to the plasma membrane (modified from **ref. 7**).

4. Incubate the membrane with a 2000-fold dilution of HRP-conjugated rabbit anti-goat IgG in Tris-buffered saline containing 4% (w/v) fat-free dried milk and 0.1% (v/v) Tween-20 for 1–2 h with constant agitation at 4°C.
5. Wash membranes and visualize and record protein bands using a chemiluminescent reagent as decribed in **Subheading 3.2.1.1., steps 4–6** (*see* **Notes 6,7**).

3.3. Protein Kinase B

Protein kinase B (PKB, also known as RAC or Akt kinase) is a 57-kDa, phospholipid-dependent serine/threonine kinase and a product of the oncogene v-akt of the acutely transforming retrovirus Akt8 **(Fig. 3)**. It was first isolated from a rodent T-cell lymphoma and had the capacity to induce cell transformation. Subsequently, PKB was reported to prevent apoptosis, to regulate glyconeogenesis by phosphorylation of glycogen synthase kinase 3 (GSK3), and to control glucose uptake by inducing translocation of the glucose transporter GLUT-4 to the plasma membrane. In platelets, protein kinase B is activated by α-thrombin, thrombopoietin, and collagen via pathways involving members of the PtdIns-3 kinase family and Ca^{2+}-dependent protein kinase C isoforms. The downstream effects in platelets are yet to be identified.

The human genome encodes for at least three different PKB genes, which display more than 80% sequence homology and are named PKBα, -β, and -γ. Human platelets contain PKBα and, to a lesser extent, the PKBβ isoform, but not PKBγ *(2)*. The protein kinase Bα and -β isoforms differ in the position of the phosphorylation sites. Both contain a pleckstrin homology (PH) domain for binding to the plasma membrane. Protein kinase Bα is activated by dual phosphorylation on Thr[308] and Ser[473] by phosphatidylinositol-dependent kinases **(Fig. 4)**. Protein kinase C contributes to the phosphorylation of Ser[473] but not to the phosphorylation of Thr[308]. The extent by which both sites are phosphorylated is proportional to the catalytic activity *(2)*.

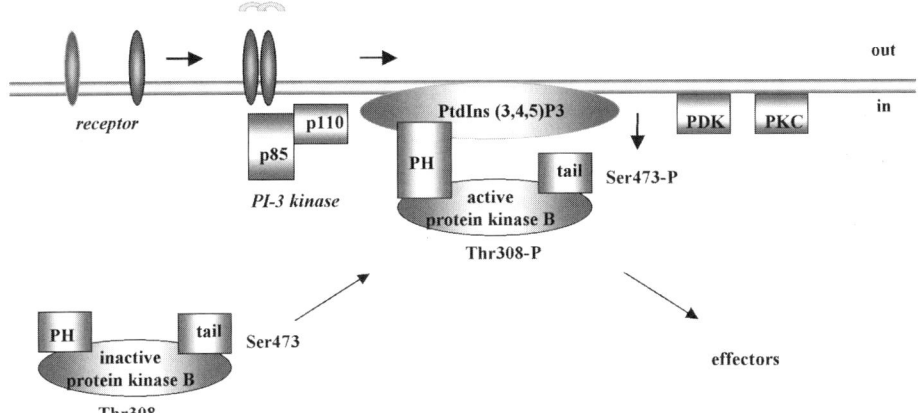

Fig. 4. Activation of protein kinase Bα. A strong activator of platelet protein kinase B is thrombopoietin. Binding to its receptor, the cMpl-receptor, induces dimerization and signaling to PtdIns-3 kinase. PtdIns $(3,4,5)P_3$ is formed which serves as a docking site for the PH domains of protein kinase B. The cytosolic enzyme is inactive. Docking to PtdIns $(3,4,5)P_3$ brings the enzyme together with PtdIns-dependent kinases (PDKs), which phosphorylate and thereby activate protein kinase B. A second activator of protein kinase B is α-thrombin, which signals to Ser473 via Ca^{2+}-dependent isoforms of protein kinase C.

3.3.1. Measurement of Protein Kinase Bα and Protein Kinase Bβ Expression

This method is based on detection of protein kinase B subtypes by Western blotting.

1. Prepare and separate platelet proteins by SDS-PAGE, and transfer onto nitrocellulose membranes in preparation for immunoblotting as described in **Subheading 3.2.1., steps 1–8**.
2. Incubate the membrane with a 7000-fold dilution of the anti-protein kinase Bα IgG or the anti-protein kinase Bβ IgG in Tris-buffered saline containing 5% (w/v) fat-free dried milk and 0.1% (v/v) Tween-20 for 2 h or, if more convenient, overnight with constant agitation at 4°C.
3. Continue with six washing steps, each consisting of a 10-min incubation with 10 mL Tris-buffered saline containing 1% (w/v) fat-free dried milk and 0.1% (v/v) Tween-20 with constant agitation at 4°C.
4. Incubate the membrane with a 2000-fold dilution of HRP-labeled rabbit, anti-goat, and rabbit anti-sheep IgG in Tris-buffered saline containing 5% (w/v) fat-free dried milk and 0.1% (v/v) Tween-20 for 1–2 h with constant agitation at 4°C.
5. Wash membranes and visualize and record protein bands using a chemiluminescent reagent as decribed in **Subheading 3.2.1.1., steps 4–6**.

3.3.2. Measurement of Protein Kinase Bα Phosphorylation

Since protein kinase Bα is activated by phosphorylation of Thr[308] and Ser[473], detection of these phosphorylated residues with phosphospecific antibodies serves to evaluate the activation state of the enzyme (*see* **Note 8**). The method is based on protein separation by electrophoresis followed by Western blotting with appropriate antibodies. A control on total (phosphorylated and nonphosphorylated) protein kinase B obtained by stripping and reprobing with an anti-total protein kinase B antibody should accompany the blots of phosphoprotein kinase B.

1. Isolated platelets suspended in HEPES-Tyrode's are incubated at 20 or 37°C for 10 min and stimulated with the desired agonist, for example 0.2 U/mL α-thrombin (final conc.), without stirring.
2. At 0, 1, 2, 5, and 10 min a 100-μL sample is collected and mixed with 50 μL 3X concentrated Laemmli buffer. The use of screwcap or safe-lock microtubes is recommended.

Prepare and separate platelet proteins by SDS-PAGE, and transfer onto nitrocellulose membranes in preparation for immunoblotting as described in **Subheading 3.2.1., steps 2–8**.

The following procedures depend on the properties of protein kinase Bα that are of interest.

3.3.2.1. ANALYSIS OF PHOSPHORYLATED THR[308] ON PROTEIN KINASE Bα

1. Incubate the membrane with a 1500-fold dilution of anti-phospho-Akt Thr[308] antibody in Tris-buffered saline containing 5% (w/v) fat-free dried milk and 0.1% (v/v) Tween-20 for 4 h or, if more convenient, overnight with constant agitation at 4°C.
2. Continue with six washing steps, each consisting of a 10-min incubation with 10 mL Tris-buffered saline containing 1% (w/v) fat-free dried milk and 0.1% (v/v) Tween-20 with constant agitation at 4°C.
3. Incubate the membrane with a 2000-fold dilution of HRP-labeled anti-rabbit antibody in Tris-buffered saline containing 5% (w/v) fat-free dried milk and 0.1% (v/v) Tween-20 for 1–2 h with constant agitation at 4°C.
4. Wash membranes and visualize and record protein bands using a chemiluminescent HRP substrate as decribed in **Subheading 3.2.1.1., steps 4–6 (Fig. 5)** (*see* **Note 6**).

3.3.2.2. ANALYSIS OF PHOSPHORYLATED SER[473] ON PROTEIN KINASE Bα

1. Incubate the membrane with a 1500-fold dilution of anti-phospho-Akt Ser[473] antibody in Tris-buffered saline containing 5% (w/v) fat-free dried milk and 0.1% (v/v) Tween-20 for 4 h or, if more convenient, overnight with constant agitation at 4°C.
2. Continue with six washing steps, each consisting of a 10-min incubation with 10 mL Tris-buffered saline containing 1% (w/v) fat-free dried milk and 0.1% (v/v) Tween-20 with constant agitation at 4°C.
3. Incubate the membrane with a 2000-fold dilution of HRP-labeled anti-rabbit IgG in Tris-buffered saline containing 0.1% (v/v) Tween-20 and 5% (w/v) fat-free dried milk for 1–2 h with constant agitation at 4°C.
4. Wash membranes and visualize and record protein bands using a chemiluminescent reagent as decribed in **Subheading 3.2.1.1., steps 4–6** (*see* **Note 6**).

3.3.2.3. ANALYSIS OF TOTAL PROTEIN KINASE Bα,β

1. Incubate the membrane with a 3000-fold dilution of anti-total protein kinase Bα,β antibody in Tris-buffered saline containing 5% (w/v) fat-free dried milk and 0.1% (v/v) Tween-20 for 4 h or, if more convenient, overnight with constant agitation at 4°C.

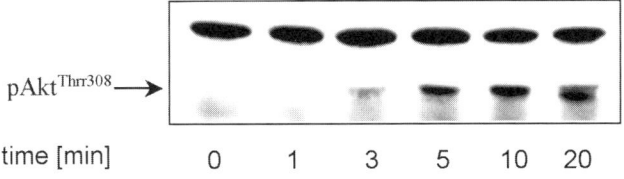

$pAkt^{Thr308} \longrightarrow$

time [min]   0   1   3   5   10   20

Fig. 5. Phosphorylation of protein kinase B (Akt) on Thr[308]. Platelets were stimulated with thrombopoietin and the phosphorylation of Thr[308] was measured. A nonspecific band of about 70 kDa served as a control for lane loading (from **ref. 2**).

2. Continue with six washing steps, each consisting of a 10-min incubation with 10 mL Tris-buffered saline containing 1% (w/v) fat-free dried milk and 0.1% (v/v) Tween-20 with constant agitation at 4°C.
3. Incubate the membrane with a 5000-fold dilution of HRP-conjugated rabbit anti-goat IgG (RAGPO) in Tris-buffered saline containing 5% (w/v) fat-free dried milk and 0.1% (v/v) Tween-20 for 1–2 h with constant agitation at 4°C.
4. Wash membranes and visualize and record protein bands using a chemiluminescent reagent as decribed in **Subheading 3.2.1.1., steps 4–6**.

3.3.2.4. MEASUREMENT OF PROTEIN KINASE Bα CATALYTIC ACTIVITY

Because detection of activation-specific phosphorylation sites gives only a semi-quantitative indication of the activity of the enzyme, it is often necessary to combine these assays with a test on the catalytic activity. The method is based on the isolation of activated protein kinase B from a platelet lysate by immunoprecipitation and analysis of $[\gamma^{32}P]$ATP incorporation in an artificial substrate of protein kinase B *(2)*. (This procedure should be carried out under the appropriate safety regulations for the safe use of radioisotopes.)

1. Stimulate 250 µL of washed platelets with a platelet agonist for 10 min at 20°C.
2. Stop the reaction with by adding 250 µL 2X concentrated lysis buffer.
3. Snap-freeze the samples in liquid nitrogen and store at –70°C.
4. Thaw samples and add 4 µg anti-total protein kinase B antibody coupled to protein A-sepharose beads (Amersham Pharmacia Biotech, Uppsala, Sweden); incubate with constant agitation overnight at 4°C to immunoprecipitate the enzyme.
5. Wash pellet three times with 500 µL lysis buffer containing 0.5 M NaCl; twice with 500 µL washing buffer and twice with 100 µL assay buffer.
6. Resuspend pellet in 10 µL ice-cold assay dilution buffer, 10 µL of 40 µM cAMP-dependent protein kinase A inhibitor peptide, and 100 µM protein kinase B substrate peptide.
7. Start the reaction by adding 10 µCi (370 kBα) $[\gamma^{-32}P]$ATP (NEN PerkinElmer Life Science Products, Boston, MA) to a final concentration of 1 mCi/mL and incubate for 10 min at 30°C under continuous shaking.
8. Precipitate the proteins with 40% (v/v) trichloroacetic acid and spot 40 µL samples onto phosphocellulose paper.
9. Wash the paper three times in 0.75% (v/v) phosphoric acid and once in acetone.
10. Measure the radioactivity.

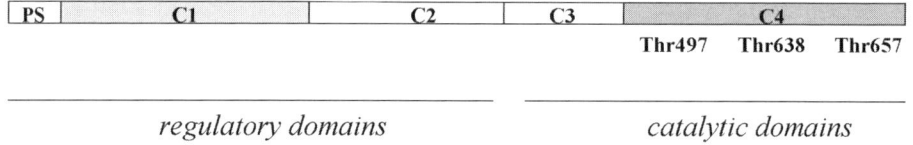

<table>
<tr><td>PS</td><td>C1</td><td>C2</td><td>C3</td><td>C4</td></tr>
</table>

Thr497   Thr638   Thr657

regulatory domains   *catalytic domains*

Fig. 6. Structure of protein kinase Cα. Platelets contain the classical protein kinase C isoforms α, β_1, and β_{11}, which are regulated by diacylglycerol (DAG), phosphatidylserine, and Ca^{2+},; the novel protein kinase C isoforms δ, ϵ, θ, and η, which are regulated by DAG and phosphatidylserine; and the atypical ζ isoform, which is regulated by phosphatidylserine. The classical isoforms contain a PS domain, which inhibits the catalytic domains; the C1 domains with binding sites for DAG and phorbolester; and the C2 domain, which binds Ca^{2+} and acidic phospholipids. The catalytic domains C3 and C4 bind to the substrates and possess the kinase activity.

3.4. Protein Kinase C

Protein kinase C is a group of serine-threonine kinases with molecular weights between 67 and 83 kDa that play a key role in signal transduction pathways in platelets. The members of this family of kinases differ in their activation mechanism (**Fig. 6**). A central step in the activation of protein kinase C is its translocation from the cytosol to specific sites at the inner leaflet of the plasma membrane that are rich in phosphatidylserine and diacylglycerol, a product of the phospholipase C pathway (**Fig. 7**). Platelets contain the classical protein kinase C isoforms α, β_1, and β_{11}, which are regulated by diacylglycerol (DAG), phosphatidylserine and Ca^{2+}, the novel protein kinase C isoforms δ, ϵ, θ, and η, which are regulated by DAG and phosphatidylserine; and the atypical ζ isoform, which is regulated by phosphatidylserine. Ca^{2+} ions are important cofactors for the so-called classical (or conventional) isotypes, but not for the novel and atypical isotypes. The downstream effects of activated protein kinase C are numerous, including regulation of integrin $\alpha_{IIb}\beta_3$ and aggregation, secretion of dense-, α-, and lysosomal granule contents, cytoskeletal reassembly, and clot retraction (*see* **Note 9**).

3.4.1. Measurement of Protein Kinase Cα and Protein Kinase CβII Expression

The method is based on detection of these conventional protein kinase C subtypes by Western blotting.

1. Prepare and separate platelet proteins by SDS-PAGE as described in **Subheading 3.2.1., steps 1–3**.
2. Prepare a 7.5% SDS-polyacrylamide gel with a thickness of 1.5 mm; acrylamide/bisacrylamide 37.5/1 (w/w).
3. Separate platelet proteins by SDS-PAGE, and transfer onto nitrocellulose membranes in preparation for immunoblotting as described in **Subheading 3.2.1., steps 5–8**.

The following procedures depend on the properties of protein kinase C that are of interest.

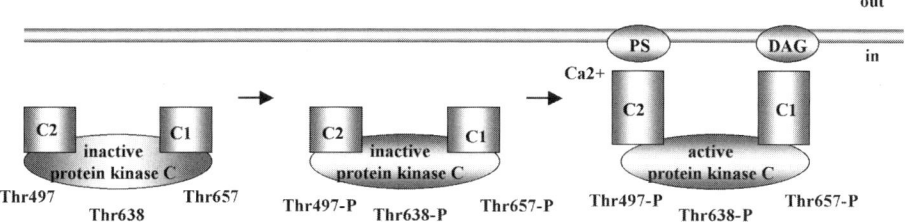

Fig. 7. Activation of protein kinase Cα. A strong activator of the protein kinase Cα is α-thrombin. Binding to its receptors (PAR1,3,4) starts signaling to phospholipase Cβ. Diacylglycerol (DAG) is formed, which together with phosphatidylserine (PS) serves as docking site for the cytosolic, inactive protein kinase C. Thr^{497} is the initial phosphorylation site of protein kinase C and is phosphorylated by a protein kinase C-kinase. Thr^{638} and Thr^{657} serve as autophosphorylation sites. Phosphorylated protein kinase C binds to phosphatidylserine via its C2 domain in a Ca^{2+}-dependent manner, bringing the enzyme in close proximity of the activator DAG. DAG then binds to the C1 domain and relieves the inhibition of the catalytic domains.

3.4.2. Analysis of Protein Kinase Cα

1. Incubate the membrane with a 7000-fold dilution of anti-protein kinase Cα antibody in Tris-buffered saline containing 5% (w/v) fat-free dried milk and 0.1% (v/v) Tween-20 for 2 h or, if more convenient, overnight with constant agitation at 4°C.
2. Continue with six washing steps, each consisting of a 10-min incubation with 10 mL Tris-buffered saline containing 1% (w/v) fat-free dried milk and 0.1% (v/v) Tween-20 with constant agitation at 4°C.
3. Incubate the membrane with a 10,000-fold dilution of HRP-conjugated swine anti-rabbit IgG (SWARPO) in Tris-buffered saline containing 5% (w/v) fat-free dried milk and 0.1% (v/v) Tween-20 for 1–2 h with constant agitation at 4°C.
4. Wash membranes and visualize and record protein bands using a chemiluminescent reagent as decribed in **Subheading 3.2.1.1., steps 4–6**.

3.4.3. Analysis of Protein Kinase CβII

1. Incubate the membrane with a 7000-fold dilution of antiprotein kinase CβII antibody in Tris-buffered saline containing 5% (w/v) fat-free dried milk and 0.1% (v/v) Tween-20 for 4 h or, if more convenient, overnight with constant agitation at 4°C.
2. Continue with six washing steps, each consisting of a 10-min incubation with 10 mL Tris-buffered saline containing 1% fat-free dried milk and 0.1% Tween-20 and with constant agitation at 4°C.
3. Incubate the membrane with a 10,000-fold dilution of HRP-conjugated swine anti-rabbit IgG in Tris-buffered saline containing 5% (w/v) fat-free dried milk and 0.1% (v/v) Tween-20 for 1–2 h with constant agitation at 4°C.
4. Wash membranes and visualize and record protein bands using a chemiluminescent reagent as decribed in **Subheading 3.2.1.1., steps 4–6**.

3.4.4. Measurement of Thr⁶³⁸/Thr⁶⁴¹-Phosphorylation of Protein Kinase Cα/βII

3.4.4. Measurement of Thr^{638}/Thr^{641}-Phosphorylation of Protein Kinase Cα/βII

Phosphorylation of Thr residues is a key step in the activation of protein kinase Cα/βII. The method is based on protein separation by electrophoresis followed by Western blotting with appropriate antibodies directed against the phosphorylated Thr residues.

1. Incubate isolated platelets suspended in HEPES-Tyrode's at 20 or 37°C for 10 min and stimulated with the desired agonist, for example 0.2 U/mL α-thrombin (final conc) without stirring.
2. At 0, 1, 2, 5, and 10 min, collect a 100-μL sample and mix with 50 μL 3X concentrated Laemmli buffer. The use of screwcap or safe-lock microtubes is recommended.
3. Separate platelet proteins on a 7.5% SDS-polyacrylamide gel and transfer proteins onto a nitrocellulose membrane as described in **Subheading 3.2.1., steps 2–8**, in preparation for immunoblotting.
4. Incubate the membrane with a 10,000-fold dilution of the anti-phosphoprotein kinase Cα/βII antibody in Tris-buffered saline containing 0.1% (v/v) Tween-20 and 5% (w/v) fat-free dried milk for 2 h or, if more convenient, overnight with constant agitation at 4°C.
5. Continue with six washing steps, each consisting of a 10-min incubation with 10 mL Tris-buffered saline containing 0.1% (v/v) Tween-20 and 1% (w/v) fat-free dried milk with constant agitation at 4°C.
6. Incubate the membrane with a 10,000-fold dilution of HRP-labeled anti-rabbit IgG antibody in Tris-buffered saline containing 0.1% (v/v) Tween-20 and 5% (w/v) fat-free dried milk for 1–2 h with constant agitation at 4°C.
7. Wash membranes and visualize and record protein bands using a chemiluminescent reagent as decribed in **Subheading 3.2.1.1., steps 4–6** (*see* **Note 6**).

3.4.5. Measurement of Catalytic Activity of Protein Kinase C

One of the major substrates of protein kinase C in platelets is the 47-kDa protein pleckstrin. Since it is phosphorylated much more strongly than other substrates of this kinase and is easily separated by electrophoresis, its phosphorylation has become a popular marker to assess the overall activity of protein kinase C family members *(3)*. The assay is based on ^{32}P incorporation in pleckstrin, protein separation by electrophoresis, and analysis of radioactivity in the 47-kDa protein band detected using X-ray film **(Fig. 8)**.

3.4.5.1. PREPARATION OF ^{32}P-LABELED PLATELETS

This procedure should be carried out under the appropriate regulations for the safe use of radioisotopes.

1. Prepare PRP as described in **Subheading 3.1.**
2. Add to PRP [^{32}P]orthophosphoric acid to a final concentration of 3.7 MBq/mL.
3. Mix gently without turning the sample upside down.
4. Incubate for 60 min at 37°C without stirring.
5. Let the sample stand at 20°C until the temperature of the sample is adjusted.
6. Add 0.1 volume ACD.

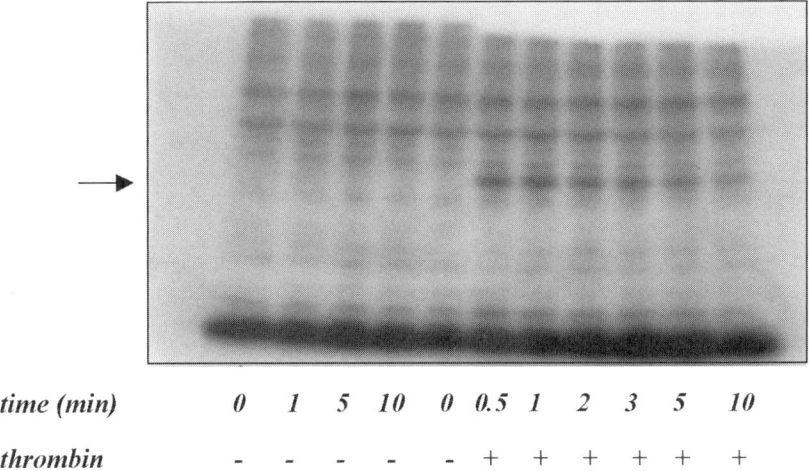

time (min)	0	1	5	10	0	0.5	1	2	3	5	10
thrombin	-	-	-	-	-	+	+	+	+	+	+

Fig. 8. Phosphorylation of pleckstrin. Pleckstrin phosphorylaton induced by 0.1 U/mL α-thrombin. The arrow indicates the 47-kDa band.

7. Centrifuge at 330*g* for 20 min at 20°C.
8. Remove the supernatant completely (use caution, as this contains >90% of the total ^{32}P).
9. Resuspend the platelet pellet carefully in HEPES-Tyrode's buffer, pH 7.25, to a final platelet density of 0.2×10^9 platelets/mL.
10. Allow 30 min at 20°C, without stirring, to restore platelets to a resting state.

3.4.5.2. MEASUREMENT OF PLECKSTRIN PHOSPHORYLATION

1. ^{32}P-labeled platelets are incubated at 20 or 37°C and stimulated with the desired agonist, for example 0.2 U/mL α-thrombin (final conc.), without stirring.
2. At 0, 1, 2, 5, and 10 min, collect a 100-μL sample and mix with 50 μL 3X concentrated Laemmli buffer. The use of screwcap or safe-lock microtubes is recommended.
3. Heat samples for 10 min at 100°C, cool to 20°C, and centrifuge at 10,000*g* for 1 min at 20°C in a microcentrifuge.
4. Collect the supernatant for analysis by electrophoresis, for example using the Bio-Rad Miniprotean II system (Bio-Rad, Hercules, CA).
5. Prepare 11% SDS-polyacrylamide gels with a thickness of 1.5 mm (acrylamide/bisacrylamide 37.5/1 [w/w]).
6. Apply samples of 30 or 40 μL each and apply prestained molecular mass protein standards (e.g., from Bio-Rad).
7. Separate the proteins by electrophoresis (30 mA/gel, maximal voltage, 20°C).
8. Collect the gels and visualize the separated proteins by staining with Coomassie brilliant blue for 30 min with mild shaking at 20°C.
9. Destain the gels with destaining solutions 1, 2, and 3 for 90, 60, and 60 min respectively (for practical reasons, the last washing step is usually an overnight incubation) with mild shaking at 20°C.
10. Dry the gel on a Geldryer (Bio-Rad) between two sheets of cellophane.
11. Visualize the radioactive proteins by exposure to X-ray film, e.g., Kodak X-Omat blue XB-1 film (Eastman Kodak, Rochestor, NY) (*see* **Note 6**).

3.5. Protein Kinase G

Protein kinase G or AMP-dependent protein kinase is a homodimer, with each subunit containing a catalytic and a regulatory domain. Two isozymes have been identified (types I and II) but platelets contain only type I. A major substrate of protein kinase G is the protein VASP, which can be used for the measurement of protein kinase G activity (*see* **Subheading 3.2.3.**).

3.6. p38-MAPKinase

p38 MAPkinase (p38[MAPK]) is a member of the family of mitogen-activated protein kinases (MAP kinases) which are proline-directed serine-threonine kinases. MAPkinases are activated in response to their simultaneous phosphorylation on threonine and tyrosine in a Thr-X-Tyr motif by dual-specificity mapkinase-kinases. For p38[MAPK] this motif is Thr[180]-Gly[181]-Tyr[182] located in a regulatory loop between subdomains VII and VIII. In turn, these MAPkinase-kinases are activated by phosphorylation on serine and threonine by upstream MAPkinase-kinase-kinases. Downstream targets of activated p38[MAPK] are cytosolic phospholipase A_2 and the mechanisms that control F-actin polymerization. Platelet p38[MAPK] is activated by α-thrombin, collagen, thromboxane A_2-analog, and low-density lipoprotein (LDL), among other agonists *(4)*. There are four subtypes, named p38α (38 kDa), p38β (39 kDa), p38γ (43 kDa), and p38δ (40 kDa). The measurement is based on detection of phosphorylated Thr[180]/Tyr[182] **(Fig. 9)**. Since stripping-reblotting often leads to erroneous results, a second electrophoresis step is used to detect total p38[MAPK] (*see* **Notes 10,11**).

1. Collect a 100-µL sample of platelet suspension in 15 µL 1.0 *M* formaldehyde in 154 m*M* NaCl in a microtube.
2. Leave the sample on ice for 30 min.
3. Centrifuge at 10,000*g* for 1 min at 4°C in a microfuge and remove the complete supernatant.
4. Resupend the pellet in 30 µL 3X concentrated Lammli buffer.
5. Clear the sample by pressing it through a needle (0.33 mm diameter).
6. Heat the sample for 10 min at 100°C.
7. Cool the sample to 20°C and store the sample at –20°C until further analysis.
8. Centrifuge the samples at 10,000*g* for 1 min for 20°C and collect the supernatant for analysis by SDS-PAGE.
9. Prepare two 12% SDS-polyacrylamide gels with a thickness of 1.5 mm (acrylamide/bisacrylamide 37.5/1 [w/w]).
10. Apply 15 µL sample/lane on each of the gels together with prestained molecular mass protein standards.
11. Separate the proteins by electrophoresis (30 mA/gel, maximal voltage, 20°C).
12. Transfer the proteins to a nitrocellulose membrane for example using a Transblot system (Bio-Rad; 1 h, 125 V, 4°C) with stirring to prevent local heating.
13. Block membranes using 10 mL of blocking buffer to diminish nonspecific binding of the antibody used for detection of p38[MAPK] and incubate 1–2 h at 4°C with constant agitation.
14. Incubate one of the membranes with a 2000-fold dilution of the antiphospho-p38[MAPK] antibody in phosphate-buffered saline containing 1% (w/v) fat-free dried milk and 0.1% (v/v) Tween-20 for 2 h or, if more convenient, overnight with constant agitation at 4°C. To measure total p38[MAPK] incubate another membrane containing samples treated in an iden-

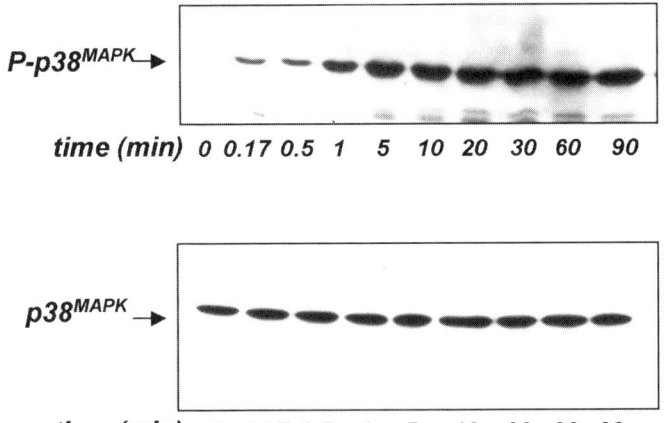

Fig. 9. Phosphorylation of p38^{MAPK}. Phosphorylation of p38^{MAPK} by 1 g/L low-density lipoprotein, LDL (from **ref. 4**).

tical manner with a 2000-fold dilution of anti-total p38^{MAPK} antibody in phosphate-buffered saline containing 1% (w/v) fat-free dried milk and 0.1% (v/v) Tween-20 for 2 h or, if more convenient, overnight with constant agitation at 4°C.

15. Wash membranes and incubate the membrane with a 10,000-fold dilution of HRP-labeled anti-rabbit IgG antibody in Tris-buffered saline containing 0.1% (v/v) Tween-20 and 1% (w/v) fat-free dried milk for 1–2 h with constant agitation at 4°C.

16. Wash membranes and visualize and record protein bands using a chemiluminescent HRP substrate as decribed in **Subheading 3.2.1.1., steps 4–6** (*see* **Note 6**).

3.7. p42-MAPKinase

p42^{MAPK} is also a member of the family of mitogen-activated protein kinases; it is also known as extracellular signal-regulated kinase, or ERK2. Members of the p42^{MAPK} family contain the Thr-Glu-Tyr motif *(5)*. The detection of phosphorylated p42^{MAPK} is based on the mobility shift on SDS-polyacrylamide electrophoresis that accompanies the phosphorylation of p42^{MAPK} (**Fig. 10**) (*see* **Note 12**).

1. Incubate platelets at 20 or 37°C and stimulate with an agonist, for example, 0.2 U/mL α-thrombin (final conc), without stirring.

2. At 0, 1, 2, 5, and 10 min, collect a 100-μL sample and mix with 50 μL 3X concentrated Laemmli buffer. The use of screwcap or safe-lock microtubes is recommended.

3. Heat samples for 10 min at 100°C, cool to 20°C, and centrifuge at 10,000g for 1 min at 20°C in a microcentrifuge.

4. Collect the supernatant for analysis by electrophoresis, for example using the Bio-Rad Protean II system.

5. Prepare a 12.8% SDS polyacrylamide gel with a thickness of 0.75 mm (acrylamide/bisacrylamide 173/1 [w/w]).

6. Apply samples of 30 or 40 μL each containing about 25 μg protein per well. Also apply prestained molecular mass protein standards (e.g., from Bio-Rad).

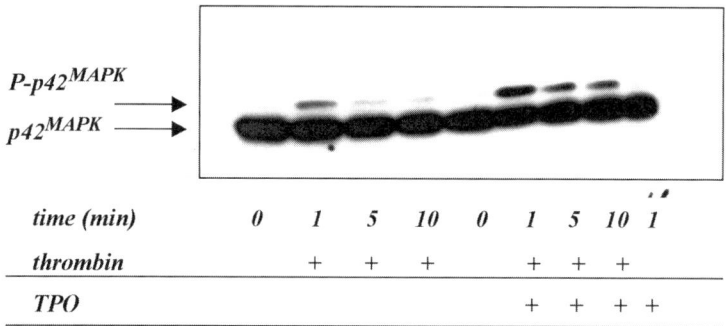

Fig. 10. Phosphorylation of p42^{MAPK}. Activation of p42^{MAPK} by 0.1 U/mL α-thrombin, a combination of α-thrombin and 20 ng/mL thrombopoietin (TPO, 5 min preincubation) and by thrombopoietin alone (1 min).

7. Separate the proteins by overnight electrophoresis (5 mA/gel, maximal voltage, 20°C) until the 30-kDa marker has reached the bottom of the gel.
8. Collect the gels and transfer the proteins electrophoretically to a polyvinylidene difluoride (PVDF) membrane (Immobilon-P Millipore, Millipore, Bedford, MA), for example, using a Transblot system (BioRad) for 90 min at 125 V. The transfer procedure is carried out at 4°C and the system is placed on a magnetic stirrer to prevent local heating. If necessary, the upper part of the gel is cut off to adapt the gel to the size of the blotting system.
9. Collect the membrane and place it against the inner surface of a 50-mL tube with the proteins exposed.
10. Block the membranes by adding 10 mL of the blocking buffer to diminish nonspecific binding of the antibody used for detection of p42^{MAPK} and incubate 1–2 h at 20°C with constant agitation.
11. Incubate the membrane with a 10,000-fold dilution of anti p42^{MAPK} antibody in phosphate-buffered saline containing 0.1% (v/v) Tween-20 for 2 h or, if more convenient, overnight with constant agitation at 4°C.
12. Continue with six washing steps, each consisting of a 10-min incubation with 10 mL phosphate-buffered saline containing 0.1% (v/v) Tween-20 with constant agitation at 4°C.
13. Incubate the membrane with a 2000-fold dilution of HRP-linked protein A in phosphate-buffered saline containing 0.1% (v/v) Tween-20 for 1–2 h with constant agitation at 4°C.
14. Wash membranes and visualize and record protein bands using a chemiluminescent reagent as decribed in **Subheading 3.2.1.1., steps 4–6** (*see* **Note 6**).

4. Notes

1. Store platelet suspensions in plastic tubes at 20°C, without stirring, and close the tubes to prevent pH changes. Do not cool the platelet suspensions and avoid rapid changes in temperature. Keep the time period between blood collection and the execution of the experiments within 2–3 h, as platelets rapidly lose their responsiveness after blood collection.
2. The activity of serine-threonine kinases in platelets is best studied in platelets suspended in buffer. Most buffers contain 2 g/L albumin, which is one of the factors that preserves platelet integrity and responsiveness. Unfortunately, this 65-kDa molecular-weight protein

disturbs the electrophoresis patterns of proteins in the molecular-weight range between 50 and 80 kDa by overshadowing other proteins present in lesser quantities and interfering with the proper positioning of proteins within—and often beyond—this molecular-weight range. Hence, studies on these enzymes are best carried out in albumin-free medium.

3. When, for special reasons, platelet serine-threonine kinases must be studied in a plasma environment, collect 100-μL samples of the platelet suspension and centrifuge immediately (5 s at 10,000g at 20°C) as a "quick pulse." Remove the supernatant completely. Resuspend the pellet in 50 μL 3X concentrated Laemmli buffer (*see* below) and keep the samples on ice. Heat the samples at 100°C for 10 min and store them frozen until further analysis or centrifuge (1 min at 14,000g at 20°C) in a microcentrifuge. Use the supernatants for SDS-PAGE and immunoblot analysis.

4. In most experimental designs, regulation of serine-threonine kinases is studied in comparison with one or more functional responses, most notably platelet aggregation. This poses a problem. The formation of aggregates makes it impossible to collect equal quantities of platelets. Thus, the regulation of these kinases is best studied in unstirred suspensions to avoid the formation of aggregates, with gentle mixing after additions.

5. An alternative method to prevent aggregation is the use of inhibitors of fibrinogen binding to integrin $\alpha_{IIb}\beta_3$ such as antibodies, interfering peptides (GRGDS), or therapeutically used $\alpha_{IIb}\beta_3$ blockers such as abciximab. Keep in mind that these treatments also interfere with ligand-induced signal generation through this integrin, which is an important mechanism to support platelet-platelet interaction and platelet adhesion to surface-bound ligands of this integrin (so-called outside-in signaling).

6. Many serine-threonine kinases are activated by phosphorylation. The standard approach is to detect phosphorylated residues by SDS-PAGE to separate the protein of interest, Western blotting with a primary antibody directed against the phosphorylated enzyme followed by incubation with a secondary antibody, which is directed against the primary antibody and coupled to peroxidase (**Table 1**). A final incubation with chemiluminescence reagent converts peroxidase activity to a dark spot on a film. This is a sensitive technique provided that each lane in the electrophoresis gel contains exactly the same amount of protein, transfer of the protein of interest to the blot membrane is complete, and exposure to the film is in the range (in terms of time and quantity) where the intensity of the spot is proportional to the amount of peroxidase. Pitfalls in this procedure are unequal lane loading, incomplete protein transfer, and, especially, overexposure of the film. The obvious basis for equal lane loading is the determination and adjustment of protein content or platelet count. Furthermore, it is now common to repeat a first blotting procedure to analyze the phosphorylated enzyme by a stripping procedure and a second blotting step with an antibody against the total enzyme (independent of the phosphorylation state) again followed by a secondary antibody directed against the first and coupled to peroxidase. Blots of phosphorylated enzyme can then be compared with blots of total enzyme content. In some assays, the stripping procedure leads to erroneous results. A second sample run in parallel with the first is then the basis for the control on total enzyme content. A third possibility is to use as an internal control for lane loading.

7. VASP phosphorylation is inhibited by 10 μM H89, 15 min preincubation at 37°C (H89-dihydrochloride, Alexis Biochemicals, San Diego, CA).

8. The corresponding antibodies also recognize phosphorylated Thr[309] and Ser[474] on protein kinase Bβ.

9. Protein kinase C is inhibited by a 1-min preincubation (37°C) of platelets with 5 μM bisindolylmaleimide (Roche, Basel, Switzerland). Ca^{2+}-dependent protein kinase C subtypes

are inhibited by a 20-min incubation (37°C) of platelets with 5 μ*M* Gö 6976 (Calbiochem, La Jolla, CA).

10. Platelet p38^{MAPK} is strongly inhibited by slight increases in cAMP. Thus, the use of cAMP-elevating agents such as prostacyclin to prevent platelet activation during isolation of the cells is potentially harmful. Platelets should be given ample time (30 min or more) to restore the normal cAMP level of the resting state (4–6 nmol/10^{11} platelets) before the start of the experiments.

11. A possible role of p38^{MAPK} in platelet signaling sequences is often inferred from blockade with 10 μ*M* SB203580 (preincubation 30 min, 37°C). It should be noted that this agent is known to inhibit p38^{MAPK}-α and -β, but not the other subtypes.

12. P42^{MAPK} is inhibited following a 10-min incubation (37°C) with 20 μ*M* MAPkinase-kinase (MEK) inhibitor PD98059 (Calbiochem).

References

1. Den Dekker, E., Gorter, G., Heemskerk, J. W. M., and Akkerman, J. W. N. (2002) Development of platelet inhibition by cAMP during megakaryocytopoiesis. *J. Biol. Chem.* **277,** 29,321–29,329.

2. Kroner, C., Eybrechts, K., and Akkerman, J. W. N. (2000) Dual regulation of platelet protein kinase B. *J. Biol. Chem.* **275,** 27,790–27,798.

3. Van Willigen, G. and Akkerman, J. W. N. (1992) Regulation of Glycoprotein IIB/IIIA exposure on platelets stimulated with α-thrombin. *Blood* **79,** 82–90.

4. Hackeng, C. M., Relou, I., Pladet, M. W., van Rijn, H. J. M., and Akkerman, J. W. N. (1999) Early platelet activation by low density lipoprotein via p38 mapkinase. *Thromb. Haemost.* **82,** 1749–1756.

5. Van Willigen, G., Gorter, G., and Akkerman, J. W. N. (2000) Thrombopoietin increases platelet sensitivity to α-thrombin via activation of the ERK2-cPLA2 pathway. *Thromb. Haemost.* **83,** 610–616.

6. Den Dekker, E., Heemskerk, J. W. M., Gorter, G., van der Vuurst, H. de Jong-Donath, J., Kroner, C., et al. (2002) cAMP raises Ca^{2+} in human megakaryocytes independent of protein kinase A. *Arterioscler. Thromb. Vasc. Biol.* **22,** 179–186.

7. Coffer, P. J., Jin, J., and Woodgett, J. R. (1998) Protein kinase B (c-Akt): a multifunctional mediator of phosphatidylinositol 3-kinase activation. *Biochem. J.* **335,** 1–13.

Table 1
Antibodies for Analysis of Serine-Threonine Kinases in Platelets

Enzyme/substrate		Antibody
Protein kinase A-C	1st	rabbit polyclonal anti-protein kinase A-C IgG, Cat. no. SC-903 (Santa Cruz Biotechnology)
	2nd	horseradish peroxidase-labeled goat anti-rabbit IgG, Cat. no. 7071-1 (Cell Signaling Technology)
Protein kinase A-RI α/β	1st	mouse anti R-I IgG2b, Cat. no. P19920 (Transduction Laboratories)
	2nd	peroxidase-conjugated goat anti-mouse IgG (GAMPO; DAKO A/S)
Protein kinase A-RIIα	1st	mouse anti-R-IIa IgG1, Cat. no. P55120 (Transduction Laboratories)
	2nd	peroxidase-conjugated goat anti-mouse IgG (GAMPO; DAKO A/S)
Protein kinase A-RIIβ	1st	mouse anti-R-IIb IgG1, Cat. no. P54720 (Transduction Laboratories)
	2nd	peroxidase-conjugated goat anti-mouse IgG (GAMPO; DAKO A/S)
VASP	1st	goat anti-VASP IgG Cat. no. SC-1950 (Santa Cruz Biotechnology)
	2nd	peroxidase-conjugated rabbit anti-goat immunoglobulin RAGPO, Cat. no. P0449 (DAKO A/S)
Protein kinase Bα	1st	goat polyclonal anti-Akt1 IgG Cat. no. SC-7126 (Santa Cruz Biotechnology)
	2nd	rabbit anti-goat peroxidase-conjugated IgG RAGPO, Cat. no. P0449 (DAKO A/S)
Protein kinase Bβ	1st	sheep polyclonal anti-Akt 1/2 IgG Cat. no. 06-606 (Upstate Biotechnology)
	2nd	horseradish peroxidase labeled rabbit anti-sheep IgG (RASPO, DAKO A/S)
Protein kinase B total	1st	goat polyclonal anti-Akt IgG, Cat. no. SC-1618 (Santa Cruz Biotechnology)
	2nd	rabbit anti-goat peroxidase-conjugated RAGPO, Cat. no. P0449 (DAKO A/S)
Protein kinase Bα-Thr308-P	1st	rabbit polyclonal phospho-Akt Thr308 Cat. no. 9275 (Cell Signaling Technology)
	2nd	horseradish peroxidase-labeled goat anti-rabbit IgG Cat. no. 7071-1 (Cell Signaling Technology)
Protein kinase Bα-Ser473-P	1st	rabbit polyclonal phospho-Akt Ser473 Cat. no. 9271 (Cell Signaling Technology)
	2nd	horseradish peroxidase-labeled goat anti-rabbit IgG Cat. no. 7071-1 (Cell Signaling Technology)
Protein kinase Cα	1st	rabbit polyclonal IgG Cat. no. SC-208 (Santa Cruz Biotechnology)
	2nd	swine anti-rabbit peroxidase-conjugated IgG (SWARPO, DAKO A/S)
Protein kinase CβII	1st	rabbit polyclonal IgG Cat. no. SC-210208 (Santa Cruz Biotechnology)
	2nd	swine anti-rabbit peroxidase-conjugated IgG SWARPO (DAKO A/S)
Protein kinase Cα/βII-Thr-638-P/Thr641-P	1st	polyclonal rabbit anti-phosphoprotein kinase Cα/β IgG (Thr^{638}/Thr^{641}) Cat. no. 9375 (Cell Signaling Technology)
	2nd	horseradish peroxidase-labeled goat anti-rabbit IgG, Cat. no. 7071-1 (Cell Signaling Technology)
p38-MAPkinase	1st	polyclonal rabbit anti-p38MAPK IgG Cat. no. 9212 (Cell Signaling Technology)
	2nd	horseradish peroxidase-labeled goat anti-rabbit IgG Cat. no. 7071-1 (Cell Signaling Technology)
p38-MAPkinase-P	1st	polyclonal rabbit anti-phospho-p38MAPK IgG (Thr^{180}/Thr^{182}) Cat. no. 9211 (Cell Signaling Technology)
	2nd	horseradish peroxidase-labeled goat anti-rabbit IgG Cat. no. 7071-1 (Cell Signaling Technology)
p42-MAPkinase	1st	rabbit polyclonal anti-p42MAPK IgG Cat. no. C-154 (Santa Cruz Biotechnology)
	2nd	horseradish peroxidase-linked protein A Cat. no. NA 9120 (Amersham Biosciences)

13

Phosphoinositides

Lipid Kinases and Phosphatases

Bernard Payrastre

1. Introduction

Phosphoinositides (PIs) are a family of eight quantitatively minor membrane lipids playing important roles in the control of a variety of intracellular signaling mechanisms in eukaryotic cells *(1–4)*. The metabolism of these peculiar lipids is highly controlled by a set of enzymes such as kinases, phosphatases, and phospholipases. Phosphatidylinositol (PtdIns) is the quantitatively major PI and is sequentially phosphorylated by specific kinases to produce the different polyPIs that are biologically active compounds *(1,2)*. The so-called canonical pathway involves 4- and 5-kinases and leads to the production of PtdIns(4,5)P$_2$, which can be hydrolyzed by phospholipase C (PLC) generating diacylglycerol (DAG), an activator of protein kinase C (PKC), and inositol 1,4,5 *tris*-phosphate (InsP$_3$), stimulating calcium release from the endoplasmic reticulum *(5)*. DAG can be degraded by lipases or, as in blood platelets, it can be rapidly phosphorylated by DAG-kinases to phosphatidic acid (PtdOH), a lipid that might have important biological roles *(6)*.

Besides this classical pathway, it is now becoming clear that polyPIs can act as signaling molecules on their own by directly and specifically interacting with a number of proteins involved in signal transduction *(1,2,7,8)*. They can be rapidly synthesized and degraded in discrete membrane domains through different metabolic pathways involving specific 3-, 4-, or 5-kinases and phosphatases *(1,2,9)*. Several of these kinases, particularly PI 3-kinases, and phosphatases are regulated and/or relocated by cell surface receptors for extracellular ligands. Several PI-binding domains (e.g., PH, FYVE, PX, ENTH) have been identified recently *(1–3)*. They allow specific interactions between PIs and proteins, leading to their relocalization, activation, or changes in their conformation. Thus, PIs strongly contribute to the spatial and temporal organization of key signaling pathways and are implicated in the rearrangement of the actin cytoskeleton and in the intracellular vesicle trafficking *(1–4)*. Recent discoveries have brought PI-metabolizing

From: *Methods in Molecular Biology, vol. 273:*
Platelets and Megakaryocytes, Vol. 2: Perspectives and Techniques
Edited by: J. M. Gibbins and M. P. Mahaut-Smith © Humana Press Inc., Totowa, NJ

enzymes to the forefront of biomedical research and suggest that probably all PIs have distinct biological roles. Several PI kinases and phosphatases have been cloned and some of them are clearly implicated in human diseases *(1,9)*. Therefore, the measure of the level and/or turnover of the different PIs in vivo is an important source of information. It is also of interest to follow the specific activity of the major PI kinases and phosphatases during cell stimulation or under pathological situations.

In platelets, the PI metabolism is particularly active and generates a number of second messenger molecules upon activation *(2)*. Some of these molecules such as InsP$_3$, DAG, phosphatidylinositol 3,4,5-*tris*-phosphate (PtdIns(3,4,5)P$_3$) (*see* **Fig. 1**), or phosphatidylinositol 3,4-*bis*-phosphate (PtdIns(3,4)P$_2$) have been shown to play a key role in platelet-activation processes *(2,10–13)* and in megakaryocyte functions *(14)*.

This chapter focuses on several techniques currently used to analyze PI metabolism. We will describe procedures for in vitro assay of the major PI kinases, including PI 3-kinase, and phosphatases, such as SHIP or myotubularin. A method allowing in vivo labeling and quantification of the various PIs, including PI 3-kinase products, will also be described. This method is based on lipid extraction and analysis from ortho[^{32}P]phosphate or [^{3}H]inositol-labeled cells by a combination of thin-layer chromatography (TLC) and high-pressure liquid chromatography (HPLC) techniques, allowing an accurate separation of the various PIs and their quantification. Finally, a simple method will be described to quantify the intracellular phosphatidic acid (PtdOH) production as a reflection of PLC activation in platelets.

2. Materials

2.1. In Vitro Lipid Kinase Assays

1. Lipid stocks of the PIs to be used as substrates such as PtdIns, PtdIns(4)P$_1$ and PtdIns(4,5)P$_2$ and phosphatidylserine (PS) (Sigma, St Louis, MO; Avanti Polar Lipids, Alabaster, AL; or Echelon Research Laboratories, Salt Lake City, UT). Stored in CHCl$_3$/CH$_3$OH (v/v) at –20°C under an N$_2$ atmosphere.
2. [γ-^{32}P]ATP (3000 Ci/mmol, 370 MBq/mL) stored at 4°C.
3. TLC plates: silica gel 60 (20 × 20 and 0.2-mm thickness) (Merck, Nogent sur Marne, France). Prior to use, plates are first premigrated once in a mixture containing 90 mL H$_2$O, 1.5 g potassium oxalate, 3 mL 100 m*M* EDTA, and 60 mL CH$_3$OH and then placed in a drying oven at 100°C for 20 min.
4. Solvent mixture for separation of PIs comprising: CHCl$_3$/CH$_3$COCH$_3$/CH$_3$OH/CH$_3$ COOH/H$_2$O (80/30/26/24/14, v/v).
5. A nitrogen stream for evaporation of the solvents in which lipids are resuspended.
6. Iodine vapor for the visualization of PI standards.
7. Suitable antibody to immunoprecipitate the kinase of interest. For example, anti-p85 (UBI, Lake Placid, NY, ref. 06-195).

2.2. PI Phosphatase Assays

1. Lipids, thin-layer chromatography (TLC) plates, and the solvent for the separation of PIs are as described in **Subheading 2.1.**
2. di-C8-NBD6-PIs (Echelon Research Laboratories) resuspended in CHCl$_3$/CH$_3$OH (v/v) and stored at –20°C for a few weeks (*see* **Note 1**).

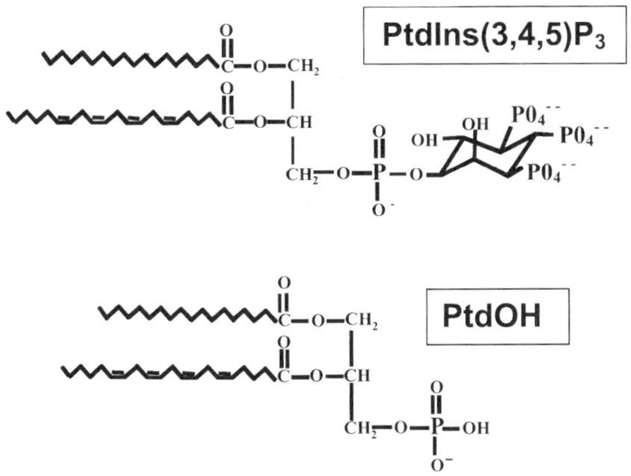

Fig. 1. PtdIns(3,4,5)P$_3$ and PtdOH are two important molecules produced during platelet activation and derived from the phosphoinositide metabolism. PtdIns(3,4,5)P$_3$ is hardly detectable in resting platelets but is rapidly and transiently produced upon platelet activation by a variety of physiological agonists. The level of PtdOH is very weak in resting platelets but rises rapidly upon platelet activation. In these cells, PtdOH is mainly produced through the action of a PLC-generating DAG, which is rapidly phosphorylated by a DAG-kinase into PtdOH.

3. A solvent mixture used for the separation of di-C8-NBD6-PIs, comprising CHCl$_3$/CH$_3$OH/CH$_3$COCH$_3$/CH$_3$COOH/H$_2$O (70/50/20/20/20, v/v).
4. Antibodies for p85 (as described in **Subheading 2.1.**) and the phosphatase of interest to enable the immunoprecipitation of PI3-kinase and the phosphatase.

2.3. Identification and Quantification of Intracellular PIs

1. TLC plates, lipid stocks, and the solvent for PI separation as described in **Subheading 2.1.**
2. Ortho[^{32}P]phosphate (370 MBq/mL) or, alternatively, *myo*-[^{3}H]inositol (with stabilizer, 370-740 GBq/mmol) for cell labeling.
3. Deacylation reagent: 26.8% (v/v) of 40% (v/v) methylamine, 45.7% (v/v) CH$_3$OH, 11.4% (v/v) n-butanol, and 16% (v/v) H$_2$O.
4. 0.2-μm filters for sample filtration before HPLC analysis.
5. An HPLC system with an online continuous-flow liquid-scintillation detector is required.
6. HPLC column: Whatman Partisphere 5 SAX column 4.6 mm × 125 mm (Ref. 4621-0505, Whatman International Ltd., Maidstone, UK) with guard-cartridge anion exchanger units (Ref. 4641 0005, Whatman International Ltd.).
7. HPLC elution solution 1 *M* (NH$_4$)$_2$HPO$_4$ (pH 3.8) and bi-distilled water.

2.4. Quantification of Intracellular PtdOH Production as a Reflection of PLC Activation in Platelets

1. Ortho-[^{32}P]phosphate (370 MBq/mL) for cell labeling.
2. TLC plates: silica gel 60 (20 × 20 and 0.2-mm thickness) (Merck). Plates are pre-migrated once in a mixture containing 30 mL H$_2$O, 3.15 g oxalic acid, and 70 mL CH$_3$OH and placed in a drying oven at 70°C for 20 min prior to use.
3. Solvent mixture for PtdOH separation comprising CHCl$_3$/CH$_3$OH/10 *N* HCl (87/13/0.5, v/v).

Note that plexiglass screens must be used for protection against radiation and that all the solvents must be manipulated under a hood ($CHCl_3$, chloroform; CH_3OH, methanol; CH_3COCH_3, acetone; CH_3COOH: acetic acid).

3. Methods

3.1. In Vitro Lipid Kinase Assays

The method described is usually performed to measure the activity of type IA PI 3-kinase (p85α-p110), a key enzyme in cell signaling *(4)*. However, as mentioned below, this method can be applied to other PI kinases.

1. Platelets are lysed by adding to the suspension 1 volume of twice-concentrated ice-cold lysis buffer containing 80 mM Tris-HCl (pH 7.4), 200 mM NaCl, 200 mM NaF, 20 mM EDTA, 80 mM $Na_4P_2O_7$, 1 mM Na_3VO_4, 2% (v/v) Triton X-100, 1 mM phenylmethyl-sulfonyl fluoride, and 10 µg/mL each of aprotinin and leupeptin.
2. After gentle shaking for 20 min at 4°C and centrifugation (12,000g for 10 min at 4°C), the soluble fraction is collected and precleared by mixing for 30 min with protein A-Sepharose CL4B (10% [w/v]).
3. Following removal of the protein A-Sepharose CL4B, the precleared suspensions is then incubated 2 h at 4°C with the anti-p85 antibody (UBI, ref. 06-195).
4. Immune complexes are precipitated by addition of 10% (w/v) protein A-Sepharose and mixing for 1 h at 4°C followed by centrifugation (6000g for 5 min at 4°C).
5. The immunoprecipitate is washed once in lysis buffer and twice in washing buffer containing 10 mM Tris-HCl (pH 7.4), 100 mM NaCl, 200 µM Na_3VO_4, 1 mM EDTA, and 1 µg/mL each of aprotinin and leupeptin.
6. To prepare the lipid substrate, PtdIns and PS (1:2; w/w) are dried under a nitrogen stream, resuspended in 50 mM Tris-HCl (pH 7.4), and sonicated (20 kHz, three times 1 min at 4°C) in order to make lipid vesicles containing 1 µg PtdIns/µL. The final concentration of PtdIns in the assay is 0.14 µg/µL and ATP is 50 µM.
7. The immunoprecipitate (about 15 µL of protein A-Sepharose beads plus 10 µL of washing buffer) is left in a microcentrifuge tube and the kinase assay reaction is started by addition of 10 µL of the lipid substrate and 35 µL of kinase buffer containing 50 mM Tris-HCl, 100 mM NaCl, 20 mM $MgCl_2$, 100 µM ATP, and 20–30 µCi [γ-^{32}P]ATP (pH 7.4). The reaction is performed at 37°C with shaking for 15 min (*see* **Notes 2** and **3**).
8. The reaction is stopped by addition of 200 µL of $CHCl_3/CH_3OH$ (v/v) and 30 µL of HCl (3 N) and the sample is vortexed thoroughly for 1–2 min at room temperature.
9. After centrifugation (7000g for 5 min) the organic phase (lower phase) is collected and dried under a nitrogen stream.
10. Dried lipid samples are resuspended in a minimal volume (60 µL) of $CHCl_3/CH_3OH$ (v/v) and spotted onto a TLC plate. Authentic PIs (5–10 µg) used as standards are also spotted on the plate. The solvent system used for analysis of [^{32}P]PIs is a mixture of $CHCl_3/CH_3COCH_3/CH_3OH/CH_3COOH/H_2O$ (80/30/26/24/14, v/v) (*see* **Note 4**).
11. The solvent is allowed to migrate to 2 cm from the top of the plate (approx 1 h 30 min).
12. Authentic PI standards are visualized using iodine vapor, whereas radioactive spots are visualized by autoradiography or with a PhosphorImager (*see* **Fig. 2**).
13. After scraping silica from the plates, the radioactivity incorporated in the product of the reaction is quantified by scintillation counting in order to calculate the specific activity of the kinase.

TLC

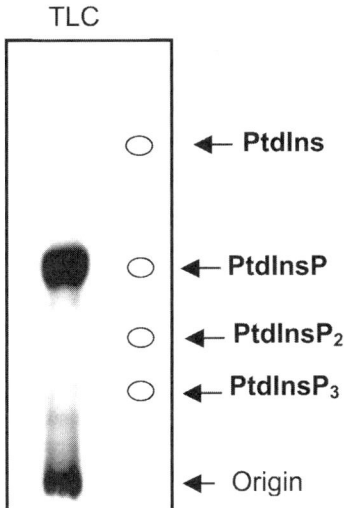

Fig. 2. Separation of the products of the in vitro kinase assay. The PI 3-kinase assay was performed as indicated in **Subheading 3.1.** PtdIns was used as a substrate and the production of [^{32}P]PtdIns(3)P was visualized by TLC. The solvent system used was a mixture of CHCl$_3$/CH$_3$COCH$_3$/CH$_3$OH/CH$_3$COOH/H$_2$O (80/30/26/24/14, v/v). The migration of authentic standards visualized by iodine vapor is shown on the right. Only the lower part of the TLC plate is shown here.

3.2. PI Phosphatase Assays

3.2.1. Assay Utilizing Radioactive Substrate

This method can be used to measure the activity of PI phosphatases in general *(15)*. Described here is an assay used for the SH2 domain containing inositol 5-phosphatase (SHIP1), one of the major PtdIns(3,4,5)P$_3$ 5-phosphatases in platelets *(16,17)*.

1. The radioactive substrate, [^{32}P]PtdIns(3,4,5)P$_3$, is prepared using PtdIns(4,5)P$_2$/PS vesicles, [γ-^{32}P]ATP, and immunopurified PI 3-kinase as described in **Subheading 3.1.**
2. [^{32}P]PtdIns(3,4,5)P$_3$ is scraped off the TLC plate and the silica is resuspended in 500 μL of HCl (2 *N*).
3. One mL of CHCl$_3$/CH$_3$OH (v/v) is rapidly added and the sample is vortexed thoroughly for 2 min at room temperature.
4. After centrifugation (7000*g* for 5 min) the organic phase is collected.
5. For one assay, about 30,000 dpm of purified [^{32}P]PtdIns(3,4,5)P$_3$ together with 50 μg of PS in CHCl$_3$/CH$_3$OH (v/v) are dried under a nitrogen stream, resuspended in a minimal volume (30 μL) of 50 m*M* Tris-HCl (pH 7.5), and sonicated (20 kHz, three times 1 min at 4°C).
6. The reaction mixture (50 μL) containing 20 μL of immunoprecipitated SHIP1 (immunoprecipitated using the protocol described in **Subheadings 3.1., steps 1–5**) in 50 m*M* Tris-HCl (pH 7.5), 10 m*M* MgCl$_2$, and lipid vesicles (30 μL) is incubated for 30 min at 37°C with shaking.

7. The reaction is stopped by addition of 200 µL of CHCl$_3$/CH$_3$OH (v/v) and 50 µL of HCl (3 *N*) and the sample is vortexed thoroughly for 2 min at room temperature.

8. After centrifugation (10,000*g* for 5 min) the organic phase is collected and dried under a nitrogen stream.

9. Dried lipid samples are resuspended in a minimal volume (60 µL) of CHCl$_3$/CH$_3$OH (v/v) and spotted onto a TLC plate.

10. [^{32}P]PtdIns(3,4,5)P$_3$ and its degradation product [^{32}P]PtdIns(3,4)P$_2$ are separated by TLC using CHCl$_3$/CH$_3$COCH$_3$/CH$_3$OH/CH$_3$COOH/H$_2$O (80/30/26/24/14, v/v) as a solvent system (*see* **Note 4**).

11. The radioactive spots are visualized by a PhosphorImager or by autoradiography.

12. Authentic PtdIns(3,4,5)P$_3$ and PtdIns(3,4)P$_2$ standards resolved on the same TLC are visualized using iodine vapor. To quantify the specific activity of the phosphatase, a known amount of PtdIns(3,4,5)P$_3$ can be added to [^{32}P]PtdIns(3,4,5)P$_3$ (which is in trace amount)/PS in the assay.

3.2.2. Assay Utilizing Fluorescent Substrates

This simple procedure, recently developed by Taylor and Dixon *(18)*, can be used to investigate the activity and specificity of any PI phosphatase, either recombinant or immunoprecipitated from platelets or megakaryocytes. This assay has been used for determining the specificity of PTEN, SHIP, and myotubularin *(18–20)*. This method is useful for determination of a phosphatase activity in a fraction and for the identification of its substrate specificity. However, the quantification of the phosphatase activity is only approximate and the radioactive assay or the malachite green-based assay *(21)* are more appropriate for an accurate quantification. As an example, an assay of the ubiquitous PtdIns(3)P 3-phosphatase myotubularin (MTM1) is described using di-C$_8$-NBD6-PtdIns(3)P as a substrate *(18,19)*.

1. One µg of di-C$_8$-NBD6-PtdIns(3)P is dried under a nitrogen stream in a microcentrifuge tube and resuspended in 30 µL of 50 m*M* ammonium acetate (pH 6.0) containing 2 m*M* dithiothreitol.

2. Immunoprecipitated MTM1 (15 µL of beads) is added and the mixture is incubated for 30 min at 37°C under shaking (*see* **Subheading 3.1., steps 3–5** for protocol for immunoprecipitation).

3. The reaction is stopped by centrifugation (10,000*g*, for 2 min) to pellet the beads, and 20 µL of supernatant is removed.

4. After addition of 100 µL of CH$_3$COCH$_3$ the supernatant is dried under a nitrogen stream at 37°C and resuspended in 10 µL of CH$_3$OH/isopropyl alcohol (CH$_3$)$_2$CHOH/CH$_3$COOH (5/5/2, v/v) and spotted onto a TLC plate.

5. The solvent system used is a mixture of CHCl$_3$/CH$_3$OH/CH$_3$COCH$_3$/CH$_3$COOH/H$_2$O (70/50/20/20/20, v/v) and is allowed to migrate to 2 cm from the top of the plate (approx 1 h) (*see* **Note 4**).

6. Fluorescent lipids are then visualized under UV light (*see* **Fig. 3**). This method can be applied to other phosphatases provided that appropriate fluorescent substrate is used (*see* **Note 5**).

3.3. Identification and Quantification of Intracellular PIs

A combination TLC and HPLC techniques allows the separation and detection of the eight PIs extracted either from *myo*-[^{3}H]inositol- or from ortho[^{32}P]phosphate-labeled

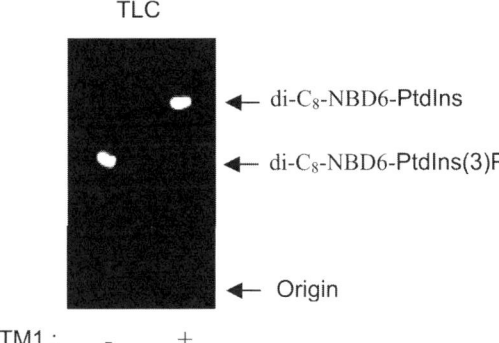

← di-C$_8$-NBD6-PtdIns

← di-C$_8$-NBD6-PtdIns(3)P

← Origin

MTM1 :   -   +

Fig. 3. Separation of the products of the in vitro phosphatase assay using fluorescent substrates. MTM1 was immunoprecipitated as described *(19)* and phosphatase assay was performed as indicated in **Subheading 3.2.2.** Di-C$_8$-NBD6-PtdIns(3)P was used as a substrate and the production of di-C$_8$-NBD6-PtdIns was visualized by TLC using a mixture of CHCl$_3$/CH$_3$OH/CH$_3$COCH$_3$/CH$_3$COOH/H$_2$O (70/50/20/20/20, v/v) as a solvent. The migration of authentic di-C8-NBD6-standards is shown on the right.

cells. This accurate method developed by Auger et al. *(22)* is necessary to unambiguously demonstrate the synthesis of PI 3-kinase products in vivo in radiolabeled cells *(23)* and can be applied successfully for PI analysis in platelets *(11,16)* (*see* **Note 6**).

After extraction and deacylation, PIs can be analyzed directly by HPLC. For a better separation, they can be first separated on TLC, scraped off, deacylated, and then analyzed by HPLC (**Fig. 4**).

1. Platelets are washed in a buffer (pH 6.5) containing 140 m*M* NaCl, 5 m*M* KCl, 5 m*M* KH$_2$PO$_4$, 1 m*M* MgSO$_4$, 10 m*M* HEPES, 5 m*M* glucose, and 0.35% BSA (w/v).
2. Platelets are labeled with 0.5 mCi/mL ortho[^{32}P]phosphate by incubation for 60 min in phosphate-free washing buffer (pH 6.5) at 37°C (*see* **Notes 7** and **8**).
3. After one washing step, platelets are resuspended in the same buffer containing 1 m*M* CaCl$_2$ and pH is adjusted to 7.4.
4. [^{32}P]-labeled platelets (~1.5 × 10^9 in 0.5 mL) are stimulated with the appropriate agonist and the reaction is stopped by addition of 1 mL CHCl$_3$/CH$_3$OH (v/v).
5. HCl is then added to obtain a final concentration of 0.4 *N* and the sample is vortexed for 5 min at room temperature.
6. After centrifugation (3000*g*, 5 min) the organic phase is collected and dried under a nitrogen stream at 37°C. Dried lipids are resuspended in a minimal volume (<100 µL) of CHCl$_3$/CH$_3$OH (v/v) and resolved by TLC using CHCl$_3$/CH$_3$COCH$_3$/CH$_3$OH/CH$_3$COOH/ H$_2$O (80/30/26/24/14, v/v) as a solvent (time of migration ~1 h 30 min) (*see* **Note 4**).
7. Spots corresponding to [^{32}P]-PtdInsP, [^{32}P]-PtdInsP$_2$, and [^{32}P]-PtdIns(3,4,5)P$_3$, identified with authentic standards, are visualized by autoradiography or with a PhosphorImager (**Fig. 4**) and scraped off. A mixture of crude phospholipids (~4 µg) can be added to the silica, before scraping, as a carrier to enhance the recovery.
8. PIs are then deacylated by adding to the silica powder 1 mL of methylamine reagent composed of 26.8% (v/v) of 40% (v/v) methylamine, 45.7% (v/v) CH$_3$OH, 11.4% (v/v)

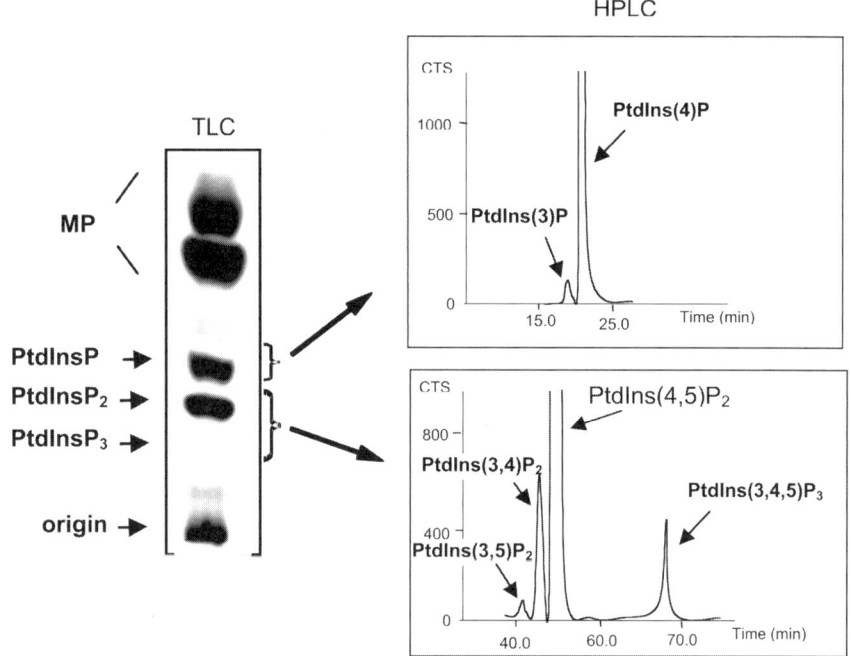

Fig. 4. Analysis of the different polyPIs extracted from ^{32}P-labeled platelets. Phospho-inositides are extracted from ^{32}P-labeled platelets stimulated by collagen (10 μg/mL, 1 min) and separated by TLC (left panel). The radioactive spots corresponding to PtdInsP and PtdInsP$_2$ + PtdInsP$_3$ are scraped off, deacylated, and analyzed by HPLC (right panel). MP stands for major phospholipids (phosphatidylserine, phosphatidylcholine, and phosphatidylethanolamine). The position of authentic PI standards is shown. Note that a specific HPLC gradient must be used to separate PtdIns(4)P and PtdIns(5)P *(2)*.

n-butanol, and 16% (v/v) H$_2$O. After incubation at 53°C for 50 min (vials must be tightly capped), the methylamine reagent is completely evaporated under a nitrogen stream at 37°C.

9. Bi-distillated water (1.2 mL) is added and the sample is vortexed thoroughly (*see* **Note 9**).

10. After filtration through a 0.2-μ filter, the samples are analyzed by HPLC on a Whatman Partisphere 5 SAX column (**Fig. 4**). The compounds are eluted with 1 *M* (NH$_4$)$_2$HPO$_4$ (pH 3.8) and H$_2$O at a flow rate of 1 mL/min using the following gradient : 0% 1 *M* (NH$_4$)$_2$HPO$_4$ for 5 min; 0–22% 1 *M* (NH$_4$)$_2$HPO$_4$ for 55 min; 22–100% 1 *M* (NH$_4$)$_2$HPO$_4$ for 15 min; 100% 1 *M* (NH$_4$)$_2$HPO$_4$ for 5 min; 100–0% 1 *M* (NH$_4$)$_2$HPO$_4$ for 15 min. Eluate from the HPLC column flows into an online continuous-flow liquid-scintillation detector (**Fig. 4**).

3.4. Quantification of Intracellular Phosphatidic Acid (PtdOH) Production as a Reflection of PLC Activation in Platelets

In [32P]-labeled platelets, the main part of DAG produced by PLC is rapidly converted into [32P]PtdOH by a DAG-kinase. In this model, the contribution of phospholipase D to the production of [32P]PtdOH is relatively minor *(11,24)*. Moreover, [32P]PtdOH is hardly detectable in resting platelets. Thus, the formation of [32P]PtdOH is a good reflection of PLC activation in platelets.

1. Platelets are labeled as described in **Subheading 3.3., steps 1–3**, although lower concentrations of ortho[32P]phosphate can be used (0.2 mCi/mL).
2. [32P]-labeled platelets (~10^9 cells in 0.5 mL) are stimulated and the reaction is stopped by addition of 1 mL of $CHCl_3/CH_3OH$ (v/v). HCl is added to obtain a final concentration of 0.1 N and the sample is vortexed thoroughly for 2 min at room temperature.
3. After centrifugation (3000g, 5 min) the organic phase is collected and dried under a nitrogen stream at 37°C.
4. Dried lipids are resuspended in a minimal volume (50 µL) of $CHCl_3/CH_3OH$ (v/v) and spotted onto a TLC plate. Before use, the TLC plate is first pre-migrated once in a mixture containing H_2O (30 mL), oxalic acid (3.15 g), and CH_3OH (70 mL) and placed in a drying oven at 70°C for 20 min prior to use. The solvent system used for analysis of [32P]PtdOH is a mixture of $CHCl_3/CH_3OH/10$ N HCl (87/13/0.5, v/v) as described previously *(11)* (*see* **Note 4**).
5. The solvent is allowed to migrate through two thirds of the plate (approx 25 min).
6. The spot corresponding to [32P]PtdOH is visualized by autoradiography or by Phosphor-Imager (**Fig. 5**), scraped off, and the radioactivity incorporated in this lipid is counted by scintillation spectrometry. This solvent system resolves PtdOH but other major phospholipids remain at the origin.

4. Notes

1. di-C_8-NBD6-PIs are light-sensitive and it is worth checking by TLC analysis that they are not degraded prior to use.
2. Mixed lipid compositions of PtdIns(4,5)P_2 and PS (1,2; w/w) or PtdIns, PtdIns(4)P, PtdIns(4,5)P_2 and PS (1,1,1,1; w/w/w/w) can also be used as a substrate for PI 3-kinase assay. Moreover, this assay can be adapted to measure the activity of other PI kinases (4- or 5-kinases) provided that the appropriate substrate is used. DAG-kinase can also be assayed by this method using a mixture of 1,2-dioleyl-sn-glycerol and PS (1,2; w/w) as a substrate and the [32P]PtdOH formed can be analyzed as described in **Subheading 3.4.**
3. Immunoprecipitates (to check potential co-immunoprecipitation) or subcellular fractions (such as isolated cytoskeleton or membrane rafts) can be used as a source of the kinase to be measured.
4. It is important to note that all the TLC solvents must form a single phase when mixed. Addition of $CHCl_3$ as the last constituent will help this. The TLC solvents should be prepared two days before the migration and can be used for three to four different migrations. Authentic lipid standards are always spotted on the TLC plate and visualized using iodine vapor.
5. The optimization of pH of the reaction and the concentration of potential cofactors will be required when applying this protocol to other phosphatases and substrates.

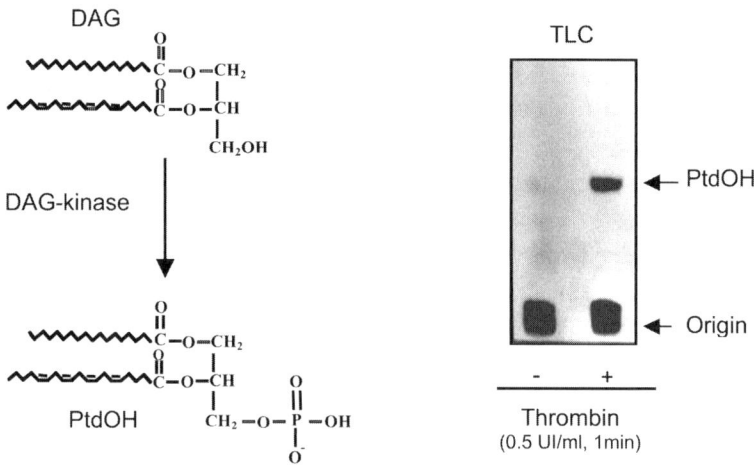

Fig. 5. Production of PtdOH in thrombin-stimulated platelets. In thrombin-stimulated [^{32}P]-labeled platelets, DAG is rapidly transformed into [^{32}P]PtdOH by a DAG-kinase (left panel). The production of [^{32}P]PtdOH in [^{32}P]-labeled human platelets stimulated by thrombin is analyzed by a specific TLC/solvent system that resolves PtdOH but other major phospholipids remains at the origin (right panel).

6. Another method exists to quantify intracellular PtdIns(3,4,5)P$_3$ level *(25)*. Briefly, lipids are extracted from cells and hydrolyzed by KOH to obtain Ins(1,3,4,5)P$_4$ from PtdIns(3,4,5)P$_3$. A competition with radio-labeled Ins(1,3,4,5)P$_4$ vs an Ins(1,3,4,5)P$_4$-binding protein from cerebellum is then performed.

7. In ortho[^{32}P]phosphate-labeled cells, changes in the labeling of PtdIns(4)P and PtdIns(4,5)P$_2$ may reflect an increased turnover and not necessarily and increased level. Labeling with [^{3}H]-*myo*inositol is more appropriate for quantification of these "house-keeping" PIs, but its incorporation in isolated platelets is not very efficient.

8. Recently, a mass assay for PtdIns(5)P has been developed based on the use of recombinant PIP 4-kinase α *(26,27)*.

9. Detection *in cellulo* of PIs can be performed by transfection of GFP-PH domains (Btk, ARNO, GRP1 for PtdIns(3,4,5)P$_3$; GFP-PH domain of PLCδ for PtdIns(4,5)P$_2$) or GFP-FYVE domains (EE1 for PtdIns(3)P) *(2)*. However, this method is possible only in living cells, since fixative methods generally do not prevent lipid relocation. Finally, the analysis of Akt phosphorylation by Western blotting is a good reflection of Type I PI 3-kinase activation *(4)*.

References

1. Toker, A. (2002) Phosphoinositides and signal transduction. *CMLS Cell. Mol. Life Sci.* **59,** 761–779.
2. Payrastre, B., Missy, K., Giuriato, S., Bodin, S., Plantavid, M., and Gratacap, M. P. (2001) Phosphoinositides—Key players in cell signalling, in time and space. *Cell Signal.* **13,** 377–387.

3. Rameh, L. E. and Cantley, L. C. (1999) The role of phosphoinositide 3-kinase lipid products in cell function. *J. Biol. Chem.* **274,** 8347–8350.

4. Katso, R., Okkenhaug, K., Ahmadi, K., White, S., Timms, J., and Waterfield, M. D. (2001) Cellular function of phosphoinositide 3-kinase: implication for development, immunity, homeostasis and cancer. *Annu. Rev. Cell. Dev. Biol.* **17,** 615–675.

5. Michell, R. H. (1975) Inositol phopholipids and cell surface receptor function. *Biochim. Biophys. Acta* **415,** 81–147.

6. Topham, M. K. and Prescott, S. M. (2002) Diacylglycerol kinase and signaling roles. *Thromb. Haemost.* **88,** 912–918.

7. Hinchliffe, K. (2000) Intracellular signalling: Is PIP_2 a messenger too? *Curr. Biol.* **10,** R104–R105.

8. Czech, M. P. (2000) PIP_2 and PIP_3: complex roles at cell surface. *Cell* **100,** 603–606.

9. Maehama, T., Taylor, G., and Dixon, J. E. (2001) PTEN and myotubularin: novel phospho-inositide phosphatases. *Annu. Rev. Biochem.* **70,** 247–279.

10. Trumel, C., Payrastre, B., Plantavid, M., Hechler, B., Viala, C., Presek, P., et al. (1999) A key role of ADP in the irreversible platelet aggregation induced the PAR1-activating peptide through the late activation of phosphoinositide 3-kinase. *Blood* **94,** 4156–4165.

11. Gratacap, M. P., Payrastre, B., Viala, C., Mauco, G., Plantavid, M., and Chap, H. (1998) Phosphatidylinositol 3,4,5-trisphosphate-dependent stimulation of phospholipase C-γ2 is an early key event in FcγRIIA-mediated activation of human platelets. *J. Biol. Chem.* **273,** 24,314–24,321.

12. Pasquet, J. M., Gross, B. S., Gratacap, M. P., Quek, L., Pasquet, S., Payrastre, B., et al. (2000) Thrombopoietin potentiates collagen receptor signalling in platelets through a phosphatidylinositol 3-kinase-dependent pathway. *Blood* **95,** 3429–3434.

13. Selheim, F., Holmsen, H., and Vassbotn, F. S. (2000) PI 3-kinase signalling in platelets: the significance of synergistic, autocrine stimulation. *Platelets* **11,** 69–82.

14. Rojnuckarin, P., Miyakawa, Y., Fox, N. E., Deou, J., Daum, G., and Kaushansky, K. (2001) The roles of phosphatidylinositol 3-kinase and protein kinase Czeta for thrombopoietin-induced mitogen-activated protein kinase activation in primary murine megakaryocytes. *J. Biol. Chem.* **276,** 41,014–41,022.

15. Payrastre, B., Gironcel, D., Plantavid, M., Mauco, G., Breton, M., and Chap, H. (1994) Phosphoinositide 3-phosphatase segregates from phosphatidylinositol 3-kinase in EGF-stimulated A431 cells and fails to hydrolyse *in vitro* phosphatidylinositol (3,4,5) trisphosphate. *FEBS Lett.* **341,** 113–118.

16. Giuriato, S., Payrastre, B., Drayer, A. L., Plantavid, M., Woscholski, R., Parker, P., et al. (1997) Tyrosine phosphorylation and relocation of SHIP are integrin-mediated in thrombin-stimulated human blood platelets. *J. Biol. Chem.* **272,** 26,857–26,863.

17. Pasquet, J. M., Queck, L., Stevens, C., Bobe, R., Hubert, M., Duronio, V., et al. (2000) Phosphatidylinositol 3,4,5-trisphosphate regulates Ca^{2+} entry via btk in platelets and megakaryocytes without increasing phospholipase C activity. *EMBO J.* **19,** 2793–2802.

18. Taylor, G. S. and Dixon, J. E. (2001) An assay for phosphoinositide phosphatases utilizing fluorescent substrates. *Anal. Biochem.* **295,** 122–126.

19. Laporte, J., Liaubet, L., Blondeau, F., Tronchère, H., Mandel, J.-L., and Payrastre, B. (2002) Functional redundancy in the myotubularin family. *Biochem. Biophys. Res. Commun.* **291,** 305–312.

20. Giuriato, S., Blero, D., Robaye, B., Bruyns, C., Payrastre, B., and Erneux, C. (2002) SHIP2 overexpression strongly reduces the proliferation rate of K562 erythroleukemia cell line. *Biochem. Biophys. Res. Commun.* **296,** 106–110.

21. Maehama, T., Taylor, G. S., Slama, J. T., and Dixon, J. E. (2000) A sensitive assay for phosphoinositide phosphatases. *Anal. Biochem.* **279,** 248–250.

22. Auger, K. R., Serunian, L. A., Soltoff, P., Libby, P., and Cantley, L. C. (1989) PDGF-dependent tyrosine phosphorylation stimulates production of novel phosphoinositides in intact cells. *Cell* **57,** 167–175.

23. Soltoff, S. P., Kaplan, D. R., and Cantley, L. C. (1993) Phosphatidylinositol 3-kinase. *Meth. Neurosci.* **18,** 100–113.

24. Huang, R., Kucera, G. L., and Rittenhouse, S. E. (1991) Elevated cytosolic Ca2+ activates phospholipase D in human platelets. *J. Biol. Chem.* **266,** 1652–1655.

25. Van der Kaay, J., Batty, I. H., Cross, A. E. D., Watt, P. W., and Downes, C. P. (1997) A novel, rapid, and highly sensitive mass array for phosphatidylinositol 3,4,5-triphosphate (Ptd Ins (3,4,5) P_3) and its application to measure insulin-stimulated Ptd Ins (3,4,5)P_3 production in rat skeletal muscle in vivo. *J. Biol. Chem.* **272,** 5477–5481.

26. Morris, J. B., Hinchliffe, K. A., Ciruela, A., Letcher, A. J., and Irvine, R. F. (2000) Thrombin stimulation of platelets causes an increase in phosphatidylinositol 5-phosphate revealed by mass assay. *FEBS Lett.* **475,** 57–60.

27. Niebuhr, K., Giuriato, S., Pedron, T., Philpott, D. J., Gaits, F., Sable, J., et al. (2002) Conversion of PtdIns(4,5)P_2 into PtdIns(5)P by the *Shigella flexneri* effector IpgD reorganizes host cell morphology. *EMBO. J.* **21,** 5069–5078.

14

Isolation and Analysis of Platelet Lipid Rafts

Corie N. Shrimpton, Karine Gousset, Fern Tablin, and José A. López

1. Introduction

The cell membrane can no longer be viewed as a homogenous fluid bilayer; instead, it is now known to contain discrete lateral microdomains with characteristic subsets of lipids and proteins. One such microdomain, termed the "lipid raft," has received much attention over the past few years. Several excellent reviews of lipid rafts (1–3) and of related membrane structures called caveolae (4,5) have been published recently. Lipid rafts are enriched in cholesterol and sphingolipids, producing islands in the plasma membrane in which the lipids exist in the liquid-ordered state, discrete from the bulk of the lipids existing in the liquid-disordered state. As a consequence, lipid rafts can be isolated as a low-density, insoluble fraction after low-temperature nonionic detergent extraction (6). These properties give rise to their alternate names, detergent-resistant membranes (DRMs), detergent-insoluble glycolipid-rich domains (DIGs), Triton-insoluble floating fraction (TIFF), and glycolipid-enriched membranes (GEMs).

Lipid rafts are localized mainly within the plasma membrane, but can also be found in the late secretory and endocytic pathways, where they have been implicated in protein and lipid sorting (7–10). On the plasma membrane, lipid rafts are believed to act as platforms for signal transduction and have been shown to play a critical role in post-receptor signaling, especially in hematopoietic cells (11–13). One advantage of the organization of the plasma membrane into these microdomains is that proteins involved in signal transduction can be confined to discrete microenvironments and thus increase their local concentration, thereby facilitating signal transduction. Indeed, one of the most important features of lipid rafts is that they selectively attract certain proteins while excluding others. Proteins with affinity for lipid rafts include doubly acylated proteins, such as the Src-family kinases (14), glycosylphatidylinositol (GPI)-anchored proteins (6), and certain transmembrane proteins, especially those modified by palmitoylation. It appears that acylation—in particular, double-acylation—may be a general mechanism for targeting proteins to lipid rafts (15,16).

From: *Methods in Molecular Biology, vol. 273:*
Platelets and Megakaryocytes, Vol. 2: Perspectives and Techniques
Edited by: J. M. Gibbins and M. P. Mahaut-Smith © Humana Press Inc., Totowa, NJ

Lipid rafts have been implicated in platelet activation, although this role in is not well developed, with only a handful of studies published to date. Rafts isolated from platelets lack caveolin and are enriched in CD36 and Src family kinases *(17,18)*. Lipid rafts have been implicated as having important roles in signaling through the collagen receptor GPVI *(19,20)* and the von Willebrand factor receptor GPIb-IX-V *(21)*, in the cold activation of platelets *(22)*, and as sites for the production of phosphoinositide second messengers *(23)*.

Several approaches can be taken to investigate whether the function of a particular protein involves its localization to lipid rafts. A good starting point is to show that lipid rafts are enriched in the protein in question, based on detergent insolubility and buoyancy on sucrose gradients, and further, to examine whether the protein of interest may associate with effector proteins within these domains. Demonstration of either condition should be considered consistent with a role for lipid rafts in that protein's function, but does not constitute definitive proof.

A second approach is to disrupt the lipid rafts via the manipulation of their constituents. Cholesterol depletion has been shown to disrupt raft structural integrity and result in the loss of the function of raft-associated proteins. One must, however, be cautious when interpreting such results, as the manipulation of cholesterol levels may have pleiotropic effects on cell function independent of lipid raft disruption. Stronger evidence for a role for lipid rafts in the function of a given protein is provided by disrupting the association between that protein and the lipid raft. For example, palmitoylation of both the adapter molecule LAT *(24)* and the Src-family kinase Lck *(25)* have been shown to be imperative for the partitioning of these molecules to lipid rafts. Mutation of the palmitoylation sites results in the loss of the partitioning of these proteins to lipid rafts and a concomitant loss of their function. Such an approach is technically difficult in the platelet and highlights the need for good cell models when studying this system. The best evidence for the role of lipid rafts in the function of a given protein is provided when several approaches are employed.

2. Materials

2.1. Equipment

1. Ultracentrifuge for centrifugation at 200,000*g* (e.g., Beckman Optima LE-80K, Beckman Coulter, Fullerton, CA) and rotor (e.g., SW40, Beckman Coulter).
2. 14-mL ultracentrifuge tubes.
3. Screw-top test tubes and minivials with Teflon-lined lids.
4. Dot-blot apparatus (e.g., Bio-Rad Biodot).
5. Apparatus for SDS-PAGE and Western transfer.
6. Nitrocellulose.
7. Scintillation counter.
8. Scintillation vials and cocktail.
9. Confocal or fluorescence microscope.
10. Glass slides and coverslips.

2.2. Reagents and Buffers

1. Buffer 1: 10X stock, 1.34 *M* NaCl, 120 m*M* NaHCO$_3$, 29 m*M* KCl, 3.4 m*M* Na$_2$HPO$_4$, 10 m*M* MgCl$_2$, and 100 m*M* HEPES, pH 7.4. Store at 4°C. Dilute 10-fold with deionized H$_2$O, add 5 m*M* glucose, 0.3 g BSA/100 mL, and warm to 37°C prior to use.

2. Mes-buffered saline (MBS): 25 mM Mes, pH 6.5, 150 mM NaCl. Store at 4°C.
3. Sucrose solutions: 80% (w/v) sucrose (80 grams sucrose into a final volume of 100 mL MBS). Heat 60 mL MBS, with stirring, and gradually add sucrose until dissolved. Do not boil. Increase volume to 100 mL with MBS. 30% (w/v) and 5% (w/v) sucrose solutions can be made by dilution of the 80% (w/v) stock. Store at 4°C.
4. Triton X-100 lysis buffer: 1% (v/v) Triton X-100 and 5% (w/v) protease inhibitor cocktail (Sigma P-8340) in MBS. For tyrosine phosphorylation studies, include 5 mM sodium orthovanadate and 5 mM sodium fluoride.
5. Tris-buffered saline (TBS): 10 mM Tris-HCl, pH 7.4, 150 mM NaCl.
6. Phosphate-buffered Saline (PBS): 10X stock, 1.37 M NaCl, 26.8 mM KCl, 43 mM NaH$_2$PO$_4$, 14.7 mM KH$_2$PO$_4$.
7. Methyl-β-cyclodextrin (Sigma C-4555) and cholesterol complexed to Methyl-β-cyclodextrin (Sigma C-4951). Stock solutions made in deionized H$_2$O and stored at –20°C.
8. Cholera toxin B subunit (CTxB) peroxidase conjugate (Sigma C-4672).
9. *n*-octyl-β-glucopyranoside (Sigma O-9882).
10. [9,10-^{3}H-(N)]palmitic acid (Cat. no. NET043, NEN PerkinElmer Life Science, Boston, MA).
11. Prostacyclin I$_2$ (PGI$_2$): Dissolve 1 mg PGI$_2$ (Sigma P-6188) in 2 mL 0.1 M NaCl, 0.05 M Tris-HCl, pH 12.0. Aliquot immediately (25 µL) on dry ice and store at –70°C. For use, add 225 µL of the above buffer to a 25-µL aliquot for a final concentration of 50 µg/mL. Keep on ice and discard after use.
12. diI-C$_{18}$(1,1′-dioctadecyl-3,3,3′,3′-tetramethyl-indocarbocyanine perchlorate) (Cat. no. D-3911, Molecular Probes, Leiden, The Netherlands).
13. 2% (w/v) paraformaldehyde: For 100 mL final volume, add 2 g paraformaldehyde to 80 mL deionized H$_2$O. Heat with stirring, until the solution reaches 56–70°C. Do not boil. Add 10 N NaOH dropwise until the solution clears. Remove from the heat and allow to cool: add 10 mL of 10X PBS and H$_2$O to 100 mL. Adjust the pH to 7.4, filter the solution, and store, wrapped in foil, at 4°C. The solution will be stable for approximately one month. **Note:** Wear a protective mask to avoid inhalation of paraformaldehyde.
14. 0.01% (w/v) poly-L-lysine (Sigma P-4707).
15. Trichloroacetic acid solution (100% w/v; Sigma 490-10).

3. Methods
3.1. Lipid Raft Isolation From Human Platelets

Most researchers operationally define lipid rafts by insolubility in nonionic detergent and subsequent flotation on sucrose gradients. Despite the convenience and the usefulness of this method, it is not without limitation. First, this method cannot be used to quantitate the fraction of a given molecule present in rafts in the intact cell or provide information as to the original subcellular localization of the lipid rafts. Second, a raft-associated protein may also be connected to the cytoskeleton, so that it will not float after detergent extraction. Alternatively, its association with lipid rafts may be so weak that it will be solubilized by the detergent used. Results also often vary depending on the concentration and chemical structure of the detergent used. Consistent results may be obtained by using standard protocols, by examining the detergent concentration dependence of a given protein's association with lipid rafts, and most importantly, by paying particular attention to the lipid composition of the rafts isolated under a particular condition and cell type. Nevertheless, this approach provides a powerful tool for identifying molecules that are likely to exist in lipid rafts in the living cell. The following protocol is depicted schematically in **Fig. 1** (*see* **Note 1**).

1. Pre-cool ultracentrifuge and rotor to 4°C.
2. Lyse platelet pellet (1×10^9 platelets) in 2 mL Triton X-100 lysis buffer and incubate on ice for 30 min. The platelet pellet is obtained by centrifugation of the platelet suspension (800g for 10 min at 25°C; if required, include 1 µg/mL PGI$_2$ to prevent activation), either untreated, following agonist stimulation, or manipulation as described in this chapter. Washed platelet suspensions can be directly lysed in 2X Triton X-100 lysis buffer (*see* **Note 2**).
3. In a 14-mL ultracentrifuge tube, adjust the 2-mL platelet lysate to 40% (w/v) sucrose by the addition of 2 mL 80% (w/v) sucrose. Mix well.
4. Using a syringe or pipet, slowly layer 4 mL of the 30% (w/v) sucrose solution upon the 40% (w/v) layer. Hold the pipet at a 45° angle to the side of the centrifugation tube and allow the solution to run gently down the side of the tube so as not to disturb the gradient.
5. Repeat this method to layer 2 mL of the 5% (w/v) sucrose solution upon the 30% layer.
6. Balance the samples by the careful dropwise addition of the 5% (w/v) sucrose solution.
7. Centrifuge samples at 200,000g for 18 h at 4°C.
8. By pipet, carefully take 12 equal fractions (830 µL each) from the top of the gradient.

3.2. Localization of the Lipid Raft Fraction Within the Sucrose Gradient

3.2.1. GM1 Ganglioside

GM1 ganglioside is highly enriched in lipid rafts and thus considered a marker of these domains **(Fig. 2)**. Co-localization of a given protein with GM1 ganglioside (usually detected using cholera toxin B subunit, CTxB) is a widely used approach for identifying lipid raft association of a given protein (refer also to **Subheading 3.7.**). It is important to note, however, that while the GM1 ganglioside is enriched in lipid rafts, it is not exclusive to them nor are its buoyant properties affected by cholesterol depletion *(26)*. As such, one must be cautious about relying solely on GM1 as a marker of lipid rafts.

1. Assemble Biodot apparatus and perform dot-blotting according to manufacturer's instructions, using 50–100 µL of each of the 12 fractions taken from the sucrose gradient. Analyze an equal volume from each fraction.
2. Block nonspecific sites on the nitrocellulose membrane by incubation with 5% (w/v) skim milk in TBS for 1 h at room temperature.
3. Wash the membrane three times for 5 min with TBS.
4. Incubate membrane overnight at 4°C with 5 µg/mL peroxidase-conjugated CTxB in TBS containing 0.5% (w/v) skim milk.
5. Wash the membrane three times for 5 min with TBS.
6. Develop using chemiluminescence/ECL according to manufacturer's instructions (*see* **Note 3**).

Fig. 1. *(opposite)* A nonionic detergent extract is prepared at 4°C and centrifuged to equilibrium in a discontinous sucrose density gradient. Under these conditions, the fully solubilized proteins remain in the 40% (w/v) sucrose sample in which they were loaded, while the insoluble lipid rafts float as an opaque band at low density.

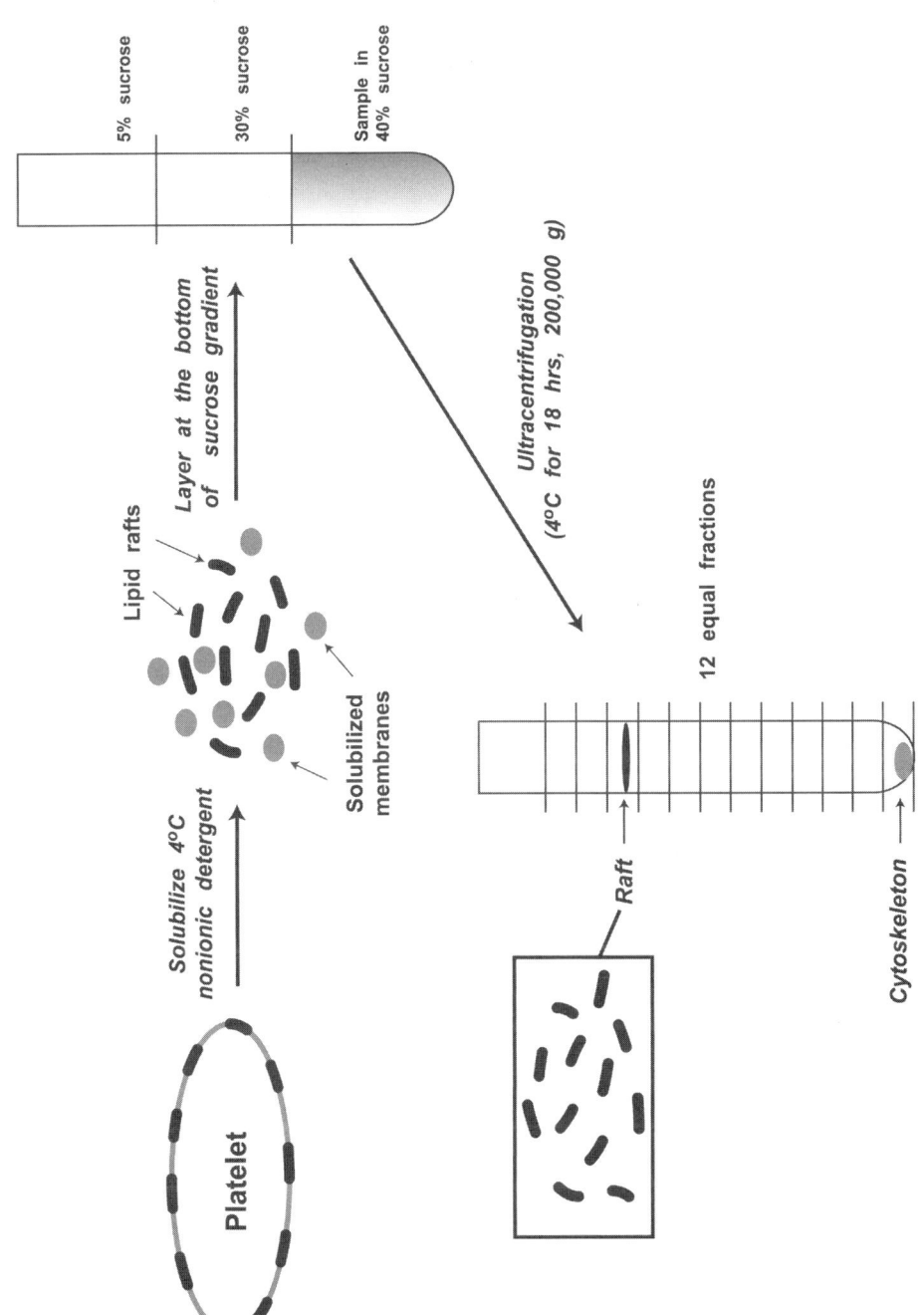

5% sucrose

30% sucrose

Sample in 40% sucrose

Layer at the bottom of sucrose gradient

Lipid rafts

Solubilized membranes

Ultracentrifugation (4°C for 18 hrs, 200,000 g)

12 equal fractions

Solubilize 4°C nonionic detergent

Platelet

Raft

Cytoskeleton

217

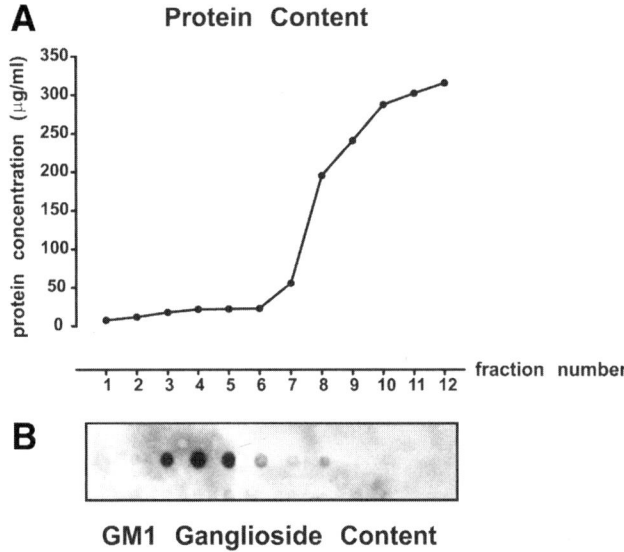

Fig. 2. **(A)** The protein content of each of the 12 fractions taken across the gradient was determined using the BCA assay. Lipid rafts are extremely protein-poor, containing less than 3% of the total cellular protein. **(B)** The position of the lipid rafts within the gradient was revealed by the presence of ganglioside GM1, which was detected by dot-blotting with HRP-conjugated CTxB. The lipid raft fraction floats within sucrose gradient fractions 3, 4, and 5.

3.2.2. Extraction of Lipids From Platelet Sucrose Gradient Fractions

Lipids from each gradient fraction can be extracted following the procedure of Bligh and Dyer *(27)*. If gradient fractions aren't to be used immediately they may be stored at –80°C until required.

For each 0.8-mL gradient fraction in a screw-top glass test tube with Teflon-coated lid:

1. Add 3 mL chloroform/methanol (1 : 2 v/v) to the tube, cap, and vortex for 60 s. The final proportions are chloroform/methanol/water: 1 : 2 : 0.8.
2. Add 1 mL chloroform to the tube, cap, and vortex for 60 s.
3. Add 1 mL water to the tube, cap, and vortex for 60 s. The final proportions are now chloroform/methanol/water: 2 : 2 : 1.8. The addition of 1 mL of either 150 mM KCl or 100 mM Na$_2$EDTA in place of the water at this stage will help prevent the lipids from smearing on thin-layer chromatography (TLC).
4. Centrifuge the tube for 5 min at 15,000g to allow the two phases to separate. Be very careful when removing the tubes from the centrifuge so as not to disturb the two layers.
5. Using a Pasteur pipet with bulb, draw up the lower, chloroform phase (which contains the lipids), being careful not to draw up any of the methanol/water phase. Transfer the layer into a clean glass vial with a Teflon-coated lid.
6. Evaporate the chloroform in a fume hood by using a gentle stream of nitrogen.

7. Store the samples at –20°C until lipid analysis. Lipid analysis can be done using TLC or high-performance liquid chromatography (HPLC). Cholesterol can be quantitated either by TLC or cholesterol oxidase or by using a commercial diagnostic kit.

3.3. Identification of Lipid Raft-Associated Proteins

Each fraction taken from the sucrose gradient should be analyzed for the protein of interest by SDS-PAGE and immunoblotting following either trichloroacetic acid (TCA) precipitation of all proteins or immunoprecipitation of a given protein. Examining the distribution of the protein across the gradient in both unactivated and activated platelets is a good starting point, as platelet activation may affect the distribution of proteins to lipid rafts (refer to **Subheading 3.5.**). Protein association with lipid rafts can also be analyzed by comparing pooled raft fractions (i.e., fraction numbers 3–5, as determined in **Subheading 3.2.**) with pooled non-raft fractions (i.e., fraction numbers 9–12).

3.3.1. Trichloroacetic Acid (TCA) Precipitation of Protein

1. Add TCA to equal volumes of each gradient fraction. The final concentration of TCA should be 10% (v/v).
2. Vortex and incubate on ice for 30 min.
3. Centrifuge at 4°C for 10 min at 13,000g. Carefully remove supernatant.
4. Add 300 µL cold acetone.
5. Centrifuge at 4°C for 10 min at 13,000g. Carefully remove supernatant without disturbing the pellet. **Note:** Pellet may not be visible at this point.
6. Air-dry pellet for 5 min at room temperature.
7. Resuspend the pellet in SDS-PAGE loading buffer. (For further details of SDS-PAGE analysis, refer to Chapter 9, vol. 2.)

3.3.2. Immunoprecipitation

The protein of interest can be immunoprecipitated from equal volumes of each fraction as described in Chapter 10, vol. 2. It cannot be assumed that two proteins are in contact with each other solely on the basis of their concurrent localization into lipid rafts. So when investigating the association of proteins within lipid rafts, the lipid rafts must be solubilized by the addition of *n*-octyl-β-glucopyranoside (60 m*M*) to each fraction, prior to the immunoprecipitation procedure, to ensure the complete solubilization of the lipid rafts. The protein–protein interaction can then be investigated.

3.4. Disruption of Lipid Raft Structural Integrity

Manipulation of lipid raft constituents may lead to the perturbation of microdomain structure, the dissociation of component proteins, and a loss of function. Given that cholesterol plays a pivotal role in stabilizing lipid rafts, reversible depletion or enrichment of plasma membrane cholesterol are widely employed methods of altering lipid raft structure. The water-soluble carrier molecule, methyl-β-cyclodextrin (MβCD), is most commonly used since it neither inserts into nor binds to the membrane. In the cholesterol-free state, MβCD effectively removes cholesterol from membranes, the extent of removal depending on both MβCD concentration and incubation time (**Fig. 3**). On the other hand, MβCD that has been pre-equilibrated with cholesterol can effectively deliver

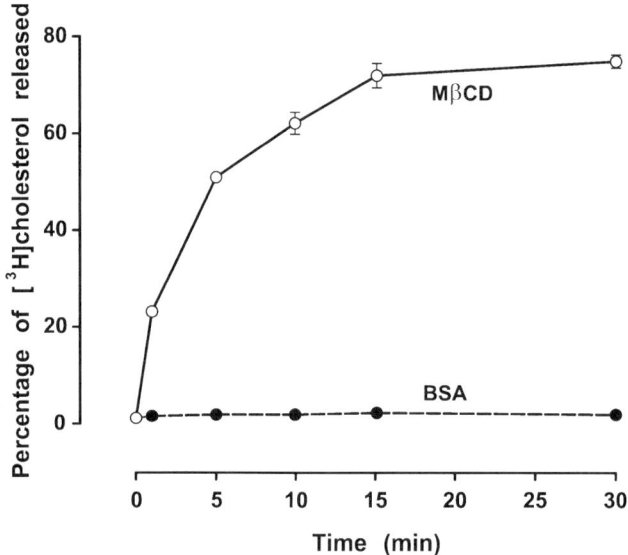

Fig. 3. A [^{3}H]cholesterol emulsion was prepared as previously described *(36)*. Platelets were pelleted from prostacyclin-treated PRP, resuspended in the [^{3}H]cholesterol-albumin emulsion, and incubated for 1 h at 37°C. Platelets were washed to remove unincorporated [^{3}H]cholesterol and resuspended with BSA-free buffer 1 to the original PRP volume. Samples were incubated with 0.03% (w/v) BSA (as a control) or with 10 mM MβCD and, at defined time periods (to 30 min), 300-μL aliquots were taken and rapidly centrifuged (20,000g, 2 min). The [^{3}H]cholesterol released into the supernatant was measured by liquid scintillation counting. At time zero, an aliquot of total platelet suspension was collected, dissolved in 1 M NaOH, and similarly counted.

cholesterol back to the membrane. Care must be taken when using MβCD to remove cholesterol, especially at high concentrations (>10 mM) or after long periods of incubation (>30 min), which may perforate or lyse the membrane, making interpretation of data problematic, especially when investigating calcium-induced signaling events. Other agents that have been used to disrupt raft structure are outlined in **Table 1**.

Following cholesterol depletion or depletion/repletion, platelets can be used in functional studies such as platelet aggregation, adhesion, or signaling studies (e.g., tyrosine phosphorylation, calcium flux). Alternatively, platelets can be used for microscopic studies (*see* **Subheading 3.7.**) or be subjected to detergent extraction and sucrose-gradient centrifugation to investigate the effect of cholesterol on protein localization to the lipid raft fraction.

3.4.1. Plasma-Membrane Cholesterol Depletion

This procedure can be performed using platelet-rich plasma (PRP) or washed platelets, depending on the platelet function to be examined. We routinely perform

Table 1
Agents That Disrupt Lipid Rafts

Agent	Action
Filipin	Cholesterol sequestration
Nystatin	
Digitonin	
Saponin	
Streptolysin	
Methyl-β-cyclodextrin	
Compactin	Inhibition of cellular cholesterol
Lovastatin	biosynthesis
Exogenous ganglioside	Perturbation of lipid raft integrity
Exogenous polyunsaturated fatty acids	

cholesterol-depletion studies using PRP that is then used immediately in functional assays. In our experience, it is not necessary to remove the MβCD from the system prior to performing functional studies, since in our experiments we have found that replacement of the MβCD containing plasma with fresh plasma produced identical results as when the MβCD was not removed. The inhibitory effects of cholesterol depletion on platelet function are both time- and dose-dependent, so it is good practice to examine a range of concentrations and incubation times to determine the optimal conditions required.

1. Incubate PRP with 10 mM final concentration of MβCD at 37°C for 30 min.
2. Platelets can now be used directly in functional studies, or washed, pelleted, lysed, and subjected to sucrose-gradient fractionation.

Note: If PRP is to be used in functional studies, it must be used immediately following incubation with MβCD, as this agent will continue to deplete the membrane of cholesterol.

3.4.2. Plasma-Membrane Cholesterol Repletion

When precomplexed with cholesterol, cyclodextrins can deliver cholesterol back to the plasma membrane and hence restore cellular functions interrupted by cholesterol depletion.

1. Wash cholesterol-depleted platelets once to remove MβCD.
2. Resuspend platelets in buffer 1 containing 3 mM MβCD:cholesterol and incubate at 37°C for 30 min.
3. Wash platelets once and use in subsequent experiments.

3.5. Receptor Redistribution Into or Out of Microdomains

Proteins may be constitutively associated with lipid rafts or recruited there upon platelet activation. Indeed, protein distribution can change dramatically upon receptor clustering. In this section, we describe the use of a radiolabeled antibody or ligand to monitor the membrane distribution of a given receptor protein in the resting or clustered state.

1. Incubate washed platelets with radiolabeled antibody or ligand (1–10 µg/mL) for 30 min to 1 h at room temperature (RT).
2. Separate platelets from unbound antibody or ligand by fivefold dilution in buffer 1 containing 1 µg/mL PGI$_2$ and centrifugation at 800g for 10 min.
3. Lyse platelets and isolate lipid rafts as described in **Subheading 3.1.** If platelet activation is desired, platelets can be resuspended in buffer 1 and stimulated with the appropriate agonist before lysis.
4. Count an equal volume of each fraction (by scintillation or γ-counting, depending on the label used) for associated radioactivity.

Note: One caveat to this approach is the cross-linking of receptors as a consequence of antibody valency. This can be overcome by the use of monoclonal Fabs.

3.6. [³H]Palmitate Labeling of Platelet Proteins

1. Incubate washed platelets with 10 µCi/mL [³H]palmitic acid for 45 min at 37°C. Prior to use, dry the [³H]palmitic acid under a gentle stream of nitrogen and resuspend by gentle vortexing in buffer 1. **Note:** Given that fatty acids stick to plastic, if possible use glass pipets and vials while preparing the [³H]palmitate.
2. Separate platelets from unincorporated [³H]palmitic acid by fivefold dilution in buffer 1 containing 1 µg/mL PGI$_2$ and centrifugation at 800g for 10 min.
3. Lyse platelets and isolate lipid rafts as described in **Subheading 3.1.** If platelet activation is desired, platelets can be resuspended in buffer 1 and stimulated with the appropriate agonist prior to the lysis step.
4. The protein of interest can be immunoprecipitated from equal volumes of each fraction. It is important to solubilize the lipid rafts by adding 60 mM n-octyl-β-glucopyranoside to each fraction prior to immunoprecipitation to ensure that the immunoprecipitated radioactivity is associated with the protein of interest rather than with other, coprecipitating lipid raft-associated proteins or with noncovalently associated free palmitate.
5. Immunoprecipitates can be transferred to scintillation tubes containing scintillation fluid and the associated radioactivity counted. Alternatively, immunoprecipitated proteins can be resuspended in 2X SDS sample buffer containing 0.1 M DTT and analyzed by SDS-PAGE and fluorography (*see* **Note 4**).

3.7. Fluorescence Microscopy

While visualization of individual lipid rafts, which are less than 100 nm in size *(28,29)*, is beyond the resolution of the fluorescence microscope, various methods can be used to detect visually both lipid raft aggregates and lipid raft-associated proteins. Such studies can provide important information regarding the co-localization of liquid-ordered domains with lipid raft-specific proteins and should be used to confirm biochemical data based on detergent insolubility.

Fluorescent-membrane lipid probes such as the lipophilic carbocyanines (diI and its derivatives) can be used to label both living and fixed cells. These dyes can mimic the behavior of lipids found in biological membranes because their head groups are simi-

Table 2
Microscopy Techniques Used for the Studies of Lipid Rafts in Live Cells

Technique	Information	Key reference
Fluorescence resonance energy transfer (FRET)	Used to detect whether two raft components are spatially close (<10 nM) e.g., GM1 ganglioside labeled with CT-B and antibody-labeled protein in question.	*28*
Single fluorophore tracking microscopy	Used to measure the diffusion and dynamics of individual raft proteins or lipids	*37*
Photonic force microscopy	Used to determine the size and dynamics of individual rafts	*29*

lar to those of lipids normally associated with the lipid bilayer *(30)*. Depending on their chain length, diI analogs can distinguish the fluid from the ordered phase in membranes. For example, diI-C 14 and shorter chains will prefer a fluid environment, whereas diI-C 18 and longer chain lengths will partition preferentially into ordered domains *(31)*. Because of these properties, diI can be used to identify the presence of liquid-ordered domains in the plasma membrane *(32)*.

Other microscopic techniques relevant to platelets that have been extensively used to study lipid rafts are outlined in **Table 2**. These methods may be more sensitive in detecting weak associations of proteins with lipid rafts than are the studies of detergent insolubility (*see* **Note 5**).

3.7.1. diI Labeling of Platelets

1. Incubate washed platelets (1×10^8/mL) with 2.5 µg/mL diI-C$_{18}$(1,1′-dioctadecyl-3,3,3′,3′-tetramethyl-indocarbocyanine perchlorate) for 5–10 min in the dark at 37°C. Gently rotate samples to ensure even distribution of the dye. **Note:** If cholesterol-depletion is required, incubate platelets with MβCD prior to diI-C$_{18}$ labeling.
2. If desired, expose the platelets to agonist for 5 min.
3. Add an equal volume of 2% (w/v) paraformaldehyde and fix for 1 h. **Note:** It is important to fix the platelets at the temperature at which they were incubated with the diI.
4. Wash platelets with buffer 1 containing 1 µg/mL PGI$_2$ to remove the paraformaldehyde.
5. If desired, label the platelets with antibodies to lipid raft-specific transmembrane proteins, previously identified by detergent insolubility studies. A protein consistently found in platelet lipid rafts is CD36 *(17,22)*. Label for 1 h in the dark. A good starting concentration of antibody is 10 µg/mL. As diI lipophilic dyes have very broad excitation and emission spectra, antibodies should be labeled with either cy5 or allophycocyanin and cells examined at far-red frequencies.
6. Place an aliquot of the platelet suspension on a poly-L-lysine-coated coverslip and allow the platelets settle in darkness for 1 h. Mount coverslips on glass slides with Aquapolymount.
7. If internalized proteins are to be detected, adherent fixed platelets should be permeabilized

Human Platelets

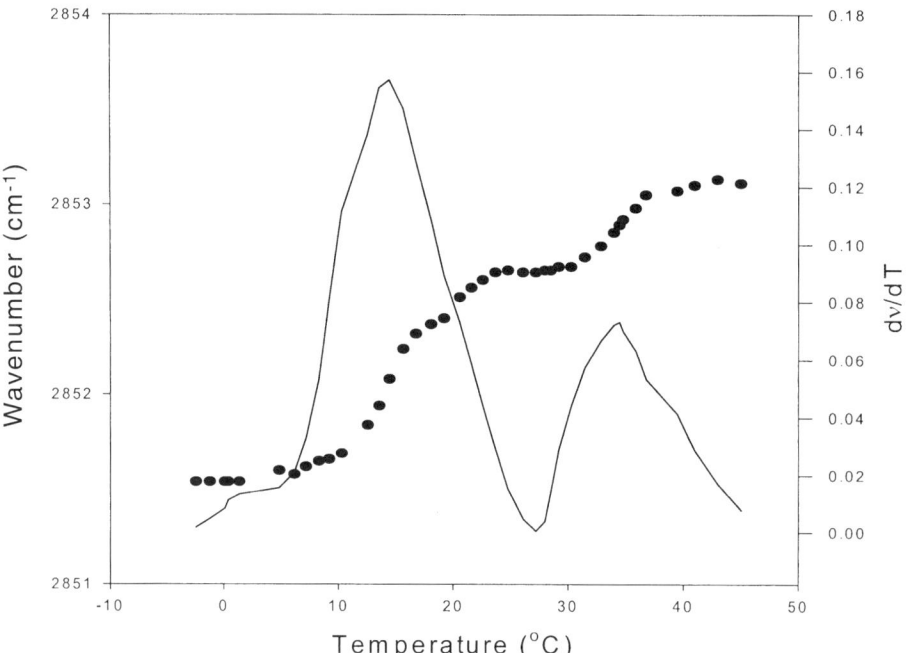

Fig. 4. When platelets are studied using Fourier transform infrared spectroscopy, examination of the CH2 stretching vibration (dotted line) reveals information regarding the wave number, and thus the phase behavior, of lipids. A high wave number indicates increased vibrational frequency (liquid-crystalline) and a low wave number indicates decreased vibrational frequency (gel phase). Also shown are the first derivatives of the data (solid line), which more easily demonstrate the two major phase transitions. Above 22°C, platelet membrane phospholipids are in the liquid-crystalline state. However, chilling the platelets to 4°C results in passage of the phospholipids through the main phase transition (15–18°C) and movement of phospholipids into the gel state. An additional phase transition can be seen at 35°C, representing the sphingomyelin phase transition indicating the presence of lipid rafts even at physiological temperatures. From the unpublished studies of K. Gousset, W. F. Wolkers, J. H. Crowe, and F. Tablin.

with the same type and concentration of nonionic detergent as used in the detergent-resistant preparation. For this, add a drop of detergent to the coverslip and incubate at RT for 10 min, then rinse the slide three times with 1X PBS. Add antibody to the coverslip and incubate for 1 h in the dark, rinse with 1X PBS, and prepare for visualization as described. Again, labels should be in the far-red range to prevent accidental fluorescence resonance energy transfer (FRET) due to the broad emissions of diI.
8. Visualize either by confocal microscopy (using a Kr/Ag laser) or by a fluorescence microscope fitted with an argon laser. Visualize within 24 h of preparation. Store in the dark at

room temperature; do not store at 4°C, as chilling may induce the artificial clustering of lipid rafts (*see* **Note 6**).

9. If there are no microscopes available to detect emissions in the far-red, then platelet rafts can be identified by labeling with antibodies against CD36, which is exclusively localized to the raft fraction. Other membrane or cytosolic proteins of interest can then be labeled using standard antibodies conjugated either to fluorescence or rhodamine (TRITC) dyes.

10. Alternatively, platelets can be labeled with either cholera toxin B (10 µg/mL per 10^8/mL platelets) or antibodies to the glycosylphosphatidylinositol (GPI)-anchored protein CD59. It should be noted that while both GM1 ganglioside and the GPI-linked proteins are found in raft fractions, they are not exclusively localized to the raft domains and care should be taken in interpreting the microscopy results. Labeling of either of these markers is best correlated with diI-C18 or CD36 labeling. Cy3 and Cy5-CTxB can be prepared from succinimidyl ester derivatives according to manufacturer's instructions (Fluorolink Reactive Dye; Amersham, Arlington Heights, IL).

3.7.2. Co-Patching Studies

An additional microscopic method for visualizing lipid rafts is to artificially crosslink and cluster lipid rafts with antibodies against resident receptors or lipids *(33)*. For example, crosslinking the raft component GM1 with fluorescent CTxB and an anti-CTxB antibody results in the formation of discrete fluorescent patches. Such studies were originally undertaken to understand the biological functions of membrane patching, but they can also be used to aggregate lipid rafts. Cross-linking studies have most commonly been used in studies of cells of the immune system, in which case these more closely mirror physiological events, where the cross-linking of immunoglobulins is key to immune cells signaling. However, such patching may well be artifactual in the platelet system.

4. Notes

1. The method for the isolation of lipid rafts described in this chapter is one we have used with success in our laboratory. However, there are many variations to this method in the literature in terms of volume, gradient composition (e.g., 40% [w/v] sucrose, followed by a stepwise gradient of 30–35% [w/v] sucrose), and number of fractions taken. The choice of the specific method will be influenced by the equipment available and by personal preference. It is important, however, to first determine the position of the lipid rafts within the sucrose gradient by GM1 ganglioside positioning and/or by analysis of the lipid composition. Rafts may also be identified by the exclusive presence of the transmembrane protein CD36.

2. Given that the type and concentration of detergent used to extract lipid rafts can affect the recovery of lipid raft-associated proteins, it is worthwhile to examine raft localization using several different nonionic detergents (e.g., Brij series, NP-40, etc.) and at various detergent concentrations (0.1–1.0%). Again, it is necessary to confirm that the relevant fractions identified have the characteristics of a liquid ordered domain, i.e., that they show an enrichment of cholesterol and sphingomyelin.

3. GM1 ganglioside may not localize exclusively to the lipid raft fraction and may be detectable in the heavier fractions. However, a discrete population should be evident with the lighter "lipid raft" fractions.

4. Do not boil [^{3}H]palmitate-containing samples. Boiling may result in the significant loss of label as a consequence of the labile nature of the thioester bond.

5. An alternate method of identifying lipid rafts employs a cholesterol-binding toxin to examine the organization of membrane cholesterol. A protease-nicked biotinylated derivative of θ-toxin (Perfringolysin O), a pore-forming cytolysin of *Clostridium perfringens*, has been demonstrated to bind selectively to the lipid raft domains of intact cells *(34)*. Hence, this agent, known as BCθ, can be employed in several ways to study rafts. It can be preincubated (5–10 μg/mL, 20 min at RT) with platelets prior to fractionation by sucrose gradient ultra-centrifugation. The position of the BCθ can be detected by immunoblotting with an anti-θ-toxin antibody following SDS-PAGE or dot-blotting. Platelets incubated with BCθ can also be used for microscopy studies, either in electron microscopy studies employing an anti-biotin antibody and 10 nM protein-A gold, or in fluorescent microscopy studies using Cy3-avidin.

6. It is important to note that lipid raft preparations are generally produced by incubation at 4°C for 3–60 min. Such cold incubation results in cold-induced platelet activation and correlates with a phase transition of membrane lipids *(35)*. Thus it is critical that platelets be activated or subjected to other treatments while at physiological temperatures prior to being incubated in the cold. If at all possible, it is best to accomplish platelet manipulations at physiological temperatures and to subject them to cold only during detergent lysis, so as to avoid the introduction of cold-induced artifact. If cold-induced activation is desired, then incubation of cells at 4°C for up to 1 h will yield raft aggregation, as assessed by DRMs, diI, and CD36 labeling, and analysis of lipid composition.

References

1. Simons, K. and Toomre, D. (2000) Lipid rafts and signal transduction. *Nat. Rev. Mol. Cell Biol.* **1,** 31–39.
2. Galbiati, F., Razani, B., and Lisanti, M. P. (2001) Emerging themes in lipid rafts and caveolae. *Cell* **106,** 403–411.
3. Brown, D. A. and London, E. (2000) Structure and function of sphingolipid- and cholesterol-rich membrane rafts. *J. Biol. Chem.* **275,** 17,221–17,224.
4. Harris, J., Werling, D., Hope, J. C., Taylor, G., and Howard, C. J. (2002) Caveolae and caveolin in immune cells: distribution and functions. *Trends Immunol.* **23,** 158–164.
5. Anderson, R. G. (1998) The caveolae membrane system. *Annu. Rev. Biochem.* **67,** 199–225.
6. Brown, D. A. and Rose, J. K. (1992) Sorting of GPI-anchored proteins to glycolipid-enriched membrane subdomains during transport to the apical cell-surface. *Cell* **68,** 533–544.
7. Simons, K. and Ikonen, E. (1997) Functional rafts in cell membranes. *Nature* **387,** 569–572.
8. Bretscher, M. S. and Munro, S. (1993) Cholesterol and the golgi-apparatus. *Science* **261,** 1280–1281.
9. Ledesma, M. D., Simons, K., and Dotti, C. G. (1998) Neuronal polarity: Essential, role of protein-lipid complexes in axonal sorting. *Proc. Natl. Acad. Sci. USA* **95,** 3966-3971.
10. Mayor, S., Sabharanjak, S., and Maxfield, F. R. (1998) Cholesterol-dependent retention of GPI-anchored proteins in endosomes. *EMBO J.* **17,** 4626–4638.
11. Janes, P. W., Ley, S. C., Magee, A. I., and Kabouridis, P. S. (2000) The role of lipid rafts in T cell antigen receptor (TCR) signalling. *Semin Immunol.* **12,** 23–34.
12. Cheng, P. C., Dykstra, M. L., Mitchell, R. N., and Pierce, S. K. (1999) A role for lipid rafts in B cell antigen receptor signaling and antigen targeting. *J. Exp. Med.* **190,** 1549–1560.
13. Sheets, E. D., Holowka, D., and Baird, B. (1999) Membrane organization in immunoglobulin E receptor signaling. *Curr. Opin. Chem. Biol.* **3,** 95–99.
14. Shenoy-Scaria, A. M., Dietzen, D. J., Kwong, J., Link, D. C., and Lublin, D. M. (1994) Cysteine(3) of Src family protein-tyrosine kinases determines palmitoylation and localization in caveolae. *J. Cell Biol.* **126,** 353–363.

15. Melkonian, K. A., Ostermeyer, A. G., Chen, J. Z., Roth, M. G., and Brown, D. A. (1999) Role of lipid modifications in targeting proteins to detergent-resistant membrane rafts—Many raft proteins are acylated, while few are prenylated. *J. Biol. Chem.* **274,** 3910–3917.

16. Webb, Y., Hermida-Matsumoto, L., and Resh, M. D. (2000) Inhibition of protein palmitoylation, raft localization, and T cell signaling by 2-bromopalmitate and polyunsaturated fatty acids. *J. Biol. Chem.* **275,** 261–270.

17. Dorahy, D. J., Lincz, L. F., Meldrum, C. J., and Burns, G. F. (1996) Biochemical isolation of a membrane microdomain from resting platelets highly enriched in the plasma membrane glycoprotein CD36. *Biochem. J.* **319,** 67–72.

18. Dorahy, D. J. and Burns, G. F. (1998) Active Lyn protein tyrosine kinase is selectively enriched within membrane microdomains of resting platelets. *Biochem. J.* **333,** 373–379.

19. Ezumi, Y., Kodama, K., Uchiyama, T., and Takayama, H. (2002) Constitutive and functional association of the platelet collagen receptor glycoprotein VI-Fc receptor gamma-chain complex with membrane rafts. *Blood* **99,** 3250–3255.

20. Locke, D., Chen, H., Liu, Y., Liu, C., and Kahn, M. L. (2002) Lipid rafts orchestrate signaling by the platelet receptor glycoprotein VI. *J. Biol. Chem.* **277,** 18,801–18,809.

21. Shrimpton, C. N., Borthakur, G., Larrucea, S., Cruz, M. A., Dong, J. F., and Lopez, J. A. (2002) Localization of the adhesion receptor glycoprotein Ib-IX-V complex to lipid rafts is required for platelet adhesion and activation. *J. Exp. Med.* **196,** 1057–1066.

22. Gousset, K., Wolkers, W. F., Tsvetkova, N. M., Oliver, A. E., Field, C. L., Walker, N. J., et al. (2002) Evidence for a physiological role for membrane rafts in human platelets. *J. Cell Physiol.* **190,** 117–128.

23. Bodin, S., Giurato, S., Ragab, J., Humbel, B. M., Viala, C., Vieu, C., et al. (2001) Production of phosphatidylinositol 3,4,5-trisphosphate and phosphatidic acid in platelet rafts: Evidence for a critical role of cholesterol-enriched domains in human platelet activation. *Biochemistry* **40,** 15,290–15,299.

24. Zhang, W., Trible, R. P., and Samelson, L. E. (1998) LAT palmitoylation: Its essential role in membrane microdomain targeting and tyrosine phosphorylation during T cell activation. *Immunity* **9,** 239–246

25. Kosugi, A., Hayashi, F., Liddicoat, D. R., Yasuda, K., Saitoh, S., and Hamaoka, T. (2001) A pivotal role of cysteine 3 of Lck tyrosine kinase for localization to glycolipid-enriched microdomains and T cell activation. *Immunol. Lett.* **76,** 133–138.

26. Ilangumaran, S. and Hoessli, D. C. (1998) Effects of cholesterol depletion by cyclodextrin on the sphingolipid microdomains of the plasma membrane. *Biochem. J.* **335,** 433–440.

27. Bligh, E. G. and Dyer, W. J. (1959) A rapid method of total lipid extraction and purification. *Can. J. Biochem. Physiol.* **37,** 911–917.

28. Varma, R. and Mayor, S. (1998) GPI-anchored proteins are organized in submicron domains at the cell surface. *Nature* **394,** 798–801.

29. Pralle, A., Keller, P., Florin, E. L., Simons, K., and Horber, J. K. (2000) Sphingolipid-cholesterol rafts diffuse as small entities in the plasma membrane of mammalian cells. *J. Cell Biol.* **148,** 997–1008.

30. Bergelson, L. D., Molotkovsky, J. G., and Manevich, Y. M. (1985) Lipid-specific fluorescent-probes in studies of biological-membranes. *Chem. Phys. Lipids* **37,** 165–195.

31. Spink, C. H., Yeager, M. D., and Feigenson, G. W. (1990) Partitioning behavior indocarbocyanine probes between coexisting gel and fluid phases in model membranes. *Biochim. Biophys. Acta* **1023,** 25–33.

32. Thomas, J. L., Holowka, D., Baird, B., and Webb, W. W. (1994) Large-scale coaggregation of fluorescent lipid probes with cell-surface proteins. *J. Cell Biol.* **125,** 795–802.

33. Mayor, S., Rothberg, K. G., and Maxfield, F. R. (1994) Sequestration of GPI-anchored proteins in caveolae triggered by crosslinking. *Science* **264,** 1948–1951.
34. Waheed, A. A., Shimada, Y., Heijnen, H. F., Nakamura, M., Inomata, M., Hayashi, M., et al. (2001) Selective binding of perfringolysin O derivative to cholesterol-rich membrane microdomains (rafts). *Proc. Natl. Acad. Sci. USA* **98,** 4926–4931.
35. Tablin, F., Oliver, A. E., Walker, N. J., Crowe, L. M., and Crowe, J. H. (1996) Membrane phase transition of intact human platelets: Correlation with cold-induced activation. *J. Cell Physiol.* **168,** 305–313.
36. Gillett, M. P. T. and Owen, J. S. (1992) Cholesterol esterifying enzymes-lethicin: cholesterol acyltransferase (LCAT) and acyltransferase (ACAT), in *Lipoprotein Analysis. A Practical Approach* (Converse, C. A. and Skinner, E. R., eds.), Oxford Press, Oxford.
37. Schutz, G. J., Kada, G., Pastushenko, V. P., and Schindler, H. (2000) Properties of lipid microdomains in a muscle cell membrane visualized by single molecule microscopy. *EMBO J.* **19,** 892–901.

15

Measurement and Manipulation of $[Ca^{2+}]_i$ in Suspensions of Platelets and Cell Cultures

Philippe Ohlmann, Béatrice Hechler, Jean-Pierre Cazenave, and Christian Gachet

1. Introduction
1.1. Platelet Ca^{2+} Signaling

The importance of cytosolic calcium (Ca^{2+}) elevation in the regulation of platelet functions (shape change, aggregation, and secretion) has been widely acknowledged. Therefore its concentration must be tightly regulated, with most of it being sequestrated in intracellular organelles such as the dense tubular system (the releasable intracellular Ca^{2+} store) and mitochondria, or bound to membranes and cytoplasmic proteins. Only a small fraction is freely available in the ionized form. The maintenance of low cytosolic calcium concentration (around 100 nM), necessary to keep platelets in a resting state or to reestablish a resting state after activation, is achieved by a combination of Ca^{2+}/ATPases and a Na^+/Ca^{2+} exchanger. Platelets contain two types of Ca^{2+}/ATPases: the plasma membrane Ca^{2+}/ATPase (PMCA types 1–4) *(1)* and the sarco-endoplasmic reticulum Ca^{2+}/ATPase (SERCA) present on intracellular stores *(2)*. Two types of Na^+/Ca^{2+} exchangers have been identified; the platelets have been suggested to contain the retinal type, which is K^+-dependent and located in the plasma membrane *(3,4)*.

When platelets are activated by agonists, Ca^{2+} elevation in the cytosol occurs via both the mobilization of stored Ca^{2+} from the dense tubular system (corresponding to the endoplasmic reticulum of nucleated cells) and an influx of extracellular Ca^{2+} across the plasma membrane *(5–7)*. The rise in cytosolic Ca^{2+} levels can vary from basal levels of approximately 50–100 n*M* to levels during activation of 1–2 µ*M*, depending on the agonist *(8)*. In the absence of external Ca^{2+}, the maximum Ca^{2+} rise is much smaller, about 200–300 nM, and results from Ca^{2+} release from the internal stores *(5,9,10)*. The binding of agonists to their specific surface receptors leads to the activation of phospholipase C (PLC), which catalyzes the hydrolysis of membrane phosphatidylinositol

From: *Methods in Molecular Biology, vol. 273:*
Platelets and Megakaryocytes, Vol. 2: Perspectives and Techniques
Edited by: J. M. Gibbins and M. P. Mahaut-Smith © Humana Press Inc., Totowa, NJ

4,5-bisphosphate to produce the Ca^{2+}-releasing messenger inositol 1,4,5-triphosphate (IP_3). Platelets contain all three subtypes of IP_3 receptors. Type I and type II receptors are predominant in intracellular membranes of human platelets, while type III is found in plasma membranes *(11)*. The release of Ca^{2+} from the intracellular stores when mediated by the activation of G protein-coupled receptors requires about 200 ms, which corresponds to the metabolic events generating IP_3. Many platelet agonists, including adenosine 5'-diphosphate (ADP), platelet-activating factor (PAF), thrombin, vasopressin, or thromboxane A2 (TxA_2), induce the generation of IP_3 through activation of PLC-β via the heterotrimeric GTP-binding protein G_q *(12)*. PAF also activates PLC-γ_1 *(7)*, while collagen activates PLC-γ_2 via tyrosine phosphorylation *(13)*, and thrombin activates PLC-γ_1 and PLC-γ_2 *(14)*.

Release of Ca^{2+} from the finite intracellular store is supplemented by Ca^{2+} entry from the external medium *(9)*. In platelets, Ca^{2+} influx is achieved at least in part by store-operated Ca^{2+} channels (SOCC) activated in response to depletion of the intracellular Ca^{2+} store, also termed "capacitative Ca^{2+} entry" *(15)*. A number of hypotheses have been put forward to explain how depletion of the intracellular Ca^{2+} stores might be signaled to the plasma membrane to activate Ca^{2+} entry. Two of the most prominent schemes involve either a form of conformational coupling between the Ca^{2+} entry channels on the plasma membrane (PM) and Ca^{2+} release channels on the endoplasmic reticulum (ER) *(16,17)*, or a diffusible calcium-factor messenger that is synthesized by depleted-Ca^{2+} stores and activates Ca^{2+} channels on the PM *(18)*. In addition to store-dependent Ca^{2+} entry, platelets also possess a noncapacitative Ca^{2+} influx pathway, for which one candidate is TRPC6 *(19)*. Platelets also express a receptor-operated channel (ROC) identified as the ligand-gated ionotropic $P2X_1$ receptor that forms nonselective cation channels *(20)*. As compared to G protein-coupled receptors, ATP-evoked calcium influx through the ionotropic $P2X_1$ receptor is of very short latency (less then 20 ms) *(6,21,22)*.

Finally, two glycoprotein complexes may be involved in platelet Ca^{2+} signaling, the $\alpha_{IIb}\beta_3$ integrin (the fibrinogen receptor), and the GPIb-V-IX complex, of which GPIbα is the receptor for the von Willebrand factor (vWF) *(7)*. Earlier studies showed that plasma-membrane Ca^{2+} exchange was decreased in platelets in which the $\alpha_{IIb}\beta_3$ integrin was forced to dissociate and in platelets drawn from thrombasthenic patients *(23,24)*. More recently, it was reported that $\alpha_{IIb}\beta_3$-dependent calcium flux involves a PI 3-kinase mechanism linked to intracellular Ca^{2+} mobilization and subsequent transmembrane Ca^{2+} influx *(25)*. On the other hand, GPIb engagement on immobilized vWF was found to elicit transient calcium spikes mediated through release of intracellular Ca^{2+} stores in a signaling mechanism independent of PI 3-kinase *(25,26)*.

In this chapter we will describe how we measure the intracellular calcium concentration ($[Ca^{2+}]_i$) variations in human and rodent blood platelets as well as in cell cultures using fluorescent Ca^{2+}-sensing indicators.

1.2. Measurement of [Ca²⁺]ᵢ Using Fluorescent Indicators

Fluorescent Ca^{2+}-sensing indicators became very popular mainly because they can be easily incorporated into cells and produce signals that can be monitored with fairly

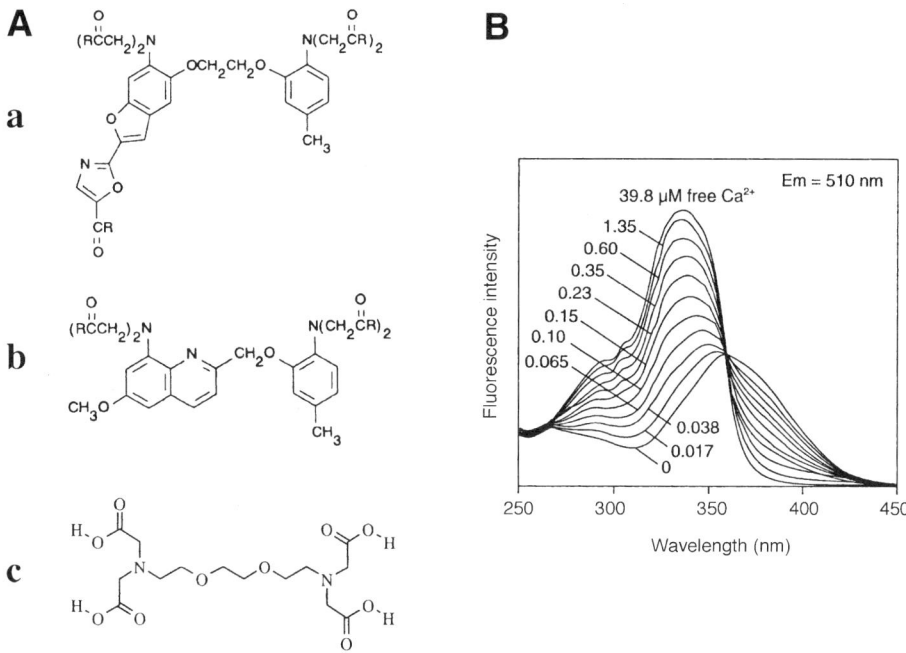

Fig. 1. Properties of the fluorescent Ca^{2+} indicator fura-2. (**A**) Chemical structures of fura-2 (a), quin-2 (b) and EGTA (c). R = •ONa$^+$ or •OK$^+$ are the sodium or potassium salt of the fluorescent probes, which are cell-impermeant. R = •OCH$_2$OCOCH$_3$ is the acetoxymethyl form of the fluorescent probe, which is cell-permeant. (**B**) Fluorescence excitation spectra of fura-2 in solutions containing zero to 39.8 μM free Ca^{2+}.

simple and widely available instrumentation. Fluorescent probes show a spectral response on binding Ca^{2+}. The first direct measurement of [Ca^{2+}]$_i$ in platelets was made in 1982 using the Ca^{2+}-sensitive fluorophore quin-2 (*8,27*). In 1985, Tsien and coworkers reported the development of several new Ca^{2+}-sensitive fluorophores (fura-2 and indo-1) with a significant number of advantages over quin-2 (*28*). Fura-2 is currently the most widely used of the fluorescent Ca^{2+} indicator dyes, especially in platelets (*29–31*). Like quin-2, fura-2 binds Ca^{2+} in a one-to-one stoichiometry (**Fig. 1A**) and responds with a change in fluorescence intensity. The fluorescence excitation maximum of fura-2 also shifts to a lower wavelength on Ca^{2+} binding, with negligible shift in the emission maximum. This allows fura-2 to be utilized as a dual excitation indicator. Fura-2 emission at 510 nm is usually monitored following excitation at both 380 nm, where the signal decreases upon binding of Ca^{2+}, and 340 nm, where the signal increases with Ca^{2+} binding (**Figs. 1B** and **3A**) (*32,33*). Ion-sensitive indicators such as fura-2 that exhibit spectral shifts upon ion binding can be used for ratiometric measurements of Ca^{2+} concentration, which are essentially independent of uneven dye loading, cell thickness, photobleaching effects, and dye leakage.

Fura-2 is approx 30-fold more fluorescent than its widely used predecessor, quin-2 *(28)*. Thus, useful signals can be achieved with much lower levels of dye loaded into the cytoplasm which allows for less disturbance (i.e., buffering) of the biological Ca^{2+} signals. In addition, fura-2 has a lower affinity for Ca^{2+} than does quin-2 (K_D = 224 n*M* and 115 n*M*, respectively), which along with reduced Mg^{2+} sensitivity *(33)*, allows one to detect $[Ca^{2+}]_i$ levels greater than 1 μ*M*.

Fura-2 is a polar molecule that does not readily cross the lipid-rich plasma membrane; however, the carboxylic acid groups of fura-2 can be blocked by esterification with acetoxymethyl groups, thus creating the lipid-soluble parent molecule fura-2/AM and thereby facilitating entry into the cell. Upon exposure of intact platelets to fura-2/AM in micromolar extracellular concentrations, rapid transmembrane passage of the molecule is followed by intracellular hydrolysis of the blocking groups by endogenous esterases to yield cell-impermeant and Ca^{2+}-sensitive fura-2.

2. Materials

2.1. Blood Collection and Preparation of Washed Platelet Suspension

The reagents used for blood collection or for the washing procedure of platelets are fully described in Chapter 2, vol. 1. These include acid citrate dextrose (ACD) anticoagulant, prostacyclin (PGI_2), heparin, apyrase, human serum albumin (HSA), or fatty-acid-free human serum albumin (FAF-HSA) when required. Anticoagulated blood, as well as platelet-rich plasma (PRP) and washed platelet suspensions, must be kept in plastic tubes (polypropylene, ref. 430291, Corning Inc., Corning, NY) to avoid platelet activation and blood coagulation. Human blood is collected with a 16-gauge needle mounted on a short length (10–20 cm) of plastic tubing (ref. HC-15R, Nissho Nipro Europe N.V., Zaventem, Belgium). 18-Gauge and 25-gauge needles are used for collection of rat and mouse blood, respectively. Appropriate ethical permission should be obtained.

2.2. Cell Line

Cultured cells are useful models for studying the Ca^{2+} signaling pathways coupled to platelet receptors, particularly since these systems are amenable to molecular alteration of specific proteins (for example, using antisense techniques, *see* Chapter 24, vol. 2). For example, cells in suspension such as Jurkat E6.1 cells (ECACC No. 88042803, Cerdic, France) stably expressing the $P2Y_1$ receptor or MEG-01 cells expressing endogenous P2Y receptors are grown in RPMI-1640 medium supplemented with 10% (v/v) heat-inactivated fetal calf serum, 2 m*M* glutamine, 100 U/mL penicillin, 0.1 mg/mL streptomycin, and 1 mg/mL geneticin *(34,35)*. Cultures are kept at 37°C in a humidified atmosphere containing 5% CO_2 and cells are subcultured every 3 days so as to maintain a density of approx 5×10^5 cells/mL. Adherent cells such as 1321 N1 human astrocytoma cells stably expressing the human $P2Y_1$ receptor *(36)* or B10 clone cells from rat brain capillary endothelium *(37)* are grown on sterile glass coverslips (13 mm × 27 mm) coated with fibronectin (*see* **Subheading 3.3.2., step 2**) in Dulbecco's modified Eagle's

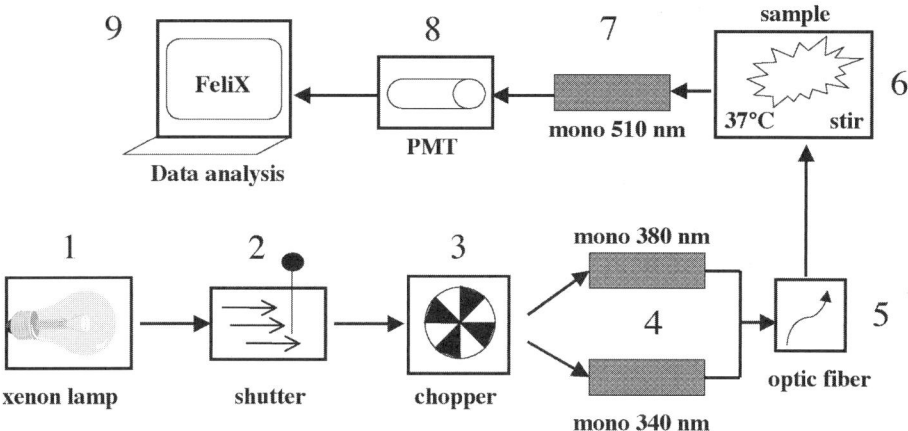

Fig. 2. Spectrofluorometer hardware system with dual excitation wavelengths. The illumination produced by the xenon arc lamp (1) first encounters a rotating chopper disk (3), which alternates the direction of the incident beam to one of two monochromators, one set at 340 nm, the other at 380 nm (4), via a focusing mirror. A shutter (2) is included to protect the sensitive PMT detector (8) from excessive light. The two separate monochromator outputs are collected by a quartz fiberoptic bundle that focuses the excitation light onto the sample (5). The sample holder is designed to accommodate a 10-mm-wide square cuvet and has a provision for sample stirring and for variable temperature (6). The sample compartment is attached to the emission monochromator (510 nm) and the photon counts are measured using a PMT (8). The fluorescence spectrometer is linked to a computer equipped with an acquisition system and appropriate software (9), for example FeliX (Photon Technology International Inc. Princeton, NJ).

medium supplemented with 10% (v/v) heat-inactivated fetal calf serum, 2 mM glutamine, 100 U/mL penicillin, 0.1 mg/mL streptomycin for 2 or 3 d to a cell density of approx 80% confluent.

2.3. Calcium Measurement

1. *Spectrofluorometer:* The essential elements of a fluorometer are a light source (PowerArc 75 W xenon arc lamp), one or two monochromators (or interference filters) for selecting the excitation wavelengths, a sample compartment (including the variable-temperature sample holder and a small magnetic stirrer to keep cells in suspension), an emission light collector and wavelength selector, a detector (usually a photomultiplier tube, PMT, mounted at 90° to the excitation light source), and an amplification and recording system (*see* **Fig. 2** and legend for further details; *see also* **Note 1**). Ratio-fluorescence techniques utilize dual wavelength fluorescence probes such as fura-2 or indo-1. Excitation-shifted probes require a special excitation light source that can rapidly alternate between two excitation wavelengths; for example, fura-2 needs excitation wavelengths at 340 nm and 380 nm. The fluorescence spectrofluorometer is linked to a computer equipped with custom-designed

software for measuring fura-2 signals at 340 and 380 nm excitation, calculating ratios (340/380), and converting them to Ca^{2+} concentration using the Grynkiewicz equation (*see* **Subheading 3.5.**) *(38)* (**Fig. 2**).

2. *Cuvets:* It is well known that the near-UV wavelengths required for fura-2 excitation are poorly transmitted by conventional glass; thus, glass cuvets should be made of quartz. Recently developed acryl cuvets are good alternatives, as they display a high transmittance of near-UV wavelengths (88% transmission at 334 nm and 89% transmission at 366 nm) (ref. 67755, Sarstedt Inc., Newton, NC). Acryl cuvets are far cheaper than quartz cuvets and can therefore be considered as disposable.

3. *Preparation and handling of fura-2/AM:* The acetoxymethyl ester form of fura-2 (fura-2/AM) is available from several commercial suppliers (Molecular Probes Inc., Eugene, OR; Sigma Inc., St Louis, MO; Calbiochem Inc., San Diego, CA) and is easily stored in solvent-tight tubes as a 10 mM solution in anhydrous dimethylsulfoxide (DMSO) at $-20°C$. This procedure will curtail the spontaneous ester hydrolysis that can occur in moist environments. This solution should be totally protected from light by using dark-colored tubes or aluminum foil. It is very stable under these conditions (several months), although repeated freeze-thawing should be avoided by dividing the stock into appropriately sized aliquots.

4. *Tyrode's buffer without calcium:* Stock solutions for preparing Tyrode's buffer without calcium are as follows:

   a. Stock I: 2.73 M NaCl (160 g), 53.6 mM KCl (4 g), 238 mM NaHCO$_3$ (20 g), 8.4 mM NaH$_2$PO$_4$, H$_2$O (1.16 g) made up to 1 L with distilled water and stored at 4°C.

   b. Stock II: 0.1 M MgCl$_2$•6 H$_2$O (20.33 g) made up to 1 L with distilled water and stored at 4°C.

   c. HEPES stock: 0.5 M (N-[2-Hydroxyethyl]piperazine-N'-[2-ethanesulfonic acid]) sodium salt (119 g) made up to 1 L with distilled water and stored at 4°C.

   Preparation of Tyrode's buffer for washed platelets: add 150 mL of distilled water to 10 mL of stock I, add 0.2 g of dextrose, 2 mL of stock II, and 2 mL of HEPES stock, 0.35 % (w/v) HSA and bring to a total volume of 200 mL with distilled water. Adjust pH to 7.35 (with NaOH/HCl) and the osmolarity to 295 mOsm/liter (with NaCl, from powder or 1 M stock) (*see* **Note 2**).

5. *Triton X-100:* A solution of Triton X-100 is prepared at 10% (v/v) in distilled water and stored at 4°C. This lysis buffer is used at a final concentration of 0.1% to release loaded dye for determining the calibration constants (*see* **Subheading 3.5.**).

6. *EGTA:* 0.8 M stock solution of ethylene glycol-*bis*(2-aminoethylether)-N,N,N',N'-tetraacetic acid (EGTA). This is made using the free acid form (Sigma) in distilled water. The pH is adjusted to 8.8 with NaOH (alkaline pH required for EGTA to dissolve) and the stock is filtered through a 0.22-μm membrane before storage at 4°C. EGTA (8 mM) is used to chelate the Ca^{2+} bound to fura-2 after lysis with Triton X-100 and give the minimum fluorescence ratio (R_{min}) and maximum 380 nm fluorescence (Sf2) in calibration experiments (*see* **Note 3** and **Subheading 3.5.**). EGTA (0.2 mM) is also used to chelate extracellular traces of Ca^{2+} in nominally Ca^{2+}-free salines, before adding an agonist in order to evaluate the $[Ca^{2+}]_i$ increase coming only from the internal stores.

7. *Calcium chloride:* 100 mM stock solution of CaCl$_2$, prepared by dissolving 2.19 g of CaCl$_2$•6 H$_2$O powder in 100 mL of distilled water. It is stored at 4°C after filtration through a 0.22-μm membrane. This solution is used to adjust the extracellular concentration of calcium, normally to a final concentration of 2 mM before addition of an agonist.

8. *Basic salt solution (BSS):* 125 mM NaCl, 5 mM KCl, 1 mM MgCl$_2$, 5 mM glucose, 25 mM HEPES, pH 7.3 (NaOH). This is supplemented with 0.1% (w/v) FAF-HSA, apyrase (0.02 U/mL), EGTA (0.1 mM) or CaCl$_2$ (2 mM) as required.

9. *Temperature-controlled centrifuge* (Sorvall RC3BP, Kendro Laboratory Products, Newtown, CT).
10. Platelets were counted in an automatic hematology analyzer for human blood (Sysmex, K-1000, Merck, Darmstadt, Germany) or for rodent blood (Coulter, Miami, FL).

3. Methods
3.1. Preparation of Fura-2-Loaded Human Washed Platelets

In this method platelets are loaded with fura-2 in a washed suspension. Further details of platelet preparation are given in Chapter 2, vol. 1 (*see also* **ref. 39**) which is a modification of the Mustard technique *(40)*.

1. Collect blood from a forearm vein with a 16-gauge needle mounted on a short length (10–20 cm) of plastic tubing (*see* **Note 4**). Discard the first 10 mL of blood, contaminated by tissue factor (TF) and containing trace amounts of thrombin. Then collect the blood directly into a conical 50-mL centrifuge tube containing 1 volume acid citrate dextrose (ACD) for every 6 volumes of blood (final pH 6.5 and citrate concentration 22 mM).
2. Immediately after collection, gently mix the blood with the anticoagulant and place in a water bath at 37°C for a maximum storage period of 15 min.
3. Prewarm the centrifuge to 37°C and spin the anticoagulated blood at 250g for a period of time depending on the quantity of blood. For a 50-mL centrifuge tube containing 50 mL of blood, the centrifugation time is 16 min; for 40 mL use 15 min and for 25 mL use 13 min.
4. Carefully collect the PRP using a 10-mL syringe and transfer to a new 50-mL centrifuge tube (*see* **Note 5**).
5. After 10 min incubation at 37°C, centrifuge the PRP at 2200g for a period of time depending on the quantity of plasma. For a 50-mL centrifuge tube containing 15 mL of plasma, the centrifugation time is 10 min; for 20 mL use 12 min, for 30 mL use 14 min, and for 40 mL use 16 min.
6. Discard the supernatant consisting of platelet-poor plasma (PPP) using a Pasteur pipet connected to a vacuum pump, in order to remove all traces of plasma from the tube walls or near the platelet pellet to prevent generation of thrombin during the subsequent washing steps.
7. Gently resuspend the platelet pellet in Tyrode's solution containing heparin (10 U/mL), HSA 0.35% (w/v), and PGI$_2$ (0.5 µM) but no Ca^{2+} and no apyrase.
8. Adjust the platelet count to 600,000/µL and incubate the suspension with fura-2/AM (2 µM) for 45 min at 37°C in the dark (*see* **Note 6**). Throughout the fura-2/AM-loading procedure and the subsequent washing steps and storage, platelets should be protected from light.
9. At the end of the incubation period, add PGI$_2$ (0.5 µM) and centrifuge the platelets at 1900g for 8 min (30 or 40 mL platelet suspension volume).
10. Resuspend the platelet pellet in the second washing solution containing HSA 0.35% (w/v) and PGI$_2$ (0.5 µM) but no Ca^{2+} and no apyrase. After incubation for 10 min at 37°C, again add PGI$_2$ (0.5 µM) and centrifuge the platelet suspension once more for 8 min at 1900g (30 or 40 mL platelet suspension volume).
11. Finally, resuspend platelets in Tyrode's buffer containing FAF-HSA 0.1% (w/v) and apyrase (0.02 U/mL; *see* **Note 7**), but no Ca^{2+}, at a density of 3×10^5 platelets/µL. Under these conditions, the fura-2-loaded platelet suspension, protected from light and kept at 37°C, is stable for 2 h (*see* **Note 8**).

3.2. Fura-2/AM Loading of Rodent Platelets

Intracellular accumulation of the calcium probe fura-2/AM is less effective in rat or mouse platelets as compared to human platelets, owing to two concomitant phenomena: difficulty of the dye to penetrate in platelets and high loss of the dye after loading *(41,42)*. There are several hypotheses to explain the difficulty of the fura-2 probe to accumulate into the cytosol, for example: (1) hydrolysis of the acetoxymethyl ester part of the fura-2 moiety via cytosolic esterases may not be as efficient in rodent compared to human platelets; (2) leak of the fluorescent dye from rodent platelets at 37°C occurs at a much greater rate than in human platelets. To overcome the problems, the rodent platelet-loading procedure is slightly different from human: the dye concentration used is 10-fold higher, the fura-2/AM loading is performed in the second washing step and the final suspension of fura-2-loaded platelets is stored at 20°C.

1. Collect blood from the abdominal aorta of a rat or mouse using an 18-gauge needle (rat) or a 25-gauge needle (mouse) mounted on a syringe containing 1 volume ACD for 6 volumes of blood. The volume of blood collected from a rat weighing 200 g is about 10 mL, while the volume of blood drawn from a mouse weighing 20 g is only 1 mL.
2. As compared to human, rodent blood is centrifuged at a higher force ($2300g$) to obtain PRP. The time of centrifugation depends on the volume of blood (10 s per mL of blood).
3. Collect the PRP and centrifuge at $2200g$ for a period of time depending on the volume of PRP. For a 15 mL centrifuge tube containing 1 mL of PRP, the centrifugation time is 4 min; for 2 mL 5 min, for 3 mL 6 min, for 4 mL 8 min, and for 5 mL 10 min.
4. Discard the supernatant consisting of platelet-poor plasma (PPP) using a Pasteur pipet connected to a vacuum pump, in order to remove all traces of plasma from the tube walls or near the platelet pellet to prevent generation of thrombin during the subsequent washing steps.
5. Resuspend the rodent platelets in Tyrode's buffer without calcium, containing HSA 0.35% (w/v), heparin (10 U/mL), and PGI_2 (0.5 μM). After incubation for 10 min at 37°C, add PGI_2 (0.5 μM) and centrifuge the platelets at $1900g$ for a period of time depending on the platelet suspension volume. For a 15 mL centrifuge tube containing 2–3 mL of platelet suspension, the centrifugation time is 3 min; for 4–5 mL 4 min, for 6–7 mL 5 min, for 8–9 mL 6 min, and for 10 mL 7 min.
6. Resuspend the rodent platelets in Tyrode's buffer without calcium, containing HSA 0.35% (w/v), PGI_2 (0.5 μM) and apyrase (0.02 U/mL) and adjust the platelet density to 7.5×10^5 platelets/μL.
7. Incubate the rodent platelets with 15 μM fura-2/AM for 45 min at 37°C in the dark (*see* **Note 9**). Again add PGI_2 (0.5 μM) and centrifuge as detailed in **step 5**.
8. After centrifugation, resuspend platelets at a density of approx 100,000 platelets/μL in Tyrode's buffer containing FAF-HSA 0.1% (w/v), apyrase (0.02 U/mL) but no added calcium (*see* **Note 7**). For a blood volume of 10 mL (one rat), 30 mL of washed platelets at 1×10^5 platelets/μL is obtained, corresponding to 15 experimental points/rat. For a mouse blood volume of 5 mL (1 mL/mouse, blood pooled from 5 mice), 30 mL of washed platelets at 1×10^5 platelets/μL is obtained. In this case, working below 5 mL of blood does not lead to a good yield. This final platelet suspension is stored at room temperature (*see* **Note 6**) and each sample is prewarmed at 37°C for 2 min before starting the calcium measurements.

3.3. Fura-2/AM-Loading of Nucleated Cell Lines

The conditions for measurement of $[Ca^{2+}]_i$ vary between different cell lines and the following methods should be considered as a basic guideline based on our experience with Jurkat E6.1 cells and MEG-01 cells, which grow in suspension, and with 1321 N1 human astrocytoma and B10 cells, which are adherent (*see* **Note 10**).

3.3.1. Cells in Suspension

1. Cells cultured in suspension are centrifuged (100*g*, 5 min) and resuspended in BSS supplemented with 2 m*M* CaCl₂, HSA 0.35% (v/v), and apyrase (0.02 U/mL).
2. After a second centrifugation (100*g*, 5 min), the cells are resuspended in BSS without calcium at a concentration of 15×10^6 cells/mL and incubated with fura-2/AM (5 μ*M*; *see* **Note 6**) at 37°C for 30 min in the dark.
3. The cells are then pelleted and resuspended at a density of 1×10^6 cells/mL in BSS containing no calcium and stored at 37°C (*see* **Note 8**).
4. Aliquots of fura-2-loaded cells (2 mL) are transferred to an acryl cuvet maintained at 37°C and fluorescence measurements are performed as described in **Subheading 3.4.**

3.3.2. Adherent Cells

1. Adherent cells are cultured until confluent on glass coverslips coated with fibronectin (60 μL of human fibronectin 500 μg/mL, purified from plasma as described by Ruoslahti et al. *(43)*, is coated on a glass coverslips (13 mm × 27 mm) and incubated at RT for 15 min).
2. The glass coverslips are directly immersed in BSS buffer without calcium, containing fura-2/AM (5 μ*M*), HSA 0.1% (w/v) and apyrase (0.02 U/mL) and incubated at 37°C for 30 min in the dark.
3. After washing with BSS, cover slips coated with cells are transferred to an acryl cuvet.
4. In order to achieve stirring, a drop of silicone grease is applied 3 mm above the bottom of the cuvet and the coverslip is inserted into the cuvet. Fluorescence measurements are performed at 37°C as described in **Subheading 3.4.**

3.4. Measurement of Fura-2 Signals

1. Prewarm the xenon arc lamp for 15 min before starting measurements and adjust the instrument setup as follows: thermostat-controlled to 37°C, excitation wavelengths adjusted to 340 nm and 380 nm (bandwidth = 5 nm) and fluorescent emission intensity monitored continuously at 510 nm (bandwidth = 5 nm) (*see* **Note 11**).
2. Pipet 2 mL of the fura-2-loaded platelet suspension into an acryl cuvet, insert into the spectrofluorometer holder, and start the recording.
3. Record the signals for 340 and 380 nm in arbitrary units (a.u.). Add CaCl₂ or agonists to the cuvet directly by dilution from stocks, as required for the specific experiment. **Fig. 3A** shows the raw traces expected in an experiment in which 2 m*M* CaCl₂ is added, followed by 5 μ*M* ADP. When 2 m*M* CaCl₂ is added, the signal at 380 nm excitation decreases, while the signal at 340 nm excitation increases and both stabilize in approx 10 to 20 s (**Fig. 3A**). Addition of the agonist (5 μ*M* ADP in this case) induces a very rapid and reversible increase in the signal recorded at 340 nm while the signal recorded at 380 nm reversibly decreases.

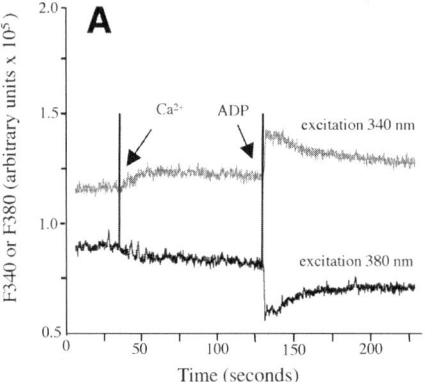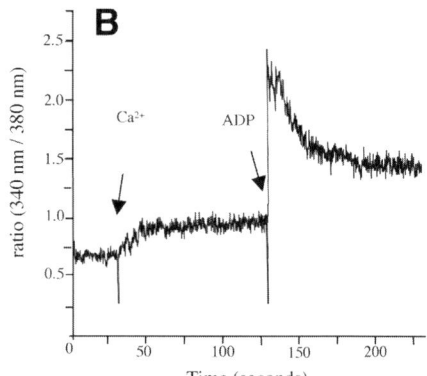

Fig. 3. ADP-evoked Ca^{2+} signals recorded by fura-2 in a platelet suspension. Fluorescence recorded at 510 nm (5 nm bandwidth) from a 2 mL suspension containing 6×10^8 fura-2-loaded platelets (2 μM fura-2AM) activated with ADP in the presence of external calcium. **(A)** Fluorescence resulting from excitation at 340 nm (gray) and 380 nm (black). Results are expressed in arbitrary units. **(B)** Ratio of the two fluorescence signals (340 nm/380 nm) after subtraction of autofluorescence (*see* text for further details). Calcium (2 mM) and ADP (5 μM), were added where indicated.

4. After the specific test experiment, calibrate the fura-2 signal by measuring the maximal and minimum 340/380 nm ratios as described in **Subheading 3.5.**
5. In addition to measuring R_{min} and R_{max}, measure the autofluorescence from the 340 and 380 nm fluorescence of unloaded platelets (*see* **Note 12**), which is then subtracted from all raw signals before calculation of the 340/380 nm ratio and calibration (*see* **Subheading 3.5.**). Addition of 5 μM ADP induces a very rapid and reversible increase in ratio signal (**Fig. 3B**) and values are converted to $[Ca^{2+}]_i$ levels in nanomolar, using the Grynkiewicz formula (*see* **Subheading 3.5.**). Agonist-evoked changes can be expressed as the peak Ca^{2+} increase (nM) or total Ca^{2+} mobilized from the integral of the increase (nM.s).

3.5. Determination of Cytoplasmic Calcium Concentration

1. At the end of an experiment in the presence of 2 mM external Ca^{2+}, lyse the platelets by addition of 0.1% Triton X-100 to obtain the fluorescence signals at maximal Ca^{2+} saturation of the dye.
2. Add 8 mM EGTA to obtain the fluorescent signals at essentially zero levels of free Ca^{2+} (**Fig. 4A**) (*see* **Note 3**).
3. After subtraction of autofluorescence (measured in **Subheading 3.4., step 5**), convert the 340 and 380 nm signals to ratios and measure the R_{max}, in 2 mM external Ca^{2+}, and R_{min} in 8 mM EGTA (*see* **Fig. 3A** and **Note 13**). The range between R_{min} and R_{max} can be used to indicate the extent of dye ester hydrolysis, as the presence of nonhydrolyzed or incompletely hydrolyzed Ca^{2+}-insensitive fura-2/AM decreases the dynamic fluorescence range.
4. Use the following equation to convert fura-2 ratios to $[Ca^{2+}]_i$ (*see* Grynkiewicz et al. *[28]* for the derivation of this equation):

$$[Ca^{2+}]_i = K_D \times [(R - R_{min}) / (R_{max} - R)] \times (Sf_2 / Sb_2)$$

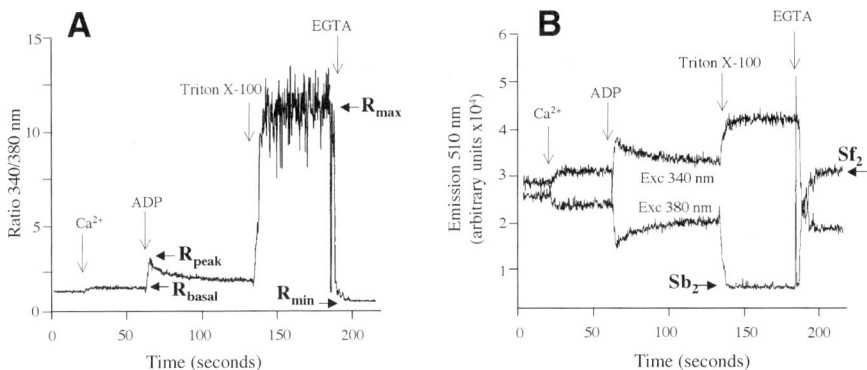

Fig. 4. Calibration of fura-2 signals using Triton X-100 and EGTA. (**A**) Ratio of fluorescence at the two wavelengths (340 nm/380 nm). Addition of 0.1% Triton X-100 (v/v) in the presence of 2 mM Ca²⁺ is used to determine R_{max}; R_{min} is obtained by adding 8 mM EGTA (*see* **Note 3**). R_{basal} is the ratio value under resting conditions and R_{peak} the maximum value after stimulation with the agonist. (**B**) The fluorescence resulting from excitation at 340 nm and 380 nm are the upper and lower traces, respectively. The Sb_2 value is the minimum 380 nm fluorescence, measured in the presence of Triton X-100 and 2 mM Ca²⁺, and the Sf_2 value is the maximum 380 nm fluorescence, measured in the presence of EGTA. All 340 and 380 nm signals are corrected for autofluorescence prior to measurement of calibration constants.

K_D is the dissociation constant of the dye for Ca²⁺, under the chosen experimental conditions: in the presence of 1 mM Mg²⁺, at 37°C, and at physiological ionic strength, and its value is 224 nM. R is the ratio of fluorescence intensities measured at two wavelengths (340 nm/380 nm) under experimental conditions. R_{max} is the ratio obtained after addition of Triton X-100 (0.1%) in the presence of 2 mM Ca²⁺, and R_{min} is the ratio value obtained after addition of EGTA (8 mM) (*see* **Note 3**) (**Fig. 4A**) (*see* **Note 2**). Sf_2/Sb_2 is a fluorescence proportionality constant that takes into account the excitation coefficient at 380 nm, path length, quantum efficiency of the dye, and the instrument's collection efficiency (*28*). Sf_2 is the maximum fluorescence at 380 nm, which corresponds to the Ca²⁺-free form of the dye (measured in the presence of EGTA). Sb2 is the minimum fluorescence at 380 nm, representing the bound form of the dye (measured in the presence of 2 mM Ca²⁺) (**Fig. 4B**).

5. After calibrating, the agonist-evoked Ca²⁺ responses are most commonly quantified as the increase above resting levels. For example, in **Fig. 4A**, the concentration of Ca²⁺ in human platelet cytosol in response to 5 µM ADP in the presence of 2 mM of external Ca²⁺ (**Fig. 4A,B**) can be calculated from the ratios prior to addition of agonist (R_{basal}) and at the peak of the response (R_{peak}) as follows (*see* **Note 14**):

$$R_{basal} = 1.322 \quad R_{peak} = 2.897$$
$$R_{min} = 0.5668 \quad R_{max} = 11.45$$
$$Sf_2 = 31730 \quad Sb_2 = 3885$$

$$[Ca^{2+}]_{basal} = 224 \times [(1.322 - 0.5668) / (11.45 - 1.322)] \times (31730 / 3885) = 136 \text{ n}M$$
$$[Ca^{2+}]_{peak} = 224 \times [(2.897 - 0.5668) / (11.45 - 2.897)] \times (31730 / 3885) = 498 \text{ n}M$$
$$ADP\text{-}evoked\ [Ca^{2+}]_i\ increase = [Ca^{2+}]_{peak} - [Ca^{2+}]_{basal} = 498 - 136 = 362 \text{ n}M$$

3.6. Cytosolic Calcium Control

Tools exist that allow the control and modification of $[Ca^{2+}]_i$. The most commonly used are thapsigargin or t-BuBHQ [2-5-di-(tert-butyl)-1,4-hydroquinone], which are both SERCA inhibitors, and BAPTA/AM [1,2-*bis*(2-aminophenoxy)-ethane-*N,N,N′,N′*-tetraacetic acid, acetoxymethylester], a calcium chelator. These compounds have the advantage that they are cell-permeant. In platelets, as well as in other cells, SERCA inhibitors like thapsigargin (500 n*M*) evoke Ca^{2+} store depletion and thus a $[Ca^{2+}]_i$ increase, which is independent of receptor activation mechanisms. BAPTA is a nonfluo-rescent calcium chelator with a chemical structure resembling that of the fura-2 molecule. The parent molecule BAPTA/AM is cell-permeant and is used to chelate the cytosolic Ca^{2+} (to virtually eliminate the Ca^{2+} responses to agonists, incubate platelets with 30 µ*M* BAPTA-AM for 45 min at 37°C *[17]*) The advantage of BAPTA as compared to EGTA or EDTA is its high selectivity and affinity for calcium ions. BAPTA has been exten-sively used to clamp $[Ca^{2+}]_i$, providing insights on the role of free cytosolic Ca^{2+}.

3.7. Calcium Signals in Washed Platelet Suspensions

3.7.1. Main Platelet Agonists

In this section, we will give several examples of agonist-induced Ca^{2+} movements in washed platelets. Human platelet activation by either serotonin, ADP, or arachidonic acid produces an immediate and reversible rise in cytosolic Ca^{2+}, peaking at 150, 400, and 500 n*M* respectively above basal levels, in the presence of external Ca^{2+} (**Fig. 5**, upper trace). When platelets are activated by thrombin, the rapid and strong rise in cytosolic Ca^{2+} peaks at 800 n*M* depending on the thrombin concentration and the signal is not reversible. To discriminate between extracellular Ca^{2+} influx and mobilization from the intracellular stores, fura-2-loaded platelets are activated in the absence of external Ca^{2+} (presence of 0.2 m*M* EGTA). This signal is reversible with all agonists and culminates around 300 n*M* when platelets are stimulated with thrombin (**Fig. 5**, lower trace), serotonin being the weakest platelet agonist. Problems in recording Ca^{2+} signals can be encountered when platelets are stimulated with fibrillar collagen where the size of the aggregates alter the signal. In such cases, platelet aggregation is pre-vented by adding integrin $\alpha_{IIb}\beta_3$ inhibitors to allow measurement of the calcium signal.

3.7.2. Control of Calcium Mobilization
Through Metabotropic P2Y₁ and P2Y₁₂ Receptors

A classical example of the role of a specific receptor for platelet agonists is shown in **Fig. 6**, where ADP-induced calcium signaling is illustrated. ADP induces platelet

---

Fig. 5. *(see facing page)* $[Ca^{2+}]_i$ increase in fura-2 loaded-platelets after stimulation with the main platelet agonists. Responses are shown for serotonin, ADP, arachidonic acid, and thrombin in the presence of 2 m*M* external Ca^{2+} (black trace) or in the absence of Ca^{2+} (0.2 m*M* EGTA) (gray trace). $[Ca^{2+}]_i$ is expressed in nanomolar concentrations.

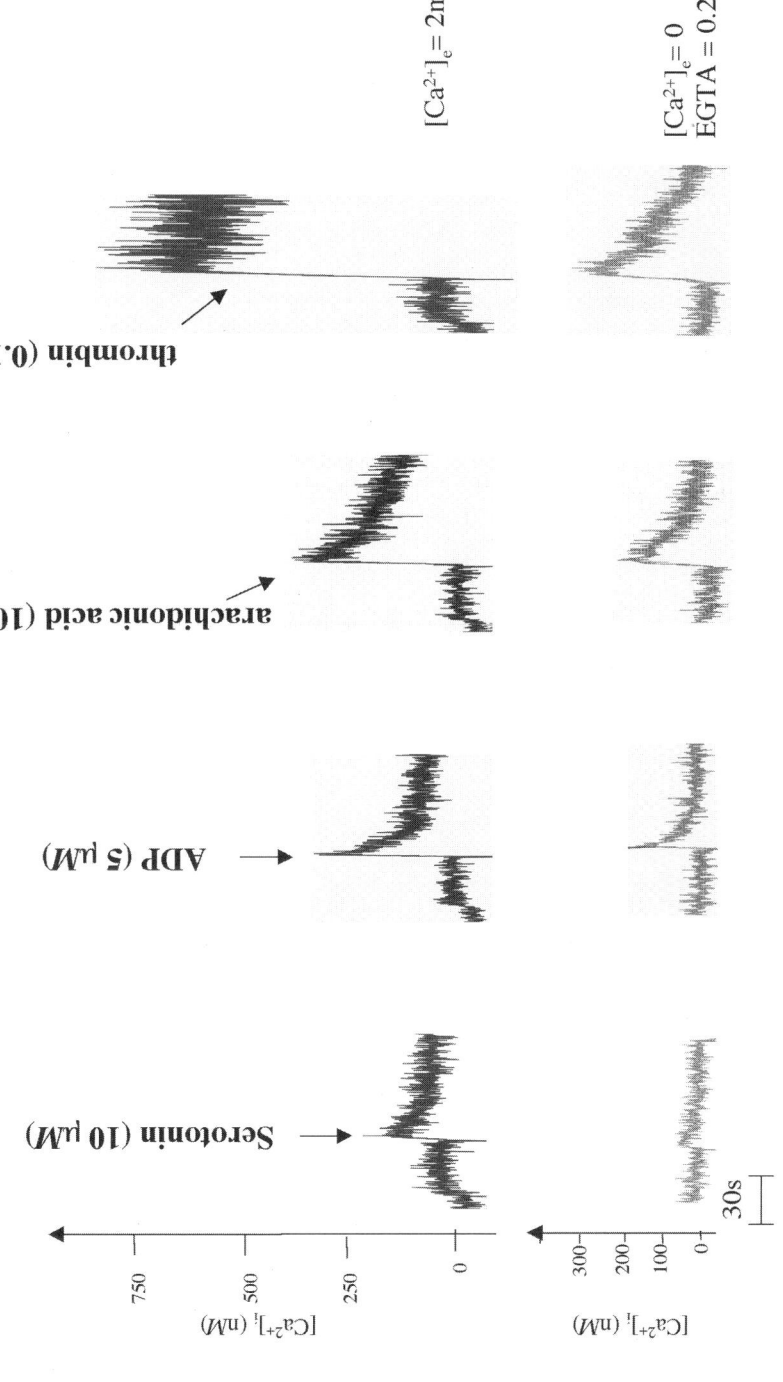

Serotonin (10 μ*M*)

ADP (5 μ*M*)

arachidonic acid (100 μ*M*)

thrombin (0.1 U/mL)

$[Ca^{2+}]_e = 2mM$

$[Ca^{2+}]_e = 0$
EGTA = 0.2 m*M*

$[Ca^{2+}]_i$ (n*M*)

750

500

250

0

$[Ca^{2+}]_i$ (n*M*)

300

200

100

0

30s

aggregation through activation of two P2Y receptors: the $P2Y_1$ receptor coupled to mobilization of intracellular calcium stores and the $P2Y_{12}$ receptor coupled to inhibition of adenylyl cyclase activity and responsible for amplification of aggregation *(44)*. As shown in **Fig. 6A**, in washed human platelets the calcium signal induced by ADP (5 μM) is completely abolished in the presence of the selective $P2Y_1$ receptor antagonist MRS2179 (100 μM). Similarly, ADP-induced calcium signaling is completely abolished in platelets derived from $P2Y_1$ receptor knockout mice **(Fig. 6B)**. Inhibition of the $P2Y_{12}$ receptor using the selective antagonist AR-C69931MX (10 μM) or the thienopyridine compounds ticlopidine and clopidogrel *(44)* has no direct effect on calcium mobilization induced by ADP as well as in patients lacking the $P2Y_{12}$ receptor **(Fig. 6C)** *(45,46)*. However, the reversal to baseline is accelerated, most probably due to the role of the $P2Y_{12}$ receptor in lowering cAMP production and thus interfering with the rate of store refilling and Ca^{2+} extrusion from the cytoplasm *(47)*.

3.7.3. Calcium Influx Through Ionotropic P2X₁ Receptor

The $P2X_1$ receptor is the major ionotropic purinoceptor expressed in platelets *(20,48,49)*. Stimulation of human platelet $P2X_1$ receptors evokes a rapid transient inward current carried by Na^+ and Ca^{2+} through nonselective cation channels *(6,20)*. The $P2X_1$-induced $[Ca^{2+}]_i$ response measured in platelet suspension reflects the time-course of activation of the underlying ion channels, peaking at 1–2 s with a lag phase of 20 ms and returning toward baseline within a few tens of seconds *(20)*. The main problem with studies of $P2X_1$ receptors is the difficulty of preserving $P2X_1$ responses during preparation of washed platelets. This is at least partly due to leakage of nucleotides and the slow time-course of recovery of $P2X_1$ receptors from desensitization *(50)*. To reduce $P2X_1$ receptor desensitization during the isolation of platelets, the ADP-removing enzyme apyrase (0.9 U/mL) is added during the washing procedure. Under these conditions, $\alpha\beta$MeATP (100 μM), a specific $P2X_1$ receptor agonist, induces a transient rise in intracellular calcium concentration that is abolished in the absence of external Ca^{2+}, indicating that it is entirely due to the entry of calcium from the external medium **(Fig. 7)**.

3.7.4. Effect of Platelet Function Inhibitors

The effect of two main platelet function inhibitors is illustrated in **Fig. 8**. Prostacyclin (PGI_2) is a strong vasodilator and a potent inhibitor of platelet aggregation, which

---

Fig. 6. *(see facing page)* Characterization of P2Y receptors responsible for ADP-evoked platelet Ca^{2+} signals. **(A)** Effect of MRS2179 and AR-C69931, the selective antagonists of the $P2Y_1$ and $P2Y_{12}$ receptors, respectively, on calcium signals induced by ADP in human platelets. Arrows indicate addition of the antagonist or vehicle control followed by ADP (5 μM). Extracellular Ca^{2+} concentration ($[Ca^{2+}]_e$) was 2 mM. **(B)** Measurement of $[Ca^{2+}]_i$ increase in WT and $P2Y_1$ knockout mice *(56)*. **(C)** Measurement of $[Ca^{2+}]_i$ increase in human platelets from a patient lacking the $P2Y_{12}$ receptor compared to the normal response *(57)*.

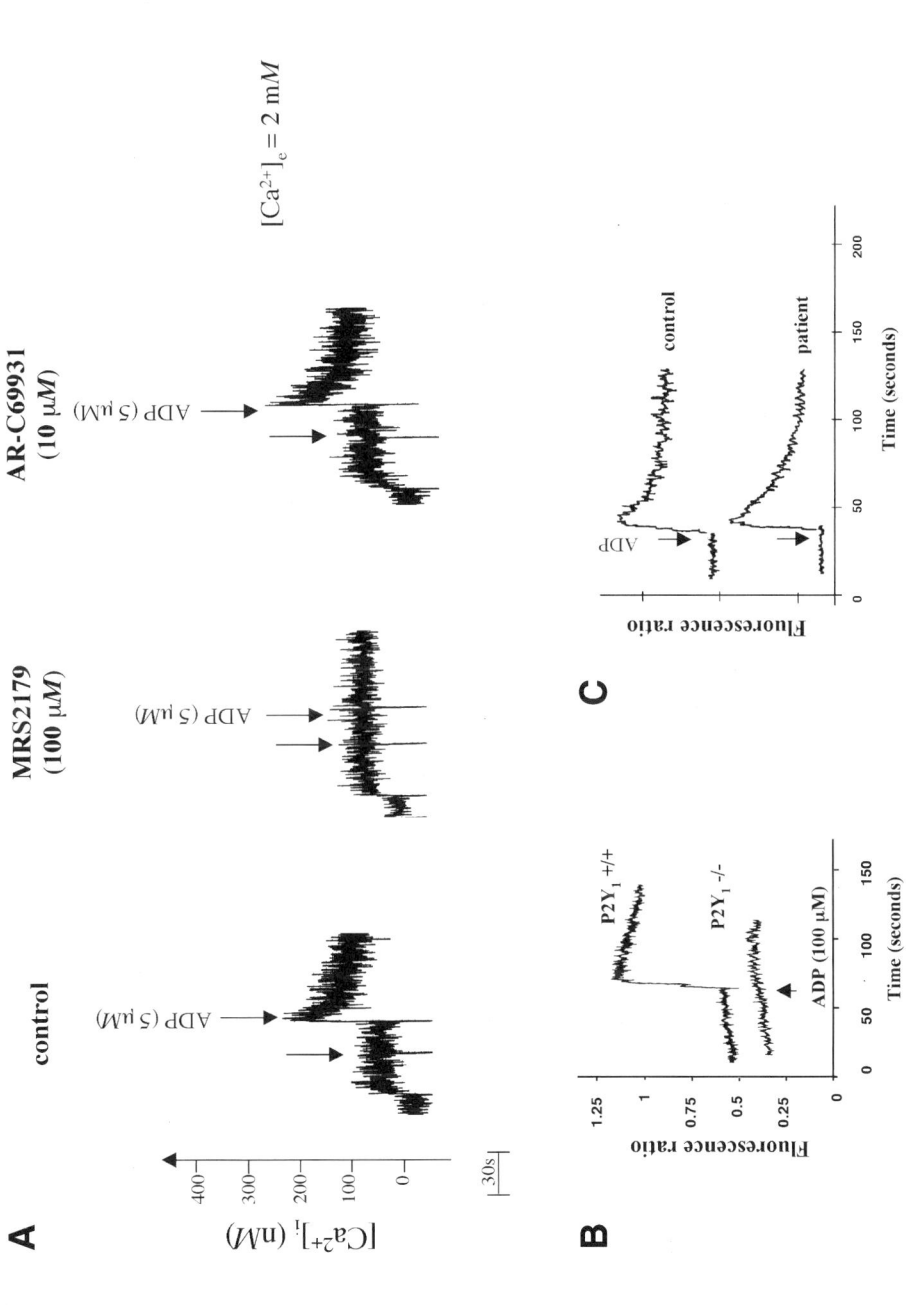

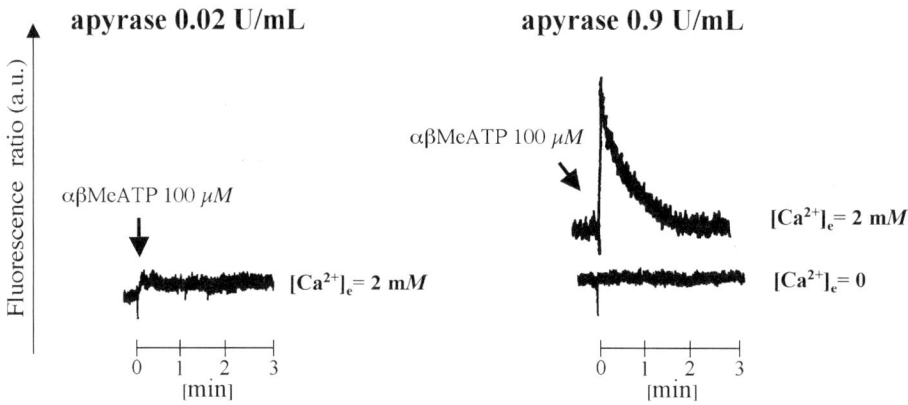

Fig. 7. Measurement of the calcium influx through the P2X$_1$ receptor activated by αβMeATP in human fura-2-loaded platelets. Reduced levels of the ectonucleotidase apyrase during platelet preparation result in loss of the Ca^{2+} influx via P2X$_1$ receptors, which can be selectively activated with the ATP analog αβmeATP.

induces increase of intracellular cAMP by activation of the prostacyclin receptor coupled to the α subunit of Gs. As such it completely inhibits the calcium mobilization and subsequent Ca^{2+} entry induced by most of the platelet agonists **(Fig. 8A)**. On the other hand, aspirin, one of the main platelet-function inhibitors, has no direct effect on calcium signaling. Acting as an inhibitor of cyclo-oxygenase activity, it abolishes the production of TXA$_2$ from arachidonic acid. Therefore, ADP- as well as U46619-induced calcium signaling is not affected by aspirin while, in contrast, arachidonic acid-induced calcium signaling is abolished. Because of this, aspirin partially inhibits calcium mobilization induced by agonists such as collagen or thrombin, which induce TXA$_2$ generation *(51)*. Platelets are treated with aspirin (1 mM) for 15 min at 37°C for complete cyclo-oxygenase activity inhibition. PGI$_2$ (10 μM) is added to the platelets 30 s before the agonist and in contrast to aspirin, its inhibitory effect is transient, while the effect of PGE$_1$ is irreversible.

3.8. Conclusion

Measurement of calcium signaling in human and rodent platelets as well as in nucleated cells, either in suspension or adherent, has been greatly facilitated by the use of dyes such as fura-2. The methods described here are easy to perform provided care is taken to have good sampling procedures and handling. As illustrated, subtle changes in the signals can be observed and interpreted that help in the understanding of the molecular mechanisms of platelet-activation processes.

4. Notes

1. A variety spectrofluorometer systems for measurement of [Ca^{2+}]$_i$ are available. The system used in our laboratory is a PTI DeltaScan RatioMaster™ and was purchased from Photon Technology International Inc. (Princeton, NJ). Another example that has been widely used

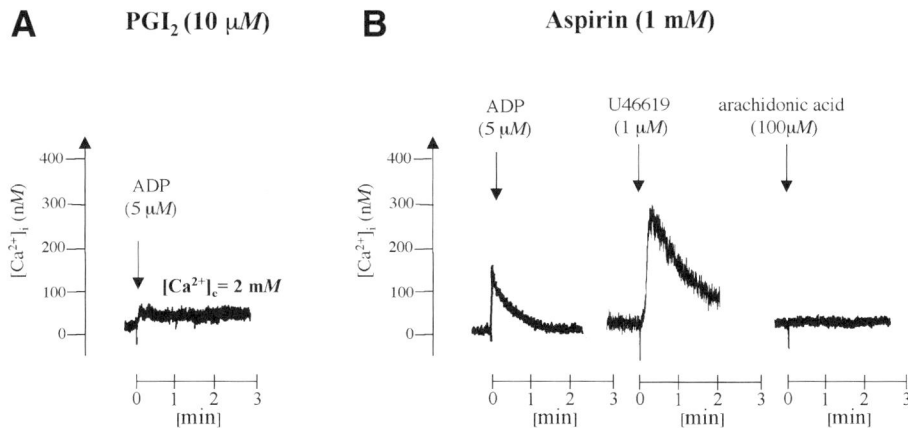

Fig. 8. Effect of platelet function inhibitors on calcium signaling in fura-2-loaded platelets. The effects of two main platelet-function inhibitors PGI₂ (**A**) and aspirin (**B**) are shown on agonist-evoked Ca²⁺ responses in human platelets. 10 μ*M* PGI₂ was added 30 s before the addition of agonist. Platelet suspension was treated with 1 m*M* aspirin in the first washing buffer for 15 min and ADP, U46619, or arachidonic acid were added as indicated. Experiments were performed in the presence of 2 m*M* external Ca²⁺.

for platelet work is the modular system from Cairn Research Ltd. (Faversham, Kent, UK; *see also* Chapter 16, vol. 2), which uses either a spinning filter wheel or monochromator as the light source and is coupled to a cuvet holder via a liquid light guide. The cuvet can be easily modified to take a second photomultiplier, which allows simultaneous measurement of transmitted light for functional measurements of shape change and aggregation *(50)*.

2. The final suspension medium is Tyrode's buffer containing 0.1% FAF-HSA. Fatty-acid-free albumin is preferred to normal albumin and is used at a lower concentration in order to minimize autofluorescence.

3. When EGTA binds Ca²⁺, protons are released. If insufficient buffer is present in the medium, the solution pH will fall and reduce the affinity of EGTA for Ca²⁺. An indication that this is happening is if the R_{min} is greater than the 340:380 nm ratio measured intra-cellularly in unstimulated cells (although *see* **Note 12** for an alternative explanation). The pH should be raised to above 8.0, for example by addition of Tris base, to lower the K_D of EGTA to the nanomolar range.

4. Since many drugs affect platelet reactivity (particularly aspirin and other nonsteroidal anti-inflammatory drugs), blood donors should be carefully questioned about the drugs they have taken during the previous 2 wk.

5. Other groups routinely load human platelets with fura-2 while still in plasma. For example following preparation of PRP with ACD anticoagulant, approx 2 μ*M* fura-2AM is added for 45 min at 37°C, followed by the preparation of a washed suspension of platelets *(50)*. As described by Cazenave et al. *(39)* and Mustard et al. *(40)*, PRP obtained from blood anti-coagulated with ACD or citrate is not stable. For this reason we never load platelets in PRP and we wait no longer than 10–15 min before centrifugation of the PRP. Working with platelet suspensions rather than PRP has the advantage that the extracellular medium can easily be controlled, particularly the level of external Ca²⁺.

6. Different incubation times and concentrations of fura-2/AM have been tried by other groups. It may also be necessary to adjust the concentration of fura-2/AM for different batches of the ester and for different cell lines. High concentrations and thus loading will result in over-buffering of the Ca^{2+} response and will affect the function of the platelets. For example, platelets loaded with higher concentrations of fura-2 (50 μM), exhibit diminished aggregation responses to all aggregating agents *(52)*. Interestingly, the integrity of fura-2 (1 μM)-loaded platelets is also modified, resulting in the potentiation of weak agonist- or low agonist concentration-induced platelet aggregation (50% increase with ADP 5 μM and 97% increase with PAF 1 μM) and secretion (3% in control vs 24% under PAF stimulation). In contrast, it has been shown that stimulation of fura-2-loaded platelets with high agonist concentrations (thrombin 0.25 U/mL, collagen 2.5 μg/mL, U-46619 5 μM, or A23187 5 μM) show a similar extent of aggregation and secretion as compared to unloaded platelet samples.

7. For measurement of $P2X_1$ receptor responses, apyrase should be present at a high concentration (0.9 U/mL) during each step of the entire platelet preparation as well as during measurements (*see* **Fig. 7**).

8. If leakage of fura-2 from the cells and platelets is a problem, this can be minimized by storing at room temperature. This is particularly a problem with cell lines and rodent platelets. The presence of extracellular dye is indicated by an immediate increase in fura-2 ratio when adding Ca^{2+} or an immediate quench of both signals when adding Mn^{2+} (100 μM).

9. Since AM esters have low aqueous solubility, dispersing agents such as the nonionic and non denaturing detergent Pluronic F-127, are often used to increase or facilitate the loading of fura-2/AM into the cells, which thus allows the use of fura-2/AM concentrations less than 15 μM *(53)*.

10. Cell lines are well known to lose Ca^{2+} from their internal stores in Ca^{2+}-free salines. Caution should therefore be taken when interpreting the contribution of influx and release to receptor-dependent Ca^{2+} mobilization. Ca^{2+} stores can be recharged by replacing external Ca^{2+} and then excess EGTA used to chelate the divalent cation immediately prior to agonist (but *see* **Note 3**).

11. If the fura-2 signal is very limited, for example by reduced platelet densities, the bandwidth of the emission wavelength can be increased as significant fura-2 emission from UV excitation occurs outside the commonly used emission bandwidth of 510 ± 5 nm. Wavelengths below approx 490 nm should be avoided due to signals from autofluorescent molecules such as NADH (peak emission 450 nm with UV excitation). In all systems, particularly if increasing the bandwidth, control experiments in unloaded cells should be conducted to check for possible changes in cell autofluorescence. A greater fura-2 signal can also be achieved by increasing the amount of excitation light, but this is at the risk of enhanced photobleaching and phototoxicity. Increasing the excitation bandwidth is not as useful, as it will reduce the dynamic range of the fura-2 signal.

12. The autofluorescence for fura-2 may also be measured by adding $MnCl_2$ excess to the external $CaCl_2$ following Triton X-100 permeabilization. Mn^{2+} binds to fura-2 with a much higher affinity than Ca^{2+} and quenches its fluorescence to negligible levels. Ideally this should be in the absence of EGTA, which also binds Mn^{2+}. Add Mn^{2+} in excess to the Ca^{2+} and increase its concentration until no further quenching is seen.

13. In some cell types, particularly cell lines, the greater viscosity of the cytoplasm compared to the extracellular environment result in a shift of R_{min} and R_{max} due to altered properties of fura-2 at its longer excitation wavelength *(54)*. An indication that this is a problem is if the R_{min} measured extracellularly is higher than the lowest ratio measured from the cells. In these cases a viscosity correction factor (commonly 0.85 for myeloid lines and

megakaryocytes; *see also* Chapter 16, vol. 2) is applied to the extracellularly derived R_{min} and R_{max} before use in the calibration equation. The viscosity correction factor can be estimated by comparing the ratio of the dye under extreme Ca^{2+}-lowering conditions (e.g., ionomycin in Ca^{2+}-free salines) with R_{min} measured extracellularly after Triton X-100 permeabilization. Viscosity correction does not seem necessary for human platelet fura-2 measurements in suspension at 37°C *(55)*.

14. The computer software provided with many spectrophotometers includes algorithms to automatically subtract the autofluorescence and use the calibration constants to convert ratios to $[Ca^{2+}]_i$. However, it is always advisable to perform manual checks of the system as described in this example.

References

1. Carafoli, E. (1994) Biogenesis: plasma membrane calcium ATPase: 15 years of work on the purified enzyme. *Faseb. J.* **8,** 993–1002.
2. Burk, S. E., Lytton, J., MacLennan, D. H., and Shull, G. E. (1989) cDNA cloning, functional expression, and mRNA tissue distribution of a third organellar Ca^{2+} pump. *J. Biol. Chem.* **264,** 18,561–18,568.
3. Kimura, M., Aviv, A., and Reeves, J. P. (1993) K^+-dependent Na^+/Ca^{2+} exchange in human platelets. *J. Biol. Chem.* **268,** 6874–6877.
4. Kimura, M., Jeanclos, E. M., Donnelly, R. J., Lytton, J., Reeves, J. P., and Aviv, A. (1999) Physiological and molecular characterization of the Na+/Ca2+ exchanger in human platelets. *Am. J. Physiol.* **277,** H911–H917.
5. Rink, T. J. and Sage, S. O. (1990) Calcium signaling in human platelets. *Annu. Rev. Physiol.* **52,** 431–449.
6. Sage, S. O., MacKenzie, A. B., Jenner, S., and Mahaut-Smith, M. P. (1997) Purinoceptor-evoked calcium signalling in human platelets. *Prostaglandins Leukot. Essent. Fatty Acids* **57,** 435–438.
7. Heemskerk, J. W. M. and Sage, S. O. (1994) Calcium signalling in platelets and other cells. *Platelets* **5,** 295–316.
8. Tsien, R. Y., Pozzan, T., and Rink, T. J. (1982) Calcium homeostasis in intact lymphocytes: cytoplasmic free calcium monitored with a new, intracellularly trapped fluorescent indicator. *J. Cell. Biol.* **94,** 325–334.
9. Sage, S. O. (1997) The Wellcome Prize Lecture. Calcium entry mechanisms in human platelets. *Exp. Physiol.* **82,** 807–823.
10. Authi, K. S. (1997) Ca^{2+} homeostasis in human platelets, in *Platelets and their Factors* (von Bruchhausen, W. U., ed.), Springer, Berlin, Germany, pp. 325–370.
11. El-Daher, S. S., Patel, Y., Siddiqua, A., Hassock, S., Edmunds, S., Maddison, B., et al. (2000) Distinct localization and function of (1,4,5)IP(3) receptor subtypes and the (1,3,4,5)IP₄ receptor GAP1_{IP4BP} in highly purified human platelet membranes. *Blood* **95,** 3412–3422.
12. Offermanns, S. (2000) The role of heterotrimeric G proteins in platelet activation. *Biol. Chem.* **381,** 389–396.
13. Watson, S. P., Asazuma, N., Atkinson, B., Berlanga, O., Best, D., Bobe, R., et al. (2001) The role of ITAM- and ITIM-coupled receptors in platelet activation by collagen. *Thromb. Haemost.* **86,** 276–288.
14. Daniel, J. L., Dangelmaier, C., and Smith, J. B. (1994) Evidence for a role for tyrosine phosphorylation of phospholipase C γ 2 in collagen-induced platelet cytosolic calcium mobilization. *Biochem. J.* **302,** 617–622.

15. Bootman, M. D., Berridge, M. J., and Roderick, H. L. (2002) Calcium signalling: more messengers, more channels, more complexity. *Curr. Biol.* **12**, R563–R565.
16. Berridge, M. J., Lipp, P., and Bootman, M. D. (2000) The versatility and universality of calcium signalling. *Nat. Rev. Mol. Cell Biol.* **1**, 11–21.
17. Rosado, J. A., and Sage, S. O. (2000) A role for the actin cytoskeleton in the initiation and maintenance of store-mediated calcium entry in human platelets. *Trends Cardiovasc. Med.* **10**, 327–332.
18. Trepakova, E. S., Csutora, P., Hunton, D. L., Marchase, R. B., Cohen, R. A., and Bolotina, V. M. (2000) Calcium influx factor directly activates store-operated cation channels in vascular smooth muscle cells. *J. Biol. Chem.* **275**, 26,158–26,163.
19. Hassock, S. R., Zhu, M. X., Trost, C., Flockerzi, V., and Authi, K. S. (2002) Expression and role of TRPC proteins in human platelets: evidence that TRPC6 forms the store-independent calcium entry channel. *Blood* **100**, 2801–2811.
20. MacKenzie, A. B., Mahaut-Smith, M. P., and Sage, S. O. (1996) Activation of receptor-operated cation channels via P_{2X1} not P_{2T} purinoceptors in human platelets. *J. Biol. Chem.* **271**, 2879–2881.
21. Mahaut-Smith, M. P., Sage, S. O., and Rink, T. J. (1990) Receptor-activated single channels in intact human platelets. *J. Biol. Chem.* **265**, 10,479–10,483.
22. Mahaut-Smith, M. P., Sage, S. O., and Rink, T. J. (1992) Rapid ADP-evoked currents in human platelets recorded with the nystatin permeabilized patch technique. *J. Biol. Chem.* **267**, 3060–3065.
23. Brass, L. F. (1985) Ca^{2+} transport across the platelet plasma membrane. A role for membrane glycoproteins IIB and IIIA. *J. Biol. Chem.* **260**, 2231–2236.
24. Fujimoto, T., Fujimura, K., and Kuramoto, A. (1991) Electrophysiological evidence that glycoprotein IIb-IIIa complex is involved in calcium channel activation on human platelet plasma membrane. *J. Biol. Chem.* **266**, 16,370–16,375.
25. Nesbitt, W. S., Kulkarni, S., Giuliano, S., Goncalves, I., Dopheide, S. M., Yap, C. L., et al. (2002) Distinct glycoprotein Ib/V/IX and integrin alpha IIbbeta 3-dependent calcium signals cooperatively regulate platelet adhesion under flow. *J. Biol. Chem.* **277**, 2965–2972.
26. Mazzucato, M., Pradella, P., Cozzi, M. R., De Marco, L., and Ruggeri, Z. M. (2002) Sequential cytoplasmic calcium signals in a 2-stage platelet activation process induced by the glycoprotein Ibalpha mechanoreceptor. *Blood* **100**, 2793–2800.
27. Rink, T. J., Smith, S. W., and Tsien, R. Y. (1982) Cytoplasmic free Ca^{2+} in human platelets: Ca^{2+} thresholds and Ca-independent activation for shape-change and secretion. *FEBS Lett.* **148**, 21–26.
28. Grynkiewicz, G., Poenie, M., and Tsien, R. Y. (1985) A new generation of Ca2+ indicators with greatly improved fluorescence properties. *J. Biol. Chem.* **260**, 3440–3450.
29. Rao, G. H., Peller, J. D., and White, J. G. (1985) Measurement of ionized calcium in blood platelets with a new generation calcium indicator. *Biochem. Biophys. Res. Commun.* **132**, 652–657.
30. Sage, S. O. and Rink, T. J. (1986) Kinetic differences between thrombin-induced and ADP-induced calcium influx and release from internal stores in fura-2-loaded human platelets. *Biochem. Biophys. Res. Commun.* **136**, 1124–1129.
31. Sage, S. O. and Rink, T. J. (1987) The kinetics of changes in intracellular calcium concentration in fura- 2-loaded human platelets. *J. Biol. Chem.* **262**, 16,364–16,369.
32. Rao, G. H. (1988) Measurement of ionized calcium in normal human blood platelets. *Anal. Biochem.* **169**, 400–404.

33. Sage, S. O., Merritt, J. E., Hallam, T. J., and Rink, T. J. (1989) Receptor-mediated calcium entry in fura-2-loaded human platelets stimulated with ADP and thrombin. Dual-wavelengths studies with Mn^{2+}. *Biochem. J.* **258,** 923–926.

34. Hechler, B., Cazenave, J. P., Hanau, D., and Gachet, C. (1995) Presence of functional P2T and P2U purinoceptors on the human megakaryoblastic cell line, Meg-01 characterization by functional and binding studies. *Nouv. Rev. Fr. Hematol.* **37,** 231–240.

35. Leon, C., Hechler, B., Vial, C., Leray, C., Cazenave, J. P., and Gachet, C. (1997) The P2Y1 receptor is an ADP receptor antagonized by ATP and expressed in platelets and megakaryoblastic cells. *FEBS Lett.* **403,** 26–30.

36. Schachter, J. B., Li, Q., Boyer, J. L., Nicholas, R. A., and Harden, T. K. (1996) Second messenger cascade specificity and pharmacological selectivity of the human P2Y1-purinoceptor. *Br. J. Pharmacol.* **118,** 167–173.

37. Hechler, B., Vigne, P., Leon, C., Breittmayer, J. P., Gachet, C., and Frelin, C. (1998) ATP derivatives are antagonists of the P2Y1 receptor: similarities to the platelet ADP receptor. *Mol. Pharmacol.* **53,** 727–733.

38. Harris, D. A., and Bashford, C. L. (1987) *Spectrophotometry and Spectrofluorimetry, A Practical Approach*, IRL Press, Oxford.

39. Cazenave, J. P., Hemmendinger, S., Beretz, A., Sutter-Bay, A., and Launay, J. (1983) L'agrégation plaquettaire: outil d'investigation clinique et d'étude pharmacologique. Méthodologie. *Ann. Biol. Clin.* **41,** 167–179.

40. Mustard, J. F., Perry, D. W., Ardlie, N. G., and Packham, M. A. (1972) Preparation of suspensions of washed platelets from humans. *Br. J. Haematol.* **22,** 193–204.

41. Heemskerk, J. W., Feijge, M. A., Rietman, E., and Hornstra, G. (1991) Rat platelets are deficient in internal Ca2+ release and require influx of extracellular Ca2+ for activation. *FEBS Lett.* **284,** 223–226.

42. Cavallini, L., Francesconi, M. A., Ruzzene, M., Valente, M., and Deana, R. (1991) A procedure allowing measurement of cytosolic Ca^{2+} in rat platelets. Inhibition of a plasma lipoprotein on fura 2-AM loading. *Thromb. Res.* **63,** 47–57.

43. Ruoslahti, E., Pierschbacher, M., Engvall, E., Oldberg, A., and Hayman, E. (1982) Molecular and biological interactions of fibronectin. *J. Invest. Dermatol.* **79,** 65–68.

44. Gachet, C. (2001) Identification, characterization, and inhibition of the platelet ADP receptors. *Int. J. Hematol.* **74,** 375–381.

45. Cattaneo, M., Lecchi, A., Randi, A. M., McGregor, J. L., and Mannucci, P. M. (1992) Identification of a new congenital defect of platelet function characterized by severe impairment of platelet responses to adenosine diphosphate. *Blood* **80,** 2787–2796.

46. Nurden, P., Savi, P., Heilmann, E., Bihour, C., Herbert, J. M., Maffrand, J. P., et al. (1995) An inherited bleeding disorder linked to a defective interaction between ADP and its receptor on platelets. Its influence on glycoprotein IIb-IIIa complex function. *J. Clin. Invest.* **95,** 1612–1622.

47. Storey, R. F., Sanderson, H. M., White, A. E., May, J. A., Cameron, K. E., and Heptinstall, S. (2000) The central role of the P_{2T} receptor in amplification of human platelet activation, aggregation, secretion and procoagulant activity. *Br. J. Haematol.* **110,** 925–934.

48. Vulchanova, L., Arvidsson, U., Riedl, M., Wang, J., Buell, G., Surprenant, A., et al. (1996) Differential distribution of two ATP-gated channels (P2X receptors) determined by immunocytochemistry. *Proc. Natl. Acad. Sci. USA* **93,** 8063–8067.

49. Vial, C., Hechler, B., Leon, C., Cazenave, J. P., and Gachet, C. (1997) Presence of P2X1 purinoceptors in human platelets and megakaryoblastic cell lines. *Thromb. Haemost.* **78,** 1500–1504.

50. Rolf, M. G., Brearley, C. A., and Mahaut-Smith, M. P. (2001) Platelet shape change evoked by selective activation of P2X1 purinoceptors with alpha,beta-methylene ATP. *Thromb. Haemost.* **85,** 303–308.

51. Siess, W. (1989) Molecular mechanisms of platelet activation. *Physiol. Rev.* **69,** 58–178.

52. Lanza, F., Beretz, A., Kubina, M., and Cazenave, J. P. (1987) Increased aggregation and secretion responses of human platelets when loaded with the calcium fluorescent probes quin2 and fura-2. *Thromb. Haemost.* **58,** 737–743.

53. Maruyama, I., Hasegawa, T., Yamamoto, T., and Momose, K. (1989) Effects of pluronic F-127 on loading of fura 2/AM into single smooth muscle cells isolated from guinea pig taenia coli. *J. Toxicol. Sci.* **14,** 153–163.

54. Poenie, M. (1990) Alteration of intracellular fura-2 fluorescence by viscosity: a simple correction. *Cell Calcium* **11,** 85–91.

55. Mahaut-Smith, M. P., Ennion, S. J., Rolf, M. G., and Evans, R. J. (2000) ADP is not an agonist at P2X(1) receptors: evidence for separate receptors stimulated by ATP and ADP on human platelets. *Br. J. Pharmacol.* **131,** 108–114.

56. Leon, C., Hechler, B., Freund, M., Eckly, A., Vial, C., Ohlmann, P., et al. (1999) Defective platelet aggregation and increased resistance to thrombosis in purinergic P2Y(1) receptor-null mice. *J. Clin. Invest.* **104,** 1731–1737.

57. Leon, C., Vial, C., Gachet, C., Ohlmann, P., Hechler, B., Cazenave, J. P., et al. (1999) The P2Y1 receptor is normal in a patient presenting a severe deficiency of ADP-induced platelet aggregation. *Thromb. Haemost.* **81,** 775–781.

16

Measurement and Manipulation of Intracellular Ca²⁺ in Single Platelets and Megakaryocytes

Michael J. Mason and Martyn P. Mahaut-Smith

1. Introduction

Intracellular free Ca^{2+} is a key second messenger in virtually all cells. In unstimulated platelets and megakaryocytes, $[Ca^{2+}]_i$ is approx 100 nM and can be rapidly elevated by a variety of different agonists that activate phospholipase-C via heterotrimeric G proteins or receptor kinases *(1–4)*. Phospholipase-C (PLC) generates cytosolic IP_3, which releases Ca^{2+} by opening cation channels on the intracellular stores. Agonists also evoke Ca^{2+} influx across the plasma membrane, although the exact nature of these pathways and their mechanism of activation and contribution under physiological conditions remain poorly understood and, in fact, highly controversial. Release of Ca^{2+} from the stores activates a Ca^{2+} influx pathway in both the platelet and megakaryocyte *(5–7)*; electrophysiological measurements have shown that ICRAC (calcium-release activated calcium current), one of the main candidates for the current underlying store-dependent Ca^{2+} influx, is present in the precursor cell *(3,8)*. Non-store-dependent Ca^{2+} influx also seems to exist in the platelet *(9)* and TRPC6, a nonselective cation channel known to be activated by the PLC product diacylglycerol (DAG), has recently been detected in platelets and megakaryocytic cell lines *(10)*. Finally, platelets possess at least one type of receptor-operated Ca^{2+}-permeable channel, the ATP-gated $P2X_1$ receptor, which allows rapid Ca^{2+} entry independently of store release *(11,12)*.

In the platelet, experiments with Ca^{2+} ionophores such as ionomycin and A23187 have demonstrated that a $[Ca^{2+}]_i$ increase can directly stimulate a number of functional responses, including shape change, secretion, and aggregation *(13–15)*. The $[Ca^{2+}]_i$ increase also synergizes with other signals to amplify functional responses during receptor activation of platelets *(16–18)*. The functional relevance of Ca^{2+} responses in the megakaryocyte is an important question currently under investigation by our laboratory and others. In many respects the megakaryocyte can be considered a giant platelet and thus a model for studies of platelet signaling, as the precursor cell is

From: *Methods in Molecular Biology, vol. 273:*
Platelets and Megakaryocytes, Vol. 2: Perspectives and Techniques
Edited by: J. M. Gibbins and M. P. Mahaut-Smith © Humana Press Inc., Totowa, NJ

responsible for generating most, if not all, proteins in the platelet. In addition, the megakaryocyte shows platelet functional responses such as exposure of glycoprotein IIbIIIa receptors (19), and it is believed that the plasma membrane of the precursor cell represents that of the future platelet (20–22). Indeed, Ca^{2+} responses to the majority of platelet agonists, including ADP, thrombin, U46619, and collagen, have been reported in the megakaryocyte (4,7,23,24). These responses may reflect the development of pathways for use in the platelet. However, the possibility that they have further relevance in megakaryocytopoesis and thombopoiesis, including, for example, gene expression, have not been explored.

The vast majority of platelet $[Ca^{2+}]_i$ studies have been performed in stirred suspensions where the signal represents the average from a cell population (usually more than a million). This approach does not therefore reveal the complex spatial and temporal patterns of Ca^{2+} increase, such as Ca^{2+} oscillations and waves, that are now known to occur in a variety of cell types. Evidence is now accumulating to suggest that these complex patterns of Ca^{2+} increase allow this single ubiquitous second messenger to encode important information such as agonist concentration (25,26). $[Ca^{2+}]_i$ in single platelets attached to coverslips or isolated individually also shows repetitive increases in response to agonists (2,27–31). Different classes of Ca^{2+} spikes have been shown to result from glycoprotein Ib versus IIbIIIa receptor signaling following binding by von Willebrand factor (32). In the megakaryocyte, single-cell $[Ca^{2+}]_i$ undergoes oscillations that are more regular than in platelets (23,33,34) and unlike in the platelet, each increase in Ca^{2+} can be observed as a wave propogating across the cell (35,36). The lack of a clear wave underlying each transient probably reflects the fact that the platelet is of a similar size to the smallest unitary IP_3-evoked Ca^{2+} events that can be detected, that is blips and puffs, which result from the opening of single or clusters of IP_3 receptors (37,38). The properties of the Ca^{2+} signals in the platelet and megakaryocyte are similar in many other respects, including, for example, IP_3-induced responses that can be potently inhibited by cAMP via activation of protein kinase A (7,39). It is likely, given the large size of the megakaryocyte and its ability to be patch-clamped and to be molecularly transfected, that the precursor cell will continue to significantly contribute to our understanding of platelet Ca^{2+} signaling, particularly in studies at the single-cell level.

Currently, measurement of intracellular Ca^{2+} in the platelet and megakaryocyte relies almost exclusively on the use of fluorescent indicators such as fura-2, indo-1, and fluo-3. Alternatives include aequorin, a bioluminescent Ca^{2+} indicator, which has been used in platelets but requires permeation of the plasma membrane (for example, with EGTA and ATP, [40]) and molecularly encoded Ca^{2+} indicators (41,42). The latter may be particularly useful in megakaryocytic cell lines where transfection is relatively easy, but in the future may also be applied to megakaryocytes and ultimately platelets generated in vitro as culture systems have now been established (see Chapters 22 and 23, vol. 1). Fura-2 or indo-1 are the indicators of choice in many laboratories for nonconfocal measurements, as they allow ratiometric measurements and thereby greatly reduce the problems caused by the effective concentration of indicator, indicator leakage, or dialysis during the experiment and cell movement. These two dyes have relative strengths and weaknesses. Indo-1 has slightly faster kinetics of Ca^{2+} binding than fura-2

(43) but shows more significant photobleaching (to a fluorescent but non-Ca^{2+}-dependent form), compared to fura-2. Indo-1 measurements require one source of light and a means of discriminating between emitted light of two wavelengths, as the predominant Ca^{2+}-dependent shift is in its emission spectrum. Fura-2, on the other hand, shows its main Ca^{2+}-dependent shift in the excitation spectrum and therefore requires a means of alternating between excitation light of two wavelengths (e.g., using a monochromator, a filter wheel, or alternative selection from two light sources). These factors play a role in the choice of ratiometric indicator for your application. Due to the the technical problems of using dual excitation within confocal microscopy, the high cost of UV lasers and UV-compatible microscopes, and the high photobleach rate of indo-1, confocal measurements of cytosolic Ca^{2+} normally use an indicator excited by visible wavelenths, such as fluo-3, fluo-4, or Oregon Green 488 BAPTA-1. These are single-wavelength indicators, excited using the 488 nm line of an argon laser, with the Ca^{2+}-dependent emission normally captured at >505 nm. Pseudoratioing of signals is performed offline to partially correct for shifts in baseline fluorescence.

The most useful means of manipulating $[Ca^{2+}]_i$ are ionophores and caged compounds. The latter represent a more controlled method to elevate Ca^{2+}. The caged, inactive form of the compound (e.g., caged IP_3) is delivered into the cytoplasm; exposure to ultraviolet light results in photolysis of the compound and release of the active biological compound. This can be used to stimulate repeated bursts of IP_3 and thus also Ca^{2+} in a highly controllable manner. The number of available caged compounds is growing and caged IP_3, cAMP, Ca^{2+}, and EGTA have all been applied to megakaryocytes or platelets *(6,33,36)* (Mahaut-Smith and Mason, unpublished observations). This chapter will describe the measurement of Ca^{2+} in single cells using fura-2, fluo-3, and Oregon Green 488 BAPTA-1, and its manipulation using caged IP_3 and caged cAMP.

2. Materials

2.1. Equipment

2.1.1. Photometric Measurements of Ca²⁺ Using Fura-2

Numerous manufacturers provide turnkey systems that can be linked to your microscope for intracellular measurements of $[Ca^{2+}]_i$ in platelets and megakaryocytes using fura-2. In this laboratory we routinely use instrumentation manufactured by Cairn Research Ltd. (Faversham, UK) (*see* **Note 1**). Unless otherwise described, Cairn can supply all items listed.

1. Inverted microscope with epifluorescence port, side port for fluorescence detection, UV-transmitting optics and high-power UV-transmitting fluorescence objectives (e.g., 40× or 63× for megakaryocytes and cell lines and 100× for platelets [*see* **Note 2**]). Measurements in this laboratory on platelets and megakaryocytes have used a Nikon Diaphot inverted microscope (*see* **Note 3**) equipped with 40× and 100× Fluor lenses, NA 1.3 (Nikon UK Ltd., Kingston upon Thames, UK).
2. 32 × 32 mm No. 1 coverslips (BDH, UK, Cat. no. 406/0187), which form the base of the Perspex experimental chamber (*see* Chapter 17, vol. 2 for further details). Thicker coverslips reduce the light transmission and reduce working distance, while thinner coverslips are very fragile.

3. Silicone grease for adhering the coverslip base to the experimental chamber (High Vacuum Grease, Dow Corning (Allesley, Coventry, UK) or RS Components (Corby, Northants, UK) UK Cat. no. 494-124).

4. A xenon arc lamp (75 W) and power supply. Xenon arc lamps are recommended for fura-2 excitation, as 100 W Hg lamps have a large spectral peak close to the isosbestic point of Fura 2 (*see* **Note 4**). An electronic shutter and adjustable diaphragm are positioned in front of the lamp.

5. Filter wheel or monochromator capable of switching between 340 and 380 nm light at a rate of at least 10 Hz. This is positioned immediately after the electronic shutter and diaphragm. We use 10-nm bandpass interference filters set in weight-balancing adapter rings (Cairn Research) or monochromator input and exit slit bandwidths of approx 5 nm. The Cairn filter wheel has 6 positions, into which we insert 4×340 and 2×380 nm filters (to balance the poorer 340 nm transmission through the lens) and the signals are combined for each wavelength to improve signal to noise (with a monochromator, the signals can be balanced using a longer integration time for 340 nm compared with 380 nm).

6. Liquid light guide and microscope couplings to deliver excitation light to the epifluorescence port of the microscope (*see* **Note 5**).

7. Electronic hardware for controlling the rotor wheel or monochromator positions (*see* **Note 1**). This may also be under computer software control.

8. An analog-to-digital acquistion system for data collection and analysis. Most manufacturers of single-cell fluorescence spectrophotometer systems provide proprietary software and/or hardware that controls both the filter wheel or monochromator and the timing of emission detection from the photomultiplier tube. Refer to **Note 1** for a more detailed discussion of this topic.

9. Photomultiplier tube and power supply. Although other sensitive fluorescence-detector devices exist (e.g., CCD or photodiode), photomultiplier tubes are still the usual choice for nonspatial Ca^{2+} indicator measurements. They are sensitive and inexpensive and have a high dynamic range (particularly important for fluorescence measurements where signal levels can vary greatly both within and between experiments). For optimal detection of fura-2 emission we use tubes with a rubidium bialkali photocathode.

10. Optical filters. 400-nm dichroic mirror mounted in a cube or adapter to suit the epifluorescence port of your microscope and 480-nm long-pass gelatin filter. The dichroic ensures that the 340- and 380-nm excitation wavelengths are reflected up through the objective to illuminate the platelet or megakaryocyte, yet allows transmission of emissions greater than approx 400 nm to the microscope sideport. **Fig. 1** schematically presents the components of the excitation and emission light path in our microscope system for fura-2 measurements of $[Ca^{2+}]_i$. A 480-nm long-pass gelatin filter in front of the photomultiplier tube and a 600-nm dichroic (*see* **item 11**) result in a bandpass for emission collection of approx 480–600 nm (*see* **item 12**) (*see* **Note 6**).

11. Region-of-interest controller and connector box to couple the microscope side port to the photomultiplier tube. The Cairn region-of-interest controller consists of four metal leaves that are positioned around a single cell to restrict the emitted light reaching the photomultiplier tube to a single cell of choice (*see* **Note 7** for more information on region-of-interest controllers). The connector box places the photomultiplier tube parfocal with the eyepieces of the microscope. Depending on the type of microscope, additional focusing lenses and different optical tube lengths may be required between the side port of the microscope and the photomultiplier tube to obtain parfocality (*see* **Note 8**).

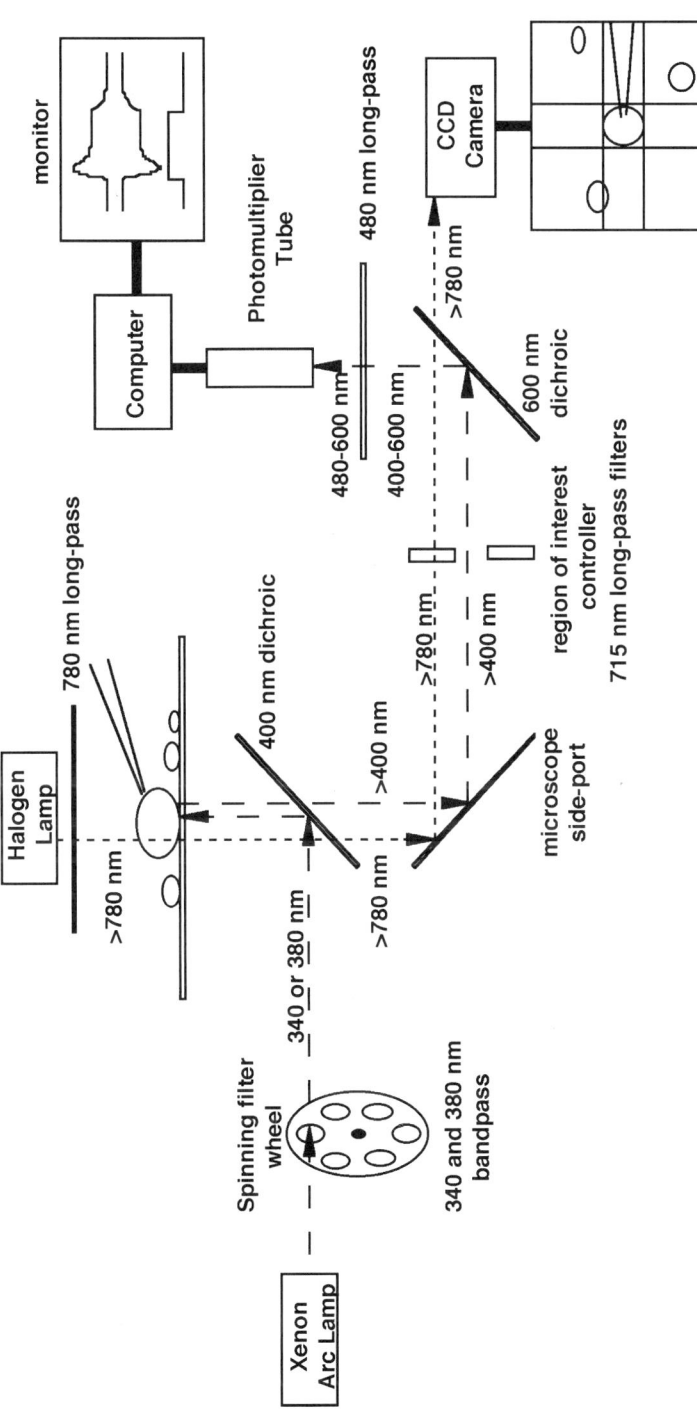

Fig. 1. Schematic representation of the filters and wavelengths within a single cell photometry setup using fura-2 as the Ca²⁺ indicator. The long dashed lines illustrate the wavelengths for excitation of fura-2 and detection of its emission by a photomultiplier tube. The short dashed lines show the path and wavelengths for simultaneous monitoring of a transmitted light image by a CCD camera. A custom region-of-interest controller with 715-nm long-pass filters permits visualization of the entire field of view. Redrawn from reference (*48*), with permission.

12. Infrared (IR) imaging camera or other method of viewing the region of interest. This is required for accurate positioning of the cell in the region of interest. The preferred system exploits the sensitivity of inexpensive CCD cameras to IR illumination, which can be viewed on an inexpensive monochrome monitor. We simultaneously illuminate our preparation with long-wavelength IR light, which is achieved by positioning a 780-nm-long pass filter in front of the standard halogen light used on microscopes for bright field illumination (*see* **Note 7**). The preparation is simultaneously illuminated with IR light greater than 780 nm and the 340- and 380-nm wavelengths required to excite fura-2. Fura-2 emission is separated from the long-wavelength IR greater than 780 nm using a 600-nm dichroic mirror to reflect light <600 nm to the photomultiplier tube and transmit light >600 nm to the CCD camera (*see* **Fig. 1**). This system allows for continual visual monitoring of the region of interest during the experiment while ensuring that no cross-talk exists. It takes advantage of the fact that the rubidium bialkali photocathodes are relatively insensitive to the long IR wavelengths. A simpler but less convenient method of viewing the region of interest is to position a parfocal monocular immediately after the region of interest controller at 90° to the light path. A movable mirror is positioned in the light path at an angle of 45° to view the region-of-interest controller and removed from the light path during photometric recordings. This type of system is incorporated into systems manufactured by Photon Technology International (Lawrenceville, NJ).

13. Agonist perfusion system. A method of changing the composition of the experimental chamber and for applying agonists is required. We have used two methods: (a) A homemade gravity-driven bath superfusion system controlled by three-way stopcocks attached to 60-mL syringes acting as storage vessels and a smaller syringe to remove bubbles from the tubes at the start of the day. The output from each syringe is positioned in a Perspex manifold containing a common outflow port, which is attached to the inflow tube of the chamber. The flow rate can be controlled by adjusting the height of the syringes or by modulating a small device that controls the diameter of the chamber inflow port. Flow rates in our system are about 5–10 mL per min through a chamber of about 300–500 μL. (b) A second method is to apply agonist from a large-bore puffer pipet positioned downstream of the megakaryocyte or platelet. Application of a regulated positive pressure (e.g., using a PLI-100 Picolitre Injector, Digitimer Ltd., [Welwyn Garden City, UK]) will eject a steady stream of saline-containing agonist onto the cell. A large bore ensures that the entire cell sees the application with minimal time delay; however, leakage of agonist can be a problem and thus counterperfusion of the chamber is normally required. Movement of the puffer pipet in the chamber is made using a micromanipulator.

2.1.2. Fluorescence Imaging of Platelets and Megakaryocytes

Fluorescence imaging can be considered as a modification of the basic methodology described here for single-cell photometry except that (1) one requires a method of imaging rather than averaging the emitted fluorescence and (2) a region-of-interest controller is not required. There are several options when selecting the imaging system. In reality, one would probably approach an imaging company, and therefore the following is a basic outline of the setup for platelet and megakaryocyte Ca^{2+} imaging.

Option 1: Confocal microscope. Virtually all confocal microscopes use a laser to provide the excitation light. Compared to filters and monochromators, the choice of wavelengths is limited; however, lasers generate light of higher intensity and narrower bandwidth. Due to the high cost of lasers and optics for UV excitation, most confocals operate at visible wavelengths. For Ca^{2+} imaging with fluo-3 or Oregon Green 488 BAPTA-1, an argon laser emitting at 488 nm is required. Our confocal is a Zeiss LSM 510 coupled to an Axiovert 100M inverted microscope (Carl Zeiss Ltd., Welwyn Garden City, UK), which scans the laser across the selected image then uses a digital processor and software to decode the signal at the photomultiplier and generate an image. The speed of acquisition is a trade-off between spatial and temporal resolution *(36)*. At least 10 Hz is required to image Ca^{2+} waves in megakaryocytes; this speed may require selection of a lower spatial resolution (companies are always improving their systems; therefore it is not appropriate to give limiting speeds here). One also has control over the "pinhole," which sets the optical slice. Increasing this allows acquisition of more signal, but loss of resolution in the *z* axis. A second type of confocal uses a spinning disk (*see* Chapter 25, vol. 2 for a more complete description) to illuminate the sample with a series of simultaneous excitation beams, and a highly sensitive camera captures the image. The two types of confocals present different advantages. The laser scanning system is more flexible in terms of controlling both the excitation region and the pinhole, and is highly compatible with UV excitation. The spinning disk confocals tend to bleach the sample less and at present can acquire more rapidly than many laser-scanning confocals.

Option 2: Nonconfocal imaging system. This is very much a modification of the photometry system described in **Subheading 2.1.1.** A highly sensitive camera replaces the photomultiplier tube and the region-of-interest controller is removed. It should be noted that the clear benefit of spatial resolution provided by using a camera as a detection device will normally come at the expense of both temporal resolution and ease of integration with other synchronous techniques (e.g., patch-clamping or flash photolysis). A vast number of cameras are available to choose from, but they must be compatible with your imaging software/excitation controlling system—and experience has taught us that it is better to purchase a complete system (e.g., from Cairn Research Ltd. or Universal Imaging Corporation). For real-time fluorescence imaging there is a choice between using a cooled digital CCD camera or an intensified camera (combined intensified *and* cooled CCDs are available, but they are very expensive). Monochrome digital CCD cameras are available with high quantum-efficiency (QE, >70%), resolution (1300 × 1000), and dynamic range (12-bit), and are ideal for fluorescence applications where signals are either relatively bright or slow-changing, or both. For fast, low-light studies there are often insufficient photons reaching the detector (within the required integration period, e.g., 50 ms) to overcome the readout noise of the CCD chip and hence produce useful images. The detection limit is improved by binning blocks of pixels on the CCD chip to increase the individual detector area at the expense of resolution; however, this is not always sufficient. Intensified cameras amplify the signal before it reaches the camera chip and thereby can ensure that there are enough photons to produce an image. This allows shorter integration times and hence faster imaging rates; however, intensifiers typically have lower QE and higher dark noise than CCD chips, so the resultant image (although acquired more quickly) will be noisier.

Two further considerations are that intensifier photocathodes are larger than CCD chips, so care must be taken to ensure appropriate magnification, and that they are also vulnerable to accidental damage by exposure to excessive light.

Since megakaryocyte Ca^{2+} waves are extremely fast, and signals from single platelets are very low, an intensified system is probably more useful overall for Ca^{2+} imaging in these cells. One flexible solution is to fit a cooled digital camera with a lens-coupled modular intensifier, which can be used when necessary and removed from the light path when not required. Unfortunately the coupling efficiency achieved by lenses is lower than that achieved in a direct-coupled (tapered fiberoptic) camera, so there is a trade-off between versatility and ultimate sensitivity. It is hoped that new CCD chips with on-chip electron multiplication (currently being integrated into cameras by Andor Technology, Photometrics, and Hamamatsu, among others) will allow cooled digital cameras to match the time response of intensified systems, but with the higher QE and small pixel size of CCDs compared to intensifiers. Whatever your final choice of components, the supplied imaging system should be capable of conducting ratiometric excitation imaging (for fura-2) at rates of at least 10 Hz. Single-wavelength measurements at 490-nm excitation are also useful and should be achievable a higher capture rates (>20 Hz). Dual-excitation wavelengths for fura-2 can be provided by a monochromator or alternative fast-stepping device.

2.1.3. Flash Photolysis of Caged Compounds While Measuring Ca^{2+} Levels With Long-Wavelength Fluorescent Indicators

Various manufacturers supply flash-photolysis systems compatible with a single-cell fluorescence system. We use the system from Cairn Research Ltd., Faversham, UK, coupled through the microscope epifluorescence port and objective lens to focus high-intensity uncaging UV light on the preparation. The UV light can also be simply directed at the preparation from above, independent of the microscope optics. A through-the-lens system results in more efficient focusing of the uncaging energy on the cell; however, it must be designed carefully to allow delivery of enough UV light via the microscope optics and to conduct simultaneous Ca^{2+} measurements. Long-wavelength fluorescence indicators for Ca^{2+} detection (e.g., fluo-3 or Oregon Green 488 BAPTA-1) are used, as they are not bleached by the high-intensity UV irradiation required to uncage. The use of these indicators requires only minor changes (**items 1–3** below) to the fura-2 fluorescence system described in **Subheading 2.1.1.**.

1. 490-nm excitation light (achieved with bandpass interference filters or resetting of the monochromator). We normally use 3×490 nm, 10-nm bandpass interference filters in the filter-wheel system (*see* **Fig. 2**).
2. 528-nm emission filter (Comar Instruments, Cambridge, UK) immediately in front of the photomultiplier tube, thereby providing an emission bandpass of 528 to 600 nm (*see* **Note 9**).
3. 503-nm dichroic mirror with extended UV reflectance (Chroma Technology Corp., Brattleboro, VT). This is installed in place of the 400-nm dichroic mirror mounted in the epifluorescence port under the objective. It ensures that both the flash light and the excitation light are directed to the sample (*see* **Note 10**).
4. Flash-photolysis system (Cairn Research Ltd.), consisting of controller unit, flash-lamp housing, and coupling to send the light through a liquid light guide. The light guide is

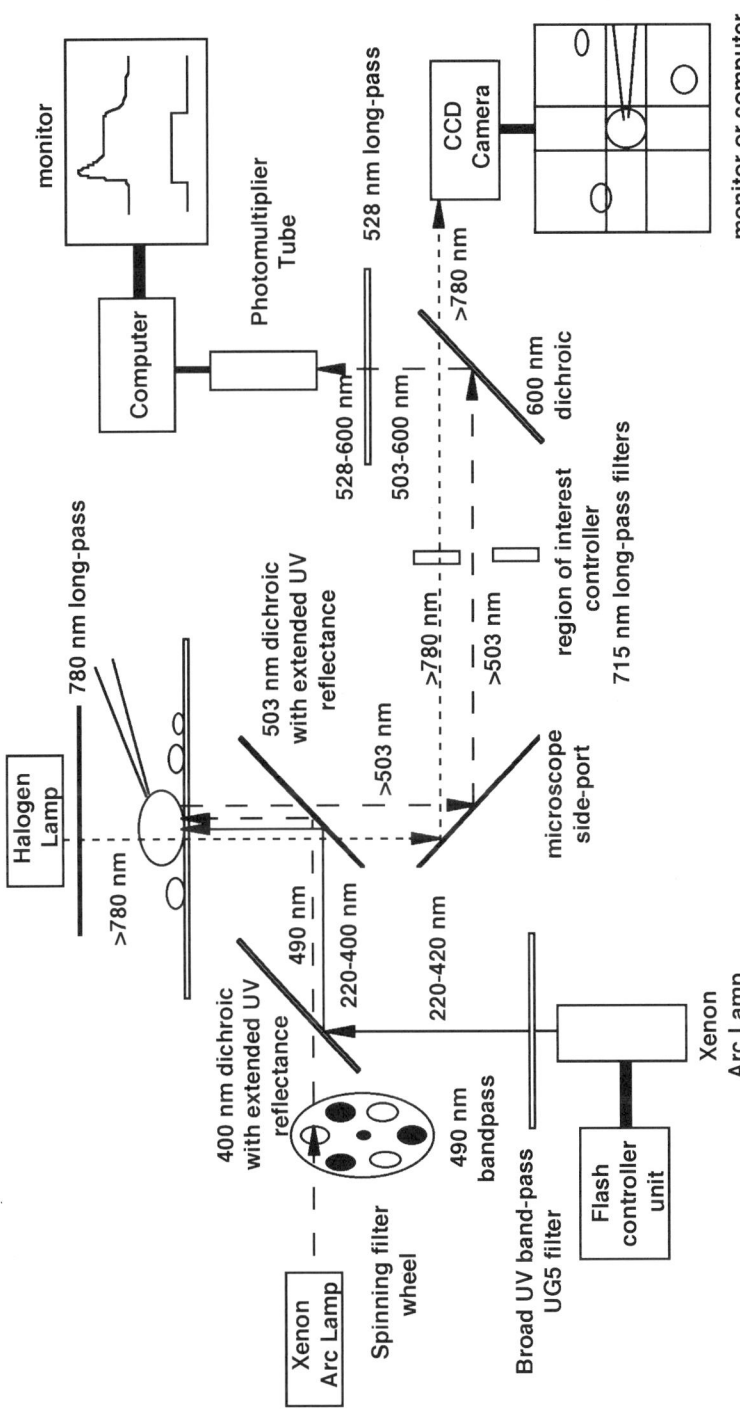

Fig. 2. Schematic representation of the filters and wavelengths within a combined flash photolysis and single-cell photometric setup. The long dashed lines illustrate the wavelengths for excitation of fluo-3 or Oregon Green 488 BAPTA-1 and detection of its emission by a photomultiplier tube. The short dashed lines show the path and wavelengths for simultaneous monitoring of a transmitted light image. The solid lines show the excitation wavelengths for flash photolysis of caged compounds.

identical to that used for the Ca^{2+} indicator excitation. The amount of uncaging light is controlled by the capacitance (up to ~4000 μF) and voltage (0–400 V) settings.

5. High-intensity xenon arc lamp (Advance Radiation Inc., Santa Clara, CA) (*see* **Note 11**).
6. Coaxial lead connecting the TTL output of the flash-photolysis unit to the "gate" of the photomultiplier housing. A key feature of the Cairn photolysis unit is a TTL output pulse during the flash, which transiently inactivates the early stages of the photomultiplier, thus ensuring that the tube is not saturated by the high-intensity flash.
7. A broad-bandpass UV filter (UG5, Comar Instruments) is positioned in front of the arc lamp to restrict the illumination to the desired UV or near-UV uncaging wavelengths.
8. Two-way liquid light guide coupling (Cairn Research Ltd.) with custom 400-nm extended UV reflectance dichroic (Chroma Technology Corp.; *see* **Note 10**). The light guides are attached at 90° to each other and the dichroic is positioned at 45° incidence to the caged light input. This allows the output from the flash photolysis arc lamp and 490-nm excitation light to be simultaneously delivered to the epifluorescence port of the microscope.

2.1.4. Patch-Clamp Measurements in Megakaryocytes

Chapter 17, vol. 2 describes whole-cell recordings from megakaryocytes and discusses the requirements of internal pipet solutions. The most common internal saline we have used consists of (in mM): 150 KCl, 2 $MgCl_2$, 0.1 EGTA, 0.05 Na_2GTP, 10 HEPES, pH 7.2 (with KOH), to which the Ca^{2+} indicator is added. An interesting paradox arises when including fluorescent indicators in the internal solutions. One wishes to add enough fluorescent indicator to improve the signal-to-noise ratio. However, these indicators are Ca^{2+} buffers that can change the spatiotemporal pattern of Ca^{2+} changes if added in too high a concentration. Experience has shown us that it is important to not exceed 100 μM indicator concentration in whole-cell patch-clamp experiments. We routinely use 50 μM.

2.2. Isolation of Rat Megakaryocytes and Human Platelets

The isolation and preparation of rat megakaryocytes and human platelets is described in Chapter 17, vol. 2.

2.3. Chemicals and Reagents

1. Standard external saline: 145 mM NaCl, 5 mM KCl, 1 mM $CaCl_2$, 1 mM $MgCl_2$, 10 mM HEPES, 10 mM D-glucose, pH 7.35 (NaOH). The $CaCl_2$ is omitted for nominally Ca^{2+}-free saline (replaced by 1 mM $MgCl_2$ to maintain a constant divalent cation concentration).
2. Fluorescent Ca^{2+} indicators. The majority of work in this laboratory has used fura-2 for quantitative measurements of $[Ca^{2+}]_i$ and either fluo-3 or Oregon Green 488 BAPTA-1 for flash photolysis. These indicators can be purchased from a variety of sources. We routinely purchase our fluorescent indicators from Molecular Probes (Leiden, the Netherlands) or Teflabs (Austin, TX). The impermeant salt is used for whole-cell patch-clamp. It is purchased as the penta potassium or sodium salts of fura-2 and the penta ammonium or potassium salts of fluo-3 and Oregon Green, dissolved in water as a 10 mM stock and stored in 5–10-μL aliquots at –20°C. Oregon Green exhibits a higher fluorescence at resting levels of cytosolic Ca^{2+} compared to fluo-3 and is preferred for imaging studies where we like to see that the cell has been adequately dialyzed with indicator before applying a stimulus. For measurements in intact cells the membrane-permeant acetoxymethylester

(AM) derivative of fura-2 is used. It is dissolved in anhydrous dimethylsulfoxide as a 5 mM stock and stored in 10–20-µL aliquots. Avoid multiple freeze-thawing of these stocks.

3. Caged IP$_3$. We use the 2-nitrophenyl ethyl ester of D-myo-inositol 1,4,5-trisphosphate (NPE-caged IP$_3$) (Calbiochem-Novachem, Nottingham, UK). For caged IP$_3$ experiments the pipet internal contains 100 µM NPE-caged IP$_3$ and dialysis is allowed to proceed following formation of the whole-cell configuration. Due to the high cost of the caged compounds, pipets are filled with the smallest volume possible, which can be <50 µL if the Ag/AgCl wire of the patch pipet holder is lengthened.

4. Caged cyclic AMP. We use 4,5 dimethoxy-2-nitrobenzyl caged cAMP, a membrane-permeant version of caged cAMP (DMNB-caged cAMP) (Molecular Probes, Eugene, OR). Cells are loaded with this derivative by incubating the marrow preparation for 2 h in standard external saline supplemented with 200 µM DMNB-caged cAMP. Loading by pre-incubation, rather than simply including in the patch pipet, is preferred owing to the possible presence of free cAMP in the purchased caged compound. Caged cAMP is membrane-permeant, whereas the free cAMP is not.

3. Methods

3.1. Simultaneous Whole-Cell Patch-Clamp Recordings and Photometric Fluorescence Measurement of Agonist-Evoked Ca²⁺ Reponses in Single Megakaryocytes Using Fura-2

It is difficult to load megakaryocytes with fluorescent Ca²⁺ indicators by incubation in acetoxymethyl derivatives without over-buffering the agonist-evoked responses *(35)*. Single-cell Ca²⁺ measurements in megakaryocytes are performed under whole-cell patch-clamp conditions, thereby allowing the megakaryocyte to be loaded by dialysis from the patch pipet.

1. A small-volume Perspex chamber (~300–500 µL, *see* Chapter 17, vol. 2 for details) is mounted on the stage of the microscope. The bottom of the chamber is formed by attaching a coverslip to the Perspex chamber with silcone grease. The chamber is fitted with a stainless steel tube for solution inflow and a tube for drawing off the solution by suction generated from a small vacuum pump. For patch-clamp experiments the chamber must be grounded, usually by connecting an Ag/AgCl grounding wire directly to the ground connection on the headstage (*see* **Note 12** and Chapter 17, vol. 2).

2. A 20–30-µL aliquot of marrow suspension is added to the chamber and allowed to briefly settle to the bottom of the chamber (*see* **Note 13**).

3. Quickly find a megakaryocyte (*see* **Note 14**) and while repeatedly applying a small square voltage step, lower the patch pipet into position, maintaining positive pressure while moving the pipet through the air-water interface.

4. Apply suction to the pipet to form a high-resistance gigaohm seal between the pipet and the cell membrane (*see* Chapter 17, vol. 2 for further details). Raise the cell off the bottom of the chamber with a pronounced upward movement of the *z* axis of the micromanipulator. Slight movements in the *x* and *y* axes may allow weakly adherant megakaryocytes to be detached from the coverslip without damage to the cell or gigaseal.

5. Position the 780-nm long-pass filter in front of the halogen light of the transillumination system of the microscope and direct the emission light path to the side port of the microscope. The IR image detected by the CCD camera should be displayed on the camera monitor. Adjust the intensity of the halogen light to give good illumination and contrast.

6. Position the cell in the center of the field of view and close in the leaves of the region of interest controller to confine the emission to the megakaryocyte. As described in **Note 7**, IR-transmitting filters forming the four leaves in the region-of-interest controller will allow you to view the entire field of view (*see* **Subheading 2.1.1.** above).

7. Open the shutter to allow the excitation light to illuminate the preparation, turn on the photomultiplier tube, and adjust the magnitude of the photomultiplier tube output by controlling the voltage to the photomultiplier tube so that a signal of appropriate magnitude is acquired by the data acquistion system. It is often recommended that photomultiplier tubes be run at a reasonably high voltage (>700 V) for optimal performance, so alter gains elsewhere in the system to allow this. Only experience with your acquistion system will allow you to determine the settings required to achieve this (*see* **Note 15**). Ideally, we attempt to acquire signals at as fast a rate as the signal allows without excessive noise. In practice, we normally sample the fluorescence signals at up to 60 Hz, with further averaging to generate final signals at 15–20 Hz.

8. While still applying the small periodic square voltage step (5–10 mV), slowly apply negative pressure to the patch pipet under voltage clamp and monitor the current using an oscilloscope or electrophysiological software. The appearance of whole-cell capacitative transients at the start and end of the test pulse will indicate that the whole-cell configuration has been acheieved. Proceed only if the capacitative transient decays with a single exponential (*see* Chapter 17, vol. 2 for further details). Neutralize the capacitative transients using the amplifier capacitance and series resistance controls and proceed if the series resistance (Rs) is low (we normally accept starting Rs values of ≤ 5 MΩ). If large-membrane currents are also activated by the experimental protocol, series resistance compensation may also be required (*see* Chapter 17, vol. 2).

9. As the cell dialyzes with the pipet solution containing fura-2, both the 340- and 380-nm fluorescence signals will increase (*see* **Fig. 3**). The lower the series resistance, the more rapid the dialysis *(44)*. As a result of the large size of the megakaryocyte, many minutes may be required until a steady-state fluorescence value is achieved; however, experiments can be started prior to reaching steady-state once the 340/380 ratio is stable.

10. Application of agonists such as ADP can be made by rapid superfusion through the bath or by a puffer pipet as detailed in **Subheading 2.1.1.** In the case of superfusion, it is useful to estimate the time for the new solution to reach the chamber and cell following a solution change. This will enable you to correct for the time lag in your responses. One easy method is to measure the potential change between two salines of different chloride concentration using a patch pipet filled with conventional internal saline.

11. Immediately after termination of the experiment, the cell is removed far from the region of interest and the emission is recorded. This provides a measure of the background fluorescence at 340 and 380 nm excitation and is used for background correcting the fura-2 ratio and for converting the fluorescence signals into estimates of $[Ca^{2+}]_i$ (*see* **Note 16** and **Subheading 3.5.**).

3.2. Photometric Measurement of Agonist-Evoked Ca²⁺ Reponses in Single Human Platelets

1. 0.5–1-mL aliquots of platelet-rich plasma (PRP), treated with apyrase and aspirin, are prepared as described in Chapter 17, vol. 2. Aspirin can be omitted if studying the effects of thromboxane generation. These are kept at room temperature in Eppendorf tubes on a rotor turning at approx 0.2 Hz.

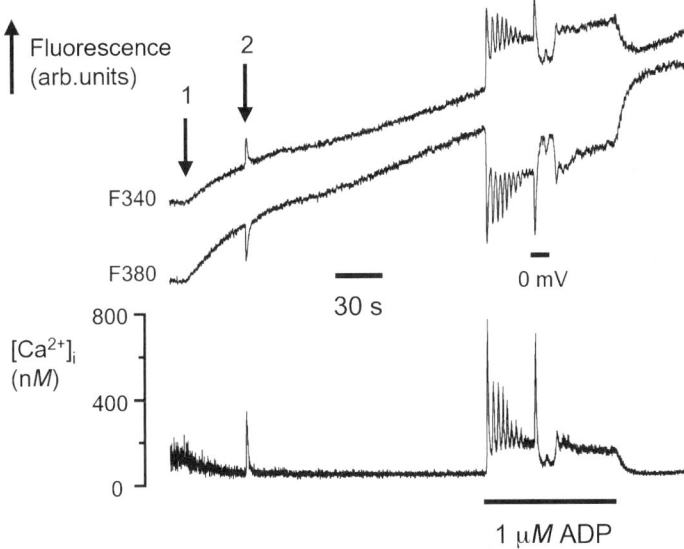

Fig. 3. Fura-2 fluorescence recording of intracellular Ca²⁺ in a megakaryocyte under whole-cell patch clamp. Emitted signals at 340- and 380-nm excitation are shown in the upper panel and the intracellular Ca²⁺ concentration calculated from the background-corrected 340/380nm ratio (*see* **Subheading 3.5.**) is shown in the lower panel. The recording starts with the cell positioned in the region of interest in cell-attached patch-clamp configuration and whole-cell recording is achieved at arrow 1. Agonist (1 µ*M* ADP) is superfused over the cell once the ratio (and thus indicated Ca²⁺ level) are stable. Spontaneous Ca²⁺ increases (arrow 2) are frequently observed during the early stages of this type of recording. The 340- and 380-nm fluorescence records have been separated for clarity, but start at the same value in the cell-attached mode.

2. Incubate an aliquot of PRP with 4 µ*M* fura-2 AM at 37°C, for 45 min. This concentration is higher than used for cuvet measurements of Ca²⁺ (usually 2 µ*M* *[45]*), in order to improve the signal-to-noise ratio.

3. Spin the platelets in a microcentrifuge for approx 20–30 s or until the platelets have formed a soft pellet (we advise that each lab assess the speed empirically, as the aim is to spin the platelets with the minimum *g* force and time to allow removal of the plasma). Remove the PRP and resuspend in nominally Ca²⁺-free saline with 0.32 U/mL apyrase. It should also be noted here that if the platelets are to be used immediately, it is possible to omit this spin step since the platelets can be washed following adherence to a fibrinogen-coated coverslip (*see* Chapter 12, vol. 1) (*see* **Note 17**) or to a pipet *(29)*.

4. Add 10–50 µL of cell suspension to the small Perspex chamber containing standard external saline, mounted on an inverted microscope, and allow the platelets to settle. Perfusion should initally be switched off. For experiments where a single platelet is to be studied independently of its neighbors (*see* **step 6**), spread platelets over only one area of the cov-erslip, towards the perfusion outflow. Perfuse with saline and adjust the flow rate and dura-

tion until a reasonable density of platelets is left floating above the uncoated (or Sylgard®-coated [*see* **Note 13**]) coverslip.

5. Fill a glass pipet, as used for platelet patch-clamp (*see* Chapter 17, vol. 2), with normal external saline and mount in a patch pipet holder (or other holder allowing pressure to be applied to the pipet interior). Using a micromanipulator, bring the tip to within 5 μm of a platelet with nonactivated appearance (discoid, lacking filopodia). Apply the lightest possible suction to draw the platelet onto the pipet. (Since the pipet is not being used for electrical recordings in this situation, a high-resistance seal is not required.)

6. To reduce the risk of paracrine influences, move the chamber so no other platelets are near the pipet tip.

7. Redirect the emision path of the microscope to the sideport and thus to the photomultiplier tube.

8. Adjust the region of interest to collect fluorescence only from an area slightly larger than the platelet.

9. Open the illumination shutter within the excitation light path and adjust the gain of the photomultiplier and input amplifier (*see* **Note 15**).

10. Acquire the fluorescence signals at a rate of ≥8 Hz (the faster the better, but the low-fluorescence signal from a single platelet normally limits the acquisition rate) and monitor the 340/380 nm ratio for several minutes to assess whether spontaneous Ca^{2+} increases are present.

11. Apply agonists/antagonists for specific experiments. During the experiment, return the platelet to the center of the ROI and focal plane using the manipulator if any movement occurs. A typical experiment showing repeated ADP-evoked Ca^{2+} transients in a single human platelet is shown in **Fig. 4**. We are unsure at present whether the heterogeneity in responses, from single to multiple spikes of increase in $[Ca^{2+}]_i$, is a result of inter-platelet variation or different extents of dye loading (i.e., different introduced buffering levels).

12. At the end of the experiment, move the platelet out of the window to measure the background fluorescence levels. The background-corrected 340/380 nm ratio can be used as an indication of intracellular Ca^{2+} levels or can be converted to Ca^{2+} as described in **Subheading 3.5.** Note that because of the very low fluorescence signals, the background-corrected 380-nm level can approach zero and thus the ratio will be extremely high. To overcome this problem, the averaging can be increased or the signal-to-noise ratio can be improved using more excitation light or slightly longer dye-loading times. However, it may ultimately be necessary to offset the fluorescence traces by one or two bit values, so the 380-nm signal is above zero, and present as raw ratios without converting to Ca^{2+}.

3.3. Fluorescence Imaging of Megakaryocyte and Platelet Ca²⁺ Responses

1. *Megakaryocytes:* The procedure for monitoring Ca^{2+} waves and other spatial information from megakaryocytes under whole-cell patch clamp is identical to that described above for photometry except that the signal from the cell is imaged rather than averaged. We use an LSM 510 confocal and either fluo-3 or Oregon Green 488 BAPTA-1 as the indicators, with excitation at 488 nm and emission collected at >505 nm. Another group *(35)* has used fura-2 and conventional imaging. Megakaryocyte waves are among the fastest recorded *(35)*; therefore, acquistion rates of at least 10 Hz are required—the faster the better, but signal or system acquisition rates may be limiting (*see* **Note 18**). Background fluorescence is measured from an area outside the cell and subtracted from all cell fluorescence signals. The signal from the indicator will increase as the patch solution dialyzes into the cell, so

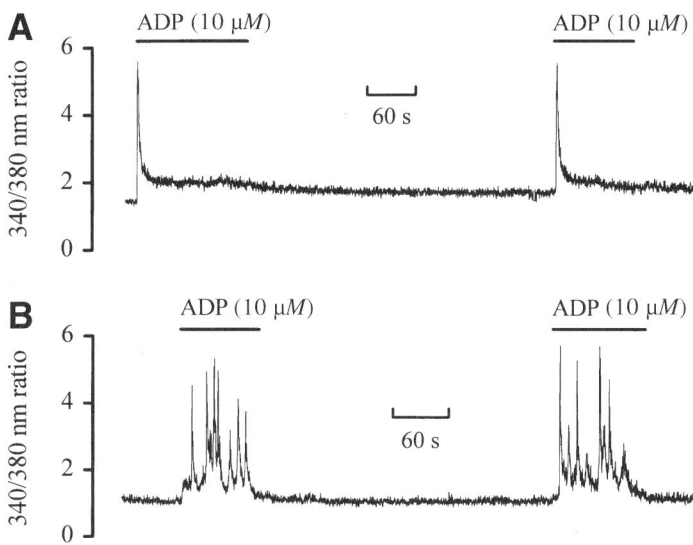

Fig. 4. Fura-2 fluorescence recording of intracellular Ca²⁺ in an individual human platelet. A platelet was loaded with fura-2 by incubation with fura-2 AM and gently held at the tip of a glass ("patch") pipet. The background-corrected 340/380 nm ratio is used as an indication of intracellular Ca²⁺. Two different platelets are shown (**A** and **B**), which illustrate the range of responses to 10 μ*M* ADP. Reproduced from reference *(29)*, with permission.

    for single-wavelength measurements, the fluorescence signal (F) over short sections of recording is normalized, by expressing as the F/F_0 ratio, where F_0 is the fluorescence at the start of the section.

2. *Platelets:* The approach is the same as in **Subheading 3.2.**, but platelets are adhered to coverslips (e.g., using fibrinogen, *see* Chapter 12, vol. 1) and images acquired at the fastest rate that the signal allows (at least 5 Hz). Offline analysis can be used to draw regions of interest around each platelet (*see* **Note 19**). Fura-2 signals can be calibrated (*see* **Subheading 3.5.**) or the background-corrected 340/380 nm ratios used to indicate Ca²⁺ changes (*see* **Fig. 4**). For single-wavelength indicators, the raw fluorescence or background-corrected F/F_0 can be used to indicate Ca²⁺ changes (*see* **Note 19**).

    Ca²⁺ responses to physiological agonists such as ADP, collagen, or thrombin can be studied at the single-platelet level following superfusion of the chamber or by pressure application from a nearby pipet (*see* **Subheading 2.1.1., item 12**). Alternatively, pharmacological tools that control elements of the Ca²⁺ signaling pathways can be applied. Useful examples include (a) ionomycin (100 n*M*–1 μ*M*), an electroneutral Ca²⁺/H⁺ exchanger that allows Ca²⁺ to flow freely across membranes driven by its chemical gradient; (b) thapsigargin (200 n*M*–1 μ*M*), which blocks endomembrane CaATPases and thus depletes internal stores to activate store-dependent Ca²⁺ influx. A significant advantage of the single-platelet approach over conventional cuvet measurements is that reagents can be washed off and thus recovery, an understudied area of platelet Ca^{2+} signaling, can be monitored. In this respect, thapsigargin is irreversible in most cells, but other, more reversible endomembrane CaATPase inhibitors exist, such as 2-5-di-(*tert*-butyl)-1,4-hydroquinone or cyclopiazonic acid.

3.4. Controlled Release of Caged Compounds by Flash Photolysis in Megakaryocytes

Release of cytosolic messengers from their caged, inactive derivatives is routinely used to study Ca^{2+} signaling in our laboratory. The use of UV-resistant long-excitation-wavelength Ca^{2+} indicators such as fluo-3 or Oregon green 488 BAPTA-1 require that the experimental setup be modified as outlined in **Subheading 2.1.3.** Following these hardware modifications a similar experimental protocol to that described in **Subheading 3.1.** can be used to monitor Ca^{2+} except that the free Ca^{2+} level is monitored from the background-corrected fluorescence at 490 nm. Ratioing of the F490 signal against the prestimulus fluorescence level can also be used (F/F$_0$, *see* **Subheading 3.3.**) to normalize short sections of data. The addition of caged compound and the procedure used to uncage is outlined below.

1. For flash release of IP$_3$ under whole-cell patch clamp, we supplement the required pipet solution with 100 μ*M* 2-nitrophenyl ethyl ester of D-myo-inositol 1,4,5-trisphosphate (NPE-caged IP$_3$) as noted in **Subheading 2.3.** In the case of flash release of cAMP it is not necessary to add caged cAMP to the pipet solution, as the cells are loaded with the membrane-permeant, 4,5 dimethoxy-2-nitrobenzyl-caged cAMP (DMNB-caged cAMP) (*see* **Subheading 2.3.**). This provides a limited amount of caged compound compared to the addition to the pipet, but reduces problems due to free cAMP and provides for at least two to three rapid elevations of enough cAMP to completely inhibit ADP-evoked Ca^{2+} release.

2. Dialysis of the cytosol with the caged compound is dependent on the series resistance, as is the dialysis of all the components of the pipet solution *(44)*. We have found that series resistances of ≤8 *M*Ω in our megakaryocyte experiments do not significantly limit dialysis of caged compounds into the cell.

3. Following adequate dialysis of fluorescent indicator into the cell, uncaging can be performed (*see* **Note 20**). Control over the amount of IP$_3$ or cAMP released during each flash is controlled by three variables: (a) the concentration of caged compound in the cytosol, (b) the intensity of the flash (capacitance × voltage; *see* **Fig. 5**) and (c) the efficiency of transmittance of the flash to the cell (*see* **Note 21**). The concentration of caged compound is governed by the concentration in the pipet or, in the case of DMNB-caged cAMP, the concentration in the incubation medium and the incubation time. The intensity of the flash is governed by both the amount of capacitance and the charging voltage set by the Cairn controller unit (charge across capacitor = capacitance × voltage). The Cairn unit is capable of providing up to approx 4 mF and 400 V. The experimenter has complete control over these variables and trial and error will determine the minimum flash intensity required to induce a measurable effect. We tend to use the highest settings for our through-the-lens system (*see* **Fig. 5**). Start with low-intensity flashes and increase the flash intensity until the desired release and effect are detected. Remember that high-intensity UV irradiation is detrimental to most cells so lower flash intensities are preferable and important controls are required (*see* **Note 22**).

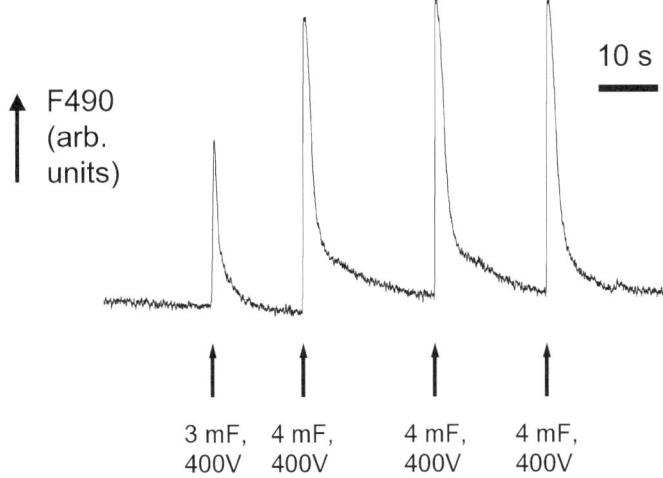

Fig. 5. Use of flash photolysis to repeatedly release IP$_3$ in a rat megakaryocyte. A rat megakaryocyte was held under whole-cell patch clamp with 100 μM caged IP$_3$ in the pipet saline. Repeated delivery of an uncaging light (small arrows) released IP$_3$ and generated a brief Ca²⁺ transient, detected with the Ca²⁺ indicator fluo-3 (50 μM in the pipet). The effect of increasing the energy of the flash, by increasing the capacitance setting from 3 to 4 mF, is illustrated.

4. Flashes can be triggered directly from the control unit or via a TTL pulse delivered to the control unit from another instrument. This is particularly convenient if correlating the time of flash delivery with electrophysiolgical protocols or rapid agonist superfusion from puffer pipets.
5. If using flash photolysis to repeatedly generate the same amount of messenger (e.g., IP$_3$), flashes should be given at equally spaced intervals to ensure that adequate and comparable dialysis of uncaged compound into the cell has occurred after each flash. In our experience, multiple flashes can be delivered with intervals of only 20–30 s with reproducible effects, provided the series resistance is low ($\leq$8 MΩ). The maximum flash within our system will leave sufficient caged compound to allow further release; thus multiple flashes can be used to study repeated (and cumulative, depending on metabolism) delivery of a messenger. The minimum interval between flashes is limited only by the rate at which the controller unit recharges.

3.5. Calibration of the Fura-2 Ratio in Platelets and Megakaryocytes

The [Ca²⁺]$_i$ indicated by the fura-2 ratio (R) is defined by the following equation *(46)*:

$$[Ca^{2+}]_i = K_d \times \frac{(R - R_{min})}{(R_{max} - R)} \cdot \frac{S_f}{S_b}$$

where K_d is the fura-2 dissociation constant, S_f is the maximum value of fluorescence at 380 nm, and S_b is the minimum value of fluorescence at 380 nm. R_{max} and R_{min} are the maximum and minimum values of the 340/380 nm ratio under saturating Ca^{2+} and Ca^{2+}-free conditions. All fluorescence values must be corrected for background fluorescence (*see* **Note 23**).

We use an extracellular procedure for the calibration of fura-2 (*see* **Note 24**). Since our megakaryocyte experiments are performed under whole-cell patch clamp at room temperature, extracellular calibration is done using the same pipet solution at a similar temperature. In most experiments we use a high-K^+ pipet solution that mimics the ionic composition of the cytosol of platelets and megakaryocytes. Therefore, this calibration procedure is equally valid for both megakaryocytes and platelets. However, if the experiments being calibrated were performed using an alternative pipet solution, this should be used in the calibration procedure. Calibration is performed on the microscope using conditions that closely match those used during megakaryocyte and platelet experiments including objective magnifications and the size of the region of interest as set by the region of interest controller as outlined in **Subheading 2.1.1.** The procedure is as follows:

1. 2.5 mL of patch solution is supplemented with 1 μM K_5fura-2 (the signal from the normal pipet concentration of 50 μM is too large).
2. The solution is then divided into two 1.25-mL aliquots. 1 mM K_4EGTA is added to one aliquot, while 1 mM $CaCl_2$ is added to the other. The former solution is Ca^{2+}-free and is used for determination of R_{min} and S_f, while the latter contains a saturating $[Ca^{2+}]$ and is used for determination of R_{max} and S_b.
3. The experimental chamber is mounted on the microscope stage and a small line is drawn on the upper face of the glass coverslip with a marker pen.
4. The appropriate oil-immersion objective (40× used in megakaryocyte experiments or 100× used in platelet experiments) is focused slightly above the permanent marker line, thus ensuring that the focal plane is in the chamber and close to the plane of focus used during the experiments.
5. The region-of-interest controller is set to the size approximating that used during the experiments.
6. The chamber is first filled with the fura-2 saline with saturating levels of Ca^{2+}. Ideally, measurements of the 340- and 380-nm signals should be recorded under similar conditions of photomultiplier voltage, amplification, and excitation intensity to ensure that conditions are as close as possible to those used during the acquisition of megakaryocyte data. It may be necessary to adjust the region of interest to allow this.
7. The chamber is then thoroughly washed with Ca^{2+}-free saline and the fura-2-containing Ca^{2+}-free saline added and 340- and 380-nm signals measured. It is necessary to ensure that the chamber is adequately rinsed between solutions to ensure that no cross-contamination occurs.
8. The chamber is again thoroughly washed with saline lacking fura-2 and background fluorescence measurements are then made with Ca^{2+}-free saline. Our data have shown that the level of background fluorescence does not vary between Ca^{2+}-free and saturating Ca^{2+} solution, but it is wise to check this for yourself.
9. **Steps 6–8** are repeated until three consecutive fluorescence measurements under each condition are consistent.
10. The only value not determined during this calibration procedure is the fura-2 dissociation constant, K_d. Most investigators use a K_d value previously published in the literature (*see* **Note 25**). For accuracy, the value should have been calculated under conditions similar to those used in your experiments.

The characteristics of fluorescent indicators varies with the conditions under which they are used. Employing an extracellular calibration means that the calibration variables may be different from those calculated in the cytosol. One variable that has been reported to contribute to these changes is the viscosity of the medium, and a correction is frequently made to take into account cytosolic viscosity *(47)*. This correction is applied to the calibration equation to give the following modified equation that can be used to convert fura-2 fluorescence ratios to estimates of $[Ca^{2+}]_i$ using the extracellular calibration procedure:

$$[Ca^{2+}]_i = K_d \times \frac{[R - (R_{min} - VCF)]}{[(R_{max} - VCF) - R]} \times \frac{S_f}{S_b}$$

VCF is the viscosity correction factor, which we take as 0.85 *(47)* for megakaryocytes (and in our work on megakaryocyte cell lines). From our work in cell suspensions, no correction appears to be necessary for platelets at normal temperatures (M. G. Rolf and M. P. Mahaut-Smith, unpublished).

4. Notes

1. The Cairn Research fluorescence spectrophotometer is based around a rack system that controls the excitation filter wheel (or monochromator) either independently or from computer software. The main advantage of the Cairn system is its modular nature such that it can also be used for cuvet and single-cell studies. The emission output from the photomultiplier tube is decoded and processed by Cairn proprietary software. Alternatively, the system can output to other acquisition systems.

2. Light transmission through the objective varies with the fourth power of the numerical objective (NA). Therefore, it is crucial for single-cell fluorescence studies to purchase an objective with a high NA (but *see also* **Note 19**). For work with fura-2, indo-1, and caged compounds, the lens must also have high transmittance in the ultraviolet part of the spectrum. We have found fluor oil-immersion objectives (e.g., 40× and 100×) from Nikon (NA 1.3) to be suitable for these purposes.

3. In order to monitor Ca^{2+} events at the highest spatial and temporal resolution, the entire fluorescence signal from the cell should be sent to the detector. For this reason, it is not recommended to use microscopes with split prisms mounted under the objective that divide the emitted light between the eyepieces and one or two side ports.

4. **Safety note:** The xenon arc lamps used for excitation of fluorescent indicators emit a very-high-intensity light beam with a high ultraviolet (UV) component that can permanently damage eyesight. Wear appropriate UV-protective goggles when setting up or adjusting the excitation lamp. To avoid risk to the user, we tend to adopt the factory settings initially and optimally focus the beam using the emitted signals while exciting a low concentration of fura-2 (~0.1 μ*M*) in the chamber. In addition, never look directly at the excitation light as it emerges from the lens. Some microscope manufacturers provide a UV-absorbing safety shield, which is mounted on the eyepiece turret.

5. Fused silica (quartz) fiberoptic guides are frequently used to couple the excitation source to the miscroscope. Liquid light guides are a less-expensive alternative. While the percentage transmission of near-UV light is lower than quartz, it is sufficient for most applications.

6. We have chosen a wide emission bandpass because the signal is often limiting during single cell measurements, particularly platelets. Caution should be taken with this approach as background signals increase, and cell autofluorescence can contribute and change during

activation. For each system and type of experiment, it is therefore important to test for autofluorescence changes in nonloaded cells at the same system gains.

7. A modification of the region-of-interest controller has been made in our laboratory that enables viewing of the complete field of view with IR wavelengths while restricting the fura-2 emission light to a single cell as defined by the controller *(48)*. This system replaces the opaque leaves of a standard controller with 715-nm long-pass filters, thus allowing the long-wavelength IR illumination to be transmitted through while rejecting the shorter emission wavelengths of fura-2. Transmission of fura-2 emission is restricted to the area defined by the border of the filters.

   Some microscopes have IR filters permanently mounted in the transillumination light path to filter out IR before it reaches the preparation. These filters need to be taken into account when developing an IR viewing system.

8. Most manufacturers of single-cell instrumentation work closely with microscope manufacturers to ensure that their instrumentation can be easily adapted to the port options of the variety of microscopes on the market.

9. With the filter-based system that uses 490-nm interference filters for fluo-3 and Oregon Green 488 BAPTA-1 excitation, we found that within the series of Wratten gelatin filters (Comar Instruments), far too much excitation light was detected with a 519-nm long-pass and therefore we use a 528-nm long-pass filter.

10. The dichroic mirrors referred to in this chapter are set at 45° to the incident light; wavelengths immediately below their described cut-off are reflected and longer wavelengths are transmitted. However, standard dichroics will also transmit wavelengths much shorter than this cut-off wavelength. For flash photolysis, it is therefore imperative that the dichroics have an additional custom coating that provides extended reflectance into the UV range.

11. Extreme caution should be taken with the uncaging light, which has a very high intensity (far greater than used for indicator excitation; *see* **Note 4**) and can cause permanent eye damage, even after transmission through the microscope optics. The system should not have light leaks and the operator should be protected at all times from the uncaging light. We operate the uncaging experiments inside a Faraday cage with solid sides and front. The photolysis controller unit is always switched off when the front of the cage is open.

12. Details of the Ag/AgCl grounds are given in Chapter 17, vol. 2. This type of grounding provides a low-resistance ground pathway that is ideal for experiments in which extracellular chloride level is not changed. To reduce the affect of junction potentials on your ground wire during changes in extracellular chloride, the chamber should be grounded using an agar bridge filled with a high concentration of KCl.

13. We have found that coating the coverslip with a thin layer of Sylgard 184 elastomer (Dow Corning) reduces the tendency of the cells to stick. The thinnest possible layer of Sylgard mixture is spread on the coverslip and then cured overnight or more rapidly with a hotplate or flame.

14. Megakaryocytes are clearly visible as large-diameter (15–50 μm) cells that can be easily distinguished from other marrow cells (*see* **Fig. 3**, Chapter 17, vol. 2).

15. It is important that the amplification allows as much of the range of the analog to digital data acquistion board to be used as possible, although one must allow for a steady increase in 340- and 380-nm excitation signals as fura-2 dialyzes into the cell during whole-cell recordings (*see* **Fig. 3**). Too often, inexperienced users run their systems using dynamic ranges of tens of millivolts rather than volts. This can result in the signal being buried in the bit noise of the acquistion system. It should also be noted that not all turnkey systems provide for variable photomultiplier voltage control, with many manufacturers running

their photomultiplier tubes at a fixed, high voltage and thus gain is controlled electronically elsewhere.

16. Some researchers use the background fluorescence of the megakaryocyte immediately prior to initiating the whole-cell configuration for background corrections. However, this value can be contaminated by small amounts of stray fluorescence originating from the patch pipet. In addition, the whole-cell configuration can often occur spontaneously before commencing acquisition of fluorescence signals. We continue to debate the pros and cons of the types of backgrounds, but adhere to the end-of-experiment cell-independent background for consistency.

17. Previous work has used fibrinogen as a method of adhering platelets to coverslips, (*see* Chapter 12, vol. 1) and certainly a method of immobilization is required for basic studies of Ca^{2+} signaling. Ca^{2+} can also be studied in nonadherent platelets or following perfusion over prothrombotic surfaces; however, movement artifacts are then a problem and thus it is important to use the highest acquisition rates possible.

18. Due to the marked morphological changes of both the platelet and megakaryocyte, it is useful to capture the transmitted light image if possible. This is normally available as an extra channel on most laser-scanning confocals.

19. One artifact to be aware of when imaging, which is not a problem in photometric recordings, is uneven fluorescence illumination. This is mainly a problem with filter- and monochromator-based systems (rather than with confocals) and can be assessed using a sample with uniform fluorescence, for example indicator in solution. The problem can be removed by reference to a sample of even fluorescence *(49)* but this is laborious and itself prone to artifacts, and therefore it is preferable to arrange for even illumination of the preparation. Uneven illumination may result from the properties of the lens (choose a lens that focuses evenly throughout the preparation) or the properties of the excitation system (discuss with the supplier).

20. We have experienced significant activation of Ca^{2+} oscillations with some batches of caged IP_3, presumably due to free IP_3 in the caged compound. These events do disappear with time during a whole-cell recording, but can be reduced by using a lower concentration of caged compound if sufficient uncaging can still be achieved.

21. Objective lenses used for fluorescence measurements have very limited transmission at the short UV wavelengths required for photolysis of some caged compounds (particularly the NPE derivatives). Nevertheless, adequate uncaging can be achieved with these compounds using a through-the-lens system by increasing the intensity of the flash and relying upon absorbance of the less-than-optimal longer UV wavelengths.

22. Important controls should be carried out to test for nonspecific effects of the uncaging light and side-products of the uncaging process, which includes protons and nitroso or related by-products (*see* **ref. 50** for full discussion). The best control is to test uncaging of a compound that is known to have no biological effect, attached to the same cage moiety. Caged phosphate is often used as a control. Another control is simply to test the same flash intensity in the absence of caged compound: any side effects indicate direct effects of the UV light on the cell. To overcome this, decrease the illumination intensity and, if possible, increase the concentration of the caged compound.

23. Background fluorescence is the contribution of the emission signal recorded at 340 and 380 nm that is not related to the fluorescence of fura-2. This includes the autofluorescence of the cell and experimental solutions and the fluorescence associated with the components of the optical path. Cell autofluorescence in our system is low, but should be checked on all new systems as filters and optical paths vary.

24. Accurate calibration of fura-2 requires that the calibration variables K_d, R_{max}, R_{min}, S_f, and S_b be determined under conditions that mimic the experimental conditions. In some cell types an *in situ* procedure has been used to determine R_{max}, R_{min}, S_f, and S_b. Ca^{2+}-transporting ionophores are used to transport Ca^{2+} across the membrane and saturate fura-2, thus providing an estimate of R_{max} *in situ*. Ca^{2+} is then removed from the extracellular solution by superfusing the cell with a Ca^{2+}-free solution containing high concentrations of a Ca^{2+} buffer such as EGTA. In the sustained presence of ionophore, Ca^{2+} is transported out of the cell and given sufficient time an estimate of R_{min} is achieved. Alternatively, patch-pipet salines with high EGTA or high Ca^{2+} levels can be used to measure calibration constants. Due to problems we experienced when trying to clamp cytosolic Ca^{2+} at high levels in the megakaryocyte (cell activation and requirement for long dialysis times), we have opted for extracellular calibration of our fluorescence signals.

25. While it is a common practice for investigators to use a dissociation constant taken from the literature, it must be noted that the K_d varies with temperature, ionic strength, and pH. As a result it is wise to calculate the K_d under conditions that match your experimental situation as closely as possible. Calculation of K_d requires the collection of fluorescence at 340- and 380-nm excitation for a range of different Ca^{2+} concentrations. Put simply, a calibration curve must be created that spans the gap between Ca^{2+}-free and saturating Ca^{2+} solutions. Solutions of different Ca^{2+} concentration can be made up using a Ca^{2+} buffer such as EGTA and varying the total Ca^{2+} added. MaxChelator is a computer program written by Dr. C. Patton and available on the Web (www.stanford.edu/~cpatton/maxc.html) that facilitates the calculation of free Ca^{2+} under defined conditions as set by your experimental solution. Alternatively, a calibration kit is available from Molecular Probes. The advantage of the calibration kit is that these kits provide accurate buffer reagents and detailed protocols including methods for determining the K_d of fluorescent Ca^{2+} indicators.

Acknowledgments

These studies were supported by the British Heart Foundation, Royal Society and Medical Research Council. We thank Jeremy Graham for helpful comments, particularly regarding fluorescence-imaging hardware.

References

1. Rink, T. J. and Sage, S. O. (1990) Calcium signaling in human platelets. *Annu. Rev. Physiol.* **52,** 431–449.
2. Siess, W. (1989) Molecular mechanisms of platelet activation. *Physiol. Rev.* **69,** 58–178.
3. Somasundaram, B. and Mahaut-Smith, M. P. (1994) Three cation influx currents activated by purinergic receptor stimulation in rat megakaryocytes. *J. Physiol.* **480,** 225–231.
4. Briddon, S. J., Melford, S. K., Turner, M., Tybulewicz, V., and Watson, S. P. (1999) Collagen mediates changes in intracellular calcium in primary mouse megakaryocytes through syk-dependent and -independent pathways. *Blood* **93,** 3847–3855.
5. Sage, S. O., Reast, R., and Rink, T. J. (1990) ADP evokes biphasic Ca^{2+} influx in fura-2-loaded human platelets. Evidence for Ca^{2+} entry regulated by the intracellular Ca^{2+} store. *Biochem. J.* **265,** 675–680.
6. Somasundaram, B. and Mahaut-Smith, M. P. (1995) A novel monovalent cation channel activated by inositol trisphosphate in the plasma membrane of rat megakaryocytes. *J. Biol. Chem.* **270,** 16,638–16,644.

7. Mason, M. J. and Mahaut-Smith, M. P. (2001) Voltage-dependent Ca²⁺ release in rat megakaryocytes requires functional IP₃ receptors. *J. Physiol.* **533,** 175–183.

8. Somasundaram, B., Norman, J. C., and Mahaut-Smith, M. P. (1995) Primaquine, an inhibitor of vesicular transport, blocks the calcium-release-activated current in rat megakaryocytes. *Biochem. J.* **309,** 725–729.

9. Rosado, J. A. and Sage, S. O. (2000) Protein kinase C activates non-capacitative calcium entry in human platelets. *J. Physiol.* **529,** 159–169.

10. Hassock, S. R., Zhu, M. X., Trost, C., Flockerzi, V., and Authi, K. S. (2002) Expression and role of TRPC proteins in human platelets: evidence that TRPC6 forms the store-independent calcium entry channel. *Blood* **100,** 2801–2811.

11. MacKenzie, A. B., Mahaut-Smith, M. P., and Sage, S. O. (1996) Activation of receptor-operated cation channels via P₂ₓ₁ not P₂ₜ purinoceptors in human platelets. *J. Biol. Chem.* **271,** 2879–2881.

12. Mahaut-Smith, M. P., Ennion, S. J., Rolf, M. G., and Evans, R. J. (2000) ADP is not an agonist at P2X₁ receptors: evidence for separate receptors stimulated by ATP and ADP on human platelets. *Br. J. Pharmacol.* **131,** 108–114.

13. Rink, T. J., Smith, S. W., and Tsien, R. Y. (1982) Cytoplasmic free Ca²⁺ in human platelets: Ca²⁺ thresholds and Ca-independent activation for shape-change and secretion. *FEBS Lett.* **148,** 21–26.

14. Hallam, T. J. and Rink, T. J. (1985) Responses to adenosine diphosphate in human platelets loaded with the fluorescent calcium indicator quin2. *J. Physiol.* **368,** 131–146.

15. Hallam, T. J., Daniel, J. L., Kendrick-Jones, J., and Rink, T. J. (1985) Relationship between cytoplasmic free calcium and myosin light chain phosphorylation in intact platelets. *Biochem. J.* **232,** 373–377.

16. Jin, J. and Kunapuli, S. P. (1998) Coactivation of two different G protein-coupled receptors is essential for ADP-induced platelet aggregation. *Proc. Natl. Acad. Sci. USA* **95,** 8070–8074.

17. Hechler, B., Eckly, A., Ohlmann, P., Cazenave, J. P., and Gachet, C. (1998) The P2Y₁ receptor, necessary but not sufficient to support full ADP-induced platelet aggregation, is not the target of the drug clopidogrel. *Br. J. Haematol.* **103,** 858–866.

18. Paul, B. Z., Daniel, J. L., and Kunapuli, S. P. (1999) Platelet shape change is mediated by both calcium-dependent and -independent signaling pathways. Role of p160 Rho-associated coiled-coil-containing protein kinase in platelet shape change. *J. Biol. Chem.* **274,** 28,293–28,300.

19. Shattil, S. J. and Leavitt, A. D. (2001) All in the family: primary megakaryocytes for studies of platelet αᵢᵢᵦβ₃ signaling. *Thromb. Haemost.* **86,** 259–265.

20. Behnke, O. (1968) An electron microscope study of the megacaryocyte of the rat bone marrow. I. The development of the demarcation membrane system and the platelet surface coat. *J. Ultrastruct. Res.* **24,** 412–433.

21. Behnke, O. (1969) An electron microscope study of the rat megacaryocyte. II. Some aspects of platelet release and microtubules. *J. Ultrastruct. Res.* **26,** 111–129.

22. Leven, R. M. (1987) Megakaryocyte motility and platelet formation. *Scanning Microsc.* **1,** 1701–1709.

23. Uneyama, C., Uneyama, H., and Akaike, N. (1993) Cytoplasmic Ca²⁺ oscillation in rat megakaryocytes evoked by a novel type of purinoceptor. *J. Physiol.* **470,** 731–749.

24. Ikeda, M., Kurokawa, K., and Maruyama, Y. (1992) Cyclic nucleotide-dependent regulation of agonist-induced calcium increases in mouse megakaryocytes. *J. Physiol. (Lond.)* **447,** 711–728.

25. Bootman, M. D., Lipp, P., and Berridge, M. J. (2001) The organisation and functions of local Ca²⁺ signals. *J. Cell Sci.* **114,** 2213–2222.
26. Berridge, M. J. (1997) The AM and FM of calcium signalling. *Nature* **386,** 759–760.
27. Nishio, H., Ikegami, Y., and Segawa, T. (1991) Fluorescence digital image analysis of serotonin-induced calcium oscillations in single blood platelets. *Cell Calcium* **12,** 177–184.
28. Heemskerk, J. W., Hoyland, J., Mason, W. T., and Sage, S. O. (1992) Spiking in cytosolic calcium concentration in single fibrinogen-bound fura-2-loaded human platelets. *Biochem. J.* **283,** 379–383.
29. Hussain, J. F. and Mahaut-Smith, M. P. (1999) Reversible and irreversible intracellular Ca²⁺ spiking in single isolated human platelets. *J. Physiol.* **514,** 713–718.
30. Mahaut-Smith, M. P. (1995) Calcium-activated potassium channels in human platelets. *J. Physiol.* **484,** 15–24.
31. van Gorp, R. M., Feijge, M. A., Vuist, W. M., Rook, M. B., and Heemskerk, J. W. (2002) Irregular spiking in free calcium concentration in single, human platelets. Regulation by modulation of the inositol trisphosphate receptors. *Eur. J. Biochem.* **269,** 1543–1552.
32. Nesbitt, W. S., Kulkarni, S., Giuliano, S., Goncalves, I., Dopheide, S. M., Yap, C. L., et al. (2002) Distinct glycoprotein Ib/V/IX and integrin $\alpha_{IIb}\beta_3$-dependent calcium signals cooperatively regulate platelet adhesion under flow. *J. Biol. Chem.* **277,** 2965–2972.
33. Heemskerk, J. W., Willems, G. M., Rook, M. B., and Sage, S. O. (2001) Ragged spiking of free calcium in ADP-stimulated human platelets: regulation of puff-like calcium signals in vitro and ex vivo. *J. Physiol.* **535,** 625–635.
34. Hussain, J. F. and Mahaut-Smith, M. P. (1998) ADP and inositol trisphosphate evoke oscillations of a monovalent cation conductance in rat megakaryocytes. *J. Physiol.* **511,** 791–801.
35. Tertyshnikova, S. and Fein, A. (1997) [Ca²⁺]ᵢ oscillations and [Ca²⁺]ᵢ waves in rat megakaryocytes. *Cell Calcium* **21,** 331–344.
36. Thomas, D., Mason, M. J., and Mahaut-Smith, M. P. (2001) Depolarisation-evoked Ca²⁺ waves in the non-excitable rat megakaryocyte. *J. Physiol.* **537,** 371–378.
37. Bootman, M. D., Berridge, M. J., and Lipp, P. (1997) Cooking with calcium: the recipes for composing global signals from elementary events. *Cell* **91,** 367–373.
38. Sun, X. P., Callamaras, N., Marchant, J. S., and Parker, I. (1998) A continuum of InsP₃-mediated elementary Ca²⁺ signalling events in Xenopus oocytes. *J. Physiol.* **509,** 67–80.
39. Tertyshnikova, S. and Fein, A. (1998) Inhibition of inositol 1,4,5-trisphosphate-induced Ca²⁺ release by cAMP-dependent protein kinase in a living cell. *Proc. Natl. Acad. Sci. USA* **95,** 1613–1617.
40. Johnson, P. C., Ware, J. A., Cliveden, P. B., Smith, M., Dvorak, A. M., and Salzman, E. W. (1985) Measurement of ionized calcium in blood platelets with the photoprotein aequorin. Comparison with Quin 2. *J. Biol. Chem.* **260,** 2069–2076.
41. Miyawaki, A., Llopis, J., Heim, R., McCaffery, J. M., Adams, J. A., Ikura, M., and Tsien, R. Y. (1997) Fluorescent indicators for Ca²⁺ based on green fluorescent proteins and calmodulin. *Nature* **388,** 882–887.
42. Nagai, T., Sawano, A., Park, E. S., and Miyawaki, A. (2001) Circularly permuted green fluorescent proteins engineered to sense Ca²⁺. *Proc. Natl. Acad. Sci. USA* **98,** 3197–3202.
43. Jackson, A. P., Timmerman, M. P., Bagshaw, C. R., and Ashley, C. C. (1987) The kinetics of calcium binding to fura-2 and indo-1. *FEBS Lett.* **216,** 35–39.
44. Pusch, M. and Neher, E. (1988) Rates of diffusional exchange between small cells and a measuring patch pipette. *Pflugers Arch.* **411,** 204–211.

45. Rolf, M. G., Brearley, C. A., and Mahaut-Smith, M. P. (2001) Platelet shape change evoked by selective activation of P2X$_1$ purinoceptors with α,β-methylene ATP. *Thromb. Haemost.* **85,** 303–308.

46. Grynkiewicz, G., Poenie, M., and Tsien, R. Y. (1985) A new generation of Ca²⁺ indicators with greatly improved fluorescence properties. *J. Biol. Chem.* **260,** 3440–3450.

47. Poenie, M. (1990) Alteration of intracellular Fura-2 fluorescence by viscosity: a simple correction. *Cell Calcium* **11,** 85–91.

48. Mahaut-Smith, M. P. (1998) An infra-red light-transmitting aperture controller for use in single-cell fluorescence photometry. *J. Microsc.* **191,** 60–66.

49. Tsien, R. Y. and Harootunian, A. T. (1990) Practical design criteria for a dynamic ratio imaging system. *Cell Calcium* **11,** 93–109.

50. Gurney, A. M. (1994) Flash photolysis of caged compounds, in *Microelectrode Techniques. The Plymouth Workshop Handbook* (Ogden, D. A., ed.), The Company of Biologists, pp. 389–406.

17

Patch-Clamp Recordings of Electrophysiological Events in the Platelet and Megakaryocyte

Martyn P. Mahaut-Smith

1. Introduction

Ion channels have fundamental roles in all cell types. Although fluorescent indicators can provide an indirect assessment of membrane conductances—for example, by studies of ion concentrations or membrane potential—our understanding of ion-channel activity ultimately relies on data from electrophysiological techniques. The first electrophysiological recordings from the megakaryocyte/platelet lineage used conventional microelectrode impalements of guinea pig marrow megakaryocytes (*1*). Membrane potentials in normal extracellular levels of Ca^{2+} were low, in the order of -14 mV, which probably reflects damage caused by microelectrode impalement since later estimates from patch clamp recordings are far more negative: approx -45 mV at rest up to approx -80 mV during activation (*2,3*). Nevertheless, this first electrophysiological study by Miller et al. (*1*) provided evidence for K^+ conductances that lead to membrane hyperpolarizations and thereby increase the driving force for Ca^{2+} entry during cell signaling. Since its advent in the late 1970s, the patch clamp technique (*4,5*) has revolutionized our ability to directly record electrophysiological events, particularly in small cells. The initial feeling of many scientists in the 1980s was that meaningful patch-clamp recordings from mammalian platelets could never be achieved due to platelets' small and fragile nature. This led to several alternative approaches. For example, thrombocytes from lower species such as the newt, *Triturus pyrrhogaster*, are larger in diameter and were used by Kawa (*6*) to demonstrate the existence of a voltage-dependent outward K^+ conductance. Single-channel recordings from lipid bilayers following insertion of platelet membranes have also provided evidence for chloride channels (*7*) and thrombin-activated cation channels (*8*) in the platelet plasma membrane. In 1987, Maruyama (*9*) surprised many in the field by reporting the first patch-clamp recordings from rat, rabbit, and human platelets. In this seminal study, he described the passive electrophysiological properties of the platelet and characterized a voltage-dependent outward K^+ conductance that was

From: *Methods in Molecular Biology, vol. 273:*
Platelets and Megakaryocytes, Vol. 2: Perspectives and Techniques
Edited by: J. M. Gibbins and M. P. Mahaut-Smith © Humana Press Inc., Totowa, NJ

later shown to be the main determinant of the negative platelet resting membrane potential *(10)*. Agonist stimulation of platelets in the study by Maruyama led to a loss of stable recording *(9)*; however, in 1990, agonist-evoked membrane currents were reported in human platelets following activation by commercial samples of ADP *(11)*. The nonselective cation channels underlying this response were later identified as $P2X_1$ receptors and are in fact stimulated by ATP (or diadenosine polyphosphates) rather than by ADP *(12,13)*.

Patch clamp recordings have also been reported from mouse, rat, guinea pig, and human primary marrow megakaryocytes *(12,14–18)*. Although a number of megakaryocytic cell lines have been recorded from with this technique, it is clear that the ion-channel phenotype of continuous cell lines is substantially different from that of the primary cell *(14,19–22)*. As a consequence of their much larger size, megakaryocytes have taken over from platelets as the choice for electrophysiological studies in this cell lineage. Furthermore, specific experiments that cannot be performed in the platelet are quite feasible in the megakaryocyte. A good example is the direct study of I_{CRAC} (Ca^{2+}-release-activated Ca^{2+} current), one candidate for the Ca^{2+} current underlying store-dependent Ca^{2+} influx. This current is at or below the limit of resolution in the platelet (M. P. Mahaut Smith, unpublished observations), but can be easily resolved in the megakaryocyte *(17,23)*. This raises an important question about the limitations of the precursor cell as a surrogate for studies of signaling in the platelet. More comparisons of ion-channel activity in platelets and megakaryocytes are required to fully address this issue. However, all platelet channels should be present in the mature megakaryocyte, as the anuclear platelet relies on its precursor cell for the production of most, if not all of its proteins.

The patch-clamp technique depends upon the formation of a very-high-resistance ($>10^9$ gigaohm [GΩ]) seal between a glass pipet and the cell membrane. The high ratio of seal-to-pipet resistance results in virtually all of the ionic current through the accessed cell membrane being faithfully recorded by the amplifier under voltage clamp mode. The high resistance also reduces thermal (Johnson) noise, which is a major source of noise in these types of recordings, and thereby allows single-channel events to be resolved. Patch-clamp amplifiers operate in essentially two modes: voltage clamp and current clamp. In the vast majority of patch-clamp experiments, voltage clamp is used to hold the pipet potential constant and measure the current flowing across the membrane. Most commercial amplifiers can also operate in "current clamp" mode, in which the potential at the tip of the pipet is monitored. The cell membrane potential is measured when no current flows from the amplifier, although responses to injected current can also be assessed. For excellent theoretical explanations of the patch-clamp technique and further discussion of equipment, see **refs. *24–26***.

A variety of different patch-clamp configurations have been developed over the last 20 or so years (*see* **Fig. 1**); each has distinct advantages and disadvantages. All recordings commence in the cell-attached configuration in which the pipet tip isolates a small patch of membrane on an intact cell. This is clearly the most noninvasive configuration; however, it has the least control over the transmembrane potential and composition of internal and external salines. Single-channel events can be studied in this mode,

Rpip : 1-10 MΩ

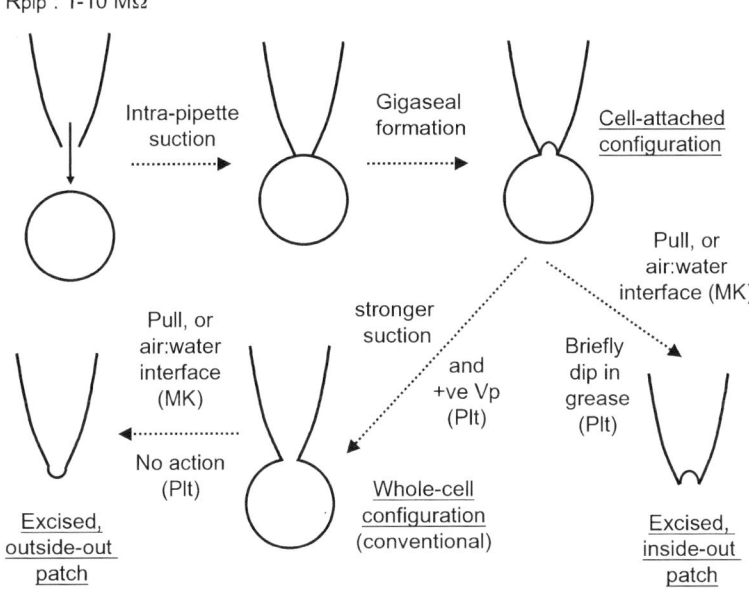

Fig. 1. Conventional patch-clamp configurations. Text beside the dotted lines indicates actions to progress between configurations, with particular requirements for megakaryocytes (MK) or platelets (Plt). As a consequence of the small size of the mammalian platelet, a whole-cell recording is essentially an excised outside-out patch. Rpip, pipet resistance. Vp, pipet voltage.

although the channel of interest may not always be present in the patch. It should also be noted that in small cells with high input resistance such as the platelet, the selective opening of a single channel in the cell-attached membrane patch can significantly alter the membrane potential in the remainder of the cell, leading to non-steady-state single-channel currents *(11,27)*. The membrane patch under the pipet can be physically ruptured or chemically permeabilized, resulting in the whole-cell configuration in which the pipet has access to the intracellular side of the plasma membrane. In megakaryocytes, suction or the pore-forming antibiotic nystatin have been used as techniques to routinely generate whole-cell recordings *(2,18,28)*. Nystatin has the advantage that it reduces or even eliminates rundown of ionic currents since the pores are permeable only to monovalent cations or uncharged molecules below approx 200 molecular weight, but has the disadvantage that it cannot be used to introduce many compounds into the cell *(see* **Fig. 2***)*. In platelets, formation of whole-cell recordings presents particular difficulties due to the small size of the pipet and probably also the interaction of the cytoskeleton with plasma membrane proteins. Conventional breakthrough by suction has been achieved, but resealing was a problem *(10)*; thus, nystatin-perforated patch-clamp recordings are preferable *(29,30)*, although a suggested alternative is addition of EGTA

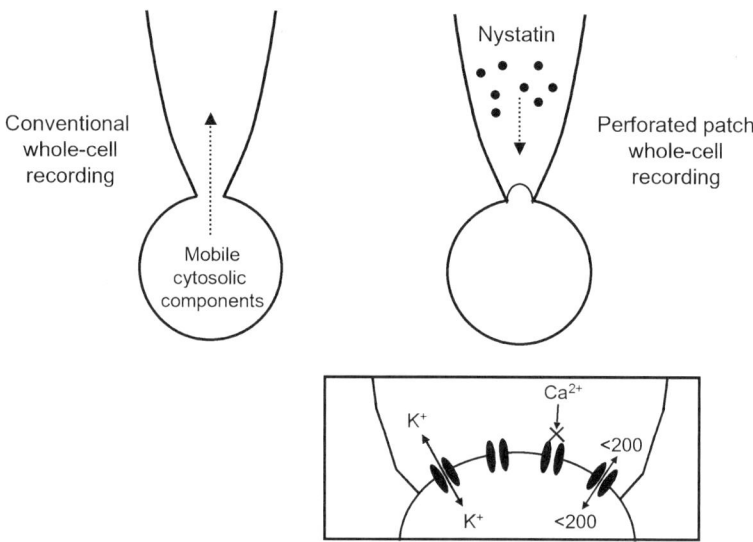

Fig. 2. Comparison of conventional and nystatin perforated patch whole-cell configu-rations. Physical disruption of the patch in conventional whole-cell mode results in loss of mobile cytosolic components into the pipet. Including a high concentration of nystatin in the pipet allows electrical access to the cell but limits dialysis of cytosolic components. The nystatin pores are permeable to monovalent cations (and to a lesser extent Cl⁻ *[40]*) but not to divalent cations or uncharged molecules with a molecular weight above approx 200 Da.

(no added Ca^{2+}) and ATP to the pipet *(9)*. Exactly how the latter promotes formation of whole-cell recordings is unknown. The whole-cell mode has the advantage that ion chan-nels through virtually the entire cell can be monitored, and agonists can be reversibly superfused over the extracellular surface. The conventional whole-cell recording mode represents an extremely useful way of manipulating the intracellular environment, for example using modulators of cell signaling. Larger-molecular-weight compounds can be introduced (e.g., peptides *[31]* and heparin *[3]*); however, dialysis depends on the access resistance, determined mainly by the size of the pipet tip and the molecular weight *(32)*. Whole-cell patch-clamp also provides an important method of introducing fluorescent Ca^{2+} indicators into the megakaryocyte as incubation with the acetoxymethylester deriv-atives often leads to overbuffering of Ca^{2+} responses *(33)*.

If the cell-attached patch is pulled away from the cell, this results in the excised inside-out patch configuration, allowing control over the intracellular environment, and is particularly useful for studies of the second messenger-dependence to channel activ-ity. The conventional method of forming excised inside-out patches is to pull the pipet away from a cell-attached recording while the cell is adhered to the coverslip, or to briefly pass through the air : saline interface. These techniques may work for megakaryocytes but to date have not been reported. However, in platelets, which must be recorded from in suspension, brief exposure to the air:saline interface destroyed the recording. Therefore, Mackenzie and Mahaut-Smith *(34)* developed the silicon-grease-trap technique, in which the platelet in cell-attached mode is briefly inserted into a small drop of silicon grease at the base of the chamber under low-power magnification. The platelet is trapped within the grease, yielding an excised patch when the pipet is withdrawn. This approach has been used to record Cl⁻ and K⁺ channels in excised platelet membrane patches *(34,35)*. The main disadvantage to the excised-patch configurations in many preparations is irreversible rundown or alteration of channel activity.

In many cell types, pulling the pipet away from the cell after formation of a whole-cell recording results in formation of an outside-out patch. This is the equivalent of a reversed inside-out patch, that is, the extracellular surface of the patch faces the bath saline. It is particularly useful for studying agonist-evoked single-channel events, but suffers from the same rundown problems as inside-out recordings. This mode has not been described in the megakaryocyte and is of no great advantage in the platelet (if it could ever be achieved), since the small size of the latter means that single-channel events can be resolved in whole-cell recordings *(30)*. Further, specialized modes of patch-clamp recording have been reported from other cell types, such as nucleated excised-patch recordings, nystatin-permeabilized vesicles, patch amperometry, and single-channel recordings from organellar membranes. The field of electrophysiology in platelets and megakaryocytes is still very much in its infancy. The size extremes of these two cells and their specialized nature provides some difficulties and yet some unique situations. Recently, the entire platelet-forming demarcation membrane system of the megakaryocyte has been shown to be electrically contiguous with the plasma membrane such that whole-cell capacitance can be used to quantify this important platelet-forming membrane system *(36)*.

2. Materials
2.1. Salines and Reagents

1. Acid citrate dextrose (ACD) anticoagulant: 85 mM trisodium citrate, 78 mM citric acid, 111 mM glucose.
2. Apyrase (Sigma, St. Louis, MO type VII) at a stock of 320 U/mL in H_2O; aliquot and store at –20°C.
3. Aspirin (acetylsalicylic acid, Sigma, A-5376), stock 100 mM in ethanol.
4. Agonists (e.g., ATP, ADP, thrombin). Commercial samples of nucleotides are normally contaminated by small amounts of related compounds (e.g., 1–2% ATP is present in ADP from Sigma). If these contaminants are of concern, they can be removed by HPLC or, in some situations, reduced to insignificant levels by enzymatic treatment. For example, ATP can be removed by incubation of a 10-mM stock of ADP with 3 U/mL hexokinase (Boehringer Mannheim, hexokinase from yeast) in SES (*see* **item 5**) containing 22 mM glucose and retitrated to pH 8.0 (*see* **Note 1**). The use of creatine phosphate/creatine phosphokinase to reduce levels of ADP within commercial ATP has also been reported *(37)*.
5. Standard extracellular saline (SES): 145 mM NaCl, 5 mM KCl, 1 mM CaCl$_2$, 1 mM MgCl$_2$, 10 mM D-glucose, 10 mM HEPES, titrated to pH 7.35 with NaOH. The CaCl$_2$ is omitted to yield nominally Ca^{2+}-free SES. For ionic selectivity experiments, use equimolar substi-

tution of salts. For example, to assess monovalent cation permeability, replace Na+ and K+ by an equal concentration of the large impermeant cation *n*-methyl-D-glucamine (NMDG). NMDG is added from the free base and solution pH titrated with HCl to yield NMDGCl. NMDGCl-based saline is also used to record I_{CRAC} (150 m*M* NMDGCl, 2 m*M* CaCl$_2$, 1 m*M* MgCl$_2$, 10 m*M* HEPES, pH 7.35 with NMDG base or HCl) as a monovalent conductance is activated alongside this store-dependent current in megakaryocytes *(17,23)*. Gluconate⁻ is a reasonable substitute for Cl⁻ but due to contaminants in commercial gluconate solutions, use gluconic acid lactone. Mix in solution with the required base (e.g., NaOH to yield Nagluconate) and stir for 48 h to ensure that the species is fully dissolved as gluconate. Gluconate displays signficant Ca^{2+}-buffering capacity; thus additional CaCl$_2$ should be added if a constant free Ca^{2+} concentration is required (*see* **Note 2**).

6. Standard pipet saline (SpipS) for conventional whole-cell recordings: 150 m*M* KCl, 2 m*M* MgCl$_2$, 0.1 m*M* EGTA, 0.05 m*M* Na$_2$GTP, 10 m*M* HEPES, titrated to pH 7.2 with KOH (*see* **Note 3**). This is filtered through 0.22-μm filters, aliquoted, and stored at –20°C. The Na$_2$ GTP is usually added to a thawed aliquot at the start of a period of recording. For recording of nonselective cation currents (via P2X$_1$ or the P2Y receptor-associated current in megakaryocytes *[17,28]*), use equimolar substitution of CsCl for KCl to block K+ currents (titrated to pH 7.2 with CsOH) . For recording of I_{CRAC}, the pipet saline contains 100 m*M* CsCl, 10 m*M* Cs$_4$BAPTA, 2 m*M* CaCl$_2$, 1 m*M* MgCl$_2$, 10 m*M* HEPES, titrated to pH 7.2 with CsOH (*see* **Note 4**). Aliquots of this solution are then stored at –20°C.

7. Nystatin (Sigma, N 3503) at a stock of 50 mg/mL in DMSO, for perforated whole-cell patch clamp recordings. Nystatin is highly light-sensitive and should be kept in the dark. Fresh aliquots are made about every 2 h to maximize nystatin activity. It is diluted into the internal saline 500X immediately before attempting to patch.

2.2. Cell Preparation

2.2.1. Human Platelets

1. Sample of fresh human blood (*see* Chapters 1 and 2, vol. 1 for advice on phlebotomy techniques designed to limit platelet activation).
2. 15-mL plastic centrifuge tubes.
3. Plastic transfer pipets.
4. 1.5-mL Eppendorf tubes.

2.2.2. Rodent Megakaryocytes (see **Note 5**)

1. Dissection instruments, including bone-cutting forceps and dental excavator.
2. For murine megakaryocytes, a "flushing device" consisting of a short section of 1-mm-bore silicone tubing attached via a shortened yellow pipettor tip to a 2- or 5-mL plastic syringe.
3. Plastic transfer pipets.
4. 1.5-mL Eppendorf tubes.
5. Plastic Petri dishes (mouse: 35-mm; rat: 50-mm diameter).

2.3. Electrophysiological Equipment

1. Patch-clamp amplifier, supplied with pipet holder and model cell (e.g., Axopatch 200B, Axon Instruments, Foster City, CA) (*see* **Notes 6,7**). The pipet holder has a side port that is connected to a length of silicone tubing, then to a three-way tap and 1-mm syringe used for pressure application. We discard the syringe plunger and apply suction by mouth (*see* **Note 8**).

A chlorided silver wire electrically couples the gold pin at the back of the pipet holder to the pipet saline and can be easily chlorided by dipping in bleach for an hour or more.

2. Computer acquisition system (e.g., Digidata series of digitizers and pClamp acquisition software, Axon Instruments) (*see* **Note 9**).

3. Oscilloscope (basic 20-MHz digital storage model is fine). Almost made obsolete by **item 2**, but it can be useful.

4. Inverted microscope (e.g., Nikon Diaphot) with remote dc light power supply, movable stage, 40× and 100× objectives, 10× eyepieces, and ideally a 4× magnifier lens in the eyepiece turret.

5. Air-damped antivibration table for microscope. A variety of these are available. Ours are from Ealing Electro-Optics and have both horizontal and vertical antivibration characteristics. They are also supplied with a 60-mm-thick "breadboard," which has additional antivibration properties.

6. Variable 8-pole low-pass Bessel electronic filter (most commercial types cover the required range, approx 0.4–10 kHz; e.g., from Frequency Devices Inc., Haverhill, MA). This is a useful addition to the limited range of the internal filters of most patch-clamp amplifiers. *See* **Note 10** for guidance on correct settings.

7. Micromanipulator with coarse and fine controls in three dimensions. The fine controls must be capable of being operated remotely (e.g., a hydraulic manipulator from Narishige Group (Japan) or piezoelectric manipulator from Luigs & Neumann, Ratingen, Germany). The manipulator holds the headstage of the patch-clamp amplifier, and thus also the patch pipet, and is fixed to the body of the microscope or the antivibration table, rather than the microscope movable stage, to allow movement independently of the electrophysiological chamber.

8. Pipet puller: The pipet puller should be capable of at least two stages of pull (e.g., Narishige PP-series of pullers, Narishige Group, or a Flaming/Brown type, Sutter Instrument Co., Novato, CA) allowing a pipet with short shank and tips of 0.5–3 µm diameter.

9. Microforge for fire-polishing of glass tips (e.g., Narishige Group). A homemade microforge can be constructed from a platinum-iridium wire connected to a variable 10-V 10-A power supply and mounted on an inverted microscope with long working distance objectives *(24)* (*see* **Note 11**).

10. Sylgard® curing rig. The power supply described in **item 6** can also be used to heat a coil of metal wire for curing of Sylgard-coated pipets.

11. Electrophysiological chamber and perfusion system. A variety of chambers with perfusion and temperature control are available commercially (Digitimer Ltd., Welwyn Garden City, Herts, UK, or Warner Instruments Inc., Hamden, CT). Our chambers are built in-house by milling a 15 × 8 mm diameter hole through a 3-mm Perspex plate with holes drilled in the side of the plate for ground (*see* **Note 12**) and solution inflow. Inflow is gravity-fed via a 6-into-1 Perspex interface, and switching is controlled manually by taps at the solution reservoirs (*see* **Note 13**). Outflow is achieved via a tube fixed to the top edge of the hole, connected to a vacuum pump to remove saline. 32 × 32 mm square No 1 glass coverslips are sealed to the underside of the Perspex plate with silicone grease to form the base of the chamber. Most patch-clamp experiments on platelets and megakaryocytes are conducted at the ambient temperature. Although far from the mammalian physiological temperature, this is a common occurrence for patch-clamp studies, as recordings are more stable at normal room temperatures compared to 35–37°C. Recently, we have achieved stable recordings from rat megakaryocytes at 36°C (J. Martinez-Pinna and M. P. Mahaut-Smith, unpublished).

12. Grounding and general connections: To shield the recording from electrical noise, a stand-alone Faraday cage surrounds the microscope and antivibration table. Its sides can be made from aluminium mesh, although ours are made of aluminium sheet, painted black, as this allows the preparation to also be shielded from light for simultaneous fluorescence recordings (*see* Chapter 16, vol. 2. The Faraday cage, solution lines, microscope, manipulators, and other peripheral equipment inside the cage are electrically connected to the main ground (we use the amplifier signal ground) (*see* **Note 14**).

13. Bath ground lead: A chlorided silver wire or silver/silver chloride pellet (Harvard Apparatus Ltd., Edenbridge, Kent, UK) is used for the ground connection and is connected to the headstage ground (*see* **Note 15**). Regular chloride-coating of the pellet or wire (*see* **item 1**) is recommended for maintaining a good junction (*see* **Note 16**).

2.4. Patch-Clamp Accessories

1. 1.5-mm outer diameter, 0.86-mm inner diameter filamented borosilicate glass capillaries (GC150F-10; Harvard Apparatus Ltd.). Prior to use, heat the ends of each piece of glass tubing in a flame to eliminate sharp edges. This facilitates insertion of the pulled pipet into the headstage holder and prevents removal of the chloride coating on the silver wire.

2. Pipet storage box with lid for dust protection (e.g., food storage box with pipets held by a strip of Blutack®).

3. "Bubble number tester" (*see* **Note 17**): consisting of a 10-cm³ syringe and an arrangement of tubing (~5 cm total length) that allows pressure to be applied to the unpulled end of a patch pipet. There should be no air leaks or volume expansion of the tubing when pressure is applied.

4. 28-gauge Microfil® nonmetallic syringe needles for filling pipets (MF-28G, World Precision Instruments, Stevenage, Herts, UK).

5. Dow Corning Sylgard 184 Silicone Elastomer kit (Farnell, Leeds, UK) or wax candles. Sylgard can be made in advance and stored at –20°C for several weeks. After mixing elastomer and curing agent at a ratio of 10:1 (v/v), and allowing bubbles to escape, aliquot into Eppendorf tubes.

6. High-vacuum grease or silicone grease (Dow Corning [Allesley, Coventry, UK] or RS Components [Corby, Northants, UK]).

3. Methods
3.1. Cell Preparation (see Note 18)
3.1.1. Human Platelets

1. Anticoagulate a fresh sample of venous blood by gently mixing with ACD (6 parts blood to 1 part ACD) and spin immediately at 700*g* for 5 min. Ideally blood should be drawn directly into the anticoagulant. Five to 10 mL of blood provides adequate platelets for a day of experimentation.

2. Remove the platelet-rich plasma (PRP; upper layer) using a plastic Pasteur pipet and transfer to a plastic container.

3. Add aspirin (100 µ*M*) and apyrase (0.32 U/mL) and mix by gentle inversion. Aspirin would be omitted for studies into the effects of thromboxane generation.

4. Transfer 0.5–1 mL aliquots of PRP into 1.5-mL Eppendorf tubes and place on a rotor running at approx 0.2 revolutions per second, which prevents the platelets from settling and thus reduces the risk of spontaneous activation.

5. Prepare a washed platelet suspension for patch-clamp by spinning at low speed in a microcentrifuge for approx 20 s. It is important to use the minimum velocity and duration of spin required to pellet the cells and allow removal of plasma. These settings are best derived empirically for each centrifuge. Remove as much plasma as possible (*see* **Note 19**) and gently resuspend the platelets in nominally Ca^{2+}-free SES with 0.32 U/mL apyrase (*see* **Note 20**).

3.1.2. Rodent Megakaryocytes

1. Euthanize adult rats or mice according to national and local regulations.
2. Dissect out the femoral and tibial bones intact, removing as much attached muscle or other tissue as possible, and immerse in a Petri dish containing SES with 0.32 U/mL apyrase (SES:apyrase).
3. After approx 15 min, wipe the bones with dry tissue to further remove attached tissues.
4. For rat marrow, split open longitudinally using a pair of bone cutters into a new Petri dish with 10 mL SES:apyrase. Separate the marrow from the bone using a blunt-ended dental excavator, removing as many bone fragments as possible.
5. For mouse marrow, cut the ends off the bones and use a syringe filled with approx 2 mL SES:apyrase, connected to a short length of silicone tubing, to flush out the marrow. Use the minimum amount of saline for each bone.
6. Aliquot clumps of marrow and cell suspension into 1.5-mL Eppendorf tubes and place on a rotor at 0.2 Hz. Normally the cell suspension initially contains mainly activated megakaryocytes. Nonactivated megakaryocytes with intact plasma membranes appear 1–3 h later (*see* **Fig. 3**), possibly following further separation of cells from the marrow. Therefore, ensure that each aliquot contains at least one clump of undissociated marrow. Megakaryocytes prepared in this way are amenable for patch-clamp experiments up to 16 h after marrow isolation. Although megakaryocytes can be grown in culture, more sterile conditions are required during isolation, and marked morphological events such as proplatelet formation can occur.

3.2. Patch-Pipet Manufacture

1. Pull pipets with an external tip diameter and filled internal resistance suitable for either platelets or megakaryocytes. As a rough guide, platelets require tips with internal diameters of about 0.5 μm (resistance 5–10 MΩ; *see* **Note 17**) and whole-cell recordings from megakaryocytes require tips of 2–3 μm (resistance 1–2 MΩ). On a two-stage Narishige puller, a first step at a high coil heat is used to thin the glass over a length of approx 8 mm, after which the pipet is allowed to cool and the coil moved to a position over the thinned section that generates two pipets of the same shank length following the second pull. The second heat is lower than the first, and is adjusted to alter the tip diameter. A higher coil temperature also increases the length of the shank and therefore it may be necessary to increase the length of the first pull if finer tips are required, or use a multistage puller (e.g., Flaming Brown type from Sutter Instruments).
2. (Optional) To improve noise levels and reduce stray capacitance, coat the pipet shank with Sylgard or wax. The settings of some amplifiers (e.g., Axopatch 200 series) are sufficient to eliminate pipet and stray capacitance without such coatings. Reduced noise levels are helpful when recording single-channel events of low conductance or for kinetic analysis of brief events, since the lower noise levels allow the recording bandwidth to be increased. Spread the Sylgard over the entire pipet shank as close to the tip as possible, using a hypodermic needle as an applicator, with the aid of a dissecting microscope. The Sylgard is cured by passing through a small (5 mm diameter) heated wire coil, prior to fire polishing.

Molten wax from a candle is applied in a similar way but hardens on its own and is applied after fire-polishing (i.e., after **step 3**) as the filament heat will remelt the wax.

3. Fire-polish pipet tips using a microforge. Fire-polishing is essential for formation of high-resistance seals on megakaryocyte membranes, but is not necessary for human platelets. Under 150–400× magnification, bring the pipet tip to 20–50 μm from the filament. Heat the filament for 2–3 s by passing current so it approaches the pipet and polishes the glass. "Light" polishing will not produce marked changes in the appearance of the tip unless viewed under very high magnification and good optics. "Heavy" polishing will physically reduce the internal tip diameter. For megakaryocytes we use medium to heavy polishing. It is best to determine the settings for each fire-polishing rig empirically from (a) the bubble number before and after polishing and (b) the ease with which gigaseals are formed.

3.3. Cell-Attached Patch-Clamp Recordings from Human Platelets

1. Evenly distribute 10–20 μL of washed platelet suspension throughout the electrophysiology chamber filled with nominally Ca^{2+}-free platelet saline.

2. Monitor the platelets at ×400 magnification as they settle towards the base of the chamber. Once a reasonable density of platelets is observed above the coverslip, perfuse nominally Ca^{2+}-free platelet saline for a period of 30 s to a few minutes. The perfusion rate at the base of the chamber is slowest; therefore by adjusting the flow rate, the number of platelets added, and the duration of perfusion, a layer of platelets can be left floating just above the coverslip (*see* **Note 21**).

3. Half-fill a patch pipet with saline (composition depending on type of experiment) with the aid of a Microfil® needle, insert into the holder attached to the headstage and lower the tip into the chamber. The patch clamp amplifier should be under voltage clamp mode and at ≥5 kHz bandwidth, with an output gain of approx 1 mV/pA (if using an Axopatch 200 series amplifier, the amplifier β gain should be "1"). All other options, such as whole-cell capacitance, series resistance correction, leak correction, and holding potential commands, should be off. Adjust the pipet offset so the output membrane current is zero. At this stage, the combination of all voltage offsets in the circuit is zero; thus no current is required to hold the pipet voltage at 0 mV. Apply a 5- to 10-mV square wave pulse, approx 20 ms duration at a frequency of approx 10 Hz. The resistance (in ohms) of the pipet can be calculated from Ohm's law ($R = V/I$), where I (in amperes) is the amount of current required to set the test voltage, V (in volts). As a rough guide, it should be 5–10 MΩ for platelets, but it is probably more useful to use the bubble number as a guide (number range 5–5.5; *see* **Note 17**). Most electrophysiological software packages can apply a test pulse and calculate the resistance automatically, since the acquisition system and amplifier have telegraph connections for detection of amplifier gain settings. A high pipet resistance may be due to bubbles trapped in the shank, which, can be dislodged by removal of the pipet and light filliping of the glass. Large drifts in the current level suggest an unstable ground and/or pipet connector, which should be replaced or rechlorided.

4. Position the pipet so its tip is 25–50 μm above the layer of platelets. A floating, nonactivated platelet is selected and the tip of the pipet maneuvered to within approx 5–10 μm at the same focal plane (*see* **Note 22**). Gentle positive pressure is used to prevent other platelets and cells from approaching the pipet tip. The chances of seal formation are greatly reduced if any cell or other particulate matter even briefly touch the pipet tip prior to attaching the selected platelet. Once the tip is directly opposite the selected platelet, suction is applied. The platelet is drawn onto the tip of the pipet and the resistance of the recording increases. Suction is released when the resistance has reached the multigigaohm range.

For unknown reasons, application of a negative holding potential up to -70 mV can aid formation of a gigaseal. This potential is usually applied after an initial increase in resistance to at least 100 MΩ. Seals can form quite slowly (over a matter of minutes); however, by far the best recordings are obtained when the gigaseal forms instantly and rapidly (*see* **Note 23**). At the normal starting gain of the amplifier, the current record will be a flat line with small capacitative transients at the start and end of the voltage test pulse. Once a gigaohm resistance seal has formed, the amplifier gain can be increased and the quality of the gigaseal measured more accurately (*see* **Note 24**). The capacitative transients are due to a combination of pipet and stray capacitance and can be cancelled using the pipet fast and slow capacitance dials (see also comment above about Sylgard or wax coating to reduce stray capacitance). To record single-channel events, the gain should be $\geq$50 mV/pA for a good signal:noise ratio. Some amplifiers (e.g., List EPC series) have a larger feedback resistor, and thus lower noise, at gains of $\geq$50 mV/pA; therefore, this range should be selected for recording single-channel events.

5. The cell-attached mode of recording can easily be used to monitor channels activated by changes in membrane voltage or in response to agonists and agents that generate diffusible second messengers *(10,30)*. It is less straightforward to measure ionotropic events, e.g., P2X receptors. However, P2X$_1$ receptor-evoked events were recorded in cell-attached patches from platelets by filling the pipet tip with normal saline and adding the agonist (usually at a high concentration) to the back of the pipet *(11)*. The pipet tip is filled simply by dipping in agonist-free saline for 20–30 s.

3.4. Excised Inside-Out Patch-Clamp Recordings from Platelets

1. Prior to addition of saline to the chamber, place a small (~2 mm diameter) drop of silicone grease onto the upper surface of the coverslip.
2. After formation of a cell-attached patch, switch the objective to lower power (10×). Bring the drop of silicone grease into the field of view and position the pipet directly above or opposite the center of the grease.
3. In one continuous movement, manipulate the pipet tip into the grease and then back into the saline.
4. Switch back to high magnification and if the platelet has not become detached from the membrane patch, repeat the previous step. Ideally, the resistance to ground should not deteriorate significantly, although some decrease in seal resistance can be tolerated. Record single-channel events at a gain of $\geq$50mV/pA. This mode of patch clamp is particularly useful for assessing the effects of second messengers; however, the channel of interest may not be present in the patch. The extent to which the open canalicular system is accessed during platelet electrophysiological recordings is unknown.

3.5. Conventional Whole-Cell Recordings from Platelets

1. After formation of a cell-attached patch, hold the pipet potential positive ($+40$ to $+80$ mV) and apply further gentle suction. With standard internal and external salines, a large outward current will appear as the whole-cell mode is achieved, due to activation of the many voltage-dependent K$^+$ channels in the platelet.
2. To study depolarization-activated voltage-dependent K$^+$ channels, the resting potential is initially held at potentials of -80 mV or more hyperpolarized and then depolarizing ramps or steps used to activate this conductance *(9,10)*. However, in normal salines, the membrane patch may show a tendency to reseal. This resealing can be prevented by brief exposures

to a positive potential (e.g., 500 ms duration ramps) at regular intervals (every 30 s–1 min) *(10)*. Maruyama *(9)* has also suggested that stable whole-platelet recordings can be obtained by adding EGTA (no added Ca^{2+}) and ATP to the pipet saline.

3. Record at a gain suitable for the experiment, attempting to fill as much of the digitizer voltage range (usually ±10 volts) as possible. If the gain is too low, single-channel events can be hidden within the digital bit noise; if too high, large conductances may saturate the A/D board.

4. This mode of patch clamp can be used to study whole-cell K^+ currents following their activation by depolarizing ramps or voltage steps *(9,10)*, normally with high K^+ in the internal saline. For unknown reasons, the pipet filling solutions commonly used to record single-channel currents through divalent cation conductances in the cell-attached or excised mode (e.g., 110 mM $BaCl_2$ (or $MgCl_2$), 10 mM HEPES, pH 7.35) allow relatively easy and stable transition to the whole-platelet patch mode *(38)* and lead to activation of chloride channels.

3.6. Gigaseal Formation and Conventional Whole-Cell Recordings From Primary Rat Megakaryocytes

1. Evenly distribute 10–50 μL of marrow suspension over the base of the electrophysiological chamber filled with SES (adjust volume of suspension depending on density of megakaryo cytes and other cells).

2. As soon as a reasonable number of cells has settled, superfuse at a low rate and adjust the flow rate and time to obtain a suitable cell density above the coverslip. Do not leave for a long time as megakaryocytes will stick to the coverslip, which results in an increased chance of activation. Coating the glass with Sylgard, as for platelets, can reduce adherence to the coverslip (*see* **Note 21**).

3. Scan the cells and select a megakaryocyte that appears healthy. Megakaryocytes are clearly distinguishable from other marrow cells by their large size **(Fig. 3)** (*see* **Note 25**).

4. Half-fill a pipet with SpipS, attach to the holder and thus to the headstage, apply a 5-mV square wave test pulse under voltage clamp at an amplifier gain of 0.2–0.5 mV/pA, and lower into the chamber (if using an Axopatch amplifier, the β gain should be "0.1"). Pipets for whole-cell recordings should have resistances of 1–2 MΩ.

5. Adjust the current level at 0 mV to be zero (i.e., the same as "ground") using the pipet offset.

6. Manipulate the pipet to within 10 μm of the selected megakaryocyte.

7. Move the pipet closer to the megakaryocyte and at the same time apply a very gentle suction (*see* **Note 26**); if it is not attached to the coverslip, the megakaryocyte will leap onto the tip of the pipet. If it is attached, push the tip lightly onto the megakaryocyte plasma membrane.

8. Often the resistance to ground will instantaneously reach a value of >10 GΩ. If not, continue to apply sustained gentle suction, starting with the lightest possible suction and increasing very gradually. A negative holding potential (up to –70 mV) can promote gigaseal formation.

9. After formation of the gigaseal, raise the megakaryocyte off the coverslip. (If the cell is firmly attached, it can be left there, but the manipulators must be drift-free and the set-up vibration-free. This is also not ideal as megakaryocytes can spread on the coverslip when activated.)

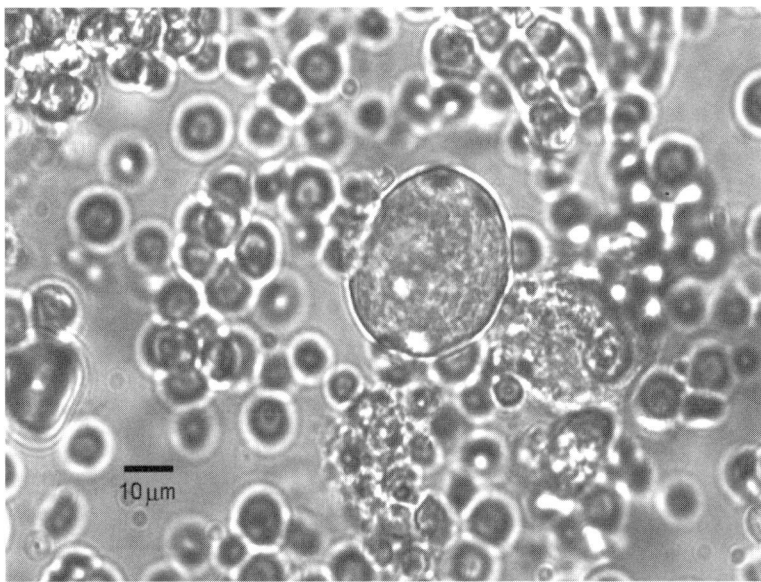

Fig. 3. Typical appearance of cells from rat marrow. A large megakaryocyte in good condition can be seen in the center. To its right is an "activated" megakaryocyte, typical of those frequently seen immediately after marrow dissociation.

10. The megakaryocyte is now in the cell-attached configuration. Cancel stray capacitance transients (*see* **Subheading 3.3.**). Apply a holding potential of −70 to −80 mV to the pipet and apply further suction to the pipet interior. Increase the suction until the patch ruptures, as indicated by the appearance of large capacitative transients (*see* **Fig 4A**). This transition to whole-cell mode can also occur spontaneously after gigaseal formation. Continue with the recording of membrane currents in the whole-cell mode if the decay of the transients follows a single exponential (*see* **Note 27**).

11. Use the whole-cell capacitance and series resistance dials of the amplifier to electronically cancel the capacitative transients. This is best achieved by iteratively increasing the capacitance and series resistance settings. The aim is to leave a rectangular-shaped current step of small magnitude, which is due to the current required to set the voltage step across the cell resistance (in the megakaryocyte this is about 1 GΩ, so only 10 pA is required per 10 mV; note that this assumes the seal resistance is much higher). It may be necessary to further alter the pipet capacitance settings. The dial settings indicate the series resistance and whole-cell capacitance. Megakaryocytes have large capacitances (70–700 pF, mean ~240 pF; *see* **Subheading 3.8.** for full details of measurement).

12. If series resistance compensation is required (*see* **Note 28**), increase the percentage compensation until the current record starts to display oscillations, which is the limit of stability, then reduce it slightly. Each amplifier varies in the exact manner in which the series resistance compensation is applied, so consult the manufacturer's manual. Further, iterative adjustments of the pipet capacitance compensation and series resistance settings are often required to restore the best capacitance transient cancellation (*see* **Note 29**).

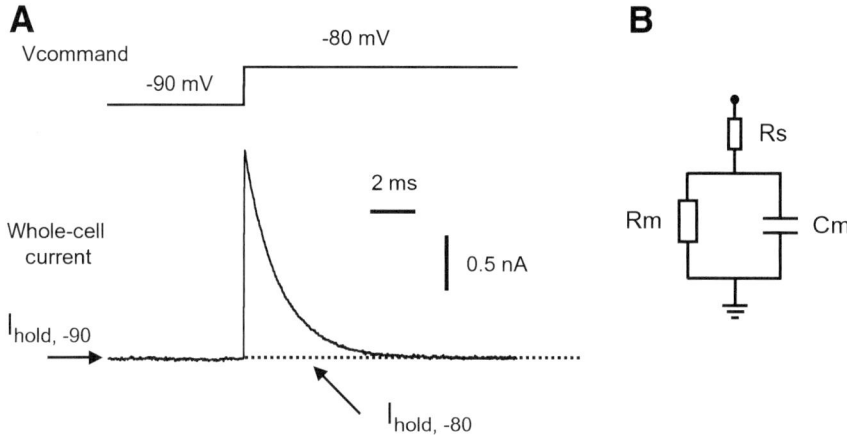

Fig. 4. Measurement of whole-cell capacitance and thus demarcation membrane system development in the megakaryocyte. (**A**) Recording of whole-cell membrane current from a rat megakaryocyte, as the command potential (Vcommand) is stepped from –90 to –80 mV. The large transient represents the current required to charge the membrane capacitance to the new potential. I_{hold} is the steady current level required to set the potential across the cell resistance at either –80 or –90 mV. The charge (Q) that flowed during the transient, above the constant level of $I_{hold,-80}$, was 3.38 pC and the cell capacitance was 338 pF (see text for full details). (**B**) Equivalent circuit for a megakaryocyte in which the capacitative transient decays with a single exponential. Cm is the cell membrane capacitance, Rm the cell resistance, and Rs the series resistance.

3.7. Nystatin Perforated-Patch Whole-Cell Clamp Configuration (see Note 30)

1. Add 2 µL of the stock (50 mg/mL) nystatin to a 1-mL aliquot of SpipS and vortex to disperse the nystatin. The solution will appear slightly cloudy (*see* **Note 31**).
2. Dip the tip of the pipet in normal SpipS for 20–30 s. Back-fill with the nystatin-containing SpipS.
3. As quickly as possible, form a gigaohm seal on a platelet or megakaryocyte as described in **Subheadings 3.3.** and **3.6.** (*see* **Note 32**).
4. After gigaseal formation, apply a holding potential of –70 to –80 mV.
5. In megakaryocytes, apply 5-mV test pulses to assess the size of the capacitative transients. As nystatin molecules are inserted into the membrane patch, they decrease the access resistance; the transients increase in peak amplitude and their decay constant decreases. Use the whole-cell capacitance cancellation dials (*see* **Subheading 3.6.**) to assess the series resistance. The series resistances tend to be higher (5–20-fold) than in conventional whole-cell recordings; thus particular care should be taken over series resistance errors (*see* **Note 28**).
6. In the platelet, whole-cell formation is best assessed by applying approx 200-ms duration depolarizing ramps from –90 to 0 mV. With good voltage control, the voltage-dependent K^+ channels start to be activated at about –60 mV, and in normal KCl internal/NaCl external salines (SpipS and SES) will appear as an outward current on top of a linear component.

The linear component is the current required to establish the potential across the seal and platelet input resistances *(10)*.

3.8. Membrane Capacitance Measurements of the Demarcation Membrane System

It has recently been shown that the demarcation membrane system of the rat megakaryocyte is accessed through only a single series resistance in the majority of whole-cell patch-clamp recordings *(36)*. Therefore, this platelet-generating system can be quantified from the whole-cell capacitative current transients following a voltage step as described below.

1. After formation of a cell-attached patch, carefully eliminate the capacitative transients with the pipet capacitance dials. Sylgarding of the pipet or reduction of saline levels in the bath may be needed to achive this.
2. Apply further suction to break the patch and obtain a conventional whole-cell recording (*see* **Subheading 3.6.**). The whole-cell capacitance neutralization facility of the amplifier should be switched off.
3. Examine the capacitative current following a small voltage step (5–10 mV) over a range of potentials that does not activate voltage-gated K^+ channels (–100 to –80 mV). In the majority of recordings, the decay of the current will follow a single exponential (*see* **Fig. 4A**). In a proportion of recordings (varies daily, but is normally 5% or less), the current will decay multiexponentially.
4. In those cells showing a single exponential decay, acquire the current transients to computer disk at 50 kHz (low-pass filtered at 10 kHz) during the 5 or 10 mV voltage step. The duration of the voltage step should be long enough to allow the current to reach a steady, flat value ($I_{hold,-80}$ in **Fig. 4A**). This duration increases with the series resistance, but for most recordings a pulse of 20 ms should be long enough. $I_{hold,-90}$ and $I_{hold,-80}$ represent the current levels required to set the potentials of –90 and –80 mV, respectively, across the combined series resistance (R_s) and cell membrane resistance (R_m) (*see* **Fig. 4B** for the equivalent circuit; note that this assumes that the seal resistance, in parallel with Rm, is much greater than R_m).
5. Using computer software (we have used Microsoft Excel or Microcal Origin, although it is also possible to analyze within pClamp software), offset the capacitative transient so that the steady level during the pulse ($I_{hold,-80}$) is zero. Then integrate the area under the transient to give the amount of charge, Q, in coulombs ($Q = I \times t$, where I is the current in amperes and t is the time in seconds). If integration is not available on your choice of software package, simply add all the data points together during the offset transient and multiply by the sampling interval (*see* **Note 33**).
6. Calculate the whole-cell capacitance, Cm, in farads from the equation: $Cm = Q/V$, where Q is the charge in coulombs, and V the voltage in volts. The average capacitance of megakaryocytes from marrow is about 240 pF, but can show a large variation (70–700 pF in **ref. 36**). The lab record currently stands at 770 pF (*see* **Note 34**).
7. The extent of development of the demarcation membrane system can be further assessed from the ratio of the measured capacitance to the predicted capacitance for a cell with no invaginations. The latter is calculated from the cell surface area and the membrane capacitance of a biological membrane (1 $\mu F/cm^2$). In our previous work *(36)*, we measured the cell diameter at two perpendicular planes and assumed a spherical geometry. The ratio of measured to predicted specific capacitance varied from 4 to 14 (note that this is also the

specific membrane capacitance for the "spherical megakaryocyte," in units of μF/cm²). Volume rendering of three-dimensional images may provide a further, more accurate measurement if the cells do not move during acquisition.

3.9. Further Comments and Correction for Errors

1. *Filtering:* Correct setting of the low-pass filter is important. Usually, the amplifier internal filter is set at 5 or 10 kHz during gigaseal formation and cancellation of the capacitative transients. The filter is then often set lower during actual experiments, at 1–2 kHz, as this greatly reduces the high-frequency noise. However, since filtering can distort or even remove real biological events, the bandwidth should be as high as possible. The data can always be further filtered or averaged off-line. On the other hand, since the sampling rate of the storage device needs to be at least twice that of the highest frequency within the digitized signal (*see* **Note 10**), a high frequency can result in an excessive computer file size.

2. *Voltage offsets:* These can occur at the junction between the silver/silver chloride electrodes and salines, or between salines of different composition, including between an agar bridge and the bath saline. At the start of an experiment, a pipet offset is added by the amplifier so that the combination of all offsets is zero. If any of these change, the pipet voltage will be shifted from the command potential and must be corrected for, or if small, its value stated as a constant offset. For example, if SES and SpipS are used, the sealing process removes the SES : SpipS junction, resulting in an offset of about –3 mV. These offsets are measured by reference to 3 M KCl, which has a large but constant offset against standard patch-clamp salines. A junctional potential calculator (based on original software by Dr. P. H. Barry, University of New South Wales) is also available with pClamp software (Axon Instruments). Offsets can be corrected during an experiment or later during analysis. *See* **ref. 39** for further discussion.

4. Notes

1. Hexokinase is highly pH-sensitive, and for preparation of purified ADP, the pH must be >8.0–8.5. If titrating at room temperature, the temperature dependence of a pH buffer should be taken into account (e.g., –0.014 pH U/°C for HEPES).

2. We do not normally check the osmolarity of salines each time they are made. However, it is important to assess the osmolarity of newly designed salines. The difference between internal and external salines should not be more than a few percent and if necessary should be adjusted by altering ionic composition or using e.g., sucrose.

3. In perforated patch-clamp recordings, there has been some concern about large Donnan potentials across the patch and thus cell swelling, with the use of a high Cl⁻ in the pipet saline *(40)*. Therefore some studies have used Cl⁻ levels in this saline that are closer to intracellular levels *(30,41)*.

4. The high BAPTA in this saline is required to amplify I_{CRAC} *(17,42)*, probably by reducing Ca^{2+}-dependent inactivation. The added Ca^{2+} results in a free Ca^{2+} concentration close to the normal intracellular level (~100 nM) in order to prevent store depletion and spontaneous activation of I_{CRAC}. The required $CaCl_2$ addition to the internal saline is derived empirically from the level at which I_{CRAC} does not develop spontaneously in whole-cell recordings. The easiest and most reliable way to activate I_{CRAC} is with superfusion of 1 μM ionomycin in external saline (Calbiochem-Novabiochem UK Ltd., Nottingham, UK; 0.2 μL/mL of a 5 mM stock in DMSO).

5. For megakaryocyte electrophysiology, we predominantly use rat (adult male Wistar) megakaryocytes, as the yield is good and responses to many platelet agonists can be

recorded. Murine (adult male, MF1 or C57BL/6) show similar responses but the megakaryocyte yield is lower. We have recorded from Hartley guinea pig megakaryocytes but the $[Ca^{2+}]_i$ responses to ADP were very poor (J. M. Martinez-Pinna and M. P. Mahaut-Smith, unpublished observations).

6. A number of different commercial patch-clamp amplifiers are available. The most important feature of an amplifier for use with megakaryocytes is a high level of capacitance cancellation. Megakaryocytes possess extensive plasma-membrane invaginations (the demarcation membrane system) and consequently have capacitances in the range of 70 to 700 pF, with an average of approx 240 pF *(36)*. Amplifiers achieve the higher range of capacitance cancellation by using a lower-feedback resistance, which increases background noise. Therefore if the same amplifier is to be used for recordings from platelets or for single-channel events, the headstage must be capable of switching to a higher-value feedback resistor.

7. The "whole-cell" circuit of the model cell supplied with an Axopatch amplifier usually has a capacitor of about 30 pF and resistor of 0.5 GΩ. For megakaryocytes, the resistor is fine, but the capacitance is too low. Replacement with a 300-pF capacitor allows testing for most megakaryocytes, although in the rat, up to 770 pF has been seen. A model cell for platelet whole-cell recordings (input resistance ~60 GΩ and capacitance approx 130 fF *[9]*) is not useful as the normal stray capacitance exceeds that of a mammalian platelet. To test the noise levels following formation of a "gigaseal," cure a drop of Sylgard to the coverslip in a chamber. Pushing the pipet tip into the Sylgard from a saline-filled chamber reversibly mimics the formation of a gigaseal.

8. Patch clampers apply pressure to the pipet either by mouth or using a 1-mL syringe. If the former method is used, a 0.22-μm pore syringe filter is attached at the end of the line, particularly if several persons use the same rig. For platelet patching, a 1-mL syringe cannot sufficiently control the pressure; however, either method can be used for megakaryocytes.

9. Choosing an acquisition system and software that allows "telegraph" connections and thus automated detection of amplifier gain (and other parameters such as capacitance and filter settings) greatly simplifies data analysis. pClamp software can support specific amplifiers other than Axon's own series. Before computer hard disks became sufficiently large to continuously record an entire experiment, the current was acquired to videotape via a digital interface (e.g., Instrutech, VR-series of digital recorders, Digitimer Ltd., Welwyn Garden City, UK), then acquired in segments on the computer for analysis. A backup system is still useful for platelet experiments, as their technically demanding nature means that lost data are especially disheartening.

10. To avoid aliasing, the data sample rate must be at least twice that of its highest component frequency. Since the filter setting on the front panel usually refers to its –3 dB frequency, sample at ≥ fourfold greater frequency than this value for most recordings and at least 10-fold higher for noise analysis.

11. Concerns about effects of metal fumes given off when heating the filament have led many workers to coat the tip of the filament with a bead of glass.

12. If the composition of the external medium is changed, particularly to a low-chloride-containing saline, large voltage offsets can occur. These can be limited by the use of a high-chloride-containing agar bridge (e.g., 3 *M* KCl in 1–2% agarose, but ensure that the cell is not close to the bridge or the local KCl concentration will rise).

13. This "slow" perfusion system is sufficient for responses via G protein-coupled receptors, but a pressure injector (e.g., PLI-100 injector, Digitimer Ltd., UK/Medical Systems Corp.) coupled to an agonist-injection "puffer" pipet *(17,29)* or Y- or U-tube system *(15,43)* is

needed for ionotropic receptor responses. The puffer pipet is placed 50–200 μm from the cell, but the Y/U-tube can be placed further away. Megakaryocytes may need to be stabilized by pushing against the coverslip when using a rapid U-tube system, to avoid movement artifacts.

14. One of the most frustrating tasks during the setting up of an electrophysiology rig is to minimize interfering sources of noise. The Faraday cage is grounded via a thick, low-resistance copper braid to the signal ground of the amplifier, or alternatively to a "clean" ground connection if provided by the building. The Faraday cage and all equipment are then connected to this main ground. The chamber ground lead is usually connected to a ground pin on the headstage unless the amplifier provides separate bath headstages or a "driven" ground is used.

15. Silver/silver chloride connectors: The nonchamber end of the silver/silver chloride ground is soldered onto a standard insulated copper wire, which then connects to the headstage. Fragile areas of the lead can be reinforced using a combination of Teflon tubing, epoxy glue, and heat-shrink tubing.

16. Problems with the ground lead are indicated by a drift of holding current with the pipet tip in the chamber (although this can also happen when new grounds are "settling in") and if the current due to a 5-mV square wave test pulse is not completely square. The first step would be to rechloride or replace the ground connector.

17. Platelets are not easily deformed; therefore, a larger tip diameter:cell diameter ratio can be used than expected from other small cells. This inability to be deformed is the main advantage (perhaps the only advantage) that the platelet presents to the technique of patch-clamp. Internal tip diameters can be assessed from electrical resistance after filling; however, both shank geometry and tip diameter contribute to the resistance. Relative internal tip diameter is easily assessed from the "bubble number" technique *(44)* since the pressure required to force air through the tip of a pipet immersed in ethanol depends on the internal tip diameter. Simply attach tubing from a 10-cm^3 syringe, with the plunger set at 10 cm^3, to the back of the pulled pipet and immerse its tip in ethanol. Reduce the volume of the syringe, watching the tip carefully. The volume at which bubbles first appear from the tip is the "bubble number." For our bubble tester, a range of 5.0–5.5 is good for platelets and around 7.3–7.8 for megakaryocyte whole-cell recordings. For the latter, slightly larger tips (up to 8.5 bubble number) can be reduced by heavily fire-polishing. Bubble numbers can be remeasured after fire-polishing. A word of caution: The bubble measuring device needs to be carefully constructed as at high pressures (small bubble numbers) the pipet can be jettisoned from the end. There are no commercially available devices. Some modern, expensive pullers (e.g., from Sutter Instruments) claim high reliability and thus may not require the bubble test.

18. Work with animal or human tissue should be carried out only with appropriate authorization. Consult national and local guidelines.

19. It is essential to remove as much plasma as possible since proteins interfere with formation of the glass:membrane gigaseal. This is best achieved by removing the final volume of plasma using a 100–200-μL pipettor with a narrow-ended yellow tip while tilting the Eppendorf tube toward a horizontal position.

20. Addition of 0.1–0.2% bovine serum albumin to the SES can also reduce spontaneous activation of platelets in washed suspension. The platelets are perfused with non-albumin-containing medium prior to attempting gigaseal formation.

21. Platelets will eventually stick to the glass and activate; thus, there is a narrow time window for selecting and sealing onto a floating platelet. This time window may be increased by coating the glass with the siliconizing agent Sigmacote® (Sigma) or Sylgard®. Sylgard is

spread as thinly as possible and cured on a hotplate, or overnight at room temperature, before sealing to the chamber.

22. The pipet tip should be directly opposite the selected platelet so that when suction is applied, the platelet is attracted without interference from other cells or particulate matter. Some Nikon inverted microscopes have a 4× magnifying lens in the turret, which is extremely useful for increasing the overall magnification during final positioning of the pipet.

23. If seal formation does not occur, try the following: (a) Filter all external salines through a 0.22-μm pore filter. (b) For megakaryocyte recordings, fire-polish the pipets more extensively. (c) To reduce the chance of particulate matter adhering to the pipet tip, perfuse the chamber with more saline prior to lowering the pipet into the chamber and/or apply positive pressure to the pipet as it is lowered into the chamber, particularly through the air:saline interface. (d) Slightly altering the osmolarity of either the pipet or bath saline has been reported to improve seals. (e) The best seals are achieved with high-divalent cations in the pipet, so try, e.g., isotonic $BaCl_2$ or $MgCl_2$ (110 mM $MgCl_2$, 10 mM HEPES, pH 7.35 with NMDG base). It should always be recognized that the seal success rate varies from day to day. This innate variability has yet to be fully explained, and some days you are better off giving up.

24. The term "gigaseal" is used to indicate that the resistance of the seal should be >10^9 ohms. Although 1 GΩ is sufficient to allow virtually all of the membrane current to be recorded by the amplifier under voltage-clamp mode, in practice it should be much higher for a useful, stable recording. The resistance to ground measured in the cell-attached configuration is complex as it also includes the membrane patch, in series with the resistance of the remainder of the cell *(45)*. As a guideline, the resistance to ground should be 10 GΩ or more at the start of a cell-attached recording from a platelet or megakaryocyte.

25. The main issue with the use of size as the only means for identifying megakaryocytes is the risk of selecting osteoclasts. However, there is little risk of this if adult rats are used and bone fragments are removed. Histological staining of our preparation has confirmed that osteoclasts are very rare. This problem may be more of an issue with neonatal tissue. In the early stages (1–2 h) after marrow removal, a large number of "activated" megakaryocytes are present in the preparation, in which the integrity of the plasma membrane appears to compromised, possibly destroyed, and one can clearly see the multilobular nucleus. Later in the preparation, megakaryocytes with healthy, intact plasma membranes appear, which are selected for patching (*see* **Fig. 3**).

26. The ideal suction and overall "technique" for reliably obtaining a gigaseal varies among different cell types. For all cells, suction should be as gentle as possible to avoid activation and to preserve membrane integrity. In practice, suction for platelets tends to be in the high range, and for megakaryocytes in the low range. A manometer can be used, but the ideal amount of suction is best judged empirically. As a guide, lightest suction is "almost no suction at all" and large suction can be compared to drinking a thick shake through a straw.

27. The variability in the shape of the capacitative transients, from single to multiple exponential decays, is interesting as it almost certainly reflects differences in the type of access to the demarcation membrane system during the whole-cell recording. We have experienced a higher percentage of multiexponential transients on some days and are unsure whether this reflects changes in pipet geometry or activation status of the cell.

28. Voltage errors across the series resistance *(Rs)* result from the fact that the current passed under voltage clamp has to pass through a resistance in series with the cell input resistance (*see* **Fig. 4B**). The magnitude of the voltage error (*Vs*, in volts) can be calculated

from Ohm's law ($Vs = Rs \times I$, where I is the current in amperes, and Rs is in ohms). This error is small enough to be ignored for cell-attached recordings and in most platelet recordings, where few channels are active; however, it is a major problem for recording some conductances in the megakaryocyte (e.g., voltage-dependent K^+ currents can easily exceed 10 nA). To reduce the error, the pipet resistance (which is the major determinant of the series resistance) should be kept to a minimum. Most amplifiers are able to compensate for the series resistance error by up to 95%, but the value that can be achieved in practice is often lower. The maximum value can also vary with the amplitude of the step with some amplifier designs. A time-lag setting is also provided with some amplifiers to limit the bandwidth and reduce the risk of oscillations. In a "good" recording from a rat megakaryocyte using an Axopatch 200B with a time lag setting of 10 μs, the maximum Rs compensation without oscillations is 70–75%. Appearance of oscillations at lower values of compensation usually results from an inability to neutralize the capacitative transient because it does not decay with a single exponential.

29. The series resistance frequently changes during a whole-cell recording in the megakaryocyte. It is almost certainly due to the dynamic nature of the demarcation membrane system. Frequently, Rs will increase during a recording. It is therefore essential to regularly reapply the 5–10 mV test pulse and readjust the Rs and Cm dial settings to reneutralize the capacitative transients.

30. Some workers use amphotericin B instead of nystatin as the conductance of the channels it forms is slightly higher. However, it is more expensive. Use amphotericin in an identical manner to nystatin.

31. Nystatin can be made at slightly higher stock concentration, but precipitates out upon dilution into an aqueous medium. 100 μg/mL is close to the highest nystatin concentration that can be achieved in aqueous medium. Some workers have sonicated the aqueous mixture in an attempt to improve solubility, but owing to the rapid loss of activity in aqueous medium, it is probably better to patch immediately after vigorous vortexing.

32. Nystatin markedly inhibits gigaseal formation. If this occurs, increase the amount of backfilling to allow extra time to generate a seal. Dialysis of the pipet with nystatin after seal formation has been suggested *(40)*, but is very difficult.

33. During the development of this method, we had to decide which parts of the transient to include. We included all data points above $I_{hold,-80}$, throughout the transient, which incorporates the rising as well as falling parts of the current transient. This makes sense as the clamp will not be instantaneous and one is attempting to include as much charge required to set the new potential across the membrane as possible.

34. This approach can also be used to calibrate the capacitance dial settings of the amplifier in "open circuit" and generate a calibration curve to convert dial settings to capacitance for each cell during a real recording.

Acknowledgments

Work within the author's laboratory is funded by grants from the British Heart Foundation (in particular, Science Lectureship, BS/10), Medical Research Council, and Royal Society. Dr. Juan Martinez-Pinna is gratefully acknowledged for helpful discussions.

References

1. Miller, J. L., Sheridan, J. D., and White, J. G. (1978) Electrical responses by guinea pig megakaryocytes. *Nature* **272,** 643–645.

2. Somasundaram, B. and Mahaut-Smith, M. P. (1995) A novel monovalent cation channel activated by inositol trisphosphate in the plasma membrane of rat megakaryocytes. *J. Biol. Chem.* **270,** 16,638–16,644.

3. Mahaut-Smith, M. P., Hussain, J. F., and Mason, M. J. (1999) Depolarization-evoked Ca^{2+} release in a non-excitable cell, the rat megakaryocyte. *J. Physiol.* **515,** 385–390.

4. Neher, E., Sakmann, B., and Steinbach, J. H. (1978) The extracellular patch clamp: a method for resolving currents through individual open channels in biological membranes. *Pflugers Arch.* **375,** 219–228.

5. Neher, E. and Sakmann, B. (1992) The patch clamp technique. *Sci. Am.* **266,** 44–51.

6. Kawa, K. (1987) Transient outward currents and changes of their gating properties after cell activation in thrombocytes of the newt. *J. Physiol.* **385,** 189–205.

7. Manning, S. D. and Williams, A. J. (1989) Conduction and blocking properties of a predominantly anion-selective channel from human platelet surface membrane reconstituted into planar phospholipid bilayers. *J. Membr. Biol.* **109,** 113–122.

8. Zschauer, A., van Breemen, C., Buhler, F. R., and Nelson, M. T. (1988) Calcium channels in thrombin-activated human platelet membrane. *Nature* **334,** 703–705.

9. Maruyama, Y. (1987) A patch-clamp study of mammalian platelets and their voltage-gated potassium current. *J. Physiol.* **391,** 467–485.

10. Mahaut-Smith, M. P., Rink, T. J., Collins, S. C., and Sage, S. O. (1990) Voltage-gated potassium channels and the control of membrane potential in human platelets. *J. Physiol.* **428,** 723–735.

11. Mahaut-Smith, M. P., Sage, S. O., and Rink, T. J. (1990) Receptor-activated single channels in intact human platelets. *J. Biol. Chem.* **265,** 10,479–10,483.

12. Mahaut-Smith, M. P., Ennion, S. J., Rolf, M. G., and Evans, R. J. (2000) ADP is not an agonist at $P2X_1$ receptors: evidence for separate receptors stimulated by ATP and ADP on human platelets. *Br. J. Pharmacol.* **131,** 108–114.

13. Sage, S. O., MacKenzie, A. B., Jenner, S., and Mahaut-Smith, M. P. (1997) Purinoceptor-evoked calcium signaling in human platelets. *Prostaglandins Leukot. Essent. Fatty Acids* **57,** 435–438.

14. Kapural, L., Feinstein, M. B., O'Rourke, F., and Fein, A. (1995) Suppression of the delayed rectifier type of voltage gated K^+ outward current in megakaryocytes from patients with myelogenous leukemias. *Blood* **86,** 1043–1055.

15. Kawa, K. (1996) ADP-induced rapid inward currents through Ca^{2+}-permeable cation channels in mouse, rat and guinea-pig megakaryocytes: a patch-clamp study. *J. Physiol.* **495,** 339–352.

16. Kawa, K. (1990) Guinea-pig megakaryocytes can respond to external ADP by activating Ca^{2+}-dependent potassium conductance. *J. Physiol.* **431,** 207–224.

17. Somasundaram, B. and Mahaut-Smith, M. P. (1994) Three cation influx currents activated by purinergic receptor stimulation in rat megakaryocytes. *J. Physiol.* **480,** 225–231.

18. Uneyama, C., Uneyama, H., and Akaike, N. (1993) Cytoplasmic Ca^{2+} oscillation in rat megakaryocytes evoked by a novel type of purinoceptor. *J. Physiol.* **470,** 731–749.

19. Somasundaram, B., Mason, M. J., and Mahaut-Smith, M. P. (1997) Thrombin-dependent calcium signaling in single human erythroleukaemia cells. *J. Physiol.* **501,** 485–495.

20. Lu, X., Fein, A., Feinstein, M. B., and O'Rourke, F. A. (1999) Antisense knock out of the inositol 1,3,4,5-tetrakisphosphate receptor $GAP1_{IP4BP}$ in the human erythroleukemia cell line leads to the appearance of intermediate conductance K_{Ca} channels that hyperpolarize the membrane and enhance calcium influx. *J. Gen. Physiol.* **113,** 81–96.

21. Sullivan, R., Koliwad, S. K., and Kunze, D. L. (1998) Analysis of a Ca^{2+}-activated K^+ channel that mediates hyperpolarization via the thrombin receptor pathway. *Am. J. Physiol.* **275,** C1342–C1348.

22. Sullivan, R., Kunze, D. L., and Kroll, M. H. (1996) Thrombin receptors activate potassium and chloride channels. *Blood* **87,** 648–656.

23. Somasundaram, B., Norman, J. C., and Mahaut-Smith, M. P. (1995) Primaquine, an inhibitor of vesicular transport, blocks the calcium-release-activated current in rat megakaryocytes. *Biochem. J.* **309,** 725–729.

24. Sakmann, B. and Neher, E. (1995) *Single Channel Recording.* 2nd edition. Plenum Press, New York.

25. Ogden, D. A. (1994) *Microelectrode Techniques. The Plymouth Workshop Handbook.* The Company of Biologists, Cambridge, UK.

26. Sherman-Gold, R. (1993) *The Axon Guide for Electrophysiology and Biophysics.* Axon Instruments, Inc., Union City, CA.

27. Barry, P. H. and Lynch, J. W. (1991) Liquid junction potentials and small cell effects in patch-clamp analysis. *J. Membr. Biol.* **121,** 101–117.

28. Hussain, J. F. and Mahaut-Smith, M. P. (1998) ADP and inositol trisphosphate evoke oscillations of a monovalent cation conductance in rat megakaryocytes. *J. Physiol.* **511,** 791–801.

29. Mahaut-Smith, M. P., Sage, S. O., and Rink, T. J. (1992) Rapid ADP-evoked currents in human platelets recorded with the nystatin permeabilized patch technique. *J. Biol. Chem.* **267,** 3060–3065.

30. Mahaut-Smith, M. P. (1995) Calcium-activated potassium channels in human platelets. *J. Physiol.* **484,** 15–24.

31. Mason, M. J. and Mahaut-Smith, M. P. (2001) Voltage-dependent Ca^{2+} release in rat megakaryocytes requires functional IP_3 receptors. *J. Physiol.* **533,** 175–183.

32. Pusch, M. and Neher, E. (1988) Rates of diffusional exchange between small cells and a measuring patch pipet. *Pflugers Arch.* **411,** 204–211.

33. Tertyshnikova, S. and Fein, A. (1997) $[Ca^{2+}]_i$ oscillations and $[Ca^{2+}]_i$ waves in rat megakaryocytes. *Cell Calcium* **21,** 331–344.

34. MacKenzie, A. B. and Mahaut-Smith, M. P. (1996) Chloride channels in excised membrane patches from human platelets: effect of intracellular calcium. *Biochim. Biophys. Acta* **1278,** 131–136.

35. MacKenzie, A. B. (1996) Ion channels in human platelets. *PhD Thesis, University of Cambridge*, Cambridge, UK.

36. Mahaut-Smith, M. P., Thomas, D., Higham, A. B., Usher-Smith, J. A., Hussain, J. F., Martinez-Pinna, J., et al. (2003) Properties of the demarcation membrane system in living rat megakaryocytes. *Biophys. J.* **84,** 2646–2654.

37. Hechler, B., Vigne, P., Leon, C., Breittmayer, J. P., Gachet, C., and Frelin, C. (1998) ATP derivatives are antagonists of the P2Y1 receptor: similarities to the platelet ADP receptor. *Mol. Pharmacol.* **53,** 727–733.

38. Mahaut-Smith M. P. (1990) Chloride channels in human platelets: evidence for activation by internal calcium. *J. Membr. Biol.* **118,** 69–75.

39. Neher, E. (1995) Voltage offsets in patch clamp experiments, in *Single Channel Recording,* 2nd ed. (Sakmann. B. and Neher, E., eds.), Plenum, New York, pp. 147–153.

40. Horn, R. and Marty, A. (1988) Muscarinic activation of ionic currents measured by a new whole-cell recording method. *J. Gen. Physiol.* **92,** 145–159.

41. Korn, S. J. and Horn, R. (1989) Influence of sodium-calcium exchange on calcium current rundown and the duration of calcium-dependent chloride currents in pituitary cells, studied with whole cell and perforated patch recording. *J. Gen. Physiol.* **94,** 789–812.

42. Hoth, M. and Penner, R. (1992) Depletion of intracellular calcium stores activates a calcium current in mast cells. *Nature* **355,** 353–356.

43. Vial, C., Rolf, M. G., Mahaut-Smith, M. P., and Evans, R. J. (2002) A study of P2X$_1$ receptor function in murine megakaryocytes and human platelets reveals synergy with P2Y receptors. *Br. J. Pharmacol.* **135,** 363–372.

44. Corey, D. P. and Stevens, C. F. (1983) Science and technology of patch-recording electrodes, in *Single Channel Recording*, 1st ed. (Sakmann, B. and Neher, E., eds.), Plenum, New York, pp. 53–68.

45. Fenwick, E. M., Marty, A., and Neher, E. (1982) A patch-clamp study of bovine chromaffin cells and of their sensitivity to acetylcholine. *J. Physiol.* **331,** 577–597.

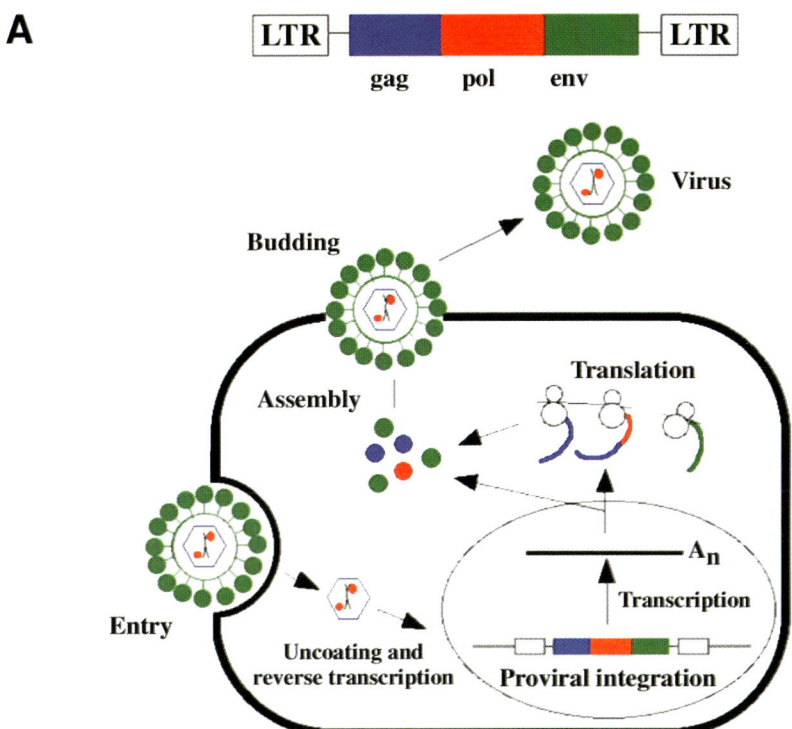

Color Plate 1, Fig. 1A (*see* discussion and full caption in Chapter 2, p 37.) **(A)** Schematic representation of basic retroviral structure and the process of infection.

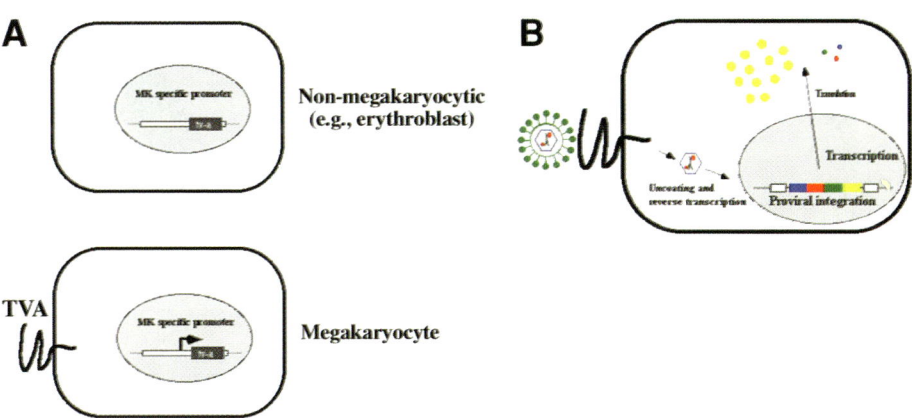

Color Plate 2, Fig. 2 (*see* discussion and full caption in Chapter 2, p 39.) Megakaryocyte-specific retroviral infection through the TVA receptor.

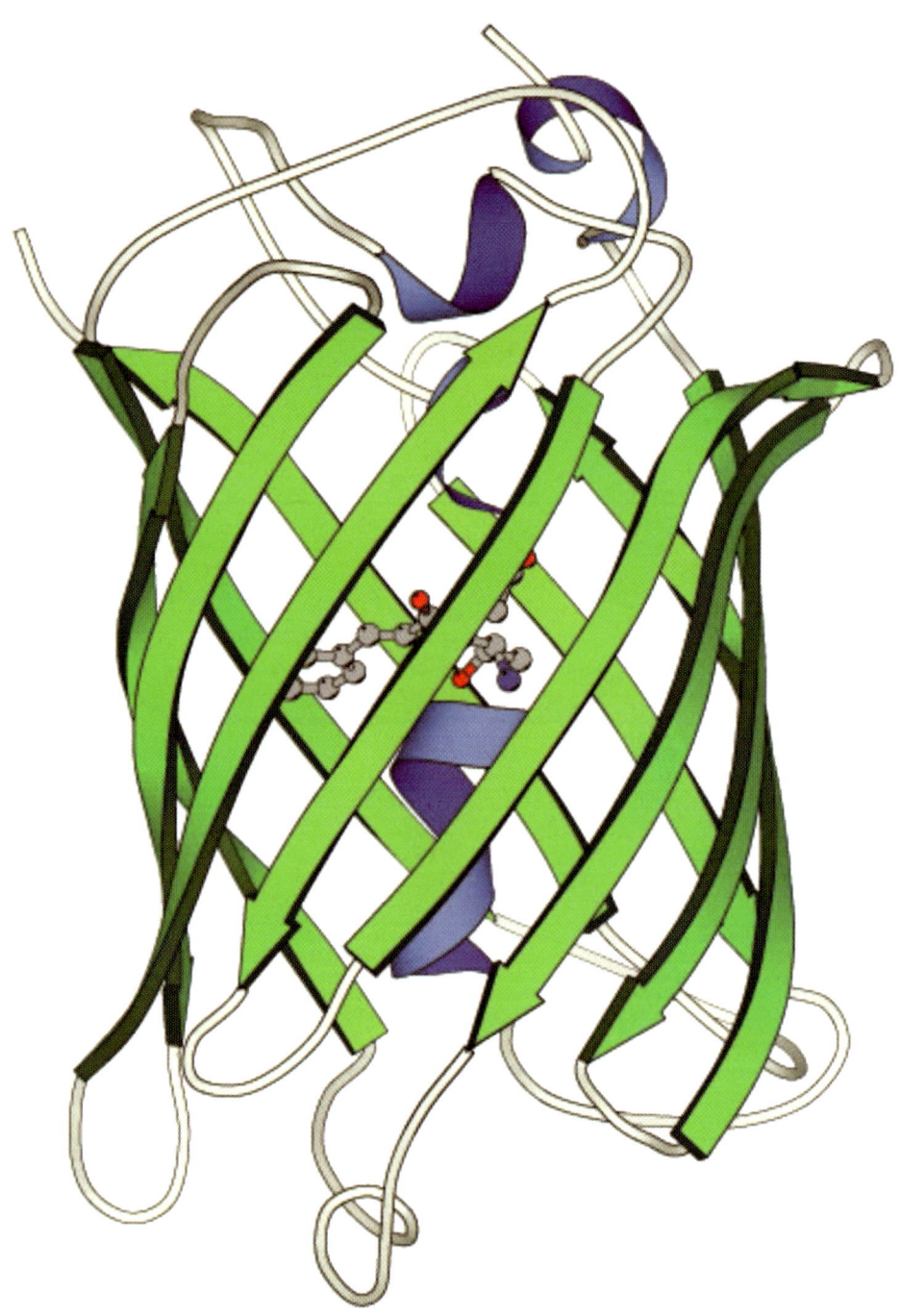

Color Plate 3, Fig. 1 (*see* discussion and full caption in Chapter 25, p 408.) Ribbon diagram representation of the crystal structure of wild-type green fluorescent protein from *Aequoria victoria.*

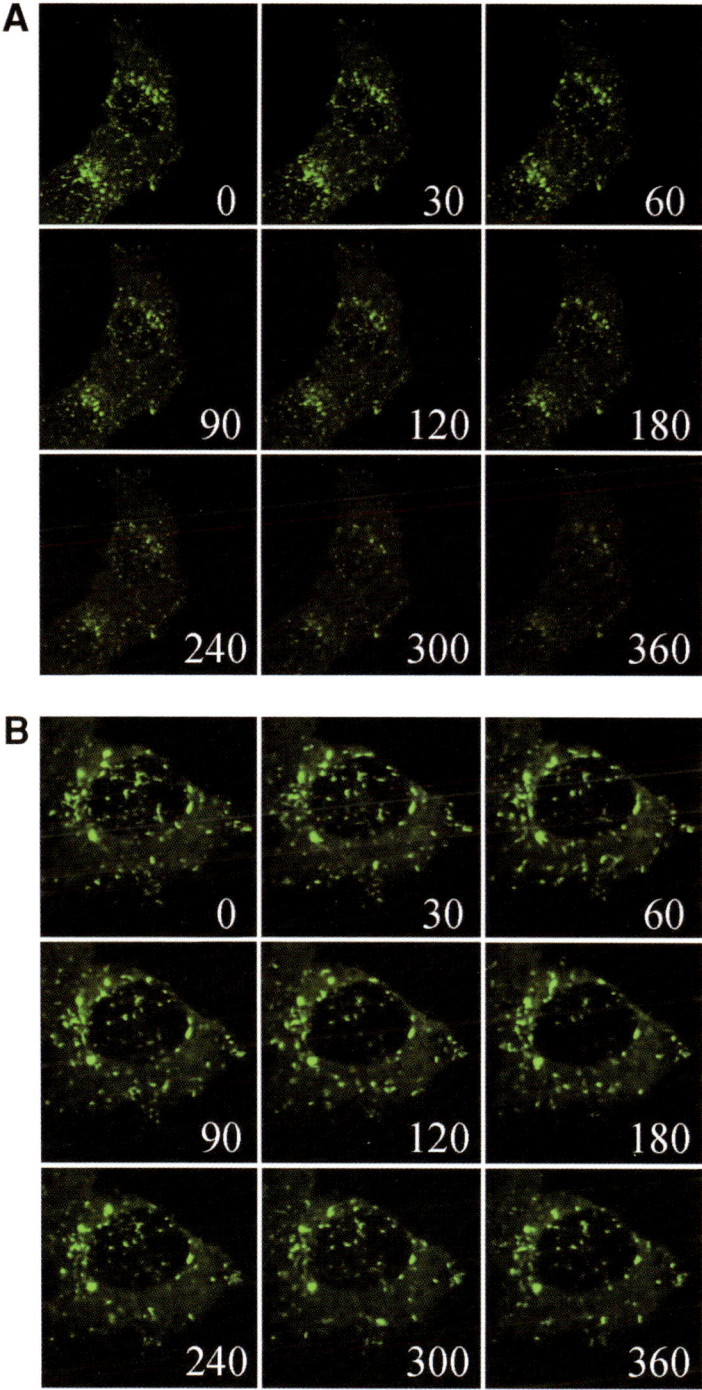

Color Plate 4, Fig. 2 (*see* discussion and full caption in Chapter 25, p 413.) Comparison of GFP images obtained with laser scanning and spinning disk confocal microscopes.

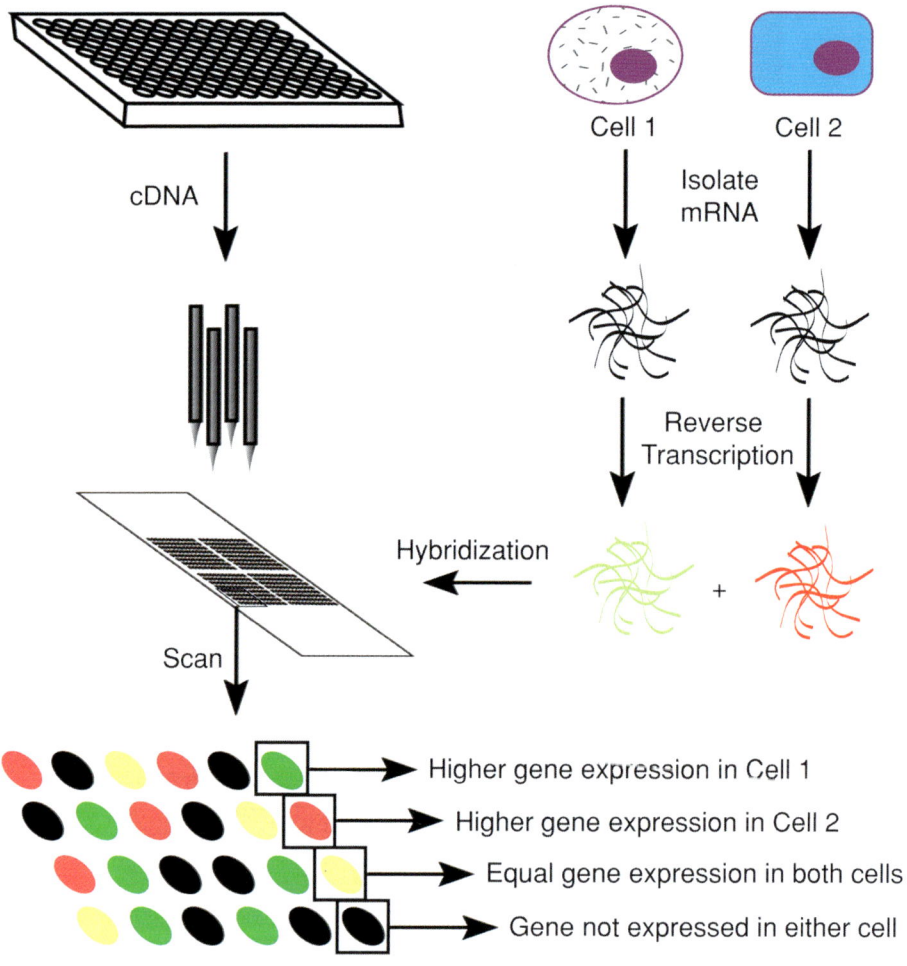

cDNA

Cell 1　　Cell 2

Isolate mRNA

Reverse Transcription

Hybridization

+

Scan

Higher gene expression in Cell 1

Higher gene expression in Cell 2

Equal gene expression in both cells

Gene not expressed in either cell

Color Plate 5, Fig. 1 (*see* discussion and full caption in Chapter 29, p 481.) Summary of the microarray process.

18

Evaluation of Nitrotyrosine-Containing Proteins in Blood Platelets

Khalid M. Naseem, Rocio Riba, and Max Troxler

1. Introduction

In this chapter we present two methods available to study the formation of nitrated proteins in blood platelets. We have used the methods in conjunction to first, identify proteins that become nitrated and second, to quantify the amounts of nitrated proteins present in platelets under a variety of different conditions. Since protein nitration is a new aspect of platelet biology we discuss some of the important findings in this field before outlining the methods.

1.1. Formation of Nitrated Proteins

Nitric oxide (NO) is a ubiquitous gaseous messenger synthesized from the amino acid L-arginine by a group of enzymes termed the nitric oxide synthases (NOSs) *(1)*. The endothelial form of this enzyme (eNOS) is responsible for cardiovascular homeo-stasis, including the control of blood pressure, vessel modeling, and regulation of platelet function. NO is released continually by the endothelium in response to the pulsatile flow of blood and acts to regulate platelet reactivity by preventing platelet adhesion to the subendothelial matrix and inhibiting platelet aggregation *(2)*. NO is an extremely lipophilic molecule and diffuses across platelet membranes. Inside the platelet NO activates the enzyme soluble guanylyl cyclase, leading to increased synthesis of cGMP. The subsequent activation of cGMP-dependent kinases leads to a reduction in $[Ca^{2+}]_i$ and inhibition of Ca^{2+}-dependent activation processes *(3)*. Platelets also possess a Ca^{2+}-dependent NOS, which has been putatively identified as an eNOS isoform *(4)*. This NOS becomes activated during platelet aggregation, releasing NO in a very local-ized environment, and acts to retard excessive aggregate formation *(5)*.

As the understanding of NO biology has increased, it has become clear that NO under-goes a plethora of different reactions beyond its activation of guanylyl cyclase. One key reaction of NO to emerge in the past 10 yr is an interaction with the superoxide anion

From: *Methods in Molecular Biology, vol. 273:*
Platelets and Megakaryocytes, Vol. 2: Perspectives and Techniques
Edited by: J. M. Gibbins and M. P. Mahaut-Smith © Humana Press Inc., Totowa, NJ

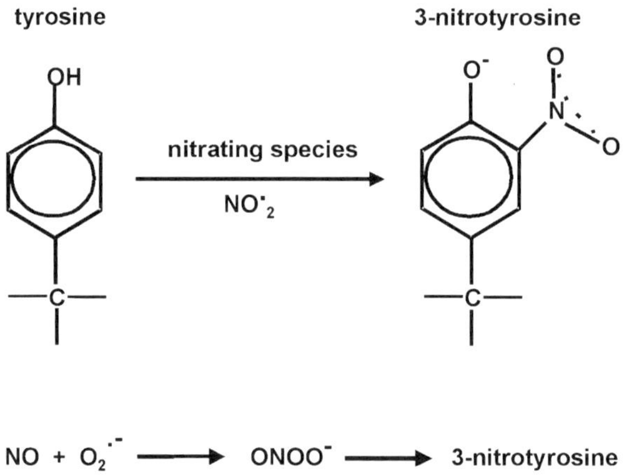

tyrosine   3-nitrotyrosine

OH   O⁻   O

nitrating species
NO·₂

—C—   —C—

NO + O₂·⁻ ⟶ ONOO⁻ ⟶ 3-nitrotyrosine

Fig. 1. Structure of nitrotyrosine. Upper panel: The nitrating species that is thought to be the nitrogen dioxide radical (NO·₂) attacks the ortho position of the phenol ring within the amino acid tyrosine to produce 3-nitrotyrosine. This can be achieved in vitro by exposing cells to chemically synthesized peroxynitrite. Lower panel: The formation of peroxynitrite from nitric oxide and superoxide anion, which leads to the production of 3-nitrotyrosine.

($O_2^{\cdot-}$). NO reacts with $O_2^{\cdot-}$ in a reaction that is diffusion-limited for NO to produce the potent cell-permeable oxidant peroxynitrite (6). Once formed, peroxynitrite can chemically modify amino acids, nucleic acids, and thiol-containing proteins and peptides. Peroxynitrite, via peroxynitrous acid, decomposes to form the nitrogen dioxide radical ($NO_2^{\cdot}$) and a species similar in reactivity to the hydroxyl radical (·OH). The $NO_2^{\cdot}$ attacks phenol groups to produce nitrophenols (7), a process termed nitration. In biological systems this leads to modification of tyrosine residues to produce 3-nitrotyrosine (**Fig. 1**). The formation of 3-nitrotyrosine can be thought of as a stable biological marker for the formation of peroxynitrite. Additional mechanisms for tyrosine nitration have also been proposed, such as myeloperoxidase-dependent oxidation of nitrite by neutrophils (8) or NO trapping of the tyrosyl radical of prostaglandin-H-synthase (9), but nitration by peroxynitrite remains the most likely mechanism.

For those new to NO biology, it is important not to confuse the process of nitration with nitrosation. Nitrosation is the addition of NO, via N_2O_3, to a reduced thiol group (-SH) of amino acids such as cysteine, resulting in the formation of a nitrosothiol (-SNO) (10). This NO can subsequently be released either enzymatically or by transition-metal ions (11). Nitrosothiols have been detected in blood plasma and may represent a circulating pool of NO donors. However, the influence of protein nitrosation on platelet function is beyond the scope of this chapter.

1.2. Exposure of Platelets to Peroxynitrite

Platelets in vivo may be exposed to peroxynitrite from several different cellular origins. However, the two primary sources are probably the vascular endothelium and platelets themselves. The endothelium is known to release NO and $O_2^{\cdot-}$ anions, with the release of both radicals increased significantly if the cells are activated *(12)* or in regions of arterial disease *(13)*. Since the reactants required to produce peroxynitrite are present, it is highly likely that peroxynitrite is formed at the luminal surface of the endothelium. Indeed, cultured aortic endothelial cells, when activated by bradykinin or calcium ionophore, have been demonstrated to release peroxynitrite as measured by luminol-enhanced chemiluminescence *(14)*. Thus, as platelets in circulation brush along the endothelial surface, they probably encounter a mixture of NO and peroxynitrite, depending on the physiological situation. Platelets also produce significant amounts of both NO and $O_2^{\cdot-}$ when activated, although the production of peroxynitrite by platelets has not been demonstrated. However, we have observed that during collagen-induced platelet aggregation the levels of protein nitration are increased *(15)*. This endogenous nitration is dependent on platelet-derived NO and is indicative of peroxynitrite formation.

1.3. Protein Nitration and Platelet Function

The nitration of proteins could have potentially important consequences for cellular function. The nitration of key tyrosine residues has been shown to alter protein conformation, leading to both inhibition and activation of enzymes. Glutathione *S*-transferase *(16)*, NADH:ubiquinone reductase *(17)*, succinyl CoA transferase *(18)*, and protein kinase C *(19)* are inhibited through nitration of important tyrosine residues. Conversely, nitrating species activate protein kinase B *(20)* and extracellular signal-regulated kinase *(21)*. These effects have not been demonstrated in platelets. In addition, peroxynitrite-induced nitration of tyrosine-containing peptides reduces their phosphorylation when exposed to specific protein kinases *(22)*. The findings suggest strongly that nitration may interfere with protein phosphorylation signaling pathways, by both altering enzyme activity and reducing substrate availability.

Platelet proteins are targets for peroxynitrite-induced nitration *(23,24)*. Exposure of platelets to peroxynitrite results in the nitration of both cytosolic and membrane proteins, although cytosolic proteins are preferentially targeted *(12)*. Peroxynitrite probably diffuses into the platelet via an anion exchanger *(25)*. The targets for nitration in platelets are widespread and have yet to be identified. Interestingly, our work has demonstrated the presence of small amounts of nitrated proteins in unstimulated platelets, though it is unclear if this is due to exposure to nitrating agents in vivo or as a result of cell stress during isolation *(26)*. The levels of nitrated proteins increase in platelets undergoing collagen-induced aggregation *(15)* in the absence of peroxynitrite. Moreover several proteins are specifically targeted by this endogenous nitration. These observations indicate that nitration may play a key role in platelet function.

2. Materials

2.1. Peroxynitrite Synthesis

1. 1.5 *M* Sodium hydroxide.
2. 1.2 *M* Sodium nitrite.
3. 1 *M* Nitric acid.
4. 30% Hydrogen peroxide.

5. Manganese oxide.
6. Glass or plastic 20-mL syringes.
7. Plastic tubing (30 cm).
8. Y-connector.

All of the above reagents can be purchased from Sigma (Poole, UK).

2.2. SDS-PAGE and Western Blotting

1. SDS (Laemlli) lysis buffer: 500 mM Tris-HCl, pH 6.8, SDS (4% [w/v]), mercaptoethanol (10% [v/v]), glycerol (20% [v/v]), and brilliant blue dye (trace).
2. Running buffer: 192 mM glycine, 25 mM Tris base and 0.1% (w/v) SDS, pH 8.3, stored at 4°C.
3. Transfer buffer: 25 mM Tris base, 192 mM glycine, in 800 mL H$_2$O; this should be made up to 1 L with methanol (200 mL), stored at 4°C.
4. Tris-buffered saline (TBS$_T$): 150 mM NaCl, 20 mM Tris-HCl, 0.047% (v/v) Tween-20 (0.470 mL/L), pH 7.4
5. Blocking buffer: 3 g BSA (Sigma) in 100 mL in TBS$_T$ (3% BSA). This can be prepared in bulk and stored at 4°C for 1 mo in the presence of 0.01% (w/v) sodium azide.
6. Rabbit anti-human anti-nitrotyrosine polyclonal antibody, diluted 1 in 800 in blocking buffer.
7. Donkey anti-rabbit HRP-linked antibody (Amersham, Bucks, UK), diluted 1 in 4000 in blocking buffer.
8. Bio-Rad Miniblot system or equivalent.
9. Photographic film (Kodak).
10. Enhanced chemiluminescence (ECL) detection reagents (Amersham).
11. Nitrocellulose membrane (Amersham).

For further details on SDS-PAGE and Western blotting, please refer to Chapter 9, vol. 2.

2.3. Materials for Nitrotyrosine ELISA

1. Carbonate buffer: (a) 50 mM Na$_2$CO$_3$, (b) 50 mM NaHCO$_3$. Mix in the ratio of 1 part (a) to 9 parts (b) should yield a pH of 10.0. This buffer can be stored for up to 1 wk at 4°C.
2. Phosphate-buffered saline (PBS): 8.3 mM Na$_2$HPO$_4$, 1.5 mM KH$_2$PO$_4$, 1.54 mM NaCl, 2.7 mM KCl (PBS packets can be used instead of making up the above mixture), pH 7.4.
3. PBS Tween-20 (0.05% [w/v]; PBS$_T$) (*see* **Note 1**).
4. Blocking buffer: PBS$_T$ with 0.5% ovalbumin (w/v; PBS$_T$/OAA). OAA is purchased from Sigma. Approximately 35 mL per plate is required.
5. Fatty-acid-free bovine serum albumin (BSA) (Boehringer Mannheim, Indianapolis, IN).
6. Rabbit IgG anti-nitrotyrosine antibody (primary antibody) diluted 1:200 in PBS$_T$/OAA and stored in 100-µL aliquots at –20°C (Upstate Biotech, Milton Keynes, UK, RPN1004). Then, 100-µL aliquot diluted to 10 mL total with PBS$_T$/OAA to give a final dilution of 1:20,000.
7. Biotinylated donkey anti-rabbit IgG antibody stored at 4°C (Amersham Biosciences).
8. Avidin biotin complex conjugated with horseradish peroxidase (ABC-HRP) (Dako, Ely, UK). Prepared by adding 1 drop of reagent A and 1 drop of reagent B to 5 mL of PBS and left at room temperature for at least 30 min. This complex is then further diluted 1:50 in

PBS$_T$/OAA before use. The activity of the ABC complex can be tested at this point (*see* **Note 2**).

9. *O*-phenylenediamine (substrate) in phosphate/citrate/perborate buffer. This is prepared by the addition of one capsule of phosphate/citrate/perborate buffer to 100 mL distilled water. One 20-mg *O*-phenylenediamine tablet is dissolved in 50 mL of the buffer (both purchased from Sigma). This should be prepared immediately before use.
10. 4 *M* sulfuric acid.
11. Maxisorp plates (Nunc).

3. Methods

Although our work has shown that several platelet proteins are both nitrated at rest and endogenously increased upon platelet activation, the most common method for studying the effects of nitration is to induce nitrotyrosine formation by exposing platelets to peroxynitrite.

3.1. Synthesis of Peroxynitrite

Peroxynitrite can be synthesized easily and inexpensively using standard laboratory equipment and chemicals. The oxidant is stable in its anionic form, but decomposes very rapidly at physiological pH.

1. In a beaker mix 6 mL of 1 *M* HNO$_3$, 0.85 mL of 30% H$_2$O$_2$, and 3.15 mL of H$_2$O. Mix and place in 20-mL syringe.
2. Place 10 mL 1.2 *M* sodium nitrite in a second 20 mL syringe.
3. Plastic tubing (10 cm) is connected to each syringe and then to the Y-connector. The final piece of tubing is connected to the end of the Y-connector.
4. Even pressure is applied to each syringe so that the two solutions will mix evenly in the Y-connector. The resulting yellow peroxynitrite solution is passed immediately into 1.5 *M* NaOH (peroxynitrite is stable in an anionic form).
5. The peroxynitrite should then be passed down a MnO column to remove any excess hydrogen peroxide. This procedure should be repeated several times until the effervescence stops. The presence of residual H$_2$O$_2$ can be checked by reading the absorbance at 220 nm.
6. To prepare control or decomposed peroxynitrite the above procedure is repeated, except that the solution is passed into water instead of sodium hydroxide. This leads to immediate decomposition of the oxidant. NaOH is then added to restore the pH to 12.0.
7. Concentrations of peroxynitrite solutions are calculated using the molar extinction coefficient (λ_{302nm} = 1670/M/cm). Dilute 10 µL of peroxynitrite in 990 µL 1 *M* NaOH and place in a cuvet. Blank the spectrophotometer using NaOH, and then scan peroxynitrite in the range of 280–320 nm. A distinct peak should be observed at 302 nm (*see* **Note 3**).
8. Aliquoted peroxynitrite can be stored for up to two months at –70°C, but the concentration does decrease with time. On the day of experiment thaw the aliquot and check concentration before use. Once thawed, do not refreeze and discard after use.
9. Working solutions can be prepared using 1 m*M* NaOH, but should be prepared on the day of use and discarded after use.

3.2. Exposure of Platelets to Peroxynitrite

The major limitations of working with peroxynitrite in cell systems are the pH of the peroxynitrite solution and its short half-life at physiological pH; both these factors need to be accounted for within the experimental design. We have found that working solutions of peroxynitrite can be prepared in 1 mM NaOH without any appreciable changes in peroxynitrite concentration. This will act to reduce the pH of the peroxynitrite added to the platelets. However, to be completely sure that the pH of the peroxynitrite is not altering cell function, it is always important to use decomposed peroxynitrite as a control. This has the added advantage of testing the effects of the decomposition products—for example, nitrate.

Peroxynitrite decays very rapidly at physiological pH, and it is therefore important to take steps to ensure that as many of the platelets are exposed to the oxidant as possible before it decomposes. A platelet aggregometer is an ideal tool with which to carry out the nitration procedure.

1. Platelets should be left to stir for one minute to allow for temperature equilibration. Peroxynitrite is added directly to the suspension, using a long thin pipet tip. The nitration reaction is almost instantaneous, although we leave the platelets for one minute before lysis (*see* **Notes 4–7**).
2. For Western blotting samples, lyse the cells with an equal volume of standard Laemmli sample buffer (*see* Chapter 9, vol. 2). For the ELISA samples, add half the volume of ice-cold PBS and place cells on ice immediately (e.g., if using 100 µL of platelets add 50 µL buffer).

3.3. Detection of Nitrated Proteins by SDS-PAGE and Western Blotting

We use a standard Western blotting technique to identify which proteins become nitrated. A polyclonal anti-human anti-nitrotyrosine antibody is commercially available (Upstate Biotech), but the protocol outlined below uses a similar antibody raised our laboratory. Our polyclonal anti-nitrotyrosine antibody was raised against nitrated keyhole limpet hemocyanin and purified using standard affinity purification techniques. Extensive characterization was performed using nitrated peptides and a variety of tyrosine derivatives. The performance of this antibody in our ELISA (**Subheading 3.4.**) was very similar to that of the commercial polyclonal antibody, but gave clearer resolution on immunoblots *(27)*.

1. Platelet lysates are separated by SDS-PAGE on 10% gels for 2 h at 100 V, using a minigel system. Each well is loaded with 30 µg of protein (we normally load nitrated BSA as a positive control; for preparation *see* **Subheading 3.4.1.**) (for further details on SDS-PAGE please refer to Chapter 9, vol. 2).
2. Proteins are then transferred to nitrocellulose membranes using standard wet-blotting techniques. The transfer is allowed to run for 2 h at 100 V, with cooling (for further details on Western blotting please refer to Chapter 9, vol. 2).
3. The membrane is blocked to prevent nonspecific binding of antibodies using blocking buffer. Membrane blocking can be performed overnight at 4°C or for 4 h at room temperature on a platform rotator.
4. Membrane is washed four times for 15 min in TBS$_T$ on a platform rotator, using fresh buffer for each wash.
5. Membrane is now incubated with primary antibody (diluted 1:800 in blocking buffer) for 2 h at room temperature on a platform rotator.
6. Membrane is washed four times for 15 min in TBS$_T$, with fresh buffer used for each wash.

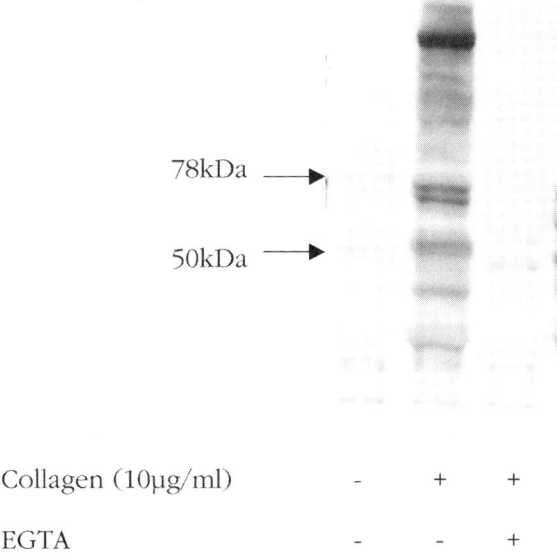

78kDa ──→

50kDa ──→

Collagen (10μg/ml)   -   +   +

EGTA   -   -   +

Fig. 2. Nitration of platelet proteins by collagen. Washed platelets were stimulated with collagen (10 μg/mL) for 1 min in the presence and absence of EGTA (1 m*M*). Cells were lysed during standard Laemmli buffer and blotted using the procedure outlined in **Subheading 3.3.**

7. Membrane is incubated with secondary antibody (diluted 1:4000 in blocking buffer) for 1 h on a platform rotator.
8. The four 15-min washing procedures are then repeated.
9. After the final wash, the membrane is incubated for 90 s with ECL reagents. The ECL reagents are poured off, the membrane covered in cling film. Signals are recorded on X-ray film and processed in a dark room (*see* **Notes 8** and **9**).

An example of results obtained using this method is shown in **Fig. 2**. In this particular example, platelets have been stimulated with collagen in the presence and absence of EGTA.

3.4. Quantification of Nitrated Proteins by ELISA

ELISA allows the semiquantitative measurement of protein bound nitrotyrosine present in a variety of biological samples *(28)*, and is based on competition between free and bound nitrated bovine serum albumin (BSA) for a limited amount of antibody. The assay uses nitrated BSA (NT-BSA) as a standard, prepared by exposure of BSA to peroxynitrite, and a commercially available rabbit anti-human polyclonal anti-nitrotyrosine antibody to detect nitrated proteins. The assay is only semiquantitative, as the antibody may react with other nitrated proteins differently from its interaction with nitrated BSA; nevertheless, it is a very useful tool to quantify the amount of nitrated proteins (*see* **Note 11**). The results are normally expressed as nmol nitrated BSA equivalents/mg protein.

3.4.1. Preparation of Nitrated BSA Standard for ELISA

1. BSA is dissolved at 2 mg/mL in PBS (**Subheading 2.3.**).
2. An aliquot of peroxynitrite is thawed and concentration determined by a wavelength scan (**Subheading 3.1.**).
3. BSA is nitrated by three additions of 1 mM peroxynitrite (final conc. 3 mM) at 37°C (*see* **Notes 12** and **13**). The BSA solution is incubated at 37°C for 5 min prior to each addition and 5 min after each addition in a shaking water bath.
4. Nitrated BSA is then dialyzed overnight against 5 L of PBS. This is followed by two more 3-h dialyses, using fresh buffer each time.
5. The protein concentration should be determined, for example, by a Bradford assay, and the extent of nitration measured by a wavelength scan. The spectrophotometer is blanked using a cuvet containing a 1:1 mixture of carbonate buffer (pH 10.0) and PBS. The nitrated protein is diluted 1:1 in carbonate buffer and the absorbance peak at 438 nm is noted. The molar extinction co-efficient for nitrated proteins at pH 10.0 at 438 nm is 4300$M^{-1}s^{-1}$. Stoichiometry should be between 3–6 mol nitrotyrosine/mol BSA.
6. The nitrated protein can be aliquoted (100 µL) and stored at –70°C for up to 6 mo.

3.4.2. Coating the ELISA Plate

1. An aliquot of frozen standard nitrated BSA (NT-BSA) is thawed and diluted to 5 µg/mL in carbonate buffer within a sialinized vessel to prevent protein loss; approx 10 mL of diluted NT-BSA is required for each plate.
2. Diluted NT-BSA (100 µL) is added to every well of the 96-well ELISA plate, and incubated at 4°C overnight.
3. The plate is washed in PBS by the following method: all wells are filled with PBS and the plate inverted and shaken to expel the PBS. This process is repeated four times. On the fourth occasion the PBS is left in the wells for approx 5 min. The plate is then inverted and the plate shaken to expel the PBS and the plate is tapped dry.
4. Nonspecific binding sites are blocked by the addition of PBS$_T$/OAA (350 µL) to every well. The plate is then covered using a plastic plate sealer and incubated at 37°C for 2 h.
5. The plate is now washed four times with PBS$_T$ (rather than just PBS). After thorough drying, it is covered and can be stored at 4°C for up to 4 wk before use.

3.4.3. Performing the Assay (see **Note 14**)

1. Standard NT-BSA is diluted in a sialinized vessel to 100 µg/mL in PBS$_T$/OAA. Approximately 10 mL of solution is required for each plate.
2. The standard curve is generated by the serial dilution of the 100 µg/mL solution of NT-BSA. NT-BSA (150 µL) is added to wells B1–B3 (*see* diagram). PBS$_T$/OAA (100 µL) is added to C1–C3, D1–D3, and so on to wells C4–C6. 50 µL of NT-BSA is aspirated from B1–B3, added to C1–C3, and mixed thoroughly with 16 up/down syringeings. The process is repeated by removal of 50 µL from C1–C3 and adding to D1–D3 and mixing again. This is repeated to produce a series of triplicate wells with serial threefold dilutions of the 100 µg/mL solution (from 100 µg/mL to 0.005 µg/mL). Finally, 50 µL is aspirated from C4–C6 after mixing and discarded to leave 100 µL in all standard wells.

	1	2	3	4	5	6	7	8	9	10	11	12
A	200 µL	200 µL	200 µL	100 µL	100 µL	100 µL						
B	150 µL	150 µL	150 µL	100 µL	100 µL	100 µL						
C	100 µL	100 µL	100 µL	100 µL	100 µL	100 µL						
D	100 µL	100 µL	100 µL									
E	100 µL	100 µL	100 µL									
F	100 µL	100 µL	100 µL									
G	100 µL	100 µL	100 µL									
H	100 µL	100 µL	100 µL							100 µL	100 µL	100 µL

3. Wells A1–A3 contain 200 µL of PBS$_T$/OAA and are a control to which the primary antibody is not added. This will demonstrate any nonspecific binding by the secondary antibody.

4. Wells H10–12 contain 100 µL of PBS$_T$/OAA, to which the primary antibody will be added; this will show the absorbance given by 100% antibody binding.

5. An optional third control consists of the addition of 10 µL of 3 nitro-tyrosine (200 mM) to a set of wells containing 90 µL of PBS$_T$/OAA, to which primary antibody will be added.

6. The other sets of wells not used in the diagram contain 100 µL of sample in triplicate.

7. Primary antibody is diluted 1:20,000 in PBS$_T$/OAA to give a final dilution of 1:40,000 (*see* **Note 15**) when 100 µL is added to all ELISA wells except A1–A3. Plates are then incubated for 2 h on a plate shaker at 37°C.

8. Plates are washed four times using PBS$_T$.

9. Secondary antibody is diluted 1:1000 in PBS$_T$/OAA and 100 µL added to all wells and then incubated for 1 h on a plate shaker at 37°C.

10. While plates are incubating, the ABC-HRP complex is prepared.

11. Plates are then washed four times with PBS$_T$.

12. ABC-HRP complex (100 µL) is added to all wells and the plate incubated again for 1 h on a plate shaker at 37°C.

13. Plates are washed four times with PBS$_T$ and then developed by the addition of 100-µL of substrate to all wells.

14. Assay is allowed to develop for a maximum of 30 min at room temperature. Assay is halted by the addition of 50-µL of 4 M H$_2$SO$_4$ and absorbance read at 490 nm using a plate reader.

15. A standard curve is constructed using values obtained for NT-BSA standards; this is used to determine the concentration of nitrated proteins in samples.

4. Notes

1. When preparing peroxynitrite we find that higher concentrations can be yielded by ensuring that all solutions and syringes are cooled on ice immediately before use and that fresh hydrogen peroxide is used. We aim to produce peroxynitrite at a concentration of 80–120 mM for BSA nitration.

2. Before addition of the substrate, remove 500 µL of substrate solution and add to 500 µL of ABC-HRP complex retained in **Subheading 2.3., step 8**. The color should quickly develop to a deep red/brown. This step is important, as it is the only one where a quality control procedure can be used in the assay. If the color does not develop, it could be that the ABC-HRP complex has not been prepared properly and the plate will fail. Alternatively, it could be that the chromagen has not been properly prepared and the plate will similarly fail. If the test solution does not change color, then firstly prepare a fresh chromagen

solution and re-test with some more ABC. If this does not produce a color, it must be assumed that the ABC-HRP complex is faulty and must be prepared again.

3. When studying the effects of protein nitration on platelet function, it is preferable to use washed platelets. PRP should be avoided since plasma contains many protein targets, particularly albumin, which will react with the peroxynitrite before the oxidant reaches the platelets.

4. In the literature groups use up to 1 mM peroxynitrite, the justification being that since this is such an unstable molecule at physiological pH the majority of peroxynitrite will decompose before entering the cell. However, the use of such high concentrations of peroxynitrite leads to massive nitration and probably damages the cell structure. We can reproducibly induce significant protein nitration with 10 μM peroxynitrite, and would not advise going above 150 μM in cellular studies.

5. If the experimenter is investigating the effects of peroxynitrite on protein nitration then using platelets suspensions at 3×10^8 platelets/mL is sufficient.

6. When assessing the influence of endogenous protein nitration on platelet signaling cascades, we stimulate platelets in the absence of EGTA. This seems to reduce agonist-induced nitration (**Fig. 2**).

7. The protocol is designed to for optimal use with our own nitrotyrosine antibody, although the commercial antibody from Upstate Biotech can be used. In our hands the commercial antibody was diluted 1:1000 and gave reasonably clear resolution. The procedure is identical to the one outlined here.

8. When using nitrated BSA as a positive control, 7 μg of protein is sufficient to give a clear band at 66 kDa.

9. It has been observed that after the developing process a high background could appear. To clear the background it is recommended to wash the membrane for 5 min with TBS$_T$ after transfer.

10. If the NT-BSA has been prepared and the plates coated and washed in advance, the entire procedure takes 1 d.

11. When measuring endogenously synthesized proteins by ELISA, there can be problems with detection, since the levels of protein modification produced endogenously are far less than those induced by peroxynitrite treatment. If the experimenter anticipates that there may be low levels of nitration then it is be better to suspend the platelets at a greater concentration, for example, 5×10^8 platelets/mL. This increases the amount of nitrated proteins into the linear section of the standard curve and will give more accurate measurements.

12. When preparing BSA solutions, mix gently to avoid frothing and subsequent loss of protein.

13. To prepare the NT-BSA we place 20 mL of BSA solution in to a 50-mL Falcon tube and after incubation at 37°C in a shaking water bath, vortex until the liquid rises three quarters of the way up the tube. Peroxynitrite (thawed on ice) is then added slowly, drop by drop, with a 200-μL pipet. When this is repeated three times we find that this produces the highest levels of nitration.

14. The concentrations of antigen and dilutions of antibodies given in the above protocols are the optimums under our experimental conditions. However, it is advisable when setting up the assay for the first time that full optimization of antibodies and antigen should be performed.

15. We use "packet" PBS for preparation and dilution of reagents and use "homemade" PBS only for plate washing.

References

1. Alderton, W. K., Cooper, C. E., and Knowles, R. G. (2001) Nitric oxide synthases: structure, function and inhibition. *Biochem. J.* **357,** 593–615.

2. de Graaf, J. C., Banga, J. D., Moncada, S., Palmer, R. M., de Groot, P. G., and Sixma, J. (1992) Nitric oxide functions as an inhibitor of platelet adhesion under flow conditions. *Circulation* **85,** 2284–2290.

3. Haynes, D. H. (1993) Effects of cyclic nucleotides and protein kinases on platelet calcium homeostasis and mobilisation. *Platelets* **4,** 231–242.

4. Sase, K. and Michel, T. (1995) Expression of constitutive endothelial nitric oxide synthase in human blood platelets. *Life Science* **57,** 2049–2055.

5. Freedman, J. E., Loscalzo, J., Barnard, M. R., Alpert, C., Keaney, J. F., and Michelson, A. D. (1997) Nitric oxide released from activated platelets inhibits platelet recruitment. *J. Clin. Invest.* **100,** 350–356.

6. Beckman, J. S., Beckman, T. W., Chen, J., Marshall, P. A., and Freeman, B. A. (1990) Apparent hydroxyl radical production by peroxynitrite: Implications for endothelial cell injury form nitric oxide and superoxide. *Proc. Natl. Acad. Sci. USA* **87,** 1620–1624.

7. Beckman, J. S., Ischiropoulos, H., Zhu, L., van der Woerd, M., Smith, C. D., Chen, J., et al. (1992) Kinetics of superoxide dismutase- and iron-catalysed nitration of phenolics by peroxynitrite. *Arch. Biochem. Biophys.* **298,** 438–445.

8. Eiserich, J. P., Cross, C. E., Jones, A. D., Halliwell, B., and van der Vliet, A. (1996) Formation of nitrating and chlorinating species by reaction of nitrite with hypochlorous acid. A novel mechanism for nitric oxide-mediated protein modification. *J. Biol. Chem.* **271,** 19,199–19,208.

9. Gunther, M. R., Hsi, L. C., Curtis, J. F., Gierse, J. K., Marnett, M. J., Eling, T. E., et al. (1997) NO trapping of the tyrosyl radical of prostaglandin-H-synthase 2 leads tyrosyl immoxyl radical and nitrotyrosine formation. *J. Biol. Chem.* **272,** 17,086–17,090.

10. Akaike, T. (2000) Mechanisms of biological S-nitrosation and its measurement. *Free Radic. Res.* **33,** 461–469.

11. Stubauer, G., Giuffre, A., and Sarti, P. (1999) Mechanism of S-nitrosothiol formation and degradation mediated by copper ions. *J. Biol. Chem.* **274,** 28,128–28,133.

12. Matsuba, T. and Ziff, M. (1986) Increased superoxide release from human endothelial cells in response to cytokines. *J. Immunol.* **137,** 3295–3298.

13. Mugge, A., Blandes, R. B., Boger, R. H., Dwenger, A., Bode-Boger, S., Folich, J. C., et al. (1994) Vascular release of superoxide radicals is enhanced in hypercholesterolaemic rabbits. *J. Cardiovasc. Pharmacol.* **24,** 994–998.

14. Kooy, N. W. and Royall, J. A. (1994) Agonist-induced peroxynitrite production from endothelial cells. *Arch. Biochem. Biophys.* **310,** 352–359.

15. Naseem, K. M., Low, S., Sabetkar, M., Bradley, N. J., Khan, J., Jacobs, M., et al. (2000) The nitration of cytosolic proteins during agonist-induced activation of platelets. *FEBS Lett.* **473,** 119–122.

16. Wong, P. S., Eiserich, J. P., Reddy, S., Lopez, C. L., Cross, C. E., and van der Vliet, A. (2001) Inactivation of glutathione S-transferases by nitric oxide-derived oxidants: exploring a role for tyrosine nitration. *Arch. Biochem. Biophys.* **394,** 216–228.

17. Riobo, N. A., Clementi, E., Melani, M., Boveris, A., Cadenas, E., Moncada, S., et al. (2001) Nitric oxide inhibits mitochondrial NADH:ubiquinone reductase activity through peroxynitrite formation. *Biochem. J.* **359,** 139–145.

18. Turko, I. V., Marcondes, S., and Murad, F. (2001) Diabetes-associated nitration o tyrosine and inactivation of succinyl-CoA:3-oxoacid CoA transferase. *Am. J. Physiol.* **281,** H2289–H2294.

19. Knapp. L. T., Kanterewicz, B. I., Hayes, E. L., and Klann, E. (2001) Peroxynitrite induced tyrosine nitration and inhibition of protein kinase C. *Biochem. Biophys. Res. Comm.* **286,** 764–770.

20. Koltz, L. O., Schieke, S. M., Sies, H., and Holbrook, N. J. (2000) Peroxynitrite activates the phosphoinositide 3-kinase/Akt pathway in human fibroblasts. *Biochem. J.* **352,** 219–225.

21. Zhang, P., Wang, Y. Z., Kagan, E., and Bonner, J. C. (2000) Peroxynitrite targets the epidermal growth factor receptor, Raf-1 and MEK independently to activate MAPK. *J. Biol. Chem.* **275,** 22,479–22,786.

22. Gow, A., Duran, D., Malcolm, S., and Ischiropoulos, H. (1996) Effects of peroxynitrite-induced protein modifications on tyrosine phosphorylation and degradation. *FEBS Letts.* **385,** 63–66.

23. Low, S. Y., Sabetkar, M., Bruckdorfer, K. R., and Naseem, K. M. (2002) The role of protein nitration in the inhibition of platelet activation by peroxynitrite. *FEBS Lett.* **511,** 59–64.

24. Mondoro, T. H., Shafer, B. C., and Vostal, J. G. (1997) Peroxynitrite induced tyrosine nitration and phosphorylation in human platelets. *Free Rad. Med. Biol.* **22,** 1055–1063.

25. Boulos, C., Jiang, H., and Balazy, M. (2000) Diffusion of peroxynitrite into the human platelet inhibits cyclooxygenase via nitration of tyrosine residues. *J. Pharm. Exp. Therap.* **293,** 222–229.

26. Sabetkar, M., Low, S. Y., Naseem, K. M., and Bruckdorfer, K. R. (2002) The nitration of proteins in platelets: significance in platelet activation. *Free Rad. Med. Biol.* **33,** 728–736.

27. Hughes, M. N. and Nicklin, H. G. (1970) A possible role for the species peroxonitrate in nitrification. *Biochem. Biophys. Acta* **222,** 660–661.

28. Khan, J., Brennan, D. M., Bradley, N. J., Gao, B., Bruckdorfer, K. R., and Jacobs, M. (1998) 3-nitrotyrosine in the proteins of human plasma determined by an ELISA method. *Biochem. J.* **330,** 795–801.

19

Nitric Oxide Signaling in Platelets

Sylvia Y. Low and K. Richard Bruckdorfer

1. Introduction

Nitric oxide was recognized fifteen years ago to be an endothelium-dependent relaxing factor with an important role in vasomotor control through its actions on vascular smooth muscle *(1)*. Shortly after this discovery it was also demonstrated that nitric oxide (NO) is an inhibitor of platelet function and plays a physiological role in the reduction in platelet activation *(2)*. This is achieved because platelets while circulating, being the smallest of the blood cells, are closest to the endothelium, which is considered to be the most important source of NO in the vasculature. However, it was soon realized that platelets themselves are capable of biosynthesising NO when they are activated *(3)*.

1.1. Nitric Oxide Biosynthesis

The biosynthesis of NO is now well understood, if not completely. It is formed by a series of oxidation-reduction mechanisms from the amino acid L-arginine *(4)*, which is normally present in high concentrations in the plasma (80 μM) and at even higher concentrations intracellularly. These mechanisms yield NO and citrulline, which is also a well-known metabolite associated with the urea cycle (**Fig. 1**). The enzyme that catalyzes this process is NO synthase, which has a number of different isoforms, but requires a large number of cofactors appropriate to the oxidation/reductions and also calmodulin to bind calcium, another requirement for activity *(5)*.

In the endothelium (and platelets) there is a low-output, constitutive enzyme that responds rapidly to the presence of selected agonists or mechanical stimulation and a slightly different enzyme in nerves (**Table 1**). In contrast, macrophages and some other cells have an inducible form, which, after induction for 18–24 h, produces large amounts of NO, mainly as a front-line defense mechanism against microorganisms.

It is now widely accepted that NO and another endothelium-derived platelet inhibitor, prostacyclin, are important inhibitory regulators of platelet activation and act synergistically. Whereas prostacyclin is not biosynthesized in platelets, which actually

From: *Methods in Molecular Biology, vol. 273:*
Platelets and Megakaryocytes, Vol. 2: Perspectives and Techniques
Edited by: J. M. Gibbins and M. P. Mahaut-Smith © Humana Press Inc., Totowa, NJ

Overall reaction:

L-arginine + 2 NADPH + O$_2$ ⟶ L-citrulline + NO + 2NADP$^+$

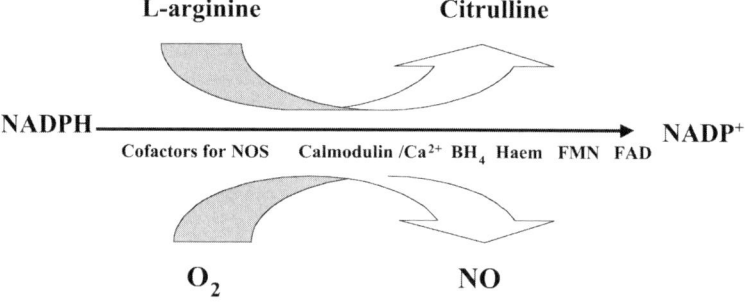

Fig. 1. Biosynthesis of nitric oxide.

Table 1
Isoforms of Nitric Oxide Synthases and Their Functions

eNOS (NOS III)	nNOS (NOS I)	iNOS (NOS II)
Cardiovascular system (platelets)	Central nervous system	Nonspecific immune system
Relaxation of smooth muscle	Neurotransmitter	Kills bacteria and other microorganisms
Regulates blood flow and blood pressure	Peripheral nonadrenergic nerves	Macrophages
Inhibits platelet activation	Important in penile erection, GI tract, and bladder	Seen in inflammatory conditions

produce the pro-aggregatory eicosanoid thromboxane A$_2$, NO is biosynthesized by platelets, but mainly when they are activated, particularly by collagen. This may have the function of self-limiting the extent of the activation of platelets *(6)*, as inhibitors of NO synthase actually increase the responsiveness of platelets to collagen. Furthermore, evidence is emerging that there is increased NO synthesis in response to the platelet inhibitor adenosine *(7,8)*.

1.2. Soluble Guanylyl Cyclase

NO is different from other regulators in that it has no specific receptor on the plasma membrane, but directly modifies intracellular target proteins to regulate their function. As a gas, it can rapidly penetrate hydrophobic membranes and diffuse to its target protein. One of these proteins is soluble guanylyl cyclase (sGC), which is abundant in platelets and, when activated, forms cyclic GMP from GTP. The activation takes place by the formation of a five-coordinate complex with the heme (ferro-protoporphyrin IX)

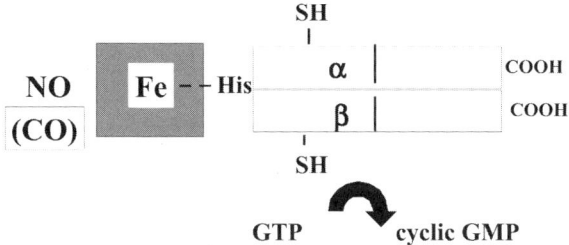

Fig. 2. The activation of soluble guanylyl cyclase by nitric oxide (and carbon monoxide).

group of sGC (**Fig. 2**). This is not quite unique to NO because carbon monoxide forms a six-coordinate complex with this heme, but conveys much lower activity on the enzyme. CO is therefore, also, a weak activator of sGC and a very weak inhibitor of blood platelets. Purification of sGC revealed that the enzyme is a heterodimer consisting of an α and a β subunit and contains the axial ligand *(9,10)*. The active parts of the enzyme (α and β subunits, which exhibit two isoforms) have little activity if the ferro-protoporphyrin IX is removed. Soluble guanylyl cyclase may be activated by a variety of NO donors, such as sodium nitroprusside and glyceryltrinitrate, which release NO either slowly or more rapidly in the presence of platelets. It may also be worth noting that there is another agent ([3-(5′-hydroxymethyl-2′-furyl)-1-benzyl indazole]), abbreviated to YC-1, which is a nonphysiological, but very potent, activator of this enzyme *(11)*. It appears to act directly on the α1 subunit in a manner analogous to the effects of forskolin on adenylyl cyclase activity.

1.2.1. Phosphodiesterases

Platelets contain a range of active phosphodiesterases (PDE). This includes the cyclic GMP-specific PDE V. This means that platelets can be inhibited by sildenafil (Viagra).

1.3. Targets for NO and Cyclic GMP

NO prevents shape change and secretion in platelets. One of the best understood of the mechanisms of action relates to its effects on the platelet cytoskeleton. The key event is the polymerization of actin and the formation of focal adhesions. As may be expected, there are many different proteins involved in this process, but the activity of one of them is regulated through cyclic nucleotides. Cyclic GMP activates the cyclic GMP-dependent protein kinase, cGPK; there is an equivalent kinase cAPK, which is activated by cyclic AMP. One of the target proteins (substrates) for both these enzymes, vasodilator-stimulated phosphoprotein (VASP), is important to the polymerization of actin (**Fig. 3**) and was discovered through the pioneering work of Halbrugge and colleagues in Germany *(12)*. This protein is a member of the ancient Enabled (Ena) protein family of *Drosophila* and probably has several members of the family represented in mammals, certainly the mouse variant Mena and the Ena-VASP like protein Evl *(13)*. They all

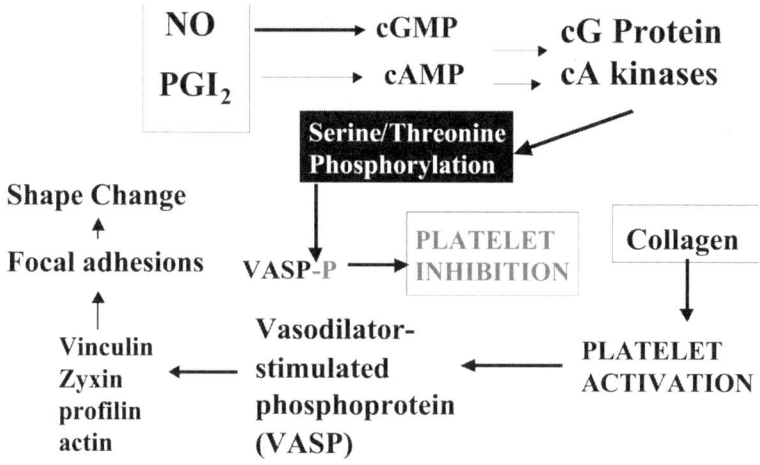

Fig. 3. Regulation of platelet activation through VASP.

share a common domain structure comprising a central proline-rich core flanked by two highly conserved Ena-VASP homology domains, EVH1 and EVH2. These domains target to focal adhesions and are recognised by VASP ligands such as zyxin, vinculin, or profilin. VASP is also thought to bind directly to F-actin.

VASP is highly enriched in platelets, and mouse VASP knockouts have shown that platelet activation is impaired in its absence *(14)*. In contrast, there were no significant changes in the activity of smooth muscle, where VASP is also expressed, presumably because of redundancy caused by the presence of other VASP family members, e.g., Mena. Therefore, platelets are ideal cells in which to study this protein in the absence of others with a similar action. VASP is phosphorylated through the activities of cyclic nucleotide-dependent protein kinases, which prevent the interactions described above and leads to the inhibition of cytoskeletal activation in platelets.

There are three sites of phosphorylation that may be phosphorylated to different extents by cAPK and cGPK as shown in **Fig. 4**. The phosphorylation of VASP is achieved at different sites by the different kinases. Serine[157] is most strongly activated by cAMP protein kinase; this brings about an apparent molecular weight shift for the protein from 46 to 50 kDa, which is not evident with the other phosphorylation sites. There are a number of tyrosines associated with sites of interaction between other proteins, but there is no strong evidence that these may be regulated by phosphorylation. This represents a least one mechanism for synergism between prostacyclin and NO where cyclic nucleotides, produced by these agents, interact to modify VASP and prevent it from participating in the reorganization of the cytoskeleton.

1.4. Nitrosation of Proteins

In addition to its regulatory mechanisms through cyclic nucleotides, NO may influence the activities of proteins by different mechanisms. These involve direct covalent

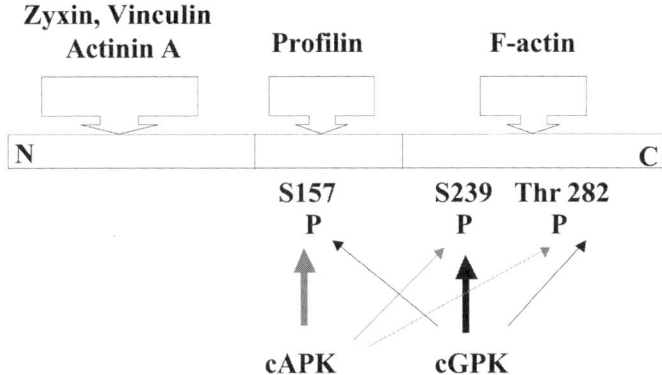

Fig. 4. Organization of VASP and its interaction with other proteins of the cytoskeleton.

modifications of proteins by nitration of the proteins (for details on the study of protein nitration, refer to Chapter 18, vol. 2) or by nitrosation of proteins. Nitration is the substitution of phenolic amino acids such as tyrosine by a nitro group, whereas nitrosation happens uniquely to sulphydryl groups of cysteine (also iron-sulfur clusters, which we will not deal with here). The nitrosations are not always very stable and may readily dissociate. They are thought to arise as the result of further chemistry involving NO.

$$NO_2 + NO \rightarrow N_2O_3$$
$$\text{Protein-SH} + N_2O_3 \rightarrow \text{Protein-SNO} + NO_2^- + H^+$$

Indeed, low levels of nitrosated proteins have been detected in plasma in the region of 20–30 nM, a great deal less than suggested in previous reports. The nitrosation of proteins has been suggested as a major method of regulation of protein activity and signal-transduction pathways *(15)*. Indeed, a list of over a hundred proteins and transcription factors, the activities of which can be modified by nitrosation, is available at the following Internet site: www.cell.com/supplemental/106/6/675/DC1 (subscription required). These include signaling proteins involved in platelet regulation, e.g., MAP kinase and JNK.

There may be other mechanisms that are likely to lead to nitrosation of proteins. These include actions of peroxynitrite on thiols, but these may not be very efficient. More importantly, proteins can be nitrosated by other nitrosated proteins or peptides by a process known as transnitrosation.

$$\text{Protein X-SH} + \text{Protein Y-S-NO} \rightarrow \text{Protein X-SNO} + \text{Protein Y-NO}$$

This process may be catalyzed enzymatically. The existence of an enzyme for denitrosation of *S*-nitrosoglutathione and also nitrosated proteins, protein disulfide isomerases, has been found in yeast and mouse macrophages *(16)*, which may terminate any signals induced by nitrosation and prevent "nitrosative stress." The presence of high concentrations of glutathione within platelets suggests that nitrosation of small peptides and proteins may be an important event in platelets, although, to date, little work has been done in this area. There is evidence that S-nitrosothiols are stored by platelets and released during platelet-neutrophil interactions *(17)*. Indeed, NO can be readily released from S-nitrosothiols such as S-nitrosoglutathione and S-nitrosoalbumin, leading to inhibition of platelet activation. This process is accelerated by glutathione peroxidase *(18)*, which is found abundantly in platelets.

NO has actions on a diverse range of cellular activity including activation of K$^+$ channels, probably the Ca^{2+}-activated channels, which are known to be redox-sensitive because of the presence of key sulfydryl groups. There may be a role here for S-nitrosation of these groups *(19)*, although there is also evidence that cyclic GMP may help regulate certain types of K$^+$ channels, most of which are represented in platelets.

The aim of this chapter is to describe a number of techniques that may be used to study nitric oxide signaling in platelets. These include analysis of nitric oxide synthesis, measurement of NO-stimulated guanyl cyclase, and cGMP-dependent kinase activity and nitrosation of proteins.

2. Materials

All materials are from Sigma (Poole, UK) or BDH (Merck, Poole, UK) unless stated otherwise.

2.1. Platelet Isolation

1. Acid citrate dextrose anticoagulant: 72.6 mM NaCl, 113.8 mM glucose, 29.0 mM trisodium citrate, and 2.8 mM citric acid, pH 6.4; *see* **Note 1**.
2. Tyrode's buffer: 137 mM NaCl, 11.9 mM NaHCO$_3$, 4.2 mM NaH$_2$PO$_4$•2 H$_2$O, and 2.7 mM KCl, pH 7.4, or a HEPES low-phosphate buffer: 150 mM NaCl, 5 mM HEPES, 0.55 mM NaH$_2$PO$_4$•2 H$_2$O, 7 mM NaHCO$_3$ (anhydrous), 2.7 mM KCl, 0.5 mM MgCl$_2$, and 5.6 mM glucose, pH 7.4, as resuspension buffers when preparing washed platelets.
3. Prostacyclin: prepared by first dissolving in ethanol (1 mg/mL), followed by division into aliquots (e.g., 10 µg) and then evaporation of the ethanol. These aliquots must be stored at –20°C under dry conditions. An aliquot, when required, is resuspended in ice-cold 1 M Tris-HCl buffer (pH 9.9) and used immediately.
4. 3.8% Sodium citrate.
5. 0.3 mM citric acid.
6. Sterile 0.2-µm filters (Sarstedt, Leicester, UK).
7. Sterile wide-gauge needles and syringes.

2.2. Nitric Oxide Preparation and Nitric Oxide Donors

1. To prepare NO solutions custom-made glass tubes are required, which are sealed with Teflon rubber septums.
2. Nitrogen and NO gases (from BOC, Luton, UK and Linde Gas UK Ltd., Stoke-on-Trent, UK. respectively).
3. Gas-tight glass syringes and tubes.
4. NO donor: A range of NO donors can be found in **Table 2** (Calbiochem, Nottingham, UK has a wide range of them; *see* **Note 2**). NO solutions can be used only on the day of preparation, and it is recommended that they be used as soon as possible after preparation.

Table 2
Range of NO Donors and Their Properties

NO Donor	Solubility	Stability	NO Release
DEA NONOate	Distilled water	Prepare fresh, keep on ice when using	$t_{1/2}$ = 16 min
NOC-5	Distilled water	Prepare fresh and keep on ice; relatively stable in alkaline solutions	$t_{1/2}$ = 93 min
NOC-7	Distilled water	Prepare fresh and keep on ice; relatively stable in alkaline solutions	$t_{1/2}$ = 10 min
MAHMA NONOate (NOC-9)	Distilled water	Prepare fresh and keep on ice; relatively stable in alkaline solutions	$t_{1/2}$ = 3 min
Spermine NONOate	Distilled water	Prepare fresh and keep on ice; relatively stable in alkaline solutions	$t_{1/2}$ = 230 min
GSNO/SNOG (*S*-nitrosoglutathione)	Distilled water (use ice-cold and free of divalent cations except Ca^{2+})	Protect from light; prepare fresh, keep on ice, and use within 2 h.	
GSNO-MEE (*S*-nitrosoglutathione monoethyl ester)	Distilled water (use ice-cold and free of divalent cations except Ca^{2+})	Protect from light; prepare fresh, keep on ice, and use within 2 h	
PROLI-NO	Distilled water	Prepare fresh and keep on ice	$t_{1/2}$ = 1.8 s at 37°C Rapidly dissociates to 1 mol Pro and 2 mol NO
SIN-1 (3-morpholino-sydnonimine) HCl	Distilled water, DMSO (*see* **Note 16**) or ethanol (*see* **Note 17**)	Prepare fresh and keep on ice	Superoxide also released on decomposition.
SNAP (*S*-nitroso-*N*-acetylpenicillamine)	DMSO, ethanol, and methanol	Protect from light; prepare fresh and keep on ice	EC_{50} = 130 nM
Sodium nitroprusside (SNP)	Distilled water	Protect from light; prepare fresh and keep on ice	

2.3. Measurement of NO Synthesis and NO Synthase Activity

1. Tritiated arginine (Amersham Biosciences, Bucks, UK).
2. Reaction buffer: 1 mM NADPH (nicotinamide adenine dinucleotide phosphate, reduced form), 6 μM tetrahydrobiopterin, 2 μM FAD (flavin adenine dinucleotide), 2 μM FMN (flavin mononucleotide), 0.2 μM calmodulin, 1.2 mM CaCl$_2$ in 50 mM Tris-HCl, pH 7.4. This is best prepared on the day of use.
3. Reaction termination buffer: 50 mM HEPES, pH 5.5, 5 mM EDTA, and a "slurry" of cationic resin (Dowex AF 50W-X8, Sigma).

Table 3
Phosphodiesterase (PDE) Inhibitors for Use in Platelets

Compound	IC_{50}	Solubility	Stability
Dipyridamole	0.9 µM, inhibits PDE type V	DMSO	Prepare fresh
IBMX (3-isobutyl-1-methylxanthine)	2–50 µM, nonselective	DMSO	Prepare fresh
4-([3′,4′-{methylenedioxy} benzyl] amino)-6-methoxyquinazoline	0.23 µM, inhibits PDE type V, no effect on other PDE isoenzymes	DMSO	Prepare fresh
Zaprinast	0.45 µM, inhibits PDE type V but can inhibit PDE IX (at higher concentration, 35 µM)	DMSO	Prepare fresh

This table gives a selection of PDE inhibitors that are relevant to platelets.

4. Platelet lysis buffer used for measuring NO synthase activity; 250 mM Tris-HCl, pH 7.4, containing 10 mM EDTA and 10 mM EGTA.
5. Liquid nitrogen or saponin (10 mg/mL) for platelet lysis.

2.4. Soluble Guanylyl Cyclase

1. Reaction buffer: 3 mM Mg^{2+}, 3 mM DTT, 0.5 mg/mL bovine serum albumin, 300 µM GTP, and 50 mM triethylamine hydrochloride.
2. The phosphodiesterase inhibitor isobutylmethylxanthine (IBMX, Calbiochem). For alternatives *see* **Table 3**.
3. Trichloroacetic acid to terminate the reaction.
4. To measure the cGMP concentrations, a commercial ELISA is suitable (Cayman, Alexis, Nottingham, UK).
5. [α-^{32}P] GTP (Amersham Biosciences) can also be used to measure the activity of this enzyme, especially for purified or partially purified preparations of the enzyme.

2.5. Other Useful Reagents

Table 4 lists useful reagents for studying NO signaling in platelets (obtainable from Calbiochem or Tocris-Cookson Chemicals, Southampton, UK; *see* **Note 2**).

2.6. cGMP Protein Targets

2.6.1. Cyclic GMP Kinases

For measuring the activity of cyclic GMP kinase, peptide substrates (e.g., RKISASEF, obtained from Promega, Southampton, UK) derived from known target proteins for cyclic GMP-dependent kinases (e.g., cyclic GMP-binding phosphodiesterase or histone H2.B) are used to determine the extent of serine phosphorylation from ^{32}P-ATP (Amersham Biosciences).

Table 4
Other Useful Compounds for Studying NO Signaling

Compound	Function	Solubility & Stability	Incubation Conditions
L-NAME (N^G-nitro-L-arginine methyl ester)	Reversible inhibitor of eNOS (IC_{50} = 500 nM).	Distilled water Protect from light Prepare fresh and keep on ice	30 min, 100 μM
L-NMMA (N^G-monomethyl-L-arginine)	Competitive inhibitor of eNOS (IC_{50} = 700 nM), iNOS (IC_{50} = 3.9 μM) and nNOS (IC_{50} = 650 nM).	Distilled water, methanol Prepare and keep on ice	30 min, 100 μM
NS 2028 (4H-8-bromo-1,2,4-oxadiazolo(3,4-d)benz(b)(1,4)oxazin-1-one	Irreversible inhibitor of soluble guanylyl cyclase (IC_{50} = 30 nM for basal and 200 nM for NO-stimulated enzyme activity).	DMSO Prepare fresh	15 min, 10 μM
ODQ (1H-[1,2,4] oxadiazolo[4,3-a] quinooxalin-1-one)	Potent, selective inhibitor of NO-sensitive soluble guanylyl cyclase (IC_{50} = 20 nM).	DMSO Prepare fresh	15 min, 10 μM
Carboxy-PTIO ([2-(4-carboxyphenyl)-4,4,5,5-tetramethylimidazoline-1-oxyl-3-oxide])	NO scavenger at pH 7.4, resulting in the generation of NO_2^-/NO_3^-.	Distilled water Prepare fresh and keep on ice	1 min, 1 μM

2.6.2. VASP

1. SDS-PAGE apparatus and reagents to run 10% polyacrylamide gels (for further details refer to Chapter 9, vol. 2).
2. Western blotting electrotransfer apparatus and reagents (for further details on Western blotting refer to Chapter 9, vol. 2).
3. Blocking and antibody-incubation buffer (PBS, pH 7.4, containing 0.5% bovine serum albumin, 1% polyethylene glycol [PEG], 1% polyvinylpyrrilidone [PVP-10], 0.2% Tween-20 [polyoxyethylene sorbitan] and 10 mM NaF) are prepared fresh and kept at 4°C until required.
4. Membrane-washing buffer contains PBS, pH 7.4, containing 0.05% Tween-20. A 10X stock solution of PBS can be made in advance and kept in the refrigerator for use in preparing these PBS buffers.
5. Antibodies: anti-VASP and anti-phosphoserine (Alexis, Nottingham, UK) anti-Ser[239] VASP (Calbiochem).
6. Horseradish peroxidase (HRP)-linked anti-mouse IgG antibody.
7. Streptavidin-HRP conjugate.
8. Nitrocellulose (pore size 0.45 μm) from Amersham Biosciences.
9. Chemiluminescence immunoblot detection reagent (e.g., West Dura chemiluminescent substrate from Perbio (Cheshire, UK).

2.7. Nitrosation of Proteins

1. Pre-cast NuPAGE 4–12% gradient gels and NuPAGE sample, running and transfer buffers (Invitrogen Life Technologies, Paisley, UK). A booklet is supplied with the NuPAGE gels that has the constituents of all the buffers used.
2. Biotinylated molecular weight markers (e.g., from Bio-Rad, Hemel Hempstead, UK).
3. Blocking and antibody incubation buffer: phosphate buffered saline, pH 7.4, Tween-20 0.05%, 10% powdered milk, and 0.1 mM diethylentriaminepentaacetic acid (*see* **Note 3**).
4. Washing buffer: phosphate-buffered saline (pH 7.4), Tween-20 0.05%, and 0.1 mM diethyl-entriaminepentaacetic acid should be made when required.
5. BCA protein assay (Perbio).
6. Chemiluminescent substrate (e.g., West Dura from Perbio).
7. Polyclonal anti-nitrosocysteine (Calbiochem).
8. Horseradish peroxidase (HRP)-linked anti-rabbit IgG antibodies.
9. Streptavidin-HRP conjugate (Amersham Biosciences).
10. Nitrocellulose (pore size 0.45 µm, Amersham Biosciences).

3. Methods
3.1. Isolation of Platelets for NO Research

The normal methods of platelet research include platelet-rich plasma, washed (isolated) platelets, and also (less commonly) whole blood. These can all be used for work with NO, but there are a few points worth noting (*see* **Notes 4,5**).

There are also other difficulties according to the methods of isolation used. One of the most popular techniques is the isolation of platelets in the presence of prostacyclin, developed by Moncada and coworkers *(20)*. The platelets are isolated in a relatively inactivated "resting" state, due to the high level of inhibition caused by the transitory elevation of cyclic AMP (*see* **Note 6**).

An alternative to this method is to lower the pH of the PRP, using citrate, to pH 6.4, which also prevents activation of the platelets during centrifugation, and restoration to pH 7.4 afterward by resuspension in buffer. It is also possible to prepare platelets by addition of NO itself during the centrifugation steps, but this may not be ideal if NO is the topic of the investigation.

1. Venous blood is collected from the antecubital vein of healthy donors or patients who deny taking medication (particularly aspirin) in the previous 14 days (*see* **Note 7**).
2. Use as wide a gauge of needle as possible (at least 21); never use vacuum methods for blood collection. The blood is put into acid citrate dextrose anticoagulant (72.6 mM NaCl, 113.8 mM glucose, 29.0 mM trisodium citrate, and 2.8 mM citric acid, pH 6.4) at ratio of 1 vol anticoagulant to 4 vol blood.
3. Centrifuge in universal tubes for 20 min at 120 g and room temperature (no brake). The upper third of the tube is carefully removed with a wide-bore pipet tip (if hemoglobin-free).
4. Prostacyclin (prepared as described in **Subheading 2.1.**) is added to a final concentration of 30 nM and the tube centrifuged for 10 min at 800g at room temperature (no cooling or brake).
5. Gently discard the supernatant and wipe the lip of the tube. Resuspend the platelets *gently* in buffer, e.g., Tyrode's buffer, pH 7.4, without disturbing any traces of red cells in the cone of the tube. They should show the characteristic swirl of platelets.

6. Adjust to the required cell count after counting in a thombocounter (typically 3×10^8 platelets/mL). After 90–120 min, the platelets recover from inhibition and will respond to agonists.

7. A variation on this is to avoid using prostacyclin (and alter cAMP), but add sodium citrate 3.8% (ratio 1:9 blood) and spin to obtain PRP ($130g$ for 15 min at 20°C). Add 0.3 mM citric acid to lower the pH to 6.4, and centrifuge at $800g$ for 10 min. The pellet is resuspended and recentrifuged, and then resuspended in appropriate buffer, e.g., Tyrode's. The platelets should be available for use after 30 min.

3.2. Preparation of Solutions of Nitric Oxide

Although this is the most laborious way of delivering NO to platelets, this is the authentic form of NO. NO donors break down to release not only NO, but also other products that may influence the processes that are of interest to the researcher (*see* **Note 8**). To produce solutions of NO, several hours need to be set aside (*see* **Note 9**). The original method for this was as described by Palmer et al. *(1)*.

1. All glassware must be cleaned in concentrated nitric acid and washed thoroughly in double-distilled water (DDW).

2. DDW is boiled for 10 min and cooled to 60°C before 100 mL is pulled under vacuum in a glass gas-sampling tube until it is full. One end is then sealed with a Teflon rubber septum.

3. The water is bubbled with a stream of N_2 for 45 min. The other end is then sealed with another septum to prevent reoxygenation. The glass tubes are not usually commercially available, but can be made by a competent glassblower (volume 100 mL).

4. Nitric oxide gas can be bought commercially and can be flushed into another gas-sampling tube until all the brown gas (NO_2) is no longer visible within it, then a Teflon septum can be placed in either end.

5. Using a gas-tight syringe, a known volume of gas is taken and injected into the deoxygenated water to make a solution of known concentrations. If a wide range of concentrations is required, e.g., to construct a dose-response curve, then several sampling tubes should be made and different volumes of gas injected.

6. The maximum amount that will dissolve is approx 5 mL of NO per 100 mL. The concentration in moles can be calculated from Avogadro's Law: 1 mole of gas at standard temperature and pressure occupies 22.4 L. If oxygen has entered the tube, the solution will turn brown (visible only at the highest concentrations) and must be discarded. The concentration of NO can be measured directly using a NO electrode (World Precision Instruments, Stevenage, UK) or by chemiluminescence after mixing with ozone using a Sievers NO detector (Sievers International, Analytix, Newcastle-upon-Tyne, UK) (*see* **Note 10**).

7. The NO solutions are then delivered in gas-tight syringes to the platelet cuvet over a range of volumes to construct a dose-response curve. The extent of inhibition will depend on the concentration of the agonists used. It may also depend on the interval between addition of antagonist and agonists. A separation of 1 min has been found to be appropriate. For an example of a typical curve, *see* **Fig. 5**. The resultant curve is sigmoid, and it is possible calculate the IC_{50} for NO and GSNO using a specific agonist of known concentration and fixed conditions; here it is 0.02 U/mL of thrombin at 37°C after 1 min *(21)*.

3.2.1. NO Donors

Clearly, the procedures for making authentic NO solutions require time and care. Consequently many researchers opt to use donors for NO that spontaneously release the

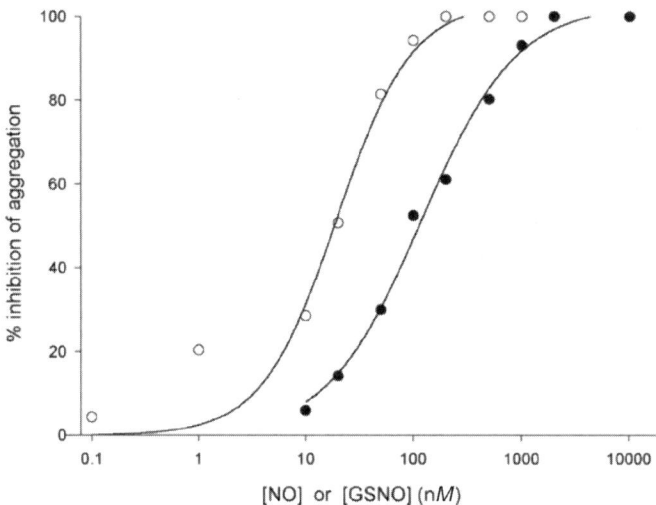

Fig. 5. The inhibition of platelet activation by NO and GSNO (dose–response). Washed platelets were incubated with a range of either NO concentrations (0–1 μM; open circles) or GSNO concentrations (0–200 nM; filled circles) for 1 min prior to addition of 0.02 U/mL thrombin, and aggregation was measured over 3 min at 37°C with stirring. The IC_{50} for NO and GSNO were determined as 110 ± 16 nM and 19.3 ± 3.5 nM respectively. Results are the mean ± SD for 3 independent experiments.

gas, each at a different rate (**Table 2** and **Note 8**). In some cases the release of NO may be accelerated by the presence of the platelets, e.g., GSNO, so that the inhibition of platelets is at least as effective as NO. An example of this is given in **Fig. 5**. S-nitrosoglutathione has a lower IC_{50} than authentic NO because most of the NO is released within the platelet.

1. Platelets are isolated as described in **Subheading 3.1.**
2. Following selection of an appropriate NO donor platelet samples are pre-incubated for a defined time. This is dependent on the rate of NO release by the NO donor used (*see* **Table 2**). An example of the range of concentrations and incubation times used for NO and NO donors are given in **Fig. 5** (*see* **Note 11**).
3. After treatment platelets can be used for aggregation experiments or for other measurements as detailed in this chapter.

3.3. Measurement of NO Synthesis

The direct way of achieving this is by measuring the conversion of radioactively labeled L-arginine to citrulline, which is formed directly in proportion to the release of NO in the reaction described in the introduction. It does require the addition of a number of cofactors (*22*), but this is simplified by the existence of kits, which can be purchased from Calbiochem.

3.3.1. Preparation of Platelet Lysates
for Measurement of NO Synthase Activity

1. Platelets are isolated as described in **Subheading 3.1.** and concentrated to 1×10^9 platelets/mL in Tyrode's buffer; then 100 μL of these platelets are treated with the reagents of interest.
2. After additions of a 10-fold concentration of lysis buffer (250 mM Tris-HCl, pH 7.4, containing 10 mM EDTA and 10 mM EGTA), the platelets are pelleted by spinning in a microfuge for 1 min.
3. Next they are lysed either with two 5-s bursts with ultrasound or with two cycles of freezing and thawing in liquid nitrogen or by addition of saponin (10 mg/mL) for 15 min at 4°C.
4. Finally the lysates are resuspended in 100 μL buffer. Lysates may also be centrifuged for 60 min at 100,000g to separate the cytosolic and membrane fractions.
5. Add to 25 μL of platelet lysate an equal volume of reaction mix containing L-^{3}H arginine (200,000 dpm), and 1 mM NADPH, 6 μM tetrahydrobiopterin, 2 μM FAD, 2 μM FMN, 0.2 μM calmodulin, 1.2 mM CaCl$_2$ in 50 mM Tris-HCl, pH 7.4, and incubate at 37°C for 1 h.
6. Terminate the reaction by addition of 400 μL of buffer containing 50 mM HEPES, pH 5.5, and 5 mM EDTA, and 100 μL of a slurry of cationic resin (Dowex AF 50W-X8).
7. Transfer the mix to a spin filter and microcentrifuge for 30 s; the eluate is then counted by liquid scintillation. Any nonenzymatic formation of citrulline can be determined by addition of the specific NO synthase inhibitor L-NAME to the mix in control tubes.
8. The rate of formation of NO (citrulline) can be determined and expressed as pmol/h/10^9 platelets.

3.4. Assay of Soluble Guanylyl Cyclase Activity

The assay of enzyme activity may be determined in intact platelets, lysates, and cytosol. There are different methods, some of which involve the use of radioactive substrate, e.g., ^{32}P-ATP or by the assay of the product cyclic GMP, using either radioimmunoassay or ELISA techniques. Typical values for cyclic GMP concentrations after exposure of platelets to authentic NO at a range of concentrations are shown in **Fig. 6**.

1. To 100 μL aliquots of the platelet preparation a final concentration of the following was added: 3 mM Mg^{2+}, 3 mM DTT, 0.5 mg bovine serum albumin/mL, 300 μM GTP, and 50 mM triethylamine hydrochloride.
2. 200 μM isobutylmethylxanthine (IBMX), a phosphodiesterase inhibitor, is added to the platelet preparations for 30 min prior to experimentation.
3. Take baseline samples before addition of a range of concentrations of NO or NO donors and then take samples after incubation for 10 min with them. The reaction is terminated by the addition of trichloroacetic acid (final concentration 7.5%). The concentrations of cyclic GMP are then measured using a commercial ELISA method (Cayman).
4. By subtracting the baseline concentrations, the rate of formation of cyclic GMP/min can be determined. Alternatively a radioimmunoassay for cyclic GMP be used.
5. The activity may also measured from the conversion of [α-^{32}P] GTP to [^{32}P] cyclic GMP; for details *see* **ref. 29**.

3.4.1. Phosphodiesterase Inhibitors

In addition, it is necessary to inhibit the breakdown of cyclic GMP when measuring soluble guanylyl cyclase activity. There is a choice of inhibitors that are specific for

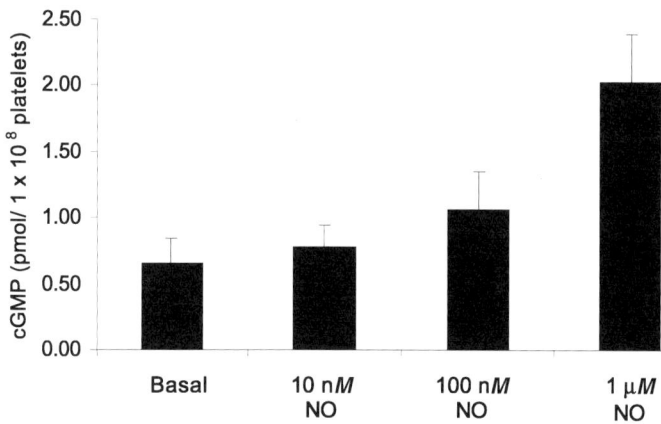

Fig. 6. Effect of NO on cyclic GMP levels in activated platelets. Platelets were exposed to NO (0–1 μ*M*) for 1 min prior to incubation with 0.02 U/mL thrombin (1 min). The reaction was terminated by addition of HCl and the samples were assayed for cGMP as described in **Subheading 3.4.** The results are means ± SD for 3 independent measurements.

cyclic GMP phosphodiesterase, or some universal ones (**Table 3**). If it is necessary to measure ambient levels of cyclic nucleotides in the platelets, the inhibitors clearly should not be used.

3.5. Other Useful Reagents for NO Research

There are other ways of regulating the activity of the NO pathway (**Table 4**), such as by inhibiting the formation of NO with specific inhibitors such as L-NAME and L-NMMA. Alternatively, all NO released can be complexed by the presence of carboxy-PTIO. Also, the stimulation of soluble guanylyl cyclase by NO may be blocked with the agents ODQ and NS 2028.

1. Platelets are isolated as in **Subheading 3.1.**
2. Prior to aggregation or other experiments, pre-incubation of platelets as indicated in **Table 4** with NOS inhibitors, soluble guanylate cyclase inhibitors, and NO scavengers can be carried out.
3. Platelet aggregation can be carried out or the platelets can be prepared for the other types of measurements described here.

3.6. Targets for Cyclic GMP

3.6.1. Cyclic GMP Kinases

The prime site of action of cyclic GMP is the activation of the specific kinase cyclic GMP-dependent protein kinase, found in membrane fractions. This enzyme is responsible for the phosphorylation of a number of target proteins that, in some cases, leads

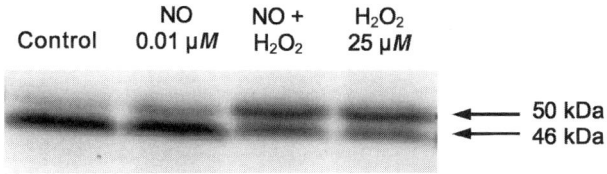

Fig. 7. Effects of NO and H_2O_2 on VASP in platelets. Platelet proteins were separated on a 10% SDS-PAGE gel, after treatment with either NO (0.01 μ*M*) or hydrogen peroxide (25 μ*M*) for 1 min. The shift in apparent molecular weight of VASP was analyzed, following immunoblotting, with a commercially available antibody against VASP (1/2000). The data are typical of at least 5 independent experiments.

to the inhibition of enzymes involved in platelet activation. Apart from VASP, which was discussed in the introduction, the list includes thromboxane A_2 receptor, heat shock protein 27, and the small GTPase Rho *(23–25)*. Platelets are particularly rich in the cyclic GMP protein kinase-1 isoform.

1. Activity can be determined either by observing changes in phosphorylation of the putative target protein or by using a known substrate (usually peptides) that it will phosphorylate. Such peptides are commercially available (e.g., from Promega, RKISASEF) and may be used, for example, to determine the extent of serine phosphorylation from ^{32}P-ATP.
2. In addition, the kinase itself (which is commercially available) can be used to investigate the phosphorylation of target proteins in platelets. Since some of these are already known, the downstream effects of NO action via cyclic GMP can be determined from the modifications observed in these proteins.
3. Specific inhibitors of these kinases exist. However, one popular inhibitor (KT5823) is reported to be ineffective in intact human platelets *(27)*.

3.6.2. VASP

The best-known substrate for cyclic GMP-dependent kinases is VASP, although it is equally regulated through cyclic AMP kinases. All three phosphorylation sites (*see* introduction) seem to be significant for phosphorylation and inactivation of the protein leading to inhibition of platelet activation. However, the phosphorylation at serine[157] is particularly significant for activation by cyclic AMP-dependent kinases and is accompanied by a specific shift in the apparent molecular weight from 46 kDa to 50 kDa (**Fig. 7**). Whereas this may give useful information, an alternative approach is to measure phosphorylation, particularly of serine. This can be achieved using anti-phosphoserine antibodies, but these are sometimes difficult to use (**Fig. 8**).

1. Washed platelets are isolated as described previously and incubated for example with NO and H_2O_2 (**Figs. 7** and **8**).
2. To stop the reaction, sample buffer (NuPAGE LDS sample buffer) with reducing agent (i.e., β-mercaptoethanol or dithiothreitol) is added. Samples and biotinylated molecular weight markers (broad range) are prepared for gel electrophoresis by incubation in a water bath at 70°C for 10 min.

NO (μ*M*)

0   0.01   1   10

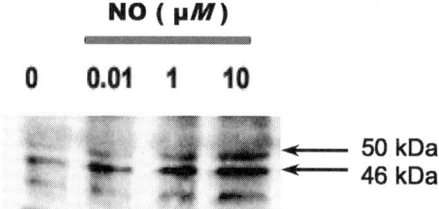

←— 50 kDa
←— 46 kDa

Fig. 8. Effect of NO on the serine phosphorylation of VASP in platelets. Platelet proteins were separated on a 15% SDS-PAGE gel, after treatment with NO (0–10 μ*M*) for 1 min. The serine phosphorylation of VASP was investigated, using immunoblotting with a commercially available antibody to phosphoserine (1/1000). Note that at high concentrations of NO (10 μ*M*) the change in apparent molecular weight is not complete and that the major part of the serine phosphorylation is in the 46-kDa band. The data are typical of at least 5 independent experiments.

3. Platelet proteins are then separated using 15% SDS-PAGE gels run under reducing conditions in standard tris-glycine SDS running buffer.
4. Following this the proteins are transferred onto nitrocellulose membrane—for example, using the NOVEX wet-transfer system with NuPAGE transfer buffer containing methanol (30 V, 1 h).
5. The membrane is blocked using the blocking buffer (*see* **Subheading 2.6.2.**) for 1 h with shaking at room temperature.
6. Next the membrane is incubated, with anti-VASP (1/2000) or anti-phosphoserine (1/1000) antibody (*see* **Note 12**) diluted in the buffer given in **step 5**, overnight with shaking at 4°C.
7. After overnight incubation, the membrane is washed several times with PBS Tween-20 0.05% (3 × 5 min and 1 × 15 min washes).
8. The membrane is next incubated with horseradish peroxidase (HRP)-linked secondary antibody (1/10,000 anti-rabbit IgG) and streptavidin-HRP conjugate (1/20,000 for detecting the biotinylated molecular-weight markers) diluted in PBS Tween-20 (as in **step 7**; *see* **Note 12**) then incubated for 1 h at room temperature with shaking. Next, washes are repeated as in **step 7**.
9. To detect proteins that have bound the antibody, the membrane is incubated with chemiluminescent reagent (e.g., West Dura) for 5 min following the manufacturer's instructions. After this the membrane is exposed to a digital camera as part of an electronic imaging system to visualize the proteins bound to the antibody.
10. Alternatively, the membrane can be exposed to photographic film after incubation of the membrane. For either of these methods the time of exposure depends on the strength of the signal.
11. Recently an antibody against Ser[239] VASP (Calbiochem) has become commercially available, which may simplify some of these procedures.

3.7. Nitrosation of Proteins

This is a relatively new area of research and little of it has yet been applied to platelet research. Nevertheless, there may be a rich future awaiting those wishing to

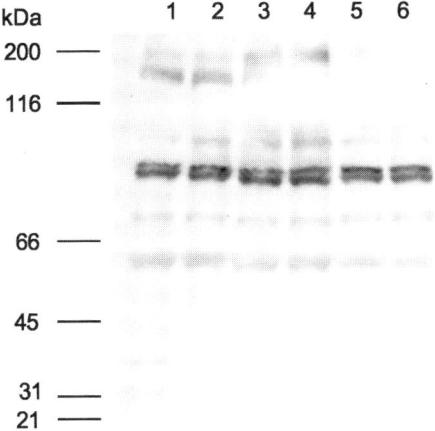

Fig. 9. Nitrosation of platelet proteins. Washed human platelet samples were removed immediately after isolation and incubated with 1 m*M* DEANO (Lane 3) or 1 m*M* GSNO (Lane 5) for 15 min. The incubation was stopped by the addition of gel sample buffer as described above. A control sample was also removed and sample buffer added (Lane 1). The same incubations were carried out 2 h later, lane 2: control, lane 4: + 1 m*M* DEANO and lane 6: + 1 m*M* GSNO. Platelet proteins were then separated on a 4–12% gradient gel and then transferred onto nitrocellulose membrane. The membrane was incubated with anti-nitrosocysteine antibody (1/100) as described above. 30 µg platelet protein loaded per lane.

invest some effort here. The principal difficulty is that nitrosation may be rapidly reversible, especially with low-molecular-weight thiols. However, protein nitrosation (at least of some proteins) may be sufficiently stable in order to identify and measure the extent of the nitrosation (*see* **Note 13**). Identification of nitrated proteins using antibodies to *S*-nitrosocysteine is also possible (*see* **Fig. 9**).

1. Washed platelets are isolated as described previously in this chapter (*see* **Note 14**). An example of a typical experiment is described in **Fig. 9**.
2. To stop the reaction, sample buffer (NuPAGE LDS sample buffer) with no reducing agent (i.e., β-mercaptoethanol or dithiothreitol) is added. Samples and biotinylated molecular-weight markers (broad range) are prepared for gel electrophoresis by incubation in a water bath at 70°C for 10 min.
3. Platelet proteins are then separated using precast 4–12% NuPAGE bis-tris gels run under nonreducing conditions in NuPAGE MOPs running buffer (200 V, 50 min).
4. Following this the proteins are transferred onto nitrocellulose membrane (e.g., using the NOVEX wet-transfer system) with NuPAGE transfer buffer containing methanol (30 V, 1 h) (*see* **Note 15**).
5. The membrane is blocked using PBS Tween-20 0.05% containing 10% powdered milk and 0.1 m*M* diethylentriaminepentaacetic acid for 1 h with shaking at room temperature.
6. The membrane is then incubated, with anti-nitrosocysteine antibody diluted 1/100 (*see* **Note 12**) in the buffer given in **step 5**, overnight with shaking at 4°C.

7. After overnight incubation, the membrane is washed several times with PBS Tween-20 0.05% containing 0.1 m*M* diethylentriaminepentaacetic acid (3 × 5 min and 1 × 15 min washes).

8. The membrane is then incubated with horseradish peroxidase (HRP)-linked secondary antibody (1/10,000 anti-rabbit IgG) and streptavidin-HRP conjugate (1/20,000 for detecting the biotinylated molecular-weight markers) diluted in PBS Tween-20 (as in **step 7**, *see* **Note 12**) then incubated for 1 h at room temperature with shaking. Next, washes are repeated as in **step 7**.

9. To detect any proteins that have bound the antibody, the membrane is incubated with chemiluminescence substrate as described for VASP experiment (*see* **Subheading 3.6.2., step 9**).

4. Notes

1. All buffers are stored at 4°C and warmed to room temperature prior to use and filtered using 0.2 µm filters.

2. NO solutions can be used only on the day of preparation, and it is recommended that they be used as soon as possible after preparation. This also applies to NO donors.

3. This buffer should be made up on the day it is required, but a 10X stock solution of PBS can be made in advance and kept refrigerated for use in preparing this buffer and the washing buffer.

4. Clearly, in whole blood, a very large proportion of any exogenous NO that may be added would be neutralized by binding to the overwhelming amounts of hemoglobin present in the erythrocytes. This may be important to the researcher, who is trying to determine the requirements for inhibition of NO or NO donors on platelets under physiological conditions. There may also be similar problems working with platelet-rich plasma, as the presence of traces of hemoglobin resulting from the lysis of erythrocytes during phlebotomy may strongly influence the dosage of NO or NO donors required for inhibition. There are some precautions that can be taken. Discard the first 2 mL of the blood sample and do not centrifuge too vigorously. Ensure that the PRP is taken well above the buffy coat and discard any "pink"-looking samples. To be absolutely sure, run a test for hemoglobin using a standard detection kit in a sample of unused PRP and accept only levels that are below the level of detection.

5. These problems can also arise with washed platelets and often require at least two washes to remove all the plasma. The presence of unlysed erythrocytes is often evident in the platelet pellet and these may have to be discarded.

6. This is also attributable to the apparent cytoprotective action of prostacyclin and, after the recovery time of approx 90 min, leaves them responsive to a wide range of platelet agonists for some hours. One disadvantage, especially if one is interested in the cyclic nucleotides, is that often cyclic AMP concentrations remain about twice as high as the normal basal levels.

7. It may also be better if the volunteers have not drunk alcohol in the past 24 h, since this also inhibits platelets, or if they have not smoked cigarettes: smoke contains large amounts of NO (up to 800 ppm). Ethics committee permission is required for healthy volunteers.

8. For example, sodium nitroprusside will release cyanide ions. Less worrisome are the products of the *S*-nitroso compounds such as *S*-nitrosoglutathione, which will also yield another product—i.e., free glutathione—which could have its own effects. This may also be a factor, as well as the rate of release, in considering which to use, depending on the nature of the planned experiments.

9. The main problems are to create oxygen-free solvents to prevent the reaction of NO with O_2 to produce NO_2 and then to handle the gas in making the solutions.

10. Testing the amount of nitrite by the Griess reaction will not indicate the actual concentration of available NO. Use of the Griess reaction is not recommended since low background levels of nitrite are required and it is sensitive only to 1 μM.

    Even more useful would be antibodies for phosphoserine, which show positional specificity in VASP. These are not commercially available, but an antibody for the serine[157] site, which recognises only the phosphorylated form, has been developed in the laboratory of U Walter.

11. If using PRP, higher concentrations of NO and NO donors are required (*see* **Fig. 5**).

12. The dilutions of primary and secondary antibodies used were optimized for the experiments described here. It is important to do this for your own experimental conditions and for new batches of antibodies. This applies to all antibodies that may be used.

13. Identification and measurement of the extent of nitrosation can be achieved in two ways. One is to use recently available antibodies against *S*-nitrosothiols on proteins separated on gels. The other is to determine the extent of nitrosation directly on whole platelets. The method of detection of *S*-nitrosothiols has been described elsewhere, but generally involves complex methodologies such as chemiluminescence *(27)*, after removal of contaminating nitrite and nitrate, as well as discontinuous fluorescence, electrochemical, and colorimetric assays. Applications of these techniques to the measurement of individual proteins isolated from cells will require a high level of sophistication, but proteomic methodology may also be applicable in the future.

14. In order to prevent the loss of nitrosylation from platelet proteins, the platelets were protected from light during incubations. All platelet incubations were carried out at room temperature.

15. All subsequent steps are carried out with the membrane being protected from light.

16. DMSO must be used with caution in platelet studies because DMSO can affect platelet function. Where possible, prepare stocks at least 100-fold more concentrated than required and in less than 100% DMSO. It is important that control incubations with DMSO at the same concentration used when preparing the compound of interest be carried out.

17. Avoid using methanol or ethanol for preparing compounds for platelet incubations. When using NO donors it is also important to consider that in some cases the breakdown products, other than NO, of these donors can also have effects on platelet function.

References

1. Palmer, R. M. J, Ferridge, A. G., and Moncada, S. (1987) Nitric oxide release accounts for the biological activity of endothelium-derived relaxing factor. *Nature (London)* **327**, 524–526.

2. Radomski, M. W., Palmer, R. M., and Moncada, S. (1987) Endogenous nitric oxide inhibits human platelet adhesion to vascular endothelium. *Lancet* **2**, 1057–1058.

3. Radomski, M. W., Palmer, R. M. J., and Moncada, S. (1990) Characterization of the L-arginine:nitric oxide pathway in human platelets. *Br. J. Pharmacol.* **101**, 325–328.

4. Palmer, R. M. J, Ashton, D. S., and Moncada, S. (1988) Vascular endothelial cells biosynthesise nitric oxide from L-arginine. *Nature (London)* **333**, 664–666.

5. Alderton, W. K., Cooper, C. E., and Knowles, R. G. (2001) Nitric oxide synthases: structure, function and inhibition. *Biochem. J.* **357**, 593–615.

6. Freedman, J. E., Loscalzo, J., Barnard, M. R., Alpert, C., Keaney, J. F., and Michelson, A. D. (1997) Nitric oxide released from activated platelets inhibits platelet recruitment. *J. Clin. Invest.* **100**, 350–356.

7. Anfossi, G., Russo, I., Massucco, P., Mattiello, L., Cavalot, F., Balbo, A., et al. (2002) Adenosine increases human platelet levels of cGMP through nitric oxide—possible role in its antiaggregating effect. *Thromb. Res.* **105**, 71–78.

8. Anfossi, G., Russo, I., Massucco, P., Mattiello, L., and Trovati, M. (2002) Catecholamines, via β-adrenoreceptors, increase intracellular concentrations of 3′,5′-cyclic guanosine monophosphate (cGMP) through nitric oxide in human platelets. *Thromb. Haemost.* **87,** 539–540.

9. Ignarro, L. J. (1989) Heme-dependent activation of soluble guanylate cyclase by nitric oxide: regulation of enzyme activity by porphyrins and metalloporphyrins. *Seminars in Haematol.* **26,** 63–76.

10. Wedel, B., Harteneck, C., Foerster, J., Friebe, A., Schultz, A., Schultz, G., et al. (1995) Functional domains of soluble guanylyl cyclase. *J. Biol. Chem.* **270,** 24,871–24,875.

11. Friebe, A., Schultz, G., and Koesling, D. (1998) Sensitizing soluble guanylyl cyclase to become a highly co-sensitive enzyme. *EMBO J.* **15,** 6863–6868.

12. Halbrugge, M., Friedrich, C., Eigenthaler, M., Schanzenbacher, P., and Walter, U. (1990) Stoichiometric and reversible phosphorylation of 46 kDa protein in human platelets in response to cGMP and cAMP elevation. *J. Biol. Chem.* **265,** 3088–3093.

13. Prehoda, K. E., Lee, D. H., and Lim, W. A. (1999) Structure of the Enabled/VASP homology 1 domain—peptide complex: a key componenet in the spatial control of actin assembly. *Cell* **97,** 471–480.

14. Aszodi, A., Pfeifer, A., Ahmad, M., Glauner, M., Zhou, X.-H., Ny, L., et al. (1999) The vasodilator-stimulated phosphoprotein is involved in cGMP and cAMP-mediated inhibition of agonist induced platelet aggregation, but is dispensable for smooth muscle function. *EMBO J.* **18,** 37–48.

15. Stamler, J. S., Lamas, S., and Fang, F. C. (2001) Nitrosylation: the prototypic redox-based signaling mechanism. *Cell* **106,** 675–683.

16. Liu, L., Hausladen, A., Zeng, M., Que, L., Heitman, J., and Stamler, J. S. (2001) A metabolic enzyme for S-nitrosothiol conserved from bacteria to humans. *Nature* **410,** 490–494.

17. Hiriyama, A., Noronha-Dutra, A. A., Gordge, M. P., Neild, G. H., and Hothersall, J. S. (1999) S-nitrosothiols are stored by platelets and released during platelet-neutrophil interactions. *Nitric Oxide: Biology and Chemistry* **3,** 95–104.

18. Waldron, G. J. and Cole, W. C. (1999) Activation of vascular smooth muscle K^+ channels by endothelium-derived relaxing factors. *Clin. Pharm. and Physiol.* **26,** 180–184.

19. Freedman, J. E., Frei, B., Welch, G. N., and Loscalzo, J. (1995) Glutathione peroxidase potentiates the inhibition of platelet function by S-nitrosothiols. *J. Clin. Invest.* **96,** 394–400.

20. Vargas, J. R., Radomski, M., Moncada, S. (1982) The use of prostaglandin in the separation from plasma and washing of human platelets. *Prostaglandins* **23,** 929–945.

21. Naseem, K. M. and Bruckdorfer, K. R. (1995) Hydrogen peroxide at low concentrations strongly enhances the inhibitory effect of nitric oxide on platelets. *Biochem. J.* **310,** 149–153.

22. Riddell, D. R., Graham, A., and Owen, J. S. (1997) Apolipoprotein E inhibits platelet aggregation through the L-arginine:nitric oxide pathway. *J. Biol. Chem.* **272,** 89–95.

23. Wang, G. R., Zhu, Y., Halushka, P. V., Lincoln, T. M., and Mendelsohn, M. E. (1998) Mechanism of platelet inhibition by nitric oxide: in vivo phosphorylation of thromboxane receptor by cyclic-GMP dependent protein kinase. *Proc. Natl. Acad. Sci. USA* **95,** 4888–4893.

24. Butt, E., ImmLer, D., Meyer, H. E., Kotlyarov, A., Laass, K., and Gaestel, M. (2001) Heat shock protein 27 is a substrate of cyclic GMP-dependent kinase in intact human platelets. *J. Biol. Chem.* **276,** 7108–7113.

25. Sawada, N., Itoh, H., Yamashita, J., Doi, K., Inoue, M., Masatasugu, K., et al. (2001) cGMP-dependent protein kinase phosphorylates and inactivates RhoA. *Biochem. Biophys. Res. Comm.* **280,** 798–805.

26. Colbran, J. L., Francis, S. H., Leach, A. B., Thomas, M. K., Jiang, H., McAllister, L. M., et al. (1992) A phenylalanine in peptide substrates provides for selectivity between cGMP and cAMP-dependent protein kinases. *J. Biol. Chem.* **287,** 9589–9594.
27. Burkhard, M., Glazova, M., Gambaryan, S., Vollkommer, T., Butt, E., Bader, B., et al. (2000) KT5823 inhibits cGMP-dependent protein kinase activity in vitro but not in intact human platelets and rat mesangial cells. *J. Biol. Chem.* **275,** 33,536–33,541.
28. Marley, R., Patel, R. P., Orie, N., Ceaser, E., Darley-Usmar, V., and Moore, K. P. (2001) Formation of nanomolar concentrations of S-nitroso-albumin in human plasma by nitric oxide. *Free Radical Biology and Medicine* **31,** 688–696.
29. Humbert, P., Niroomand, F., Fischer, G., Mayer, B., Koesling, D., Hinsch, K. D., et al. (1990) Purification of soluble guanylyl cyclase from bovine lung by a new immunoaffinity chromatographic method. *Eur. J. Biochem.* **190,** 273–278.

20

Snake Venom Toxins Affecting Platelet Function

Robert K. Andrews, Elizabeth E. Gardiner, and Michael C. Berndt

1. Introduction

Botrocetin from the South American pit viper *Bothrops jararaca* was described as an activator of von Willebrand factor-dependent platelet aggregation by Read, Shermer, and Brinkhous in 1978 *(1)*. Subsequently, botrocetin has been widely used as an important in vitro modulator in the analysis of von Willebrand factor and platelet aggregation. Botrocetin has since been identified as a heterodimer of the C-type lectin family of snake venom proteins (~25 kDa nonreduced, ~14 kDa reduced), the primary sequence and crystal structure have been determined, and specific binding sites within the A1 domain of von Willebrand factor have been identified *(2–8)*. Interestingly, members of the metalloproteinase-disintegrin family of snake venom proteins—jararhagin, jaracetin, and one-chain botrocetin—are functionally related to two-chain botrocetin, and also interact with the von Willebrand factor A1 domain *(4,9,10)*. Jaracetin and one-chain botrocetin are variably processed forms of jararhagin and are found in the same viper species as the C-type lectin family, two-chain botrocetin. In this regard, recent evidence suggests that C-type lectin proteins and metalloproteinase-disintegrins may be derived from a common gene encoding a much larger precursor protein *(11)*.

An increasing number of C-type lectin proteins and metalloproteinases from cobra or viper venoms have been reported that selectively target either von Willebrand factor or its platelet receptor, glycoprotein (GP) Ibα of the GPIb-IX-V complex (some examples are shown in **Table 1**). These include the cobra venom metalloproteinase-disintegrin, mocarhagin from the Mozambiquan spitting cobra *Naja mocambique mocambique (Naja mossambica mossambica)* that cleaves GP Ibα within an anionic sequence containing three sulfated tyrosines *(19,24,25)*. Mocarhagin also cleaves the neutrophil receptor PSGL-1 (P-selectin glycoprotein ligand-1) within an analogous sequence, and therefore has antiinflammatory activity in addition to being antithrombotic *(18)*. In contrast, kaouthiagin from *Naja kaouthia* cleaves von Willebrand factor *(21)*. A variety of C-type lectin proteins including alboaggregin-B (**Table 1**) bind to GPIbα and inhibit von

From: *Methods in Molecular Biology, vol. 273:*
Platelets and Megakaryocytes, Vol. 2: Perspectives and Techniques
Edited by: J. M. Gibbins and M. P. Mahaut-Smith © Humana Press Inc., Totowa, NJ

Table 1
Examples of Snake Toxins that Target Vascular Cell Adhesion Receptors or Ligands[a]

Protein *snake species*	Molecular mass	Vascular target(s)	Typical yield (mg/g venom)	Ref.
C-type lectin family				
Two-chain botrocetin	~25 kDa NR	von Willebrand factor	5–15	*2,3*
Bothrops jararaca	~14 kDa R	(A1 domain)		
Bitiscetin	~25 kDa NR	von Willebrand factor		*12*
Bitis arietans	~14 kDa R	(A3 domain)		
Alboaggregin-A	~50 kDa NR	GP Ibα (His1-Glu282),	2–5	*13*
Trimeresurus albolabris	~14 kDa R	and GP VI		
Alboaggregin-B	~25 kDa NR	GP Ibα (His1-Glu282)	2–5	*13*
Trimeresurus albolabris	~14 kDa R			
Convulxin	~85 kDa NR	GP VI		*14,15*
Crotalus durissus terrificus	~14 kDa R			
Ophioluxin	~85 kDa NR	GP VI		*16*
Ophiophagus hannah	~16 kDa R			
Metalloproteinases				
Alborhagin	~55 kDa NR/R	GP VI, Fibrinogen	2–5	*17*
Trimeresurus albolabris				
Mocarhagin	~55 kDa NR/R	GP Ibα (Glu282/Asp283),	2–5	*17,18*
Naja m. mocambique		Fibrinogen		
Jararhagin	~55 kDa NR/R	Binds von Willebrand factor	2–5	*9,20*
Bothrops jararaca		and $\alpha_2\beta_1$ A-type domains		
Kaouthiagin	~50 kDa NR/R	von Willebrand factor		*21*
Naja kaouthia		(cleaves 708/709)		
Crotalin	~25 kDa NR/R	GP Ibα; von Willebrand		*22*
Crotalus atrox		factor		
Catrocollastatin	~50 kDa NR/R	Collagen		*23*
Crotalus atrox				

[a]Molecular mass in kiloDaltons (kDa) based on SDS-PAGE (NR, nonreduced; R, reduced). Yields shown are for proteins purified in our laboratory. For additional examples of structurally/functionally related venom proteins, *see* **refs. 24–26**.

Willebrand factor binding *(13,24,26)*. Related venom proteins target collagen or its receptors, $\alpha_2\beta_1$ or GPVI on platelets. For example, convulxin and ophioluxin of the C-type lectin family *(14–16)* and the metalloproteinase, alborhagin *(17)*, are agonists targeting GPVI. Interestingly, the 50-kDa heterotetrameric C-type lectin protein, alboaggregin-A, interacts with both GPIbα and GPVI *(27,28)*. Together, these snake toxins provide both prothrombotic and antithrombotic reagents, and have proven invaluable for analysis of molecular mechanisms underlying platelet function. Although less well characterized, snake toxins such as botrocetin and convulxin have also been used to study megakaryocyte function *(29,30)*.

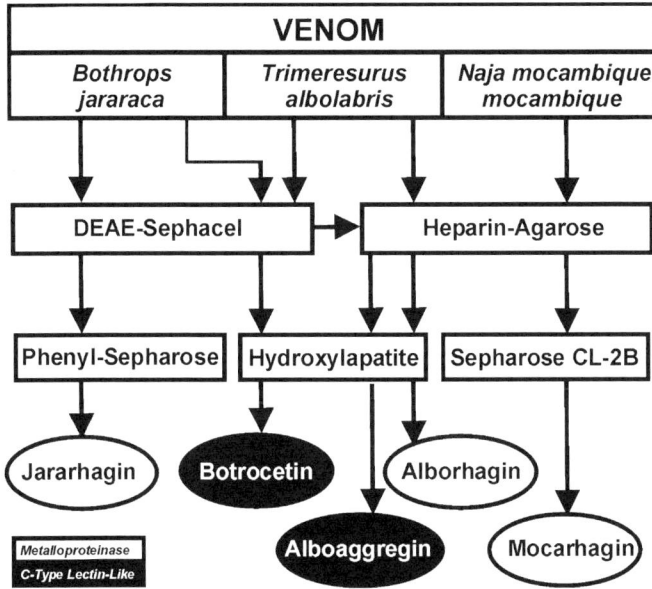

Fig. 1. Algorithm for purification of snake venom proteins.

We have used common strategies for isolation of either C-type lectin or metalloproteinase-disintegrin proteins from various species. An algorithm for purification protocols used in our laboratory is summarized in **Fig. 1**. In addition, many pitfalls and other points to consider when purifying venom proteins are relevant to a variety of snake toxins from different species. This chapter will describe the source, isolation, and use of venom proteins. We will focus on two specific examples: botrocetin of the C-type lectin family that activates von Willebrand factor, and alborhagin that targets GPVI. Finally, imminent new directions in venom protein analysis, such as isolation of selective probes using constitutively activated forms of receptor or ligand as affinity steps, or venomic (venom proteomic) approaches, will be discussed briefly.

2. Materials

2.1. Purification of Venom Proteins

Lyophilized crude venom for isolation of alborhagin *(Trimeresurus albolabris)* or botrocetin *(Bothrops jararaca)* was from Venom Supplies, Tanunda, South Australia or Sigma, St. Louis, MO, respectively. Regulations pertaining to transportation of products from endangered species may limit acquisition of some venoms internationally. Different suppliers may not always guarantee authenticity of particular species, depending on collection procedures used. In this regard, venoms from snakes and other animals utilize a conserved structural framework such as a C-type lectin, with primary sequence

Table 2
Chromatography Media and Buffers for Snake Toxin Purification[a]

Column	Column/wash buffer	Elution buffer	Ref.
Heparin-agarose (1.5×30 cm)	$0.01\ M$ Tris-HCl, $0.15\ M$ NaCl, pH 7.4 (TS buffer)	Linear $0.15–1.0\ M$ NaCl gradient in $0.01\ M$ Tris-HCl, pH 7.4 (total volume, 400 mL)	*13*
Hydroxylapatite (1.5×20 cm)	5 mM Na_2HPO_4, pH 6.8	Linear $5–200$ mM Na_2HPO_4, pH 6.8 gradient (total volume, 200 mL)	*2*
DEAE-sephacel (2.5×40 cm)	$0.01\ M$ Tris-HCl, $0.15\ M$ NaCl, pH 7.4 (TS buffer)	Linear $0.15–1.0\ M$ NaCl gradient in $0.01\ M$ Tris-HCl, pH 7.4 (total volume, 500 mL)	*2*
Phenyl-sepharose (1.5×30 cm)	$0.01\ M$ Tris-HCl, $1.2\ M$ $(NH_4)_2SO_4$, pH 7.4	Linear $1.2\ M$ to zero $(NH_4)_2SO_4$ gradient in $0.01\ M$ Tris-HCl, pH 7.4 (total volume, 400 mL)	*3*
Sepharose CL-2B (2.5×40 cm)	$0.01\ M$ Tris-HCl, $0.5\ M$ NaCl, pH 7.4	(Gel filtration)	*17*

[a]Columns run at 30 mL/h, with samples loaded in the column buffer, and washed until eluate (monitored by A_{280}) returns to baseline before applying the gradient.

differences and/or posttranslational modification leading to enormous functional variability *(24–26)*. The large number of isoforms has the advantage of providing extensive functional probes, but at the same time may complicate purification and characterization of particular forms. Batch-to-batch variation of venom profiles may also adversely affect yield. Metalloproteinases are prone to post-translational processing or autodigestion in the venom (during purification or storage), so that variant forms of the same protein may exist in different batches of venom. For example, jaracetin is a disulfide-bridged homodimer composed of one-chain botrocetin subunits *(9)*. In general, it is recommended to monitor fractions throughout the purification, under nonreducing and reducing conditions, and start again using a different batch of venom if the preparation is atypical (*see* **Note 1**).

2.1.1. Metalloproteinases: Purification of Alborhagin

The steps used for isolation of alborhagin are shown in **Fig. 1**. Resins, column buffers, and elution conditions are summarized in **Table 2**.

1. Heparin-agarose resin (Bio-Rad, Richmond, CA) packed into a 1.5×20-cm glass column, and equilibrated by washing with at least 10 bed volumes of column buffer (**Table 2**).
2. Hydroxylapatite (Bio-Rad) washed several times before use by suspending in column buffer in a beaker, and aspirating the upper layer containing fine particles as the bulk of the material settles on standing (*see* **Note 2**). It is then packed into a 1.5×20-cm column, and washed with 10 bed volumes of column buffer (**Table 2**).
3. Dialysis tubing (Union Carbide, Chicago, IL) with <10,000 mol wt pore size may be used for all dialysis steps.

4. Ultrafiltration device (10- or 50-mL capacity) fitted with a YM30 membrane (Amicon, Danvers, MA).
5. BCA (bicinchoninic acid) protein quantitation kit and standard bovine serum albumin solution (Pierce, Rockford, IL).

2.1.2. C-Type Lectin Proteins: Purification of Botrocetin

The steps used for isolation of botrocetin are shown in **Fig. 1**. Resins, column buffers, and elution conditions are summarized in **Table 2**.

1. Ammonium sulfate (solid).
2. DEAE-Sephacel (Pharmacia, Uppsala, Sweden) packed into a 2.5 × 40-cm glass column and equilibrated by washing with 10 bed volumes of column buffer. The pH of the eluate should be 7.4 when the column is equilibrated.
3. Hydroxylapatite (Bio-Rad) washed several times before use by suspending in column buffer in a beaker, and aspirating the upper layer containing fine particles as the bulk of the material settles on standing (*see* **Note 2**). It is then packed into a 1.5 × 20-cm column, and washed with 10 bed volumes of column buffer **(Table 2)**.
4. Dialysis tubing (Union Carbide) with <10,000 mol wt pore size may be used for all dialysis steps.
5. Ultrafiltration device (10- or 50-mL capacity) fitted with a YM10 membrane (Amicon, Danvers, MA).

2.2. Functional Analysis of Venom Proteins

The method used to test alborhagin or botrocetin for induction of platelet aggregation (**Subheading 2.2.1.**), may also be used for routine screening of venom fractions for activators or inhibitors targeting von Willebrand factor, GPIbα, or other ligand-receptor interactions. Alborhagin induces aggregation of citrated platelet-rich plasma by a mechanism involving binding to GPVI *(17)*. This aggregation involves activation of GPIIb-IIIa ($\alpha_{IIb}\beta_3$), but is independent of GPIbα. Botrocetin induces aggregation of citrated platelet-rich plasma due to activation of plasma von Willebrand factor, and aggregation is GPIbα-dependent *(2–5)*. These pathways are distinguishable using GPIbα antagonists, such as inhibitory anti-GPIbα monoclonal antibodies *(3)* or mocarhagin *(19)*. Finally, metalloproteinase activity of alborhagin (or many other venom metalloproteinases) may be evaluated using cleavage of fibrinogen *(31)*, a common venom metalloproteinase substrate (**Subheading 3.2.2.**).

2.2.1. Platelet Aggregation

1. Winged infusion kit with 19-gauge needle (Baxter Healthcare, Old Toongabbie, N.S.W., Australia). A 19-gauge needle should be used to minimize platelet activation during blood collection.
2. Trisodium citrate stock solution (3.2 g trisodium citrate per 100 mL distilled water, without pH adjustment).
3. EDTA: 1 *M* stock solution in distilled water, titrated to pH 7.0 with NaOH.
4. Anti-$\alpha_{IIb}\beta_3$ monoclonal antibody, CRC64, which blocks alborhagin-dependent platelet aggregation *(17)* (Dr. A. Mazurov, Moscow, Russia). Similar inhibitory anti-$\alpha_{IIb}\beta_3$ mono-

clonal antibodies are commercially available (e.g., Reopro [abciximab] from Eli Lilly, Indianapolis, IN).

5. Ristocetin A sulfate (Paesel and Lorei, Duisburg, Germany).
6. Inhibitory anti-GPIbα monoclonal antibody, AK2 *(3)* (Dako, Carpinteria, CA).
7. Dual beam lumiaggregometer, sample and reference cuvets (0.5-mL capacity), and siliconized disposable stir bars (Chronolog, Havertown, PA).

2.2.2. Digestion of Fibrinogen

Human fibrinogen is available from Kabivitrum, Stockholm, Sweden *(31)*.

3. Methods
3.1. Purification of Venom Proteins

Routine purification of platelet-targeting proteins from crude snake venom (**Subheadings 3.1.1.** and **3.1.2.**) involves separation of proteins (**Fig. 1**), then analysis of fractions on SDS-polyacrylamide gels and testing in platelet functional assays (**Subheading 3.2.**). SDS-polyacrylamide gel electrophoresis (SDS-PAGE) utilizes exponential 5–20% gradient resolving gels under nonreducing and reducing conditions, to distinguish C-type lectin family proteins (~25 kDa or higher nonreduced, 14-kDa subunits reduced) and metalloproteinases (typically 50–60 kDa, nonreduced and reduced). C-type lectin family proteins of higher multimer size may be approx 14 kDa reduced, and approx 50 kDa or approx 85 kDa nonreduced *(13–16)*. For further details on SDS-PAGE techniques, *see* Chapter 9, vol. 2.

3.1.1. Metalloproteinases: Purification of Alborhagin

1. Dissolve 0.2 g lyophilized *Trimeresurus albolabris* venom in 10 mL of TS buffer and load at 30 mL/h onto a 1.5 × 20-cm heparin-agarose column equilibrated with TS buffer. Chromatography can be performed at 22°C. Monitor the eluate by absorbance at 280 nm (*see* **Note 3**).
2. Wash with TS buffer until A_{280} returns to baseline, and elute with a linear 0.15–1.0 M NaCl gradient in 0.01 M Tris-HCl, pH 7.4. After loading, collect approx 5-mL fractions of flowthrough, wash, and eluted protein, and analyze on 5–20% SDS-polyacrylamide gels under nonreducing and reducing conditions (*see* Chapter 9, vol. 2 for further details).
3. Pool fractions containing alborhagin (~60 kDa, reduced and nonreduced) that elute with >0.5 M NaCl, and dialyze (4 × 4 L changes at 4°C) into 5 mM Na$_2$HPO$_4$, pH 6.8 (*see* **Note 2**).
4. Load at approx 30 mL/h onto 1.5 × 20-cm column of hydroxylapatite in 5 mM Na$_2$HPO$_4$, pH 6.8. Wash with this buffer, then elute with a linear gradient of 5–200 mM Na$_2$HPO$_4$, pH 6.8. Analyze fractions (~5 mL each) on SDS-polyacrylamide gels.
5. Purified protein is pooled, concentrated using an Amicon ultrafiltration device fitted with a YM30 membrane, and dialyzed into TS buffer (4 × 4 L at 4°C) (*see* **Notes 4,5**).

3.1.2. C-Type Lectin Proteins: Purification of Botrocetin

1. Dissolve 5 g lyophilized *Bothrops jararaca* venom in 100 mL TS buffer. Fractionate by adding solid ammonium sulfate (39 g) a little at a time over approx 10 min with stirring, then stir for a further 20 min at 22°C. This gives a final ammonium sulfate concentration of 60% saturation, and the precipitate represents a "0–60% cut." Centrifuge at 10,000g for 20 min. Decant the supernatant, and resuspend the pellet in 100 mL TS buffer and dialyze against the same buffer (4 × 5 L changes at 4°C).

2. Load at 30 mL/h onto a 2.5 × 40-cm column of DEAE-Sephacel equilibrated with TS buffer at 22°C. After loading, wash the column with TS buffer until the A_{280} of the eluate returns to baseline.
3. Elute with a 400-mL linear 0.15–1.0 M NaCl gradient in 0.01 M Tris-HCl, pH 7.4. Collect approx 10-mL fractions and analyze on 5–20% SDS-polyacrylamide gels (*see* **Note 6**).
4. Pool fractions containing botrocetin and dialyze into 5 mM Na$_2$HPO$_4$, pH 6.8 (4 × 4 L changes at 4°C) (*see* **Note 2**).
5. Load at 30 mL/h onto a 1.5 × 20-cm column of hydroxylapatite in 5 mM Na$_2$HPO$_4$, pH 6.8, at 22°C. Wash with the column buffer until the A_{280} of the eluate returns to baseline, and elute with a linear 200-mL gradient of 5–200 mM Na$_2$HPO$_4$, pH 6.8. Botrocetin elutes late in the gradient (>100 mM Na$_2$HPO$_4$).
6. Dialyze botrocetin into TS buffer (3 × 2 L), and concentrate using an Amicon ultrafiltration cell fitted with a YM10 membrane (*see* **Notes 7–9**).

3.2. Functional Analysis of Venom Proteins

3.2.1. Platelet Aggregation

1. Use a 19-gauge winged infusion kit to collect 27 mL blood into a syringe containing 3 mL trisodium citrate anticoagulant, and mix by gentle inversion. The final concentration of trisodium citrate in the blood is 0.32% (w/v). This concentration of citrate prevents coagulation, but there remains sufficient Ca^{2+} to enable $\alpha_{IIb}\beta_3$-dependent platelet aggregation (refer to **step 7** in this section).
2. Spin blood at 100g for 20 min at 22°C in 10-mL plastic centrifuge tubes.
3. Remove citrated platelet-rich plasma (PRP) using a plastic disposable transfer pipet, leaving approx 0.5 cm of PRP above the packed red cells since this layer contains contaminating white cells. There should be minimal turbulent shear stress applied to the platelet sample during these manipulations to minimize the possibility of platelet activation. Pool the PRP into one tube.
4. To prepare platelet-poor plasma (PPP), transfer 0.5–1 mL PRP into an Eppendorf tube, spin in a microcentrifuge for 1 min (10,000g, 22°C) to pellet the platelets, and retain the supernatant.
5. Take 400 µL PPP plus 100 µL TS buffer into an aggregation cuvet, mix, and use this for the "reference" sample in a dual-beam lumiaggregometer.
6. For aggregation measurements, use 400 µL PRP containing a stir bar in the "sample" cuvet; add TS buffer and other additions to make a total volume of 500 µL. Routinely, samples are equilibrated at 37°C, and stirred at 900 rpm. After establishing a baseline, commence the aggregation by addition of agonist. For further details of turbidimetric methods for studying platelet aggregation, *see* **refs. *17,19***.
7. For alborhagin purified as described in **Subheading 3.1.1.**, a maximal rate of aggregation is achieved at approx 7.5 µg/mL final concentration. This response is initiated via signaling pathways coupled to GP VI, followed by activation of the integrin $\alpha_{IIb}\beta_3$, which binds fibrinogen or von Willebrand factor in a Ca^{2+}-dependent manner. Alborhagin-dependent aggregation of PRP, therefore, is inhibitable by anti-$\alpha_{IIb}\beta_3$ monoclonal antibodies, such as CRC64 (10 µg/mL final concentration), or by addition of EDTA (final concentration, 10 mM) prior to alborhagin (*17*). Alborhagin also induces a platelet shape change prior to the aggregation response (indicated by *decreased* light transmission through the sample owing to platelets changing from a discoid to a spherical shape). The shape change occurs in the presence of EDTA where aggregation is blocked, suggesting that external Ca^{2+} is not

required for this response to alborhagin. If aggregation occurs in the presence of anti-$\alpha_{IIb}\beta_3$ antibody or EDTA, then the "alborhagin" may be an analog functionally related to jararhagin, and may induce plasma von Willebrand factor binding to platelet GP Ibα, a pathway which may be independent of $\alpha_{IIb}\beta_3$ and external Ca^{2+}.

8. For botrocetin purified as described in **Subheading 3.1.2.**, a maximal rate of aggregation is achieved at approx 2.5 μg/mL final concentration. Botrocetin-dependent aggregation under these conditions is strictly dependent on von Willebrand factor binding to platelet GP Ibα, and may agglutinate platelets without the requirement for platelet activation and $\alpha_{IIb}\beta_3$ activity. Botrocetin activity, therefore, may not be blocked by anti-$\alpha_{IIb}\beta_3$ antibodies or EDTA, but may be confirmed by pretreating PRP with mocarhagin (*see* **Note 3**) at a final concentration of 10 μg/mL for 6 min at 37°C prior to the addition of botrocetin. The mocarhagin used for this test should be confirmed as active by showing that it inhibits aggregation of PRP induced by ristocetin (final concentration, 1.5 mg/mL). As an alternative to mocarhagin, inhibitory anti-GP Ibα monoclonal antibodies (for example, AK2) can be pre-equilibrated with PRP at a final concentration of 10 μg/mL prior to the addition of botrocetin. AK2 can also be confirmed as functional by showing that it inhibits ristocetin-dependent aggregation. If botrocetin does not induce aggregation of PRP at 2.5–5 μg/mL, then the concentration should be increased to 50 μg/mL to assess whether it is an active form of botrocetin with lower activity (*see* **Note 7**). If there is still no aggregation, then the "botrocetin" may be inactive (*see* **Note 6**). On the other hand, if botrocetin induces aggregation of PRP that is not inhibitable by mocarhagin and/or AK2, then the "botrocetin" is not typical, and the aggregation could be owing to a contaminating thrombin-like serine protease, for example, or may represent an analog that targets another platelet receptor, rather than activating von Willebrand factor to bind GP Ibα (refer to **Subheading 2.1.**).

9. Finally, specific methods suitable for more detailed analysis of alborhagin or other venom proteins *(13–17,27,28)* involving measuring of platelet activation events such as secretion, determining expression of surface activation markers such as P-selectin, or protein kinase-dependent phosphorylation of cytosolic signaling molecules are described elsewhere in this book (*see* Chapters 7–9, 19, vol. 1 and 10 and 12, vol. 2).

3.2.2. Digestion of Fibrinogen

1. Dissolve fibrinogen in TS buffer at a stock concentration of approx 300 μg/mL.
2. Microcentrifuge the sample for 2 min at 10,000g, if necessary, to remove any insoluble precipitate.
3. To confirm metalloproteinase activity of alborhagin, digest fibrinogen at approx 100 μg/mL (final concentration) in TS buffer with a final concentration of 10 μg/mL alborhagin for 5–30 min at 22°C. Digestion should be carried out in the presence of either EDTA (10 mM, final concentration) or $CaCl_2$ (10 mM, final concentration), in a total volume of 200 μL.
4. Stop the reaction at various times ranging from 0–80 min by adding an equal volume of SDS-PAGE loading buffer.
5. Analyze samples of fibrinogen without added alborhagin, alborhagin alone, and the digests in adjacent lanes on SDS-7.5%-polyacrylamide gels to reveal cleavage of alpha, beta, and/or gamma chains that run as a closely spaced triplet at about the midpoint of the gel under reducing conditions. For alborhagin, there should be time-dependent cleavage of the alpha chain in the presence but not the absence of Ca^{2+}, and appearance of a major digestion fragment that runs below the gamma chain on SDS-PAGE. Maximum cleavage should occur after approx 80 min at 22°C *(17)*.

3.3. Future Directions

The general approach to venom protein purification by a series of simple separation techniques (**Fig. 1**) has realized numerous useful probes, as well as significant new findings concerning the nature of venoms, and the function of vascular adhesion receptor targets. It would be expected in the future, however, that more selective affinity isolation procedures will play a much greater part in screening for, and isolating, venom proteins with a particular desirable function. For instance, for antithrombotics, it may be preferable to isolate venom proteins that selectively inhibit shear-induced thrombosis. Recent evidence suggests specific sites on activated GPIbα or von Willebrand factor may be critical for initiating shear-induced platelet aggregation *(33)*. It could be envisaged that functional domains of GPIbα or the von Willebrand factor A1 domain containing gain-of-function mutations may be useful for affinity purification of venom proteins that selectively recognize these active conformers. Second, in combination with such separation methods, the occurrence of multiple, closely related isoforms in venoms would suggest that they could be analyzed using proteomic approaches. Although full genomic sequences are lacking, the viability of "venomics" will continue to be enhanced as structure-function relationships of functionally different isoforms are deciphered, together with increased availability of crystal and model structures for C-type lectin-like proteins and metalloproteinase-disintegrins *(25)*.

In summary, the plethora of functional isoforms in snake venoms, and considerable batch-to-batch variability, is a complexity when it comes to isolation of particular forms of metalloproteinase-disintegrins and C-type lectin-like proteins. However, investigation of unexpected or unusual fractions has frequently led to discovery of novel probes for analysis of vascular cell adhesion receptors or ligands. New technologies promise exciting new developments in venom protein analysis.

4. Notes

1. Toxicity of crude lyophilized venoms toward humans may not have been established. They should therefore be considered potentially hazardous and handled with care, avoiding inhalation, especially on opening vials or containers. Dispose of contaminated materials and side fractions carefully to avoid the possibility of accidental envenomation.
2. Using sodium phosphate rather than potassium phosphate avoids precipitation when adding to SDS-PAGE sample buffer. This is particularly a problem for samples eluted later in the gradient, when the potassium concentration would be relatively high.
3. For isolation of mocarhagin, dissolve 0.5 g *Naja mocambique mocambique* venom (Sigma) in 10 mL TS buffer. Load at approx 30 mL/h onto a 1.5 × 20-cm heparin-agarose column, and wash with TS buffer. Elute with a linear 200-mL 0.15–1 *M* NaCl gradient in 0.01 *M* Tris-HCl, pH 7.4. Analyze approx 5-mL fractions on 5–20% SDS-PAGE under nonreduced and reduced conditions and pool fractions containing mocarhagin (~55 kDa, nonreduced and reduced). The main contaminant is a smeary band at approx 20 kDa that resembles an upside-down "U." Concentrate to approx 5 mL by Amicon ultrafiltration using a YM30 membrane, and load at approx 30 mL/h onto a Sepharose CL-6B column in 0.01 *M* Tris-HCl, 0.5 *M* NaCl, pH 7.4. Collect approx 5-mL fractions and analyze on 5–20% SDS-PAGE, nonreduced and reduced. Pool fractions containing mocarhagin, and dialyze into TS buffer.
4. Purified venom proteins may be stable at 4°C for several months. For long-term storage, freeze aliquots at –70°C to avoid repetitive freeze-thawing. Metalloproteinases such as

alborhagin or mocarhagin are susceptible to autodigestion on storage *(17)*, and if storing at 4°C for >1 wk, 1 m*M* EDTA (ethylenediaminetetraacetic acid, tetrasodium salt) should be included in the storage buffer. Depending on the type of functional assay, it may be necessary that Ca^{2+} in the sample buffer be increased in order to compensate for Ca^{2+} binding to EDTA added along with the metalloproteinase. Changes of pH resulting from EDTA:Ca^{2+} binding when adding Ca^{2+} to PRP are possible, but usually do not affect platelet aggregation under these conditions.

5. The concentration of purified protein is estimated using the BCA (bicinchoninic acid) protein quantitation method with bovine serum albumin (BSA) as standard according to the manufacturer's instructions (Pierce).

6. An inactive analog of botrocetin elutes early in the gradient (<50 m*M* Na_2HPO_4).

7. This form of two-chain botrocetin *(3)* differs from that previously purified in our laboratory *(2)* based on its activity in platelet-rich plasma (half-maximal aggregation at approx 1 µg/mL), and is comparable to that reported by Fujimura et al. *(4)*. In our previously reported method *(3)*, the DEAE-Sephacel eluted fractions were further chromatographed on phenyl-sepharose as part of the co-purification of jararhagin and jaracetin *(9)*. This step may be omitted for botrocetin purification (unpublished results). However, to purify jararhagin (**Fig. 1**) and/or jaracetin, the fractions eluted from DEAE-Sephacel should be pooled and made up to 1.2 *M* ammonium sulfate by dropwise addition of one-half volume of 3.6 *M* ammonium sulfate in 0.01 *M* Tris-HCl, pH 7.4. Load at 30 mL/h onto a 1.5 × 20-cm column of phenyl-Sepharose (Pharmacia), wash with 0.01 *M* Tris-HCl, 1.2 *M* ammonium sulfate, pH 7.4. Elute bound protein with a linear 400–mL 1.2 *M* to zero ammonium sulfate gradient in 0.01 *M* Tris-HCl, pH 7.4. Peak fractions containing jararhagin (~55 kDa, nonreduced and reduced) or jaracetin (~60 kDa non-reduced, ~30 kDa reduced) can be pooled, dialyzed into TS buffer (4 × 4 L changes at 4°C), and concentrated where necessary with an Amicon ultrafiltration device fitted with a YM30 membrane.

8. For isolating alboaggregin-A and alboaggregin-B, dissolve 0.1 g *Trimeresurus albolabris* venom in 10 mL TS buffer, and load at 30 mL/h onto a DEAE-Sephacel column equilibrated with TS buffer at 22°C. Wash the column with TS buffer, and elute bound protein with a linear 400-mL 0.15–0.5 NaCl gradient in 0.01 *M* Tris-HCl, pH 7.4. Eluted fractions (~5 mL) containing alboaggregin-A (~50 kDa nonreduced, ~14 kDa reduced) and alboaggregin-B (~25 kDa nonreduced, ~14 kDa reduced) that co-eluted from the DEAE-Sephacel column were distinguished by SDS-PAGE under nonreducing vs reducing conditions. Pooled fractions were dialyzed into TS buffer and loaded at 25 mL/h onto a 1 × 10-cm heparin-agarose column. After washing with TS buffer until the A_{280} of the eluate returned to baseline, bound protein is eluted using a linear 200-mL 0.15–1.0 *M* NaCl gradient in 10 m*M* Tris-HCl, pH 7.4. Collect approx 5-mL fractions of flowthrough, wash, and eluate for analysis. Alboaggregin-A eluted after the flowthrough peak, alboaggregin-B eluted at approx 0.5 *M* NaCl. Pooled fractions of alboaggregin-A and alboaggregin-B were dialyzed separately into 5 m*M* Na_2HPO_4, pH 6.8. The samples were then chromatographed individually on hydroxylapatite (**Table 2**).

9. An analog of botrocetin termed aspercetin from the venom of *Bothrops asper* was isolated by chromatography on DEAE-sepharose and affinity chromatography on Affi-Gel Blue *(32)*.

Acknowledgments

We thank the National Health & Medical Research Council of Australia for financial support.

References

1. Read, M. S., Shermer, R. W., and Brinkhous, K. M. (1978) Venom coagglutinin: an activator of platelet aggregation dependent on von Willebrand Factor. *Proc. Natl. Acad. Sci. USA* **83,** 4514–4518.
2. Andrews, R. K., Booth, W. J., Gorman, J. J., Castaldi, P. A., and Berndt, M. C. (1989) Purification of botrocetin from *Bothrops jararaca* venom. Analysis of the botrocetin-mediated interaction between von Willebrand factor and the human platelet membrane glycoprotein Ib-IX complex. *Biochemistry* **28,** 8317–8326.
3. Shen, Y., Romo, G. M., Dong, J.-F., Schade, A., McIntire, L. V., Kenny, D., et al. (2000) Requirement of leucine-rich repeats of GP Ibα for shear-dependent and static binding of von Willebrand factor to the platelet membrane GP Ib-IX-V complex. *Blood* **95,** 903–910.
4. Fujimura, Y., Titani, K., Usami, Y., Suzuki, M., Oyama, R., Matsui, T., et al. (1991) Isolation and chemical characterization of two structurally and functionally distinct forms of botrocetin, the platelet coagglutinin isolated from the venom of *Bothrops jararaca*. *Biochemistry* **30,** 1957–1964.
5. Berndt, M. C., Ward, C. M., Booth, W. J., Castaldi, P. A., Mazurov, A. V., and Andrews, R. K. (1992) Identification of aspartic acid 514 through glutamic acid 542 as a glycoprotein Ib-IX complex receptor recognition sequence in von Willebrand factor. Mechanism of modulation of von Willebrand factor by ristocetin and botrocetin. *Biochemistry* **31,** 11,144–11,151.
6. Usami, Y., Fujimura, Y., Suzuki, M., Ozeki, Y., Nishio, K., Fukui, H., et al. (1993) Primary structure of two-chain botrocetin; a von Willebrand factor modulator purified from the venom of *Bothrops jararaca*. *Proc. Natl. Acad. Sci. USA* **90,** 928–932.
7. Sen, U., Vasudevan, S., Subbarao, G., McClintock, R. A., Celikel, R., Ruggeri, Z. M., et al. (2001) Crystal structure of the von Willebrand factor modulator botrocetin. *Biochemistry* **40,** 345–352.
8. Fukuda, K., Doggett, T. A., Bankston, L. A., Cruz, M. A., Diacovo, T. G., and Liddington, R. C. (2002) Structural basis of von Willebrand factor activation by the snake toxin botrocetin. *Structure (Camb)* **10,** 943–950.
9. De Luca, M., Ward, C. M., Ohmori, K., Andrews, R. K., and Berndt, M. C. (1995) Jararhagin and jaracetin: novel snake venom inhibitors of the integrin collagen receptor, $\alpha_2\beta_1$. *Biochem. Biophys. Res. Commun.* **206,** 570–576.
10. Paine, M. J. I., Desmond, H. P., Theakston, R. D. G., and Crampton, J. M. (1992) Purification, cloning, and molecular characterization of a high molecular weight hemorrhagic metalloprotease, jararhagin, from *Bothrops jararaca* venom: insights into the disintegrin gene family. *J. Biol. Chem.* **267,** 22,869–22,876.
11. Kini, R. M. (1996) Are C-type lectin-related proteins derived by proteolysis of metalloproteinase/disintegrin precursor proteins? *Toxicon* **34,** 1287–1294.
12. Obert, B., Houllier, A., Meyer, D., and Girma, J.-P. (1999) Conformational changes in the A3 domain of von Willebrand factor modulate the interaction of the A1 domain with platelet glycoprotein Ib. *Blood* **93,** 1959–1968.
13. Andrews, R. K., Kroll, M. H., Ward, C. M., Rose, J. W., Scarborough, R. M., Smith, A. I., et al. (1996) Binding of a novel 50-kilodalton alboaggregin from *Trimeresurus albolabris* and related viper venom proteins to the platelet membrane glycoprotein Ib-IX-V complex. Effect on platelet aggregation and glycoprotein Ib-mediated platelet activation. *Biochemistry* **35,** 12,629–12,639.
14. Jandrot-Perrus, M., Lagrue, A. H., Okuma, M., and Bon, C. (1997) Adhesion and activation of human platelets induced by convulxin involve glycoprotein VI and integrin $\alpha_2\beta_1$. *J. Biol. Chem.* **272,** 27,035–27,041.

15. Polgar, J., Clemetson, J. M., Kehrel, B. E., Weidemann, M., Magnenat, E. M., Wells, T. N. C., et al. (1997) Platelet activation and signal transduction by convulxin, a C-type lectin from *Crotalus durissus terrificus* (tropical rattlesnake) venom via the p62/GPVI collagen receptor. *J. Biol. Chem.* **272**, 13,576–13,583.

16. Du, X.-Y., Clemetson, J. M., Navdaev, A., Magnenat, E. M., Wells, T. N. C., and Clemetson, K. J. (2002) Ophioluxin, a convulxin-like C-type lectin from *Ophiophagus hannah* (King cobra) is a powerful platelet activator via glycoprotein VI. *J. Biol. Chem.* **277**, 35,124–35,132.

17. Andrews, R. K., Gardiner, E. E., Asazuma, N., Berlanga, O., Tulasne, D., Nieswandt, B., et al. (2001) A novel viper venom metalloproteinase, alborhagin, is an agonist at the platelet collagen receptor GPVI. *J. Biol. Chem.* **276**, 28,092–28,097.

18. De Luca, M., Dunlop, L. C., Andrews, R. K., Flannery, J. V., Ettling, R., Cumming, D. A., et al. (1995) A novel cobra venom metalloproteinase, mocarhagin, cleaves a ten amino acid peptide from the mature N-terminus of P-selectin glycoprotein ligand receptor, PSGL-1, and abolishes P-selectin binding. *J. Biol. Chem.* **270**, 26,734–26,737.

19. Ward, C. M., Andrews, R. K., Smith, A. I., and Berndt, M. C. (1996) Mocarhagin, a novel cobra venom metalloproteinase, cleaves the platelet von Willebrand factor receptor glyco-protein Ibα. Identification of the sulfated tyrosine/anionic sequence Tyr-276-Glu-282 of glycoprotein Ibα as a binding site for von Willebrand factor and α-thrombin. *Biochemistry* **35**, 4929–4938.

20. Kamiguti, A. S., Hay, C. R. M., and Zuzel, M. (1996) Inhibition of collagen-induced platelet aggregation as the result of cleavage of $\alpha_2\beta_1$-integrin by the snake venom metalloproteinase jararhagin. *Biochem. J.* **320**, 635–641.

21. Hamako, J., Matsui, T., Nishida, S., Nomura, S., Fujimura, Y., Ito, M., et al. (1998) Purifi-cation and characterization of kaouthiagin, a von Willebrand factor-binding and -cleaving metalloproteinase from *Naja kaouthia* cobra venom. *Thromb. Haemost.* **80**, 499–505.

22. Wu, W. B., Peng, H. C., and Huang, T.-F. (2001) Crotalin, a vWF and GP Ib cleaving metalloproteinase from venom of *Crotalus atrox*. *Thromb. Haemost.* **86**, 1501–1511.

23. Zhou, Q., Dangelmaier, C., and Smith, J. B. (1996) The hemorrhagin catrocollastatin inhibits collagen-induced platelet aggregation by binding to collagen via its disintegrin-like domain. *Biochem. Biophys. Res. Commun.* **219**, 720–726.

24. Andrews, R. K. and Berndt, M. C. (2000) Snake venom modulators of platelet adhesion receptors and their ligands. *Toxicon* **38**, 775–791.

25. Andrews, R. K., Kamiguti, A., Berlanga, O., Leduc, M., Theakston, R. D. G., and Watson, S. P. (2001) The use of snake venom toxins as tools to study the platelet receptors for collagen and von Willebrand factor. *Haemostasis* **31**, 155–172.

26. Fujimura, Y., Kawasaki, T., and Titani, K. (1996) Snake venom proteins modulating the interaction between von Willebrand factor and platelet glycoprotein Ib. *Thromb. Haemost.* **76**, 633–639.

27. Dormann, D., Clemetson, J. M., Navdaev, A., Kehrel, B. E., and Clemetson, K. J. (2001) Alboaggregin A activates platelets by a mechanism involving glycoprotein VI as well as glycoprotein Ib. *Blood* **97**, 929–936.

28. Asazuma, N., Marshall, S., Berlanga, O., Snell, D., Poole, A., Berndt, M. C., et al. (2001) The snake venom toxin alboaggregin-A activates GPVI. *Blood* **97**, 3989–3991

29. Kuter, D. J., Gminski, D., and Rosenberg, R. D. (1992) Botrocetin agglutination of rat megakaryocytes: a rapid method for megakaryocyte isolation. *Exp. Hematol.* **20**, 1085–1089.

30. Berlanga, O., Bobe, R., Becker, M., Murphy, G., Leduc, M., Bon, C., et al. (2000) Expres-sion of the collagen receptor glycoprotein VI during megakaryocyte differentiation. *Blood* **96**, 2740–2745.

31. Ward, C. M., Vinogradov, D. V., Andrews, R. K., and Berndt, M. C. (1996) Characterization of mocarhagin, a cobra venom metalloproteinase from *Naja mocambique mocambique*, and related proteins from other *Elapidae* venoms. *Toxicon* **34,** 1203–1206.
32. Rucavado, A., Soto, M., Kamiguti, A. S., Theakston, R. D., Fox, J. W., Escalante, T., et al. (2001) Characterization of aspercetin, a platelet aggregating component from the venom of the snake *Bothrops asper* which induces thrombocytopenia and potentiates metalloproteinase-induced hemorrhage. *Thromb. Haemost.* **85,** 710–715.
33. Dong, J.-F., Berndt, M. C., Schade, A., McIntire, L. V., Andrews, R. K., and López, J. A. (2001) Ristocetin- but not botrocetin-dependent binding of von Willebrand factor to the platelet membrane glycoprotein Ib-IX-V complex correlates with shear-dependent interactions. *Blood* **97,** 162–168.

21

Peptide Synthesis in the Study
of Collagen–Platelet Interactions

C. Graham Knight, Catherine M. Onley, and Richard W. Farndale

1. Introduction
1.1. Collagen

Collagen is the most abundant protein in vertebrates and has an essential structural role. Thus, the tensile strength of the fibrous collagens, types I, II, III, V, and XI, is fundamental to the function of bone, skin, tendons, cartilage, and blood vessel walls. Similarly, the network-forming collagens, such as types IV and VI, are key components of the extracellular matrix of endothelial and epithelial tissues. Collagens interact with cells both directly, through receptors that may modulate cell function, and indirectly, through the numerous proteins and other constituents of connective tissues that bind simultaneously to both collagen and their own specific cell surface receptors. Collagen plays a crucial role in diverse biological processes, either as a mechanical support or as an active ligand. For this reason, collagen preparations are widely used as a cell growth matrix or as a means of activating cells. The capacity of collagens of the blood vessel wall to activate platelets represents a very accessible example of this principle in operation, utilizing direct interaction with integrin $\alpha_2\beta_1$ and glycoprotein (GP) VI, and indirect interaction with GPIb via von Willebrand factor. Binding sites within collagen and their complementary receptors on the cell surface or within the matrix are potential targets, as yet unexploited, for the treatment of several disease processes, such as arterial thrombosis. For this potential to be realized, each interaction must be understood in detail.

Until recently, platelet research employed soluble collagens from connective tissues, typically skin or tendon or placenta, purified either by acid extraction or by pepsin digestion. The extraction process may yield monomeric or partially denatured collagens whose capacity to assemble as fibers is diminished, along with loss of the telopeptide extensions that form part of the fibrillar structure of native collagen. These preparations often prove unstable in use, with fiber formation either failing to occur or proceeding

From: *Methods in Molecular Biology, vol. 273:*
Platelets and Megakaryocytes, Vol. 2: Perspectives and Techniques
Edited by: J. M. Gibbins and M. P. Mahaut-Smith © Humana Press Inc., Totowa, NJ

in an uncontrolled manner so that the proper size distribution of fibers is lost. Where intact fibers are extracted from connective tissues by nonproteolytic methods, fiber preparations may be highly active, but noncollagenous components may confound attempts to identify collagen-specific interactions.

1.2. Collagen-Related Peptides

Peptides mimicking the structure of native collagen serve as defined, specific ligands with which to modulate platelet function and also as valuable tools with which to identify receptor-binding sites within collagen. Thus, such peptides can assist in gaining understanding of the structural aspects of the interaction as well as the subsequent intracellular events.

Collagens are defined by their common primary structure, a sequence of repeating Gly-Xaa-Yaa triplets, where Xaa is often proline (P) and Yaa is often hydroxyproline (O). Collagen I consists of two $\alpha_1(I)$-chains and one $\alpha_2(I)$-chain, whereas collagens II and III are disulfide-bonded homotrimers of $\alpha_1(II)$ or $\alpha_1(III)$ chains. Each of the α-chains forms a left-handed polyproline II-like helix, of which a Gly-Xaa-Yaa triplet represents about one turn, and together they comprise a right-handed superhelix. In collagens I, II, and III, each chain of the triple-helical domain is approx 1000 amino acids in length (300 nm, with a diameter of 1.5 nm).

The extreme length of the collagen molecule presents a problem when one wishes to determine those regions of the structure responsible for particular interactions. Although collagen can be fragmented by cyanogen bromide (CNBr), the fragments range from about 40 to 280 residues in length (1). Since the receptor-binding motifs are unlikely to span more than a few triplets of the collagen sequence (2), these fragments are too long to allow precise definition of the motifs. Smaller collagen fragments may lose the ability to form the native triple helical structure at 37°C (3) and so physiologically relevant cell-reactive sites may be masked. CNBr cleavage at methionine residues may similarly disrupt binding motifs. All these limitations are overcome by the use of synthetic peptides, provided steps are taken to maintain the native triple helical structure.

Identification of specific collagen-receptor-binding motifs plays a key role in establishing the function of individual receptors and examining the functional interplay between receptors. For example, collagen-related peptide (CRP) is a specific ligand for GPVI and induces platelet activation through GPVI (4). Specific triple-helical peptide ligands for the other major platelet collagen-receptor integrin $\alpha_2\beta_1$ have also been developed (5) and have facilitated production of high-resolution structures detailing the receptor–ligand interaction (2). Peptides have a wider application in studying platelet signaling, including generation of peptide inhibitors conjugated to membrane-permeant sequences, such as the calpain inhibitor calpastatin (6). Similarly, we have used biotinyl peptides corresponding to collagen-receptor cytoplasmic domains to identify and localize binding sites of downstream signaling proteins (C. M. Onley et al., unpublished work).

1.3. Synthetic Strategies for Collagen-Like Peptides

The assembly of the three individual collagen chains from random coils into the rigid triple-helical structure is a highly cooperative process (7), but the native structure is thermally labile and rapidly dissociates above a critical "melting" temperature, T_m.

Indeed, it has been demonstrated recently that type I collagen has a T_m *below* body temperature and remains triple-helical only because the rate of dissociation is very slow *(8)*. Short collagen peptides with native sequence, as would be necessary to locate receptor-binding motifs, are likely to be denatured at ambient temperatures.

Various strategies have been devised to stabilize synthetic peptides in the triple helical conformation *(9)*. Thus, tethering either the C- or the N-terminal residue of each chain to a tripodal linker confers stability. Although these methods undoubtedly have been successful, they require highly optimized synthesis conditions *(10)*, so that the proportion of deletion sequences (those with one or more residues missing) is essentially zero; otherwise the covalently linked chains are out of register. In our view, these methods are more suited to the synthesis of short triple-helical peptides.

The strategy that we have found most useful employs single chain "host–guest" peptides *(11)*, in which a stretch of native collagen sequence is inserted within an extended number of Gly-Pro-Hyp or Gly-Pro-Pro triplets, which have a natural tendency to assemble into triple helixes. For collagen-like peptides, the validity of this approach has been amply demonstrated *(12)*. Single chains are more readily purified to homogeneity and then spontaneously associate into collagen-like structures. Triple-helicity must be confirmed experimentally, as different sequences can have widely different melting points *(5)*. A variety of methods is available to determine T_m, but we favor polarimetry for its simplicity. Using the procedures described below, we routinely prepare peptides between 54 and 66 residues in length, containing up to 10 triplets of native collagen sequence *(5)*.

2. Materials (*see* Note 1)

Specific recommendations and comments for many of the materials are given in the notes. Where a supplier is not given, the material is widely available. To avoid confusing chemical nomenclature, the protecting groups and some reagents will be referred to by their standard abbreviations. These are listed by Chan and White *(13)* and in the Novabiochem catalog.

2.1. Solid-Phase Peptide Synthesis (see Note 2)

1. Peptide synthesizer (e.g., Applied Biosystems Pioneer®, Foster City, CA or Protein Technologies Symphony®, Woburn, MA). This allows the automated synthesis of peptides up to 70 or more residues in length, provided the synthetic procedures are optimized, as discussed below. The synthesis proceeds from the C- to the N-terminus and the synthetic method of choice uses Fmoc-amino acids.
2. Fmoc-amino acids are available commercially (*see* **Note 3**). The amino terminus is protected with the base-labile Fmoc group and a reactive side chain, if present, is blocked with an acid-labile protective group (*see* **Note 4**).
3. Solid-phase support (resin). The type of support depends on whether the C-terminus of the peptide is an amide ($CONH_2$) or a carboxyl group (COOH) (*see* **Note 5**).
4. Dimethylformamide (DMF).
5. 20% (v/v) Piperidine in DMF or 2% (v/v) DBU + 2% (v/v) piperidine in DMF (*see* **Note 4b**).

6. Reagent for the formation of peptide bonds. A wide variety of uronium (e.g., HBTU or TBTU) or phosphonium (e.g., PyBOP) coupling reagents is available (*see* **Note 6**). The acylation catalyst 1-hydroxy-7-azabenzotriazole (HOAt) is available from Applied Biosystems.
7. Di-isopropylethylamine (DIPEA).
8. Methanol.
9. Peroxide-free ether, prepared by filtration through basic aluminium oxide.

2.2. Post-Synthetic Tagging With Biotin or Fluorescein

1. Biotin.
2. Dimethylformamide (DMF).
3. 1-Hydroxy-7-azabenzotriazole (HOAt).
4. Di-isopropylcarbodiimide (DIC).
5. 5(6)-Carboxyfluorescein.
6. Dichloromethane (DCM) containing 10% (v/v) *N,N′*-dimethylpropyleneurea (DMPU).
7. Small ceramic filter funnel and filter paper disks.

2.3. Peptide-Resin Cleavage

1. Trifluoroacetic acid (TFA) (*see* **Notes 2** and **7**).
2. Cleavage cocktail: typically, 9.5 mL trifluoroacetic acid, 0.5 mL thioanisole, 0.25 mL triisopropylsilane, and 0.25 g dithiothreitol (*see* **Notes 7–9**).
3. Polypropylene or polytetrafluoroethylene (PTFE) flask (50 mL). Resins tend to stick firmly to glass, so a TFA-resistant plastic vessel is preferred.
4. Peroxide-free diethyl ether.
5. Sintered-glass funnel (porosity 1).
6. Phosphorus pentoxide (P_2O_5).
7. Falcon polypropylene tubes (50 mL).
8. Centrifuge with sealed rotor or explosion-proof centrifuge.

2.4. Peptide Purification

1. High-performance liquid chromatography (HPLC) system (e.g., PerkinElmer LC200, Boston, MA). Ideally, the system will comprise a four-channel pump capable of delivering 10 mL/min or more, a vacuum degasser, an autosampler, a column oven, and a diode-array detector. Data handling and storage are under computer control.
2. Analytical diphenyl-silica or octadecyl (C18)-silica HPLC column: 5-μm particles, diameter 4.6 mm, length 250 mm (*see* **Note 10**). Both Vydac (Hesperia, CA; www.vydac.com) and ACE (Berkshire, UK; www.hichrom.co.uk) columns have given good results.
3. Preparative diphenyl-silica or octadecyl (C18)-silica HPLC columns: 10-μm particles, diameter 22 mm, length 250 mm. Again, we have used Vydac and ACE columns.
4. HPLC-grade water, containing 0.1% (v/v) TFA.
5. HPLC-grade acetonitrile (CH_3CN), containing 0.08% TFA.
6. Mass spectrometer: access to this is essential to confirm the identity of the product by determining the molecular mass. This will commonly be done by a central facility with matrix-assisted laser desorption ionization = time of flight (MALDI-TOF) or electrospray ionization (ESI) instruments (*see* **Note 11**).
7. Acetic acid.

8. 0.45-µm filters (e.g., Millipore type HVLP for solvents and Whatman Puradisc 25 PP for peptide solutions).

2.5. Cross-Linking

Peptides containing both cysteine and free amino groups are cross-linked with *N*-succinimidyl 3-(2-pyridyldithio)propionate (SPDP); those containing only amino groups can be cross-linked with disuccinimidyl glutarate (DSG). Both cross-linkers are widely available (e.g., Sigma and Perbio).

1. 50 m*M* SPDP in dry ethanol.
2. 1 *M* NaHCO$_3$ solution in water.
3. QuixSep microdialyzer (0.5 mL capacity), available from Perbio.
4. Dialysis membrane having a nominal M_r cut-off of 1000 (Perbio).
5. 0.01 *M* Acetic acid in water.
6. 100 m*M* DSG in dry dimethylsulfoxide (DMSO). Prepare the DSG stock immediately before use.
7. 1 *M* Glycine in water.

3. Methods
3.1. Overview of Solid-Phase Peptide Synthesis

In the sections that follow, we assume that peptides are made by automated solid-phase synthesis, employing the 9-fluorenylmethoxycarbonyl (Fmoc) method. Our discussion is necessarily simplified, as we assume further that the reader is most likely to have peptides made by a colleague or by a central facility. For more comprehensive overviews of peptide synthesis, the books by Chan and White *(13)* and Grant *(14)* are highly recommended. Detailed synthetic methodology will be specific to the type of synthesizer and will not be presented; however, the principles will be discussed.

The steps involved in the automated synthesis of peptides are typically as follows:

1. Place the resin support (commonly 0.1 mmole) in a vessel or column that allows solvents and reactants to be added and exchanged efficiently, and attach this container to the synthesizer. Peptide amides are synthesized on a resin containing an Fmoc-protected amino group. Peptide acids are made on a resin with the C-terminal amino acid already attached (*see* **Note 4**). The amino terminus is protected with the base-labile Fmoc group and a reactive side chain, if present, is blocked with an acid-labile protective group (*see* **Note 5**).
2. Wash the resin with DMF (30 mL/min, 40 s) and then with 20% (v/v) piperidine in DMF (5 mL/min, 5 min) to remove the Fmoc group and expose the amino terminus. Wash again with DMF (30 mL/min, 40 s).
3. Add a fourfold molar excess of the next Fmoc-amino acid in the presence of a fourfold excess of HATU and an eightfold excess of DIPEA. These reagents activate the carboxyl group to form the peptide bond (other coupling agents can be used; *see* **Note 6**).
4. Recycle the reagents through the column (30 mL/min, 30 min) to allow the coupling reaction to go to completion. Repeat **step 2**.
5. Continue further cycles of piperidine in DMF, DMF wash, and coupling of amino acid until the required peptide has been constructed.
6. End the synthesis with the removal of the N-terminal Fmoc group from the fully side-chain-protected peptide using 20% piperidine in DMF.

7. The peptide product remains bound to the resin matrix. This peptide-resin is quite stable and can be safely stored at 4°C, so allowing the later addition of further amino acids or labels (*see* **Subheading 3.2.**), prior to cleavage of the peptide from the resin. It is important to wash the resin efficiently with methanol and peroxide-free ether and to dry under a vacuum in a desiccator containing P_2O_5, as residual traces of DMF interfere with the TFA-mediated cleavage (*see* **Note 6**).

8. To obtain the peptide, treat the resin with TFA, which releases the peptide into solution and cleaves all the side-chain-protecting groups. Details of this resin-cleavage step are given in **Subheading 3.3.**

9. Finally, purify the product using HPLC (*see* **Subheading 3.4.**) and establish by mass spectrometry that it has the correct identity.

3.2. Post-Synthetic Modifications: Tagging With Fluorescein or Biotin

If the desired final product is a peptide labeled with biotin or fluorescein, it is necessary to attach these probes prior to resin cleavage. We find that the following procedures work reliably:

1. Wash the peptide-resin with methanol and peroxide-free ether, before drying under nitrogen and briefly under vacuum.

2. For fluorescein and biotin labeling, use a sixfold molar excess of all the reagents. For fluorescein labeling, proceed to **step 6**.

3. For biotinylation, dissolve the biotin in DMF by cautious heating (solubility is about 40 mg/mL). Cool the solution and add to the resin, with enough DMF to produce a freely mobile suspension.

4. Add a sixfold molar excess of both HOAt and the coupling reagent di-isopropyl-carbodiimide (DIC), and allow the reaction to continue overnight with gentle shaking.

5. Take a few resin beads for a ninhydrin amine test *(13)* to check for completeness of reaction. If the test is positive, wash the resin with DMF and repeat **steps 3** and **4**.

6. For fluorescein labeling, suspend the dry resin in dichloromethane (DCM) containing 10% (v/v) *N,N'*-dimethylpropyleneurea (DMPU; the solid reagents are poorly soluble in DCM alone).

7. Shake the resin gently with a sixfold molar excess of 5(6)-carboxyfluorescein and HOAt until they dissolve.

8. Add a sixfold molar excess of DIC and leave the mixture overnight.

9. Check for completeness of reaction by the ninhydrin amine test *(13)*. We find it convenient to add a few drops of piperidine before washing the resin on a filter-paper disk held in a small ceramic filter funnel to remove the excess fluorescein. The washed resin is then transferred to a TFA-resistant plastic flask for cleavage. Do not be tempted to label the peptide-resin with fluorescein isothiocyanate (FITC). The carbamate derivative thus formed decomposes and loses fluorescein during the subsequent treatment with TFA.

3.3. Peptide-Resin Cleavage

1. Transfer the peptide-resin to a TFA-resistant plastic flask and wash it thoroughly with methanol and peroxide-free diethyl ether.

2. Allow the resin to dry in air or under a gentle stream of nitrogen, before leaving overnight in a vacuum desiccator containing P_2O_5.

3. Prepare the cleavage cocktail and pass nitrogen through the cocktail for 30 s before adding it to the resin under nitrogen (*see* **Note 7**). This minimizes oxidation of sensitive residues such as methionine and biotin (*see* **Note 9**).

4. Add the cocktail to the dried peptide-resin product to scavenge the products of the protecting groups and also the resin-bound linker fragment (*see* **Notes 8,9**).

5. After cleavage for 3–4 h, filter the reaction mixture through a sintered-glass funnel (porosity 1).

6. Rinse the cleavage vessel twice with about 5 mL of TFA and use these washings to wash the resin on the filter.

7. Evaporate the combined filtrate until most of the TFA has been removed.

8. Add the oily residue dropwise to a preweighed centrifuge tube (50-mL Falcon poly-propylene tubes are suitable) containing 40 mL of rapidly stirred ice-cold peroxide-free diethyl ether.

9. Wash the filtration flask twice with 0.5 mL of TFA and add these washings to the ether.

10. Cap the centrifuge tube and spin in a sealed rotor (*important*, or use an explosion-proof centrifuge) at about 700*g* for 2 min.

11. Decant the supernatant, resuspend the peptide pellet in ether, and centrifuge the mixture. Repeat this process three times.

12. Dry the peptide under a gentle stream of nitrogen. Weigh the tube again to obtain the amount of crude peptide.

3.4. Peptide Purification

If the synthesis has been efficient, the majority of the crude product will be the desired peptide, with traces of deletion peptides, residual TFA, and scavengers. The peptide is isolated from this mixture by high-performance liquid chromatography (HPLC), employing gradient elution from a reverse-phase column. An enormous variety of chromatographs and columns is available, so we shall outline our approach to purification to illustrate the method.

1. Perform an initial HPLC run on an analytical diphenyl-silica or octadecyl (C18)-silica column (*see* **Note 10**) with about 50 µg of crude peptide. The solvent reservoirs contain water with 0.1% TFA, and CH_3CN with 0.08% TFA (this difference in TFA concentration ensures a flat baseline). Thus TFA is always present in the eluting solvents. Both solvents are filtered through a 0.45-µm filter before use.

2. Elute the column with a linear gradient of CH_3CN increasing from 5 to 95% (v/v) over 20 min at a flow rate of 1.25 mL/min. To avoid effects of fluctuating ambient temperature, maintain all columns at 40°C in an oven.

3. Monitor the eluant at 230 nm and 260 nm, to detect peptides and other constituents. The 230-nm trace should show one major peak due to the peptide, although with collagen-like peptides self-association may cause this peak to have a complex shape *(15)*. The 260-nm trace shows the location of UV-absorbing scavengers such as thioanisole.

4. Calculate the concentration of CH_3CN corresponding to the elution time of the peptide from the analytical column. In our experience, elution from preparative columns occurs at a CH_3CN concentration about half this value.

5. Preparative chromatography is done using with the same matrix chemistry (diphenyl or C18) as analytical, but in a larger column (10-µm particles, diameter 22 mm, length

250 mm) from the same manufacturer. As for analytical runs, solvents contain TFA and the column is maintained at 40°C.

6. Pre-equilibrate the column with 5% CH$_3$CN.
7. Dissolve about 200 mg of crude peptide in 20 mL of 5% CH$_3$CN in 0.1% TFA. Pass this solution through a 0.45-μm filter and load onto the column. This is done by using the fourth channel of the quaternary pump.
8. Wash with 5% CH$_3$CN until the baseline absorbance returns to zero.
9. Elute the preparative column at a flow rate of 10–20 mL/min. Elution from reverse-phase columns occurs sharply when the concentration of CH$_3$CN reaches a critical value, so a gradient spanning that estimated value (from **step 4**) by about ±3% is satisfactory. For example, if the peptide eluted from the analytical column at 50% CH$_3$CN, the preparative elution zone is assumed to be from 22 to 28%. The CH$_3$CN content is increased over 2 min from 5% to that at the start of the elution zone and then at 0.25%/min to elute the peptide.
10. Collect 5-mL fractions during this peptide elution phase.
11. When the elution is complete, increase the CH$_3$CN to 80% over 5 min and wash the column for 10 min to elute the more nonpolar contaminants.
12. Re-equilibrate the column with 5% CH$_3$CN for 15 min (*see* **Note 10**).
13. Perform another round of analytical HPLC runs to identify the fractions containing apparently homogeneous product. Inspection of the preparative elution profile usually suggests the most likely fractions and it is seldom necessary to analyze more than 5 or 6. We find it convenient to expand and superimpose the analytical chromatography traces from the fractions and crude product.
14. Combine for freeze-drying only those fractions for which the analytical HPLC results overlay very closely. On occasion, it is difficult to decide how to make the pool. In our experience, it is unwise to guess. Mass spectrometry of samples of individual fractions allows combination of those containing the desired product (*see* **Note 7**).
15. Remove the CH$_3$CN from the pooled fractions with a rotary evaporator, dilute with 2 or 3 volumes of HPLC-grade water, and add acetic acid to 5–10% (v/v).
16. After shell-freezing in dry ice/ethanol, freeze-dry the pool overnight.

3.5. Peptide Cross-Linking

3.5.1. Peptides With Terminal Cysteine Residues

Collagen-like peptides may require cross-linking to display optimal receptor-binding activity, and particularly to facilitate the receptor clustering that appears to be a prerequisite for signaling through platelet collagen receptors. Cross-linking is achieved with 1.5 molar equivalents of SPDP (*N*-succinimidyl 3-[2-pyridyldithio] propionate), a reagent with both amine- and thiol-reactive groups. To achieve this, we synthesize the peptides with a free α-amino group and a cysteine residue in both the N- and C-terminal triplets.

1. To minimize premature aggregation, dissolve the peptide in water and add 1.0 *M* NaHCO$_3$ to give a final peptide concentration of 3.0 m*M* in 0.1 *M* NaHCO$_3$.
2. To 0.33 mL of this solution (1 μmole), add 30 μL of 50 m*M* SPDP in dry ethanol (1.5 μmoles).
3. Leave the mixture for 1 h at room temperature and transfer to a Perbio QuixSep microdialyzer (0.5 mL capacity) with the dialysis membrane having a nominal M_r cut-off of 1000.

4. Stir the microdialyzer rapidly in 500 mL of 0.01 M acetic acid for 2 h.
5. Stir for a further 2 h in fresh solvent, before transferring the cross-linked peptide to a preweighed vial with two 0.1-mL washes.
6. The concentration of product is calculated assuming 100% recovery. To ensure the efficiency of cross-linking, these proportions are maintained when cross-linking larger or smaller quantities of peptide.

3.5.2. Peptides With Terminal Lysine Residues

A similar protocol to that in **Subheading 3.5.1.** is followed for peptides having lysine residues at the N- and C-termini. In this case, the peptide is treated with 1.5 equivalents of disuccinimidyl glutarate (DSG), a bifunctional amine-reactive reagent.

1. The lysine-containing peptide is dissolved in water and 1 M NaHCO$_3$ added to give 3.0 mM peptide in 0.1 M NaHCO$_3$.
2. To 0.33 mL of this solution (1 µmol) add 15 µL of 100 mM DSG, freshly prepared in dry DMSO.
3. After 75 min, quench the reaction by adding 20 µL of 1 M glycine and leave the mixture for 1 h.
4. Dialyze the solution of cross-linked peptide against 0.01 M acetic acid, as described in **Subheading 3.5.1.**

3.6. "Difficult" Peptides

From a synthetic viewpoint, the synthesis of collagen-like peptides presents few problems, other than those associated with particular amino acid residues (*see* **Note 5**). In particular, the high content of proline and hydroxyproline prevents the self-association of the growing chains into β-sheet structures, which can severely disrupt a synthesis *(16)*. Peptides that suffer these problems during synthesis are commonly described as "difficult." These potential problems cannot be ignored, however, if one wishes to synthesize peptides related, say, to the cytoplasmic tails of collagen receptors.

Simple inspection of the proposed sequence may not reveal the tendency to β-sheet formation and it is convenient to analyze the sequence with a program such as Peptide Companion (www.5z.com/csps) to reveal aggregation potential. For example, analysis of the peptide CAAARWKKAFIAVSAANRFKKIS, derived from calmodulin, revealed that all the couplings in the 8-residue stretch WKKAFIAV were unlikely to be efficient. Two approaches can be taken to overcoming such problems, both of which were used. If the difficult sequence contains Xaa-Ser or Xaa-Thr, the dipeptide can be coupled as the serine- or threonine-derived pseudoproline (available from Novabiochem and Bachem), an acid-labile cyclic derivative that disrupts the β-sheet *(17)*. Thus, here we coupled Fmoc-Val-Ser($\psi^{Me,Me}$pro)-OH at the VS site. The pseudoproline effect extends only over 5 or 6 residues, so it seemed advisable to facilitate the WKK couplings by introducing the preceding Ala as the *N*-hydroxymethylbenzyl (Hmb) derivative, Fmoc-(FmocHmb)Ala-OH (Novabiochem). Again, the disruptive effect of Hmb on β-sheet formation extends over 5 or 6 residues *(18)*. Having taken these precautions, the final product was obtained with good yield and high purity.

3.7. Applications and Perspectives

3.7.1. Integrin-Recognition Motifs

Triple-helical peptides containing overlapping sequences derived from bovine collagen I were used to identify the sequence GFOGER as a recognition motif for integrins $\alpha_1\beta_1$ and $\alpha_2\beta_1$ *(5)*. Subsequently, derivatives of this sequence were used in co-crystallization studies *(2)* and as antagonists of platelet recognition of collagen through $\alpha_2\beta_1$ *(19)* (P. R.-M. Siljander et al., unpublished work). Cross-linked GFOGER peptides have been used to explore the signaling properties of $\alpha_2\beta_1$ *[20]* (P. Sundaresan and R. W. Farndale, unpublished work).

3.7.2. Glycoprotein VI Recognition Motifs

Triple-helical peptides containing multiple repeats of the GPO sequence have been used as ligands for GPVI. Such peptides, commonly referred to as collagen-related peptides, can be immobilized to investigate adhesion of platelets through GPVI *(21)*, and to select GPVI-specific antibodies (P. Smethurst et al., unpublished work). As monomers, collagen-related peptides serve as partial agonists or antagonists of GPVI when applied to platelets in suspension *(22)*, whereas when cross-linked, they are potent platelet activators *(21,23,24)*. Recently, a set of model peptides and related sequences from collagen III has been used to establish the basic unit of interaction of collagen with GPVI (D. J. Onley et al., unpublished work).

3.7.3. Membrane-Permeant Peptides

We have used peptides corresponding to the cytoplasmic tails of GPVI and of the integrin α_2 subunit coupled to the HIV transduction (TAT) domain to modulate platelet activities associated with these receptors. In biotinylated form, the former peptide has been used to identify intracellular proteins that associate with specific domains of GPVI.

4. Notes

1. *Safety considerations:* Many of the reagents used in peptide synthesis and purification are hazardous, and a full risk assessment should be made before commencing work. Always wear a laboratory coat, safety glasses, and appropriate gloves when handling solvents. A face mask should be worn when weighing solid reagents, such as Fmoc-amino acids and HATU. Of special note are the extreme flammability of diethyl ether and the destructive effects of trifluoroacetic acid (TFA) on tissues. These reagents must be handled only in an efficient fume hood.
2. *Choosing the reagents for peptide synthesis:* The quality of the final product in any synthetic scheme depends crucially on the quality of the starting materials. Most reagents for peptide synthesis are commercially available with high purity and can be used as supplied. An exception perhaps is TFA, which is required both for peptide-resin cleavage and for high-performance liquid chromatography (HPLC), and tends to degrade on storage. We prefer to buy a less-expensive grade (Aldrich T62200) and to redistill it before use. It should be noted that the prices of reagents such as dimethylformamide (DMF) vary widely and some suppliers are significantly cheaper than others.

3. *Fmoc-amino acids:* These are available in excellent quality from companies such as Sigma-Aldrich (St. Louis, MO; www.sigma-aldrich.com), but they may also be purchased from specialist companies such as Novabiochem (Laufelfingen, Switzerland; www.novabiochem.com) and Bachem (Bubendorf, Switzerland; www.bachem.com). The specialist companies rapidly introduce useful novel compounds described in the literature or developed in their own laboratories. Novabiochem also publishes a range of free technical pamphlets, which can be accessed online. Many instrument makers provide preweighed vials of Fmoc-amino acids, but these products may prove to be expensive compared to the bulk reagents that we prefer to use.

4. *Considerations for amino acid derivatives:* During the 40 yr since the introduction of solid-phase peptide synthesis *(25)*, a prodigious number of amino acid derivatives have been invented, especially to improve side-chain protection, and many remain commercially available, although superseded by improved versions. Some derivatives that avoid particular pitfalls are recommended here (a–e).

   a. The intrinsic reactivity of the guanidino side chain of arginine needs to be suppressed, while allowing the efficient removal of the protecting group at the end of the synthesis. This was especially problematic with peptides containing multiple Arg residues protected as Arg(Mtr), but the introduction of Arg(Pmc) and especially the more recent Arg(Pbf) has solved this problem. When peptides contain both Arg and Trp, the Pbf fragment released by Arg during the final TFA cleavage can add to the indole ring. Use of Trp(Boc) avoids this problem *(26)*.

   b. Under the basic conditions of Fmoc deprotection, the side chain of aspartic acid, and on occasion asparagine, can react with the following NH group to form a cyclic aspartimide, which in turn can undergo further reactions. Aspartimide formation is highly sequence-dependent *(27)* and can be difficult to suppress completely, unless the amino acid C-terminal to Asp has been coupled as the Hmb derivative *(28)*, in which case the offending N atom is already substituted. Asp-Gly and Asn-Gly sequences are especially prone to aspartimide formation. To prevent this we use the protected dipeptide derivatives with Asp or Asn coupled to (Hmb)Gly, recently introduced by Novabiochem. In the special case of sequences containing Asp-Ser, use of the pseudoproline avoids the problem. For coupling of Asp to other amino acids, we use Asp(OMpe), which is less susceptible to aspartimide formation *(29)*. In peptides not containing Asp, Fmoc deprotection is efficiently achieved by a mixture of 2% (v/v) 1,8-diazabicyclo*(5.4.0)*undec-7-ene (DBU, a very strong base) and 2% (v/v) piperidine in DMF. This mixture should not be used with Asp(OMpe), but rather 20% (v/v) piperidine in DMF containing 0.1 M HOBt.

   c. Although cysteine, like other amino acids, can be coupled as Fmoc-Cys(Trt)-OH, this is not recommended. Experiments with dipeptides have shown that extensive racemization occurs under the basic conditions of the coupling *(30)*. This is avoided by coupling in DMF with the pentafluorophenyl ester Fmoc-Cys(Trt)-OPfp, preferably in the presence of the acylation catalyst 1-hydroxy-7-azabenzotriazole (HOAt).

   d. HOAt (or HOBt) is also an essential additive to ensure the efficient coupling of Asn(Trt) and Gln(Trt). These trityl derivatives have the advantages over Fmoc-Asn-OH and Fmoc-Gln-OH of greatly improved solubility and less susceptibility to side reactions.

   e. We routinely couple other side chain-protected amino acids as Fmoc-Glu(OtBu)-OH, Fmoc-His(Trt)-OH, Fmoc-Hyp(tBu)-OH, Fmoc-Lys(Boc)-OH, Fmoc-Ser(tBu)-OH, Fmoc-Thr(tBu)-OH, Fmoc-Trp(Boc)-OH and Fmoc-Tyr(tBu)-OH.

5. *Choice of resin:* A bewildering range of resins is available for solid-phase peptide synthesis, but by electing to synthesize peptide amides, the choices become restricted to a few.

Polyethylene glycol-grafted polystyrene is the base resin of choice, due to superior swelling and permeability properties. There is probably little to choose between polyethylene glycol-grafted polystyrene. We have used all three successfully, although TentaGel is preferred for longer peptides. Resins are available in various loadings, but we always use those with a capacity of around 0.2 mmol/g. With long peptides, the use of high-load resins could potentially lead to steric hindrance between the growing peptide chains and an increased risk of deletion sequences. All our syntheses are done on a 0.1-mmole scale. Occasionally, one requires a peptide with a free carboxyl terminus. Novabiochem provides a wide range of Fmoc-amino acids attached to NovaSyn TGA resin. Peptides with a C-terminal proline are best synthesized on Fmoc-Pro-NovaSyn TGT resin to avoid the risk of diketopiperazine formation. Should it be necessary to prepare one's own Fmoc-amino acid resin, suitable protocols can be found in Chan and White *(13)*.

6. *Reagents for formation of peptide bonds:* Although the variety is huge and new reagents frequently appear in the literature, most practitioners in the field would choose one of the uronium (e.g., HBTU or TBTU) or phosphonium (e.g., PyBOP) reagents developed in recent years. These have in common a high coupling efficiency and a relative absence of side reactions, combined with low toxicity and an extended shelf life *(31)*. There appears little to choose between HBTU, TBTU, and PyBOP, especially if coupling is enhanced by the addition of the acylation catalyst 1-hydroxybenzotriazole (HOBt). Under some circumstances, the uronium salts can terminate the growing peptide chain by guanidination, but this side reaction is rare.

   For longer peptides, it is necessary to ensure essentially complete peptide-bond formation at every step to avoid an unacceptable build-up of deletion products. For this reason, we routinely use the reagent HATU, which is superior to HBTU, TBTU, and PyBOP in terms of coupling efficiency, stability in solution, and minimizing racemization *(31)*. The reagent is expensive, but avoids the cost and inconvenience of resynthesizing a poor-quality final product. Very recently, a new reagent, HCTU, has been described (www.peptide-and-dna.com/pub.html), which appears to be comparable to HATU in coupling efficiency, but at a much lower cost.

7. *Handling of TFA and the cleavage mixture:* This must be done in an efficient fume hood.

8. *Scavengers:* The fragments released by TFA from side-chain-protecting groups are highly reactive, and it is necessary to include "scavengers" in the cleavage mixture to protect the peptide from damage. The composition of the peptide will determine which additives are necessary, and various mixtures have been proposed. An Acrobat® file discussing these is available from the technical resources pages of the Novabiochem Web site. Except for the very simplest peptides, we use a cleavage mixture containing 9.5 mL trifluoroacetic acid, 0.5 mL thioanisole, 0.25 mL triisopropylsilane, and 0.25 g dithiothreitol. This efficiently scavenges the products of the protecting groups and also the resin-bound linker fragment. Note that we prefer to substitute the essentially odorless dithiothreitol for the noxious 1,2-ethanedithiol.

9. *Oxidation:* Peptides containing methionine are uniquely susceptible to oxidation, and the same is probably true of biotinyl peptides. Although sulfoxide-containing peptides can be reduced, it is better to prevent the problem by adding 1.5% (w/v) ammonium iodide to the cleavage cocktail *(32)*. Tetrabutylammonium bromide has been proposed *(33)* as an alternative to ammonium iodide, but we find this to be carried over into the crude peptide. If the oxidation of methionine cannot be prevented, consider replacing Met by norleucine (Nle), which is closely similar in molecular size and shape.

10. *HPLC columns:* For collagen-derived peptides, we use diphenyl-silica as recommended *(15)*, but other peptides are separated on octadecyl (C18)-silica. It is recommended that after use all reverse phase columns are stored in 95% CH$_3$CN *without TFA*.

11. *Characterization:* Despite all the care taken during the synthesis and purification of a peptide, it cannot be assumed that the final product is the desired one. Recently a peptide containing four *trans*-4-fluoroproline residues displayed a single peak on HPLC, but mass spectrometry showed that the product was not the expected peptide, but rather deletion peptides missing one or two prolines. So do not accept the identity of any peptide until you have seen the mass-spectrometry results and work on the basis that "the mass spectrum doesn't lie." In matrix-assisted laser desorption time of flight mass spectrometry (MALDI-TOF MS), some peptides fly as the Na$^+$ or K$^+$ adducts, but one usually sees the expected H$^+$ species. If doubt exists, request that the sample be run by an alternative method, such as electrospray ionization mass spectrometry. In any event, be sure that the mass spectroscopist has all the facts about your sample, so that the appropriate conditions can be chosen. An estimated final purity in excess of 95% is acceptable.

References

1. Rossi, A., Zanaboni, G., Cetta, G., and Tenni, R. (1997) Stability of type I collagen CNBr peptide trimers. *J. Mol. Biol.* **269,** 488–493.
2. Emsley, J., Knight, C. G., Farndale, R. W., Barnes, M. J., and Liddington, R. C. (2000) Structural basis of collagen recognition by integrin α2β1. *Cell* **101,** 47–56.
3. Consonni, R., Zetta, L., Longhi, R., Toma, L., Zanaboni, G., and Tenni, R. (2000) Conformational analysis and stability of collagen peptides by CD and by ^{1}H- and ^{13}C-NMR spectroscopies. *Biopolymers* **53,** 99–111.
4. Knight, C. G., Morton, L. F., Onley, D. J., Peachey, A. R., Ichinohe, T., Okuma, M., et al. (1999) Collagen-platelet interaction: Gly-Pro-Hyp is uniquely specific for platelet Gp VI and mediates platelet activation by collagen. *Cardiovasc. Res.* **41,** 450–457.
5. Knight, C. G., Morton, L. F., Onley, D. J., Peachey, A. R., Messent, A. J., Smethurst, P. A., et al. (1998) Identification in collagen type I of an integrin α$_2$β$_1$-binding site containing an essential GER sequence. *J. Biol. Chem.* **273,** 33,287–33,294.
6. Croce, K., Flaumenhaft, R., Rivers, M., Furie, B., Furie, B. C., Herman, I. M., et al. (1999) Inhibition of calpain blocks platelet secretion, aggregation, and spreading. *J. Biol. Chem.* **274,** 36,321–36,327.
7. Privalov, P. L. (1982) Stability of proteins. Proteins which do not present a single cooperative system. *Adv. Protein Chem.* **35,** 1–104.
8. Leikina, E., Mertts, M. V., Kuznetsova, N., and Leikin, S. (2002) Type I collagen is thermally unstable at body temperature. *Proc. Nat. Acad. Sci. USA* **99,** 1314–1318.
9. Jenkins, C. L. and Raines, R. T. (2002) Insights on the conformational stability of collagen. *Nat. Prod. Rep.* **19,** 49–59.
10. Grab, B., Miles, A. J., Furcht, L. T., and Fields, G. B. (1996) Promotion of fibroblast adhesion by triple-helical peptide models of type I collagen-derived sequences. *J. Biol. Chem.* **271,** 12,234–12,240.
11. O'Neil, K. T. and DeGrado, W. F. (1990) A thermodynamic scale for the helix-forming tendencies of the commonly occurring amino acids. *Science* **250,** 646–651.
12. Persikov, A. V., Ramshaw, J. A. M., Kirkpatrick, A., and Brodsky, B. (2000) Amino acid propensities for the collagen triple-helix. *Biochemistry* **39,** 14,960–14,967.

13. Chan, W. C. and White, P. D. (eds.) (2000) *Fmoc Solid Phase Peptide Synthesis. A Practical Approach*, Oxford University Press, Oxford, UK.

14. Grant, G. A. (ed.) (2002) *Synthetic Peptides. A User's Guide.* 2nd ed., Oxford University Press USA, New York.

15. Fields, C. G., Grab, B., Lauer, J. L., and Fields, G. B. (1995) Purification and analysis of synthetic, triple-helical "minicollagens" by reversed-phase high-performance liquid chromatography. *Anal. Biochem.* **231**, 57–64.

16. Johnson, T., Quibell, M., Owen, D., and Sheppard, R. C. (1993) A reversible protecting group for the amide bond in peptides. Use in the synthesis of "difficult sequences." *J. Chem. Soc. Chem. Commun.* 369–372.

17 Wöhr, T., Wahl, F., Nefzi, A., Rohwedder, B., Sato, T., Sun, X., et al. (1996) Pseudo-prolines as a solubilising, structure-disrupting protection technique in peptide synthesis. *J. Amer. Chem. Soc.* **118**, 9218–9227.

18. Johnson, T., Quibell, M., and Sheppard, R. C. (1995) *N,O*-bisFmoc derivatives of *N*-(2-hydroxy-4-methoxybenzyl)-amino acids: useful intermediates in peptide synthesis. *J. Peptide Sci.* **1**, 11–25.

19. Knight, C. G., Morton, L. F., Peachey, A. R., Tuckwell, D. S., Farndale, R. W., and Barnes, M. J. (2000) The collagen-binding A-domains of integrins $\alpha_1\beta_1$ and $\alpha_2\beta_1$ recognize the same specific amino acid sequence, GFOGER, in native (triple-helical) collagens. *J. Biol. Chem.* **275**, 35–40.

20. Achison, M., Elton, C. M., Hargreaves, P. G., Knight C. G., Barnes M. J., and Farndale R. W. (2001) Integrin-independent tyrosine phosphorylation of p125[FAK] in human platelets stimulated by collagen. *J. Biol. Chem.* **276**, 3167–3174.

21. Morton, L. F., Hargreaves, P. G., Farndale, R. W., Young, R. D., and Barnes, M. J. (1995) Integrin $\alpha_2\beta_1$-independent activation of platelets by collagen: collagen tertiary (triple helical) and quaternary (polymeric) structures are sufficient alone for activity. *Biochem. J.* **306**, 337–344.

22. Asselin, J., Knight, C. G., Farndale, R. W., Barnes, M. J., and Watson, S. P. (1999) The monomeric glycine-proline-hydroxyproline (GPP*)$_{10}$ repeat sequence is a partial agonist of the collagen receptor glycoprotein VI. *Biochem. J.* **339**, .413–418

23. Gibbins, J., Okuma, M., Farndale, R. W., Barnes, M. J., and Watson S. P. (1997) Glycoprotein VI is the collagen receptor in platelets which underlies tyrosine phosphorylation of the Fc receptor γ-chain. *FEBS Lett.* **413**, 255–259.

24. Kehrel, B., Wierwille, S., Clemetson, K. J., Anders, O., Steiner, M., Knight, C. G., et al. (1998) Glycoprotein VI is a major collagen receptor, recognizing the platelet-activating quaternary structure of collagen, whereas CD36, GPIIb/IIIa and vWf do not. *Blood* **91**, 491–499.

25. Merrifield, R. B. (1963). Solid phase peptide synthesis. I. The synthesis of a tetrapeptide. *J. Amer. Chem. Soc.* **85**, 2149–2154.

26. Fields, C. G. and Fields, G. B. (1993) Minimization of tryptophan alkylation following 9-fluorenylmethoxycarbonyl solid-phase peptide-synthesis. *Tetrahedron Lett.* **34**, 6661–6664.

27. Lauer, J. L., Fields, C. G., and Fields, G. B. (1994) Sequence dependence of aspartimide formation during 9-fluorenylmethoxycarbonyl solid-phase peptide synthesis. *Lett. Peptide Sci.* **1**, 197–205.

28. Packman, L. C. (1995) N-2-Hydroxy-4-methoxybenzyl (Hmb) backbone protection prevents double aspartimide formation in a "difficult" peptide sequence. *Tetrahedron Lett.* **36**, 7523–7526.

29. Karlström, A. H. and Undén, A. E. (1996) A new protecting group for aspartic acid that minimises piperidine-catalysed aspartimide formation in Fmoc solid phase peptide synthesis. *Tetrahedron Lett.* **37,** 4243–4246.
30. Han, Y., Albericio, F., and Barany, G. (1997) Occurrence and minimization of cysteine racemization during stepwise solid-phase peptide synthesis. *J. Org. Chem.* **62,** 4307–4312.
31. Albericio, F., Bofill, J. M., El-Faham, A., and Kates, S. A. (1998) Use of onium salt-based coupling reagents in peptide synthesis. *J. Org. Chem.* **63,** 9678–9683.
32. Huang, H., and Rabenstein, D. L. (1999) A cleavage cocktail for methionine-containing peptides. *J. Peptide Res.* **53,** 548–553.
33. Taboada, L., Nicolas, E., and Giralt, E. (2001). One-pot full peptide deprotection in Fmoc-based solid-phase peptide synthesis: methionine sulfoxide reduction with Bu$_4$NBr. *Tetrahedron Lett.* **42,** 1891–1893.

22

Platelet Permeabilization

Robert Flaumenhaft

1. Introduction

Methods of platelet permeabilization have been used for more than 20 yr as a means to study the role of various intracellular molecules in platelet function (*1–3*). Initially, permeabilization of platelets was used largely to introduce cations, nucleotides, and other small molecules into platelets. Since platelets are anucleate, it is more difficult to express proteins in these cells. Thus, permeabilization represents the most convenient method by which to introduce recombinant proteins into platelets. Inhibitory antibodies and peptides have also been dialyzed into permeabilized platelets in order to assess the role of specific proteins in platelet physiology. Several methods of permeabilizing platelet membranes have been described. The decision of which method to use will depend on the application for which the permeabilization procedure is being performed. This chapter will describe the use of bacterial toxins as permeabilizing agents. Detailed descriptions of alternative methods for platelet permeabilization (e.g., by electroporation or by detergents such as saponin and digitonin) have been described elsewhere (*4*).

Pore-forming bacterial toxins have become increasingly popular as permeabilizing agents based on several characteristics. These toxins are restricted to surface-connected membranes (i.e., the platelet plasma membrane and open canalicular system) and do not permeabilize intracellular membranes. Thus, they do not cause the release of granular contents or perturb the dense tubular system. The pores formed by these reagents are reproducible and stable over time. Use of different toxins enables the investigator to form uniform pores ranging from 0.8 nm to 30 nm in size (*5*). This characteristic allows for the introduction of macromolecules into the platelet cytosol. In addition, these toxins are devoid of enzymatic activity and do to not appear to interact with platelet proteins. These reagents are easy to use. They can be stored for long periods of time when in a purified form and require no special handling or preparation prior to use. This chapter will describe the use of α-toxin and streptolysin-O (SL-O) in permeabilizing platelets.

α-Toxin is a pore-forming protein secreted by *Staphylococcus aureus*. The toxin has an apparent molecular mass of 33.4 kDa (*5*) and is secreted as a water-soluble

From: *Methods in Molecular Biology, vol. 273:*
Platelets and Megakaryocytes, Vol. 2: Perspectives and Techniques
Edited by: J. M. Gibbins and M. P. Mahaut-Smith © Humana Press Inc., Totowa, NJ

monomeric protein that forms a heptameric transmembrane β-barrel pore upon insertion into plasma membranes *(6)*. It was first used to permeabilize chromaffin cells *(7)*. Permeabilization of platelets with α-toxin results in pores of approx 0.6–1 nm that permit flux of ions and nucleotides, but not large proteins *(5)*. On the other hand, α-toxin is capable of marked permeabilization that is stable, very reproducible, and does not require reducing reagents. α-toxin is the bacterial toxin of choice under circumstances in which the investigator wishes to selectively and irreversibly perme-abilize the surface-connected membrane (i.e., the plasma membrane and membranes of the open canalicular system) without losing intracellular proteins.

SL-O is a protein with an apparent molecular mass of 61.5 kDa that is secreted in a monomeric form by streptococcal species. Like α-toxin, SL-O is water soluble and forms pores only following insertion into membranes. Unlike α-toxin, the number of monomers contributing to the pore is variable and is dependent, in part, on the concen-tration of SL-O that is used. Thus, SL-O provides the investigator with the ability to vary the size of pores. Pores are generally stable over time. A method for generating reversible pores using low concentrations of SL-O has been described *(8)*. This tech-nique, however, has not yet been used for platelets. SL-O contains a cysteine residue that is spontaneously oxidized at atmospheric oxygen, resulting in the loss of perme-abilizing activity *(5)*. The toxin regains activity upon reduction. SL-O, therefore, is used in the presence of a reducing agent. Permeabilization of platelets with SL-O enables the introduction of ions, nucleotides, peptides, recombinant and purified proteins, and antibodies into platelets. The investigator must be cognizant of the fact, however, that cytosolic proteins are lost from platelets following permeabilization. Although loss of cytosolic proteins can impair stimulation of the platelet function of interest, it also offers an opportunity to identify proteins capable of reconstituting a particular function. For example, protein kinase C_α (PKC$_\alpha$) *(9)* and type II phosphati-dylinositol-5-phosphate 4-kinase (type II PIPK) *(10)* have been found to augment granule secretion from SL-O-permeabilized platelets. Thus, both loss of function and gain of function strategies can be used with SL-O-permeabilized platelets.

2. Materials
2.1. Analysis of α-Granule Secretion From Permeabilization Platelets

1. 30-mL syringe with 18-gauge needle.
2. Platelet-rich plasma and platelet-poor plasma.
3. Sepharose 2B column (Amersham Pharmacia Biotech, Piscataway, NJ).
4. PIPES/KCl buffer: 25 mM PIPES, 2 mM EGTA, 137 mM KCl, 4 mM NaCl, 0.1% glucose, 0.1% BSA, pH 6.4.
5. 100 mM NaOH.
6. Antibody, peptide, or recombinant protein to be tested.
7. MgATP: store in frozen aliquots at 100 mM; MgATP will lose its activity over a period of days to weeks when stored at 4°C.
8. CaCl$_2$.
9. Phycoerythrin-conjugated AC1.2 anti-P-selectin antibody (Becton Dickinson, San Jose, CA).
10. SFLLRN (Bachem, Torrance, CA) (store at 4°C), PMA (Calbiochem, San Diego, CA) (store in frozen aliquots of 10 mM), or other platelet agonists.

11. Streptolysin-*O* (Sigma-Aldrich, St. Louis, MO) (store in frozen aliquots of 10,000 U/mL; use one aliquot per day) (*see* **Note 2**).
12. α-Toxin (Calbiochem) (store in frozen aliquots at 2500 U/mL; once an aliquot is open, it can be stored at 4°C and used for several days).
13. Sheath fluid (Becton Dickinson).
14. FACSCalibur flow cytometer.

2.2. Assessment of Permeabilization

1. Dextran sulfate-FITC conjugates (Sigma-Aldrich).
2. Sulforhodamine or fluorescein (Sigma-Aldrich).

2.3. Analysis of Leakage of Cytosolic Proteins

1. SDS-PAGE (sodium dodecyl-sulfate polyacrylamide gel electrophoresis) equipment.
2. Electrophoretic protein transfer apparatus.
3. PVDF membranes.
4. Standard buffers for immunoblotting.
5. Antibodies directed at the protein of interest.

3. Methods

The actual permeabilization of platelet surface-connected membranes using bacterial toxins for permeabilizing platelets is not complicated. The challenges of the technique are to maintain the platelet function of interest despite compromise of the platelet plasma membrane and to assess the degree of permeabilization. Once the permeabilization conditions for a particular function are established, however, permeabilization of platelets enables interrogation of several aspects of platelet biology and biochemistry that cannot be analyzed in intact platelets. For example, permeabilization of platelets enables the investigator to introduce highly specific reagents into the cytosol of platelets in order to establish the function of particular proteins. In addition, permeabilization allows for the determination of proteins that are cytosolic (i.e., they diffuse from permeabilized platelets) from those proteins that are associated with either the membrane or cytoskeletal structures (i.e., they remain associated with the permeabilized platelet following centrifugation). Since there is extensive experience in evaluating platelet granule secretion using these methods, examples given in this chapter will focus on the study of granule secretion in permeabilized platelets. Permeabilization with SL-O can also be used to evaluate activation-induced cytoskeletal changes *(11)*. In addition, platelet aggregation has been studied in platelet systems permeabilized with bacterial toxins *(12,13)* (*see* **Note 1**).

3.1. Platelet Permeabilization

3.1.1. Analysis of Platelet α-Granule Secretion From Permeabilization Platelets

1. Blood from healthy volunteers who have not ingested aspirin in the two weeks prior to donation is collected by venipuncture into 0.38% (w/v) sodium citrate. If freshly obtained blood from donors cannot be obtained, platelets may be isolated from platelet-rich plasma (PRP) obtained from blood banks.
2. Citrate anticoagulated blood is centrifuged at 200*g* for 20 min to prepare PRP. PRP is removed from the tube using a polypropylene pipet, leaving the buffy coat undisturbed.

3. Platelets are purified from PRP by gel filtration using Sepharose 2B prepared in a siliconized glass column containing a nylon microfilament disk. We have found that isolation of platelets using newly poured columns occasionally results in activation of platelets. To avoid activation, we equilibrate newly poured columns with platelet-poor plasma (10 vol) prior to equilibration in buffer. The column should have a flow rate of approx 1 mL/min. The column is then equilibrated in PIPES/KCl buffer (5–6 vol).

4. Platelets (3–6 mL) are applied directly onto the gel. Once the full volume of platelets has entered the gel, PIPES/KCl buffer is applied to the column. The relatively acidic pH of this buffer helps maintain the platelets in a resting state. Cloudiness of the eluate indicates that platelets are coming off the column. Final gel-filtered platelet concentrations are approx $1–2 \times 10^8$ platelets/mL.

5. Platelets (4–8 mL) are collected into a 15-mL polypropylene tube. Collection is stopped before plasma proteins are eluted from the column. The column is then washed in ddH$_2$O. (It is also possible to wash platelets by sequential centrifugation in order to obtain platelets suitable for permeabilization. It is important to recall, however, that the cytosol rapidly equilibrates with the buffer in which the platelets are suspended upon permeabilization of platelets. The relatively high concentrations of calcium in plasma can cause platelet activation upon permeabilization of platelets in PRP. Other plasma components may also affect platelet function when introduced into the cytosol. Therefore, only gel-filtered or adequately washed platelets should be used.)

6. For permeabilization experiments, 18 μL gel-filtered platelets are distributed into 1.5-mL polypropylene tubes. One μL of 100 mM MgATP (final concentration 5 mM) is then added to the platelets. If the investigator wishes to test the activity of antibodies, peptides, or recombinant proteins, these reagents can be added prior to the addition of the bacterial toxin.

7. Permeabilization with α-toxin is accomplished by adding 1 μL of 2500 U/mL α-toxin (final concentration 125 U/mL) to the platelet sample. Permeabilization using SL-O is accomplished by adding 1 μL of 10,000 U/mL SL-O (final concentration 500 U/mL) reduced with 5 mM DTT to the platelet sample. The platelet sample is adjusted to pH 6.9 using 100 mM NaOH (~0.8 μL). Samples are performed in triplicate.

8. Following a 15-min incubation at room temperature, an amount of CaCl$_2$ sufficient to achieve a final concentration of 10 μM Ca^{2+} is added to the reaction mixture. The amount of CaCl$_2$ may vary depending on the specific buffering conditions of the assay. The CaCl$_2$ concentration can be determined using algorithms such as those available at http://www.stanford.edu/~cpatton/webmaxclite11.htm. Platelets are incubated with 10 μM Ca^{2+} for 5 min.

9. P-selectin surface expression is used to monitor Ca^{2+}-triggered platelet α-granule secretion. For this assay, 10 μL platelets are incubated with 5 μL phycoerythrin-conjugated AC1.2 anti-P-selectin antibody for 20 min. 10 μL of this sample is then added to PBS (500 μL) and the platelets are analyzed immediately by flow cytometry. Fluorescent channels are set at logarithmic gain. Ten thousand particles are acquired for each sample. A 530/30 band-pass filter is used for the FL-2 fluorescence channel, in which phycoerythrin fluorescence is measured.

10. P-selectin expression from several control samples is also evaluated. Samples permeabilized in the absence of MgATP, samples not exposed to Ca^{2+}, and samples not exposed to either serve as negative controls. Intact platelets incubated in the presence or absence of 100 μM SFLLRN serve as controls for maximum granule secretion from intact platelets. P-selectin surface expression in response to other strong agonists such as thrombin or PMA may also be used with intact platelets.

11. The percent secretion of α-granules from permeabilized platelets compared with α-granule secretion from intact platelets is next determined. For this calculation, background P-selectin expression from permeabilized or intact platelets in the absence of Ca^{2+} or SFLLRN, respectively, are subtracted from all values. The value of P-selectin expression from permeabilized platelets exposed to Ca^{2+} is then divided by the value of P-selectin expression from intact platelets exposed to SFLLRN. This value is multiplied by 100 to obtain the percentage of secretion compared to intact platelets.

Several considerations should be attended to when testing the effect of reagents on the function of the permeabilized system. If a reagent directed at a specific intracellular target appears to affect secretion from permeabilized platelets, the investigator should determine the effect of the reagent on intact platelets. An effect on intact platelets suggests a nonspecific activity of the reagent. It is important that a nonimmune antibody or an irrelevant control peptide or protein be solubilized in exactly the same buffer and tested in the permeabilized system **(Fig. 1A)** before attributing an activity to a particular reagent, as permeabilized systems are very sensitive to buffering conditions (*see* **Note 2**). Whenever possible, the investigator should use both gain-of-function studies and loss of function studies to determine whether a specific molecule is involved in the secretory process **(Fig. 1B)**. In interpreting the results of a study, it is important to recall that if reagents directed at a target fail to affect a platelet function, it does not necessarily indicate that the target is not involved in that platelet function. For example, an inhibitory antibody may not inhibit because it fails to attain adequate cytosolic concentrations or it may be sterically hindered so that it cannot reach the target in the intracellular milieu. Similarly, a recombinant protein may not have access to the same compartment as the endogenous protein. With proper attention to controls and with thoughtful interpretation, however, functional experiments using permeabilization with bacterial toxins can reveal much about intracellular processes that direct platelet function.

Whether the permeabilized platelet model is designed to evaluate granule secretion (as in the protocol above) or another platelet function, it is valuable to determine the degree of functionality of a permeabilized system compared to that of the intact system. This can vary widely depending on several variables including the agonist used for stimulation (*see* **Note 3**). For example, secretion of granules from permeabilized platelets can vary from 10% to 95% of secretion in comparable intact platelet models *(14–17)*. Most assays using bacterial toxins as permeabilizing agents demonstrate ≥50% granule secretion compared with maximal secretion from intact platelets *(16,17)*. Although there are no firm guidelines in the field, it is worth further optimizing conditions if functionality from the permeabilized system is <50% that of the intact system. Optimization involves determining the stability of the permeabilized system (described in **Subheading 3.1.2.**), assessing the degree of permeabilization (as described in **Subheading 3.2.**), varying toxin concentrations (*see* **Note 4**), and altering buffering conditions.

3.1.2. Assessing Stability of Permeabilized Secretory Systems

Since many proteins diffuse out from permeabilized platelets, it is possible that the permeabilized platelet model of interest will lose activity following permeabilization.

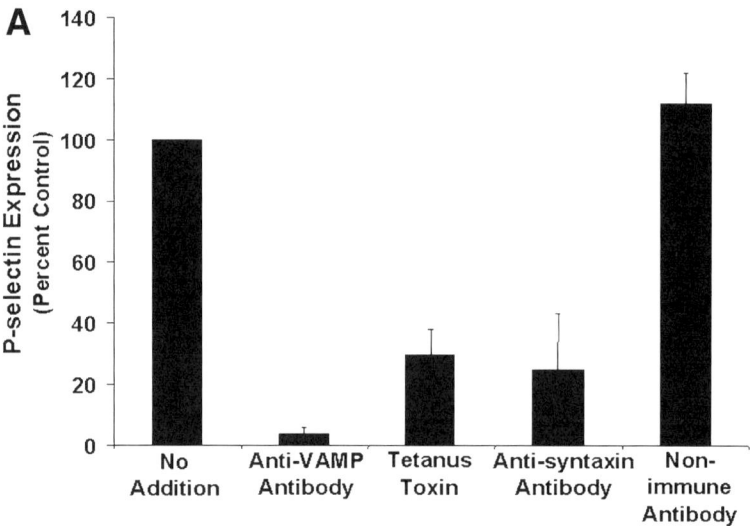

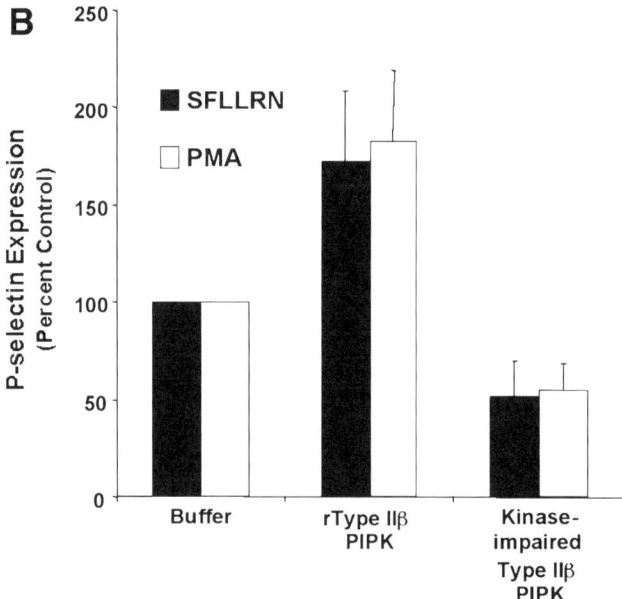

Fig. 1. Use of SL-O-permeabilized platelets to identify proteins involved in platelet α-granule secretion. (**A**) Platelets permeabilized with 2 U/mL Corgenix SL-O were incubated in the presence of the indicated reagents that target proteins that mediate α-granule secretion. α-granule secretion was then stimulated by exposure to 10 μ*M* Ca^{2+}. Decreased P-selectin expression in the presence of these reagents demonstrates that they inhibit Ca^{2+}-induced α-granule release as described in **ref. *17***. (**B**) Both

For example, loss of activity over time is characteristic of agonist-induced secretory models (**Fig. 2A**). In contrast, stimulation of secretion from MgATP-exposed platelets by Ca^{2+} is stable over time (**Fig. 2B**). Time-course experiments assessing the stability of the system may suggest whether or not proteins involved in the process of interest are leaking out of the platelet. Such experiments are also useful in optimizing the functionality of the permeabilized system compared with intact platelets.

3.1.2.1. ASSESSMENT OF STABILITY OF α-GRANULE SECRETION FROM PERMEABILIZED PLATELETS

1. Gel-filtered platelets equilibrated in PIPES/KCl buffer are permeabilized in the presence of 5 m*M* MgATP using α-toxin or SL-O as described in **steps 1–3** of **Subheading 3.1.1.**
2. To test the stability of the response of permeabilized platelets to an agonist such as SFLLRN, permeabilized platelets are exposed to agonists immediately following permeabilization and at 1, 2, 5, and 10 min after permeabilization. To test the stability of the response of permeabilized platelets exposed to 10 μ*M* Ca^{2+}, permeabilized platelets are exposed to Ca^{2+} at time 0 (prior to permeabilization) and at 15, 30, 45, and 60 min after permeabilization. Time points can be modified in subsequent experiments to more clearly define the stability of α-granule secretion over time (**Fig. 2**). Control platelets are treated with buffer alone at the indicated time points.
3. P-selectin surface expression is subsequently monitored as a measure of α-granule secretion as described in **step 9** of **Subheading 3.1.1.**

If a permeabilized system loses activity too quickly for adequate evaluation of a particular reagent, the investigator can decrease the concentration of toxin used in the system. This may enhance the stability of the system. Decreasing the toxin concentration may, however, limit the amount of reagent incorporated into the cytosol during the course of the experiment. Intracellular proteases may compromise the stability of a permeabilized platelet system. Calpain is a particularly important protease to consider in Ca^{2+}-triggered systems. Calpain has been demonstrated to cleave essential components of the secretory machinery *(18,19)* as well as numerous other intracellular platelet proteins. General inhibitors of cysteine protease (such as leupeptin) can be added to the permeabilized system to prevent proteolytic cleavage of intracellular proteins.

3.2. Assessment of Permeabilization

Assessment of platelet permeabilization is useful for several reasons. Such assessment is the first step in troubleshooting a permeabilized platelet system. For example, an investigator often wishes to test various reagents in an SL-O-permeabilized platelet model to determine whether they inhibit or augment functions. Failure of the reagent to affect the model may be because the reagent does not enter the platelet or because the reagent is ineffective despite adequate access to the cytosol. To distinguish between these two possibilities, the investigator needs to assess the degree of permeabilization. It

---

Fig. 1. *(continued)* SFLLRN- and PMA-induced α-granule secretion are augmented by recombinant type II PIPK and inhibited by a kinase-impaired form of this enzyme that acts in a dominant negative manner as described in **ref. 10**.

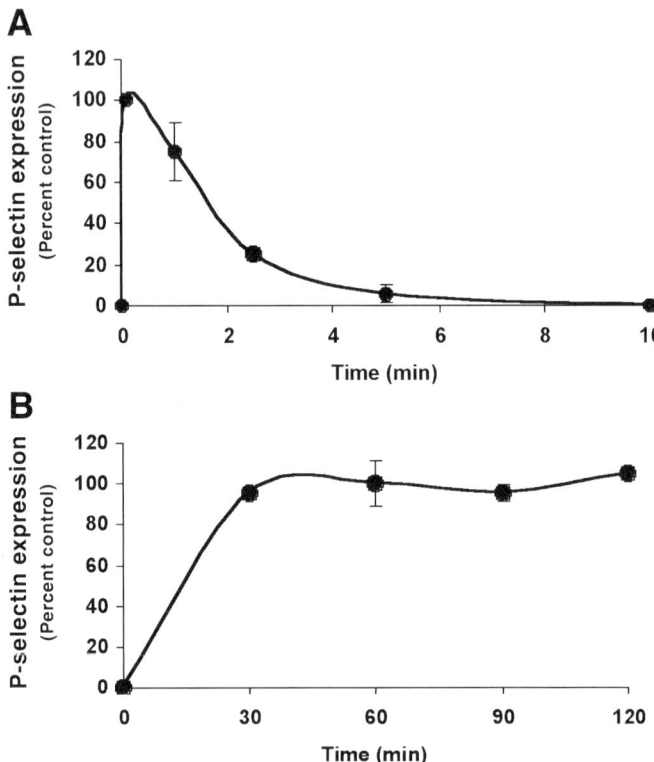

Fig. 2. Assessment of the stability of a permeabilized system. (**A**) Platelets permeabilized with 4 U/mL SL-O lose the ability to respond to SFLLRN within 10 min
following permeabilization. (**B**) In contrast, SL-O-permeabilized platelets incubated
with MgATP remain responsive to Ca^{2+} up to 120 min following permeabilization as
described in **ref. 17**.

is often useful to evaluate the ability of molecules of various sizes to enter the cytosol.
Such assessment is conveniently performed using fluorescently labeled oligopeptides or
dextran-sulfates in conjunction with flow cytometry *(17,20)*. Alternative methods, however, may also be appropriate depending on the application (*see* **Note 5**).

3.2.1. Assessment of Permeabilization Using FITC-Labeled Dextran Sulfates

1. Gel-filtered platelets are prepared as described in **Subheading 3.1.1., steps 1–3**.
2. For assessment of permeabilization using α-toxin, gel-filtered platelets are incubated in
   the presence of 25 μ*M* sulforhodamine, Oregon Green, or fluorescein. Platelets are then
   incubated in the presence or absence of the desired concentration of α-toxin (**Fig. 3**). For
   assessment of permeabilization using SL-O, gel-filtered platelets (20 μL) are incubated
   with 25 μ*M* of FITC-labeled dextran-sulfates of various molecular masses (e.g., 40, 70,

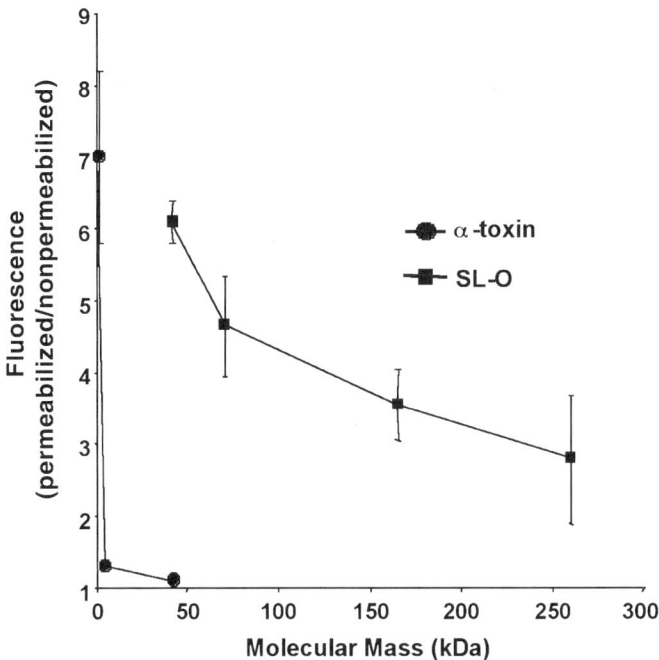

Fig. 3. Assessment of permeabilization of platelets with α-toxin and streptolysin-O. Platelets were incubated with 125 U/mL α-toxin in the presence of Oregon Green (0.5 kDa) or FITC-dextran sulfates of the indicated molecular masses (circles) or incubated with 2 U/mL Corgenix SL-O in the presence of FITC-dextran sulfates of the indicated molecular masses (squares). After a 15-min incubation, platelets were analyzed by flow cytometry. Association of the indicated compounds with platelets is expressed as the relative fluorescence of toxin-exposed platelets relative to nonpermeabilized platelets as described in **refs.** *23* and *17*.

    250 kDa) **(Fig. 3)**. Platelets are then incubated in the desired concentration of SL-O reduced with 5 m*M* DTT.
3. After a 15-min incubation, platelets are diluted in PBS (500 µL) and analyzed immediately for fluorescence by flow cytometry. Comparison of the relative fluorescence of permeabilized and unpermeabilized platelets will enable the investigator to determine whether the free fluores or FITC-dextran sulfates of a particular molecular weight are able to enter the cytosol. If there is no increase in fluorescence in the permeabilized platelet relative to the unpermeabilized platelet, then the permeabilization procedure needs to be reassessed.

    The major limitation of existing permeabilization assessment strategies is that precise intracellular concentrations of the reagent of interest cannot easily be calculated based on the data. By performing the procedure described above in the presence of varying the concentrations of bacterial toxins, however, the investigator can determine the lowest toxin concentration at which permeabilization is achieved. In addition, the

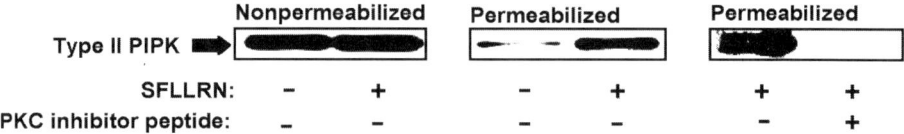

Fig. 4. Assessment of translocation of intracellular proteins in SL-O-permeabilized platelets. Platelets were exposed to BAPTA-AM and incubated in the presence or absence of 100 μ*M* SFLLRN and/or 5 μ*M* PKC inhibitor peptide as indicated. The indicated samples were then permeabilized with 4 U/mL Corgenix SL-O for 15 min. Platelets were then pelleted. Proteins from platelet pellets were assayed for type II PIPK by immunoblotting. This experiment demonstrates that type II PIPK translocates from cytosol to a platelet-associated compartment only following exposure to SFLLRN. SFLLRN-induced translocation is blocked by the PKC inhibitor peptide. This analysis demonstrates that SFLLRN-induced translocation of type II PIPK is PKC-dependent, as described in **ref. *10***.

investigator can determine the toxin concentration at which permeabilization is maximal for a FITC-dextran sulfate of a particular size.

3.3. Leakage of Cytosolic Proteins

Permeabilization of platelets with bacterial toxins allows for both the incorporation of molecules into cytosol and the diffusion of cytosolic proteins from the cytosol. Assessment of proteins that diffuse out of the permeabilized platelet can be useful in characterizing proteins required for particular platelet functions. This technique can also be used to determine conditions under which a protein translocates from the cytosol to a nondiffusable compartment such as the platelet membrane or cytoskeleton *(10,21,22)* (**Fig. 4**).

3.3.1. Detection of Leakage of Intracellular Proteins

1. Gel-filtered platelets equilibrated in PIPES/KCl buffer are permeabilized with 10,000 U/mL stock SL-O as described in **steps 6–8** of **Subheading 3.1.1.** A higher concentration of SL-O is often required to facilitate diffusion of proteins. Since several proteins translocate in an ATP-dependent manner, it is useful to assess protein translocation in the presence or absence of 5 m*M* MgATP or in the presence or absence of platelet agonists. Similarly, it is useful to assess proteins in the presence or absence of 10 μ*M* Ca^{2+}, as platelet proteins can undergo Ca^{2+}-dependent translocation or proteolysis *(18,19)*.
2. Thirty minutes following permeabilization, platelets are pelleted at 1000*g* for 10 min. The supernatants are lyophilized to recover proteins that have diffused out of the permeabilized platelet. Lyophilized proteins are then dissolved in sample buffer for SDS-PAGE. The pellets are washed three times and solubilized in sample buffer.
3. Platelet proteins from lyophilized samples and from washed pellets of permeabilized platelets are separated by SDS-PAGE and electrophoretically transferred to PVDF membranes. For further details on SDS-PAGE and Western blotting, refer to Chapter 9, vol. 2.

Table 1
Buffer Requirements for Ca^{2+}-Induced α-Granule Secretion
from α-Toxin-Permeabilized Platelets

Nucleotide	% Control[a]	Monovalent cation	% Control	Divalent cation	% Control
ATP	100 ± 16	137 mM NaCl 4 mM KCl	100 ± 2	Mg^{2+}	100 ± 5
GTP	72 ± 9	70.5 mM NaCl 70.5 mM KCl	115 ± 6	Mn^{2+}	9 ± 7
CTP	72 ± 11	4 mM NaCl 137 mM KCl	112 ± 9	No addition	15 ± 1
UTP	52 ± 28				
ITP	37 ± 5				
No addition	0 ± 0				

[a]P-selectin expression induced by 10 μM Ca^{2+} in the presence of 5 mM ATP, 5 mM MgCl$_2$, and PIPES/KCl buffer is considered 100%

4. Immunoblot analysis of PVDF membrane is performed with antibodies directed at the protein of interest using standard techniques. For further details on immunoblotting, refer to Chapters 9 and 10, vol. 2.

Proteins that are found in the lyophilized supernatants of permeabilized platelets, but not in the washed pellets, are presumed to be cytosolic proteins. Those proteins that are found only in the washed pellets are considered platelet-associated proteins. Further analysis of permeabilized platelet pellets can be performed (e.g., using a membrane-dissolving detergent such as Triton X-100) to assess the distribution of noncytosolic proteins.

4. Notes

1. Although this chapter details methods of evaluating platelet α-granule secretion, multiple platelet responses can be analyzed using platelets permeabilized by bacterial toxins. Dense granule *(26,29)* and lysosome *(27)* secretion have been analyzed using SL-O-permeabilized platelets. Platelet aggregation has been evaluated using SL-O-permeabilized platelets *(12,13)*. Effects of permeabilization on the platelet cytoskeleton have also been evaluated using SL-O *(11)*. In addition to the evaluation of multiple platelet responses, bacterial toxins can be used to permeabilize megakaryocytes (unpublished observations).

2. The buffer used in permeabilization experiments quickly equilibrates with the cytosol of the permeabilized platelet and, thus, represents a critical component of successful permeabilization studies. Several aspects of buffer conditions warrant consideration. pH is a critical determinant of stimulation-induced granule secretion from permeabilized platelets *(23)*. Thus, buffer pH must be controlled carefully when attempting to optimize a stimulation-induced function in permeabilized platelets. Another critical component of the buffer is its nucleotide content. A requirement for ATP in supporting stimulation-induced granule secre-

tion from platelets is well-established *(23,24)*. Although other nucleotides can be used, they are less effective (**Table 1**). Permeabilized platelets also demonstrated a requirement for the divalent cation Mg^{2+}, for which Mn^{2+} cannot substitute *(23)*. The ratio of monovalent cations Na^+ and K^+ does not affect the ability of permeabilized platelets to respond to stimulation *(23)*. Since K^+ is the dominant monovalent intracellular cation, it is used at a much higher concentration than Na^+ in permeabilization buffers. Buffers using glutamate as the dominant anion are also used for permeabilization assays *(16,25)*. Glucose and BSA do not appear to be critical determinants of function (unpublished results). Careful attention to buffer constituents is essential to designing a functional permeabilized platelet model and the buffer must be considered first when troubleshooting the model.

3. Platelets permeabilized with bacterial toxins can be stimulated to respond to many different types of agonists. The method described in this chapter details the use of Ca^{2+} to trigger α-granule secretion. Strong agonists, such as thrombin, SFLLRN, phorbol esters, and Ca^{2+} ionophores, capable of stimulating intact platelets, can also be used to stimulate agonist-dependent platelet functions such as granule secretion, aggregation, and shape change from permeabilized platelets. One caveat of using permeabilized platelets with these agonists is that molecules critical for signal transduction diffuse from the platelet following permeabilization with SL-O. Thus, agonist-dependent systems are typically not stable over time following permeabilization with SL-O. For this reason, platelets are usually exposed to agonists within 2 min following permeabilization. Permeabilized platelets will also respond to a variety of agonists such as stable analogs of GTP *(21,22,27)* and IP_3 *(3,28)* that are not capable of stimulating intact platelets.

4. α-Toxin obtained from Calbiochem yields very efficient and highly reproducible permeabilization of platelets. The major shortcoming of α-toxin is that the size of the pores created using this reagent limits the size of reagents that can be introduced into the cytosol. In contrast, obtaining an SL-O product with reproducible activity is somewhat more difficult. A product, originally available through Murex, Abbott, and/or Corgenix, which was quite reliable, is no longer available. SL-O can be obtained from Sigma (St. Louis, MO) and has been used successfully for platelet permeabilization. There is, however, more lot variation with this product (unpublished observation). Furthermore, this product must be reduced using DTT by the investigator prior to use. It is also important to be aware that different vendors use very different units to quantify the activity of their products. Thus, one must be cognizant of the origin of SL-O when reading the literature. (The experiments represented by the figures in this chapter were performed using SL-O from Corgenix. Corgenix SL-O is no longer available, therefore protocols are described for Sigma SL-O.)

5. An alternative to assessing permeabilization using fluorescein-labeled molecules is to evaluate the leakage of intracellular contents from the permeabilized platelet. Two cytosolic markers that have been used for this purpose are ATP and LDH *(16,26)*. Leakage of ATP from platelets is readily monitored using a luciferin-luciferase detection system. ATP leakage should occur following permeabilization with either α-toxin or SL-O. In contrast, leakage of LDH should occur following permeabilization with SL-O, but not with α-toxin. Thus, use of these two indicators of permeabilization yields some limited information regarding pore size.

Acknowledgment

The author would like to thank Nataliya Rozenvayn for technical expertise. Robert Flaumenhaft is supported by NIH grant HL63250. He is a recipient of an American

Society of Hematology Junior Faculty Scholar Award and the Burroughs Wellcome Fund Career Award.

References

1. Launay, J. M. and Alouf, J. E. (1979) Biochemical and ultrastructural study of the disruption of blood platelets by streptolysin O. *Biochim. Biophys. Acta* **556**, 278–291.
2. Haslam, R. J. and Davidson, M. M. (1984) Receptor-induced diacylglycerol formation in permeabilized platelets; possible role for a GTP-binding protein. *J. Recept. Res.* **4**, 605–629.
3. Lapetina, E. G., Watson, S. P., and Cuatrecasas, P. (1984) myo-Inositol 1,4,5-trisphosphate stimulates protein phosphorylation in saponin-permeabilized human platelets. *Proc. Natl. Acad. Sci. USA* **81**, 7431–7435.
4. Scrutton, M. C., Knight, D. E., and Authi, K. S. (1996) Preparation and uses of semi-permeabilized platelets, in *Platelets: A Practical Approach* (Watson, S. P. and Authi, K. S., eds.), Oxford University Press, New York, pp. 47–65.
5. Bhakdi, S., Bayley, H., Valeva, A., Walev, I., Walker, B., Kehoe, M., et al. (1996) Staphylococcal alpha-toxin, streptolysin-O, and Escherichia coli hemolysin: prototypes of pore-forming bacterial cytolysins. *Arch. Microbiol.* **165**, 73–79.
6. Song, L., Hobaugh, M. R., Shustak, C., Cheley, S., Bayley, H., and Gouaux, J. E. (1996) Structure of staphylococcal alpha-hemolysin, a heptameric transmembrane pore. *Science* **274**, 1859–1866.
7. Ahnert-Hilger, G., Bhakdi, S., and Gratzl, M. (1985) Minimal requirements for exocytosis. A study using PC 12 cells permeabilized with staphylococcal alpha-toxin. *J. Biol. Chem.* **260**, 12,730–12,734.
8. Walev, I., Bhakdi, S. C., Hofmann, F., Djonder, N., Valeva, A., Aktories, K., et al. (2001) Delivery of proteins into living cells by reversible membrane permeabilization with streptolysin-O. *Proc. Natl. Acad. Sci. USA* **98**, 3185–3190.
9. Yoshioka, A., Shirakawa, R., Nishioka, H., Tabuchi, A., Higashi, T., Ozaki, H., et al. (2001) Identification of protein kinase C alpha as an essential, but not sufficient, cytosolic factor for Ca^{2+}-induced alpha- and dense-core granule secretion in platelets. *J. Biol. Chem.* **276**, 39,379–39,385.
10. Rozenvayn, N. and Flaumenhaft, R. (2002) Protein kinase C mediates translocation of type II phosphatidylinositol 5-phosphate 4-kinase required for platelet alpha-granule secretion. *J. Biol. Chem.* **278**, 8126–8134.
11. Walch, M., Ziegler, U., and Groscurth, P. (2000) Effect of streptolysin O on the micro-elasticity of human platelets analyzed by atomic force microscopy. *Ultramicroscopy* **82**, 259–267.
12. Nishioka, H., Horiuchi, H., Tabuchi, A., Yoshioka, A., Shirakawa, R., and Kita, T. (2001) Small GTPase Rho regulates thrombin-induced platelet aggregation. *Biochem. Biophys. Res. Commun.* **280**, 970–975.
13. Hers, I., Donath, J., Litjens, P. E., van Willigen, G., and Akkerman, J. W. (2000) Inhibition of platelet integrin $\alpha_{IIb}\beta_3$ by peptides that interfere with protein kinases and the beta(3) tail. *Arterioscler. Thromb. Vasc. Biol.* **20**, 1651–1660.
14. Watson, S. P., Ruggiero, M., Abrahams, S. L., and Lapetina, E. G. (1986) Inositol 1,4,5-trisphosphate induces aggregation and release of 5-hydroxytryptamine from saponin-permeabilized human platelets. *J. Biol. Chem.* **261**, 5368–5372.
15. Coorssen, J. R. and Haslam, R. J. (1993) GTP gamma S and phorbol ester act synergistically to stimulate both Ca^{2+}-independent secretion and phospholipase D activity in permeabilized human platelets. Inhibition by BAPTA and analogues. *FEBS Lett.* **316**, 170–174.

16. Padfield, P. J., Panesar, N., Henderson, P., and Baldassare, J. J. (1996). Differential effects of G-protein activators on 5-hydroxytryptamine and platelet-derived growth factor release from streptolysin-O- permeabilized human platelets. *Biochem J.* **314,** 123–128.

17. Flaumenhaft, R., Croce, K., Chen, E., Furie, B., and Furie, B. C. (1999) Proteins of the exocytotic core complex mediate platelet alpha-granule secretion. Roles of vesicle-associated membrane protein, SNAP-23, and syntaxin 4. *J. Biol. Chem.* **274,** 2492–2501.

18. Lai, K. C. and Flaumenhaft, R. (2003) SNARE protein degradation upon platelet activation: Calpain cleaves SNAP-23. *J. Cell Physiol.* **194,** 206–214.

19. Rutledge, T. W. and Whiteheart, S. W. (2002) SNAP-23 is a target for calpain cleavage in activated platelets. *J. Biol. Chem.* **277,** 37,009–37,015.

20. Shattil, S. J., Cunningham, M., Wiedmer, T., Zhao, J., Sims, P. J., and Brass, L. F. (1992) Regulation of glycoprotein IIb-IIIa receptor function studied with platelets permeabilized by the pore-forming complement proteins C5b–9. *J. Biol. Chem.* **267,** 18,424–18,431.

21. Sloan, D. C. and Haslam, R. J. (1997) Protein kinase C-dependent and Ca2+-dependent mechanisms of secretion from streptolysin O-permeabilized platelets: effects of leakage of cytosolic proteins. *Biochem. J.* **328,** 13–21.

22. Sloan, D. C., Wang, P., Bao, X., and Haslam, R. J. (2002) Translocation of pleckstrin requires its phosphorylation and newly formed ligands. *Biochem. Biophys. Res. Commun.* **293,** 640–646.

23. Flaumenhaft, R., Furie, B., and Furie, B. C. (1999) Alpha-granule secretion from alpha-toxin permeabilized, MgATP-exposed platelets is induced independently by H^+ and Ca^{2+}. *J. Cell Physiol.* **179,** 1–10.

24. Patel, S. and Scrutton, M. C. (1991) Ca^{2+}-driven [3]Harachidonate release in electropermeabilized human platelets shows an absolute requirement for $MgATP^{2-}$. *Biochem. J.* **273,** 561–564.

25. Bader, M. F., Thierse, D., Aunis, D., Ahnert-Hilger, G., and Gratzl, M. (1986) Characterization of hormone and protein release from alpha-toxin- permeabilized chromaffin cells in primary culture. *J. Biol. Chem.* **261,** 5777–5783.

26. Chen, D., Bernstein, A. M., Lemons, P. P., and Whiteheart, S. W. (2000) Molecular mechanisms of platelet exocytosis: role of SNAP-23 and syntaxin 2 in dense core granule release. *Blood* **95,** 921–929.

27. Chen, D., Lemons, P. P., Schraw, T., and Whiteheart, S. W. (2000) Molecular mechanisms of platelet exocytosis: role of SNAP-23 and syntaxin 2 and 4 in lysosome release. *Blood* **96,** 1782–1788.

28. Authi, K. S., Evenden, B. J., and Crawford, N. (1986) Metabolic and functional consequences of introducing inositol 1,4,5-trisphosphate into saponin-permeabilized human platelets. *Biochem. J.* **233,** 707–718.

29. Polgar, J. and Reed, G. L. (1999) A critical role for N-ethylmaleimide-sensitive fusion protein (NSF) in platelet granule secretion. *Blood* **94,** 1313–1318.

III

MOLECULAR AND POST-GENOMIC TECHNIQUES

23

Using Retroviruses to Express Genes in Primary Megakaryocyte Lineage Cells

Meenakshi Gaur, George J. Murphy, Jonathan Frampton, and Andrew D. Leavitt

1. Introduction

Megakaryocytes in adult human bone marrow are estimated to constitute approx 0.4% of the total marrow cells (*1*), and our experience suggests that fewer than 0.5% of low-density nucleated murine bone marrow cells express the megakaryocyte-lineage marker CD41 (integrin α_{IIb}). Historically, the infrequent occurrence of megakaryocyte-lineage cells in bone marrow has been a significant obstacle to the procurement of primary megakaryocyte-lineage cells for biological studies. However, currently available conditions allow one to expand these cells in culture. In this chapter we describe protocols for using retroviruses to selectively infect early megakaryocyte-lineage cells and to infect mature megakaryocytes. The protocols allow one to study the effect of specific gene products on lineage development and biological functions of these primary cells. A basic understanding of retroviruses and the retrovirus life cycle is assumed (for review, *see* **refs.** *2,3*).

Modified retroviruses are used as vectors to introduce and express genes of interest in a variety of cell types (*4*). Integration of the viral genome into the genome of the infected cell ensures retention of the vector in the infected cell and its daughter cells. Furthermore, the ability to increase viral titer (the number of infectious particles/mL) through simple centrifugation techniques (*5*), and methods to selectively infect or regulate gene expression in a subset of cells within a heterogeneous cell population, have all advanced the usefulness of retroviral vectors (*4*).

Using retroviruses to express genes in megakaryocytes typically involves an initial purification step (i.e., fluorescence-activated cell sorting or magnetic bead selection) to first isolate CD34+ and/or CD41+ hematopoietic progenitor cells. Following infection, the isolated cells are cultured under conditions that favor outgrowth of megakaryocyte-lineage cells, and gene expression is assayed in CD41+ cells. For example, a murine

From: *Methods in Molecular Biology, vol. 273:*
Platelets and Megakaryocytes, Vol. 2: Perspectives and Techniques
Edited by: J. M. Gibbins and M. P. Mahaut-Smith © Humana Press Inc., Totowa, NJ

leukemia virus (MLV) vector was used to express murine CD9 in human megakaryo-cytes cultured from infected CD34+ cells, with CD9 expression controlled by the viral long-terminal repeat (LTR) promoter and CD9 typically detected in half or more of the megakaryocytes *(6)*. Others used an MLV vector in which the LTR promoter was dismantled and an internal human integrin α_{IIb} promoter was used in an attempt to restrict gene expression to the megakaryocyte-lineage cells *(7)*. Virus-encoded β3 integrin could be detected in 7–10% of the megakaryocyte cells derived from the infected CD34+ cells obtained from patients with Glanzmann's thrombasthenia *(8)*. Still others used an MLV vector to express β1-tubulin in megakaryocytes by starting with purified, lineage-negative murine hematopoietic progenitor cells obtained from embryonic day 13.5 fetal livers *(9)*. The total number of infected progenitor cells ranged from 5 to 10%, but the number and percentage of infected megakaryocytes was not reported. Finally, the murine stem cell retrovirus (MSCV) *(10)* has been used to infect lineage-negative murine marrow cells that were then cultured in thrombopoietin (TPO) and stem cell factor (SCF). Roughly 50% of the culture-derived megakaryocytes expressed the virus-encoded gene of interest, the cell cycle inhibitor p21, allowing a study of its effects on megakaryocyte endomitosis *(11)*.

To study the early stages of megakaryocyte-lineage development, one would ideally want to use primary hematopoietic cells, have gene expression restricted to the infrequent and difficult-to-identify cells early in the megakaryocyte developmental pathway, and not require isolation of a subset of the bone marrow cells prior to infections. We used avian retroviruses to achieve this goal. Avian retroviruses fail to infect mammalian cells because the cells lack the required viral receptor, but expression of the subgroup A, avian retroviral receptor (TVA) on mammalian cells overcomes the naturally occurring block to retrovirus infection *(12–14)*. We used the human GPIb-α *(15)* and the human integrin α_{IIb} (GPIIb) *(16)* promoters to generate transgenic mice with megakaryocyte-lineage restricted TVA expression (*see* **Note 1**). Infecting total bone marrow from these mice using a subgroup A avian retrovirus restricts infection and gene expression to the megakaryocyte-lineage (TVA-expressing) cells *(15,16)*. This precludes the need to purify hematopoietic progenitor cells, overcomes the complicating factors associated with infecting non-megakaryocyte-lineage cells, and allows one to infect the infrequent, early megakaryocyte-lineage cells. With the ever-increasing number of genetic knockout mice with a megakaryocyte-lineage phenotype, crossing strains of interest with one of these two TVA-expressing strains should allow one to perform gene replacement and other functional studies in early, primary megakaryocyte-lineage cells with specific and selected genetic expression *(16)*.

Using the TVA-expressing mice described above, one can generate highly purified populations of early megakaryocyte-lineage cells by infecting with an avian vector that expresses the antibiotic-selectable marker puromycin (RCAS-PURO) *(15)*. The puromycin-resistant cells express CD41 and von Willebrand factor but do not express appreciable amounts of acetylcholinesterase, placing them somewhere in the early stage of the murine megakaryocyte lineage. Furthermore, the cells do not express markers of the erythroid (Ter119), lymphoid (B220, CD3), or macrophage (F480) lineages *(15)*. The cells can be used directly for specific analysis or the cells can be infected with a second RCAS vector that expresses a second gene of interest *(15,17)*. The ability to super-infect cells already infected by RCAS is a potentially powerful feature of avian vectors in mammalian cells. If one crosses the TVA-expressing mice to an Mpl$^{-/-}$ genetic background, infecting the total bone marrow with a vector that expresses Mpl cDNA allows one to select the infected cells by growth in TPO *(16)*. These cells have the same phenotype as the RCAS-PURO selected cells described above.

Infecting through the TVA receptor is most commonly achieved using a replication competent vector called RCAS (reviewed in **ref. 18**). The strength of this vector is that it is easy to generate high titer stocks, often $\geq 5 \times 10^6$ infectious particles/mL. Key limits of this vector are that the gene of interest can be expressed only by the viral LTR promoter and that high-titer virus is typically achieved only when expressing a cDNA less than approx 2.5–3.0 Kb in length. With longer cDNAs, titer often falls dramatically. The avian vector RCAN allows one to use alternate promoters, but one remains limited to approx 2.5–3.0 Kb for the combined length of the promoter and the cDNA to be expressed *(18)*. Alternatively, one can use a replication-incompetent avian retroviral vector that allows for larger cDNA insertions *(19)*, or one can pseudotype MLV *(15)*, FIV (G. J. M., unpublished results) or HIV *([20]* and M. G. and A. D. L., unpublished results) vectors with envA envelope protein and use the pseudotyped virus to selectively infect TVA-expressing cells. The replication-incompetent vectors are typically of lower titer but they can sometimes be concentrated (*[20]*; and M. G. and A. D. L., unpublished results).

While the TVA-expressing mice and RCAS vectors have proven useful for the study of Mpl signaling in primary, early megakaryocyte-lineage cells, the mice have proven less useful for studies of mature megakaryocytes *(16)*. We can detect RCAS-expressed gene products in TVA-expressing megakaryocytes *(15)*, but the expression levels typically require highly sensitive assays. For example, one can readily detect human placental alkaline phosphatase (AP) following infection with RCAS-AP, yet we have had no meaningful success detecting EGFP in mature megakaryocytes after RCAS-EGFP infection of TVA-expressing cells (*[15]*; and M. G. and A. D. L., unpublished observations).

We have turned to a replication-defective, self-inactivating lentiviral vector *(21)* for retroviral infection of mature megakaryocytes from unfractionated low-density bone marrow cells. The vector uses the PGK promoter and yields significantly higher levels of expression in mature megakaryocytes than does the RCAS vector, which uses the avian LTR as the promoter. The lentiviral vector allows us to readily detect EGFP expression in the megakaryocytes using both fluorescent microscopy and flow cytometry. As such, this vector allows for a simple method to determine the percentage of infected cells, and depending on the type of experiment, to "gate" on infected cells for assays that can be performed on a flow cytometer (e.g., a Becton Dickinson FACS™ analyzer). FACS analysis shows that we can consistently infect 70–75% of the mature megakaryocytes, which is substantiated by manual counting using a fluorescent microscope (M. G. and A. D. L., unpublished results).

Retroviruses have been used for years as vectors to express genes of interest in various cell types. This chapter focuses on methods for selectively infecting primary early megakaryocyte-lineage cells and for infecting primary mature megakaryocytes with simple and complex retroviruses, respectively.

2. Materials

2.1. Maintaining and Transfecting DF1 and 293T Cells

2.1.1. Cells and Media

1. DF1 cells (ATCC, Manassas, VA, Cat. no. CRL-12203, ATCC name UMNSAH/DF-1; ATCC phone numbers: 800-638-6597 or 703-365-2700) are an avian fibroblast cell line used to generate replication-competent avian virus, RCAS.
2. 293T cells (ATCC Cat. no. CRL-11268, ATCC name 293T/17) are a human kidney epithelial cell line that express the SV40 large T antigen and are used for making replication-defective lentiviral vectors. The large T antigen increases the copy number in the cell for those plasmids with an SV40 origin of replication.
3. Both cell lines are grown in Dulbecco's modified Eagle medium [DMEM] with 4.5 g/L glucose (Gibco BRL 11965-092) supplemented with 10% fetal calf serum (FCS, Gibco BRL 26140-079), and 100 U/mL penicillin and 100 µg/mL streptomycin (Gibco BRL 15140-122) (*see* **Note 2**).
4. You may find that adding 1–2.5% chicken serum (Sigma, St. Louis, MO) to the above media keeps the DF1 cells healthier longer and helps in the recovery when thawing frozen vials.
5. Adherent cells are removed from culture plates using 0.05% trypsin-EDTA (Gibco BRL-25300-054).
6. Phosphate-buffered saline without calcium or magnesium (PBS-CMF; Gibco BRL) is used for cell washes.

2.1.2. Transfection Reagents

1. 2X HBS (HEPES-buffered saline): Dissolve 16.4 g NaCl (0.28 M final), 11.9 g HEPES (N-2-hydroxyethylpiperazine-N'-2-ethanesulfonic acid [50 mM final]; Sigma H-3375), 0.21 g Na_2HPO_4 (1.5 mM final; Sigma S-7909) in 800 mL sterile deionized H_2O. Titrate to pH 7.05 (very critical) with 5 N NaOH. Bring final volume to 1 L using ddH_2O. Filter-sterilize through 0.2-µm nitrocellulose filter unit (Nalgene 156-4020). Store at –20°C in 10-mL aliquots in 15-mL polypropylene tubes. This solution can be thawed and refrozen at least four to five times.
2. $CaCl_2$ (2 M): Dissolve 29.2 g $CaCl_2$ (Sigma C-7902) in a final volume of 100 mL ddH_2O. Filter-sterilize as in **step 1**, aliquot, and store at –20°C.
3. Sterile round-bottom 17 × 100-mm snap-cap polypropylene tubes (Fisher 149561J, Pittsburgh, PA).
4. Chloroquine diphosphate (10 mM): Dissolve 51.6 mg chloroquine diphosphate (Sigma C-6628) in a final volume of 10 mL ddH_2O. Store at –20°C in 1-mL aliquots. Keep tubes protected from light by wrapping them in aluminum foil.
5. RCAS vectors can be obtained from Steve Hughes at The NCI-Frederick Cancer Research and Development Center, Frederick, MD. Other vectors for pseudotyping can be obtained from any of a number of investigators listed on the TVA web site (http://rex.nci.nih.gov/RESEARCH/basic/varmus/tva-web/tva2.html).

2.2. Storage and Titer of Retroviral Stocks

1. Nalgene 0.45-µm syringe filters (SFCA; Nalgene 190-2545).
2. Cryovials (Nunc 343958, Rochester, NY).
3. 3T3-0.8A cells, an NIH 3T3 cell line that expresses TVA. These cells are grown in the same media as described above for 293T cells. Alternatively, one can use any of a variety of other established cell lines with heterologous TVA expression (*20*).

4. 12-well tissue-culture plastic dishes.
5. Hexadimethrine bromide, also called "polybrene" (Sigma H-9258).

2.3. Harvesting Murine Bone Marrow

1. Anesthetic and materials needed to euthanize mice should be obtained in accordance with your institution's animal care facility protocols.
2. Transgenic mice with megakaryocyte-specific TVA receptor expression (via GPIb-α or GPIIb promoters) can be obtained from Andrew D. Leavitt, University of California, San Francisco.
3. Dissection instruments, 20–25 g hypodermic needles, 3-mL syringes, razor blades.
4. Red blood cell (ACK) lysis buffer: Dissolve 4.1 g NH_4Cl (0.15 M final; Sigma A-4514) and 0.5 g $KHCO_3$ (1.0 mM final; Fisher P-184) in 400 mL ddH_2O, and then add 100 µL 0.5 M EDTA, pH 8.0 (0.1 mM final). Adjust pH to 7.2–7.4 using 1 N HCl and adjust to 500 mL with ddH_2O. Filter-sterilize through 0.2-µm nitrocellulose filter unit (Nalgene 156-4020) and store at room temperature.
5. IsoPrep—specific gravity 1.072 (Matrix Technologies, Sunnyvale, CA) for generating low-density hematopoietic progenitor cells following marrow harvests.
6. CATCH buffer: PBS (pH 7.4) supplemented with 0.38% (w/v) sodium citrate tribasic dihydrate (Sigma S-4641) and 2% (wt:vol) bovine serum albumin (BSA; Sigma A-9418). Filter-sterilize as in **step 4** and store for up to 1 wk at 4°C.
7. Hanks' Balanced Salts Solution (HBSS; Gibco BRL) supplemented with 1% (w/v) BSA (HBSS-BSA).

2.4. Infecting Murine Megakaryocyte-Lineage Cells Using RCAS Vectors

1. IMDM complete marrow media: IMDM (Iscove's modified Dulbecco's medium) culture media (Gibco BRL) supplemented with 10% horse serum (Gibco BRL 16050-122), 100 U/mL of penicillin and 100 µg/mL streptomycin (Gibco BRL 15140-122). Cytokine supplementation is influenced by your experimental design and may include some combination of murine interleukin-3 (IL-3; Peprotech, Rocky Hill, NJ), murine stem cell factor (SCF; Peprotech), murine Flt-3 ligand (Peprotech), and murine thrombopoietin (TPO; R & D Systems, Inc., Minneapolis, MN) at concentrations indicated. More details are provided in **Subheading 3.6.**
2. DMEM complete media as in **Subheading 2.1.1.**
3. Puromycin (Clontech 8052-2, Palo Alto, CA). Stock solution (1.5 mg/mL) is kept at 4°C. Working concentration is typically 1.5 µg/mL. We sometimes need to use a working concentration of 2.0 µg/mL.

2.5. Infecting Murine Megakaryocytes With Lentiviral Vectors

1. Serum-free medium is used to generate megakaryocytes from low-density hematopoietic progenitors. The required reagents are listed in **Table 1**. To prepare, dissolve 0.625 g BSA, 25 mg transferrin, and 1.25 mg insulin in 21.3 mL of IMDM in a 50-mL conical tube. Mix thoroughly but gently to dissolve the BSA with minimal bubbles. Add 125 µL of 50 mM β-mercaptoethanol (Sigma [M-3148], comes as a 14.3 M stock solution. Therefore, add 7 µL of stock β-mercaptoethanol to 2 mL of PBS to give a 50 mM β-mercaptoethanol solution). Then add 0.5 mg low density lipoprotein (LDL), 25 µL each of 100 mM dNTP (dATP, dCTP, dGTP, dTTP), 25 µL each of 100 mM NTP (ATP, CTP, GTP, UTP), 1.25 mL each of 100X L-glutamine (Gibco BRL 25030-081), and 100X penicillin/

Table 1
Serum-Free Medium to Generate Megakaryocytes From Hematopoietic Progenitors

Amount	Item	Final concentration
121.3 mL	IMDM (Iscove's modified Dulbecco's medium)	
1.25 mL	100X L-glutamine (Gibco BRL 25030-081)	
1.25 mL	100X Pen/Strep (Gibco BRL 15140-122)	
0.625 mg	BSA (Sigma A-9418)	0.5%
0.5 mg	LDL (Sigma L-07914)	4 µg/mL
25 mg	Saturated human transferrin (Sigma T-4132)	200 µg/mL
1.25 mg	Human insulin (Sigma I-0259)	10 µg/mL
125 µL	50 mM β-mercaptoethanol (Sigma M-3148)	50 µM
25 µL	NTP and dNTP (100 mM stocks of each)	20 µM

      streptomycin (Gibco BRL 15140-122). Then add 100 mL of IMDM and filter-sterilize through 0.2-µm filter unit (Nalgene 157-0020). Store at 4°C without cytokines.

2. Add cytokines to the media just prior to use: murine IL-6 (Peprotech), murine IL-11 (R & D Systems, Inc.), and murine TPO (R & D Systems, Inc. or other sources) at final concentrations of 10 ng/mL, 10 ng/mL, and 50 ng/mL, respectively.

3. Methods

3.1. Generating Replication-Competent Avian Retrovirus (RCAS)

3.1.1. Maintaining DF1 Cells

    The avian fibroblast cells DF1 are maintained in DMEM complete media. Freeze cells in 10% DMSO with 90% FCS. When thawing cells, be sure to wash out the freezing media prior to plating because the cells are very intolerant of DMSO. We do not let the cells get overly confluent, and a confluent plate can be split up to 1:8 without apparent loss of doubling time. However, for most of our experiments, splits are typically 1:4–1:6. We passage the cells up to 15–20 times from a given thawed aliquot.

3.1.2. Generating RCAS Virus, a Replication-Competent Avian Retrovirus

    We use a calcium phosphate transfection protocol *(22)* and DF1 cells. Because RCAS is a replication-competent vector, virus generated by each transfected cell infects other untransfected cells in the dish. Consequently, transfection efficiency is not critical because the virus spreads throughout the dish and essentially all cells become virus-producer cells within about 6 d. This contrasts with the more commonly used replication-incompetent virus, in which virus-producing cells include only those cells directly transfected. We describe the process for RCAS-EGFP but it is the same for any of the RCAS vectors.

1. Split DF1 cells into a 10-cm culture dish the day before a planned transfection. Cells should be approx 50–60% confluent at the time of transfection, which translates into a 1:4 split from a 90% confluent plate.

2. Remove media and replace with 9 mL of prewarmed fresh media 1 h prior to transfection.
3. Thaw RCAS-EGFP plasmid ($CsCl_2$ or QIAGEN prep), 2X HBS, and 2 *M* $CaCl_2$ and have them at room temperature.
4. Combine 10 µg of the RCAS-EGFP plasmid DNA and water to a total volume of 437 µL in a 1.5-mL microcentrifuge tube; add 63 µL of the 2 *M* $CaCl_2$ stock and mix gently but thoroughly with a micropipet.
5. Slowly (~1 drop/2 s from a P-1000 pipettor) add the DNA mixture into a sterile round-bottom polypropylene tube containing 500 µL 2X HBS. Bubble air through the 2X HBS solution throughout the "dripping" procedure using a pipetaid with an attached 1-mL pipet. The pipetaid introducing air into the HBS solution is held in one hand while the P-1000 dropping the DNA solution is held in the other hand (for detailed picture *see* **ref. 22**). The tube should be supported in a rack in which you can see the pipet tips.
6. Vortex for 10 s and then use a P-1000 to add the mixture dropwise, within 1–2 min of vortexing, to the plate of DF1 cells with the fresh media. Add drops at approx 1 drop every 1–2 s while very gently swirling the plate to evenly distribute the DNA mixture. Incubate plates at 37°C for 8–12 h.
7. Change media after 8–12 h. You will see small dark precipitates in a phase-contrast microscope. This is normal (*see* **Note 3**).
8. Monitor the EGFP expression 48 h later by viewing under an inverted fluorescence microscope to confirm successful transfection. By 5–6 d post-transfection you can expect >95% of the DF1 cells to produce virus as evidenced by fluorescent microscopy or, more quantitatively, by FACS analysis (*see* **Note 4**).
9. Once you have >90–95% infected cells, which typically translates to d 5 or 6, you can harvest the virus as described in **Subheading 3.3.** One can continue to harvest virus every 2–3 d by replenishing with fresh media, as the cells will continue to produce virus for weeks. We only do this for up to 10 d (*see* **Note 5**).
10. We use these producer cells for our co-culture infections of TVA-expressing mega-karyocyte-lineage bone marrow cells as described in **Subheading 3.6.** The viral supernatant is used to determine titer.

3.2. Generating Replication-Defective Lentivirus

As discussed in the introduction, we have had little success using RCAS vectors to achieve high-level expression (i.e., detect EGFP by FACS analysis) of virus-encoded genes in mature megakaryocytes. Using self-inactivating, EGFP-expressing lentiviral vectors *(21)* pseudotyped with envA envelope protein, 20–25% of mature megakaryo-cytes express FACS-detectable EGFP. However, if we use the vesicular stomatitis virus protein G (VSVG) envelope *(23)*, 70–75% of the mature megakaryocytes express FACS-detectable EGFP. Furthermore, the level of gene expression is higher with VSVG, even when controlling for virus titer. We find that compared with the CMV promoter, the PGK promoter gives a higher average and maximal level of EGFP expression in megakaryocytes (M. G., A. D. L., unpublished observations).

Our main focus for expressing gene products in large megakaryocytes is for assays in which we can use a FACS machine to identify gene expression in only the most mature megakaryocytes using FL1 gating. As such, it is less important for these applications to have infection restricted to megakaryocyte-lineage cells. Our typical lentivirus vector is therefore pseudotyped with VSVG envelope. The *gag-pol* vector pCMVDR8.91 and self-inactivating transfer vector are described elsewhere *(21,24)* (*see* **Note 6**). The transfer vector we use contains a PGK promoter and is derived from pRRL-PGK-EGFP sin18 *(21)*. We have added an HIV-1 cppt region to improve titer via improved complete reverse transcription and nuclear entry *(25–27)*, as well as a WPRE element for mRNA stability *(28,29)*. A very similar vector is now available via MTA from D. Trono, as is a more extensive description of the lentiviral system and useful references *(24)*.

3.2.1. Maintaining 293T Cells

Virus is generated via transient transfection of 293T cells, which are grown in DMEM complete media. They are only loosely adherent to tissue-culture plastic, so be gentle when making media changes. Not surprisingly, 293T cells are quite sensitive to trypsin.

1. Passage cells by aspirating the media, gently washing with 5 mL PBS-CMF, and then adding 1 mL of 0.05% trypsin-EDTA to a 10 cm culture dish and incubating at 37°C for 60–90 s.
2. Disperse cells to single-cell suspension using a 1-mL pipet and then immediately add complete DMEM media and replate at desired "split." Be sure to disperse cells evenly without clumps. Uneven plating and clumped cells yield inferior virus titers. We have experienced reduced titers after an extended number of cell passages, and therefore suggest no more than approx 15 passages from a vial of thawed cells. Do not let the cells become overconfluent.

3.2.2. Transfecting 293T Cells to Generate Self-Inactivating Lentivirus

1. Split 293T cells the day before transfection into 10-cm culture dishes to have approx 70–80% confluent cells at the time of transfection. Typically, a 1:3 split of a 90% confluent plate in the afternoon is good for a transfection the following evening.
2. Remove media and replace with 9 mL of fresh, prewarmed DMEM complete medium 1 h prior to transfection.
3. Thaw all plasmids (envelope, *gag-pol*, and transfer vector expressing the gene of interest), chloroquine, 2X HBS, and 2 *M* CaCl$_2$ and keep them at RT (*see* **Note 7**).
4. Combine the DNA, chloroquine, and ddH$_2$O to give a total volume of 437 µL. We use 10 µg of the *gag-pol* plasmid, 10 µg of the gene transfer plasmid, 15 µg of the envelope plasmid, and 33 µL of 10 m*M* chloroquine for a typical transfection (*see* **Note 8**). Add 63 µL of 2 *M* CaCl$_2$ to give a final volume of 500 µL, and mix thoroughly using a P-1000 pipettor.
5. Follow **steps 5–6** in **Subheading 3.1.2.**
6. Change media after 8–12 h. You will see small dark precipitates in a phase-contrast microscope. This is normal. Be sure to use prewarmed medium and add it slowly. If not, many of your transfected cells will leave the dish and they will be lost.
7. Harvest supernatant containing virus after 48–72 h as described below.

3.3. Harvesting Virus Stocks

1. Collect supernatant-containing virus using a syringe.
2. Filter through a 0.45-µm syringe filter and collect in a 15-mL polypropylene tube.
3. Use a 5-mL pipet to transfer 1-mL aliquots into prelabeled cryovials and freeze immediately in a –80°C freezer. Alternatively, larger volumes can be frozen in 15-mL polypropylene tubes.

3.4. Determining Viral Titers

Viral titer, the number of infectious U/mL, is a critical characteristic of the virus stock. As such, it is important to determine the titer for each stock. The cell line you use influences titer. Establish a protocol and cell line for your lab so that you can, at a minimum, compare relative virus titers across virus stocks. We titer RCAS virus using 3T3-O.8A cells, a clone of transfected murine NIH 3T3 cells that expresses the TVA receptor. Although RCAS is a replication-competent virus, it cannot replicate in mammalian cells; therefore 3T3-O.8A cells work well for titering virus stocks. We also use 3T3-0.8A cells to titer lentivirus pseudotyped with VSVG or avian envA protein.

1. Plate 125,000 3T3-O.8A cells /well in a 12-well dish in the morning.
2. Six to eight hours later, replace the media with 500 µL of serially diluted (1X, 10X, and 100X dilutions is a good place to start) virus stock containing 1X (8 µg/mL) polybrene. Do the serial dilutions using DMEM complete media that contains 1X polybrene. The uninfected control well will have 500 µL of DMEM complete medium containing 1X polybrene.
3. Incubate overnight, aspirate virus/media, and add 2 mL of fresh media. Observe the cells under a fluorescent microscope to monitor EGFP expression. We usually see the EGFP expression by 24 h post-infection. At 72 h post-infection, trypsinize the cells and determine the percentage of EGFP-expressing cells using a FACS machine. Titer can be calculated as follows: "% of EGFP-expressing cells (from FACS analysis) × 125,000 cells in dish × dilution factor) = infectious particles/mL." This allows for determination of titers as low as approx 1×10^3. There is no upper limit to the titer one can detect because one can continue to serially dilute the input virus stock. For vectors that lack EGFP, one needs an alternate method such as an antibiotic-selectable marker or an antibody to the expressed protein to allow for staining or FACS detection.

3.5. Harvesting Total Bone Marrow

3.5.1. For Infection Using RCAS to Infect Early Megakaryocyte-Lineage Cells

1. Set up a work space with plenty of 3-mL syringes; 20-, 23-, and 25-g needles; dissection instruments, including razor blades; a small beaker with 95% ethanol; an ethanol squirt bottle, Kimwipes, gloves, and paper towels. Soak instruments in ethanol for a few minutes and then remove and dry.
2. Have a 35-mm sterile Petri dish half full with HBSS-BSA for storing bones prior to flushing out the marrow.
3. Harvest the marrow into 50-mL conical tubes pre-filled with HBSS-BSA, 25 mL for 2 mice or 15 mL for 1 mouse. Keep these on ice.
4. Sacrifice the animal under appropriate licenses and in accordance with your institution's regulations. Use scissors to make a skin incision in the low abdomen, blunt-dissect toward the chest to detach the skin, and peel skin to the sides. Do not enter the peritoneum. Use closed scissors to hook under the exposed knee to free the leg from the skin.
5. Use the razor blade to free the leg from the hip joint, being careful to not cut the bone.
6. Scrape off the soft tissue and isolate the femur and the tibia from each leg. Place in a Petri dish until you harvest marrow.
7. When all the bones are collected, two femurs and two tibias per mouse, you are ready to flush the marrow.

8. Cut the end of the bone with a razor blade, cutting off as little as you can to get the needle into the marrow space. Use a 3-mL syringe containing 2.5 mL of HBSS-BSA. We typically start with a 23-g needle, but sometimes we need to use a 25-g needle.

9. Flush bone marrow completely. This may require a syringe refill with HBSS-BSA and going from each end. Bone goes from red/pink to pale after flushing marrow. Be sure to secure the bone with the forceps or against the wall of the tube when you flush to prevent the bone from falling into the 50-mL conical tube. Using a 10-mL syringe and a 20-g needle, gently aspirate the marrow chunks and flush them gently but repeatedly in and out of the needle with the bevel gently pressed against the wall of the tube. Do this until the chunks are all gone or you stop making progress. You may have to go to a smaller-bore needle.

10. Sediment the cells at 1000 rpm (~220g) for 5 min at 4°C (*see* **Note 9**). Resuspend in 5 mL of ACK lysis buffer and mix gently but thoroughly using a 5-mL pipet. Let stand at room temperature for 1 min and spin again.

11. Wash cells once with PBS-CMF and resuspend in IMDM complete medium using 8 mL/mouse. Plate in 6-well dishes.

3.5.2. For Infection Using Lentivirus to Express Genes in Mature Megakaryocytes

1. The day before harvest, make plenty of CATCH buffer and serum-free media. Anticipate 20 mL/mouse for each.

2. Follow **steps 1–9** in **Subheading 3.5.1.** above, substituting CATCH buffer for HBSS-BSA. The cells will be in a 50-mL conical tube, one or two mice per tube.

3. For each 50-mL conical, do one low-density gradient. Put 12 mL of IsoPrep into a fresh 50-mL conical, being careful to not get it on the sides of the tube. Using a 5-mL pipet, layer the cell suspension on top of the IsoPrep. Do not exceed the 40-mL mark and be careful to not mix the two liquids. After you layer on all of the harvest CATCH buffer, wash that tube with 5 mL of fresh CATCH buffer and add it to the top of the gradient.

4. Spin at 400g (1400 rpm in a Sorvall RC5 Plus with SH-3000 rotor) at room temperature for 30 min *with the brake off*.

5. Be sure that you have a nice band of low-density cells at the interface. It is a white haze. Aspirate the upper liquid down to approximately the 17-mL mark. Then remove the cells by taking the next approx 10 mL of liquid, being careful to aspirate from the meniscus down. Transfer the cells to a new 50-mL conical tube containing 5 mL of fresh CATCH buffer, and gently mix by pipetting.

6. Spin at approx 300g (1200 rpm in Sorvall RC5 Plus with SH-3000 rotor) at room temperature for 5 min *with the brake on*.

7. Aspirate supernatant but do not disturb the pelleted cells. Resuspend the cells in 20 mL of medium and determine the cell concentration with a hemocytometer. Adjust the media volume to give approx 1×10^6 cells/mL and plate up to 5 mL/well in a 6-well dish. Check under a microscope and place in an incubator.

3.6. Infecting Early Megakaryocyte-Lineage Cells Using RCAS Virus

3.6.1. Infection Using RCAS Supernatants

1. Day 0: Harvest bone marrow as described in **Subheading 3.5.1.** and culture in complete IMDM medium plus selected cytokines (*see* **Note 10**) at a cell concentration of 5×10^6/mL in a 6-well dish.

2. Day 1: Following overnight incubation, use gentle but thorough pipetting to transfer the nonadherent cells to a new dish. Cytokine exposure from day 0 to day 2 promotes cell cycling, significantly improving infections by RCAS vectors.

3. Day 2: Harvest RCAS retroviral supernatant as discussed in **Subheading 3.3.** Timing your experiments so that fresh virus will be available on the day of infection is advised because freeze/thaw of virus stocks often reduces their titer by up to one-half log.

4. Day 2: Count cells, and plate 5×10^5 cells per well in 500 µL of fresh medium containing 2X polybrene. Immediately add 500 µL of filtered virus supernatant stock to generate 1 mL total volume per well of medium containing 1X (8 µg/mL) polybrene. Incubate the infection overnight (*see* **Note 11**).

5. Day 3: Using a P1000, collect the day 2 infected bone marrow cells, transfer to a 15-mL conical tube containing 10 mL of PBS-CMF, and sediment at 1000 rpm (~220*g*) for 5 min at 4°C. Resuspend cells in complete growth medium (500 µL per infection) and repeat the day 2 infection.

6. Day 4: Repeat steps of day 3 to complete three daily infections.

7. Day 5: Transfer cells to selective media. For RCAS-PURO, use IMDM complete medium with 1.5 µg/mL puromycin. For RCAS-MPL infections of *mpl*$^{-/-}$ cells, the IMDM "complete" medium lacked TPO and one now changes to "complete" medium with TPO to select for growth of the infected cells.

8. Day 6 onward: Add and change media as needed to maintain good culture conditions. Do not let the media get orange. Infected cells will expand following a selection-induced death of the uninfected cells. Selected cells express surface CD41 and intracellular von Willebrand factor, and they will expand for a few months before senescing. It takes typically 7–10 d of selection to achieve a homogeneous population of early megakaryocyte-lineage cells.

3.6.2. Infection Using RCAS Retrovirus-Producing DF1 Cells

1. Day 0: Harvest bone marrow as described in **Subheading 3.5.1.** and culture in cytokine-supplemented (*see* **Note 10**) complete IMDM medium at 5×10^6 cells/mL in a 6-well dish.

2. Day 1: Transfer nonadherent cells and supernatant to a fresh dish the following morning. In the afternoon, plate RCAS virus producing DF1 cells into 6-well dishes at varying density. Plate for a confluence of 50–60% in two wells (to be used for the day 2 infection), 30–40% in 2 wells (to be used for the day 3 infection), and 10–20% in two wells (to be used for the day 4 infection). This translates to plating approx 6×10^5 cells in each of two wells, 3×10^5 cells into each of two wells, and 1.5×10^5 cells into each of two wells. Plating may need to be adjusted.

3. Day 2 morning: Aspirate media from the DF1 wells to be used for day 2 infection and add 1.5 mL fresh IMDM complete media.

4. Day 2 evening: Add $1–2 \times 10^6$ cultured marrow cells/infection in 1.5 mL IMDM complete medium by gently layering on the DF1 cells in the wells prepared for day 2 infection, giving a final volume of 3 mL/well. Co-culture overnight.

5. Day 3 morning: Aspirate media from the DF1 wells to be used for day 3 infection and add 1.5-mL fresh IMDM complete medium per well. Also, using a P1000, gently collect the day 2 infected bone marrow cells, taking care not to dislodge the DF1 virus-producing cells. Transfer the bone marrow cells to a 15-mL conical tube containing 10 mL of PBS-CMF and sediment at 1000 rpm (~220*g*) for 5 min at 4°C. Resuspend cells in 3 mL of fresh IMDM complete medium and plate in a fresh 6-well dish.

6. Day 3 evening: Repeat infections as described in **step 5**, but using the wells prepared for day 3 infection.

7. Day 4: Prepare the wells for day 4 infection (the third infection of the same cells) and transfer the marrow cells, as described in **steps 5** and **6**, thus completing three infections of the same original marrow cells.

8. Day 5: Transfer the bone marrow cells to a 15-mL conical tube containing 10 mL of PBS-CMF and sediment at 1000 rpm (~220g) for 5 min at 4°C. Resuspend cells in 3 mL of fresh IMDM complete medium and plate in a fresh 6-well dish.

9. Day 6: Transfer cells to selective media and culture cells as discussed in **steps 7** and **8** of **Subheading 3.6.1.**

3.7. Lentivirus Infections to Express Genes in Mature Megakaryocytes

Total bone marrow infections are performed on day 2 and/or day 3, with the day of harvest called "Day 0." Cells are harvested and plated as described in **Subheading 3.5.2.** (*see* **Note 12**).

1. Day 2 evening: Gently but thoroughly pipet the media to be sure all suspension cells are floating. Place 1–2 × 10^6 cells/infection into a 15-mL conical tube and sediment at 1200 rpm (~310 g) for 5 min at 4°C. Resuspend cells in fresh, serum-free complete medium containing 2X polybrene at a concentration of 4–8 × 10^6 cells/mL. Transfer 250 μL (1–2 × 10^6 cells) to each well of a 12-well dish. The total number of wells used will be determined by your experimental design/plans. Add 250 μL viral supernatant to each well into which you put the marrow cells, mix gently, and incubate overnight.

2. Day 3 morning: Using a P1000, transfer the cells from each well to a 15-mL conical tube containing 6 mL of PBS-CMF, and spin at 900 rpm (175g) for 5 min at 4°C. Use one 15-mL conical tube per well of cells. After centrifugation, aspirate supernatant, add 500 μL of fresh media, and transfer the cells to a new well of a 12-well dish.

3. Day 3 evening: For a second infection of the same cells, repeat the infection procedure described above for day 2. Likewise, for a "Day 3 only" infection, perform infections as described for day 2.

4. Day 4 morning: Repeat the wash step as described for day 3 morning if a "Day 3" infection was performed.

5. Day 5: Analyze the cells for EGFP expression using a FACS machine, or use an alternate method appropriate for the gene expressed.

4. Notes

1. Integrin α$_{IIb}$ has long been a marker for megakaryocyte-lineage cells, yet some recent reports suggest that it may be expressed at low levels in avian intra-embryonic multilineage progenitors as well as in postnatal avian and murine multilineage progenitors *(30–32)*. Nevertheless, many reports demonstrate that GPIb-α promoters regulate gene products that are expressed selectively in megakaryocyte-lineage cells *(33–35)*.

2. All mammalian cell incubations described in this chapter occur at 37°C in humidified air supplemented with 5% CO_2. As such, the medium as purchased contains bicarbonate.

3. Cells can look vacuolated and "sick" when you come back to change the medium. If so, and it recurs with a given plasmid, try shortening the time between transfection and the first medium change, or use less DNA.

4. Simply observing the cells through a fluorescent microscope allows one to prove that RCAS-EGFP virus has spread throughout the plate by day 5 or 6 post-transfection. FACS analysis can provide a specific percentage of EGFP-expressing cells, but that is not typically necessary. It is not as easy to check for spread through the plate for other genes of

interest. We therefore always perform an RCAS-EGFP transfection alongside any other RCAS vector we are transfecting. This at least allows one to control for that particular batch of transfections. Once the RCAS-EGFP plate shows >90% infected cells, we wait one more day and assume the other vectors are behaving similarly. However, for new vectors, one needs to develop vector-specific methods for determining the titer of the virus supernatant. It is a good idea to check the titer for every new vector 6–10 d posttransfection to determine its kinetics, which will allow you to optimize harvests if need be.

5. Replication-competent virus can accumulate mutations with prolonged production. We chose 10 d as our maximum cut-off to prevent that problem.

6. The self-inactivation, along with the three independent vectors required for the transfection, provide a high level of safety for using the vectors, but nothing is risk-free. An appropriate understanding of the risks of these vectors is essential prior to using them, as is approval by one's institutional biological review board.

7. Our typical lentivirus is pseudotyped with the VSVG envelope protein. The VSVG and *gag-pol* vectors have been described *(21,36)*.

8. The amount of each plasmid given in the text is the level we initially use with new vectors. However, it may be necessary to adjust total and relative amounts to get maximal titers. For example, we have experienced situations in which the gene product expressed is toxic to the cells, and have had to use less transfer vector to improve titers. We have typically adjusted down the other two vectors, but trial and error is typically needed.

9. Unless stated otherwise, centrifugations are performed in either a Sorvall RT6000B or a Sorvall Legend RT.

10. The choice of added cytokines should reflect your experimental needs because the cytokine choice will influence megakaryocytopoiesis and therefore the nature of the infected cells. Although capable of infecting nondividing cells at low efficiency *(37)*, RCAS viruses are far superior at infecting dividing megakaryocyte-lineage cells than ones that have stopped dividing and begun the process of endoreduplication. Our unpublished data suggest that cells grown and selected in IMDM marrow medium (**Subheading 2.4.**) supplemented with IL-3 alone are phenotypically the most "primitive" at the RNA level, not expressing the later stage megakaryocyte developmental markers GPIbα and PF4. Cells grown in IL-3 alone also proliferate slowly, adversely affecting their infectability. Cells grown and selected in IL-3 and SCF express low levels of GPIbα and PF4 RNA and have a shorter dividing time than those grown in IL-3 only. The faster growth rate translates into more productive infections. Cells grown and selected in IL-3, SCF, and TPO grow even faster and express yet higher levels of GPIbα, PF4, and GPIIb RNA. Another point to consider is the inclusion of IL-3 in the growth medium in the culture and selection of the megakaryocyte-lineage cells. Although IL-3 expands the number of megakaryocyte-lineage cells and can improve infectivity, it also impairs megakaryocyte-lineage maturation *(15,38–40)*. To avoid this, one can combine SCF and TPO with Flt-3 ligand with trivial loss of growth rate as compared to cells grown in SCF, TPO and IL-3 *(41,42)*. One must therefore exercise judgment as to which cytokines to include, and comparing different combinations may well prove useful.

11. Perform infection using a 50:50 mix of virus and fresh, cytokine-supplemented complete growth medium in a 12-well dish. Using a 50:50 mix of virus and fresh medium is better than using pure virus supernatant because the cells remain healthier during the overnight infection.

12. Day 2 and day 3 infections each yield approx 70–75% infected mature megakaryocytes on day 5. We do not see a considerable increase in the percentage of cells infected when we do one vs two infections. However, on day 5 we see a somewhat higher level of expression

in day 2-infection cells as compared with Day 3-infected cells. However, these findings may differ somewhat between investigators and we strongly recommend that you try various approaches to see what works best in your own hands.

Acknowledgments

We thank Didier Trono for providing vectors and Tamihiro Kamata for helpful discussions during the development of these protocols. This work was supported by grants RO1-HL65198 and PL50-HL54476 from the National Institutes of Health, Bethesda, MD.

References

1. Levine, R. F. (1980) Isolation and characterization of normal human megakaryocytes. *Br. J. Haematol.* **45,** 487–497.
2. Coffin, J. M., Hughes, S. H., and Varmus, H. E., eds. (1977) *Retroviruses.* Cold Spring Harbor Laboratory Press, Plainville, NY.
3. http://www.stanford.edu/group/nolan/.
4. Miller, A. D. (1997) Retroviruses, in *Retroviruses* (Coffin, J. M., Hughes, S. H., and Varmus, H. E., eds.), Cold Spring Harbor Press, Plainville, NY, pp. 437–474.
5. O'Doherty, U., Swiggard, W. J., and Malim, M. H. (2000) Human immunodeficiency virus type 1 spinoculation enhances infection through virus binding. *J. Virol.* **74,** 10,074–10,080.
6. Burstein, S. A., Dubart, A., Norol, F., Debili, N., Friese, P., Downs, T., et al. (1999) Expression of a foreign protein in human megakaryocytes and platelets by retrovirally mediated gene transfer. *Exp. Hematol.* **27,** 110–116.
7. Wilcox, D. A., Olsen, J. C., Ishizawa, L., Griffith, M., and White, G. C., 2nd (1999) Integrin α_{IIb} promoter-targeted expression of gene products in megakaryocytes derived from retrovirus-transduced human hematopoietic cells. *Proc. Natl. Acad. Sci. USA* **96,** 9654–9659.
8. Wilcox, D. A., Olsen, J. C., Ishizawa, L., Bray, P. F., French, D. L., Steeber, D. A., et al. (2000) Megakaryocyte-targeted synthesis of the integrin β_3-subunit results in the phenotypic correction of Glanzmann thrombasthenia. *Blood* **95,** 3645–3651.
9. Lecine, P., Italiano, J. E., Jr., Kim, S. W., Villeval, J. L., and Shivdasani, R. A. (2000) Hematopoietic-specific beta 1 tubulin participates in a pathway of platelet biogenesis dependent on the transcription factor NF-E2. *Blood* **96,** 1366–1373.
10. Hawley, R. G. (1994) High-titer retroviral vectors for efficient transduction of functional genes into murine hematopoietic stem cells. *Ann. NY Acad. Sci.* **716,** 327–330.
11. Baccini, V., Roy, L., Vitrat, N., Chagraoui, H., Sabri, S., Le Couedic, J. P., et al. (2001) Role of p21(Cip1/Waf1) in cell-cycle exit of endomitotic megakaryocytes. *Blood* **98,** 3274–3282.
12. Young, J. A., Bates, P., and Varmus, H. E. (1993) Isolation of a chicken gene that confers susceptibility to infection by subgroup A avian leukosis and sarcoma viruses. *J. Virol.* **67,** 1811–1816.
13. Bates, P., Young, J. A., and Varmus, H. E. (1993) A receptor for subgroup A Rous sarcoma virus is related to the low density lipoprotein receptor. *Cell* **74,** 1043–1051.
14. Federspiel, M. J., Bates, P., Young, J. A., Varmus, H. E., and Hughes, S. H. (1994) A system for tissue-specific gene targeting: transgenic mice susceptible to subgroup A avian leukosis virus-based retroviral vectors. *Proc. Natl. Acad. Sci. USA* **91,** 11,241–11,245.
15. Murphy, G. J. and Leavitt, A. D. (1999) A model for studying megakaryocyte development and biology. *Proc. Natl. Acad. Sci. USA* **96,** 3065–3070.

16. Gaur, M., Murphy, G. J., deSauvage, F. J., and Leavitt, A. D. (2001) Characterization of Mpl mutants using primary megakaryocyte-lineage cells from *mpl*$^{-/-}$ mice: a new system for Mpl structure-function studies. *Blood* **97**, 1653–1661.

17. Quintrell, N., Hughes, S. H., Varmus, H. E., and Bishop, J. M. (1980) Structure of viral DNA and RNA in mammalian cells infected with avian sarcoma virus. *J. Mol. Biol.* **143**, 363–393.

18. Federspiel, M. J. and Hughes, S. H. (1997) Retroviral gene delivery. *Meth. Cell Biol.* **52**, 179–214.

19. Boerkoel, C. F., Federspiel, M. J., Salter, D. W., Payne, W., Crittenden, L. B., Kung, H. J., et al. (1993) A new defective retroviral vector system based on the Bryan strain of Rous sarcoma virus. *Virology* **195**, 669–679.

20. Lewis, B. C., Chinnasamy, N., Morgan, R. A., and Varmus, H. E. (2001) Development of an avian leukosis-sarcoma virus subgroup A pseudotyped lentiviral vector. *J. Virol.* **75**, 9339–9344.

21. Zufferey, R., Dull, T., Mandel, R. J., Bukovsky, A., Quiroz, D., Naldini, L., et al. (1998) Self-inactivating lentivirus vector for safe and efficient in vivo gene delivery. *J. Virol.* **72**, 9873–9880.

22. Kingston, R. E., Chen, C. A., Okayama, H., and Rose, J. K. (2003) Transfection of DNA into mammalian cells, in *Current Protocols in Molecular Biology* (Ausubel, F. M., Brent, R., Kingston, R. E., Moore, D. D., Seidman, J. G., Smith, J. A., et al., eds.), Vol. 2. John Wiley & Sons, Inc., New York, pp. 9.1.1–9.1.11.

23. Yee, J. K., Friedmann, T., and Burns, J. C. (1994) Generation of high-titer pseudotyped retroviral vectors with very broad host range. *Methods Cell Biol.* **43**, 99–112.

24. http://tronolab.unige.ch/.

25. Zennou, V., Serguera, C., Sarkis, C., Colin, P., Perret, E., Mallet, J., et al. (2001) The HIV-1 DNA flap stimulates HIV vector-mediated cell transduction in the brain. *Nat. Biotechnol.* **19**, 446–450.

26. Sirven, A., Pflumio, F., Zennou, V., Titeux, M., Vainchenker, W., Coulombel, L., et al. (2000) The human immunodeficiency virus type-1 central DNA flap is a crucial determinant for lentiviral vector nuclear import and gene transduction of human hematopoietic stem cells. *Blood* **96**, 4103–4110.

27. Follenzi, A., Ailles, L. E., Bakovic, S., Geuna, M., and Naldini, L. (2000) Gene transfer by lentiviral vectors is limited by nuclear translocation and rescued by HIV-1 pol sequences. *Nat. Genet.* **25**, 217–222.

28. Donello, J. E., Loeb, J. E., and Hope, T. J. (1998) Woodchuck hepatitis virus contains a tripartite posttranscriptional regulatory element. *J. Virol.* **72**, 5085–5092.

29. Zufferey, R., Donello, J. E., Trono, D., and Hope, T. J. (1999) Woodchuck hepatitis virus posttranscriptional regulatory element enhances expression of transgenes delivered by retroviral vectors. *J. Virol.* **73**, 2886–2892.

30. Ody, C., Vaigot, P., Quéré, P., Imhof, B. A., and Corbel, C. (1999) Glycoprotein IIb-IIIa is expressed on avian multilineage hematopoietic progenitor cells. *Blood* **93**, 2898–2906.

31. Tronik-Le Roux, D., Roullot, V., Schweitzer, A., Berthier, R., and Marguerie, G. (1995) Suppression of erythro-megakaryocytopoiesis and the induction of reversible thrombocytopenia in mice transgenic for the thymidine kinase gene targeted by the platelet glycoprotein alpha IIb promoter [published erratum appears in *J. Exp. Med.* 1995 Oct 1;182(4):1177]. *J. Exp. Med.* **181**, 2141–2151.

32. Tropel, P., Roullot, V., Vernet, M., Poujol, C., Pointu, H., Nurden, P., et al. (1997) A 2.7-kb portion of the 5′ flanking region of the murine glycoprotein alpha IIb gene is transcriptionally active in primitive hematopoietic progenitor cells. *Blood* **90**, 2995–3004.

33. Fujita, H., Hashimoto, Y., Russell, S., Zieger, B., and Ware, J. (1998) In vivo expression of murine platelet glycoprotein Ibα. *Blood* **92,** 488–495.
34. Ware, J., Hashimoto, Y., Zieger, B., and Russell, S. (1996) Controlling elements of platelet glycoprotein Ibα expression. *C. R. Acad. Sci. III* **319,** 811–817.
35. Ware, J., Russell, S. R., Marchese, P., and Ruggeri, Z. M. (1993) Expression of human platelet glycoprotein Ibα in transgenic mice. *J. Biol. Chem.* **268,** 8376–8382.
36. Naldini, L., Blomer, U., Gallay, P., Ory, D., Mulligan, R., Gage, F. H., et al. (1996) In vivo gene delivery and stable transduction of nondividing cells by a lentiviral vector. *Science* **272,** 263–267.
37. Hatziioannou, T. and Goff, S. P. (2001) Infection of nondividing cells by Rous sarcoma virus. *J. Virol.* **75,** 9526–9531.
38. Mazur, E. M., Cohen, J. L., Bogart, L., Mufson, R. A., Gesner, T. G., Yang, Y. C., et al. (1988) Recombinant gibbon interleukin-3 stimulates megakaryocyte colony growth in vitro from human peripheral blood progenitor cells. *J. Cell. Physiol.* **136,** 439–446.
39. Dolzhanskiy, A., Hirst, J., Basch, R. S., and Karpatkin, S. (1998) Complementary and antagonistic effects of IL-3 in the early development of human megakaryocytes in culture. *Br. J. Haematol.* **100,** 415–426.
40. Segal, G. M., Stueve, T., and Adamson, J. W. (1988) Analysis of murine megakaryocyte colony size and ploidy: effects of interleukin-3. *J. Cell. Physiol.* **137,** 537–544.
41. Veiby, O. P., Jacobsen, F. W., Cui, L., Lyman, S. D., and Jacobsen, S. E. (1996) The flt3 ligand promotes the survival of primitive hemopoietic progenitor cells with myeloid as well as B lymphoid potential. Suppression of apoptosis and counteraction by TNF-alpha and TGF-beta. *J. Immunol.* **157,** 2953–2960.
42. Kobari, L., Giarratana, M. C., Poloni, A., Firat, H., Labopin, M., Gorin, N. C., et al. (1998) Flt 3 ligand, MGDF, Epo and G-CSF enhance ex vivo expansion of hematopoietic cell compartments in the presence of SCF, IL-3 and IL-6. *Bone Marrow Transplant* **21,** 759–767.

24

Use of Antisense Oligonucleotide Technology to Investigate Signaling Pathways in Megakaryocytes

Hava Avraham, Shalom Avraham, and Radoslaw Zagozdzon

1. Introduction

1.1. Background Information

The use of inhibitors of signal transducers or signal modulators is essential for studies on intracellular signaling. A variety of pharmacological compounds is available to block the main pathways of signal transduction. These compounds, however, have limited specificity and considerable toxicity. To investigate the role of a single molecule within a pathway, this level of precision is usually insufficient. Antisense oligonucleotide technology provides a particularly specific and rapid method for inhibiting the expression of a particular gene, and therefore for exploring the function of the gene product. Over the past two decades, several hundred publications have documented the usefulness of this approach to inhibit gene expression at the mRNA level and, consequently, to decrease the expression level of specific proteins.

The theory behind the antisense strategy is exceptionally simple. In general, the antisense oligonucleotides are designed as short sequences complementary to specific regions within the sequence of the gene of interest. Once within the cell, the oligonucleotide binds to the target mRNA via Watson-Crick base pair interactions and blocks the translation of the specific protein. Compounds acting through antisense mechanisms include antisense oligodeoxynucleotides (ODNs) and nucleotide analogs, antisense RNA, and ribozymes. Out of these categories, oligodeoxynucleotides, particularly their phosphorothioate derivatives, are the most widely used compounds for the sequence-specific inhibition of gene expression. These ODN analogs combine a number of properties considered essential for an efficacious antisense compound. These include the reasonable stability of mRNA-ODN duplexes, resistance against nucleases, negative charge allowing delivery with cationic lipids, and the ability to enter certain cells without a carrier. Additionally, phosphorothioate ODNs similar to unmodified phosphodiester ODNs elicit efficient RNase H activity (*1*) to cleave the target mRNA, which (along

From: *Methods in Molecular Biology, vol. 273:*
Platelets and Megakaryocytes, Vol. 2: Perspectives and Techniques
Edited by: J. M. Gibbins and M. P. Mahaut-Smith © Humana Press Inc., Totowa, NJ

with the physical blocking of mRNA translation in ribosomes) provides the major mechanism of their antisense effect. On the other hand, the biggest concern while using phosphorothioate ODNs is related to their length-dependent and sequence-independent capability of binding to proteins that contain polyanion binding sites. Examples of such proteins include a large number of heparin-binding proteins, such as EGF-R, bFGF, VEGF, PDGF *(2)*, CD4 *(3)*, Mac-1 *(4)*, laminin, fibronectin, and many others *(5)*. Also, the activation of RNase H, although potentiating the anti-mRNA activity of the phosphorothioate ODNs, is partially responsible for their nonsequence-specific side effects. The specificity problem arises because DNA/RNA and phosphorothioate-DNA/RNA duplexes as short as 5 bp in length are recognized and cleaved by RNase H. As a consequence, basically every RNase H-activating antisense ODN is expected to impair the fate of at least hundreds if not thousands of different targets within the cell. In light of these facts, there is a need to discriminate between the antisense and non-antisense effects of ODNs by designing carefully controlled experiments *(1)*.

1.2. Antisense Strategy in Studies on Megakaryocytes

The use of an antisense strategy to study megakaryocytes dates back to the late 1980s. The first reports utilized unmodified phosphodiester ODNs against c-Myb *(6,7)*, proving the role of this protein in hematopoiesis and specifically in megakaryopoiesis. Unmodified phosphodiester ODNs were also used to study the role of the protooncogene, c-mpl, in megakaryopoiesis in vitro *(8)*. Further investigations utilized mainly all-phosphorothioate ODNs. Proteins that were assessed in these studies included cyclin D3 *(9)*, MATK *(10)*, STK-1 *(11)*, PTP-RO *(12)*, p95^{vav} *(13)*, and several others. Although many of these publications focused on the general effects on megakaryopoiesis of antisense ODNs against specific proteins, studies of their effects on intracellular signaling in megakaryocytes were also reported. One example is the investigation of PTP-RO protein in c-Kit signaling during megakaryocyte differentiation *(12)*. Another illustration is the study of the contribution of p95^{vav} to the anti-proliferative effect of IFN-α in megakaryocytic cell lines *(13)*. The results obtained with the antisense strategy in studies of this type may be reproduced by other methodological approaches. For instance, in the study reporting the requirement for Mdm2 protein in the survival effects of Bcr-Abl and interleukin-3 in the M07e megakaryocytic cell line, the authors confirmed the antisense effects by using the tyrosine kinase inhibitor STI571 or IL-3 deprivation, both acting upstream of Mdm2 *(14)*. The results of those investigations undoubtedly proved the usefulness of the antisense strategy in studies on megakaryopoiesis and signal transduction in megakaryocytes.

The following protocol describes the use of antisense oligonucleotides to downregulate a specific gene product, using the intracellular tyrosine kinase MATK (also termed Csk homologous kinase [CHK]) as an example. Methods to assess changes in the target mRNA and protein expression levels, which are essential initial experiments in all antisense studies, are also described.

2. Materials

1. Oligodeoxynucleotides (ODNs): custom-synthesized, 18-mer phosphorothioate-modified and HPLC-purified:

- MATK (also termed Csk homologous kinase, CHK) antisense ODN 5′-AAC CAG AGA GCC TCG CCC CGC-3′ corresponding to nucleotides +4 to +24 bp (*see* **Note 1**).
- Control ODNs: sense 5′-GCG GGG CGA GGC TCT CTG GTT-3′ and scrambled (*see* **Note 2**).

   Lyophilized ODNs should be reconstituted in sterile H_2O at a concentration of 1750 μg/mL, stored at –20°C, and further diluted before being added to cell cultures in the incubation medium (see below) to yield a final concentration of 70 μg/mL (10 μM). Sterile technique must be used during preparation of the oligodeoxynucleotide solutions.

2. Cells: Megakaryocytic cells (primary or established cell lines such as CMK, DAMI, M07e, etc.) or human CD34+ cells incubated at a concentration of 1×10^6 cells/mL in custom-enriched serum-deprived Iscove's modified Dulbecco's medium.

3. Custom-enriched serum-deprived Iscove's modified Dulbecco's medium (IMDM): IMDM with added 300 μg/mL iron-saturated human transferrin, 100 ng/mL insulin, 28 μg/mL calcium chloride, 2% deionized bovine serum albumin, 614 μg/mL oleic acid, 7.4 mg dipalmitoyl lecithin. For CD34+ cells or primary bone marrow megakaryocytes, the incubation medium is supplemented with 100 U/mL recombinant human interleukin-3.

4. RNA extraction kit (e.g., RNeasy Mini Kit, Qiagen, Valencia, CA).

5. Superscript™ First-Strand Synthesis System for RT-PCR (Invitrogen, Carlsbad, CA).

6. MATK and control β-actin primers: The nucleotide sequence of the MATK upstream primer is 5′-GCG GGG CGA GGC TCT CTG GTT-3′ (corresponding to position +265 to +285 bp). The nucleotide sequence of the downstream primer is 5′-TGC GAG CAC ACC CGC CCC AAG-3′ (corresponding to position +430 to +450 bp). Primers for the β-actin message are: upstream primer 5′-ATG GAT GAT GAT ATC GCC GCG-3′ and downstream primer 5′-CTA GAA GCA TTT GCG GTG GAC GAT GGA GGG GCC-3′.

7. Synthetic ^{32}P-γATP-labeled oligomer hybridization probes: MATK probe 5′-GCC GTC ATG ACG AAG ATG CAA-3′ and β-actin probe 5′-GAG GAG CAC CCC GTG CTG CTG A-3′. When working with radioactivity, take appropriate precautions to avoid contamination of the researcher and the surroundings. Carry out the experiment and dispose of wastes in an appropriately designated area, following the guidelines provided by your local radiation safety officer.

8. RIPA buffer: 50 mM Tris-HCl, pH 7.4, 1% NP-40, 0.25% Na-deoxycholate, 150 mM NaCl, 1 mM EDTA, 1 mM PMSF, 1 μg/mL each of aprotinin, leupeptin, pepstatin, 1 mM Na_3VO_4, 1 mM NaF.

9. Protein assay kit: e.g., DC Protein Assay Kit from Bio-Rad Laboratories, Inc., Hercules, CA.

10. 2X Laemmli sample buffer (Bio-Rad).

11. Rabbit anti-human MATK antibody (Lsk, Santa Cruz Biotechnology, Santa Cruz, CA) and HRP-conjugated anti-rabbit secondary antibody.

12. Laboratory reagents and equipment for cell culture, RT-PCR and analysis of PCR products (*see [15]*) Unit 15.5 for detailed protocol), autoradiography, and SDS-PAGE followed by Western blot analysis (*see [15]* for detailed protocols; also Chapter 9, vol. 2).

13. Renaissance Enhanced Chemiluminescence immunodetection kit (NEN Life Science Products, Boston, MA).

3. Methods
3.1. Treatment of Cells With ODNs

1. Transfer the suspension of CD34+ or megakaryocytic cells into the wells of a 12-well plate (1 mL/well).

2. Add antisense or control ODN solutions to a final concentration of 10 µ*M* (*see* **Note 3**). Include appropriate controls to test for expression levels in the absence of ODNs and to assess the effects of any lipofection agents being used.
3. Incubate the cells with ODNs for 16 h at 37°C, 5% CO_2, 95% air, in a humidified atmosphere (*see* **Note 4**).
4. Add additional ODN solutions, as in **step 2**, again to a final concentration of 10 µ*M*. This step is necessary because of ODN degradation in the cell culture medium.
5. Incubate the cells with ODNs for an additional 6 h.
6. Divide incubated cells into two pools, the first containing 1×10^5 CD34$^+$ cells (for detection of MATK mRNA level), the second 9×10^5 CD34$^+$ cells (for detection of MATK protein level).
7. Centrifuge the cells for 7 min at 500*g*.

Following **step 5**, cells can be used for practically any other desired experiment (e.g., colony formation assay, immunofluorescent staining, detection of protein phosphorylation, etc).

3.2. Detection of MATK mRNA Level

1. Extract RNA from a pellet containing 1×10^5 CD34$^+$ or megakaryocytic cells, e.g., using the Qiagen RNeasy kit, according to the manufacturer's instructions.
2. Perform reverse transcription reaction, e.g., using commercial kit (Superscript™ First-Strand Synthesis System) according to the manufacturer's instructions.
3. Add the 5′- and 3′-specific primers, for either MATK or β-actin (as a PCR control) at a final concentration of 5 ng/50 µL each (*see* **Note 5**).
4. Subject the mixture to 30 amplification cycles with settings as follows: denaturation at 94°C for 1 min, primer annealing at 55°C for 1 min, and extension at 72°C for 2 min.
5. Perform electrophoresis of the PCR product through 2% agarose gel, followed by blotting to nylon membrane.
6. Detect the amplification products by an overnight hybridization to specific probes (25–50 ng of oligomer/probe, 42°C).
7. Wash the membrane: 15 min at room temperature, then 15 min at 42°C, followed by 30–60 min at 55–60°C.
8. Expose the autoradiographs for 6–18 h at –80°C with intensifying screens.

3.3. Detection of MATK Protein Level

1. Lyse a pellet containing 9×10^5 CD34$^+$ or megakaryocytic cells by addition of 400 µL of ice-cold RIPA buffer.
2. Determine the protein concentration in each sample using a protein assay (e.g., BioRad DC Protein Assay Kit).
3. Transfer 50 µg of each protein extract to new Eppendorf tubes and add equal volumes of 2X Laemmli sample buffer. Incubate the samples at 100°C for 5 min.
4. Separate the samples electrophoretically on 10% polyacrylamide-SDS gel (90–100 V, run until front of the sample reaches bottom of the gel).
5. Transfer separated proteins to PVDF or nitrocellulose membrane (for further details of Western blotting techniques, *see* Chapter 9, vol. 2).
6. Probe the membrane with antibodies against MATK (1:1000 dilution, 1 h incubation at room temperature or 16 h in 4°C), followed by incubation with HRP-conjugated anti-rabbit secondary antibodies (1:5000, 1 h at room temperature).
7. Detect the immunoconjugates using the enhanced chemiluminescence system.

3.4. Anticipated Results

In the cells treated with the antisense ODN, both mRNA and protein levels should be markedly diminished in comparison with the nontreated or control (sense and scrambled) ODN-treated cells.

3.5. Discussion and Perspective on Future Directions

3.5.1. Oligonucleotide Modifications

The susceptibility of unmodified phosphodiester ODNs to nucleases led to the extensive search for modifications that would allow for avoidance of this problem, while also allowing the retention of the antisense mode of action of the modified oligonucleotides (oligos). Methylphosphonate-linked ODNs were the first to provide resistance to enzymatic degradation. However, poor aqueous solubility and limited effectiveness did not permit the popular use of these compounds. In an attempt to overcome such limitations, numerous other modifications of the nucleotide backbone have been developed over the past decade. Of these, phosphorothioate-linked ODNs have come to dominate the antisense field. Still, with their common usage, it is becoming evident that phosphorothioate ODNs provide sensible sequence specificity only within a narrow concentration range and that higher concentrations generate numerous sequence-independent side effects. With the goal of overcoming this remaining limitation, new modifications of nucleotides are being continually created. Out of these, one of the most promising seems to be morpholino antisense oligomers. The main feature of morpholino-oligos is the replacement of the ribose or deoxyribose rings (characteristic for RNA or DNA, respectively) with the morpholine ring (for detailed review *see* **ref. 20**). Although considerably more expensive, morpholino-oligos present several advantages over phosphorothioate ODNs. These include immunity to nucleases, predictable targeting, high specificity, minimal sequence-independent side effects, RNase H independence, high solubility, effectiveness in the presence of serum, as well as reliable activity both in cell-free systems and in cells. A disadvantage is their inability to cooperate with the cationic lipid carriers. Therefore, the most popular delivery of morpholino-oligos to the cells is scrape delivery. Unfortunately, this type of delivery cannot be applied to the nonadherent cells, and therefore is of negligible use in studies on megakaryocytes and/or platelets. Thus, delivery systems that might be of use in this case include microinjection, electroporation, or a special delivery system using ethoxylated polyethylenimine (EPEI, Gene Tools, LLC, Philomath, OR). Alas, since no reports on the use of morpholino-oligos in megakaryocyte/platelet studies have appeared to date, we therefore cannot present any previously tested protocol utilizing these compounds. Nevertheless, on a theoretical basis, one should also assume the usefulness of morpholino-oligos in signaling studies in these types of cells.

3.5.2. Antisense RNA and Ribozymes

Both the antisense RNA strategy (for review *see* **ref. 21**) and a modification of this approach, i.e., ribozymes (for review *see* **ref. 22**), offer an alternative to the classical methods utilizing antisense oligos. The principle is the use of cDNA-containing genes that express antisense RNA, which is supposed to form RNA:RNA hybrids with the specific target. Additionally, ribozymes possess catalytic activity that allows them to cleave target RNA. As this strategy has not been extensively studied in megakaryocytic systems, it is difficult to assess the practical advantages of antisense RNA technology over ODNs. Of note, the anti-PKCα ribozyme was successfully applied to study the role of PKCα protein in actin reorganization and proplatelet formation in murine megakaryocytes *(23)*.

3.5.3. Small Interfering RNA (siRNA)

RNA interference (RNAi) is a phenomenon of gene silencing at the mRNA level offering a quick, powerful, and reliable method to determine the function of a particular gene both in vivo and in vitro (for detailed review *see* **ref. 24**). RNAi was first discovered in the late 1980s in worms. This discovery was immediately followed by studies in mammals and by adoption of RNAi technology to silence genes in numerous studies. Although this approach has not yet been used to investigate signaling pathways in megakaryocytes, it must be mentioned here as a novel and extremely promising method of gene-expression knockdown. The new tool used for this purpose is called small interfering RNAs (siRNAs). Usually these are duplexes of 19 or 21 nucleotides with symmetric 2-nucleotide (preferably TT) 3' overhangs. siRNA appears to suppress gene expression without producing nonspecific cytotoxic effects. The most basic principles of siRNA design are as follows:

1. Starting from the ATG transcription initiation codon of your cDNA, scan downstream for AA dinucleotide sequences. Record the occurrence of each AA and the 3' adjacent 19 nucleotides as potential siRNA target sites.

    Usually the target sequence should be located 50–100 nucleotides downstream from the ATG start codon. Also, sequences located in the 5' or 3' UTR of the target mRNA should be avoided, as UTR binding proteins may interfere with the scanning process of the siRNA complex.
2. Check the nucleotide sequence using a gene database (e.g., BLAST, which can be found on the NCBI server at www.ncbi.nlm.nih.gov/BLAST/) to ensure that only your desired gene will be inactivated.
3. Select qualifying target sequences for synthesis.

Synthesis of large-scale siRNA might be ordered commercially—e.g., from Ambion, Inc., Austin, TX. Some companies also offer plasmid DNA vector-based kits for introduction of siRNA to mammalian cells by transfection without the use of in vitro transcribed synthetic siRNA (e.g., GeneSuppressor™ RNA Interference Kits, Imgenex, San Diego, CA).

A complete siRNA experiment should include the proper negative control. For this purpose the same types of controls as described for ODN studies can be used (*see* **Note 2**). Similarly, for siRNA sequence, all the control sequences should be scanned using a gene database to avoid potential interference with other genes.

The most limiting factor in siRNA-based experiments is the introduction of siRNA into the target cells. Most researchers use either cationic or noncationic lipid-based carriers. In the case of cells resistant to such transfection method, electroporation might be a method of choice. Recently, the use of retrovirus vector was also described as suitable for siRNA delivery *(25)*.

4. Notes

1. *Antisense oligodeoxynucleotide design–phosphorothioate ODN:* Although theoretically the design of the ODN sequence is simple, because it is complementary to a region of the gene of interest, in practice this is the most difficult and crucial step in the antisense experiment. It has been shown by a number of studies that a well-designed antisense ODN can be a highly specific and efficient inhibitor of gene expression. On the other hand, sequence-nonspecific side effects of improperly designed ODNs might result in misleading and false results. The following parameters are considered crucial for the correct design of antisense ODNs:

   a. *Length of the antisense ODN sequence:* The minimum size of the antisense ODN needed to specifically recognize the target sequence was calculated to be between 12 and 15 bases. Although longer sequences theoretically ensure higher specificity in binding, an extensive increase in sequence length also boosts the sequence-independent side effects and toxicity of ODNs. For this reason, the length of the antisense ODN must be optimized. At the present time, the optimal length of an antisense ODN appears to be approx 16–20 nucleotides.

   b. *mRNA target region:* There are no general rules for predicting which region within the gene of interest is the best target for antisense compounds. It has been shown that any region of the mRNA can theoretically be targeted using antisense phosphorothioate ODNs (5′ or 3′ untranslated regions, AUG initiation codon, splice junctions, introns, and coding sequences) *(16)*. Therefore, all antisense ODNs must be generated from a panel of potential candidates. The initial panel may consist of all-phosphorothioate ODNs, which are relatively inexpensive and can be obtained from commercial sources.

   c. *Downregulation of target gene expression:* RT-PCR, Northern, or Western blotting can be utilized to demonstrate the effects of antisense ODNs on the content of target mRNA or protein, respectively. Particularly convincing, although not necessary, seems to be demonstration of such effects in the system of artificial gene expression (e.g., plasmid transfection).

   d. *The choice of DNA backbone:* The use of unmodified all-phosphodiester ODNs should be avoided, mainly because of their susceptibility to the nucleases. Also, their degradation products (particularly dGMP) may be toxic to the cells *(17)*. However, it is acceptable in an antisense experiment to use chimeric phosphorothioate/phosphodiester ODNs. In such cases, the ODN should be protected at the 3′ and 5′ termini by at least three phosphorothioate linkages. The advantages and disadvantages of using ODNs with modifications other than those with phosphorothioate are discussed in **Subheading 4.1.** of this chapter.

   e. *GC content:* A GC content of 45% to 65% is considered to be optimal for antisense ODN design *(18)*. Higher GC content may result in the formation of hairpin structures within the antisense ODN, impairing the efficiency of binding to the target mRNA. Special attention should be drawn to avoiding the use of four contiguous guanosine residues (GGGG) within the antisense ODN. Such structures can form G-quartets and tetraplexes via Hoogsteen base-pair formation *(1)*, which can substantially decrease the specificity and efficacy of the antisense ODN.

   f. *Fluorescence staining of ODNs:* Most manufacturers of custom ODNs have in their offerings the use of a fluorescent marker, e.g., carboxyfluorescein, attached to the ODN.

This modification might be useful to assess the cytosolic delivery of ODNs. It is important to use viable cells for this purpose, as fixation of the cells can lead to the leakage of ODNs into the cells and give a false-positive signal. Only diffuse fluorescence dimly spread throughout the cytosol of the cell indicates successful delivery into the cytosol, the site of antisense action. Dotted fluorescence does not indicate delivery to the cytosol (though it does not preclude it). Of note, visible diffuse fluorescence requires an ODN concentration several times higher than the concentration needed for antisense action against most targets. It is also noteworthy that flow cytometry or FACS analyzers cannot distinguish between dotted and diffuse fluorescence, and therefore cannot easily detect delivery.

2. *Control sequences:* Because of the fairly high probability of sequence-independent effects of ODNs in cells, the use of an appropriate control ODN is necessary. Several types of control sequences have been utilized in various studies. These include:

   a. "Sense" configuration—the sequence complementary to antisense ODN. In other words, this is a sequence derived directly from the target sequence within the gene of interest.

   b. "Random" or "scrambled" configuration—the numbers of each type of base within the control ODN are the same as in the antisense ODN; however, they are placed in scrambled order.

   c. Mismatch ODN—several (usually three or four) bases within the ODN sequence are mismatched to prevent duplex formation with the target mRNA. The same idea is also utilized in different types of controls, when one or more mutations are introduced into the target region within the gene sequence and, subsequently, the antisense ODN is shown to fail to influence mRNA and/or protein expression.

   d. "Positive" control—the sequence of this oligonucleotide is identical to the primary antisense ODN; however, the backbones are different (e.g., phosphorothioate-DNA and morpholino-oligos). If the inhibition of the molecular target is identical for both, the case for an antisense mechanism is strengthened. Similarly, additional inhibition of a molecular target by antisense RNA supports the notion of the antisense mode of action.

   Most studies use either sense, scrambled, or simultaneously both types of controls. If doubt remains, additional controls should be utilized.

3. *Delivery method:* A major challenge for antisense development is the low permeability of ODNs through cellular membranes. Although there is no doubt that ODNs undergo endocytosis when incubated with most cell types, the issue as to whether they can enter the cytosol through endosomal or lysosomal membranes remains controversial. Thus, relatively high (usually more than 10 μM) concentrations of phosphorothioate ODNs are required to ensure sufficient inhibition of gene expression, when no carrier is used. In such concentrations, there is a very high likelihood of sequence-independent side effects of the ODN. For this reason, the use of a delivery system is strongly recommended. Usually, the commercially available cationic lipid delivery systems (e.g., Lipofectamine 2000 [Gibco BRL, Carlsbad, CA] and several others *[19]*) allow for substantially decreased concentrations of antisense phosphorothioate ODN. The advantage of such reagents as Lipofectamine 2000 is that there is no need to change the medium after incubation with the DNA/cationic lipid mixture, a feature that is especially useful in experiments with nonadherent cells. However, numerous other delivery systems, including physical methods such as microinjection or electroporation, have been employed to target and/or deliver the antisense compounds in vitro and in vivo. The choice of delivery system should be optimized for the type of experiment being performed.

4. *Length of exposure:* The length of exposure of the cells to the antisense ODN depends on the experimental system and the turnover rate of the protein studied. Nevertheless, in most studies on megakaryopoiesis, the incubation times are not shorter than 18 h (**steps 3** and **5** in **Subheading 3.1.**).
5. For this step a commercial kit might also be used—e.g., DyNAzyme II DNA polymerase kit from Finnzymes OY, Espoo, Finland. For detailed reaction parameters, please follow the manufacturer's instructions.

Acknowledgments

This work was supported in part by National Institutes of Health grants HL39558 (SA), HL51456 (HA), CA76226 (HA), DAMD 17-98-1-8032 (HA), DAMD 17-99-1-9078 (HA) and CA76772 (GDM), and the Jennifer Randall Breast Cancer Research Fund (R. Z). This work was created during the term of an established investigatorship from the American Heart Association (HA). In 2002 R. Z. was the recipient of a Postdoctoral Traineeship Award from the Department of Defense Breast Cancer Research Program (Grant No. DAMD 17-02-1-0302). This chapter is dedicated to Charlene Engelhard and Ronald Ansin for their continuing friendship and support for our research program.

References

1. Stein, C. A. (2001) The experimental use of antisense oligonucleotides: a guide for the perplexed. *J. Clin. Invest.* **108,** 641–644.
2. Rockwell, P., O'Connor, W. J., King, K., Goldstein, N. I., Zhang, L. M., and Stein, C. A. (1997) Cell-surface perturbations of the epidermal growth factor and vascular endothelial growth factor receptors by phosphorothioate oligodeoxynucleotides. *Proc. Natl. Acad. Sci. USA* **94,** 6523–6528.
3. Lederman, S., Sullivan, G., Benimetskaya, L., Lowy, I., Land, K., Khaled, Z., et al. (1996) Polydeoxyguanine motifs in a 12-mer phosphorothioate oligodeoxynucleotide augment binding to the v3 loop of HIV-1 gp120 and potency of HIV-1 inhibition independency of G-tetrad formation. *Antisense Nucleic Acid Drug Dev.* **6,** 281–289.
4. Benimetskaya, L., Loike, J. D., Khaled, Z., Loike, G., Silverstein, S. C., Cao, L., et al. (1997) Mac-1 (CD11b/CD18) is an oligodeoxynucleotide-binding protein. *Nat. Med.* **3,** 414–420.
5. Khaled, Z., Benimetskaya, L., Zeltser, R., Khan, T., Sharma, H. W., Narayanan, R., et al. (1996) Multiple mechanisms may contribute to the cellular anti-adhesive effects of phosphorothioate oligodeoxynucleotides. *Nucleic Acids Res.* **24,** 737–745.
6. Gewirtz, A. M. and Calabretta, B. (1988) A c-myb antisense oligodeoxynucleotide inhibits normal human hematopoiesis in vitro. *Science* **242,** 1303–1306.
7. Caracciolo, D., Venturelli, D., Valtieri, M., Peschle, C., Gewirtz, A. M., and Calabretta, B. (1990) Stage-related proliferative activity determines c-myb functional requirements during normal human hematopoiesis. *J. Clin. Invest.* **85,** 55–61.
8. Methia, N., Louache, F., Vainchenker, W., and Wendling, F. (1993) Oligodeoxynucleotides antisense to the proto-oncogene c-mpl specifically inhibit in vitro megakaryocytopoiesis. *Blood* **82,** 1395–1401.
9. Wang, Z., Zhang, Y., Kamen, D., Lees, E., and Ravid, K. (1995) Cyclin D3 is essential for megakaryocytopoiesis. *Blood* **86,** 3783–3788.

10. Avraham, S., Jiang, S., Ota, S., Fu, Y., Deng, B., Dowler, L. L., et al. (1995) Structural and functional studies of the intracellular tyrosine kinase MATK gene and its translated product. *J. Biol. Chem.* **270,** 1833–1842.

11. Ratajczak, M. Z., Ratajczak, J., Ford, J., Kregenow, R., Marlicz, W., and Gewirtz, A. M. (1996) FLT3/FLK-2 (STK-1) Ligand does not stimulate human megakaryopoiesis in vitro. *Stem Cells* **14,** 146–150.

12. Taniguchi, Y., London, R., Schinkmann, K., Jiang, S., and Avraham, H. (1999) The receptor protein tyrosine phosphatase, PTP-RO, is upregulated during megakaryocyte differentiation and Is associated with the c-Kit receptor. *Blood* **94,** 539–549.

13. Micouin, A., Wietzerbin, J., Steunou, V., and Martyre, M. C. (2000) p95(vav) associates with the type I interferon (IFN) receptor and contributes to the antiproliferative effect of IFN-alpha in megakaryocytic cell lines. *Oncogene* **19,** 387–394.

14. Goetz, A. W., van der Kuip, H., Maya, R., Oren, M., and Aulitzky, W. E. (2001) Requirement for Mdm2 in the survival effects of Bcr-Abl and interleukin 3 in hematopoietic cells. *Cancer Res.* **61,** 7635–7641.

15. Ausubel, F. M. (2001) *Current Protocols in Molecular Biology*, John Wiley & Sons, New York.

16. Flanagan, W. M. and Wagner, R. W. (1997) Potent and selective gene inhibition using antisense oligodeoxynucleotides. *Mol. Cell Biochem.* **172,** 213–225.

17. Vaerman, J. L., Moureau, P., Deldime, F., Lewalle, P., Lammineur, C., Morschhauser, F., et al. (1997) Antisense oligodeoxyribonucleotides suppress hematologic cell growth through stepwise release of deoxyribonucleotides. *Blood* **90,** 331–339.

18. Wahlestedt, C. (1994) Antisense oligonucleotide strategies in neuropharmacology. *Trends Pharmacol. Sci.* **15,** 42–46.

19. Axel, D. I., Spyridopoulos, I., Riessen, R., Runge, H., Viebahn, R., and Karsch, K. R. (2000) Toxicity, uptake kinetics and efficacy of new transfection reagents: increase of oligonucleotide uptake. *J. Vasc. Res.* **37,** 221–234; discussion 303–324.

20. Summerton, J. and Weller, D. (1997) Morpholino antisense oligomers: design, preparation, and properties. *Antisense Nucleic Acid Drug Dev.* **7,** 187–195.

21. Weiss, B., Davidkova, G., and Zhou, L. W. (1999) Antisense RNA gene therapy for studying and modulating biological processes. *Cell Mol. Life Sci.* **55,** 334–358.

22. Gibson, S. A. and Shillitoe, E. J. (1997) Ribozymes. Their functions and strategies for their use. *Mol. Biotechnol.* **7,** 125–137.

23. Rojnuckarin, P. and Kaushansky, K. (2001) Actin reorganization and proplatelet formation in murine megakaryocytes: the role of protein kinase c alpha. *Blood* **97,** 154–161.

24. McManus, M. T. and Sharp, P. A. (2002) Gene silencing in mammals by small interfering RNAs. *Nat. Rev. Genet.* **3,** 737–747.

25. Devroe, E. and Silver, P. A. (2002) Retrovirus-delivered siRNA. *BMC Biotechnol.* **2,** 15.

25

GFP Fusion Proteins to Study Signaling in Live Cells

Simon A. Walker, Gyles E. Cozier, and Peter J. Cullen

1. Introduction

In just a few years, the green fluorescent protein (GFP) from the jellyfish *Aequorea victoria* has jumped from relative obscurity to become one of the most widely studied and exploited proteins in cell biology. Discovered by Shimomura et al. *(1)*, GFP acts as a fluorophore and native companion to the famous chemiluminescent protein aequorin *(2)*, also from *Aequorea*. GFP absorbs blue light emitted by aequorin and re-emits photons at a longer wavelength, thus accounting for the green glow of the intact jellyfish *(3)*. In research terms, the crucial breakthroughs came with the cloning of the gene *(4)* and the demonstration that expression of the gene in other organisms creates green fluorescence *(5,6)*. This is possible because the gene alone contains all the information necessary for the post-translational synthesis of the chromophore, and no jellyfish-specific enzymes are required. These properties have made GFP a powerful and versatile tool for investigating virtually all fields of cell biology, including the study of membrane traffic and dynamics, organelle structure, and gene expression. By using both homologous and heterologous systems, and with the development of methods to culture megakaryocytes in vitro (*see* Chapters 22, 23, and 27, vol. 1), it is likely that the use of GFP will make a significant contributibution to our understanding of megakaryocyte and platelet signaling. In this chapter we discuss aspects of using GFP in the imaging of protein dynamics in living cells.

1.1. The Structure of GFP

Although GFP was first crystallized in 1974 *(7)* and diffraction patterns reported in 1988 *(8)*, it was not until 1996 that the actual structure was first solved *(9,10)*. GFP is an 11-stranded β-barrel threaded by an α-helix running up the axis of the cylinder (**Fig. 1**). The chromophore is attached to the α-helix and is buried almost perfectly in the center of the cylinder, which has been called a β-can *(10,11)*. Almost all of the primary sequence is used to build the β-barrel and axial helix. The chromophore is

From: *Methods in Molecular Biology, vol. 273:*
Platelets and Megakaryocytes, Vol. 2: Perspectives and Techniques
Edited by: J. M. Gibbins and M. P. Mahaut-Smith © Humana Press Inc., Totowa, NJ

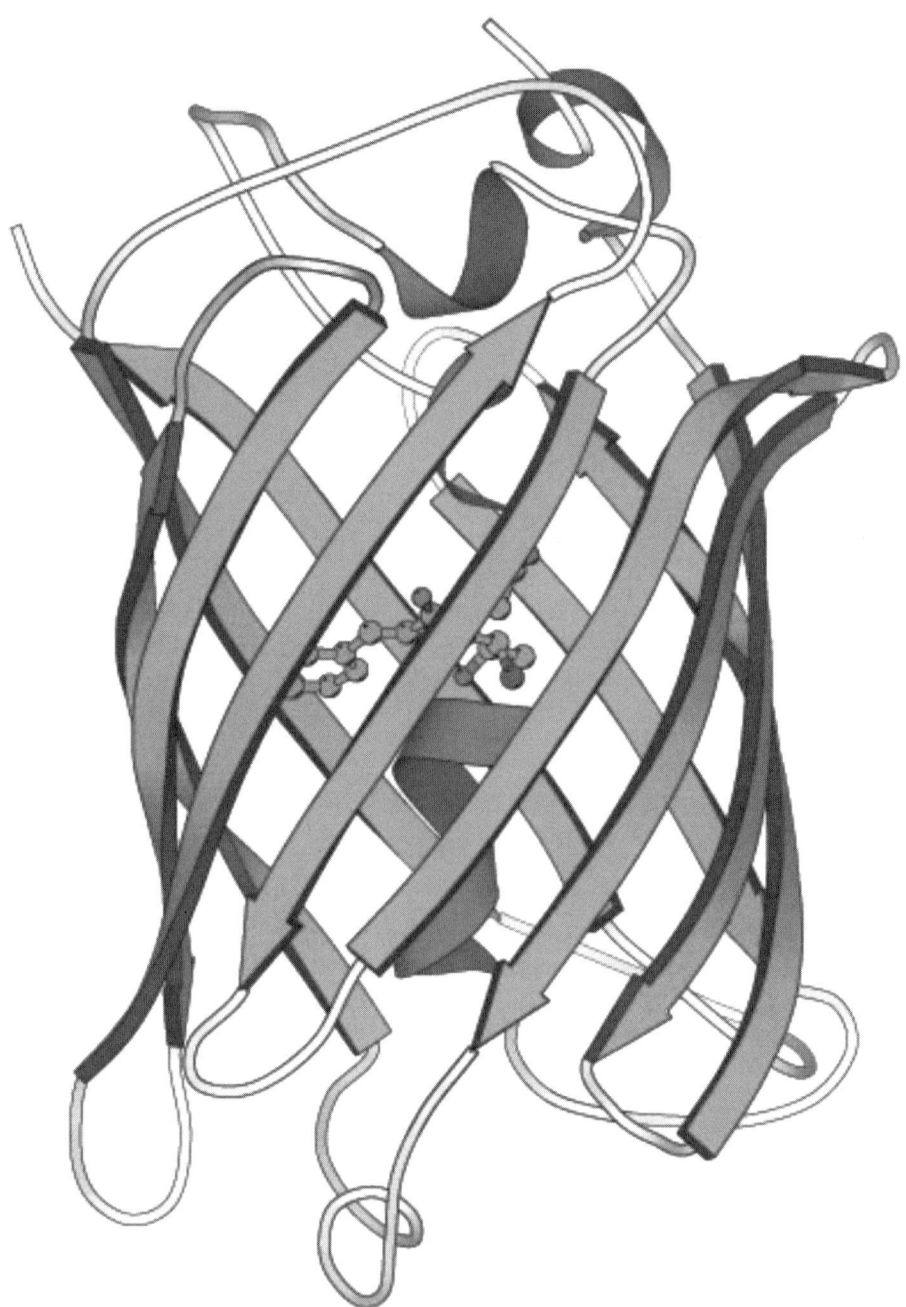

Fig. 1. Ribbon diagram representation of the crystal structure of wild-type green flu-orescent protein from *Aequoria victoria*. The β-barrel is shown in green, the α-helices in blue. The chromophore is shown in molecular detail at the center of the β-barrel. Generated using MOLSCRIPT *(31)*. (*See* color insert following p. 300.)

ρ-hydroxybenzylideneimidazolinone *(4,12)* formed from residues 65–67, which in the native protein are Ser-Tyr-Gly (for a detailed discussion of the biophysics of chromophore formation, *see* **ref. 3**). Formation of the chromophore directly follows the synthesis of the protein after a relatively brief time lag.

1.2. Molecular Engineering of GFP: Generating a "Rainbow" of Fluorescent Proteins

Perhaps one of the major reasons for the explosive success of GFP is that the fluorescent moiety is a gene product with no requirement for a cofactor. GFP is therefore amenable to molecular engineering and can be transiently or stably expressed in virtually every cell type. Targeted mutagenesis has allowed the adaptation of the fluorescent properties of GFP for particular experimental purposes. There are a number of so-called "optimizing" mutations that increase light emission by either altering the intrinsic properties of the fluorescent protein *(13,14)* or increasing its production in mammalian cells *(15)*. A number of mutations have been described that alter the stability and/or quantum efficiency of GFP upon illumination with visible light. Of these, the most useful appears to be substitution of Ser-65 by Thr (S65T). Compared to wild-type GFP, this mutant has a sixfold enhanced quantum efficiency upon excitation with blue light and is markedly less sensitive to photobleaching *(14,16,17)*. In addition, "humanized" versions of GFP are commonly used, in which silent mutations are introduced into the cDNA that convert some of the codons into the most common and efficient for translation in mammalian cells *(15)*. This results in production of more fluorescent protein for the same amount of mRNA and is an important consideration when low levels of GFP mRNA need to be detected—for example, in assays where GFP production is used as a reporter for gene promoter activity *(18)*.

Amino acid substitutions within the chromophore have generated a number of GFP variants with altered emission spectra compared to the native protein *(3)*. Of these, the most popular mutants include cyan, yellow, and blue variants of GFP (for spectral properties of these mutants, *see* **Table 1**). Attempts to isolate a truly "red" GFP variant by mutagenesis have proven unsuccessful *(19)*. However, a red companion of GFP, red fluorescent protein (RFP), has been isolated from the coral *Discosoma (20–23)*, and a far-red fluorescent protein (HcRED) generated from a native chromoprotein isolated from the coral *Heteractis crispa (24)*.

1.3. Generating Reporter-Gene Constructs and Fusion Proteins

The first proposed application of GFP was as a reporter gene for the detection of promoter activity in vivo *(5)*. This was especially useful in the nematode *Caenorhabditis elegans*, whose cuticle hinders access of the substrates required for detecting other reporter genes. However, GFP seems to require a rather strong promoter to drive sufficient expression for detection, especially in mammalian cells. Indeed, Rutter and colleagues *(18)* have imaged gene transcription at the single-cell level using both GFP and luciferase-based gene reporters, and shown the latter to be preferable. This in part stems from the fact that even for the mutant GFPs with improved extinction coefficients, 10^5 copies of matured GFP per (typical) 1–2 pL volume of a given cell are required to

Table 1
Details of Green Fluorescent Proteins and Spectral Variants

Name	Mutation	Excitation (max), nm	Emission (max), nm
EBFP	F64L, S65T, Y66H, Y145F	380	440
ECFP	F64L, S65T, Y66W, N146I, M153T, V163A, N212K	434	477
GFP	Wild-type	395, 470	509
EGFP	F64L, S65T	489	508
EYFP	S65G, S72A, T203Y	514	527
RFP	Wild-type (from *Discosoma sp.*)	558	583
HcRED	(from *Heteractis crispa*)	588	618

RFP and HcRED are isolated from different organisms and are not GFP variants. All proteins are available in a variety of expression vectors available from Clontech and Q-Biogene. Filter sets for imaging the various fluorescent proteins can be obtained directly from microscope manufacturers or specialist companies, e.g., Chroma Technology (http://www.chroma.com).
Data from **refs. *29,30***.

equal the endogenous autofluorescence—i.e., to double the fluorescence over background. This is in contrast to other reporter gene products such as luciferase that can act enzymatically on large amounts of substrate loaded into intact viable cells to generate significant luminescence.

By far the most frequent application of GFPs, however, is to use them as tags. Here, the gene encoding a GFP is fused in frame with the gene encoding the protein of interest and the resulting chimera expressed in the cell or organism of interest. For example, a recent study examined the mechanism of plasma membrane translocation of phospholipase-Cγ2 and Bruton's tyrosine kinase (important signals during platelet activation), using GFP fusion proteins *(25)*. The optimal fusion protein is one that maintains the normal function(s) and localization of the endogenous protein but is now fluorescent. In this regard, not all fusions are successful, but the failures are rarely reported, so it is difficult to assess the overall success rate. However, it is noteworthy that GFP has been targeted successfully to practically every major organelle of the cell, including the plasma membrane, nucleus, endoplasmic reticulum, Golgi apparatus, secretory and endosomal vesicles, mitochondria, peroxisomes, vacuoles, and phagosomes.

1.4. Experimental Set-Ups for Collecting and Analyzing GFP Images

The basic requirements for any GFP imaging system are the same—namely, a light source to provide the necessary excitation wavelength, a microscope fitted with an appropriate filter set for visualization of the emitted light, and a means of recording the emitted light. The choice of imaging system will depend on many factors, not least of which is the available budget. Questions to bear in mind include: Will the imaging system be used for both fixed and live cell imaging? Will the system be used for imaging other than GFP (e.g., calcium imaging)? Will the system be required to image dual or triple wavelengths from multilabeled samples? There are now literally hundreds of commercially available systems capable of imaging GFP, and readers are advised to consult individual manufacturers for further details. Conversion of existing microscopes to enable GFP visualization and recording is possible, with a variety of companies specializing in tailor-made configurations. A comprehensive list of companies involved in all aspects of imaging can be found at http://www.kaker.com/mvd/vendors.html.

Irrespective of the imaging system chosen, there tends to be a compromise, for a variety of reasons, between the rate of image capture and the resolution of the images obtained. Thus, using GFP imaging in live cells to address temporal aspects of protein function will come at the expense of high-definition spatial information. As the technology of imaging systems improves, however, this is becoming less of an issue. For example, for routine imaging of the dynamics of GFP-tagged proteins in live cells we use an Ultra*VIEW* Live Cell Imaging confocal system (PerkinElmer Life Sciences, Boston, MA), which is capable of obtaining good-quality images at high frame rates (up to 10 frames per second at maximum resolution). Our system is currently composed of a krypton/argon laser coupled through a Yokogawa CSU10 scanhead (fitted with excitation and emission filters for 488 and 568 nm excitation) to a Hamamatsu Orca ER Interline CCD camera. Cells are placed in a temperature-controlled stage of a microscope fitted with a piezoelectric objective stepper for high speed, precision z-axis control. Images are processed using Volocity 3-D rendering software.

Compared to a standard laser scanning confocal system, the Ultra*VIEW* allows us to analyze the dynamics of GFP-tagged proteins with high image quality and lower-level phototoxcity (**Fig. 2**), achieved, in part, by the Yokogawa scanhead (*see* **Note 1**). This uses dual spinning disks with microlens technology to create low-energy laser beams that illuminate the total field of view at high frequency. Instead of a single beam of high-intensity laser light delivered point by point across an image, the Ultra*VIEW* laser beam passes through 1000 individual pinholes—each with a corresponding microlens that focuses the light—before reaching the sample. An Orca ER Interline CCD camera uses On-Chip microlens technology to ensure high-efficiency light detection and shorter exposures. Overall, this provides fast three-dimensional imaging from living cells with a limited danger of phototoxicity to the cell or bleaching of the fluorophore.

The main disadvantage with our Ultra*VIEW* system is that it is currently not equipped to discriminate between GFP/YFP/CFP (although it can discriminate between GFP and RFP), making multilabeling experiments problematic. This also means that fluorescence resonance energy transfer (FRET)-based assays cannot be done with this system (*see* **ref. 27** for a discussion on using GFP molecules in FRET-based assays). For these purposes we use a Leica TCS SP2 confocal microscope. This is a more standard laser-scanning confocal microscope, which provides good-quality images, but at a much reduced frame rate compared to the Ultra*VIEW* (**Fig. 2**). However, PerkinElmer has recently launched a newer Ultra*VIEW* model with five excitation lines, which, according to the manufacturer, can be used to track CFP and YFP-labeled proteins inside cells. Intermolecular FRET measurements have yet to be done experimentally with this system although this should be possible in principle between CFP and YFP or GFP and a red fluorophore.

As mentioned earlier, the methods of imaging live cells expressing GFP chimeras are many and varied. By means of an example we will outline our protocol for imag-

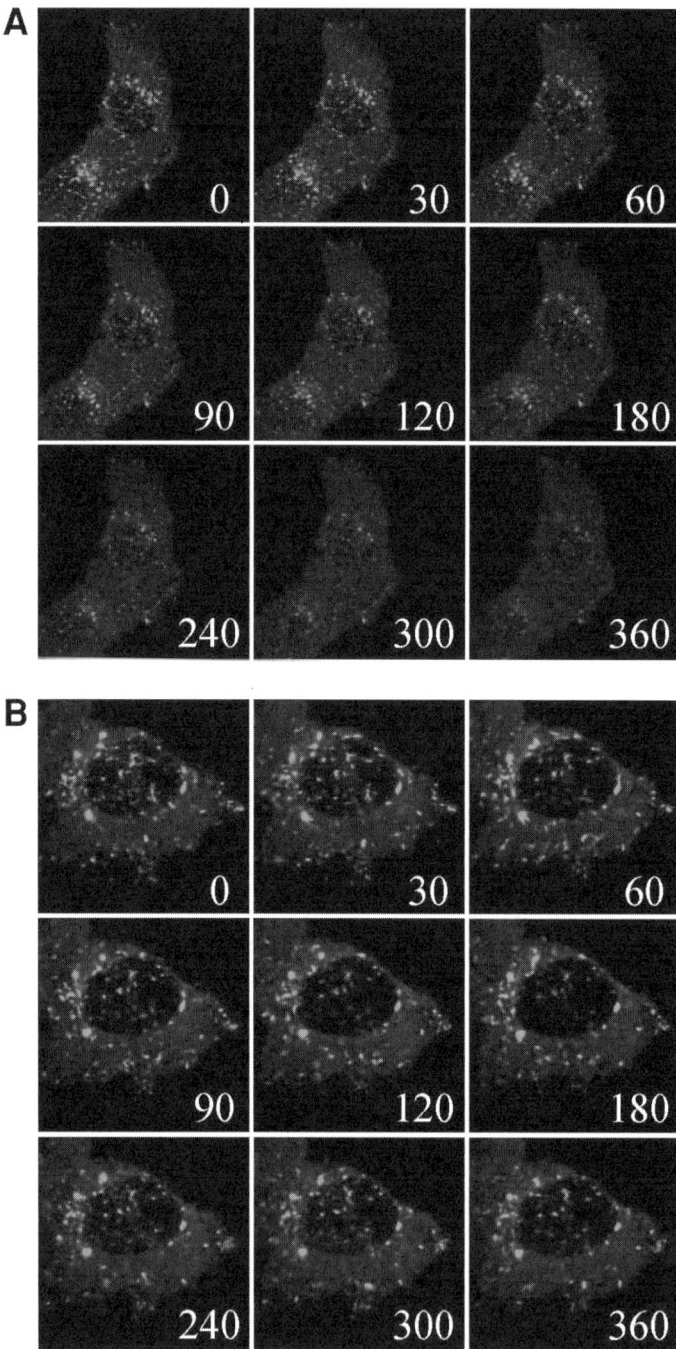

ing live cells using a PerkinElmer Ultra*VIEW* Live Cell Imager. A brief discussion on some of the important considerations to be made before making the GFP fusion protein is also included.

2. Materials
2.1. Preparation of GFP Chimera

1. Cloned gene of interest.
2. GFP expression vector (Clontech, Palo Alto, CA ; Q-Biogene, Carlsbad, CA; Stratagene, La Jolla, CA) (*see* **Note 2**).
3. Midi prep kit (Qiagen, Valencia, CA) (or other method of high-purity DNA preparation) (*see* **Note 3**).

2.2. Cell Culture and Transfection

1. Sterile 22-mm round coverslips (No. 1 thickness).
2. Six-well tissue culture plates.
3. Dulbecco's modified Eagle medium (Invitrogen, Carlsbad, CA) supplemented with 10% fetal bovine serum.
4. GeneJuice transfection reagent (Novagen, Madison WI). Other transfection reagents or methods (electroporation, calcium phosphate) can also be used, if appropriate for the cell type under study.

2.3. Imaging

1. Imaging buffer: 121 mM NaCl, 5.4 mM KCl, 1.6 mM MgCl$_2$, 6 mM NaHCO$_3$, 9 mM glucose, 25 mM HEPES, 1.3 mM CaCl$_2$; pH 7.4 with 5 M NaOH.
2. Ultra*VIEW* Live Cell Imager (PerkinElmer Life Sciences) with appropriate laser and excitation and emission filters for the GFP variant being used (*see* **Table 1**) (*see* **Note 4**). The Ultra*VIEW* is attached to a standard microscope and samples are imaged using a high-magnification objective lens with a high numerical aperture to maximize the amount of emitted light collected (*see* **Subheading 1.4.** for discussion of appropriate imaging systems). Our Ultra*VIEW* is attached to an Olympus IX70 inverted microscope, and we image using a 63× oil-immersion lens N.A. 1.25.
3. Twenty-two mm coverslip holder and heated stage (supplied by PerkinElmer with the Ultra*VIEW*). The heated stage mantains the chamber temperature at 37°C. The holder

---

Fig. 2. *(see opposite page)* Comparison of GFP images obtained with laser scanning and spinning disk confocal microscopes. Images from a time series experiment, acquired on either a laser-scanning confocal microscope (**A**) or a confocal equipped with a Yokogawa scanhead (**B**). For equivalent spatial resolution, the laser-scanning confocal microscope exposes the sample to laser light at higher power and results in more photobleaching (*see* **Note 1** for further discussion). Images show SNX1 (a member of the sorting nexin family of proteins involved in intracellular vesicular sorting) tagged at the N-terminus of GFP, imaged in live HeLa cells. Numbers indicate time in seconds after the initiation of imaging and are representative of a series of frames taken every 2–3 s during the course of the experiment. (Courtesy J. Carlton.) (*See* color insert following p. 300.)

consists of a thin plastic lower ring and a larger upper metallic ring. The two are sealed together with the coverslip sandwiched inbetween; the upper ring forms the chamber. Silicone grease is applied to the edges of the rings where they come into contact with the coverslip to form a watertight seal.

3. Methods

3.1. Considerations Before Constructing the GFP Chimera

The choice of cloning vector will depend on the nature of the protein of interest, and the system being used to detect the expression. In general, it is important to take into account known functional aspects of the target protein when choosing which termini to tag with the GFP. For example, does the host protein undergo any form of post-translational processing such as lipid modification at one of its termini, or is a terminal region required for protein:protein interactions? In cases where this is not an issue, two distinct fusion proteins, tagged either at the amino or carboxyl terminus of the host protein, should be generated. The function of these chimeras can then be compared with the localization and known biochemical aspects of the endogenous protein. For example, does the GFP signal co-localize with endogenous antibody-labeled protein in fixed cells? Does the GFP variant potentiate or antagonize pathways in the same way that the overexpressed wild-type protein does? Does the GFP variant co-immunoprecipitate with the same proteins as the wild-type? Can the GFP variant rescue the wild-type phenotype of a deletion mutant? In a few cases, intervening spacer peptides (e.g., 5–10 glycine residues) are introduced between the GFP and the host protein, but in general we have failed to detect any real effect of these linker peptides on the success of the resultant chimera. If the protein has a key regulatory region at its terminus adjacent to the GFP moiety, then a linker reigon may be more important.

3.2. Constructing a GFP Chimera

The following assumes a knowledge of basic molecular biology procedures. If further information on this is required, please refer to any standard laboratory molecular biology guide, e.g. *(26)*. (*See* **ref. 27** for a more in-depth discussion on the construction of GFP chimeras.)

1. Clone the gene of interest in-frame into the GFP expression vector such that when transcribed, the mRNA sequence will encode an uninterrupted codon sequence. The Clontech mammalian enhanced GFP expression vectors (pEGFPs) are available with GFP in all three potential reading frames (numbered 1 to 3) relative to the multiple cloning site, making this as straightforward as possible. Furthermore, these GFP expression vectors are available with the GFP gene either upstream or downstream of the multiple cloning site, allowing the gene of interest to be cloned either N-terminally (using pEGFP-N) or C-terminally (using pEGFP-C) relative to the GFP. Transferring the gene of interest from its original vector into the GFP expression vector is usually achieved by subcloning using complementary restriction sites. This will be limited by both the availability of restriction sites within the multiple cloning sites of the two vectors, and whether a certain restriction enzyme cuts within the coding sequence of the gene of interest. A useful Internet tool to check for this latter problem is called Webcutter and can be found at http://www.firstmarket.com/cutter/cut2.html. If no suitable restriction sites exist it may be necessary to use an intermediate shuttle vector or

to design PCR primers containing restriction sites and amplify the gene of interest prior to cloning. Introducing the necessary coding sequence into PCR primers will also be required if a linker peptide is to be introduced into the region between the protein and GFP moities (*see* **Subheading 3.1.**). If PCR is the preferred course of action it is highly recommended that the amplified gene product be sequenced to check for any errors introduced into the coding sequence.

2. Amplify plasmid DNA using bacterial cultures (26) and isolate using a method capable of achieving high purity DNA—for example, Midi/Maxi kits from Qiagen (*see* **Note 3**).

3.3. Cell Culture and Transfection

1. Two to three days before imaging, seed cells onto sterile 22-mm coverslips in 6-well tissue culture plates. Each well should contain one coverslip and approx 2 mL DMEM. Seed cells for an approximate confluency of 60% after 24 h (*see* **Note 5**).
2. Twenty-four hours after seeding, transfect cells with the GFP chimera plasmid DNA using an appropriate transfection method for the cell type used. We routinely use GeneJuice transfection reagent (Novagen) due to its ease of use, high transfection efficiency, and low cytotoxicity. We use 3 μL GeneJuice and 1 μg plasmid DNA per well. Contrary to the manufacturer's instructions, we usually dilute plasmid DNA in 100 μL serum-free DMEM (per well) prior to addition of GeneJuice. The DNA/transfection reagent mix is gently mixed and then incubated for 15 min at room temperature before addition to the medium in the well (*see* **Note 6**). It can then be left until the coverslip is imaged.
3. If the dynamics of the GFP fusion protein are to be studied following agonist stimulation, it is sometimes appropriate to serum-starve the cells for a period prior to imaging. This is usually done 24 h after transfection and up to 12 h before imaging. Cells are serum-starved by replacing the 10% fetal bovine serum-supplemented DMEM, with DMEM + 0.1% bovine serum albumin.
4. Depending on the levels of gene expression, cells are usually ready for imaging 24–48 h after transfection.

3.4. Imaging

1. Mount the coverslip in the custom holder supplied with the PerkinElmer Ultra*VIEW* using silicone grease applied to the edges of the rings where they come into contact with the coverslip to form a watertight seal.
2. Add 1 mL of imaging buffer to the chamber and mount the coverslip assembly into the heated stage of the inverted microscope.
3. Visualize the cells through a 63× oil-immersion lens (or other high-magnification lens) with high numerical aperture (*see* **Notes 7,8**).
4. Acquire images from cells expressing sufficient levels of GFP to allow reasonable image capture without excessive laser power; cells expressing high levels of GFP are best avoided, as these often show an abnormal physiology. We usually have our laser set to half-power or less, as this virtually eliminates bleaching but still allows sufficient illumination to capture images at a rate of up to 10 per second without the need for binning (which reduces the resolution). The optimum laser power level for any individual experiment will vary depending on the imaging system used and the levels of GFP expression. The images shown in **Fig. 2B** were taken from a medium-bright cell using maximum camera resolution (1024 × 1344 pixels), with camera sensitvity at minimum, using approximately half laser power. Our images are recorded digitally so it is usually necessary to subarray (image a region of interest from the total available image area) and/or increase binning and/or reduce the frame

capture rate to limit the size of the files generated. A single experiment can often generate tens of megabytes of data so this is an important consideration (*see* **Note 9**).

5. Perform suitable control experiments (*see* **Subheading 3.1.** for general considerations on the GFP chimera). It is very unlikely that there will be any problems with cytotoxicity caused by the GFP tag, although if this is suspected then transfect in the empty GFP vector to see if there are any adverse effects of GFP expression (the cell phenotype should remain entirely normal—except for being bright green). Expressing GFP alone is also a good initial test to check whether a particular cell type can express GFP to sufficiently high levels to image, and as a reference with which to compare tagged protein localization. We have found that the empty GFP vectors transfect into cells very readily; however, it should not be assumed that the GFP chimera will transfect with the same efficiency. Laser toxicity may become a problem at high laser power (usually seen as membrane blebbing and/or the cessation of cellular processes), although imaging sufficiently bright cells with a reasonably sensitive detection system should mean that this is not a problem. A simple test of untransfected cells at the same gain settings and laser power will show whether cellular autofluorescence (e.g., from NADH) contributes to the signals.

3.5. Post-Imaging Analysis

After images have been collected they are analyzed. For this purpose it is useful to have an "off-line" imaging system so that valuable time is not taken up from the system being used to record experimental data. Our off-line system is an identical computer to that which is attached to the Ultra*VIEW* and runs the same Ultra*VIEW* software (PerkinElmer). When analyzing the temporal aspects of a GFP-tagged protein, a sequence of images, representing a single confocal slice over a given period of time, will be played as a movie. The dynamics of the protein can then be assessed; regions of interest can be analyzed to calculate measurements of, for example, distance, surface area, minimum fluorescence intensity, maximum fluorescence intensity, mean fluorescence intensity and crucially how these parameters change over time.

Our off-line system is also equipped with Volocity software (Improvision Inc., Lexington, MA), which allows us to deconvolve images (i.e., process them to improve confocal image quality by reduction of out-of-focus fluorescence). Volocity software will reconstruct confocal *z*-sections into three-dimensional views, which can then be played as movies, making it possible to determine how objects change and develop in three dimensions over time. For these purposes it is necessary to acquire image stacks (sets of confocal images representing a section through the cell) with each stack captured as rapidly as possible so as to minimize the chance of fluorescent bodies moving between confocal sections. The actual number of confocal sections in a stack and the *z*-distance between these sections will vary depending on the experiment and the available data storage. If done correctly, a single stack of confocal images should provide a true representation of that section through the cell at that point in time. Playing back the stacks in sequence gives the three-dimensional view. Although this sounds wonderful in theory, in practice it is difficult and time-consuming to achieve. Stacks of confocal images generate large file sizes that take a long time to deconvolve and playing these stacks as movies is often a slow and laborious process. At the moment this technology seems to be pushing desktop computing power to its limits and we look forward to the days when this work can be done routinely.

4. Notes

1. The exact reason why the Ultra*VIEW* confocal microscope system is capable of generating fluorescent images with reduced phototoxicity and photobleaching compared to conventional laser scanning confocal microscopes is currently being debated (*see*, for example, http://listserv.acsu.buffalo.edu/cgi-bin/wa?S1=confocal). The microlenses have a high efficiency of light transmission and the Interline CCD cameras have a high detection efficiency; however, these are probably not the main reasons for reduced photobleach levels. Two explanations have been proposed: (a) A nonlinear relationship between photobleach/phototoxicity and energy of the exciting laser beam. With the Yokogawa spinning disk system, the fluorescence image results from multiple simultaneous points of excitation, whereas in standard confocals a single excitation point is scanned across the sample. Therefore the latter system must use a higher energy excitation beam to achieve the same average intensity over time, which may cause more photobleaching. (b) The Ultra*VIEW* confocal has less confocality compared to scanning confocals and by acquiring emitted light from a thicker optical section is able to use less excitation to achieve the same signal *(28)*.

2. When using plasmid preparations of GFP vectors that carry the kanamycin resistance gene and the f1 origin, an anomalous 500 bp has sometimes been noted. It has been suggested that this is small circular DNA, possibly single-stranded, generated as a byproduct of plasmid replication *(29)*. This does not interfere with ligating inserts into the vectors. Indeed, in our experience the Clontech GFP expression vectors are highly amenable.

3. Transfection efficiencies are greatly affected by the quality of the DNA preparation. We use DNA anion-exchange resin-based kits (Qiagen or Sigma) to ensure minimal bacterial endotoxin contamination.

4. We would highly recommend the use of a dedicated room in which to put imaging equipment. Importantly, ambient light conditions should be controllable, as should ambient air temperature. In operation, the majority of imaging systems will generate a lot of excess heat, and thus some kind of air conditioning should be in place. This is an important consideration not only for the comfort of those using the equipment, but also for the correct operation and longevity of the equipment itself.

5. Using the methods described we have imaged many cell lines, including HeLa, HEK293, PC12, CHO, COS, and MEG-01 cells. If the cells are not attaching very well to the surfaces of the coverslips, reagents such as poly-L-lysine and poly-L-ornithine (Sigma) can be used to coat the coverslip surface and enhance cell adhesion.

6. Low transfection efficiency is the bane of our lives. It is highly desirable to have a large number of cells available to image so that the best ones can be selected (you do not want to have to image a cell expressing vast amounts of GFP, blebbing at the plasma membrane, and generally looking abnormal just because it's the only one you can find). Reduced transfection efficiency can be due to a number of factors including the cell type used (for example, PC12 cells transfect at a particularly low level), the quality of plasmid DNA, the trans fection method employed, the passage number of the cell culture, and contamination of the cell culture. If low transfection efficiency is a problem, ensure that the expression vector and transfection method are appropriate for the cell type and use methods of obtaining high-quality plasmid DNA with minimal endotoxin contamination (*see* **Note 3**). If cells

become difficult to transfect after having previously transfected efficiently, start a new cell culture from a frozen stock and replace suspect media.

7. Photobleaching is always a problem when imaging with GFP and its spectral variants (particularly BFP). To minimize photobleaching, image bright cells using low laser power (though *see* **Subheading 3.4.**).

8. A problem we often encounter is focal drift, where the imaged cell gradually moves out of the focal plane during the course of an experiment. To minimize this we ensure that the coverslip assembly is firmly positioned in the heated stage. It is also useful to weigh down the heated stage, e.g. with lead pots.

9. Recording digital confocal images requires large amounts of available storage. The computer attched to our Ultra*VIEW* is equipped with a 100-GB hard disk (large by current standards). However, this storage capacity quickly fills up as experiments are recorded. We have installed a DVD rewriter to provide an inexpensive means of archiving the data (each disk has a 4.7-GB capacity). Scrupulous management of the hard disk is nonetheless required.

References

1. Shimomura, O., Johnson, F. H., and Saiga, Y. (1962) Extraction, purification and properties of aequorin, a bioluminescent protein from the luminous hydromedusan. *J. Cell. Comp. Physiol.* **59,** 223–239.

2. Blinks, J. R., Mattingly, P. H., Jewell, B. R., van Leewen, M., Marrer, G. C., and Allen, D. (1978) Practical aspects of the use of aequorin as a calcium indicator: Assay, preparation, microinjection, and interpretation of signals. *Methods Enzymol.* **57,** 292–328.

3. Tsien, R. Y. (1998) The green fluorescent protein. *Annu. Rev. Biochem.* **67,** 509–544.

4. Prasher, D. C., Eckenrode, V. K., Ward, W. W., Prendergast, F. G., and Cormier, M. J. (1992) Primary structure of the *Aequorea victoria* green-fluorescent protein. *Gene* **111,** 229–233.

5. Chalfie, M., Tu, Y., Euskierchen, G., Ward, W. W., and Prasher, D. C. (1994) Green fluorescent protein as a marker for gene expression. *Science* **263,** 802–805.

6. Inouye, S. and Tsuji, F. I. (1994) Aequorea green fluorescent protein. Expression of the gene and fluorescence characteristics of the recombinant protein. *FEBS Lett.* **341,** 277–280.

7. Morise, H., Shimomura, O., Johnson, F. H., and Winant, J. (1974) Intermolecular energy transfer in the bioluminecent system of *Aequorea. Biochemistry* **13,** 2656–2662.

8. Perozzo, M. A., Ward, K. B., Thompson, R. B., and Ward, W. W. (1988) X-ray diffraction and time-resolved fluorescence analyses of *Aequorea* green fluorescent protein crystals. *J. Biol. Chem.* **263,** 7713–7716.

9. Ormo, M., Cubitt, A. B., Kallio, K., Gross, L. A., Tsien, R. Y., and Remmington, S. J. (1996) Crystal structure of the *Aequorea victoria* green fluorescent protein. *Science* **273,** 1392–1395.

10. Yang, F., Moss, L. G., and Phillips, G. N., Jr. (1996) The molecular structure of green fluorescent protein. *Nat. Biotechnol.* **14,** 1246–1251.

11. Phillips, G. N., Jr. (1997) Structure and dynamics of green fluorescent protein. *Curr. Opin. Struct. Biol.* **7,** 821–827.

12. Cody, C. W., Prasher, D. C., Westler, W. M., Prendergast, F. G., and Ward, W. W. (1993) Chemical structure of the hexapeptide chromophore of the Aequorea green-fluorescent protein. *Biochemistry* **32,** 1212–1218.

13. Heim, R., Prasher, D. C., and Tsien, R. Y. (1994) Wavelength mutations and post-translational autoxidation of green fluorescent protein. *Proc. Natl. Acad. Sci. USA* **91,** 12,501–12,504.

14. Heim, R., Cubbit, A. B., and Tsien, R. Y. (1995) Improved green fluorescence. *Nature* **373**, 663–664.
15. Zolotukhin, S., Potter, S. M., Hauswirth, W. W., Guy, J., and Muzyczka, N. (1996) A "humanized" green fluorescent protein cDNA adapted for high-level expression in mammalian cells. *J. Virol.* **70**, 4646–4654.
16. Patterson, G. H., Knobel, S. M., Sharif, W. D., Kain, S. R., and Piston, D. W. (1997) Use of the green fluorescent protein and its mutants in quantitative fluorescence microscopy. *Biophys. J.* **73**, 2782–2790.
17. Ward, W. W. (1997) Biochemical and physical properties of green fluorescent protein, in *Green Fluorescent Protein: Properties, Applications, and Protocols* (Chalfie, M. and Kain, S., eds.), John Wiley and Sons, New York, pp. 45–75.
18. Rutter, G. A., Kennedy, H. J., Wood, C. D., White, M. R. H., and Tavare, J. M. (1998) Quantitative real-time imaging of gene expression in single cells using multiple luciferase reporters. *Chem. Biol.* **5**, R2850–R290.
19. Delgrave, S., Hawtin, R. E., Silva, C. M., Yang, M. M., and Youvan, D. C. (1995) Red-shifted excitation mutants of the green fluorescent protein. *Biol. Technol.* **13**, 151–154.
20. Matz, M. V., Fradkov, A. F., Labas, Y. A., Savitsky, A. P., Zaraisky, A. G., Markelov, M. L., et al. (1999) Fluorescent proteins from non-bioluminescent Anthozoa species. *Nat. Biotechnol.* **17**, 969–973.
21. Cotlet, M., Hofkens, J., Habuchi, S., Dirix, G., Van Guyse, M., Michiels, J., et al. (2001) Identification of different emitting species in the red fluorescent protein DsRed by means of ensemble and single-molecule spectroscopy. *Proc. Natl. Acad. Sci. USA* **98**, 14,398–14,403.
22. Garcia-Parajo, M. F., Koopman, M., van Dijk, E. M. H. P., Subramaniam, V., and van Hulst, N. F. (2001) The nature of fluorescence emission in the red fluorescent protein DsRed, revealed by single-molecule detection. *Proc. Natl. Acad. Sci. USA* **98**, 14,392–14,397.
23. Terskikh, A. V., Fradkov, A. F., Zaraisky, A. G., Kajava, A. V., and Angres, B. (2002) Analysis of DsRed Mutants. Space around the fluorophore accelerates fluorescence development. *J. Biol. Chem.* **277**, 7633–7636.
24. Gurskaya, N. G., Fradkov, A. F., Terskikh, A., Matz, M. V., Labas, Y. A., Martynov, V.I., et al. (2001) GFP-like chromoproteins as a source of far-red fluorescent proteins. *FEBS Lett.* **507**, 16–20.
25. Bobe, R., Wilde, J. I., Maschberger, P., Wenkateswarlu, K., Cullen, P. J., Siess, W., et al. (2001) Phosphatidylinositol 3-kinase-dependent translocation of phospholipase Cγ2 in mouse megakaryocytes is independent of Bruton tyrosine kinase translocation. *Blood* **97**, 678–684.
26. Sambrook, J., Fritsch, E. F., and Maniatis, T. (1989) *Molecular Cloning: A Laboratory Manual.* Cold Spring Harbor Laboratory Press, Cold Spring Harbor, NY, Vol. 1, 2, 3.
27. Wilson, L., Sullivan, K. F., and Matsudaira, P., eds. (1998) *Methods in Cell Biology: Green Fluorescent Proteins*, Academic Press, New York.
28. Reichelt, S. and Amos, W. B. (2001) SELS: a new method for laser scanning microscopy of live cells. *Microscopy and Analysis* **86**, 9–11.
29. http://www.clontech.com.
30. Tavaré, J. M., Fletcher, L. M., and Welsh, G. I. (2001) Review: Using green fluorescent protein to study intracellular signaling. *J. Endocrinol.* **170**, 297–306.
31. Kraulis, P. J. (1991) MOLSCRIPT: A program to produce both detailed and schematic plots of protein structures. *J. Appl. Crystallogr.* **24**, 946–950.

26

Two-Dimensional Polyacrylamide Gel Electrophoresis for Platelet Proteomics

Katrin Marcus and Helmut E. Meyer

1. Introduction

Blood platelets are important components of hemostasis, contributing to healing of wounds by forming thrombi and to the initiation of repair processes. They are also involved, however, in the pathogenesis of life-threatening complications such as stroke or myocardial infarction. Following injuries to blood vessels, platelets adhere to the damaged vessel wall, resulting in the formation of vascular plugs and release of intracellular substances, which initiate repair processes. Genetic defects may result in dysfunction of the platelets, inducing bleeding diseases such as Glanzmann thrombasthenia and Bernard-Soulier syndrome (1). Platelets arise as fragments of megakaryocytes, and are anucleate. Hence only restricted synthesis of proteins from residual megakaryocyte and mitochondrial mRNA may be possible, and therefore genome and transcriptome analysis of platelets is a substantial challenge. The clinical relevance of platelet dysfunctions and increased knowledge of the intracellular processes involved, make analyses of platelet proteome potentially very valuable.

The term "proteome" was first introduced in 1996 by Wilkins et al. (2). It was designed to denote the protein complement of a genome. The proteome indicates the quantitative expression profile of a cell, an organism, or a tissue under exactly defined conditions. In contrast to the temporally constant genome, the proteome, being dependent on intracellular and extracellular parameters, is dynamic and variable (3,4) (see **Fig. 1**). The analysis of a proteome represents an important supplementation to genome analysis. The human genome encodes in the order of 100,000 proteins, although the number of genes in the genome is substantially lower, with current estimates of around 38,000. It now well appreciated that one gene may produce a variety of protein products in consequence of alternative splicing of the RNA and post-translational modifications. An additional introduction of post-translational modifications, e.g., sugars and phosphates or controlled proteolysis, contributes to a complex protein pattern. Additionally,

From: *Methods in Molecular Biology, vol. 273:*
Platelets and Megakaryocytes, Vol. 2: Perspectives and Techniques
Edited by: J. M. Gibbins and M. P. Mahaut-Smith © Humana Press Inc., Totowa, NJ

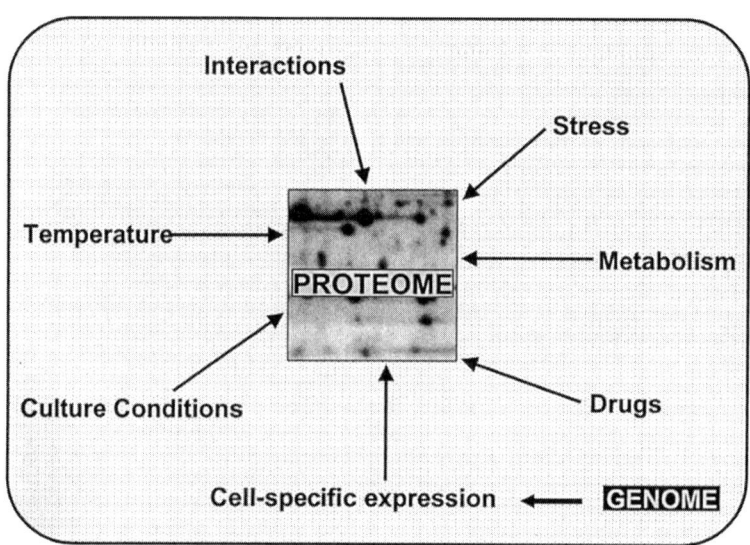

Fig. 1. Influence on the proteome. One important task of proteome analysis is the characterization and quantification of preferably all proteins expressed in a cell or tissue at a given time point. The proteome represents a dynamic and flexible variable and depends on a variety of inner and outer parameters, such as interactions, stress, metabolism, temperature, and the genome. Hence an exact definition of the starting conditions is essential.

there is only a poor correlation between mRNA and the corresponding gene product levels *(5)*. Consequently, proteome analysis must incorporate analysis of protein expression patterns, protein quantitation, and post-translational modifications.

The description of a biological system requires the characterization of a complex mixture of thousands of proteins. To get a wide and preferably complete overview of the proteins present in the respective cell system, an analysis strategy must be established. A broad consensus exists regarding the value of proteome analysis incorporating two-dimensional polyacrylamide gel electrophoresis (2D-PAGE) *(6,7)*.

The most important steps in the proteome analysis are:

1. *The design of a proteome study:* As the protein expression rate is influenced by different parameters, such as cell stress, temperature, and interactions (*see* **Fig. 1**), an exact definition of the proteome study and the starting conditions prior to the analysis is necessary.

    Generally, the aim of a proteome study includes the comparison between two different states of a specific cell type, tissue, or organism. Depending on the particular problem under investigation, differences in the protein expression profile may lead to the identification of biological mechanisms that underlie given cellular changes. Reproducible and reliable results can be obtained only when the starting conditions are well defined, and the variance in the protein pattern is due to the intended manipulation.

2. *The effective solubilization and separation of preferably all proteins of one proteome via two-dimensional polyacrylamide gel electrophoresis (2D-PAGE) and visualization of the proteins:* An efficient and reproducible sample preparation and protein separation of preferably all proteins present in the sample is essential to enable reliable analysis of results. In this regard effective and quantitative solubilization of cellular proteins is of paramount importance. Any additional manipulation during the sample preparation holds a risk of the loss of protein or artifactual modification, resulting in artificial differences in the 2D-protein pattern, and therefore great care is required during analysis. 2D-PAGE, a combination of iso-electric focusing in the first dimension and SDS-PAGE in the second dimension, represents the most established technique for the separation of complex protein mixtures. Up to 10,000 protein species may be resolved and qualitatively and quantitavely analyzed using image-analysis computer software. There are two different methods of separation technique in the first dimension (isoelectric focusing): the method of O'Farrell *(8)* and Klose *(9)*, where a pH gradient is formed via carrier ampholytes during the focusing process, and the method of Görg et al. *(10)* where immobilized pH gradients are used. Proteins separated in this way are then applied to SDS-PAGE for further separation by molecular mass.

   After separation the proteins must be visualized using dyes that are compatible with subsequent mass spectrometric analysis methods. Different dyes and staining procedures are described to be compatible with mass spectrometry, such as Coomassie R-250 *(11)*, Coomassie G-250 (colloidal Coomassie) *(12,13)*, silver *(14)*, and different fluorescent dyes *(15,16)*.

3. *The differential analysis and quantification of the proteins by image analysis systems:* For the identification and characterization of the proteins of interest, the exact analysis of the visualized 2D-protein pattern is necessary. Several commercial software packages are available that are designed to provide quantitative and qualitative information from 2D-gels. The fundamental steps of this gel analysis are the digitalization of the gels, the matching of differential protein spots, the comparison of the gels, and the data analysis *(17,18)*. The digitalization takes place via camera systems, phosphorimagers, fluorescence scanners, and densitometers, depending on the staining method used. Digitalized protein patterns are analyzed by using algorithms to detect the protein spots, calculating their intensity, and defining their coordinates in the gel. Both quantitative differences and individual displacements of the spots can be measured by comparing various gels.

4. *The identification and characterization of the proteins and detection of post-translational modifications:* After detection the proteins of interest are identified and characterized either by Edman sequencing or by mass spectrometric methods. In protein and peptide analysis, matrix-assisted laser desorption/ionization mass spectrometry (MALDI-MS) *(19)* or electrospray ionization mass spectrometry (ESI-MS) *(20)* are especially important. The main advantages of MS over other methods in protein identification lies in a short analysis time, the high sensitivity (up to the attomole range), and the potential for automation.

5. *Data analysis:* The received data of the MS-spectra are evaluated automatically via search algorithms comparing measured data with theoretically estimated data from protein and DNA databases. The data evaluation may be performed using different search programs, which are freely accessible on the Internet, e.g., ProFound (http://prowl.rockefeller.edu), Mascot (www.matrixscience.com), MS-Fit (http://prospector.ucsf.edu), or SEQUEST (distributed by Finnigan Corp.). To date only a few results of platelet proteomics have been published, giving an overview of proteins present in platelets *(21–23)*.

In this chapter, methods are described that are optimized for the separation of the platelet proteome by 2D-PAGE.

2. Materials

2.1. Equipment

2.1.1. Sample Preparation

1. Centrifuge.
2. Sonication bath.

2.1.2. 2D-PAGE

1. Horizontal isoelectric focusing apparatus (e.g., Multiphor II Electrophoresis Unit, or IPGphor, Amersham Biosciences, Freiburg, Germany).
2. Power supply (e.g., Amersham Biosciences).
3. Thermostatic circulator (e.g., Amersham Biosciences).
4. Immobiline DryStrips (IPG strips) (different pH gradients, e.g., Amersham Biosciences).
5. Reswelling cassette (e.g., Amersham Biosciences).
6. Electrode strips (e.g., Amersham Biosciences).
7. Sample cups and sample cup holders (e.g., Amersham Biosciences).
8. Glass plates and spacers.
9. Grease (e.g., Roth, Life Sciences, Karlsruhe, Germany).
10. Gradient mixer.
11. Gel multicasting chamber (e.g., Anderson-IsoDalt Electrophoresis Systems, Hoefer Pharmacia Biotech Inc., San Francisco, CA).
12. Vertical SDS-PAGE unit (e.g., Anderson-IsoDalt Electrophoresis Systems, Hoefer Pharmacia Biotech Inc.).
13. Gel dryer (e.g., Model 583 Gel Dryer, Bio-Rad Laboratories, Hercules, CA).

2.2. Buffers and Solutions

2.2.1. Platelet Preparation (see **Notes 1–3**)

1. ACD: 85 mM trisodium citrate, 70 mM citric acid, 110 mM glucose.
2. Tyrode's buffer (pH 7.4): 10 mM HEPES, 137 mM NaCl, 2.7 mM KCl, 1 mM EDTA, 5 mM glucose.
3. Washing buffer (pH 6.5): 36 mM trisodium citrate, 90 mM NaCl, 5 mM KCl, 10 mM EDTA, 5 mM glucose, 0.1 mM aspirin (see **Note 4**).
4. Lysis solution: 7 M urea, 2 M thiourea, 2% (w/v) CHAPS, 0.5% (v/v) Servalyte (3–10) (Serva Feinbiochemica GmbH & Co., Heidelberg, Germany), 100 mM dithiothreitol (DTT), 2 M sodium orthovanadate, 20 μM ZnCl$_2$, 1 μM ocadaic acid, 1/10 of a tablet Complete Mini™ protease inhibitor mixture (Roche Diagnostics GmbH, Mannheim, Germany), or equivalent. To prepare 50 mL of the solution, dissolve urea in small portions in 30 mL of deionized water, and make volume up to to 50 mL. Add 0.5 g of mixed-bed ion-exchanger resin (e.g., Amberlite IRN 150, Sigma Aldrich), stir for 2 min and filter. Add thiourea, CHAPS, and Servalyte to 48 mL of this solution and make volume up to 50 mL. Small aliquots (1 mL) can be stored at –80°C. DTT should be added on the day that the buffer is used. Solution thawed once should not be refrozen (see **Note 5**).

Table 1
Summary of Conditions for 2D-PAGE for Different pH Gradients

pH	Rehydration solution	IEF	SDS-PAGE
3–10	8 M urea 2 M thiourea 2% CHAPS 100 mM DTT 0.8% Servalyte (3–10)	Anodic sample application Cover with oil 80 kVh	Equilibration time: 15 min Gel: 11% T; 2.5% C 240 min, 350 mA
4–7	8 M urea 2 M thiourea 2% CHAPS 100 mM DTT 0.8% Servalyte (4–7)	Anodic sample application Cover with oil 80 kVh	Equilibration time: 15 min Gel: 11% T; 2.5% C 240 min, 350 mA
6-11	8 M urea 2 M thiourea 2% CHAPS 100 mM DTT 0.4% Servalyte (6–9, 9–11)	Anodic sample application Cover with oil 30 kVh	Equilibration time: 20 min Gel: 11% T; 2.5% C 240 min, 350 mA
3.5–4.5	8 M urea 2 M thiourea 4% CHAPS 100 mM DTT 0.8% Servalyte (4–7)	Cathodic sample application Cover with oil 75 kVh	Equilibration time: 15 min Gel: 11% T; 2.5% C 240 min, 350 mA
4.5–5.5	8 M urea 2 M thiourea 4% CHAPS 100 mM DTT 0.8% Servalyte (3–10)	Centric sample application Cover with oil 75 kVh	Equilibration time: 15 min Gel: 11% T; 2.5% C 240 min, 350 mA
5.5–6.7	8 M urea 2 M thiourea 2% CHAPS 100 mM DTT 0.8% Servalyte (3–10)	Anodic sample application Cover with oil 75 kVh	Equilibration time: 15 min Gel: 11% T; 2.5% C 240 min, 350 mA

T, total acrylamide concentration; C, crosslinker concentration.

2.2.2. 2D-PAGE

2.2.2.1. ISOELECTRIC FOCUSING (IEF)

Rehydration solution: Refer to **Table 1** for various buffer compositions.

2.2.2.2. SDS-PAGE

1. Equilibration buffer (pH 8.8): 6 M urea, 50 mM Tris-HCl, 2% (w/v) SDS, 30% (w/v) glycerin.
2. Running buffer (pH 8.8): 50 mM Tris-HCl, 0.1% (w/v) SDS, 1.92 M glycerin.
3. Separation buffer (pH 8.8): 1.5 M Tris-HCl, 0.4% (w/v) SDS.

4. Agarose solution: 0.3% (w/v) agarose (in running buffer), 0.1% (w/v) bromophenol blue.
5. Gel solution (for one 11% gel 25 × 20 cm): 25 mL separation buffer, 29 mL deionized water, 31 mL acrylamide (24:1, 30% T, 2.6% C), 55 μL ammonium persulfate solution (40% (w/v)), 46 μL *N,N,N',N'*-tetramethylethylenediamine (TEMED).
6. Water-saturated isopropanol.

3. Methods
3.1. Preparation of Platelets and Platelet Protein Extracts

1. Mix 80 mL of freshly taken blood with 16 mL ACD.
2. Remove remaining erythrocytes by centrifugating the concentrate for 20 min at 250*g* at room temperature (RT).
3. Platelets in the supernatant (platelet-rich plasma, PRP) are obtained by centrifugation (700*g*, 12 min, RT).
4. Resuspend the pellet gently in 2 mL Tyrode's buffer, and make volume up to 10 mL with washing buffer.
5. After 1 h at RT repeat the centrifugation step.
6. Resuspend the pellet in 4 mL Tyrode's buffer, and incubate it for 30 min at RT.
7. Spin down the platelets at 5000*g* and lyse them in 600 μL of lysis solution (to a concentration of about 15 μg/μL).
8. Complete solubilization of the proteins is achieved by sonicating the sample for 10 min at 4°C (*see* **Note 6**).
9. Freeze samples immediately and store at –80°C.
10. Before using the samples for IEF, mix an aliquot (for Coomassie-stained gels, about 500 μg of protein) with the same volume of rehydration buffer and incubate in the sonication bath at 4°C (four times for 15 s each with rest periods of 1 min) (*see* **Note 6**).

3.2. 2D-PAGE

3.2.1. IEF

The first dimension, where the proteins are separated by charge, is performed using immobilized pH gradients according to the method described by Görg et al. *(10)*. Sample application is achieved using sample application cups (*see* **Note 7**).

1. Prior to IEF, rehydrate the IPG gel strips overnight at RT in the respective rehydration solution (*see* **Table 1**). For 18-cm strips a volume of 350 μL solution is used.
2. Rinse the rehydrated strips with deionized water for a few seconds and carefully remove excess water with a filter paper.
3. Place the strips, gel side up, in the aligned tray. The basic ends of the strip must be faced toward the cathode.
4. Place two electrode strips (or filter paper, respectively) soaked with water at the ends of the strips, one at the basic and one at the acidic end.
5. Position the electrodes and press them gently on the top of the electrode strips.
6. Apply sample cup holder on top of the strips (basic, acidic end, or centric; *see* **Table 1**) depending on the pH gradient used.
7. Fix sample cups at the sample cup holders and pipet the sample into the sample cups.
8. Place the tray on the cooling plate.

9. Cover the strips and the samples with silicone oil.
10. Run the IEF with following voltage gradient: For improved sample entry, the voltage at the beginning of the gradient is limited to 200 V. The focusing time depends on the IPG strip used (*see* **Table 1**). The optimum focusing temperature is 20°C *(24)*. 1 h at 200 V; 1 h at 500 V; 1 h at 1000 V; 1 h at 2000 V; and the rest of the time at 3500 V.
11. Strips of different pH gradients can be run in parallel, keeping in mind the final kVh.
12. Following focusing, strips can be stored at –80°C up to several months.

3.2.2. SDS-PAGE

In the second dimension, the proteins are separated by exploiting their molecular mass differences. Therefore proteins must be loaded with SDS after the IEF. Additionally, disulfide bonds are reduced and alkylated by the addition of DTT and iodoacetamide to safeguard complete denaturation of the proteins. SDS-PAGE may be run horizontally or vertically *(25)*. For our application vertical gels are chosen in the second dimension (*see* **Note 8**).

3.2.2.1. PREPARING THE MULTICASTING CHAMBER

1. Prior to the assembly of the casting chamber, carefully clean the glass plates with deionized water and ethanol/water.
2. Dry the glass plates with lint-free tissue (e.g., Kim Wipe).
3. Stick two spacers (one on each side) on the plate using small amounts of grease.
4. Place the second glass plate on the spacers.
5. Place the glass plates in the casting chamber. Each set of glass plates should be separated from the next by thin plastic sheets and after each second set of sheets a thicker (about 0.5 cm) plastic plate should be introduced (*see* **Note 9**).

3.2.2.2. PREPARING THE GEL SOLUTION

Before pouring the gels, the exact gel volume for one gel must be estimated. In our experiments, for preparing $20 \times 25 \times 1$ cm gels with 11% T (total acrylamide concentration) in the Anderson-IsoDalt® multicasting chamber a volume of 76 mL gel-solution is required (*see* **Subheading 2.2.**).

1. Mix the solution thoroughly.
2. Pour the gel solution bottom-up into the chamber, using a gradient mixer to prevent the creation of air bubbles (*see* **Note 9**).
3. Allow the solution to fill the chamber until there is a gap of about 0.5 cm at the top of the glass plates.
4. Do not allow air bubbles to enter the chamber.
5. Seal the tube using a clamp and remove the tube from the gradient mixer.
6. Overlay the gel surface with water-saturated isopropanol to exclude air and ensure a level surface on the top of the gel.
7. Allow polymerization for at least 2 h at RT.
8. The gels can be stored up to 1 wk at 4°C.
9. Prior to electrophoresis, wash the gel surface carefully with deionized water.

3.2.2.3. Equilibration of IPG Strips and Electrophoresis

1. After IEF or storage at –80°C equilibrate the strips in two consecutive steps, each time for 15 or 20 min respectively (*see* **Table 1**). In the first step 65 m*M* DTT is added to the equilibration buffer (for the reduction of proteins) and in the second step 280 m*M* iodoacetamide (for the alkylation of proteins) is introduced (*see* **Note 10**).
2. Rinse the strip with deionized water and gently dry with Whatman paper to remove excess solution.
3. Fill the gap on the top of the gel with the molten agarose solution.
4. Immediately transfer the equilibrated strip onto the top of the gel using clean tweezers and spatulas (*see* **Note 11**).
5. Insert the gel cassette into the electrophoresis chamber and run the gels for 20 min at 50 mA and 3 h at 350 mA (*see* **Note 12**).
6. The proteins may be detected by any of the staining techniques applicable to SDS-PAGE analysis (*see* **Note 13**).
7. After staining, the gels should be dried to ensure improved stability of the gels and to avoid protein loss.

4. Notes

1. For the preparation of every solution and buffer, use chemicals of the highest quality. In every step clean vials and equipment should be used to prevent the contamination of the sample.
2. Gloves should be worn in every preparation step to reduce the risk of contamination by keratins.
3. Do not use glass vials when working with platelets, as this may cause unintentional activation.
4. Washing buffer contains citrate and EDTA at a higher concentration to avoid platelet activation during the centrifugation steps.
5. Never heat urea solutions above 37°C to reduce the risk of protein carbamylation.
6. Sonication steps must be performed on ice and should be as short in duration as possible to prevent the degradation of proteins.
7. Sample can be applied using sample cups or by in-gel sample rehydration. In the technique described by Goerg et al. (*10*), the sample is introduced in the rehydration solution and directly soaked into the strip during the overnight rehydration step. In our experience much better results are obtained when using sample cups (*see* **Fig. 2**). After in-gel rehydration a strong horizontal smear is observed, caused by actin. Best results were achieved when applying the sample with sample cups near the isoelectric point of the main protein component (actin). Diluting the sample with rehydration solution and covering it with silicone oil minimized protein precipitation at the application point (*see* **Fig. 3**).
8. Generally vertical gels are larger and therefore provide a better resolution. Additionally, multiple vertical gels can be run in large tanks in parallel. Parallel gel preparation and simultaneous electrophoresis ensure a better reproducibility between experiments.
9. The introduction of air bubbles can be minimized by separating the glass plates by thicker plastic sheets. Additionally, the gel solution should be poured into the chamber bottom up. The simplest way of doing this is using a gradient mixer.
10. In our experience strips of the pH gradient 6.0–11.0 should be equilibrated for 20 min in each step to ensure complete loading with SDS of the basic proteins (*see* **Fig. 4**).

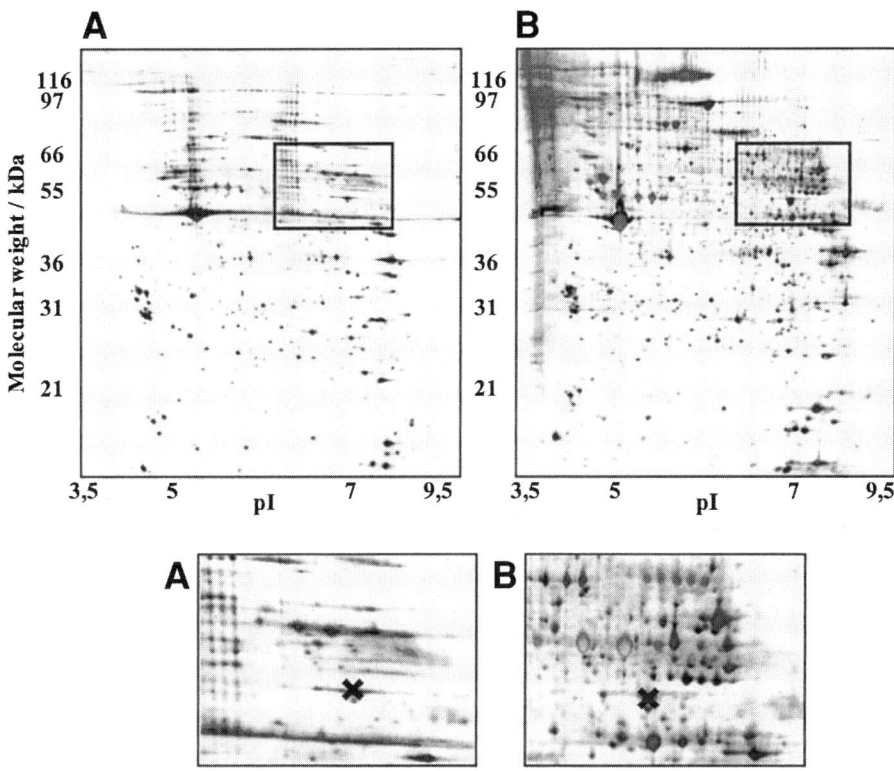

Fig. 2. Influence of sample application method on IEF. (**A**) 2D-gel (pH 3.0–10.0, 11% T, silver staining) of 150 μg platelet protein after in-gel rehydration of the sample. (**B**) 2D-gel (pH 3.0–10.0, 11% T, silver staining) of 150 μg platelet protein after sample application with sample cups. An improved separation caused by less precipitation effects was detected mainly in the upper-molecular-weight region. (**X**) Corresponding spots in both gels.

11. The agarose solution must be hot when loading; otherwise, it will set quickly.
12. The protein transfer from the IPG strip to the SDS-gel is enhanced when starting the electrophoresis with low voltage. Second-dimension gels of different dimensions will require the application of different total voltage and running times.
13. Some protocols for silver staining of SDS-gels prior to MS analyses are described *(14)*. By our experience Coomassie staining is more effective prior to MS. Coomassie-stained gels are particularly suitable for image analyses due to their high dynamic range. For both purposes fluorescence dyes are particularly suitable, but the instrumentation for visualization is very expensive. Coomassie G-250 is much more sensitive (about 10 ng) than Coomassie R-250 staining (about 25–500 ng).

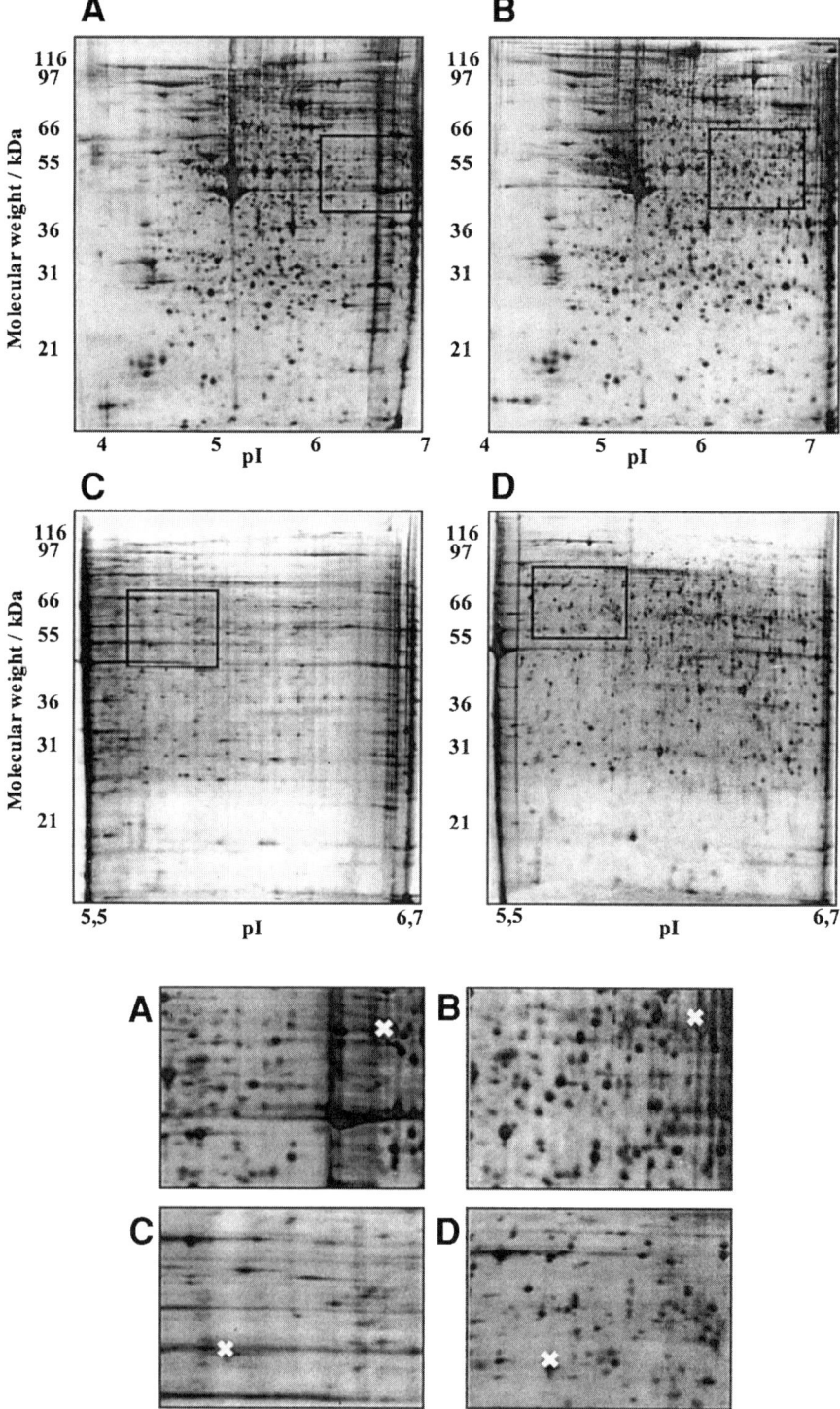

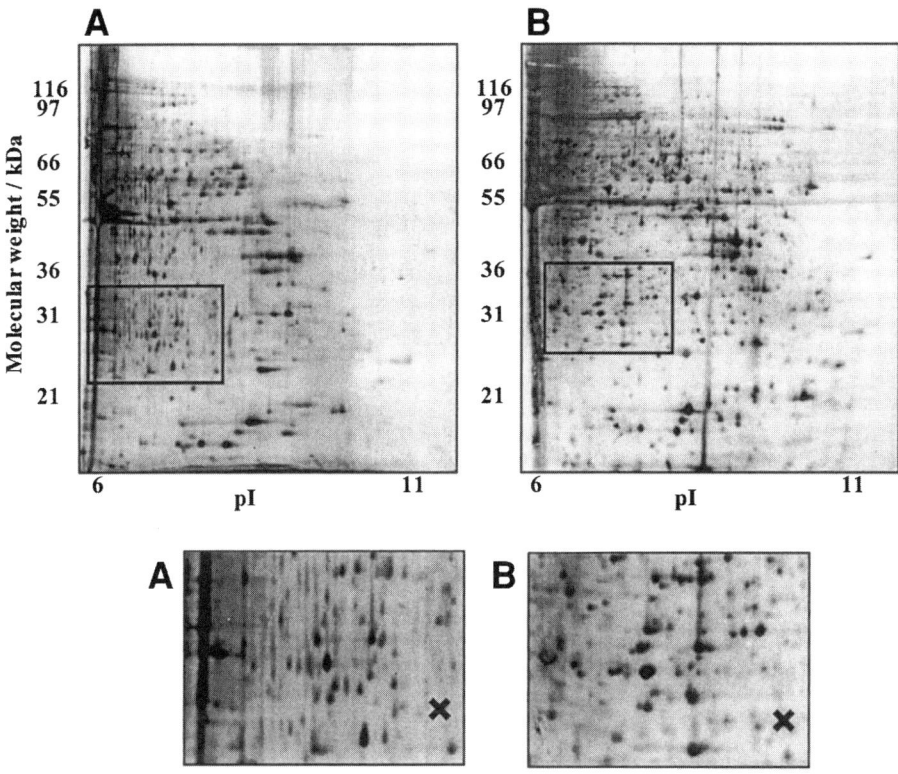

Fig. 4. Influence of equilibration time on IEF. Sample application was performed using sample cups. (**A**) 2D-gel (pH 6.0–11.0, 11% T, silver staining) of 250 μg platelet protein after 15 min equilibration time. (**B**) 2D-gel (pH 6.0–11.0, 11% T, silver staining) of 250 μg platelet protein after 20 min equilibration time. Lower panels A and B correspond to close-up views of the boxed regions in A and B above. An improved separation with sharp and clear protein spots was obtained by prolongation of the equilibration time. (**X**) Corresponding spots in both gels.

Fig. 3. *(see opposite page)* Influence of sample application point on IEF. (**A**) 2D-gel (pH 4.0–7.0, 11% T, silver staining) of 250 μg platelet protein after cathodic (at pH 6.8) sample application with sample cups. (**B**) 2D-gel (pH 4.0–7.0, 11% T, silver staining) of 250 μg platelet protein after anodic (at pH 4.4) sample application with sample cups. (**C**) 2D-gel (pH 5.5–6.7, 11% T, silver staining) of 250 μg platelet protein after cathodic (at pH 6.6) sample application with sample cups. (**D**) 2D-gel (pH 5.5–6.7, 11% T, silver staining) of 250 μg platelet protein after anodic (at pH 4.5–6.0) sample application with sample cups. Lower panels A–D correspond to close-up views of the boxed regions in A–D above. An improved separation was detected mainly in the basic and upper-molecular-weight region. (**X**) Corresponding spots in both gels.

References

1. Nurden, A. T. (1999) Inherited abnormalities of platelets. *Thromb. Haemost.* **82,** 468–480.
2. Wilkins, M., Sanchez, J. C., Gooley, A. A., Appel, R. D., Humphrey-Smith, I., Hochstrasser, D. F., et al. (1996) Progress with proteome projects: why all proteins expressed by a genome should be identified and how to do it. *Biotechnol. Genet. Eng. Rev.* **13,** 19–50.
3. Anderson, N. G. and Anderson, N. L. (1996) Twenty years of two-dimensional electrophoresis: past, present and future. *Electrophoresis* **17,** 443–453.
4. Kellner, R. (2000) Proteomics. Concepts and perspectives. *Fresenius J. Anal. Chem.* **366,** 517–524.
5. Gygi, S. P., Rochon, Y., Franza, B. R., and Aebersold, R. (1999) Correlation between protein and mRNA abundance in yeast. *Mol. Cell. Biol.* **19,** 1720–1730.
6. Lottspeich, F. (1999) Ein Genom—verschiedene Proteome. *Angew. Chemie* **111,** 2630–2647.
7. Blackstock, W. P. and Weir, M. P. (1999) Proteomics: quantitative and physical mapping of cellular proteins. *Tibtech.* **17,** 121–127.
8. O'Farrell, P. H. (1975) High resolution two-dimensional electrophoresis of proteins. *J. Biol. Chem.* **250,** 4007–4021.
9. Klose, J. (1975) Protein mapping by combined isoelectric focusing and electrophoresis of mouse tissues. A novel approach to testing for induced point mutation in mammals. *Humangenetik* **26,** 231–243.
10. Görg, A., Postel, W., and Günther, S. (1988) The current state of two-dimensional electrophoresis with immobilized pH-gradients. *Electrophoresis* **9,** 531–546.
11. Wilson, C. M. (1983) Staining of proteins on gels: comparisons of dyes and procedures. *Methods Enzymol.* **91,** 236–247.
12. Neuhoff, V., Arold, N., Taube, D., and Erhard, W. (1988) Improved staining of proteins in polyacrylamide gels including isoelectric focusing gels with clear background at nanogram sensitivity using Coomassie Brilliant Blue G-250 and R-250. *Electrophoresis* **9,** 255–262.
13. Doherty, N. S., Littman, B. H., Reilly, K., Swindell, A. C., Buss, J. M., and Anderson, N. L. (1998) Analysis of changes in acute-phase plasma proteins in an acute inflammatory response and in rheumatoid arthritis using two-dimensional gel electrophoresis. *Electrophoresis* **19,** 355–363.
14. Shevchenko, A., Wilm, M., Vorm, O., and Mann, M. (1996) Mass spectrometric sequencing of proteins silver-stained polyacrylamide gels. *Anal. Chem.* **68,** 850–858.
15. Lopez, M. F., Berggren, K., Chernokalskaya, M., Lazarev, V., Robinson, M., and Patton, W. F. (2000) A comparison of silver stain and SYPRO Ruby Protein Gel Stain with respect to protein detection in two-dimensional gels and identification by peptide mass profiling. *Electrophoresis* **17,** 3673–3683.
16. Yan, J. X., Harry, R. A., Spibey, C., and Dunn, M. J. (2000) Postelectrophoretic staining of proteins separated by two-dimensional gel electrophoresis using SYPRO dyes. *Electrophoresis* **17,** 3657–3665.
17. Dunn, M. J. (1992) The analysis of two-dimensional polyacrylamide gels for the construction of protein databases, in *Microcomputer in Biochemistry: A Practical Approach* (Bryce, C. F. A., ed.), IRL Press, Oxford, pp. 215–242.
18. Appel, R. D. and Hochstrasser, D. F. (1999) Computer analysis of 2-D images, in *2-D Proteome Analysis Protocols* (Link, A. J., ed.), Humana Press, Totowa, NJ, pp. 363–381.
19. Karas, M. and Hillenkamp, F. (1988) Laser desorption ionization of proteins with molecular mass exceeding 10,000 daltons. *Anal. Chem.* **60,** 2299–2301.
20. Fenn, J. B. (1989) Electrosprayionization for mass spectrometry of large biomolecules. *Science* **246,** 64–71.

21. Marcus, K., Immler, D., Sternberger, J., and Meyer, H. E. (2000) Identification of platelet proteins separated by two-dimensional gel electrophoresis and analyzed by matrix assisted laserdesorption/ionization-time of flight-mass spectrometry and detection of tyrosine-phosphorylated proteins. *Electrophoresis* **21,** 2622–2636.
22. O'Neill, E. E., Brock, C. J., von Kriegsheim, A. F., Pearce, A. C., Dwek, R. A., Watson, S. P., et al. (2002) Towards complete analysis of the platelet proteome. *Proteomics* **2,** 288–305.
23. Marcus, K. (2002) Analyse des Phosphoproteoms humaner Thrombin-stimulierter Thrombozyten. Dissertation, Ruhr-University Bochum.
24. Görg, A., Postel, W., Friedrich, C., Kuick, R., Strahler, J. R., and Hanash, S. M. (1991) Temperature-dependent spot positional variability in two-dimensional polypeptide patterns. *Electrophoresis* **12,** 653–658.
25. Görg, A., Boguth, G., Obermaier, C., Posch, A., and Weiss, W. (1995) Two-dimensional polyacrylamide gel electrophoresis with immobilized pH gradients in the first dimension (IPG-Dalt): The state of the art and the controversity of vertical versus horizontal systems. *Electrophoresis* **16,** 1079–1086.

27

Preparation of mRNA and cDNA Libraries From Platelets and Megakaryocytes

Benjamin Z. S. Paul, Jiango Jin, and Satya P. Kunapuli

1. Introduction

The challenge of working with platelet and megakaryocyte mRNA is contamination with leukocyte mRNA. When attempting to detect mRNA encoding platelet or megakaryocytic specific proteins using the extremely sensitive technique reverse transcription polymerase chain reaction (RT-PCR); for either mRNA analysis or cloning, the concern for a few contaminating leukocytes is greatly amplified. Platelets pose an additional challenge because they contain only small amounts of residual mRNA from the parent megakaryocyte. Therefore, molecular biology techniques that require large amounts of RNA, such as Northern hybridization, have limited application. Usually, RT-PCR is used to answer questions of gene expression or to provide material for cDNA libraries for the cloning of proteins.

The following protocols make use of magnetized polystyrene beads that are coated in a polyclonal antibody that is specific for the Fc region of a murine monoclonal IgG antibody. A murine monoclonal antibody that is specific for a cell-specific surface glycoprotein is used to target the particular cell type of interest. The magnetic polystyrene beads bind to surfaces of the target cells following the primary incubation with the murine monoclonal antibody and an additional incubation with the anti-murine-IgG coated magnetic beads. In the presence of a magnetic field all the cells that have two to four beads attached will be pulled out of suspension and held firmly to the side of the suspension vessel. This mechanism for cell selection is utilized in two ways. "Positive selection" is incorporated into a method so that the magnetic bead/target cell complex is pulled out of a population of cells (e.g., magnetic immunoselection of megakaryocytes). "Negative selection" utilizes a magnetic polystyrene bead that is coated in a mouse monoclonal IgG antibody specific for a cell surface glycoprotein. The contaminating cell binds to the magnetic beads and is subsequently removed. The

From: *Methods in Molecular Biology, vol. 273:*
Platelets and Megakaryocytes, Vol. 2: Perspectives and Techniques
Edited by: J. M. Gibbins and M. P. Mahaut-Smith © Humana Press Inc., Totowa, NJ

application of a negative selection protocol requires that the cell population of interest has already been substantially purified (e.g., magnetic immunopurification of platelets).

Following collection of platelet rich plasma (PRP), platelets are isolated using gradients prepared from Larex UF Powder (Larex Inc., St. Paul, MN, USA). The manufacturer's protocol utilizes a highly purified larch arabinogalactan to separate platelets from contaminating leukocytes based on their different densities. The remaining 1% of contaminating leukocytes (B-cell) is subsequently removed by immunopurification as recommended by the manufacturer, Dynabeads (Dynal Inc., Lake Success, NY). In order to remove these few B cells that contaminate the platelet preparation, M-450 Pan-B (CD-19) Dynabeads are used while following the manufacturer's recommended negative selection/depletion protocol. These beads are covered with mouse-anti-human IgG against the pan-B lymphocyte antigen found on both early B-cell precursors as well as pre-B cells *(1)*. Isolation of megakaryocytes from the marrow preparation utilizes a mouse monoclonal antibody to human platelet glycoprotein Ib (Dako Corporation, Santa Barbara, CA). Following the initial incubation, magnetic beads coated with antimouse IgG (Robbins Scientific, Mountain View, CA) are added to coat the megakaryocytes and permit "positive selection" in a magnetic field.

The method published by Gubler and Hoffman *(2)* detailing the production of a cDNA library from the products of the reverse transcription reaction was adapted from the method described by Okayama and Berg *(3)*. In short, the method utilizes RNase H and DNA polymerase I-mediated second-strand synthesis with subsequent cloning into a transformation vector. We recommend a number of molecular biology kits produced by Stratagene (La Jolla, CA) for the production of cDNA libraries using a phage virus. Many of the enzymes required for the production of cDNA libraries are sold with instructions and buffers. Although we provide a description of a basic method, manufacturers' recommendations are more suitable.

The following topics will be covered in this chapter: (1) collection of platelets; (2) isolation of platelets from PRP using arabinogalactan density gradient and immunopurification; (3) methods for the isolation of megakaryocytes; (4) purification of mRNA and production of cDNA; (5) assessing purity of platelet cDNA by PCR analysis; and finally, (6) Production of a cDNA library (a standard protocol for the creation of a cDNA library).

2. Materials
2.1. Purification of Platelets and Megakaryocytes

1. Acid citrate dextrose (ACD): 85 mM sodium citrate, 78.1 mM citric acid, and 111 mM glucose.
2. Buffered saline citrate-glucose (BSG): 83.84 mM NaCl, 8.45 mM Na$_2$HPO$_4$, 1.47 mM KH$_2$PO$_4$, 13.6 mM sodium citrate-dihydrate, and 11.1 mM D-glucose.
3. Stock BSA solution (23.5% (w/v)): 29 g BSA added to 100 mL of distilled water and 7.9 mL of 10X concentrated calcium- and magnesium-free Hank's balanced salt solution (CMFH); adjust pH to 7.4 by slow addition of approx 7 mL of 1 N NaOH and then 2–3 mL of 0.15 N NaOH as required.
4. Calcium- and magnesium-free Hank's balanced salt solution (CMFH) (Grand Island Biological Co., Grand Island, NY).

5. CATCH solution: CMFH with 0.38% (w/v) sodium citrate, 10 m*M* adenosine (Sigma), 20 m*M* theophylline (Sigma), and 3.5% (w/v) bovine serum albumin (BSA, Fraction V, Armour Pharmaceutical Co., Chicago, IL). The pH is adjusted to 7.4 by addition of isotonic HCl or NaOH after all other additions. Citrated solutions are made by adding 1/10 vol of 3.8% (w/v) sodium citrate.

6. Modified CATCH solution *(4)* may also be used: CMFH with 13.6 m*M* sodium citrate, 1.0 m*M* adenosine (Sigma), 1.0 m*M* theophylline (Sigma), 11.1 m*M* glucose, and 0.5% (v/v) bovine serum albumin (BSA, Path-O-Cyte 4; Miles Inc., Kankakee, IL), 10 m*M* HEPES (pH 7.25), and 0.15 U/mL apyrase (Sigma).

7. Supplemented Iscove's modified Dulbecco's medium (IMDM): serum-free Iscove's modified Dulbecco's medium (IMDM) (Sigma) with 1.5% (v/v) bovine serum albumin, sonicated lipids, insulin-transferrin-sodium selenite supplment (Boehringer Mannheim, Indianapolis, IN), nonessential amino acids, soidium pyruvite, and L-glutamine

8. Supplemented IMDM with 1% (v/v) Nutridoma (Boehringer Mannheim), penicillin, streptomycin, amphotericin B, and L-glutamine (Sigma).

9. Phosphate-buffered saline/sodium citrate/bovine serum albumin solution (PBS/sodium citrate/BSA): 1.14 m*M* NaH_2PO_4, 5.53 m*M* Na_2HPO_4, 1.38 *M* NaCl, 20.4 m*M* sodium citrate, and 0.3% (w/v) BSA.

2.2. Reverse Transcription, PCR, and Production of cDNA Library

For all solutions used to manufacture the cDNA library, use only DEPC-treated, autoclaved water and pass each solution through a sterile 0.45-μm filter. Store stock solutions at room temperature. Enzyme buffers should be aliquoted and frozen at –80°C in screw-top microcentrifuge tubes.

1. 10X annealing buffer: 100 m*M* Tris-HCl, pH 7.5, 10 m*M* EDTA, and 1.5 *M* NaCl. Final reaction concentration is 1X annealing buffer: 10 m*M* Tris-HCl at pH 7.5, 1 m*M* EDTA, and 150 m*M* NaCl.

2. 7.5 *M* ammonium acetate.

3. Buffered phenol (a yellow organic solution)—phenol without products of oxidation to prevent damage to DNA. Store at 4°C for up to 2 mo. **Caution:** Phenol causes severe burns; use caution and perform the following in a fume hood:

    a. Add 0.5 g of 8-hydroxyquinoline (antioxidant) to a 2-L glass beaker containing a stir bar.
    b. Add 500 mL of liquid phenol or melted crystals of redistilled phenol (65°C water bath).
    c. Add 500 mL of 50 m*M* Tris-HCl base.
    d. Cover with aluminum foil and stir at low speed for 10 min at room temperature.
    e. Let phases separate.
    f. Decant aqueous phase (top) into a waste receptacle and remove the rest with a pipet.
    g. Add 500 mL of 50 m*M* Tris-HCl, pH 8.0, and repeat stirring, phase separation, and decanting.
    h. Again, add 500 mL of 50 m*M* Tris-HCl, pH 8.0, and repeat stirring, phase separation, and decanting.
    i. Add 250 mL TE buffer and store at 4°C in brown glass bottle or, if using clear glass, be sure to wrap in foil to keep out light.
    j. If used for DNA purification, mix 25 volumes of phenol with 24 volumes of chloroform and 1 volume of isoamyl alcohol.

4. 10X *EcoR*1 buffer: 100 m*M* Tris-HCl, pH 7.5, 100 m*M* $MgCl_2$, 10 m*M* dithiothreitol, 1 mg/mL bovine serum albumin, 100 m*M* NaCl.

5. 10X *EcoR*1 methylase buffer: 100 mM NaCl, 100 mM Tris-HCl, pH 7.5, 1 mM EDTA.

6. 0.5 M EDTA (ethylenediamine tetraacetic acid) (stock solution; store at room temperature): 186.1 g Na$_2$EDTA•2H$_2$O in 700 mL H$_2$O; add 10 M NaOH (~50 mL) to a final pH of 8.0.

7. Luria-Bertani (LB) broth: Add to a final 1-L volume: filtered H$_2$O, 10 g bacto-tryptone, 10 g NaCl, 5 g bacto-yeast extract. Dissolve contents and adjust the pH to 7.0 with 5 N NaOH. Sterilize by autoclaving for 20 min at 15 lbs/in^2 on liquid cycle.

8. Luria-Bertani (LB) broth top agar: Prepare in 1-L volume, autoclave for 15 min to melt, cool to 50°C, swirl to mix, pour into separate 100-mL bottle, reautoclave, cool, and store at room temperature. To a final 1-L volume add filtered H$_2$O, 10 g bactotryptone, 10 g NaCl, 7 g agar.

9. LB plates, per 1 L volume, prewarm to 37°C before use: 10 g tryptone, 5 g NaCl, 5 g yeast extract, 1 mL 1 N NaOH, 15 g agar, antibiotics for selection of transfected cells.

10. 10X loading buffer (store at 4°C): 20% (v/v) Ficoll 400, 0.1 M Na$_2$EDTA, pH 8.0, 1.0% (w/v) sodium dodecyl sulfate, 0.25% Bromphenol blue.

11. 10X second-strand synthesis reaction buffer: 200 mM Tris-HCl, pH 7.5, 50 mM MgCl$_2$, 100 mM (NH$_4$)$_2$SO$_4$, 1 M KCl, 1.5 mM β-NAD, 500 µg/mL BSA.

12. 2 M sodium acetate, pH 5.5.

13. Suspension medium: 5.8 g NaCl, 2 g MgSO$_4$•7H$_2$O, 50 mL 1 M Tris-HCl, pH 7.5, 0.01% (w/v) gelatin.

14. 10X T4 DNA ligase buffer: 5 mM ATP, 0.5 M Tris-HCl, pH 7.5, 50 mM MgCl2, 50 mM dithiothreitol, 0.5 mg/mL BSA.

15. TE buffer, pH 7.5 (stock solution; store at room temperature): 10 mM Tris-HCl [tris(hydroxymethyl)aminomethane], pH 7.5, 1 mM EDTA, pH 8.0.

16. 1 M Tris-HCl, pH 7.5 (stock solution; store at room temperature): 121 g Tris base in 800 mL H$_2$O, 80.6 mL 0.1 M HCl at 25°C; if temperature is different then adjust 0.028 pH/1°C for target pH measurement. Mix and add water to 1 L final volume.

17. 1 M Tris-HCl, pH 8.0 (stock solution; store at room temperature): 121 g Tris base in 800 mL H$_2$O, 58.4 mL 0.1 M HCl at 25°C; if temperature is different, then adjust 0.028 pH/1°C for target pH measurement. Mix and add water to 1 L final volume.

2.3. Reagents

1. Anti-CD34 class II monoclonal antibodies (QBEND 10, Miltenyi Biotec, Bergisch Gladback, Germany).

2. Antimouse-IgG-coated magenetic beads (Robbins Scientific, Mountain View, CA).

3. 1 M acetylsalicylic acid (ASA) dissolved in dimethyl sulfoxide (DMSO).

4. Bovine serum albumin (BSA) (Sigma).

5. Cellsep (Larex Inc., St. Paul, MN).

6. *Escherichia coli* DNA polymerase I (Boehringer Mannheim).

7. T4 DNA Ligase (New England Biolabs Inc., Beverly, MA).

8. dNTPs (Promega Corp.).

9. dCTP (Promega Corp.).

10. dGTP (Promega Corp.).

11. *Eco* R1 (New England Biolabs Inc.).

12. *E. coli* RNase H (Gibco-BRL, Gaithersburg, MD.).

13. Calf intestine phosphatase (CIP) (New England Biolabs Inc.).

14. *EcoR1* methylase (Gibco-BRL).

15. *S*-adenosylmethionine (Promega Corp.).
16. DEPC-treated water.
17. *E. coli* C600*hflA*, or *E. coli* Y1088 (New England Biolabs Inc.).
18. λgt10 or λgt11 (New England Biolabs Inc.).
19. CL-4B column: for the removal of linkers or adapters in order to prevent subsequent interference in the cloning of a cDNA fragment.
20. Preswollen Sepharose (CL-4B, Pharmacia Corporation, Peapack, NJ.)
21. CL-4B column buffer (stock solution, filter-sterilized, store at 25°C).
    a. 5 mL 1 *M* Tris-HCl, pH 8.0.
    b. 60 mL 5 *M* NaCl.
    c. 1 mL 0.5 *M* EDTA, pH 8.0.
    d. 2.5 mL 20% N-lauroylsarcosine (Sarkosyl).
    e. 431.5 mL H_2O.
22. Silanized glass wool: submerge glass wool in 1:100 dilution of a silanizing agent (Prosil 28, VWR International, West Chester, PA) for 15 min with shaking. Rinse extensively with distilled H_2O. Autoclave for 10 min and store at room temperature.
23. New plastic tubing with a metal clamp for control of elution.
24. Disposable 5-mL plastic pipet tip.
25. Ficoll Hypaque (Sigma).
26. IL-3 (Boehringer Mannheim).
27. Iloprost (stable prostacyclin PGI_2) (Sigma).
28. Mouse monoclonal antibody to human platelet glycoprotein Ib (Dako Corporation, Santa Barbara, CA).
29. RPMI-1640 (Sigma).
30. AmpliTaq (GeneAMP RNA PCR kit, PerkinElmer, Norwalk, CT).
31. Chloroform.
32. First Strand Synthesis kit (Gibco-BRL).
33. Formamide.
34. MMLV Reverse Transcriptase (Promega Corp.).
35. oligo(dT$_{12-18}$) primers (Stratagene, La Jolla, CA).
36. Pfu DNA Polymerase (Stratagene).
37. Primer sequences as reported (**Table 1**).
38. Stem cell factor (Sigma).
39. Thrombopoeitin, murine (TPO) (Zymogenetics Inc., Seattle, WA).
40. Thrombopoeitin, human recombinant (TPO) (Sigma).
41. TRIzol reagent, total RNA isolation reagent (GIBCO-BRL).
42. Rapid mRNA Purification Kit (AMRESCO, Solon, OH).
43. RNAzol (Tel-Test, Friendswood, TX).

2.4. Equipment

1. 16-gauge needle for phlebotomy.
2. 19-gauge syringe with a 10-cm piece of Teflon tubing.
3. 50-mL centrifuge tubes (polystyrene or polypropylene).
4. CD34 progenitor cell isolation kit (Miltenyi Biotec).
5. Magnetic particle concentrator (Dynal Inc., Lake Success, NY).
6. Mini MACS system (Miltenyi Biotec).
7. Siliconized Pasteur pipets (with the pulled end removed and flamed) or plastic transfer pipets.

Table 1
PCR Primer Sequences and Product Size

Target	Primer	Sequence (5′ to 3′ direction)	Expected PCR product size (bp)
GP 1bα	Forward	GGTGCGTGCCACAAGGACTGT	282
	Reverse	TTTGGGGCGGGCTCCGGGACG	
P2Y$_{12}$ receptor	Forward	ATGCAAGCCGTCGACAATCTC	1019
	Reverse	CATTGGAGTCTCTTCATTTGG	
LTB4 receptor	Forward	GGGGCTTCCCGGCAA	821
	Reverse	GCAGCTTGGCGACGAAGC	
CD2	Forward	GGTCATCGTTCCCAGGCACCTAGT	405
	Reverse	TGGTGTGATGGAGCTCTCTGAGGA	
CD-20	Forward	CGTGCTCCAGACCCAAATCTAACA	439
	Reverse	GCGTGACAACACAAGCTGCTACAA	
HLADQb	Forward	GTCTCAATTATGTCTTGGAA	702
	Reverse	GCCACTCAGCATCTTGCT	

3. Methods
3.1. Collection of Platelets (see Note 1)

1. Human blood is to be collected from informed healthy volunteers who are not taking any medications or herbal supplements. The donated blood is collected into a one-sixth volume of Acid Citrate Dextrose (ACD) (*see* **Note 2**). Phlebotomy is performed using a 16-gauge needle with attached IV tubing (similar to a blood collection apparatus used during donation).
2. Blood is collected into 50-mL centrifuge tubes filled with 8 mL of ACD. Final volume per tube is 48 mL.
3. Platelet-rich plasma (PRP) is isolated by centrifugation of citrated blood at 180g for 15 min at room temperature.
4. Two thirds of the volume of PRP is gently collected using a large-bore transfer pipet. When collecting PRP be certain not to disturb the layer of leukocytes that lies above red blood cell layer (*see* **Note 3**). This layer of leukocytes is frequently referred to as the "buffy coat" layer.
5. Place the collected PRP into a 50-mL centrifuge tube. To prevent platelet activation, first add enough iloprost to a final concentration of 28 nM. Alternatively, to prevent platelet activation by stopping the production of thromboxane A$_2$, add a final concentration of 1 mM acetylsalicylic acid and incubate at 37°C for 30 min.

3.2. Isolation of Platelets From PRP Using Arabinogalactan Density Gradient and Immunopurification

1. To prepare the arabinogalactan gradient prepare a volume of 1:1 Cellsep and buffered saline glucose-citrate (BSG). Enough of this 1:1 mixture should be made so that 6 mL of the mixture may be added to all the 50-mL centrifuge tubes to be used.
2. Divide the amount of collected PRP by 12 to determine the number of gradient tubes to be assembled (*see* **Note 4**).
3. Insert 12 mL of Cellsep underneath using a 19-gauge syringe with a 10-cm piece of Teflon tubing attached to the tip of the needle. Run the Teflon tubing through the 1:1 Cellsep/

BSG-C solution to the bottom of the tube and deliver the denser solution (**Fig. 1**). It is critical that a sharp interface should exist between the two layers.

4. Float 12 mL of PRP (gentle drops) on top using a transfer pipet).
5. The gradient tubes undergo 20 min of centrifugation at 1450g. Following centrifugation a sharp band of platelets appears at the interface of the Cellsep gradient solutions. The plasma remains at the top and a pellet of leukocytes and other large cells forms at the bottom of the centrifuge tube (**Fig. 1**).
6. Aspirate off the plasma and 1:1 Cellsep layer. Remove the platelet layer with a plastic transfer pipet and collect in a round-bottom centrifuge tube.
7. To each collected layer add 12–30 mL of BSG-C and swirl in order to enhance the removal of Cellsep.
8. Following centrifugation at room temperature at 630g for 10 min, pour off the supernatant and gently resuspend in 1 mL of phosphate buffered saline/sodium citrate/bovine serum albumin solution (PBS/Na-citrate/BSA).
9. Combine the resuspended pellets into a 15-mL centrifuge tube with a final 10-mL volume of PBS/Na-citrate/BSA. The cell suspension should have a pH of 6.6 to 6.8. To prevent cell activation, iloprost should be added to a final concentration of 28 nM.
10. Following a modification of the procedure recommended for Dynabeads® M-450 Pan-B (CD19) for the rapid depletion of CD19-positive B cells, the platelet suspension is immunopurified. Resuspend the beads by inversion and transfer 50 µL of the supplied suspension into a centrifuge tube for cleaning. The manufacturer recommends that a final concentration of beads to target cells be 4:1 (*see* **Note 5**).
11. Place the tube on the magnetic particle concentrator (a magnet supplied by Dynal Inc., Lake Success, NY) for at least 1 min to clear the solution of magnetic particles.
12. Pipet off the fluid and remove from magnetic concentrator.
13. Clean the magnetic beads by resuspending the beads in 2 mL of PBS Na-Citrate/BSA. Remove the PBS Na-Citrate/BSA and repeat.
14. Finally, resuspend the beads in the 10 mL volume of platelets and PBS Na-citrate/BSA. The platelet and magnetic bead suspension is then incubated for at least 30 min at room temperature (*see* **Note 6**) with gentle agitation on a tilt rotator.
15. Remove the Dynabeads from the platelet suspension using the magnetic particle concentrator for at least 3 min. During this process the few contaminating B cells with beads adhered to their surface are pulled out of solution along with the unbound beads. A plastic transfer pipet is finally used to collect the immunopurified platelets into an RNase free tube.

3.3. Methods for the Isolation of Megakaryocytes

3.3.1. Guinea Pig Megakaryocytes

Levine and Fedorko (*5*) were the first to produce a method for isolation of intact megakaryocytes from guinea pig femoral marrow based on the cells' low density and large size. This method and subsequent modifications have made the in vitro study of megakaryocytes possible.

1. Femurs, tibiae, and humeri are removed, cleaned of adherent tissue, and cracked open.
2. The marrow is scooped out in large pieces using a spatula and cut up with scissors and forceps into pieces no greater than 2 mm in diameter in CATCH solution.
3. The supernatant is aspirated and the remaining tissue is minced with a razor blade. Following resuspension in 5 mL of CATCH solution the sample is transferred to a polypropylene centrifuge tube and vigorously triturated with a wide (~3 mm) bore pipet (8–10 times)

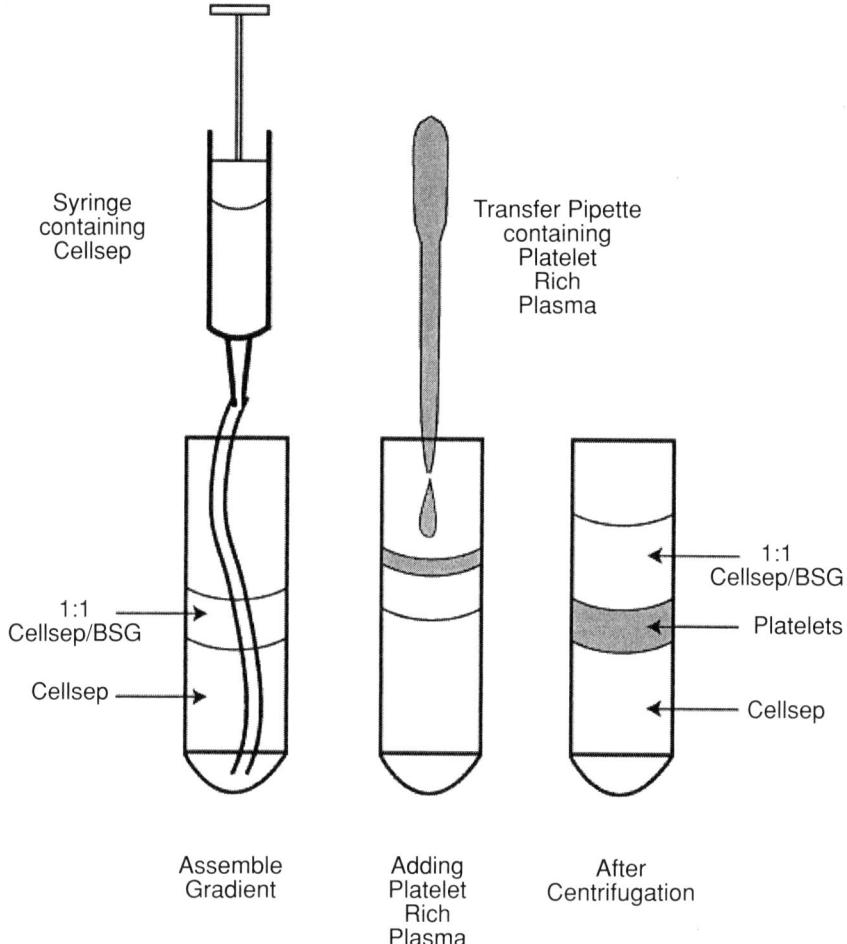

Fig. 1. Assembly of the Cellsep density gradient and collection of platelets.

in order to release single cells from the clumps of tissue. Siliconized Pasteur pipets (with the pulled end removed and flamed) or plastic transfer pipets may be used.

4. The marrow cells are then filtered through a sieve (hole diameter ~150 μM) or 100-μM pore nylon mesh into a centrifuge tube and washed twice by resuspending the cells in CATCH solution.
5. Cells are centrifuged at 250g for 10 min.
6. Resuspend marrow cells, 10^8 cells/mL.
7. Place a 1.0 mL volume of the marrow suspension over the four discontinuous density layers (1.8 mL each) of BSA dissolved in CMFH, pH 7.4. The specific gravity of each solution, starting from the top down, is 1.035, 1.040, 1.045, and 1.050. The gradient is made by serial dilutions of the 23.5% stock BSA solution. Measuring the refractive index and relating that measurement to a standard curve will establish the specific gravity of each dilution.

8. Gently layer 1.8 mL of each solution into a 15-mL centrifuge. Centrifuge the assembled gradient with the top layer of marrow cell suspension in a swinging bucket rotor at 10,000g for 30 min at 4°C. Low-density populations of cells are suspended at the four density gradient interfaces.

9. Pool the low-density populations and collect them into another centrifuge tube; the higher-density cells are collected in a pellet at the bottom of the centrifuge tube.

10. Wash the cells and resuspend in CATCH solution at a concentration of 3×10^6 cells/mL. Layer 1.0 mL of this suspension over a four-layer BSA-CMFH density gradient (1.8 mL volume in each layer) with the following specific gravities: 1.0022, 1.0045, 1.0067, and 1.0089.

11. While the tube is kept upright, allow the cells to settle at room temperature for 30 min.

12. The uppermost 3 mL (the two uppermost cloudy bands of cells at the first two interfaces) are withdrawn and discarded.

13. The remaining cells are pelleted, resuspended in CATCH solution, and pelleted again.

14. The pellet is then resuspended in 1.0 mL of CATCH solution and cell viability is assessed by combining 20 µL of the cell suspension with 20 µL of 2% trypan blue and observing the cells using a hemocytometer and phase-contrast optics. This method yields about 6×10^5 megakaryocytes from four guinea pig femurs. The collected population consists of 33% megakaryocytes with other types of cells present, including nucleated erythrocyte precursors, promyelocytic and myelocytic neutrophils, and eosinophils (*see* **Note 7**). Small numbers of erythrocytes and mononuclear cells are also present. Lymphocytes, macrophages, mature granulocytes, osteoblasts, osteoclasts, and adipocytes are absent or rare.

100% pure megakaryocytes are isolated from the above preparation using immuno-magnetic beads. However, in contrast to the previously described method for platelets, the immunomagnetic beads are used to isolate the target cells (megakaryocytes) and not the contaminating cells.

15. The resuspended pellet is mixed with 20 µL of 1 mg/mL mouse monoclonal antibody to human platelet glycoprotein Ib (Dako Corporation, Santa Barbara, CA) and mixed in a 1.5-mL plastic reaction tube on a rotary mixer for 1 h at room temperature.

16. Following the initial incubation, 5.0 mL of magnetic beads coated with antimouse IgG (Robbins Scientific, Mountain View, CA) are added and the suspension mixed for an additional 30 min (*see* **Note 6**).

17. The megakaryocytes covered in beads are then removed from suspension with a Dynal magnetic particle concentrator.

18. Four washes are performed using the magnetic particle concentrator to collect the cells.

3.3.2. Mouse Megakaryocytes

1. Drachman et al. (*6*) utilize a method for obtaining megakaryocytes from mice that first increases the number of cells through daily subcutaneous injections with pure, recombinant murine thrombopoeitin (TPO) for 5 d (10 µg/kg/d; Zymogenetics Inc., Seattle, WA).

2. Mice are sacrificed and marrow cells are obtained by flushing the contents of the femurs and tibias into serum-free IMDM containing supplements as listed in **Subheading 2.1.**

3. Vigorous trituration produces a single-cell suspension.

4. Marrow cells are resuspended at 1×10^6 cells/mL and incubated (37°C, 5% CO_2) in serum-free IMDM supplemented medium plus recombinant murine thrombopoeitin (3% (v/v) conditioned medium) for 72 to 96 h.

5. Mature megakaryocytes are obtained by centrifugation (400 rpm for 10 min) and layered on top of on a discontinuous bovine serum albumin (BSA; Sigma) density gradient (0%/1.5%/3.0% BSA in phosphate-buffered saline).
6. Megakaryocytes comprise greater than 90% of the cells that settled to the bottom within 30 min at 1*g*.
7. The cells should be observed microscopically either in culture or after Wright's stain. Cell viability can be determined using trypan blue.
8. To further purify megakaryocytes from other contaminating hematopoietic cells a second round of density separation should be performed.
9. Megakaryocytes are collected by low-speed centrifugation and resuspended in IMDM + 0.5% (w/v) BSA.

3.3.3. Human Megakaryocytes

Miyazaki et al. *(7)* have isolated human CD34+ hemapoietic progenitor cells that were subsequently cultured and stimulated with human recombinant TPO (Sigma). Human recombinant TPO stimulation causes these progenitor cells to undergo differentiation into mature polynucleated platelet-forming megakaryocytes. CD34+ cells that were greater than 80% pure were obtained after 14 d in culture. Cellular maturity was established by ploidy, expression of CD41, CD42a, and CD61, as well as proplatelet formation.

1. To obtain megakaryocytes from the human donated marrow, cells are first prepared by centrifugation through Ficoll Hypaque 1.077 g/mL (Sigma) and resuspended in RPMI-1640.
2. The hemapoietic progenitor cells are isolated using anti-CD 34 class II monoclonal antibodies (QBEND 10, Miltenyi Biotec), a CD34 progenitor cell isolation kit (Miltenyi Biotec), and repeated use (twice) of the Mini MACS system (Miltenyi Biotec) following manufacturer's instructions. This method will produce a suspension of CD34+ cells that is more than 95% pure.
3. The CD34+ cells are then cultured at low density (1×10^4 cells/mL) in supplemented serum-free IMDM. Cell cultures undergo maturation with the addition of 10 ng/mL IL-3 (Boehringer Mannheim), 10 ng/mL of stem cell factor (Sigma), and 10 ng/mL human recombinant TPO (Sigma). Cell culture is performed in 24-well plates at 37°C in a humidified atmosphere with 5% CO_2 for 7 d for maximal proliferation and maturity.

3.4. Purification of mRNA

There are many commercial kits available for the isolation of total RNA from tissue, cell culture, and suspended cells such as those found in blood. Three examples of kits that are commonly used include TRIzol Reagent (Total RNA Isolation Reagent) (GIBCO-BRL), the Rapid mRNA Purification Kit (AMRESCO), and RNAzol (Tel-Test).

1. When using TRIzol the platelets are dissolved by trituration in 500 µL of TRIzol reagent and incubated at room temperature for 5 min in a microcentrifuge tube.
2. To the TRIzol, 200 µL of chloroform is added and the mixture is gently vortexed for 15 s.
3. The mix is then incubated at room temperature for 3 min.
4. Following centrifugation (12,000*g* for 15 min at 4°C) the aqueous phase is removed and transferred to a fresh tube to which an equal volume of 100% isopropyl alcohol is added.
5. The solution is incubated at room temp for 10 min and the centrifuged (12,000*g* for 15 min at 4°C).

6. Look for a small pellet at the base of the tube. Wash the RNA pellet once in 500 μL of 75% ethanol. Vortex and then spin down the pellet (12,000g for 15 min at 4°C).

7. The ethanol is removed by air-drying for 10–15 min and the pellet is dissolved in DEPC-treated water.

8. Estimate concentration and purity of RNA by optical density (OD). The OD_{260}/OD_{280} ratio determines the purity of the preparation from protein contamination (a ratio of 1.9 to 2.0 indicates highly purified RNA). Nucleic acid concentration is determined by the equation: (40 μg/mL)(OD_{260})(dilution factor) = concentration in μg/mL.

3.5. Reverse Transcription of mRNA Into cDNA

1. First-strand cDNA can be synthesized using oligo(dT) primers (Stratagene, La Jolla, CA), 5 μg of total RNA, and MMLV Reverse Transcriptase (Promega Corp.). Following the manufacturer's protocol the cDNA is synthesized during incubation at 42°C for 60 min.

2. The MMLV reverse transcriptase is inactivated by the addition of 20 mM EDTA and the products are extracted with phenol.

3. Following the addition of 1/10 volume 2 M ammonium acetate solution, pH 5.5, the RNA/cDNA hybrid is precipitated by adding 2 volumes of absolute ethanol and incubation (~–20°C overnight).

4. The precipitate is collected by centrifugation 16,319g for 10 min, and then dissolved in a small volume of DEPC-treated water (20 to 50 μL) (*see* **Note 8**). An alternative commercially available kit is the First Strand Synthesis kit (Gibco-BRL).

3.6. Assessing Purity of Platelet cDNA by PCR Analysis

After producing platelet cDNA, the presence of leukocyte contamination is determined by PCR using leukocyte-specific targets. The presence of a PCR product indicates contamination.

3.6.1. Glycoprotein Ib

As a positive control for both platelet and megakaryocyte cDNA, PCR is performed in a final volume of 50 μL using primers specific for the α chain of the platelet membrane glycoprotein Ib (GPIbα) *(8,9)*.

1. PCR is performed using 2.5 units of the Pfu DNA Polymerase (Stratagene, La Jolla, CA), 125 ng of the forward and 125 ng of the reverse primer (**Table 1**), as well as 2.5 μL of formamide to enhance product formation.

2. A thermocycle with an initial incubation of 98°C for 45 s, followed by 40 cycles of denaturation at 98°C for 30 s, annealing at 67°C for 30 s, and extension at 72°C for 45 s, and ending with a final incubation at 72°C for 10 min will result in a 282-base-pair product *(10)*.

3. To ascertain the presence of genomic DNA contamination, the same reaction should be performed using total RNA that has not undergone reverse transcription. There should be no product from such a negative control reaction.

4. PCR products are electrophoresed on a 1.0 or 1.4% agarose gel and stained with ethidium bromide.

3.6.2. P2Y$_{12}$ Receptor

Another excellent control for the quality of platelet or megakaryocytic RNA is the P2Y$_{12}$ receptor *(11,12)*. This receptor is expressed only in the brain, megakaryocytes, and platelets, but not in the other blood cells *(12)*. Hence this receptor would be an excellent marker for the platelet or megakaryocytic mRNA.

1. PCR is performed using 2.5 units of the Pfu DNA Polymerase (Stratagene) 125 ng of the forward and 125 ng of the reverse primer (**Table 1**), as well as 2.5 μL of formamide to enhance product formation.
2. A thermocycle with an initial incubation of 98°C for 45 s, followed by 40 cycles of denaturation at 94°C for 45 s, annealing at 50°C for 45 s, and extension at 72°C for 1 min. The final cycle of an additional extension for 7 min at 72°C will result in a 1019-base-pair product *(11,12)*.
3. To ascertain the presence of genomic DNA contamination, the same reaction should be performed using total RNA that has not undergone reverse transcription. There should be no product from such a negative control reaction.
4. PCR products are electrophoresed on a 1.0 or 1.4% agarose gel and stained with ethidium bromide (**Fig. 2**).

3.6.3. LTB4 Receptor

LTB4 receptor mRNA is expressed maximally in leukocytes including polymorphonuclear cells, monocytes, and lymphocytes. but not in platelets *(13)*, and thus is a good target to assess the presence of leukocyte contamination. The following protocol will produce an 821-bp PCR product using 125 ng each of the primers specific for the LTB4 receptor (**Table 1**) *(13,14)*.

1. After an initial denaturation at 94°C for 5 min, the sequence amplification is performed through 35 cycles using 5.0 units of Pfu DNA polymerase in a final 50 μL volume using a thermocycle of 94°C for 1 min, annealing at 59°C for 1 min, and extension at 72°C for 1 min. The final cycle was followed by a final incubation at 72°C for 7 min.
2. PCR products are electrophoresed on a 1.0 or 1.4% agarose gel and stained with ethidium bromide (**Fig. 2**).

3.6.4. Glycoprotein CD2

Another negative control for the absence of T-cell mRNA in the cDNA preparation is the surface glycoprotein CD2 *(1)*.

1. The reaction is performed using 2.5 units of Pfu DNA polymerase in a final 50 μL volume with 125 ng of the forward primer and 125 ng of the reverse primer (**Table 1**).
2. A thermocycle begins with an initial incubation at 98°C for 45 s, followed by 40 cycles of denaturation at 98°C for 45 s, annealing at 69°C for 45 s, and extension at 72°C for 45 s.
3. A final incubation is performed at 72°C for 10 min. The reaction will result in a 405-bp PCR product *(10)*.
4. The B-cell surface glycoprotein, CD-20, can also be used as a PCR target for assessing the presence of leukocyte contamination *(1)*. The same reaction conditions are used as above except that the thermocycle conditions are denaturation at 98°C for 30 s, annealing at 67°C for 30 s, and extension at 72°C for 30 s. PCR performed using the forward and reverse primers for CD-20 (**Table 1**) results in a 439-bp product *(10)*.
5. PCR products are electrophoresed on a 1.0 or 1.4% agarose gel and stained with ethidium bromide (**Fig. 2**).

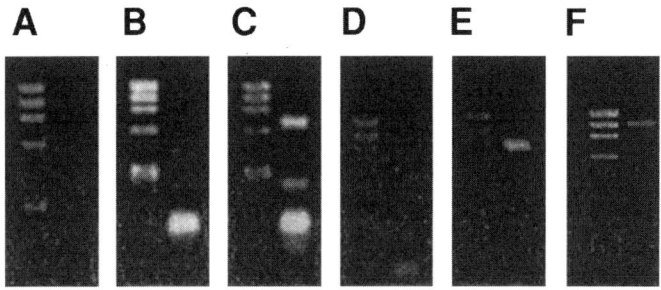

Fig. 2. Evaluation of the purity of platelet mRNA by RT-PCR. (**A**) Platelet $P2Y_{12}$ receptor primers on platelet RNA (control for genomic DNA contamination). (**B**) HLA-DQb primers on platelet cDNA. (**C**) HLA-DQb primers on leukocyte cDNA. (**D**) LTBR primers on platelet cDNA. (**E**) LTBR primers on leukocyte cDNA. (**F**) $P2Y_{12}$ receptor primers on platelet cDNA.

3.6.5. Leukocyte Marker HLADQb

The common leukocyte marker, HLADQb, has been used as a RT-PCR target to assess the purity of platelet mRNA preparations *(15)*. A 702-bp product will result from the amplification reaction using the forward and reverse primers listed in **Table 1**.

1. The mixture is heated for 2 min, and AmpliTaq (GeneAMP RNA PCR kit, PerkinElmer, Norwalk, CT) is then added.
2. The thermocycle is 35 cycles of denaturing at 95°C for 30 s, annealing at 50°C for 1 min, and extension at 60°C for 1 min.
3. PCR products are electrophoresed on a 1.0 or 1.4% agarose gel and stained with ethidium bromide (**Fig. 2**).

3.7. Production of cDNA Library

There are a number of reasons for using a lambda phage-derived cloning vector system instead of a bacterial plasmid based system *(16)*. The lambda genome has restriction sites flanking the middle third of the genome, which is not required for lytic growth. Foreign DNA can be inserted and packaged into the phage sequence in vitro. Each assembled phage containing a cloned sequence will form a plaque, whereas a plasmid at best will successfully transform 1/1000 bacterial cells. The appropriate phage must be chosen depending on the size of the DNA fragment to be cloned. Each manufacturer of phage cloning kits will provide such specifications. The level of expression of the target mRNA and the size of library required are inversely related. Low-abundance mRNA requires a large cDNA library for successful cloning. Similarly, the 5′ ends of large mRNA targets will be less frequently cloned as compared to abundance of the mRNA target. In general, lambda will permit the cloning of small fragments of DNA when the total size of the recombinant genome is at least as big as 78% of the wild-type genome and no larger than 105% *(17)*.

The formation of a platelet or megakaryocyte cDNA library involves a number of steps and follows a protocol as described by Ausubel *(17)*. First, the mRNA must be converted into double-stranded cDNA. Linkers or adapters are ligated onto the 3′- and 5′-ends of the cDNA for insertion into a chosen phage vector. A vector either undergoes restriction digestion to release the 3'- and 5'-phage arm regions or is bought in a "prepared" form by a commercial manufacturer. Following ligation of the arms to the purified DNA, the recombinant virus is assembled and amplified. Any phage that is assembled without insert will be too small to be packaged, thereby increasing efficiency.

1. The initial step is reverse transcription of the mRNA using oligo dT primers (*see* above).
2. The mRNA/cDNA hybrid products of reverse transcription are first extracted with phenol and then precipitated from the aqueous phase by the addition of 1/10 volume of 2 *M* sodium acetate, pH 5.5, and 2 volumes of absolute ethanol (–20°C overnight).
3. The precipitate is collected by centrifugation at 16,319*g* for 10 min, and then dissolved in a small volume of DEPC-treated water (20 to 50 μL).
4. The mRNA is "nicked" by the enzyme RNAse H. These fragments then serve as primers for DNA synthesis by *E. coli* DNA polymerase I.
5. Finally the 5′- and 3′- ends of the second strand are linked up by T4 DNA ligase. Suggested enzymes for second strand synthesis include *E. coli* RNase H (Gibco-BRL), DNA polymerase I (Boehringer Mannheim), and T4 DNA ligase (New England Biolabs Inc, Beverly, MA). The manufacturers' recommended reaction conditions and buffers are provided with each enzyme.
    A typical protocol would include the following:
    a. Combining reagents in a reaction volume of 100 μL with 1X second-strand synthesis reaction buffer: 500 ng of single-stranded cDNA (equal to 1 μg of hybrid), 40 μ*M* dNTPs, 8.5 U/mL of *E. coli* RNase H, 230 U/mL DNA polymerase I, and 10 U/mL *E. coli* DNA ligase.
    b. The incubation for nicking of the mRNA, thereby producing RNA fragments, is carried out at 12°C for 60 min.
    c. DNA synthesis (including translation of the nicks to remove the RNA primers) and ligation of the DNA fragments are performed at 22°C for 60 min.
    d. 20 m*M* EDTA is added to stop the reaction. The double-stranded products are extracted twice with phenol and precipitated from the aqueous phase by the addition of 1/10 volume of 2 *M* sodium acetate, pH 5.5, and 2 volumes of absolute ethanol (–20°C overnight).
    e. The precipitate is collected by centrifugation as described above, and then dissolved in a small volume of DEPC-treated water (20 μL).

Finally, the cDNA clone undergoes *EcoR1* methylation and ligation of *EcoR1* linkers, digestion of the linkers, size fractionation, and assembly of the phage for bacterial transformation. Cloning requires that only one adapter/linker be attached to each end of the cDNA sequences, which is accomplished by a restriction digest to remove any multiple linkers. Protection of the cDNA sequence is accomplished by methylation prior to the restriction digest reaction. Preparing a library using the prepared double-stranded cDNA is possible with the following kits produced by Stratagene: Lambda gt11 Vector and corresponding library production kits, the Lambda ZAP II Vector and corresponding Lambda ZAP II Vector Kits, the Lambda ZAP-CMV Vector and corresponding kits, and the ZAP Express Vector with corresponding kits. Promega Corp. also produces phage vectors for cDNA library production including: Lambda gt11 Vector, Lambda gt10 Vector, and the Packagene Lambda DNA Packaging System.

A general protocol for methylation and ligation of linkers/adapters incorporates the following: methylation by *EcoR1* methylase, ligation to EcoR1 linkers, restriction digestion with *EcoR1*, and purification.

6. To begin the methylation reaction, dissolve the blunt-ended, double-stranded cDNA pellet in 23 µL of water and add 25 µL of 2X methylase buffer (for a final 1X concentration, dilute the 10X stock by adding four volumes of H$_2$O to one volume of 10X stock) and 1 µL of 1 mg/mL *S*-adenosylmethionine.

7. Mix with a pipettor, add 1 µL (20 U) *EcoR1* methylase (final concentration 400 U/mL), and incubate the final 50 µL volume of reagents for 2 h at 37°C.

8. Add 150 µL of TE buffer and extract product by adding 200 µL buffered phenol, vortex, and microcentrifuge at room temperature for 1 min to achieve phase separation.

9. Transfer the upper aqueous phase to a new tube and add 100 µL TE buffer to the phenol layer.

10. Vortex and microcentrifuge the preparation as previously described. Collect the aqueous phase and add it to the collection in the new tube, which should have a total volume of about 300 µL. Discard the phenol appropriately.

11. Extract the 300 µL aqueous volume twice. Add 1 mL diethyl ether. Vortex and microcentrifuge this preparation as previously described. Following phase separation (the bottom layer is the aqueous phase), discard the upper ether layer using a pipet, and repeat this extraction with another 1 mL volume of diethyl ether.

12. Finally, ethanol-precipitate the methylated cDNA by adding 125 µL of 7.5 *M* ammonium acetate and 950 µL of 95% ethanol. Incubate in a dry ice/ethanol bath for 15 min, warm to 4°C, microcentrifuge for 10 min at 4°C, and remove the supernatant with a pipet.

13. Fill the tube with ice-cold 70% ethanol and microcentrifuge for 3 min at 4°C.

14. Dry the precipitated cDNA in a vacuum desiccator.

15. To ligate the *EcoR1* linkers, first dissolve the pellet in 23 µL of DEPC-treated water.

16. Add 3 µL of 10X T4 DNA ligase buffer containing 5 m*M* ATP and 2 µL of 1 mg/mL phosphorylated *EcoR1* linkers. Mix by repeat pipetting and add 2 µL (800 U) T4 DNA ligase; incubate overnight at 4°C.

17. Microcentrifuge briefly and incubate at 65°C for 10 min to inactivate the ligase.

18. After a 2 min incubation on ice add 95 µL H$_2$O and 15 µL 10X *EcoR1* buffer (1X final concentration). Pipet up and down to mix, add 10 µL of *EcoR1* restriction endonuclease (final concentration 1300 U/mL), and incubate for 4 h at 37°C.

19. Add another 3 µL (60 U) *EcoR1* to the digestion reaction and incubate for an additional hour at 37°C.

20. Inactivate the endonuclease by incubating the volume at 65°C for 10 min.

21. To purify the cDNA with attached and digested linkers by chromatography, first assemble a chromatography column. Add to a 50-mL polypropylene tube 10 mL of preswollen Sepharose CL-4B and fill the tube with CL-4B column buffer. Mix by inversion. Let the Sepharose CL-4B settle by gravity for 15 min. Aspirate off the buffer just above the level of the settled Sepharose CL-4B. Repeat the addition and removal of buffer two more times. Add 10 mL of CL-4B buffer, mix by inversion, and then incubate at 37°C for 10 min.

22. Break off the top of a 5-mL plastic pipet. Using gloved hands, use a 1-mL pipet tip to push a small piece of silanized glass wool down by the base of the pipet. Place a small

length (3–5 cm) of plastic tubing firmly onto the tip. Clamp the tubing shut and attach the column to a ring clamp and ring stand.

23. Using a pipet, fill the column with the washed Sepharose CL-4B. Release the clamp so that the buffer runs out of the column and the Sepharose settles. Clamp the tube shut and add more of the sepharose CL-4B in buffer. Again, release the clamp until the slurry settles. Repeat this until the column is filled to the 5-mL mark. Allow the level of buffer to fall in the column until it is just above the level of the sepharose and then clamp the tube shut. The column is ready to be loaded.

24. Add 2 µL of 10X loading buffer to the restriction digestion reaction and load the cDNA onto the column. Wait until the top of the gel becomes dry and then fill the column with CL-4B buffer. Allow the column to flow by gravity.

25. Collect the effluent in 200-µL volumes until the Bromphenol blue nears the bottom of the column. The Bromphenol blue indicates the position of the digested linkers.

26. Stop the collection before the dye starts to elute. Add 2.5 volumes of ethanol to each tube and incubate on dry ice for 15 min.

27. Thaw the contents at room temperature and microcentrifuge at full speed for 10 min at 4°C. Remove most of the supernatant, fill the tubes with ice-cold 70% ethanol, and repeat the centrifugation. Again, remove most of the supernatant and dry the pellet under vacuum.

28. Resuspend and collect the pellet into a total volume of 50 µL TE buffer. Measure the absorbance A_{260}/A_{280} to determine concentration and purity.

29. Run 2.5 µL on a 1% agarose minigel, stain with ethidium bromide, and determine the average cDNA fragment size. Using a marker, such as Lambda DNA/Hind III, the average size should be above 1.5 kb.

The final step in the production of a complete cDNA library is ligation of the cDNA into the phage vector followed by transfection. The double-stranded cDNA with *Eco*R1 ends is ligated into the phage vector DNA in a mixture that has a twofold molar excess of the phage vector DNA. Prior to ligation the phage vector DNA is digested with *Eco*R1 and treated with a phosphatase (to prevent self-ligation without the cDNA insert). Phosphatased phaged arms are commercially available and can be used following the manufacturer's recommendations. The final step is to package the ligation mixture into infectious phage and transform into the chosen host bacteria (e.g., *E. coli* C600hflA for λgt10, *E. coli* BB4 for λZAP, or *E. coli* Y1088 for λgt11).

30. Resuspend a twofold molar excess of the phage DNA λgt10 or λgt11 (New England Biolabs) in 40 µL H₂O and 5 µL 10X *EcoR1* buffer (1X final concentration). Pipet up and down to mix, add 5 µL of *EcoR1* restriction endonuclease (final concentration 1300 U/mL), and incubate for 4 h at 37°C.

31. Add another 3 µL (60 U) *EcoR1* to the digestion reaction and incubate for an additional hour at 37°C.

32. Inactivate the endonuclease by extracting the products with phenol and precipitating from the aqueous phase by the addition of 0.1 volume of 2 *M* sodium acetate, pH 5.5, and 2 volumes of absolute ethanol. Incubate at –20°C overnight.

33. The precipitate is collected by centrifugation as described above, and then dissolved in a small volume of DEPC-treated water (20 µL).

34. Assemble the phosphatase reaction using calf intestine phosphatase (CIP) (New England Biolabs) to remove the 3′- and 5′-end phosphates. Add the 20 µL to a final 50 µL reaction volume, which includes 20 m*M* Tris-HCl, pH 8.0, 1 m*M* MgCl₂, 1 m*M* ZnCl₂, and 0.1 U CIP.

35. Incubate for 30 min and stop the reaction by extracting with phenol and precipitating with ethanol as previously described.
36. Dissolve the pellet in a small volume of DEPC-treated water (20 μL).
37. Assemble the phage arms and cDNA inserts by combining 100 ng of the prepared cDNA, 10 μg of the digested/phosphatased λgt10 or λgt11 arms (500 μg/mL final concentration), 2 μL of 10X ligase buffer, add DEPC-treated water to a volume of 19 μL.
38. Mix by repeat pipetting and add 1 μL (400 U) of T4 DNA ligase (final concentration 20,000 U/mL). Incubate at 4°C overnight.
39. At the same time inoculate a culture of plating bacteria overnight.
40. In the morning, inoculate 50 mL broth with 0.5 mL of the overnight culture of host bacteria. Measure the OD and when approx 0.5, pellet the bacteria by centrifugation (1500*g* for 30 min at 4°C), and resuspend in cold 10 m*M* MgSO$_4$. Store at 4°C.
41. While the culture is growing, incubate the ligation at –20°C for 30 min and then thaw at room temperature and package the DNA mixture using commercially available packaging extracts following the manufacturer's instructions. Add suspension medium to the packaging mixture and collect entire 2-mL volume in one polypropylene tube. Add a few drops of chloroform and gently shake by hand for 3 s. Place the library on ice and store for no longer than 24 h at 4°C.
42. Grow a culture of *E. coli* to saturation in Luria-Bertani (LB) broth containing 0.2% maltose and 10 m*M* MgSO$_4$.
43. Melt LB top agar (used to distribute phage or cells evenly in a thin layer over the surface of a plate) by heating in a boiling water bath for 15 min followed by cooling to room temperature for 15 min. Store in a 45° to 50°C water bath before use. Cells will be killed if the agar is hotter than 65°C.
44. Add 300 μL of *E. coli* culture (OD$_{600}$ = 0.5) to five 8 × 80-mm tubes.
45. Make 100-fold serial dilutions of the phage lysate in SM (e.g., 1 : 10,000, 1 : 100,000, 1 : 1,000,000). Label the dilution tubes.
46. Add 100 μL of each dilution to a corresponding *E. coli* 8 × 80-mm tube. Label the *E. coli*/phage mixture and incubate the tubes at room temperature for 20 min. The phage particles adsorb to the bacteria during this step.
47. Heat the tubes to 37°C in a water bath for 10 min (phage will inject the cloned DNA) and label five fresh LB plates (surface should not be wet) to correspond with the labels on the dilution tubes.
48. Add 2.5 mL of top agar to a tube, vortex gently, and pour onto a plate. Spread over the entire surface by tilting the plate. Repeat for all preparations. Place the plates in a 37°C incubator.
49. Plaques of λ-derived phages will appear following 8 to 12 h of incubation. The presence of plaques indicates that production of the library was a success.
50. Estimate the titer [in plaque-forming units per milliliter (pfu/mL)] by the following formula: [{number of plaques (pfu) × dilution factor}/volume of the phage dilution plated] × 1000 μL/mL. Store the plates at 4°C.

4. Notes

1. For researchers who have access to a clinical transfusion laboratory: Hogman et al. *(18)* have described a method to produce leukocyte-free platelet concentrate (276–279 × 10^9 cells per unit). Whole-blood collection and component isolation was performed using an Opti System (Baxter, Fenwal Division, La Chatre, France). The authors utilized a leukocyte-removing filter, either a Pall PL 50 (Pall Biomedical Products Co., Glen Cove, NY) or Sepacell 5N (Asahi Biomedical Co., Tokyo, Japan) to remove the leukocytes from

a platelet-rich supernatant. The procedure is designed for platelet transfusions but could be adapted for final laboratory use.

2. The collection of unstimulated responsive platelets can be accomplished in a number of ways, including the use of a calcium chelator such as EDTA or using inhibitors of the coagulation cascade such as heparin. We have always had very reactive platelets using the low-pH ACD buffer.

3. Only removing the upper 2/3 volume of the PRP can reduce contamination from leukocytes.

4. More platelets collected from a greater volume of PRP (at least 50 to 60 mL) results in a higher yield of platelet mRNA.

5. If leukocyte contamination is occurring based on PCR control results then increase the concentration of magnetized beads.

6. Increase the time of incubation by half-hour increments to increase yield.

7. Increase purity up to 42% by discarding more of the contaminating cell volume; however, the yield of megakaryocytes will fall.

8. Gubler and Hoffman *(2)* have shown that size enrichment of the poly(A)-mRNA over a sucrose gradient will improve the yield and length of transcripts in the cDNA library (~2.4 kb maximal size). Unfractionated poly(A)-mRNA routinely results in inserts that are a maximum of approx 1.2 kb in size.

References

1. James-Yarish, M., Bradley, W. G., Emmanuel, P. J., Good, R. A., and Day, N. K. (1994) Detection of cell specific cluster determinant expression by reverse transcriptase polymerase chain reaction. *J. Immunol. Methods* **169,** 73–82.

2. Gubler, U. and Hoffman, B. J. (1983) A simple and very efficient method for generating cDNA libraries. *Gene* **25,** 263–269.

3. Okayama, H. and Berg, P. (1992) High-efficiency cloning of full-length cDNA. 1982, *Biotechnology* **24,** 210–219.

4. Miyazaki, H., Inoue, H., Yanagida, M., Horie, K., Mikayama, T., Ohashi, H., et al. (1992) Purification of rat megakaryocyte colony-forming cells using a monoclonal antibody against rat platelet glycoprotein IIb/IIIa. *Exp. Hematol.* **20,** 855–861.

5. Levine, R. F. and Fedorko, M. E. (1976) Isolation of intact megakaryocytes from guinea pig femoral marrow. Successful harvest made possible with inhibitions of platelet aggregation; enrichment achieved with a two-step separation technique. *J. Cell Biol.* **69,** 159–172.

6. Drachman, J. G., Sabath, D. F., Fox, N. E., and Kaushansky, K. (1997) Thrombopoietin signal transduction in purified murine megakaryocytes. *Blood* **89,** 483–492.

7. Miyazaki, R., Ogata, H., Iguchi, T., Sogo, S., Kushida, T., Ito, T., et al. (2000) Comparative analyses of megakaryocytes derived from cord blood and bone marrow. *Br. J. Haematol.* **108,** 602–609.

8. Weiss, H. J. (1980) Congenital disorders of platelet function. *Semin. Hematol.* **17,** 228–241.

9. Tabilio, A., Rosa, J. P., Testa, U., Kieffer, N., Nurden, A. T., Del Canizo, M. C., et al. (1984) Expression of platelet membrane glycoproteins and alpha-granule proteins by a human erythroleukemia cell line (HEL). *EMBO J. 3, 453–459.*

10. Paul, B. Z., Ashby, B., and Sheth, S. B. (1998) Distribution of prostaglandin IP and EP receptor subtypes and isoforms in platelets and human umbilical artery smooth muscle cells. *Br. J. Haematol.* **102,** 1204–1211.

11. Jin, J., Tomlinson, W., Kirk, I. P., Kim, Y. B., Humphries, R. G., and Kunapuli, S. P. (2001) The C6-2B glioma cell $P2Y_{AC}$ receptor is pharmacologically and molecularly identical to the platelet $P2Y_{12}$ receptor. *Br. J. Pharmacol.* **133,** 521–528.

12. Hollopeter, J., Jantzen, H.-M., Vincent, D., Li, G., England, L., Ramakrishnan, V., et al. (2001) Identification of the platelet ADP receptor targeted by antithrombotic drugs. *Nature* **409,** 202–207.
13. Dasari, V. R., Jin, J., and Kunapuli, S. P. (2000) Distribution of leukotriene B4 receptors in human hematopoietic cells. *Immunopharmacology* **48,** 157–163.
14. Akbar, G. K., Dasari, V. R., Webb, T. E., Ayyanathan, K., Pillarisetti, K., Sandhu, A. K., et al. (1996) Molecular cloning of a novel P2 purinoceptor from human erythroleukemia cells. *J. Biol. Chem.* **271,** 18,363–18,367.
15. van Willigen, G., Donath, J., Lapetina, E. G., and Akkerman, J. W. (1995) Identification of alpha-subunits of trimeric GTP-binding proteins in human platelets by RT-PCR. *Biochem. Biophys. Res. Commun.* **214,** 254–262.
16. Ausubel, F. M. (1987) *Current Protocols in Molecular Biology*, Greene Pub. Associates and Wiley-Interscience: John Wiley & Sons, New York.
17. Ausubel, F. M. (1988) *Current Protocols in Molecular Biology*, Greene Pub. Associates and Wiley-Interscience: John Wiley & Sons, New York.
18. Hogman, C. F., Eriksson, L., and Kristensen, J. (1993) Leukocyte-depleted platelets prepared from pooled buffy coat post-transfusion increment and "in vitro bleeding time" using the Thrombostat 4000/2. *Transfus. Sci.* **14,** 35–39.

28

Platelet Receptor Structures and Polymorphisms

Thomas J. Kunicki, Steven Head, and Daniel R. Salomon

1. Introduction
1.1. Platelet Adhesion and Thrombus Formation

Efficient platelet cohesion (i.e., platelet aggregation) is necessary for the life-saving process of hemostasis. Plaque rupture and/or endothelial damage lead to exposure of collagens, retention of von Willebrand factor (vWF), and the adhesion of circulating platelets to the damaged vessel wall through the concerted participation of a handful of important receptors. With blood flow conditions ranging from low shear ($\leq$300 s) to high shear ($\geq$1500 s), the initial *transient* arrest of platelets on collagen requires vWF acting as a bridge between collagen and the glycoprotein (GP) Ib complex *(1,2)*. Although not usually considered a collagen receptor, the GPIb complex does initiate the first adhesive platelet contact with the collagen-rich matrix. It is collagen that captures the vWF molecule by binding to its A3 domain, thereby localizing it and somehow altering its conformation to make it bind via its A1 domain with greater avidity to the GPIb complex. This interaction is relatively weak and short in duration, resulting in a slowing of platelet motion and a tethering or rolling of the platelet across the thrombogenic collagen-rich surface. Nonetheless, the GPIb complex plays an important role in signal transduction, which activates, in part, other platelet receptors, particularly the integrins $\alpha_2\beta_1$ and $\alpha_{IIb}\beta_3$. Blockade of vWF binding to the GPIb complex will inhibit this initial contact, and stable platelet monolayer formation and thrombus formation would then not ensue.

A more stable attachment of platelets is mediated by one or both of the two primary collagen receptors, the integrin $\alpha_2\beta_1$ and the platelet-specific receptor GPVI. GPVI binds to collagens, albeit weakly, and plays an important role in augmenting the transduction of signals leading to platelet activation that had already been initiated by the GPIb complex. A more avid platelet attachment is then afforded by $\alpha_2\beta_1$ leading to a stable monolayer of activated platelets that serves as a nidus for prothrombin conversion and thrombus formation. Like many integrins, $\alpha_2\beta_1$ can undergo activation-dependent

From: *Methods in Molecular Biology, vol. 273:*
Platelets and Megakaryocytes, Vol. 2: Perspectives and Techniques
Edited by: J. M. Gibbins and M. P. Mahaut-Smith © Humana Press Inc., Totowa, NJ

increases in avidity for collagens *(3)*. The contributions of both GPVI and $\alpha_2\beta_1$ to signal transduction need to be considered in a comprehensive model of platelet activation *(4–6)*. Platelet adhesion in flowing blood mediated by $\alpha_2\beta_1$ is supported by several collagen types, including types I, III, IV, and VI *(7–11)*, and the rate of platelet mono-layer formation is directly proportional to the platelet $\alpha_2\beta_1$ density *(12)*.

In the process of platelet activation, a conformational change in the second integrin $\alpha_{IIb}\beta_3$ facilitates fibrinogen binding and platelet aggregation. Thrombin generated at the blood-plaque interface converts fibrinogen to fibrin, which stabilizes thrombus growth. Therefore, any genetic differences that might alter surface expression of key platelet receptors could influence the risk for adverse outcomes such as thrombosis. In the last five years, there has been a rapid accumulation of literature concerning the relationship between genetic variations in platelet glycoproteins and risk of coronary heart disease.

Ultimately, thrombus formation ensues, mediated by activated platelet integrin $\alpha_{IIb}\beta_3$ binding to fibrinogen and/or vWF. Other platelet-derived mediators contribute substantially to the rate of thrombus formation. The arachidonate-thromboxane A_2 pathway and secreted adenosine diphosphate (ADP) enhance platelet activation and aggregation initiated by a variety of agonists, including immobilized collagen or vWF. Variability in this response follows from the heterogeneity of receptors for both ADP and thromboxane A_2 (TXA$_2$).

Platelets synthesize TXA$_2$ upon stimulation with a variety of agonists, including thrombin, collagen, or ADP. TXA$_2$ is then released from platelets, where it stimulates platelet TXA$_2$ receptors, resulting in activation of phospholipase C and an increase in cytosolic calcium. The increase in calcium then amplifies platelet aggregation and stimulates the synthesis of additional TXA$_2$ and the release of ADP. In this manner, both TXA$_2$ and ADP participate in a positive feedback loop that leads to irreversible platelet aggregation. The antiplatelet activity of aspirin is based primarily on its ability to irreversibly inhibit platelet cyclooxygenase and thereby suppress TXA$_2$ synthesis *(13)*.

Platelets express at least three purinergic receptors, two of which interact with ADP *(14)*: the P2Y$_1$ receptor responsible for mobilization of ionized calcium from internal stores, which mediates shape change and initiates aggregation induced by ADP, and the P2Y$_{12}$ receptor coupled to adenylyl cyclase inhibition, essential for the full aggregation response to ADP *(15)*. Clopidogrel is an effective inhibitor of ADP-induced platelet aggregation because its metabolite is a specific antagonist of the P2Y$_{12}$ purinogenic receptor *(13)*. Four patients with abnormalities of platelet function due to a severe defect of P2Y$_{12}$ have been described *(14,16,17)*. Each has a life-long history of mucosal bleeding, easy bruising, and/or excessive post-operative bleeding and mildly to severely prolonged bleeding times. Their platelets show only a slight, rapidly reversible primary wave of aggregation to ADP and abnormal aggregation responses induced by colla-gen, arachidonate, and thromboxane A_2 analogs, although aggregation induced by high concentrations of thrombin is normal.

Ischemic heart disease and cerebrovascular disease are the leading causes of morbid-ity and mortality among both adult males and adult females in the developed Western world *(18–21)*. Recent evidence indicates that these diseases are now the leading cause of death among American Indians/Native Alaskans *(22)*, and their incidence is steadily increasing among Asian American *(23)* and Mexican American populations *(24)*. Epidemiological studies indicate that these diseases result from complex interactions between genetic susceptibility factors, chronic environmental influences (for example, hormonal imbalance, smoking, or obesity), together with established, intercurrent disorders (such as diabetes, hypertension, dyslipidemia, or hyperhomocysteinemia). The most devastating complication of these disorders is acute myocardial infarction, resulting from the formation of an occlusive thrombus at the site of a ruptured atherosclerotic plaque. The critical role of platelets in this process is now well accepted *(25)*.

It is the rate of platelet activation that will influence these outcomes, controlled by subtle and heritable differences in the expression and avidity of platelet receptors and key plasma glycoprotein ligands.

1.2. Platelet Glycoprotein Gene Single-Nucleotide Polymorphisms (SNPs)

The density or avidity of adhesion receptors on blood platelets would be expected to influence the rate of thrombus formation and clinical risk for negative outcomes. In previous studies *(26,27)*, we showed that integrin α_2 gene (*ITGA2*) dimorphisms regulate platelet $\alpha_2\beta_1$ density, which can vary as much as fourfold from one individual to another. At the same time, the levels and function of the platelet GPIb complex are now known to vary as a result of the inheritance of different alleles of the GPIbα gene. In addition, we have determined that the binding of a specific ligand (the snake venom protein convulxin) to platelet GPVI varies by as much as fivefold between normal donors, and that this variation correlates directly with the rate of platelet-catalyzed prothrombin conversion *(28)*.

Platelet and plasma glycoprotein gene SNPs that are associated with qualitative or quantitative differences in platelet adhesion and thrombus formation are likely to also influence the extent of dysfunction that can be achieved during low-dose aspirin therapy. A number of relevant SNPs are listed in **Table 1**. The location of each nucleotide substitution is indicated with respect to the assembled gene sequence, and the appropriate GenBank Accession Number is referenced in the right hand column.

1.2.1. ITGA2

Quantitative differences in platelet $\alpha_2\beta_1$ have been correlated with inheritance of three major *ITGA2* haplotypes. Haplotype 1 (**807T**/1648G) is associated with increased levels of $\alpha_2\beta_1$, while haplotypes 2 (**807C**/1648G) and 3 (**807C**/1648A) are each associated with decreased levels of this receptor *(12)*. The haplotype frequencies in a typical white non-Hispanic population are 0.39, 0.53, and 0.08, respectively. There is a correlation between haplotype 1 (807T) (high receptor density) and risk for arterial thrombosis in younger men with a history of myocardial infarction *(29,30)*, women who are heavy smokers *(31)*, patients with diabetic retinopathy *(32)*, and younger patients with stroke *(33)*.

To readily distinguish each of the three major *ITGA2* haplotypes, we screen for the G65265A and A65986G dimorphisms (GenBank NT_025718) that define the *Bgl* II

Table 1
Platelet and Plasma Glycoprotein SNPs Relevant to Arterial Thrombosis

| Glycoprotein | Gene | Prothrombotic SNP | | | Neutral SNP | | GenBank accession Number |
		bp	a.k.a.	Freq[a]	bp	a.k.a.	
Platelet							
Integrin α_2	ITGA2	62113T	807T	0.39	62113C	807C	NT_025718
		65265A	Bgl II site		65265G		NT_025718
		65986A	Ase I site		65986G		NT_025718
Integrin β_3	ITGB3	29519C	PlA2, Pro33	0.15	29519T	PlA1, Leu33	NT_010833
GPIbα	GP1BA	757T	Met145	0.10	757C	Thr145	NT_010823
		271C	-5C	0.10	271T	-5T	NT_010823
GPVI	GP6	13039C	Pro219	0.05	13039T	Ser219	NT_011225
TXR	TXBA2R	1915C	924C	0.19	1915T	924T	NM_001060
Other							
vWF	VWF	-1793G	-1793G	0.36	-1793C	-1793C	NT_009731
Fibrinogen Bβ	FGB	-455A	-455A	0.19	-455G	-455G	NM_005141
Interleukin-6	IL-6	1510C	-174C	0.43	1510G	-174G	AF372214

[a]White non-Hispanic

and *Ase* I restriction sites within intron G previously employed in RFLP assays *(34)*. Haplotype 1 is *Bgl* II (+) *Ase* I (neg); haplotype 2 is *Bgl* II (neg) *Ase* I (+); and haplotype 3 is *Bgl* II (neg) *Ase* I (neg).

1.2.2. ITGB3

In the integrin β_3 subunit, a point mutation results in the Leu→Pro substitution at residue 33 *(35)* that is found in about 15% of the white population. A number of reports have found an association between the expression of Pro33 (PlA2 alloantigen) and risk for thrombosis, particularly among young individuals *(36,37)*, and a notable decrease in its frequency among young survivors of myocardial infarction *(38)*. A number of other studies on large patient cohorts have failed to find an association with risk for myocardial infarction *(39,40)* or stroke *(39)*. A biological basis for a qualitative effect of PlA2 on integrin function has also been debated *(41–43)*. Interestingly, one recent meta-analysis concluded that there is not an association between PlA2 and increased thrombotic risk *(44)*, while a second recent meta-analysis *(45)* concluded that there is a weak but significant correlation.

1.2.3. GP1BA

The GPIb complex is a heptamer composed of four distinct gene products: two molecules of GPIbα, two of GPIbβ, two of GPIX and one of GPV. vWF is directly bound by the GPIbα subunits, each of which is disulfide-linked to a GPIbβ subunit. Relevant polymorphisms have been associated with the GPIbβ subunit. A threonine/methionine substitution at amino acid 145 lies within the region of ligand binding *(46,47)*. The gene frequencies of the Thr and Met alleles in a typical white population are 90% and 10%, respectively *(46,47)*. An association has been found between inheritance of the Met145 allele and risk for coronary artery disease *(48,49)* or stroke *(49,50)* in younger individuals. The biological effect of the Met/Leu substitution remains to be determined, but is thought to reflect a change in the avidity of the GPIb complex for vWF.

A T/C substitution in the region of the translation start site, at a position 5 nucleotides upstream (–5) from the initiator codon (ATG) *(51,52)* influences translation efficiency. The presence of the –5C allele (gene frequency of about 0.10 in various white populations) increases the mean level of GPIbα on the platelet plasma membrane (roughly, a 50% increase in homozygous individuals and a 33% increase in heterozygous individuals) *(51)*. There is an indication of an association between –5C and the severity of negative outcomes following acute myocardial infarction in younger individuals (≤62 yr old) *(53)*. Moreover, a recent study has documented a synergistic effect of –5C and Met145 that results in an increased risk for stroke in younger individuals *(54)*.

1.2.4. GP6

Two haplotypes have been defined. Haplotype *a* (T13254) encodes a serine at residue 219, while haplotype *b* (C13254) encodes a proline *(55)*. While this nonconservative amino acid replacement may alter GPVI function, there is no direct evidence yet that this the case. This dimorphism does not influence expression of platelet GPVI (T. J. Kunicki, unpublished observations). The GPVI 13254CC (Pro219/Pro219) genotype has been associated with risk for myocardial infarction, particularly among older females (≥60 yr old) who were smokers and carried the fibrinogen Bβ-455A haplotype (higher fibrinogen level).

1.2.5. Thromboxane A_2 Receptor (TXR)

Even though there is no direct evidence yet for an association between variation in TXR expression or activity and risk for arterial thrombosis, the pivotal role of this receptor in aspirin-inhibitable platelet function makes it worthy of closer scrutiny. There are two isoforms of the thromboxane A_2 receptor (TXRα and TXRβ) that differ only in their carboxyl-termini (via alternative slicing) and show similar ligand binding properties and phospholipase C (PLC) activation, but *oppositely* regulate adenylyl cyclase activity. TXRα activates adenylyl cyclase, while TXRβ inhibits it. Both isoforms are present in platelets. Five cases of platelet dysfunction have been reported in which an Arg60→Leu mutation (G179T SNP) of TXRα results in impaired PLC and adenylyl cyclase activation *(56,57)*. This syndrome is relatively rare, and the allele frequency of 179T in Asians (Japanese) is 0.003 *(58)*.

A second SNP in the TXR gene (T924C) has been positively associated with bronchial asthma (BA), more so in adults than in children *(58)*. This more common SNP is a synonymous mutation (allele frequency = 0.19) that may directly influence the transcription or translation rate of TXR. Alternatively, variation in other genomic elements tightly linked to this polymorphism or nearby genes may influence the phenotype of BA.

The impact of the former, rare SNP (G179T) on platelet function has been established, and 179T would be considered antithrombotic. The impact of the second more common SNP (T924C) on platelet TXR levels and its prevalence in other ethnic groups remain to be investigated.

1.2.6. ADP Receptors

To date, there is no evidence of any ADP receptor gene SNP that correlates with differences in expression or activity. Nonetheless, continued scrutiny of this receptor is also warranted.

1.3. Additional Relevant Candidate Gene SNPs

1.3.1. von Willebrand Factor

Two distinct haplotypes of the *VWF* gene have been defined that influence expression: haplotype 1 (-1793G/-1234C/-1185A/-1051G) and haplotype 2 (-1793C/-1234T/-1185G/-1051A), with frequencies of 0.36 and 0.64, respectively *(59,60)*. Individuals homozygous for haplotype 1 have the highest mean vWF:Ag levels (0.962 U/mL); heterozygotes have intermediate levels (0.867 U/mL); and those homozygous for haplotype 2 have the lowest levels (0.776 U/mL) (analysis of covariance, $p = 0.008$; Kruskal-Wallis test, $p = 0.006$) *(60)*. Thus, circulating levels of plasma vWF may be determined, in part, by polymorphic variation in the promoter region of the *VWF* gene. High levels of vWF are predictive of cardiovascular mortality in survivors of myocardial infarction *(61)*.

1.3.2. Fibrinogen

Fibrinogen is a hepatically derived acute-phase protein whose levels can rise dramatically in response to infection, smoking, or trauma. Elevated levels are an independent risk factor for coronary vascular events among healthy individuals or those with coronary artery disease *(62)*. An increase in as little as 0.1 mg/mL of plasma fibrinogen will increase cardiovascular risk by about 15% *(63)*. Since production of the fibrinogen Bβ chain is the rate-limiting step in fibrinogen biosynthesis, it is not surprising that the promoter region SNP G-455A was shown to have a direct effect on transcription *(62)*. In a recent study of 250 male army recruits undergoing basic training, fibrinogen levels (mg/mL) were significantly elevated over baseline levels 2, 48, and 96 h after a strenuous 48-h final military exercise (FME), representing increases of 15.7%, 3.4%, and 7.6%, respectively. Higher levels were attained in -455A allele carriers, relative to -455GG subjects: 3.17 ± 0.05 vs 2.94 ± 0.05 ($p < 0.001$) at 2 h post-FME; 2.86 ± 0.05 vs 2.60 ± 0.05 ($p < 0.0005$) at 48 h; and 2.98 ± 0.06 vs 2.69 ± 0.06 ($p < 0.0005$) at 96 h. At the same time, there was no effect of a second SNP G-854A on fibrinogen levels. Thus, the fibrinogen Bβ chain SNP G-455A influences fibrinogen levels following exercise.

1.3.3. IL-6

Inflammation is a key component of coronary heart disease, and genes coding for cytokines are candidates for predisposing to risk for coronary heart disease. Humphries

Table 2
Glycoprotein Prothrombotic SNP Frequencies in Different Racial/Ethnic Groups

	Haplotype frequency				
Prothrombotic SNP	White Non-Hispanic	White Hispanic	Asian	Black American	American Indian
Platelet					
ITGA2 807T	0.39	**0.51**	0.30	0.34	**0.54**
ITGB3 PlA2	0.15	nd[a]	**0.005**	nd	nd
GP1BA Met145	0.10	nd	0.08	**0.178**	0.119
GP1BA-5C	0.15	nd	nd	nd	nd
GP6 Pro219	0.05	nd	nd	nd	nd
Plasma					
VWF-1793G	0.36	nd	nd	nd	nd
FGB-455A	0.19	nd	nd	nd	nd

[a]nd, not determined

et al. *(64)* examined the effect of the –174 G→C substitution in the promoter of the IL-6 gene on risk of coronary heart disease, and on intermediate risk traits including fibrinogen and systolic blood pressure, in 2751 middle-aged healthy UK men. The –174C allele (frequency 0.43) was associated with a significantly ($p = 0.007$) higher systolic blood pressure. Compared to men with genotype GG, those carrying the –174C allele had a relative risk of coronary heart disease of 1.54 ($p = 0.048$). This effect was exacerbated among heavy smokers (compared to GG nonsmokers, relative risk = 2.66). These effects remained statistically significant even after adjusting for classical risk factors, including blood pressure ($p = 0.04$). In a subset of the genotyped men ($n = 494$), carriers of the –174C allele had higher levels of C-reactive protein than noncarriers. These data confirm the importance of the inflammatory system in the development of coronary heart disease. They suggest that, at least in part, the effect of the IL-6 –174 (G/C) SNP on blood pressure is likely to be operating through inflammatory mechanisms. The molecular mechanisms underlying the effects of these genetically determined differences in plasma levels of IL-6 remain to be determined. However, the fact that IL-6 can influence the expression of other potential risk factors, such as fibrinogen and von Willebrand factor, makes it an important target in a comprehensive study of the genetics of arterial thrombosis. For the above reasons, the IL-6 –174 (G/C) SNP should be included in any list of candidate gene SNPs relevant to arterial thrombosis.

1.4. Genotype Frequencies in Different Racial/Ethnic Groups

There are substantial differences in haplotype frequencies among racial/ethnic groups, particularly in the case of the three most prothrombotic platelet glycoprotein gene SNPs, *ITGA2* 807T, *ITGB3* PlA2, and *GP1BA* Met145 *(65–68)*. The most disparate allele frequencies are indicated in bold in **Table 2**. The lack of information on haplotype frequencies in different racial/ethnic groups is cause for concern, and increased effort needs to be placed on the demographics of platelet glycoprotein SNPs and their contribution to risk for thrombosis.

1.5. Platelet Glycoprotein Immunogenetics

Membrane glycoproteins play a pivotal role in the various aspects of the hemostatic function of platelets, and our appreciation of platelet immunogenicity has generally paralleled our progress in understanding the biology of these important receptors. These glycoproteins figure prominently as alloantigens, autoantigens, and targets of drug-dependent antibodies. Consequently, the molecular nature of glycoprotein epitopes is a major focus of this review. Platelet membrane glycolipids and phospholipids can also represent autoantigenic components, but much less is known about the structure and immunogenicity of these platelet constituents.

As a point of departure, one can classify platelet-associated alloantigens into two groups: "platelet-nonspecific" alloantigens, which are shared by platelets and many diverse cell types; and "platelet-specific" alloantigens, which are uniquely expressed by these cells.

The number of serologically defined alloantigens on platelet glycoproteins has necessitated the development of a new nomenclature, by which each is prefixed as human platelet antigen (HPA-) *(69)*, as shown in **Table 3**.

Sensitization to platelet-specific alloantigens in man can result in either of two clinically significant syndromes: neonatal alloimmune thrombocytopenia (NATP) or post-transfusion purpura (PTP). Development of platelet alloantibodies can also be a major obstacle to platelet transfusion therapy in sensitized patients, such as those undergoing chemotherapy in conjunction with myelodysplastic syndromes or hematopoietic stem cell transplantation.

1.5.1. Neonatal Alloimmune Thrombocytopenia

Maternal sensitization to paternal alloantigens on fetal platelets is the cause of neonatal alloimmune thrombocytopenic purpura (NATP). In North America, the alloantigen system most often involved in NATP is HPA-1. Responsiveness to HPA-1a shows an HLA restriction *(70–72)*, so that individuals who are homozygous for Pro_{33} (homozygous HPA-1b) and responsive to the predominant HPA-1a antigen are almost exclusively HLA DRB3*0101 *(71)* or DQB1*02 *(73,74)*. In the case of DRB3*0101, the calculated risk factor is 141, a risk level equivalent to that of the hallmark of HLA-restriction in autoimmune disease, ankylosing spondylitis, and HLA B27 *(74)*. In contrast, responsiveness of homozygous HPA-1a individuals to the HPA-1b allele is not linked to HLA. *(74,75)* T cells are the likely candidates for providing HLA restriction in this case, and Maslanka et al. *(74)* have provided elegant evidence that in one case of NATP, T cells that share CDR3 motifs are stimulated by peptides that contain the same Leu_{33} polymorphism that is recognized by anti-HPA-1a alloantibodies. In the case of another less-frequent antigen, HPA-6b, there appears to be an association between responsiveness and the MHC genes HLA DRB1*1501, DQA1*0102 or DQB1*0602 *(76)*.

Table 3
Molecular Genetics of Human Platelet Alloantigens[a]

Antigen	Synonym	Glycoprotein location	Nucleotide Substitution	Amino acid substitution	Ref.
HPA-1a	Zw[a], PI[A1]	Integrin β_3	T_{196}	Leu[33]	*35*
HPA-1b	Zw[b], PI[A2]		C_{196}	Pro[33]	
HPA-2a	Ko[b]	GPIbα	C_{524}	Thr[145]	*46*
HPA-2b	Ko[a], Sib[a]		T_{524}	Met[145]	
HPA-3a	Bak[a], Lek[a]	Integrin β_{IIb}	T_{2622}	Ile[843]	*96*
HPA-3b	Bak[b]		G_{2622}	Ser[843]	
HPA-4a	Yuk[b], Pen[a]	Integrin β_3	G_{526}	Arg[143]	*97*
HPA-4b	Yuk[b], Pen[a]		A_{526}	Gln[143]	
HPA-5a	Br[b], Zav[b]	Integrin α_2	G_{1648}	Glu[505]	*98*
HPA-5b	Br[a], Zav[a], Hc[a]		A_{1648}	Lys[505]	
HPA-6bW	Ca[a], Tu[a]	Integrin β_3	A_{1564}	Gln[489]	
			G_{1564}	Arg[489]	*99*
HPA7bW	Mo[a]	Integrin β_3	G_{1317}	Ala[407]	
			C_{1317}	Pro[407]	*100*
HPA-8bW	Sr[a]	Integrin β_3	T_{2004}	Cys[636]	*101*
			C_{2004}	Arg[636]	
HPA-9bW	Max[a]	Integrin α_{IIb}	A_{2603}	Met[837]	*102*
			G_{2603}	Val[837]	
HPA-10bW	La[a]	Integrin β_3	A_{281}	Gln[62]	*103*
			G_{281}	Arg[62]	
HPA-11bW	Gro[a]	Integrin β_3	A_{1996}	His[633]	*104*
			G_{1996}	Arg[633]	
HPA-12bW	Iy[a]	GPIbβ	A_{141}	Glu[15]	*105*
			G_{141}	Gly[15]	
HPA-13bW	Sit[a]	Integrin α_2	T_{2531}	Met[799]	*106*
			C_{2531}	Thr[799]	
HPA-	Oe[a]	Integrin β_3			*107*
HPA-	Va[a]	Integrin β_3			*108*
HPA-	Pe[a]	GPIbα			*109*
HPA-	Gov[a/b]	CD109			*110*

[a]Adapted from Santoso and Kiefel *(111)*.

Substantial clinical and serologic evidence documents the most frequent antigen targets in NATP. In a large series *(77)*, 78% of serologically confirmed cases were due to anti-HPA-1a and 19% to anti-HPA-5b. All other specificities accounted for no more than 5% of cases. An association of NATP with other alloantigens, such as HPA-3a, HPA-3b, HPA-1b, or HPA-2b, has been noted, but is much less frequent *(78–81)*. Obviously, differences in allelic gene frequencies between different racial or ethnic populations will have an important impact on the frequency of responsiveness to a particular alloantigen (*see* **Table 2**). Thus, among Asians anti-HPA-1a has never been shown to be involved in NATP, and antibodies specific for HPA-4b play a dominant clinical role *(82)*. This is probably because the gene frequency for the HPA-1b allele among the Japanese (0.02) is much lower than that found in Western populations (0.15). Conversely, the gene frequency of the HPA-4b allele in Japan (0.0083) is higher than that observed in Western populations (<0.001).

1.5.2. Post-Transfusion Purpura

Following an immunogenic blood (platelet) transfusion, post-transfusion purpura (PTP) can result within 7 to 10 d. It most often affects previously nontransfused, multiparous women. As with NATP, there is an increased risk of developing PTP among HLA-DR3-positive individuals, and HPA-1a is the antigen most often implicated (in Western populations) *(70,83)*.

Even though they are antigen-negative, the recipient's platelets in PTP are cleared very rapidly from the circulation. The exact mechanism by which this occurs remains to be proved, but a number of proposals have been put forward. Firstly, it has been proposed that during the initial phase of PTP, the recipient develops antibodies that recognize "framework" determinants (conserved protein structures surrounding the specific polymorphic sites). These would then react with all allelic forms of the antigen-bearing glycoprotein, including that of the recipient. In a second proposed scenario, the recipient's antibodies form immune complexes with soluble antigens originating from the donor platelets. These complexes could then interact with recipient platelets via an Fc-receptor-dependent mechanism. In the third scenario, soluble antigen from the transfused product is adsorbed onto recipient platelets, rendering them passively positive for the antigen in question. The recipient's antibody would then bind to his or her own platelets via this passively acquired form of "autoimmunity." Platelet-membrane microparticles are known to be a constituent of fresh frozen plasma and platelet concentrates and seem to be generated in particularly high levels during platelet activation *(84)*. It is conceivable that the $\alpha_{IIb}\beta_3$ complex within these membrane microparticles could become adsorbed onto neighboring platelets. Conclusive evidence to support or refute any one of these proposals is currently lacking, however.

1.6. DNA-Array-Based Genotyping Assay

Numerous techniques for genotyping single-nucleotide polymorphisms and short insertion/deletion mutations have been developed in recent years *(85–87)*. The use of primer extension-based techniques such as minisequencing or genetic bit analysis (GBA) for resequencing and genotyping of mutations has been demonstrated in a miniaturized glass slide format *(88–90)*, gel-based formats *(91)*, and homogenous assay formats *(92)* and have gained wide acceptance as a reliable, cost-effective approach to SNP genotyping. To facilitate the genotyping of large numbers of donors, we have recently adapted a primer extension-based assay, shown schematically in **Fig. 1**, using fluorescent-labeled ddNTPs to genotype polymorphisms in template DNA. PCR-amplified template DNA is hybridized to glass slide-bound minisequencing primers. Addition of a DNA polymerase with ddNTPs labeled with detectable haptens (Bio = biotin, Dig = digoxygenin) results in single base extension of the minisequencing primers. The identification of the label tags

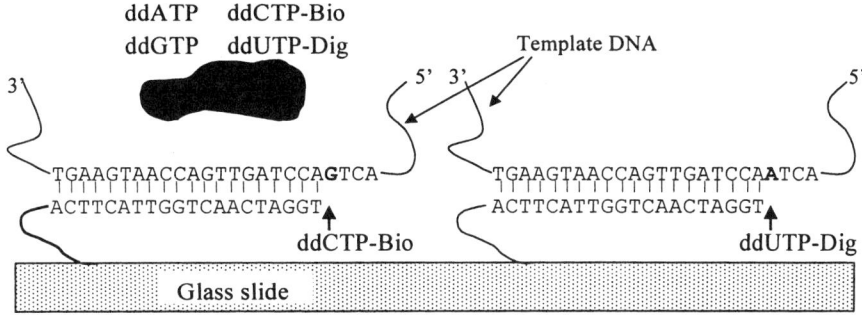

Fig. 1. Genetic bit analysis (GBA). Disulfide-tagged minisequencing primers are attached to mercaptosilane-coated glass plates by thioldisulfide exchange. PCR-amplified template DNA is hybridized to these minisequencing primers such that the polymorphic base adjacent to the 3′ end of the mini sequencing primer. Primer extension using a DNA polymerase and ddNTPs labeled with detectable haptens (Bio = biotin; Dig = digoxygenin) results in single-base extension of the minisequencing primers. The identification of the label tags indicates which nucleotide was incorporated in the extension reaction, thus identifying the complementary base on the template strand. Biotin is detected with phycoerythrin-conjugated streptavidin; digoxygenin is detected with Cy5-conjugated anti-digoxygenin.

indicates which nucleotides were incorporated in the extension reaction, thus identifying the complimentary base in the template strand. Using this approach, we have already developed the conditions to genotype for five glycoprotein gene SNPs listed in **Table 1**.

In the case of *ITGA2*, haplotype 1 bears the 807T allele (prothrombotic), while haplotypes 2 and 3 share the 807C allele. In the past, we used an *Bgl* II/*Ase* I RFLP to type for the three haplotypes *(34)*. Our DNA array-based approach utilizes the same two linked *ITGA2* loci. In **Fig. 2**, T at Locus 1 is equivalent to the previously described *Bgl*2 restriction site **(Fig. 2A)**; T at locus 2 represents the previously described *Ase* I restriction site **(Fig. 2B)**. The interpretation of the results is straightforward. For example, if a donor is homozygous for haplotype 1, then they would be homozygous TT at locus 1 and homozygous TT at locus 2. The typing for the Thr/Met-145 alleles of *GP1BA* is depicted in **Fig. 2C**. The typing of the Leu-219/Pro-219 dimorphism of GPVI is depicted in **Fig. 2D**.

The following method describes our primer extension assay, carried out on a microarray, to screen for polymorphisms within samples of DNA.

2. Materials

2.1. PCR Amplification of the Target Polymorphic DNA Fragment

1. Synthetic oligonucleotide forward and reverse primers with four phosphorothioated bases at the 5′ end of one primer of each pair (Invitrogen, Carlsbad, CA). Oligonucleotides are diluted to a 200 μM stock solution in TE (10 mM Tris-HCl pH 8.0, 1 mM EDTA).

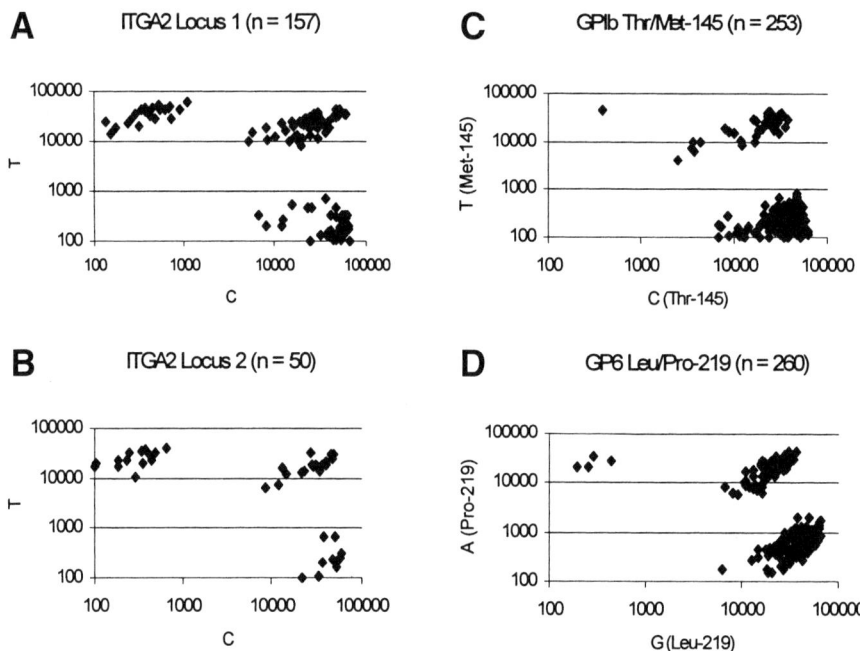

Fig. 2. Detection of platelet glycoprotein gene SNPs by GBA. The results with four representative SNPs are depicted: **(A)** Integrin α_2 *(ITGA2)* Locus 1 (*BGl* II site in intron G described in **ref. 34**); **(B)** Integrin α_2 *(ITGA2)* Locus 2 (*Ase* I site in intron G described in **ref. 34**); **(C)** Glycoprotein Ibα *(GP1BA)* C75T encoding Thr-145 versus Met-145; and **(D)** Glycoprotein VI *(GP6)* C13039T encoding Leu-219 vs Pro-219. In each case, one of the alleles is detected with Cy5-conjugated anti-digoxygenin and the other with phycoerythrin-conjugated streptavidin (*see* **Subheading 3.**) The corresponding fluorescence units are plotted on the X and Y axes (log scale). Each data point is the average of duplicate assays of single donor DNA. *n* = number of donors tested.

2. dNTPs: 100 m*M* stock each of dA, dG, dC, and dT (Invitrogen).
3. Amplitaq DNA polymerase (PerkinElmer, Foster City, CA).
4. Reaction buffer: 50 m*M* KCl, 200 m*M* Tris-HCl (pH 8.4), 1.5 m*M* MgCl$_2$ normally prepared as a 10X buffer, then master mix is made up to volume with deionized H$_2$O.
5. MicroAmp® Optical 96-Well Reaction Plates (Applied Biosystems Inc., Foster City, CA).
6. 96-Well GeneAmp® PCR System 9700 (Applied Biosystems Inc.).
7. DNA samples: genomic DNA is isolated using Puregene DNA Purification (Gentra, Minneapolis, MN) (*see* **Note 1**).

2.2. Conversion of Double-Stranded PCR Product to Single-Stranded DNA Template

1. T7 gene 6 exonuclease (Amersham, Arlington Heights, IL). Dilute the enzyme to a stock concentration of 6.25 U/µL in the buffer supplied by the manufacturer.

2. 4.5 m*M* NaCl stock.
3. Cetyltrimethylammonium bromide (Sigma, St Louis, MO).
4. 100 m*M* Ethylenediaminetetraacetic acid (Na$_2$EDTA) pH 7.5.

2.3. Preparation of Glass Slides for Minisequencing Assays

1. Glass slides (SR204; Erie Scientific Co., Portsmouth, NH).
2. 25% ammonium hydroxide (A6899; Sigma-Aldrich, St. Louis, MO).
3. 3-mercaptopropyltrimethoxysilane (M8500; United Chemical Technologies, Bristol, PA).
4. Vacuum jar.
5. Vacuum oven.

2.4. Synthesis of Disulfide Oligonucleotides

1. ABI 8909 Synthesizer with MOSS attachment (Applied Biosystems).
2. 1-*O*-dimethoxytrityl-hexyl-disulfide,1′-((2-cyanoethyl)-(*N,N*-diisopropyl)(-phosphoramidite (C6 S-S; catalog no. 10-1936-90; Glen Research, Sterling, VA).
3. Dry acetonitrile.
4. NAP10 columns (17-0854-02, Amersham Pharmacia, Piscataway, NJ).
5. Oliogonucleotide primers: All oligonucleotide synthesis reagents are purchased from Glen Research. Oligonucleotides are synthesized using standard phosphoramidite chemistry at the 0.2 µ*M* scale. Oligonucleotides are 25 bases in length and designed to be complementary to the template strand that is protected from the nuclease (in **Subheading 3.2.**) by the phosphorothioated bases introduced by the PCR primers during template amplification.
6. 10 m*M* Tris-HCl/1 m*M* EDTA, pH 8.3.

2.5. Immobilization of Oligonucleotides Onto Glass Slides

1. 0.5 *M* carbonate buffer, pH 9.6 (C3041; Sigma-Aldrich).
2. TNTw: 10 m*M* Tris-HCl (pH 7.5), 150 m*M* NaCl, and 0.05% Tween-20.
3. Dry nitrogen.
4. 16-pin robotic arrayer (Robotic Labware Designs, Encinitas, CA).

2.6. Hybridization of Template DNA to Glass Slide Oligonucleotide Arrays

1. TNTw: 10 m*M* Tris-HCl (pH 7.5), 150 m*M* NaCl, 0.05% Tween-20.

2.7. Extension Reactions and Signal Detection

1. Klenow fragment (3′→5′ exonuclease-free) of *Escherichia coli* DNA polymerase (USB, Cleveland, OH).
2. 1 m*M* ddNTPs labeled with biotin or digoxygenin (NEN, Boston, MA).
3. 200 µ*M* unlabeled ddNTPs (Amersham-Pharmacia Biotech).
4. Extension buffer: 20 m*M* Tris-HCl (pH 7.5), 10 m*M* MgCl$_2$, 25 m*M* NaCl, 10 m*M* MnCl$_2$, 15 m*M* sodium isocitrate normally prepared as a 10X buffer, then master mix is made up to volume with deionized H$_2$O.
5. Anti-digoxin antibody conjugated to cy5 (e.g., Cy-5 conjugated mouse monoclonal anti-digoxin, catalog no. 200-172-156, Jackson ImmunoResearch Laboratories, Inc., West Grove, PA).
6. Streptavidin, R-phycoerythrin conjugate (Molecular Probes, Eugene, OR).
7. 100 m*M* NaOH.
8. TNTw (*see* **Subheading 2.6.**).

9. BSA-fraction V (Sigma).
10. ScanArray 5000 (Packard Bioscience, Downers Grove, IL) confocal scanner equipped with 488 nm argon, 543 nm HeNe, 590 nm HeNe, and 632 nm HeNe lasers.
11. Imagene image analysis software (Biodiscovery, Marina Del Ray, CA).
12. Excel (Microsoft).

3. Methods

The overall scheme for genotyping by minisequencing requires several steps: (1) PCR amplification of the target polymorphism, (2) conversion of double-stranded PCR product to single stranded template, (3) hybridization of the template to the glass-slide-bound minisequencing primer, (4) reaction of the minisequencing primer and hybridized template with a polymerase and labeled ddNTPs, and (5) detection of the resulting signal for genotype calling (**Fig. 1**).

3.1. PCR Amplification of the Target Polymorphic DNA Fragment

Polymorphic DNA fragments are first amplified by PCR according to a previously published method *(93)*. Standard precautions to avoid PCR contamination should always be observed during this step, including conducting sample preparation and PCR reaction assembly in a "clean area," using dedicated pipetting tools, and wearing clean gloves at all times.

1. Design PCR primer pairs to amplify specific fragments of DNA containing each SNP listed in **Table 1**. One primer in each pair is synthesized with four phosphorothioated bases at its 5′ end, while the second primer is unmodified.
2. Set up 15-μL reaction mixtures for PCR in 96-well polypropylene MicroAmp Optical 96-well reaction plates. The final concentration of the reaction mixture is 400 μ*M* of each dNTP, 0.5 μ*M* of each primer, 2.0 ng/μL DNA sample, and 0.025 U/μL Amplitaq DNA polymerase in reaction buffer.
3. Perform PCR reactions in a 96-well GeneAmp PCR System 9700 as follows:
   a. An initial 2-min denaturation step at 94°C.
   b. Further denaturation step (1 min at 94°C).
   c. Annealing step (2 min at 55°C).
   d. Extension step (3 min at 72°C).
   e. Repeat **steps b–d** for a further 35 cycles.
   f. After final extension step, cool samples to 4°C.

3.2. Conversion of Double-Stranded PCR Product to Single-Stranded DNA Template

Double-stranded PCR products amplified with one phosphorothioated primer and one unmodified primer are converted to single-strand templates by selective exonuclease digestion, using a published method *(94)* (*see* **Note 2**).

1. Add 0.06 U/μL of T7 gene 6 exonuclease to the PCR product.
2. After a 30-min incubation at ambient temperature, add NaCl, cetyltrimethylammonium bromide, and EDTA to the single-stranded DNA template to a final concentration of 1.5 *M*, 1 m*M*, and 10 m*M*, respectively, to stop digestion and aid in the subsequent hybridization to arrayed GBA primers (described in **Subheading 3.6.**).

3.3. Preparation of Glass Slides for Minisequencing Assays

1. Immerse glass slides in 25% ammonium hydroxide for 1 h.
2. Rinse in boiling distilled water for 20 min and air-dry.
3. Place the slides in a vacuum jar along with a small beaker containing 10 mL of 3-mercaptopropyltrimethoxysilane, and then place the jar under vacuum.
4. Allow vapor deposition to proceed overnight.
5. Cure the slides in a vacuum oven for 5 h at 143°C.
6. Cool the oven to room temperature before removing the slides, which are then arrayed within 1–4 h after removal.

3.4. Synthesis of Disulfide Oligonucleotides

5′-Disulfide-modified oligonucleotides may be purchased commercially from most oligonucleotide vendors or may be synthesized in-house. We synthesize our own disulfide-modified oligonucleotides using an ABI 8909 Synthesizer with MOSS attachment employing standard phosphoramidite chemistry at the 0.2 μmole scale. The disulfide linkage is added to the 5′-terminus of the oligonucleotides using a 5′-thiol modifier, 1-*O*-dimethoxytrityl-hexyl-disulfide,1′-[(2-cyanoethyl)-(*N,N*-diisopropyl)]-phosphoramidite at a concentration of 100 m*M* in dry acetonitrile, as described *(95)*. The coupling reaction proceeds for 90 s, after which the trityl groups are removed. The oligonucleotides are deprotected using a standard protocol and desalted employing NAP10 columns. After desalting, the oligonucleotides are diluted in 10 m*M* Tris-HCl/1 m*M* EDTA, pH 8.3, to a 200 μ*M* stock solution and stored at −80°C until use.

3.5. Immobilization of Oligonucleotides onto Glass Slides

The 5′-disulfide oligonucleotides are attached to the glass surface via the intermediate mercaptosilane layer using a thiol/disulfide exchange reaction *(95)*.

1. Dilute the 5′-disulfide oligonucleotides (from 200 μ*M* stock concentrations) to 10 μ*M* in 0.5 *M* carbonate buffer, pH 9.6.
2. Array oligonucleotides onto the mercaptosilane-coated glass slides using a 16-pin robotic arrayer according to the manufacturer's instructions.
3. Immediately after printing, incubate the slides in a humid chamber for 1.5 h.
4. Rinse each slide three times in TNTw, rinse once in distilled water, air-dry, and store under dry nitrogen until use.

3.6. Hybridization of Template DNA to Glass Slide Oligonucleotide Arrays

1. Add 1 μL of the DNA template solution to each well of the glass slide array for hybridization.
2. After 30 min of hybridization at room temperature, rinse the slides in TNTw.

3.7. Extension Reactions and Signal Detection

Genotyping primers hybridized to PCR-amplified templates are extended using Klenow fragment (3′→5′ exonuclease-free) of *E. coli* DNA polymerase and ddNTPs, one or more of which are labeled with biotin or digoxygenin. Extension reactions are performed under conditions described previously *(88)*.

1. Make 600 µL of extension mix containing 1.5 µ*M* of one biotin- and one digoxygenin-labeled ddNTP, 1.5 µ*M* of each of the remaining two unlabeled ddNTPs, and 0.025 U/µL Klenow fragment in extension buffer.
2. Add 3 µL extension mix to each DNA template-hybridized array. Arrays having no hybridized DNA template are also extended to control for template independent noise (TIN; *see* **Note 3**). Extension conditions are optimized for each hybridized template.
3. Carry out extension reactions at the ambient temperature for 15 min.
4. Briefly rinse the slides in 100 m*M* NaOH, then in TNTw followed by addition of a 1:500 dilution of anti-digoxin antibody conjugated to cy5 and 1:500 dilution of streptavidin R-phycoerythrin conjugate.
5. After incubation for 15 min, rinse slides in TNTw and scan using a ScanArray 5000 confocal scanner equipped with 488 nm argon, 543 nm HeNe, 590 nm HeNe, and 632 nm HeNe lasers to detect the signals of the labeled ddNTPs.
6. Quantify fluorescent signals using Imagene image analysis software and export into Microsoft Excel for further analysis. Data collected from samples is compiled on an X-Y scatter plot along with control samples of known genotype. Genotypes are called based on data clustering (*see* **Fig. 2**, **Note 4**).

3.8. Summary

Platelet glycoprotein polymorphisms figure prominently in the efficiency of normal hemostasis, in the risk for pathological outcome of thrombosis, and in the immunogenicity of the platelet. In general, the importance of platelet glycoprotein polymorphisms as genetic risk factors for arterial thrombosis is a new area of human genomics that needs to be carefully addressed. Discrepancies in the degree to which they are reported to contribute to risk for clinical thrombosis will be resolved only once there is a universal standard for clinical study design. At the present time, there is substantial evidence that the integrin β_3 PlA2 haplotype, the GPIbα Met145 haplotype, the GPIbα -5C haplotype, and the integrin α_2 haplotype 1 (807T) can each contribute to the risk for and morbidity of thrombotic disease. There may remain dispute as to the extent of their contribution. However, well-designed, large, prospective, genetic, and epidemiologic studies are needed to clarify the role of these and other platelet-receptor polymorphisms. DNA array-based methods for identifying glycoprotein SNPs provide a rapid and reproducible means to accomplish large-scale genotyping. Most importantly, the cumulative effects of multiple platelet and plasma glycoproteins SNPs to thrombotic risk must be evaluated concurrently. Additional in vitro studies of the functional relevance underlying these polymorphisms are needed to provide a sound biological explanation for the results of clinical correlations. The opportunity now exists to make significant inroads into the development of strategies for the prevention of thrombotic disease.

4. Notes

1. Roughly 50 ng of genomic DNA is needed for each PCR reaction. The minimum amount necessary to obtain a product is probably 10 ng, but at that DNA concentration, there is an undesirable increase in the risk of PCR contamination and PCR failure.
2. Since the 5′→3′ hydrolytic activity of T7 gene 6 exonuclease is completely inhibited by the presence of four phosphorothioates, digestion by this enzyme results in the selective protection of the phosphorothioated strand.

3. *Potential problems:* Primer extension assays such as minisequencing have been well developed in multiple formats. Problems related to self-extension of minisequencing primers (template-independent noise, TIN) occasionally occur if steps are not taken to eliminate them in advance. Modification of the primer sequence through the introduction of single-base changes is usually sufficient to eliminate any self-complimentarity of the mini-sequencing primer's 3′ end without significant loss of template hybridization or polymerase extension efficiency. Occasionally, multiple priming sites may exist within a given template sequence, resulting in predictable and reproducible "template-dependent noise" or TDN. TDN can be addressed in two ways: (a) through redesign of the PCR primers so that the secondary priming site is eliminated from the template sequence or (b) genotyping of the opposite strand, eliminating the problematic sequence from the template strand.

Other potential problems have been reported with regard to allele-specific amplification due to uncharacterized polymorphisms that may occur under the PCR priming sites. While this problem rarely occurs when genotyping well-studied loci, it can occur in the development of assays for newly discovered polymorphisms where the surrounding sequence may not be sufficiently characterized. This problem can often be detected as multiple groupings of data points during cluster analysis. For example, the heterozygotes may fall into two discrete groups, both clearly distinguishable from the homozygous samples. This results from differences in amplification efficiency due to the presence or absence of a mismatch between the PCR primer and its target sequence. If the mismatch is proximal to the 3′ end of the PCR primer, one allele may remain completely unamplified, resulting in erroneous genotype calling. This type of genotyping error is best resolved by performing standard Sanger sequencing on loci yielding suspicious scatter plots or unusually high genotyping error rates.

4. Genotyping errors may be detected through two methods: (a) Mendelian errors can be readily detected if genotyping results conflict with pedigree data, and (b) random experimental error rates may be calculated by repeat genotyping of a subset of samples at each locus. To address the issue of genotyping errors, one can genotype approximately 10% of the population samples at least three times at each locus. Any locus that shows a greater than 0.54% error rate as measured by repeat genotyping or by Mendelian error calculations is flagged for further study using gel-based Sanger sequencing of selected samples. As an illustration, preliminary error analysis from repeat genotyping of the *ITGA2*-Locus no. 1 SNP indicates an overall failure rate of <4% and a mistyping rate of <1/355 (no inconsistencies in genotype calls were obtained when 92 samples were genotyped four times). These results were obtained from four replicate genotyping experiments that included all steps in the genotyping process (from PCR through data analysis).

References

1. Moroi, M., Jung, S. M., Nomura, S., Sekiguchi, S., Ordinas, A., and Diaz-Ricart, M. (1997) Analysis of the involvement of the von Willebrand factor-glycoprotein Ib interaction in platelet adhesion to a collagen-coated surface under flow conditions. *Blood* **90,** 4413–4424.
2. Savage, B., Almus-Jacobs, F., and Ruggeri, Z. M. (1998) Specific synergy of multiple substrate-receptor interactions in platelet thrombus formation under flow. *Cell* **94,** 657–666.
3. Jung, S. M. and Moroi, M. (1998) Platelets interact with soluble and insoluble collagens through characteristically different reactions. *J. Biol. Chem.* **273,** 14,827–14,837.
4. Kamiguti, A. S., Theakston, R. D., Watson, S. P., Bon, C., Laing, G. D., and Zuzel, M. (2000) Distinct contributions of glycoprotein VI and $\alpha_2\beta_1$ integrin to the induction of

platelet protein tyrosine phosphorylation and aggregation. *Arch. Biochem. Biophys.* **374,** 356–362.

5. Kehrel, B., Wierwille, S., Clemetson, K. J., Anders, O., Steiner, M., Knight, C. G., et al. (1998) Glycoprotein VI is a major collagen receptor for platelet activation: It recognizes the platelet-activating quaternary structure of collagen, whereas CD36, glycoprotein IIb/IIIa, and von Willebrand factor do not. *Blood* **91,** 491–499.

6. Ichinohe, T., Takayama, H., Ezumi, Y., Arai, M., Yamamoto, N., Takahashi, H., et al. (1997) Collagen-stimulated activation of Syk but not c-Src is severely compromised in human platelets lacking membrane glycoprotein VI. *J. Biol. Chem.* **272,** 63–68.

7. Santoro, S. A., Rajpara, S. M., Staatz, W. D., and Woods, V. L., Jr. (1988) Isolation and characterization of a platelet surface collagen binding complex related to VLA-2. *Biochem. Biophys. Res. Commun.* **153,** 217–223.

8. Kunicki, T. J., Nugent, D. J., Staats, S. J., Orchekowski, R. P., Wayner, E. A., and Carter, W. G. (1988) The human fibroblast class II extracellular matrix receptor mediates platelet adhesion to collagen and is identical to the platelet glycoprotein Ia-IIa complex. *J. Biol. Chem.* **263,** 4516–4519.

9. Pischel, K. D., Bluestein, H. G., and Woods, V. L. (1988) Platelet glycoprotein Ia,Ic, and IIa are physicochemically indistinguishable from the very late activation antigens adhesion-related proteins of lymphocytes and other cell types. *J. Clin. Invest.* **81,** 505–513.

10. Takada, Y., Wayner, E. A., Carter, W. G., and Hemler, M. E. (1988) Extracellular matrix receptors, ECMRII and ECMRI, for collagen and fibronectin correspond to VLA-2 and VLA-3 in the VLA family of heterodimers. *J. Cell. Biochem.* **37,** 385–393.

11. Saelman, E. U. M., Nieuwenhuis, H. K., Hese, K. M., De Groot, P. G., Heijnen, H. F. G., Sage, E. H., et al. (1994) Platelet adhesion to collagen types I through VIII under conditions of stasis and flow is mediated by GPIa/IIa (alpha 2 beta 1 integrin). *Blood* **83,** 1244–1250.

12. Kritzik, M., Savage, B., Nugent, D. J., Santoso, S., Ruggeri, Z. M., and Kunicki, T. J. (1998) Nucleotide polymorphisms in the alpha 2 gene define multiple alleles which are associated with differences in platelet alpha 2 beta 1. *Blood* **92,** 2382–2388.

13. Clutton, P., Folts, J. D., and Freedman, J. E. (2001) Pharmacological control of platelet function. *Pharmacol. Res.* **44,** 255–264.

14. Cattaneo, M., Lecchi, A., Randi, A. M., McGregor, J. L., and Mannucci, P. M. (1992) Identification of a new congenital defect of platelet function characterized by severe impairment of platelet responses to adenosine diphosphate. *Blood* **80,** 2787–2796.

15. Hollopeter, G., Jantzen, H. M., Vincent, D., Li, G., England, L., Ramakrishnan, V., et al. (2001) Identification of the platelet ADP receptor targeted by antithrombotic drugs. *Nature* **409,** 202–207.

16. Cattaneo, M., Lecchi, A., Lombardi, R., Gachet, C., and Zighetti, M. L. (2000) Platelets from a patient heterozygous for the defect of P2$_{CYC}$ receptors for ADP have a secretion defect despite normal thromboxane A$_2$ production and normal granule stores: further evidence that some cases of platelet "primary secretion defect" are heterozygous for a defect of P2$_{CYC}$ receptors. *Arterioscler. Thromb. Vasc. Biol.* **20,** E101–E106.

17. Nurden, P., Savi, P., Heilmann, E., Bihour, C., Herbert, J. M., Maffrand, J. P., et al. (1995) An inherited bleeding disorder linked to a defective interaction between ADP and its receptor on platelets. Its influence on glycoprotein IIb-IIIa complex function. *J. Clin. Invest.* **95,** 1612–1622.

18. (2001) Mortality from coronary heart disease and acute myocardial infarction—United States, 1998. *MMWR Morb. Mortal. Wkly. Rep.* **50,** 90–93.

19. Bedinghaus, J., Leshan, L., and Diehr, S. (2001) Coronary artery disease prevention: what's different for women? *Am. Fam. Physician* **63**, 1393–1396.
20. Brass, L. M. (2000) The impact of cerebrovascular disease. *Diabetes Obes. Metab.* **2 (Suppl.2),** S6–S10.
21. Pellicano, R., Oliaro, E., Gandolfo, N., Aruta, E., Mangiardi, L., Orzan, F., et al. (2000) Ischemic cardiovascular disease and Helicobacter pylori. Where is the link? *J. Cardiovasc. Surg. (Torino)* **41**, 829–833.
22. Harwell, T. S., Gohdes, D., Moore, K., McDowall, J. M., Smilie, J. G., and Helgerson, S. D. (2001) Cardiovascular disease and risk factors in Montana American Indians and non-Indians. *Am. J. Prev. Med.* **20**, 196–201.
23. Kauffmann-Zeh, A., Dhand, R., and Allen, L. (2000) Vascular biology. *Nature* **407**, 219.
24. Luepker, R. V. (2001) Cardiovascular disease among Mexican Americans. *Am. J. Med.* **110**, 147–148.
25. Rauch, U., Osende, J. I., Fuster, V., Badimon, J. J., Fayad, Z., and Chesebro, J. H. (2001) Thrombus formation on atherosclerotic plaques: pathogenesis and clinical consequences. *Ann. Intern. Med.* **134**, 224–238.
26. Kunicki, T. J., Orchekowski, R., Annis, D., and Honda, Y. (1993) Variability of integrin alpha 2 beta 1activity on human platelets. *Blood* **82**, 2693–2703.
27. Kunicki, T. J., Kritzik, M., Annis, D. S., and Nugent, D. J. (1997) Hereditary variation in platelet integrin $\alpha_2\beta_1$ copy number is associated with two silent polymorphisms in the α_2 gene coding sequence. *Blood* **89**, 1939–1943.
28. Furihata, K., Clemetson, K. J., Deguchi, H., and Kunicki, T. J. (2001) Variation in human platelet glycoprotein VI content modulates glycoprotein VI-specific prothrombinase activity. *Arterioscler. Thromb. Vasc. Biol.* **21**, 1857–1863.
29. Moshfegh, K., Wuillemin, W. A., Redondo, M., Lammle, B., Beer, J. H., Liechti-Gallati, S., et al. (1999) Association of two silent polymorphisms of platelet glycoprotein Ia/IIa receptor with risk of myocardial infarction: a case-control study. *Lancet* **353**, 351–354.
30. Santoso, S., Kunicki, T. J., Kroll, H., Haberbosch, W., and Gardemann, A. (1999) Association of the platelet glycoprotein Ia $C_{807}T$ gene polymorphism with myocardial infarction in younger patients. *Blood* **93**, 2449–2453.
31. Roest, M., Banga, J. D., Grobbee, D. E., De Groot, P. G., Sixma, J. J., Tempelman, M. J., et al. (2000) Homozygosity for 807 T polymorphism in α_2 subunit of platelet $\alpha_2\beta_1$ is associated with increased risk of cardiovascular mortality in high-risk women. *Circulation* **102**, 1645–1650.
32. Matsubara, Y., Murata, M., Maruyama, T., Handa, M., Yamagata, N., Watanabe, G., et al. (2000) Association between diabetic retinopathy and genetic variations in $\alpha_2\beta_1$ integrin, a platelet receptor for collagen. *Blood* **95**, 1560–1564.
33. Carlsson, L. E., Santoso, S., Spitzer, C., Kessler, C., and Greinacher, A. (1999) The alpha 2 gene coding sequences T807/A873 of the platelet collagen receptor integrin $\alpha_2\beta_1$ might be a genetic risk factor for the development of stroke in younger patients. *Blood* **93**, 3583–3586.
34. Jacquelin, B., Tarantino, M., Kritzik, M., Rozenshteyn, D., Koziol, J. A., and Kunicki, T. J. (2001) Allele-dependent transcriptional regulation of the human integrin alpha 2 gene. *Blood* **97**, 1721–1726.
35. Newman, P. J., Derbes, R. S., and Aster, R. H. (1989) The human platelet alloantigens, PLA1 and PLA2, are associated with a leucine33/proline33 amino acid polymorphism in membrane glycoprotein IIIa, and are distinguishable by DNA typing. *J. Clin. Invest.* **83**, 1778–1781.

36. Carter, A. M., Ossei-Gerning, N., and Grant, P. J. (1996) Platelet glycoprotein IIIa PI^A polymorphism in young men with myocardial infarction. *Lancet* **348**, 485–486.
37. Weiss, E. J., Bray, P. F., Tayback, M., Schulman, S. P., Dickler, T. S., Becker, L. C., et al. (1996) A polymorphism of a platelet glycoprotein receptor as an inherited risk factor for coronary thrombosis. *N. Engl. J. Med.* **334**, 1090–1094.
38. Ardissino, D., Mannucci, P. M., Merlini, P. A., Duca, F., Fetiveau, R., Tagliabue, L., et al. (1999) Prothrombotic genetic risk factors in young survivors of myocardial infarction. *Blood* **94**, 46–51.
39. Ridker, P. M., Hennekens, c. H., Schmitz, C., Stampfer, M. J., and Lindpaintner, K. (1997) PI$^{A1/A2}$ polymorphism of platelet glycoprotein IIIa and risks of myocardial infarction, stroke, and venous thrombosis. *Lancet* **349**, 385–388.
40. Herrmann, S. M., Poirier, O., Marques-Vidal, P., Evans, A., Arveiler, D., Luc, G., et al. (1977) The Leu^{33}/Pro polymorphism (PI^{A1}/PI^{A2}) of the glycoprotein IIIa (GPIIIa) receptor is not related to myocardial infarction in the ECTIM study. *Thromb. Haemost.* **77**, 1179–1181.
41. Vijayan, K. V., Goldschmidt-Clermont, P. J., Roos, C., and Bray, P. F. (2000) The PI^{A2} polymorphism of integrin β_3 enhances outdie-in signaling and adhesive functions. *J. Clin. Invest.* **105**, 793–802.
42. Michelson, A. D., Furman, M. I., Goldschmidt-Clermont, P., Mascelli, M. A., Hendrix, C., Coleman, L., et al. (2000) Platelet GP IIIa PI(A) polymorphisms display different sensitivities to agonists. *Circulation* **101**, 1013–1018.
43. Bennett, J. S., Catella-Lawson, F., Rut, A. R., Vilaire, G., Qi, W., Kapoor, S. C., et al. (2001) Effect of the PI^{A2} alloantigen on the function of β_3-integrins in platelets. *Blood* **97**, 3093–3099.
44. Zhu, M. M., Weedon, J., and Clark, L. T. (2000) Meta-analysis of the association of platelet glycoprotein IIIa PlA1/A2 polymorphism with myocardial infarction. *Am. J. Cardiol.* **86**, 1000–1005.
45. Di Castelnuovo, A., De Gaetano, G., Donati, M. B., and Iacoviello, L. (2001) Platelet glycoprotein receptor IIIa polymorphism PLA1/PLA2 and coronary risk: a meta-analysis. *Thromb. Haemost.* **85**, 626–633.
46. Kuijpers, R. W. A. M., Faber, N. M., Cuypers, H. T. M., Ouwehand, W. H., and Von dem Borne, A. E. G. K. (1992) NH$_2$-Terminal globular domain of human platelet glycoprotein Iba has a methionine^{145}/threonine^{145} amino acid polymorphism, which is associated with the HPA-2 (Ko) alloantigens. *J. Clin. Invest.* **89**, 381–384.
47. Murata, M., Furihata, K., Ishida, F., Russell, S. R., Ware, J., and Ruggeri, Z. M. (1992) Genetic and structural characterization of an amino acid dimorphism in glycoprotein Ibα involved in platelet transfusion refractoriness. *Blood* **79**, 3086–3090.
48. Murata, M., Matsubara, Y., Kawano, K., Zama, T., Aoki, N., Yoshino, H., et al. (1997) Coronary artery disease and polymorphisms in a receptor mediating shear stress-dependent platelet activation. *Circulation* **96**, 3281–3286.
49. Gonzalez-Conejero, R., Lozano, M. L., Rivera, J., Corral, J., Iniesta, J. A., Moraleda, J. M., et al. (1998) Polymorphisms of platelet membrane glycoprotein Ibα associated with arterial thrombotic disease. *Blood* **92**, 2771–2776.
50. Sonoda, A., Murata, M., Ito, D., Tanahashi, N., Ohta, A., Tada, Y., et al. (2000) Association between platelet glycoprotein Ib alpha genotype and ischemic cerebrovascular disease. *Stroke* **31**, 493–497.
51. Afshar-Kharghan, V., Li, C. Q., Khoshnevis-Asl, M., and Lopez, J. A. (1999) Kozak sequence polymorphism of the glycoprotein (GP) Ibalpha gene is a major determinant of the plasma membrane levels of the platelet GP Ib- IX-V complex. *Blood* **94**, 186–191.

52. Kaski, S., Kekomaki, R., and Partanen, J. (1996) Systematic screening for genetic polymorphism in human platelet glycoprotein Ib alpha. *Immunogenetics* **44,** 170–176.
53. Santoso, S., Zimmermann, P., Sachs, U. J., and Gardemann, A. (2002) The impact of the Kozak sequence polymorphism of the glycoprotein Ibα gene on the risk and extent of coronary heart disease. *Thromb. Haemost.* **87,** 345–346.
54. Sonoda, A., Murata, M., Ikeda, Y., Fukuuchi, Y., and Watanabe, K. (2001) Stroke and platelet glycoprotein Ibα polymorphisms. *Thromb. Haemost.* **85,** 573–574.
55. Croft, S. A., Samani, N. J., Teare, M. D., Hampton, K. K., Steeds, R. P., Channer, K. S., et al. (2001) Novel platelet membrane glycoprotein VI dimorphism is a risk factor for myocardial infarction. *Circulation* **104,** 1459–1463.
56. Higuchi, W., Fuse, I., Hattori, A., and Aizawa, Y. (1999) Mutations of the platelet thromboxane A$_2$ (TXA$_2$) receptor in patients characterized by the absence of TXA$_2$-induced platelet aggregation despite normal TXA$_2$ binding activity. *Thromb. Haemost.* **82,** 1528–1531.
57. Hirata, T., Ushikubi, F., Kakizuka, A., Okuma, M., and Narumiya, S. (1996) Two thromboxane A$_2$ receptor isoforms in human platelets. Opposite coupling to adenylyl cyclase with different sensitivity to Arg60 to Leu mutation. *J. Clin. Invest.* **97,** 949–956.
58. Unoki, M., Furuta, S., Onouchi, Y., Watanabe, O., Doi, S., Fujiwara, H., et al. (2000) Association studies of 33 single nucleotide polymorphisms (SNPs) in 29 candidate genes for bronchial asthma: positive association a T924C polymorphism in the thromboxane A2 receptor gene. *Hum. Genet.* **106,** 440–446.
59. Harvey, P. J., Keightley, A. M., Lam, Y. M., Cameron, C., and Lillicrap, D. (2000) A single nucleotide polymorphism at nucleotide -1793 in the von Willebrand factor (vWF) regulatory region is associated with plasma vWF:Ag levels. *Br. J. Haematol.* **109,** 349–353.
60. Keightley, A. M., Lam, Y. M., Brady, J. N., Cameron, C. L., and Lillicrap, D. (1999) Variation at the von Willebrand Factor (vWF) gene locus is associated with plasma vWF: Ag Levels: Identification of three novel single nucleotide polymorphisms in the vWF gene promoter. *Blood* **93,** 4277–4283.
61. Jansson, J. H., Nilsson, T. K., and Johnson, O. (1998) von Willebrand factor, tissue plasminogen activator, and dehydroepiandrosterone sulphate predict cardiovascular death in a 10 year follow up of survivors of acute myocardial infarction. *Heart* **80,** 334–337.
62. Brull, D. J., Dhamrait, S., Moulding, R., Rumley, A., Lowe, G. D., World, M. J., et al. (2002) The effect of fibrinogen genotype on fibrinogen levels after strenuous physical exercise. *Thromb. Haemost.* **87,** 37–41.
63. Ernst, E. and Resch, K. L. (1993) Fibrinogen as a cardiovascular risk factor: a meta-analysis and review of the literature. *Ann. Intern. Med.* **118,** 956–963.
64. Humphries, S. E., Luong, L. A., Ogg, M. S., Hawe, E., and Miller, G. J. (2001) The interleukin-6 -174 G/C promoter polymorphism is associated with risk of coronary heart disease and systolic blood pressure in healthy men. *Eur. Heart J.* **22,** 2243–2252.
65. Dinauer, D. M., Friedman, K. D., and Hessner, M. J. (1999) Allelic distribution of the glycoprotein Ia (α$_2$-integrin) C807T/G873A dimorphisms among caucasian venous thrombosis patients and six racial groups. *Br. J. Haematol.* **107,** 563–565.
66. Kim, H. O., Jin, Y., Kickler, T. S., Blakemore, K., Kwon, O. H., and Bray, P. F. (1995) Gene frequencies of the five major human platlet antigens in African American, white and Korean populations. *Transfusion* **35,** 863–867.
67. Seo, D. H., Park, S. S., Kim, D. W., Furihata, K., Ueno, I., and Han, K. S. (1998) Gene frequencies of eight human platelet-specific antigens in Koreans. *Transfus. Med.* **8,** 129–132.
68. Aramaki, K. M. and Reiner, A. P. (1999) A novel isoform of platelet glycoprotein Ib alpha is prevalent in African Americans. *Am. J. Hematol.* **60,** 77–79.

69. von dem Borne, A. E. G. Kr. (1990) Nomenclature of platelet antigen systems. *Br. J. Haematol.* **74,** 239–240.

70. Reznikoff-Etievant, M. F., Dangu, C., and Lobet, R. (1981) HLA-B8 antigen and anti-P1[A1] alloimmunization. *Tissue Antigens* **18,** 66–68.

71. Valentin, N., Vergracht, A., Bignon, J. D., Cheneau, M. L., Blanchard, D., Kaplan, C., et al. (1990) HLA-Drw52a is involved in alloimmunization against PL-A1 antigen. *Human Immunol.* **27,** 73–79.

72. Reznikoff-Etievant, M. F., Muller, J. Y., and Julien, F. (1981) An immune response gene linked to MHC in man. *Tissue Antigens* **22,** 312.

73. L'Abbé, D., Tremblay, L., Filion, M., Busque, L., Goldman, M., Décary, F., et al. (1992) Alloimmunization to platelet antigen HPA-1a (P1[A1]) is strongly associated with both HLA-DRB3*0101 and HLA-DQB1*0201. *Hum. Immunol.* **34,** 107–114.

74. Maslanka, K., Yassai, M., and Gorski, J. (1996) Molecular identification of T cells that respond in a primary buk culture to a peptide derived from a platelet glycoprotein implicated in neonatal alloimmune thrombocytopenia. *J. Clin. Invest.* **98,** 1802–1808.

75. Kuijpers, R. W. A. M., von dem Borne, A. E., Kifel, V., Eckhardt, C. M., Waters, A. H., Zupanska, B., et al. (1992) Leucine 33-proline 33 substitution in human platelet glycoprotein IIIa determines HLA-DRw52a (Dw24) association of the immune response against HPA-1a (Zwa/PIA1) and HPA-1b (Zwb/PIA2). *Human Immunol.* **34,** 253–256.

76. Westman, P., Hashemi-Tavoularis, S., Blanchette, V., Kekomaki, S., Laes, M., Porcelijn, L., et al. (1997) Material DRB1*1501, DQA1*0102,DQB1*0602 haplotype in fetomaternal alloimmunization against human platelet alloantigen HPA-6b (GPIIIa-Gln489). *Tissue Antigens* **50,** 113–118.

77. Mueller-Eckhardt, C., Kiefel, V., Grubert, A., Kroll, H., Weisheit, M., Schmidt, S., et al. (1989) 348 cases of suspected neonatal alloimmune thrombocytopenia. *Lancet* **1,** 363–366.

78. von dem Borne, A., von Riesz, E., Verheugt, F., ten Cate, J., Koppe, J., Englefreit, C., et al. (1980) Bak[a], a new platelet-specific antigen involved in neonatal alloimmune thrombocytopenia. *Vox Sang.* **39,** 113–120.

79. McGrath, K., Minchinton, R., Cunningham, I., and Ayberk, H. (1989) Platelet anti-Bak[b] antibody associated with neonatal alloimmune thrombocytopenia. *Vox Sang.* **57,** 182–184.

80. Mueller-Eckhardt, C., Becker, T., Weishet, M., Witz, C., and Santoso, S. (1986) Neonatal alloimmune thrombocytopenia due to fetomaternal Zw[b] in compatability. *Vox Sang.* **50,** 94–96.

81. Grenet, P., Dausset, J., Dugas, M., Petit, D., Badoual, J., and Tangun, Y. (1965) Purpura thrombopenique neonatal avec isoimmunisation foeto-maternelle anti-Ko[a]. *Arch. Fr. Pediatr.* **22,** 1165–1174.

82. Shibata, Y., Matsuda, I., Miyaji, T., and Ichikawa, Y. (1986) Yuk[a], a new platelet antigen involved in two cases of neonatal alloimmune thrombocytopenia. *Vox Sang.* **50,** 177–180.

83. Mueller-Eckhardt, C. (1982) HLA-B8 antigen and anti-P1[A1] alloimmunization. *Tissue Antigens* **19,** 154–158.

84. George, J. N., Pickett, E. B., and Heinz, R. (1987) Platelet membrane microparticles in blood bank fresh frozen plasma and cryoprecipitate. *Blood* **68,** 307–309.

85. Cotton, R. G. (1998) Mutation detection and mutation databases. *Clin. Chem. Lab. Med.* **36,** 519–522.

86. Cotton, R. G. (1997) Slowly but surely towards better scanning for mutations. *Trends Genet.* **13,** 43–46.

87. Shi, M. M. (2001) Enabling large-scale pharmacogenetic studies by high-throughput mutation detection and genotyping technologies. *Clin. Chem.* **47,** 164–172.

88. Head, S. R., Rogers, Y. H., Parikh, K., Lan, G., Anderson, S., Goelet, P., et al. (1997) Nested genetic bit analysis (N-GBA) for mutation detection in the p53 tumor suppressor gene. *Nucleic Acids Res.* **25,** 5065–5071.

89. Head, S. R., Parikh, K., Rogers, Y. H., Bishai, W., Goelet, P., and Boyce-Jacino, M. T. (1999) Solid-phase sequence scanning for drug resistance detection in tuberculosis. *Mol. Cell Probes* **13,** 81–87.

90. Raitio, M., Lindroos, K., Laukkanen, M., Pastinen, T., Sistonen, P., Sajantila, A., et al. (2001) Y-chromosomal SNPs in Finno-Ugric-speaking populations analyzed by mini-sequencing on microarrays. *Genome Res.* **11,** 471–482.

91. Makridakis, N. M. and Reichardt, J. K. (2001) Multiplex automated primer extension analysis: simultaneous genotyping of several polymorphisms. *BioTechniques* **31,** 1374–1380.

92. Ronaghi, M. (2001) Pyrosequencing sheds light on DNA sequencing. *Genome Res.* **11,** 3–11.

93. Nikiforov, T. T., Rendle, R. B., Goelet, P., Rogers, Y. H., Kotewicz, M. L., Anderson, S., et al. (1994) Genetic Bit Analysis: a solid phase method for typing single nucleotide polymorphisms. *Nucleic Acids Res.* **22,** 4167–4175.

94. Nikiforov, T. T., Rendle, R. B., Kotewicz, M. L., and Rogers, Y. H. (1994) The use of phosphorothioate primers and exonuclease hydrolysis for the preparation of single-stranded PCR products and their detection by solid-phase hybridization. *PCR Methods Appl.* **3,** 285–291.

95. Rogers, Y. H., Jiang-Baucom, P., Huang, Z. J., Bogdanov, V., Anderson, S., and Boyce-Jacino, M. T. (1999) Immobilization of oligonucleotides onto a glass support via disulfide bonds: A method for preparation of DNA microarrays. *Anal. Biochem.* **266,** 23–30.

96. Lyman, S., Aster, R. H., Visentin, G. P., and Newman, P. J. (1990) Polymorphism of human platelet membrane glycoprotein IIb associated with the Bak[a]/Bak[b] alloantigen system. *Blood* **75,** 2343–2348.

97. Wang, R., Furihata, K., McFarland, J. G., Friedman, K., Aster, R. H., and Newman, P. J. (1992) An amino acid polymorphism within the RGD binding domain of platelet membrane glycoprotein IIIa is responsible for the formation of the Pen[a]/Pen[b] alloantigen system. *J. Clin. Invest.* **90,** 2038–2043.

98. Mellins, E., Cameron, P., Amaya, M., Goodman, S., Pious, D., Smith, L., et al. (1994) A mutant human histocompatibility leukocyte antigen DR molecule associated with invariant chain peptides. *J. Exp. Med.* **179,** 541–549.

99. Wang, R., McFarland, J. G., Kekomaki, R., and Newman, P. J. (1993) Amino acid 489 is encoded by a mutational "hot spot" on the beta 3 integrin chain: the CA/TU human platelet alloantigen system. *Blood* **82,** 3386–3391.

100. Kuijpers, R. W. A. M., Simsek, S., Faber, N. M., Goldschmeding, R., van Wermerkerken, R. K. V., and Von dem Borne, A. E. G. K. (1993) Single point mutation in human glycoprotein IIIa is associated with a new platelet-specific alloantigen (Mo) involved in neonatal alloimmune thrombocytopenia. *Blood* **81,** 70–76.

101. Santoso, S., Kalb, R., Kiefel, V., Mueller-Eckhardt, C., and Newman, P. J. (1994) A point mutation leads to an unpaired cysteine residue and a molecular weight polymorphism of a functional platelet beta 3 integrin subunit. The Sra alloantigen system of GPIIIa. *J. Biol. Chem.* **269,** 8439–8444.

102. Noris, P., Simsek, S., De Bruijne-Admiraal, L. G., Porcelijn, L., Huiskes, E., van der Vlist, G. J., et al. (1995) Maxa, a new low-frequency platelet-specific antigen localized on glycoprotein IIb, is associated with neonatal alloimmune thrombocytopenia. *Blood* **86,** 1019–1026.

103. Peyruchaud, O., Bourre, F., Morel-Kopp, M.-C., Reviron, D., Mercier, P., Nurden, A., et al. (1997) HPA-10w[b] (La[a]): Genetic determination of a new platelet-specific alloantigen on glycoprotein IIIa and its expression in COS-7 cells. *Blood* **89,** 2422–2428.

104. Simsek, S., Folman, C., Van der Schoot, C. E., and Von dem Borne, A. E. G. K. (1997) The Arg633His substitution responsible for the platelet antigen Gro[a] unravelled by SSCP analysis and direct sequencing. *Br. J. Haematol.* **97,** 330–335.

105. Sachs, U. J. H., Kiefel, V., Bohringer, M., Afshar-Kharghan, V., Kroll, H., and Santoso, S. (2000) Single amino acid substitution in human platelet glycoprotein Ib beta is responsible for the formation of the platelet-specific alloantigen Iy[a]. *Blood* **95,** 1849–1855.

106. Santoso, S., Amrhein, J., Hofmann, H. A., Sachs, U. J. H., Walka, M. M., Kroll, H., et al. (1999) A point mutatin Thr[799]Met on the alpha-2 integrin leads to the formation of new human platelet alloantigen Sit[a] and affects collagen-induced aggregation. *Blood* **94,** 4103–4111.

107. Santoso, S., Pylipiw, R., Wilke, L. G., Kroll, H., and Kiefel, V. (1998) One amino acid deletion of the PI[A2] allelic form of GPIIIa leads to the formation of the new platelet alloantigen, Oe[a]. *Blood* **92,** 472a.

108. Kekomaki, R., Raivio, P., and Kero, P. (1992) A new low-frequency platelet alloantigen, Va[a], on glycoprotein IIb/IIIa associated with neonatal alloimmune thrombocytopenia. *Transf. Med.* **2,** 27–33.

109. Kekomaki, R., Partanen, J., Pitkanen, S., Ilanmaa, E., Ammala, P., and Teramo, K. (1993) Glycoprotein Ib/IX-specific alloimmunization in an HPA 2b-homozygous mother in association with neonatal thrombocytopenia. *Thromb. Haemost.* **69,** 99.

110. Kelton, J. G., Smith, J. W., Horsewood, P., Humbert, J. R., Hayward, C. P. M., and Warkentin, T. E. (1990) Gov[a/b] alloantigen system on human platelets. *Blood* **75,** 2172–2176.

111. Santoso, S. and Kiefel, V. (2001) Human platelet alloantigens. *Wien. Klin. Wochenschr.* **113,** 806–813.

29

Gene Array Technology and the Study of Platelets and Megakaryocytes

Lloyd T. Lam and Emery H. Bresnick

1. Introduction

The expression profile of the complete set of cellular genes or global gene expression provides a remarkable snapshot of physiological and pathophysiological mechanisms underlying cell regulation. Thus, accurate and precise measurements of global gene expression reveal unique insights into critical processes such as cell proliferation, differentiation, and survival. Prior to the existence of state-of-the-art approaches to analyzing global gene expression, gene expression was commonly measured one gene at a time using methods such as Northern blotting *(1)*, RNase protection *(2)*, and primer extension *(3)*. These methods rely on the ability of nucleic acid "probes" to recognize complementary sequences in hybridization reactions via base pairing. The invention of reverse transcription polymerase chain reaction (RT-PCR) *(4)* greatly improved the sensitivity of gene expression analysis, allowing for the detection of very low abundant transcripts semiquantitatively and, more recently, quantitatively via real-time RT-PCR. RNA is first converted into complementary DNA (cDNA) in a reaction catalyzed by reverse transcriptase, and the cDNA is then amplified by PCR using gene-specific primers. With various modifications, differentially expressed transcripts representing both known and novel genes can be identified by techniques called differential display (DD) *(5)* and representational difference analysis *(6)*. Although these techniques are quite powerful, their labor-intensive nature and moderate rate of false positives limits their utility somewhat. Importantly, these techniques do not allow for a facile analysis of thousands or even hundreds of gene expression changes.

Simultaneous analysis of the expression of thousands of genes is a particularly powerful approach to elucidating mechanisms that control cell differentiation. For example, growth and differentiation factors that stimulate megakaryopoiesis, such as thrombopoietin, interleukin-6, and interleukin-11, induce complex plasma membrane to nuclear signaling events, thereby establishing specific patterns of gene expression *(7,8)*.

From: *Methods in Molecular Biology, vol. 273:*
Platelets and Megakaryocytes, Vol. 2: Perspectives and Techniques
Edited by: J. M. Gibbins and M. P. Mahaut-Smith © Humana Press Inc., Totowa, NJ

Although identifying one or several target genes for such factors can lead to intriguing hypotheses concerning mechanisms controlling stem and progenitor cell proliferation and differentiation, global gene expression analysis has considerably greater potential to reveal the overall circuitry that dictates cell regulation.

One method that has been used to measure global changes in gene expression is serial analysis of gene expression (SAGE) *(9)*. Short stretches of "signature" sequences of individual genes are joined together. After sequencing this joint fragment, the frequency of occurrence of the signature sequences is compiled, yielding a global gene expression profile. Again, this technique is labor-intensive. An alternative method of measuring global changes in gene expression incorporates DNA chips or microarrays. Microarray analysis involves a hybridization-based technique that can, in principle, measure thousands of gene expression changes simultaneously, with the only limitation being the number of immobilized "probes" *(10,11)*. A microarray consists of thousands of probes, each representing a single gene, immobilized on a solid support. In contrast to Northern blotting, which allows one to measure the absolute expression of an individual gene, microarray analysis compares the relative activity of genes in two states, termed differential gene expression. Common applications of microarray analysis include comparison of gene expression between two cell types to reveal cell-specific patterns of gene expression *(12)*. In addition, microarray analysis is commonly used to measure differential gene expression within a single cell type subjected to various stimuli, such as a growth or differentiation factor *(13,14)*.

Oligonucleotide and cDNA microarrays represent the two common microarray formats. Oligonucleotide microarrays consist of high-density oligonucleotides synthesized *in situ* using photolithographic methods *(15)*. In contrast, cDNA microarrays consist of thousands of purified cDNAs arrayed on glass slides or membrane filters with a robotic arrayer. These cDNAs are amplified from templates using PCR. This chapter will focus mainly on cDNA microarrays printed on glass slides **(Fig. 1)**. The comparison of gene expression between two samples, for example a tissue or cell line under resting and stimulated conditions, is carried out by obtaining purified RNA. Fluorescently labeled cDNAs for each sample are then generated by reverse transcription using nucleotides tagged with one of two dyes, commonly Cy3 for the control and Cy5 for the stimulated sample. The cDNAs from the two samples are then combined and hybridized on the microarray. After washing, the microarray is scanned with a high-resolution laser scanner to measure the fluorescent intensity of each dye. The ratio of the two signals is represented by different colors to give a visual representation of the measurements. Finally, mathematical analysis should be done to obtain statistically significant results.

Why use microarrays? First, despite the extensive data accumulation, data acquisition is relatively easy, and results can be obtained in a short time. Second, microarrays are becoming more economical as the production of materials improves and market competition increases. Although one disadvantage of this technique is that the gene expression measurement is limited only to genes printed on the arrays, with the completion of genome sequencing projects the whole human genome will be incorporated into array sets in the near future. Microarrays have been used to accomplish diverse research goals, including disease diagnosis and prognosis *(16)*, assessment of drug response and

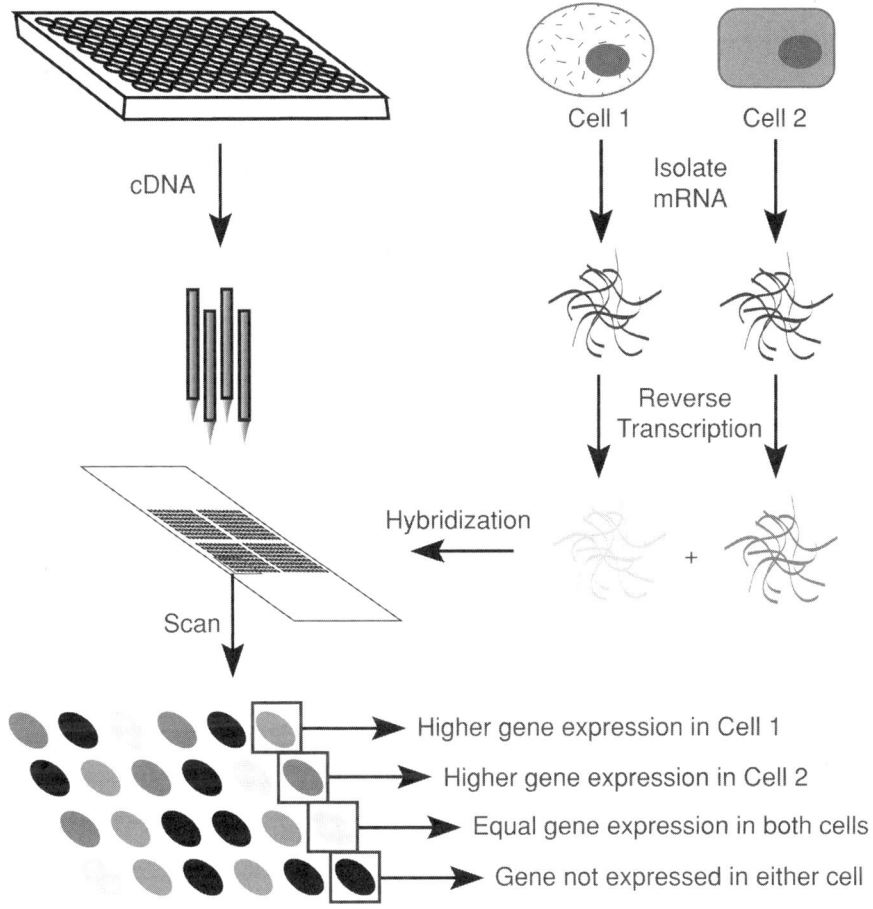

Fig. 1. Summary of the microarray process. cDNA clones are amplified in a 96- or 384-well plate and printed on glass slides using an arrayer. After purifying mRNA from two different cell samples, fluorescently labeled cDNAs are generated by reverse transcription (Cy3 for one sample and Cy5 for the other). The cDNAs from the two samples are mixed and hybridized on the microarray. The microarray is then scanned with a high-resolution laser scanner to measure the fluorescence intensity of the individual samples. The ratio of the two signals indicates the differential gene expression and is represented within analysis software by different colors to give a visual representation of the measurements. The interpretation of relative expression of one clone is shown. (*See* color insert following p. 300.)

mechanism of action *(17)*, and analysis of mechanisms of cell differentiation *(12,14)*. In this chapter, we will describe technical aspects of how microarrays are constructed and used, which can be readily adapted to the analysis of platelets and megakaryocytes

(14,18). Readers are encouraged to consult other protocols *(19)* before starting the process and choose the optimal one based on research goals and resource availability.

2. Materials
2.1. cDNA Production

1. Clones can be purchased from Research Genetics (Res Gen™, Invitrogen Corporation; www.resgen.com) or obtained from a private collection (*see* **Note 1**).
2. Luria-Bertani (LB) broth.
3. Ampicillin (100 mg/mL).
4. 96-well flat-bottom blocks (cat. no. 19579; Qiagen, Valencia, CA).
5. AirPore™ Tape Sheets (cat. no. 1006452; Qiagen).
6. Turbo 96 miniprep kit (cat. no. 27173; Qiagen) and the materials required for the kit.
7. M13 Primers-forward (5′-GTTGTAAAACGACGACGGCCAGTG-3′) and reverse (5′-CACACAGGAAACAGCTATG-3′) (stock concentration of 1 mM).
8. 96-well thermal cycler (MJ Research, Waltham, MA).
9. 96-well PCR multiplate™ (cat. no. MLP-9601; MJ Research).
10. PCR Taq polymerase (comes in 250 units/vial, 5 units/mL, cat. no. D1806; Sigma, St. Louis, MO) (–20°C).
11. 10X reaction buffer containing MgCl$_2$: 15 mM MgCl$_2$, 100 mM Tris-HCl, pH 8.3, 500 mM KCl, 0.01% gelatin (supplied as ready-made buffer with Cat. no. D1806; Sigma) (–20°C).
12. Deoxynucleotide (dNTP) mix (10 mM) (cat. No. D7295; Sigma) (–20°C).
13. 1 M MgCl$_2$.
14. Double-distilled water (ddH$_2$O).
15. 1X TAE buffer (40 mM Tris-HCl acetate, 1 mM EDTA).
16. Agarose.
17. Ethidium bromide solution (stock 10 mg/mL).
18. 96-well cell-culture cluster round-bottom microplates with polystyrene lids (cat. no. 3790; Costar, Hercules, CA).
19. 3 M sodium acetate, pH 5.2.
20. Isopropanol.
21. 100% and 70% ethanol.
22. 20X saline sodium citrate (SSC) (Gibco-BRL, Carlsbad, CA), diluted to 3X SSC and sterile-filtered.
23. 384-well plates (optional; *see* **Note 2**).
24. Biomek TM Seal & Sample aluminium foil lids for sealing 96-well plates (cat. no. 538619; Beckman, Fullerton, CA).

2.2. Coating Slides

1. Gold Seal Micro slides (cat. no. 3010; Gold Seal, Portsmouth, NH).
2. Stainless steel slide rack.
3. NaOH (cat. no. S3185; Fisher Scientific, Fairlawn, NJ).
4. 95% ethanol.
5. ddH$_2$O.
6. Poly-L-lysine solution (cat. no. P8920; Sigma).
7. Phosphate-buffered saline (PBS) (TC grade).

2.3. Printing Microarrays

1. Robotic arrayer (Gene Machine, CA).
2. Printing pins (cat. no. SMP3; Arrayit, Sunnyvale, CA).

2.4. Processing Printed Microarrays

1. Diamond-tipped scriber (VWR).
2. Rehydration chamber (for example, a plastic slide holder).
3. UV Stratalinker 1800 TM (Stratagene, Cedar Creek, TX).
4. Succinic anhydride (cat. no. 23969-0; Aldrich Chemical Company, Inc., Milwaukee, WI).
5. 1-methyl-2-pyrrolidinone (cat. no. NC988/0242; Fisher).
6. 1 M boric acid (pH 8.0 with NaOH).
7. Grain alcohol 190 proof (cat. no. 11100G190CSGP; Pharmco, Brookfield, OH). Use without dilution.

2.5. RNA Preparation

2.5.1. Total RNA Extraction

1. Trizol (cat. no. 15596-018; Gibco/BRL) (4°C).
2. Chloroform.
3. Isopropanol.
4. Diethylpyrocarbonate (DEPC)-treated H_2O (cat. no. 750023; Invitrogen) (0.1% DEPC w/v in ddH_2O, stored overnight at 37°C, then autoclaved).
5. 75% ethanol diluted with DEPC-treated H_2O.

2.5.2. mRNA Extraction

Fast-track (Cat. no. K1593-03; Invitrogen).

2.6. Labeling With Fluorescence and Hybridization

1. 500 mg/mL Oligo dT$_{12-18}$ (Operon) (–20°C).
2. SuperScript II reverse transcriptase and 5X 1st Strand Buffer (Cat. no. 18064-071; Invitrogen) (–20°C).
3. RNAsin (Cat. no. 799025; Roche) (–20°C).
4. 50X dT-dNTPs (25 mM each of dA, dG, dC, and 10 mM of dT) (Cat. no. 27-2035-01; Amersham) (–20°C).
5. Cy3-dUTP (1 mM, NEN Life Sciences, Boston, MA) (–20°C). This reagent is light-sensitive.
6. Cy5-dUTP (1 mM, NEN Life Sciences) (–20°C). This reagent is light-sensitive.
7. DEPC-treated H_2O.
8. 0.5 M EDTA, pH 8.0 (NaOH).
9. 1 M NaOH.
10. 1 M Tris-HCl, pH 7.5.
11. TE buffer: 10 mM Tris-HCl, pH 8.0, and 1 mM EDTA, pH 8.0 (NaOH).
12. Microcon YM-30 spin columns (Cat. no. 42409; Millipore).
13. Human COT-1 DNA®, 1 µg/mL (Cat. no. 15279011; Invitrogen) (–20°C) (*see* **Note 3**).
14. Yeast tRNA, 2 µg/mL (Cat. no. R8759; Sigma) (–20°C).
15. Poly dA, 8 µg/mL (Cat. no. 27,7922-01; Amersham) (–20°C).
16. 20X SSC (Gibco-BRL).
17. 0.22-µm Ultrafree-MC filter unit (Cat. no. UFC30GV00; Millipore).
18. 10% (w/v) sodium dodecyl sulfate (SDS).
19. Hybridization chamber (Cat. no. 2551; Corning, Corning, NY).
20. Glass coverslips (Corning Glass Works, Corning, NY).

2.7. Washing

1. Slide rack.
2. Glass staining dishes (five are required).
3. 2X SSC/0.1% SDS wash solution.
4. 1X SSC wash solution.
5. 0.5X SSC wash solution.
6. 0.2X SSC wash solution.

2.8. Scanning and Analysis: Fluorescence

Genepix Scanner (Axon Instruments, Union City, CA).

3. Methods

3.1. cDNA Production

The first step is to amplify the selected cDNA clones to provide enough material for generation of a microarray. Prepare as many clones as you wish to test.

1. Grow clones in LB plus ampicillin (200 µg/mL) in 96-well flat-bottom blocks overnight at 37°C. Seal blocks with AirPore tape sheets.
2. Obtain plasmid DNA according to the procedures of the Qiagen Turbo 96 miniprep kit. Store DNA at –20°C.
3. Amplify DNA in 96-well PCR multiplates by PCR. Mix in a 50-mL tube the following reagents for one 96-well plate: 1 mL 10X reaction buffer, 200 µL dNTP (10 mM), 10 µL of each M13 forward and reverse primer, 5 µL of 1 M MgCl$_2$, and 1 vial of 250 units Taq polymerase, and fill up to 10 mL with ddH$_2$O. Dispense 95 µL of mix into each well of the 96-well PCR multiplate and 5 µL of the plasmid DNA to each well.
4. Carry out PCR with the following conditions:
   a. 95°C for 3 min.
   b. 95°C for 30 s.
   c. 55°C for 30 s.
   d. 72°C for 2.5 min.
   e. Repeat **steps b–d** for 30 cycles.
   f. 72°C for 10 min.
5. Analyze 5 µL of the PCR product in each well on a 1% TAE agarose gel with 0.5 µg/mL ethidium bromide and estimate the quality of the product (*see* **Note 4**).
6. Transfer PCR products to 96-well round-bottom microplates. Add 10 µL of 3 M sodium acetate (pH 5.2) and add 100 µL of isopropanol to each well. Keep the plate at –20°C overnight.
7. Spin plate at 5900g for 45 min at 4°C. Remove supernatant and add 100 µL of 70% EtOH. Spin at 5900g for 45 min at 4°C. Remove supernatant. Invert plates and dry overnight at room temperature.
8. Resuspend the DNA in 11 µL of 3X SSC. Let the plate stand for one day at 4°C.
9. Seal with Biomek™ aluminum foil lid . The DNA is now ready for printing (enough for 10–15 prints from each well) (*see* **Note 5**).

3.2. Coating Slides

1. Place Gold Seal microscope slides into stainless steel slide racks.
2. Prepare the cleaning solution in a chemical hood as follows: Stir and add 100 mL of ddH$_2$O into 50 g of NaOH until solution becomes clear. Add 250 mL of 95% EtOH very slowly.

3. Add 350 mL of cleaning solution to each of four glass dishes in a plastic container, submerge the slide-filled racks, and shake for 2 h (*see* **Note 6**).
4. Discard cleaning solution, refill with ddH$_2$O, and shake for 2 to 5 min with vigorous plunging.
5. Refill with ddH$_2$O, shake, and plunge for four more times to remove all residual NaOH.
6. Prepare lysine coating solution by combining 280 mL of ddH$_2$O, 35 mL of poly-L-lysine solution, and 35 mL of PBS. Add this coating solution to each dish, submerge slides, and shake for 1 h.
7. Discard lysine coating solution and rinse slides for 10–15 s in ddH$_2$O.
8. Quickly spin-dry in a centrifuge at 90g for 3 min.
9. Transfer slides to a clean slide seal box and let slides age for at least two weeks at room temperature (*see* **Note 7**).

3.3. Printing Microarrays

Printing refers to the process of transferring cDNA from the 96- or 384-well plate to the glass slide. Varieties of arrayers are available in the market. Readers are encouraged to choose their arrayer based on their research needs and resource availability. Follow the manufacturer's instructions for the arrayer for printing. In addition, we recommend using printing pins from Arrayit (*see* **Note 8**). It is recommended that multiple clones (same sequence or different sequence) of the genes that are of high interest be included on the same array. This is for checking repeatability and useful for statistical analysis.

3.4. Processing Printed Microarrays

1. Mark the border, and number the slides with a diamond-tipped scriber (VWR) (*see* **Note 9**).
2. Heat a beaker of ddH$_2$O to near-boiling and pour into a rehydration chamber.
3. Place a slide, array-side down, on the basin above the water. Let the steam coat the slide, and let the spots rehydrate until each spot has absorbed water. This should take 3 to 10 s.
4. Turn the slide over, and lay flat on a heating block of approx 70°C to let all moisture evaporate. This should take 2 to 4 s. Do not heat for more than 6 s.
5. Lay slide array-side up inside a Stratalinker. Adjust the setting for 600 (×100) microjoules. Push start to crosslink DNA on the glass slide.
6. After UV crosslinking, transfer slides to a 15-slot rack.
7. Bring to a boil 1 L of ddH$_2$O in a 2.5-L beaker. This will be used for denaturing the DNA, after blocking with succinic anhydride.
8. Add 7.2 g of succinic anhydride to a 500-mL clean glass dish containing a stir bar. Pour 400 mL of 1-methyl-2-pyrrolidinone into the glass dish and stir. Once the succinic anhydride has dissolved, add 17.9 mL of 1 M boric acid to the glass dish and stir thoroughly.
9. Submerge a rack of rehydrated arrays and place on an orbital shaker for 15 min. Remove and resubmerge vigorously every 2 min (*see* **Note 10**).
10. Turn off the heat for the boiling water. Wait until bubbles subside, and transfer the rack into the beaker for 2 min.
11. Quickly remove the rack and place in a dish containing 350 mL of 95% grain EtOH. Plunge several times, and transfer to the centrifuge microplate carrier and spin dry at 90g for 2 min.
12. Transfer slides to a clean plastic slide box for storage.

3.5 RNA Preparation

For most experiments, total RNA will suffice for labeling, especially if cell lines are used. However, if tissue is used, one might consider using an mRNA isolation kit to purify poly A RNA. For most labeling, around 50 μg of total RNA or 2 μg of poly A RNA is optimal (*see* **Note 11**).

Special attention is necessary when selecting the reference sample. In general, it is easy to determine which sample should be the reference when only two samples are compared or when only one initial time is measured in a time-course study (Type I experiments). However, if multiple samples are involved (Type II experiments), a common reference is necessary. The goal is to select a reference sample that would hybridize to most genes on the array (*see* **Note 12**). After comparing all samples to the reference sample, the data can be normalized.

3.5.1. Total RNA Extraction

All the plasticware and solutions for RNA extraction should be designated for RNA work only. In addition, the work should be done on a "clean bench." Gloves should be worn at all times to minimize sample contamination.

1. Obtain 1×10^6 cells and spin down at low speed (*see* **Note 13**). Remove cell culture media. The cell pellet can be stored at –70°C or processed immediately.
2. Add 1 mL of Trizol, and resuspend the pellet with a 1-mL pipet until it completely dissolves. Add 0.2 mL of chloroform, shake tubes vigorously for 15 s, and incubate samples at room temperature for 2 to 3 min.
3. Centrifuge at 12,000*g* at 4°C for 15 min.
4. Obtain the aqueous phase and transfer into a new tube. Add 0.5 mL of isopropanol. Let the tube sit at room temperature for 10 min (*see* **Note 14**).
5. Centrifuge at 12,000*g* at 4°C for 10 min.
6. Discard the supernatant. Wash RNA pellet with 1 mL of 75% ethanol.
7. Centrifuge at 7500*g* at 4°C for 5 min.
8. Discard the wash. Air-dry the pellet for 5 min. Dissolve the RNA pellet completely in 40 μL of DEPC-treated H_2O. Store RNA at –80°C.
9. Quantitate RNA by UV spectroscopy. The goal is to obtain around 50 μg of total RNA (around 2.5 μg/mL) for each labeling reaction.

3.5.2. mRNA Extraction

Follow the protocol of the Invitrogen Fast-Track kit or other similar mRNA processing kits for the isolation of mRNA (*see* **Note 15**).

3.6. Generation of Fluorescent Probes and Hybridization

1. Make the reverse transcription reaction mix in the following order: for each reaction: 8 μL of 5X first-strand buffer, 4 μL of 0.1 *M* DTT, 1 μL of RNAsin, 2 μL of 500 μg/mL Oligo dT, and 2 μL of each of the stock (50X) dT-dNTPs.
2. Add 17 μL of this mix to 19 μL of RNA sample (a total of 40 to 60 μg total RNA or 2 to 4 μg mRNA is added into DEPC-treated H_2O to get 19 μL). Then add 4 μL of 1 m*M* Cy3-dUTP to the control and 4 μL of 1 m*M* Cy5-dUTP to the test sample. Mix the sample by vortexing or pipetting.

3. Incubate at 65°C for 5 min. Quick-spin and incubate at 42°C for 3 min.
4. Add 2 μL of Superscript II and mix with pipetting.
5. Quick-spin and incubate at 42°C for 30 min.
6. Quick-spin and add 2 μL of Superscript II. Mix with pipetting and incubate at 42°C for 30 min.
7. Add 5 μL of 0.5 *M* EDTA and mix.
8. Add 10 μL of 1 *M* NaOH and mix.
9. Spin rapidly and incubate at 65°C for 20 min.
10. Spin rapidly and add 25 μL of 1 *M* Tris-HCl, pH 7.5.
11. Add 450 μL of TE buffer to the first labeling tube, and then transfer the liquid into another labeling tube. After mixing, transfer the sample into a Microcon YM-30 column.
12. Spin at 12,000*g* for approx 7 min until approx 20 to 40 μL of probe is left (*see* **Note 16**).
13. Warm the hybridization chambers on top of the 65°C water bath.
14. Discard the buffer in the collecting tube. Add 400 μL of TE buffer to the column. Add 20 μL of Cot-1 DNA®, 2 μL of yeast tRNA, and 2 μL of poly dA.
15. Spin at 12,000*g* for 6 min until 20 μL of probe is left.
16. Invert probe into a new collection tube, and spin at 12,000*g* for 1 min.
17. Add 4.25 μL of 20X SSC and transfer sample to a 0.22-μm Ultrafree-MC filter unit. Spin at 12,000*g* for 1 min.
18. Add 0.75 μL of 10% SDS, and incubate the tube at 100°C for 2 min (*see* **Note 17**).
19. Spin at 12,000*g* for 1 min.
20. Put 25 μL of the probe onto the microarray. Add TE to achieve this volume in the tube before putting on the microarray. Cover with a coverslip, while avoiding the introduction of bubbles (*see* **Note 18**).
21. Place 20 μL of water into the water chambers. Close hybridization chamber, and place in a 65°C water bath overnight (*see* **Note 19**).

3.7. Washing

1. Carefully disassemble the hybridization chamber, and remove the slide. Place in a slide rack and submerge in staining dish with 2X SSC/0.1% SDS wash solution. Let the coverslip fall off and plunge gently (*see* **Note 20**).
2. Wash twice in staining dish with 1X SSC wash solution. Plunge gently.
3. Wash in staining dish with 0.5X SSC wash solution. Plunge gently.
4. Wash in staining dish with 0.2X SSC wash solution. Plunge gently.
5. Quickly spin-dry at 90*g* for 3 min.
6. Place in a clean slide box. Scan microarray (*see* **Note 21**).

3.8. Scanning and Analysis: Fluorescence

1. Scan the microarray with a high-resolution laser confocal scanner (*12*) such as the Axon Genepix Scanner to obtain fluorescence signals.
2. Quantitate and analyze the data using computer software for microarrays. Free software available for academic institutions include BRB-ArrayTools (http://linus.nci.nih.gov/BRB-ArrayTools.html), Cluster (http://rana.lbl.gov/EisenSoftware.htm), and TreeView (http://rana.lbl.gov/EisenSoftware.htm). Commonly used software for building databases include FileMaker Pro (Claris) and Excel (Microsoft Office).

4. Notes

1. The method described in this chapter assumes that the clones are supplied in a plasmid with M13 primer sites within the vector backbone and that the vector encodes ampicillin resistance. It is also assumed that the clones are available within transformed *E. coli* ready for plasmid amplification.

2. DNA can be stored in 96-well plates. However, if 96 well plates are used, only 4 pins can be used for printing due to the spacing between wells of the 96-well plate. Another option is to transfer DNA to flexible 384-well plates. This allows printing with 16 or even 32 pins, which would save time and storage space for DNA plates.

3. COT-1 is double-stranded DNA, predominantly 50 to 300 base pairs in length, and is used to block nonspecific hybridization, thus reducing background signals. It is also available in a form for use with murine tissue.

4. The goal is to obtain a single product (as measured by a single band) for each clone to avoid cross-contamination. If there are multiple bands for a particular clone, the clone should be restreaked and verified by sequencing.

5. There is usually evaporation of liquid on the foil seal. Quick-spin the plates before printing.

6. The slides must be completely covered in the liquids for these steps in order to get even coating.

7. This would allow the slide to become hydrophobic. To test the hydrophobicity of the slides, add a drop of water and see whether no trace of water is left behind. Also, one can breathe on the slide and see whether moisture throughout the slide evaporates equally, which would reflect the evenness of the coating.

8. Particular care should be given to the printing pins. The sharpness of the pin is the major determinant of the spot size. The pins can be tested by printing with a solution containing 100 ng/μL salmon sperm DNA, an inexpensive source of DNA (Invitrogen), in 3X SSC. This can determine the size of the spots and whether the pins yield spots of a similar size.

9. The slides should be marked so that the DNA side can be recognized, since the spots are invisible after processing.

10. Like the coating steps, the slides must be completely covered in the liquids to obtain even processing.

11. It is not encouraged to use amounts of RNA greater than the amount indicated, since this might increase background. On the other hand, if the RNA sample is limiting, the signal can be amplified (20). A commercially available kit such as the MessageAmp TM aRNA kit from Ambion can be used to amplify RNA (Cat. no. 1750; Ambion, Austin, TX). However, a random hexamer is used instead of oligo dT for the labeling reaction.

12. The reference RNA "pool" can simply be a combination of an equal amount of material from each individual sample.

13. Using too many cells would affect the RNA extraction. If a high yield of RNA is desired, do extractions in multiple tubes and combine.

14. RNA samples can be stored at −80°C in isopropanol if not used immediately.

15. To check the quality of the RNA, analyze 1 μL of the purified RNA on a 1% TAE agarose gel with 0.5 μg/mL ethidium bromide. The quantity of the RNA can be checked by spectroscopy.

16. Do not overspin to avoid drying of the membrane. Excess drying would result in significant loss of signal due to difficulty with recovery.

17. Do not use more than 0.75 μL of 10% SDS. Otherwise, small particles will be seen on the image after scanning.

18. Place the labeled sample on the center of the microarray. With the coverslip touching one side of the microarray slide, slowly lower the other side with a forceps until the drop of sample is touching most of the coverslip. Then, release the coverslip. Minimize repositioning of the coverslip once it has been placed to avoid scratching the microarray.

19. The microarray can be scanned after 12 h. Do not hybridize for more than 16 h to avoid drying of the sample, which would increase background.

20. Be extremely careful not to scratch the microarray with the coverslip.

21. Scan the microarray within 2 h to avoid loss of signal.

References

1. Alwine, J. C., Kemp, D. J., and Stark, G. R. (1977) Method for detection of specific RNAs in agarose gels by transfer to diazobenzyloxymethyl-paper and hybridization with DNA probes. *Proc. Natl. Acad. Sci. USA* **74,** 5350–5354.

2. Melton, D. A., Krieg, P. A., Rebagliati, M. R., Maniatis, T., Zinn, K., and Green, M. R. (1984) Efficient in vitro synthesis of biologically active RNA and RNA hybridization probes from plasmids containing a bacteriophage SP6 promoter. *Nucleic Acids Res.* **12,** 7035–7056.

3. Ghosh, P. K., Reddy, V. B., Swinscoe, J., Lebowitz, P., and Weissman, S. M. (1978) Heterogeneity and 5′-terminal structures of the late RNAs of simian virus 40. *J. Mol. Biol.* **126,** 813–846.

4. Saiki, R. K., Gelfand, D. H., Stoffel, S., Scharf, S. J., Higuchi, R., Horn, G. T., et al. (1988) Primer-directed enzymatic amplification of DNA with a thermostable DNA polymerase. *Science* **239,** 487–491.

5. Liang, P. and Pardee, A. B. (1992) Differential display of eukaryotic message RNA by means of the polymerase chain reaction. *Science* **257,** 967–971.

6. Hubank, M. and Schatz, D. G. (1994) Identifying differences in mRNA expression by representational difference analysis of cDNA. *Nucleic Acids Res.* **22,** 5640–5648.

7. Kaushansky, K. and Drachman, J. G. (2002) The molecular and cellular biology of thrombopoietin: the primary regulator of platelet production. *Oncogene* **21,** 3359–3367.

8. Taga, T. and Kishimoto, T. (1997) Gp130 and the interleukin-6 family of cytokines. *Ann. Rev. Immunol.* **15,** 797–819.

9. Velculescu, V. E., Zhang, L., Vogelstein, B., and Kinzler, K. W. (1995) Serial analysis of gene expression. *Science* **270,** 484–487.

10. Fodor, S. P., Rava, R. P., Huang, X. C., Pease, A. C., Holmes, C. P., and Adams, C. L. (1993) Multiplexed biochemical assays with biological chip. *Nature* **364,** 555–556.

11. Schena, M., Shalon, D., Davis, R. W., and Brown, P. O. (1995) Quantitative monitoring of gene expression patterns with a complementary DNA microarray. *Science* **270,** 467–470.

12. Phillips, R. L., Ernst, R. E., Brunk, B., Ivanova, N., Mahan, M. A., Deanehan, J. K., et al. (2000) The genetic program of hematopoietic stem cells. *Science* **288,** 1634–1640.

13. Iyer, V. R., Eisen, M. B., Ross, D. T., Schuler, G., Moore, T., Lee, J. C., et al. (1999) The transcriptional program in the response of human fibroblasts to serum. *Science* **283,** 83–87.

14. Lam, L. T., Ronchini, C., Norton, J., Capobianco, A. J., and Bresnick, E. H. (2000) Suppression of erythroid but not megakaryocytic differentiation of human K562 erythroleukemic cells by Notch-1. *J. Biol. Chem.* **275,** 19,676–19,684.

15. Cheung, V. G., Morley, M., Aguilar, F., Massimi, A., Kucherlapati, R., and Childs, G. (1999) Making and reading microarrays. *Nat. Genet.* **21,** 15–19.

16. Alizadeh, A. A., Eisen, M. B., Davis, R. E., Ma, C., Lossos, I. S., Rosenwald, A., et al. (2000) Distinct types of diffuse large B-cell lymphoma identified by gene expression profiling. *Nature* **403,** 503–511.

17. Lam, L. T., Pickeral, O., Peng, A. C., Rosenwald, A., Hurt, E. M., Giltnane, J. M., et al. (2001) Genomic-scale measurement of mRNA turnover and the mechanisms of action of the anti-cancer drug flavopiridol. *GenomeBiology*. 2/10/0041.
18. Gnatenko, D. V, Dunn, J. J., McCorkle, S. R., Weissmann, D., Perrotta, P. L., and Bahou, W. F. (2003). Transcript profiling of human platelets using microarray and serial analysis of gene expression. *Blood* **101,** 2285–2293.
19. Bowtell, D. D. (1999) Options available—from start to finish—for obtaining expression data by microarray. *Nat. Genet.* **21,** 25–32.
20. Wang, E., Miller, L. D., Ohnmacht, G. A., Liu, E. T., and Marincola, F. M. (2000) High-fidelity mRNA amplification for gene profiling. *Nat. Biotech.* **18,** 457–459.

Index